MW01079160

LINEAR ALGEBRA AND LEARNING FROM DATA

GILBERT STRANG

Massachusetts Institute of Technology

WELLESLEY - CAMBRIDGE PRESS
Box 812060 Wellesley MA 02482

Linear Algebra and Learning from Data

ISBN 978-0-692-19638-0

LaTeX typesetting by Ashley C. Fernandes (info@problemsolvingpathway.com)

Printed in the United States of America 9 8 7 6 5 4 3 2 1

Other texts from Wellesley - Cambridge Press

Introduction to Linear Algebra, 5th Edition, Gilbert Strang ISBN 978-0-9802327-7-6
Computational Science and Engineering, Gilbert Strang ISBN 978-0-9614088-1-7
Wavelets and Filter Banks, Gilbert Strang and Truong Nguyen ISBN 978-0-9614088-7-9
Introduction to Applied Mathematics, Gilbert Strang ISBN 978-0-9614088-0-0
Calculus Third edition (2017), Gilbert Strang ISBN 978-0-9802327-5-2
Algorithms for Global Positioning, Kai Borre & Gilbert Strang ISBN 978-0-9802327-3-8
Essays in Linear Algebra, Gilbert Strang ISBN 978-0-9802327-6-9
Differential Equations and Linear Algebra, Gilbert Strang ISBN 978-0-9802327-9-0
An Analysis of the Finite Element Method, 2017 edition, Gilbert Strang and George Fix ISBN 978-0-9802327-8-3

Wellesley - Cambridge Press
Box 812060
Wellesley MA 02482 USA
www.wellesleycambridge.com

math.mit.edu/weborder.php (orders)
linearalgebrabook@gmail.com
math.mit.edu/~gs
phone (781) 431-8488 fax (617) 253-4358

The website for this book is **math.mit.edu/learningfromdata**
That site will link to 18.065 course material and video lectures on YouTube and OCW

The cover photograph shows a neural net on Inle Lake. It was taken in Myanmar.
From that photograph Lois Sellers designed and created the cover.
The snapshot of **playground.tensorflow.org** was a gift from its creator Daniel Smilkov.

Linear Algebra is included in MIT's OpenCourseWare site **ocw.mit.edu**
This provides video lectures of the full linear algebra course 18.06 and 18.06 SC

Deep Learning and Neural Nets

Linear algebra and probability/statistics and optimization are the mathematical pillars of machine learning. Those chapters will come before the architecture of a neural net. But we find it helpful to start with this description of the goal: *To construct a function that classifies the training data correctly, so it can generalize to unseen test data.*

To make that statement meaningful, you need to know more about this learning function. That is the purpose of these three pages—to give direction to all that follows.

The inputs to the function F are vectors or matrices or sometimes tensors—one input $\boldsymbol{v}$ for each training sample. For the problem of identifying handwritten digits, each input sample will be an image—a matrix of pixels. We aim to classify each of those images as a number from 0 to 9. Those ten numbers are the possible outputs from the learning function. In this example, the function F learns what to look for in classifying the images.

The MNIST set contains $70,000$ handwritten digits. We train a learning function on part of that set. By assigning weights to different pixels in the image, we create the function. The big problem of optimization (the heart of the calculation) is to choose weights so that the function assigns the correct output $0, 1, 2, 3, 4, 5, 6, 7, 8$, or 9. And we don't ask for perfection! (One of the dangers in deep learning is *overfitting the data,*)

Then we validate the function by choosing unseen MNIST samples, and applying the function to classify this test data. Competitions over the years have led to major improvements in the test results. Convolutional nets now go below 1% errors. In fact it is competitions on known data like MNIST that have brought big improvements in the structure of F. That structure is based on the architecture of an underlying neural net.

Linear and Nonlinear Learning Functions

The inputs are the samples $\boldsymbol{v}$, the outputs are the computed classifications $\boldsymbol{w} = F(\boldsymbol{v})$. The simplest learning function would be linear: $\boldsymbol{w} = A\boldsymbol{v}$. The entries in the matrix A are the weights to be learned: not too difficult. Frequently the function also learns a *bias vector* $\boldsymbol{b}$, so that $F(\boldsymbol{v}) = A\boldsymbol{v} + \boldsymbol{b}$. This function is "*affine*". Affine functions can be quickly learned, but by themselves they are too simple.

More exactly, linearity is a very limiting requirement. If MNIST used Roman numerals, then II might be halfway between I and III (as linearity demands). But what would be halfway between I and XIX? Certainly affine functions $A\boldsymbol{v} + \boldsymbol{b}$ are not always sufficient.

Nonlinearity would come by squaring the components of the input vector $\boldsymbol{v}$. That step might help to separate a circle from a point inside—which linear functions cannot do. But the construction of F moved toward "sigmoidal functions" with S-shaped graphs. It is remarkable that big progress came by inserting these standard nonlinear S-shaped functions between matrices A and B to produce $A(S(B\boldsymbol{v}))$. Eventually it was discovered that the smoothly curved logistic functions S could be replaced by the extremely simple ramp function now called $\mathbf{ReLU}(\boldsymbol{x}) = \mathbf{max}\,(\mathbf{0}, \boldsymbol{x})$. The graphs of these nonlinear "*activation functions*" R are drawn in Section VII.1.

Neural Nets and the Structure of $F(v)$

The functions that yield deep learning have the form $F(\boldsymbol{v}) = L(R(L(R(\ldots(L\boldsymbol{v})))))$. This is a *composition* of affine functions $L\boldsymbol{v} = A\boldsymbol{v} + \boldsymbol{b}$ with nonlinear functions R—which act on each component of the vector $L\boldsymbol{v}$. The matrices A and the bias vectors $\boldsymbol{b}$ are the **weights in the learning function F**. It is the A's and $\boldsymbol{b}$'s that must be learned from the training data, so that the outputs $F(\boldsymbol{v})$ will be (nearly) correct. Then F can be applied to new samples from the same population. If the weights (A's and $\boldsymbol{b}$'s) are well chosen, the outputs $F(\boldsymbol{v})$ from the unseen test data should be accurate. More layers in the function F will typically produce more accuracy in $F(\boldsymbol{v})$.

Properly speaking, $F(\boldsymbol{x}, \boldsymbol{v})$ depends on the input $\boldsymbol{v}$ and the weights $\boldsymbol{x}$ (all the A's and $\boldsymbol{b}$'s). The outputs $\boldsymbol{v}_1 = \text{ReLU}(A_1\boldsymbol{v} + \boldsymbol{b}_1)$ from the first step produce the **first hidden layer in our neural net**. The complete net starts with the input layer $\boldsymbol{v}$ and ends with the output layer $\boldsymbol{w} = F(\boldsymbol{v})$. The affine part $L_k(\boldsymbol{v}_{k-1}) = A_k\boldsymbol{v}_{k-1} + \boldsymbol{b}_k$ of each step uses the computed weights A_k and $\boldsymbol{b}_k$.

All those weights together are chosen in the giant optimization of deep learning:

Choose weights A_k and b_k to minimize the total loss over all training samples.

The total loss is the sum of individual losses on each sample. The loss function for least squares has the familiar form $||F(\boldsymbol{v}) - \text{true output}||^2$. Often least squares is not the best loss function for deep learning.

One input v = 2 **One output $w = 2$**

Here is a picture of the neural net, to show the structure of $F(\boldsymbol{v})$. The input layer contains the training samples $\boldsymbol{v} = \boldsymbol{v}_0$. The output is their classification $\boldsymbol{w} = F(\boldsymbol{v})$. For perfect learning, $\boldsymbol{w}$ will be a (correct) digit from 0 to 9. **The hidden layers add depth to the network**. It is that depth which has allowed the composite function F to be so successful in deep learning. In fact the number of weights A_{ij} and b_j in the neural net is often larger than the number of inputs from the training samples $\boldsymbol{v}$.

This is a feed-forward fully connected network. For images, a *convolutional* neural net (**CNN**) is often appropriate and weights are shared—the diagonals of the matrices A are constant. Deep learning works amazingly well, when the architecture is right.

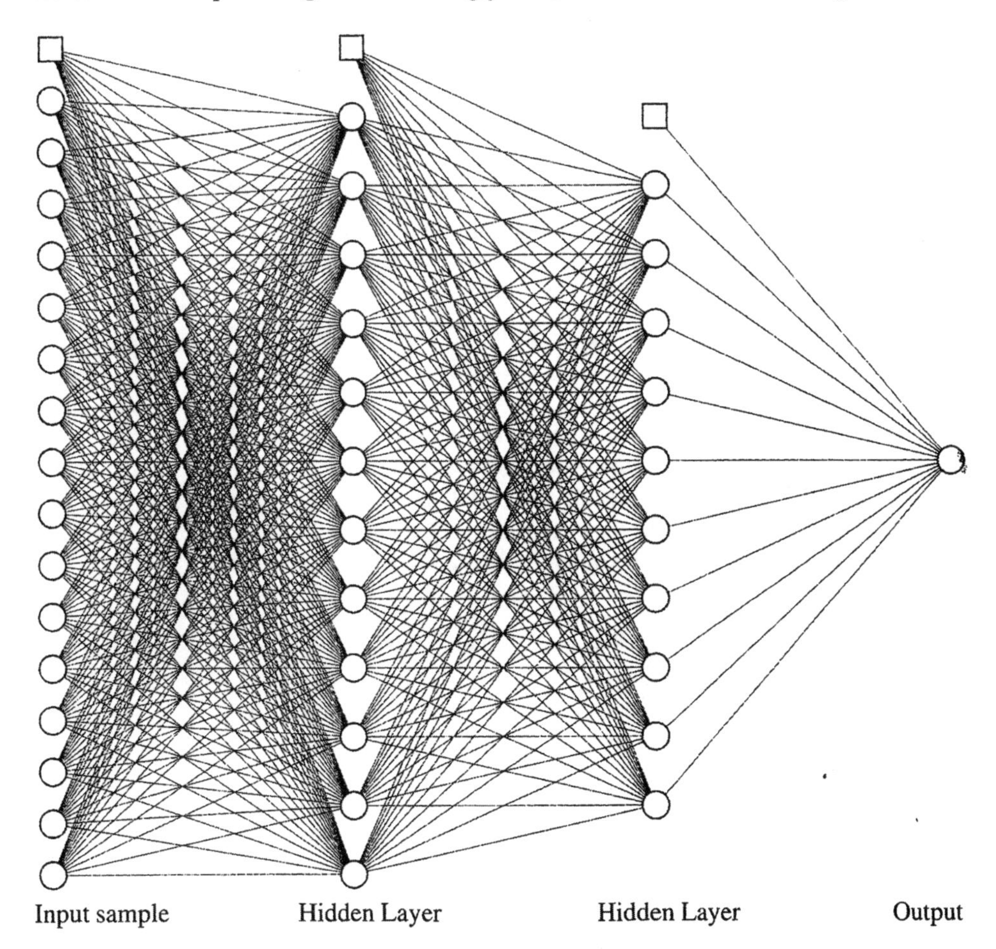

Each diagonal in this neural net represents a weight to be learned by optimization. Edges from the squares contain bias vectors $\boldsymbol{b}_1, \boldsymbol{b}_2, \boldsymbol{b}_3$. The other weights are in A_1, A_2, A_3.

Preface and Acknowledgments

My deepest gratitude goes to Professor Raj Rao Nadakuditi of the University of Michigan. On his sabbatical in 2017, Raj brought his EECS 551 course to MIT. He flew to Boston every week to teach 18.065. Thanks to Raj, the students could take a new course. He led the in-class computing, he assigned homeworks, and exams were outlawed.

This was linear algebra for signals and data, and it was alive. 140 MIT students signed up. Alan Edelman introduced the powerful language *Julia*, and I explained the four fundamental subspaces and the Singular Value Decomposition. The labs from Michigan involved rank and SVD and applications. We were asking the class for *computational thinking*.

That course worked, even the first time. It didn't touch one big topic : **Deep learning**. By this I mean the excitement of creating a learning function on a neural net, with the hidden layers and the nonlinear activation functions that make it so powerful. The system trains itself on data which has been correctly classified in advance. The optimization of weights discovers important features—the shape of a letter, the edges in an image, the syntax of a sentence, the identifying details of a signal. Those features get heavier weights—*without overfitting the data* and learning everything. Then unseen test data from a similar population can be identified by virtue of having those same features.

The algorithms to do all that are continually improving. Better if I say that they are *being improved.* This is the contribution of computer scientists and engineers and biologists and linguists and mathematicians and especially optimists—those who can optimize weights to minimize errors, and also those who believe that deep learning can help in our lives.

You can see why this book was written :

1. To organize central methods and ideas of **data science**.

2. To see how the language of **linear algebra** gives expression to those ideas.

3. Above all, to show how to **explain and teach** those ideas—to yourself or to a class.

I certainly learned that projects are far better than exams. Students ask their own questions and write their own programs. From now on, **projects !**

Linear Algebra and Calculus

The reader will have met the two central subjects of undergraduate mathematics: Linear algebra and calculus. For deep learning, it is linear algebra that matters most. We compute "weights" that pick out the important features of the training data, and those weights go into matrices. The form of the learning function is described on page iv. Then calculus shows us the *direction to move*, in order to improve the current weights $\boldsymbol{x}_k$.

From calculus, it is partial derivatives that we need (and not integrals):

$$\textbf{Reduce the error } L(\boldsymbol{x}) \textbf{ by moving from } \boldsymbol{x}_k \textbf{ to } \boldsymbol{x}_{k+1} = \boldsymbol{x}_k - s_k \boldsymbol{\nabla} \boldsymbol{L}.$$

That symbol $\boldsymbol{\nabla L}$ stands for the first derivatives of $L(\boldsymbol{x})$. Because of the minus sign, $\boldsymbol{x}_{k+1}$ *is downhill from* $\boldsymbol{x}_k$ on the graph of $L(\boldsymbol{x})$. The stepsize s_k (also called the learning rate) decides how far to move. You see the basic idea: Reduce the loss function $L(\boldsymbol{x})$ by moving in the direction of fastest decrease. $\boldsymbol{\nabla L} = \mathbf{0}$ at the best weights $\boldsymbol{x}^*$.

The complication is that the vector $\boldsymbol{x}$ represents thousands of weights. So we have to compute thousands of partial derivatives of L. And L itself is a complicated function depending on several layers of $\boldsymbol{x}$'s as well as the data. So we need the chain rule to find $\boldsymbol{\nabla L}$.

The introduction to Chapter VI will recall essential facts of multivariable calculus.

By contrast, *linear algebra is everywhere in the world of learning from data.* This is the subject to know! The first chapters of this book are essentially a course on applied linear algebra—the basic theory and its use in computations. I can try to outline how that approach (to the ideas we need) compares to earlier linear algebra courses. Those are quite different, which means that there are good things to learn.

Basic course

1. Elimination to solve $A\boldsymbol{x} = \boldsymbol{b}$
2. Matrix operations and inverses and determinants
3. Vector spaces and subspaces
4. Independence, dimension, rank of a matrix
5. Eigenvalues and eigenvectors

If a course is mostly learning definitions, that is not linear algebra in action. A stronger course puts the algebra to use. The definitions have a purpose, and so does the book.

Stronger course

1. $A\boldsymbol{x} = \boldsymbol{b}$ in all cases: square system—too many equations—too many unknowns.
2. Factor A into $\boldsymbol{LU}$ and $\boldsymbol{QR}$ and $\boldsymbol{U\Sigma V}^{\mathrm{T}}$ and $\boldsymbol{CMR}$: *Columns times rows.*
3. *Four fundamental subspaces*: dimensions and orthogonality and good bases.
4. Diagonalizing A by eigenvectors and by left and right singular vectors.
5. Applications: Graphs, convolutions, iterations, covariances, projections, filters, networks, images, matrices of data.

Linear algebra has moved to the center of machine learning, and we need to be there.

A book was needed for the 18.065 course. It was started in the original 2017 class, and a first version went out to the 2018 class. I happily acknowledge that this book owes its existence to Ashley C. Fernandes. Ashley receives pages scanned from Boston and sends back new sections from Mumbai, ready for more work. This is our seventh book together and I am extremely grateful.

Students were generous in helping with both classes, especially William Loucks and Claire Khodadad and Alex LeNail and Jack Strang. The project from Alex led to his online code **alexlenail.me/NN – SVG/** to draw neural nets (an example appears on page v). The project from Jack on **http://www.teachyourmachine.com** learns to recognize hand-written numbers and letters drawn by the user: open for experiment. See Section VII.2.

MIT's faculty and staff have given generous and much needed help:

Suvrit Sra gave a fantastic lecture on stochastic gradient descent (now an 18.065 video)

Alex Postnikov explained when matrix completion can lead to rank one (Section IV.8)

Tommy Poggio showed his class how deep learning generalizes to new data

Jonathan Harmon and Tom Mullaly and Liang Wang contributed to this book every day

Ideas arrived from all directions and gradually they filled this textbook.

The Content of the Book

This book aims to explain the mathematics on which data science depends: *Linear algebra, optimization, probability and statistics.* The weights in the learning function go into matrices. Those weights are optimized by "stochastic gradient descent". That word stochastic (= random) is a signal that success is governed by probability not certainty. The law of large numbers extends to the law of large functions: If the architecture is well designed and the parameters are well computed, there is a high probability of success.

Please note that this is not a book about computing, or coding, or software. Many books do those parts well. One of our favorites is *Hands-On Machine Learning* (2017) by Aurélien Géron (published by O'Reilly). And online help, from Tensorflow and Keras and MathWorks and Caffe and many more, is an important contribution to data science.

Linear algebra has a wonderful variety of matrices: symmetric, orthogonal, triangular, banded, permutations and projections and circulants. In my experience, *positive definite symmetric matrices* S are the aces. They have positive eigenvalues λ and orthogonal eigenvectors $\boldsymbol{q}$. They are combinations $S = \lambda_1 \boldsymbol{q}_1 \boldsymbol{q}_1^{\mathrm{T}} + \lambda_2 \boldsymbol{q}_2 \boldsymbol{q}_2^{\mathrm{T}} + \cdots$ of simple rank-one projections $\boldsymbol{q}\boldsymbol{q}^{\mathrm{T}}$ onto those eigenvectors. And if $\lambda_1 \geq \lambda_2 \geq \ldots$ then $\lambda_1 \boldsymbol{q}_1 \boldsymbol{q}_1^{\mathrm{T}}$ is the most informative part of S. For a sample covariance matrix, that part has the greatest variance.

Chapter I In our lifetimes, the most important step has been to extend those ideas from symmetric matrices to all matrices. Now we need **two sets of singular vectors, u's and v's.** Singular values σ replace eigenvalues λ. The decomposition $A = \sigma_1 u_1 v_1^T + \sigma_2 u_2 v_2^T + \cdots$ remains correct (this is the SVD). With decreasing σ's, those rank-one pieces of A still come in order of importance. That "Eckart-Young Theorem" about A complements what we have long known about the symmetric matrix $A^T A$: For rank k, stop at $\sigma_k u_k v_k^T$.

II The ideas in Chapter I become algorithms in Chapter II. For quite large matrices, the σ's and u's and v's are computable. For very large matrices, we resort to randomization: Sample the columns and the rows. For wide classes of big matrices this works well.

III-IV Chapter III focuses on low rank matrices, and Chapter IV on many important examples. We are looking for properties that make the computations especially fast (in III) or especially useful (in IV). The *Fourier matrix* is fundamental for every problem with constant coefficients (not changing with position). That discrete transform is superfast because of the **FFT**: the Fast Fourier Transform.

V Chapter V explains, as simply as possible, the statistics we need. The central ideas are always **mean and variance**: The *average* and the *spread* around that average. Usually we can reduce the mean to zero by a simple shift. Reducing the variance (the uncertainty) is the real problem. For random vectors and matrices and tensors, that problem becomes deeper. It is understood that the *linear algebra of statistics* is essential to machine learning.

VI Chapter VI presents two types of optimization problems. First come the nice problems of linear and quadratic programming and game theory. Duality and saddle points are key ideas. But the goals of deep learning and of this book are elsewhere: *Very large problems with a structure that is as simple as possible.* "Derivative equals zero" is still the fundamental equation. The second derivatives that Newton would have used are too numerous and too complicated to compute. Even using all the data (when we take a descent step to reduce the loss) is often impossible. That is why we choose only a minibatch of input data, in each step of stochastic gradient descent.

The success of large scale learning comes from the wonderful fact that *randomization often produces reliability*—when there are thousands or millions of variables.

VII Chapter VII begins with the architecture of a neural net. An input layer is connected to hidden layers and finally to the output layer. For the training data, input vectors v are known. Also the correct outputs are known (often w is the correct classification of v). **We optimize the weights x in the learning function F so that $F(x, v)$ is close to w for almost every training input v.**

Then F is applied to *test data*, drawn from the same population as the training data. If F learned what it needs (*without overfitting*: we don't want to fit 100 points by 99th degree polynomials), the test error will also be low. The system recognizes images and speech. It translates between languages. It may follow designs like ImageNet or AlexNet, winners of major competitions. A neural net defeated the world champion at Go.

The function F is often *piecewise linear*—the weights go into matrix multiplications. Every neuron on every hidden layer also has a nonlinear "activation function". The ramp function **ReLU**(x) = **(maximum of 0 and** x**)** is now the overwhelming favorite.

There is a growing world of expertise in designing the layers that make up $F(x, v)$. We start with *fully connected* layers—all neurons on layer n connected to all neurons on layer $n + 1$. Often CNN's are better—*Convolutional neural nets* repeat the same weights around all pixels in an image: a very important construction. Other layers are different. A *pooling layer* reduces the dimension. *Dropout* randomly leaves out neurons. *Batch normalization* resets the mean and variance. All these steps create a function that closely matches the training data. Then $F(x, v)$ is ready to use.

Acknowledgments

Above all, I welcome this chance to thank so many generous and encouraging friends:

Pawan Kumar and Leonard Berrada and Mike Giles and Nick Trefethen in Oxford
Ding-Xuan Zhou and Yunwen Lei in Hong Kong
Alex Townsend and Heather Wilber at Cornell
Nati Srebro and Srinadh Bhojanapalli in Chicago
Tammy Kolda and Thomas Strohmer and Trevor Hastie and Jay Kuo in California
Bill Hager and Mark Embree and Wotao Yin, for help with Chapter III
Stephen Boyd and Lieven Vandenberghe, for great books
Alex Strang, for creating the best figures, and more
Ben Recht in Berkeley, especially.

Your papers and emails and lectures and advice were wonderful.

THE MATRIX ALPHABET

A	Any Matrix	Q	Orthogonal Matrix
C	Circulant Matrix	R	Upper Triangular Matrix
C	Matrix of Columns	R	Matrix of Rows
D	Diagonal Matrix	S	Symmetric Matrix
F	Fourier Matrix	S	Sample Covariance Matrix
I	Identity Matrix	T	Tensor
L	Lower Triangular Matrix	U	Upper Triangular Matrix
L	Laplacian Matrix	U	Left Singular Vectors
M	Mixing Matrix	V	Right Singular Vectors
M	Markov Matrix	X	Eigenvector Matrix
P	Probability Matrix	Λ	Eigenvalue Matrix
P	Projection Matrix	Σ	Singular Value Matrix

Video lectures: OpenCourseWare ocw.mit.edu and YouTube (**Math 18.06 and 18.065**)
Introduction to Linear Algebra (5th ed) by Gilbert Strang, Wellesley-Cambridge Press
Book websites: **math.mit.edu/linearalgebra** and **math.mit.edu/learningfromdata**

Table of Contents

Deep Learning and Neural Nets iii

Preface and Acknowledgments vi

Part I : Highlights of Linear Algebra 1

I.1	Multiplication Ax Using Columns of A	2
I.2	Matrix-Matrix Multiplication AB	9
I.3	The Four Fundamental Subspaces	14
I.4	Elimination and $A = LU$	21
I.5	Orthogonal Matrices and Subspaces	29
I.6	Eigenvalues and Eigenvectors	36
I.7	Symmetric Positive Definite Matrices	44
I.8	Singular Values and Singular Vectors in the SVD	56
I.9	Principal Components and the Best Low Rank Matrix	71
I.10	Rayleigh Quotients and Generalized Eigenvalues	81
I.11	Norms of Vectors and Functions and Matrices	88
I.12	Factoring Matrices and Tensors : Positive and Sparse	97

Part II : Computations with Large Matrices 113

II.1	Numerical Linear Algebra	115
II.2	Least Squares : Four Ways	124
II.3	Three Bases for the Column Space	138
II.4	Randomized Linear Algebra	146

Part III: Low Rank and Compressed Sensing 159

III.1 Changes in A^{-1} from Changes in A 160
III.2 Interlacing Eigenvalues and Low Rank Signals 168
III.3 Rapidly Decaying Singular Values 178
III.4 Split Algorithms for $\ell^2 + \ell^1$ 184
III.5 Compressed Sensing and Matrix Completion 195

Part IV: Special Matrices 203

IV.1 Fourier Transforms : Discrete and Continuous 204
IV.2 Shift Matrices and Circulant Matrices 213
IV.3 The Kronecker Product $A \otimes B$ 221
IV.4 Sine and Cosine Transforms from Kronecker Sums 228
IV.5 Toeplitz Matrices and Shift Invariant Filters 232
IV.6 Graphs and Laplacians and Kirchhoff's Laws 239
IV.7 Clustering by Spectral Methods and k-means 245
IV.8 Completing Rank One Matrices 255
IV.9 The Orthogonal Procrustes Problem 257
IV.10 Distance Matrices 259

Part V: Probability and Statistics 263

V.1 Mean, Variance, and Probability 264
V.2 Probability Distributions 275
V.3 Moments, Cumulants, and Inequalities of Statistics 284
V.4 Covariance Matrices and Joint Probabilities 294
V.5 Multivariate Gaussian and Weighted Least Squares 304
V.6 Markov Chains 311

Part VI: Optimization 321

VI.1 Minimum Problems : Convexity and Newton's Method 324

VI.2 Lagrange Multipliers = Derivatives of the Cost 333

VI.3 Linear Programming, Game Theory, and Duality 338

VI.4 Gradient Descent Toward the Minimum 344

VI.5 Stochastic Gradient Descent and ADAM 359

Part VII: Learning from Data 371

VII.1 The Construction of Deep Neural Networks 375

VII.2 Convolutional Neural Nets 387

VII.3 Backpropagation and the Chain Rule 397

VII.4 Hyperparameters : The Fateful Decisions 407

VII.5 The World of Machine Learning 413

Books on Machine Learning 416

Eigenvalues and Singular Values : Rank One 417

Codes and Algorithms for Numerical Linear Algebra 418

Counting Parameters in the Basic Factorizations 419

Index of Authors 420

Index 423

Index of Symbols 432

Part I

Highlights of Linear Algebra

I.1 Multiplication Ax Using Columns of A

I.2 Matrix-Matrix Multiplication AB

I.3 The Four Fundamental Subspaces

I.4 Elimination and $A = LU$

I.5 Orthogonal Matrices and Subspaces

I.6 Eigenvalues and Eigenvectors

I.7 Symmetric Positive Definite Matrices

I.8 Singular Values and Singular Vectors in the SVD

I.9 Principal Components and the Best Low Rank Matrix

I.10 Rayleigh Quotients and Generalized Eigenvalues

I.11 Norms of Vectors and Functions and Matrices

I.12 Factoring Matrices and Tensors : Positive and Sparse

Part I : Highlights of Linear Algebra

Part I of this book is a serious introduction to applied linear algebra. If the reader's background is not great or not recent (in this important part of mathematics), please *do not rush through this part.* It starts with multiplying Ax and AB using the columns of the matrix A. That might seem only formal but in reality it is fundamental.

Let me point to five basic problems studied in this chapter.

$Ax = b$	$Ax = \lambda x$	$Av = \sigma u$	**Minimize** $\|\|Ax\|\|^2/\|\|x\|\|^2$	**Factor the matrix** A

Each of those problems looks like an ordinary computational question :

Find x Find x and λ Find v, u, and σ Factor A = **columns** times **rows**

You will see how *understanding* (even more than solving) is our goal. We want to know if $Ax = b$ has a solution x in the first place. "Is the vector b in the column space of A?" That innocent word "space" leads a long way. It will be a productive way, as you will see.

The eigenvalue equation $Ax = \lambda x$ is very different. There is no vector b—we are looking only at the matrix A. We want eigenvector directions so that Ax **keeps the same direction as** x. Then along that line all the complicated interconnections of A have gone away. The vector A^2x is just $\lambda^2 x$. The matrix e^{At} (from a differential equation) is just multiplying x by $e^{\lambda t}$. We can solve anything linear when we know every x and λ.

The equation $Av = \sigma u$ is close but different. Now we have **two vectors** v **and** u. Our matrix A is probably rectangular, and full of data. What part of that data matrix is important? The *Singular Value Decomposition* (SVD) finds its simplest pieces $\sigma u v^{\mathrm{T}}$. Those pieces are matrices (column u times row v^{T}). Every matrix is built from these orthogonal pieces. **Data science meets linear algebra in the SVD.**

Finding those pieces $\sigma u v^{\mathrm{T}}$ is the object of Principal Component Analysis (PCA).

Minimization and factorization express fundamental applied problems. They lead to those singular vectors v and u. Computing the best $\widehat{x}$ in least squares and the principal component v_1 in PCA is the algebra problem that *fits the data.* We won't give codes—those belong online—we are working to explain ideas.

When you understand column spaces and nullspaces and eigenvectors and singular vectors, you are ready for applications of all kinds : Least squares, Fourier transforms, LASSO in statistics, and stochastic gradient descent in deep learning with neural nets.

I.1 Multiplication Ax Using Columns of A

We hope you already know some linear algebra. It is a beautiful subject—more useful to more people than calculus (in our quiet opinion). But even old-style linear algebra courses miss basic and important facts. This first section of the book is about *matrix-vector* multiplication Ax and the column space of a matrix and the rank.

We always use examples to make our point clear.

Example 1 Multiply A times x using the three rows of A. Then use the two columns:

By rows $$\begin{bmatrix} 2 & 3 \\ 2 & 4 \\ 3 & 7 \end{bmatrix}\begin{bmatrix} x_1 \\ x_2 \end{bmatrix} = \begin{bmatrix} 2x_1 + 3x_2 \\ 2x_1 + 4x_2 \\ 3x_1 + 7x_2 \end{bmatrix} = \begin{matrix}\text{inner products} \\ \text{of the rows} \\ \text{with } \boldsymbol{x} = (x_1, x_2)\end{matrix}$$

By columns $$\begin{bmatrix} 2 & 3 \\ 2 & 4 \\ 3 & 7 \end{bmatrix}\begin{bmatrix} x_1 \\ x_2 \end{bmatrix} = x_1\begin{bmatrix} 2 \\ 2 \\ 3 \end{bmatrix} + x_2\begin{bmatrix} 3 \\ 4 \\ 7 \end{bmatrix} = \begin{matrix}\text{combination} \\ \text{of the columns} \\ \boldsymbol{a}_1 \text{ and } \boldsymbol{a}_2\end{matrix}$$

You see that both ways give the same result. The first way (a row at a time) produces three inner products. Those are also known as "dot products" because of the dot notation:

$$\textbf{row} \cdot \textbf{column} = (2,\ 3) \cdot (x_1,\ x_2) = 2x_1 + 3x_2 \qquad (1)$$

This is the way to find the three separate components of Ax. We use this for computing—but not for understanding. It is low level. Understanding is higher level, using vectors.

The vector approach sees Ax as a "linear combination" of a_1 and a_2. This is the fundamental operation of linear algebra! A linear combination of a_1 and a_2 includes two steps:

(1) Multiply the columns a_1 and a_2 by "scalars" x_1 and x_2

(2) Add vectors $x_1 a_1 + x_2 a_2 = Ax$.

Thus Ax is a linear combination of the columns of A. This is fundamental.

This thinking leads us to the **column space** of A. The key idea is to take **all combinations** of the columns. All real numbers x_1 and x_2 are allowed—the space includes Ax for all vectors x. In this way we get infinitely many output vectors Ax. And we can see those outputs geometrically.

In our example, each Ax is a vector in 3-dimensional space. That 3D space is called $\mathbf{R}^3$. (The $\mathbf{R}$ indicates real numbers. Vectors with three complex components lie in the space $\mathbf{C}^3$.) We stay with real vectors and we ask this key question:

All combinations $Ax = x_1 a_1 + x_2 a_2$ produce what part of the full 3D space?

Answer: Those vectors produce a **plane**. The plane contains the complete line in the direction of $a_1 = (2, 2, 3)$, since every vector $x_1 a_1$ is included. The plane also includes the line of all vectors $x_2 a_2$ in the direction of a_2. And it includes the *sum* of any vector on one line plus any vector on the other line. **This addition fills out an infinite plane containing the two lines**. But it does not fill out the whole 3-dimensional space $\mathbf{R}^3$.

Definition **The combinations of the columns** fill out the **column space** of A.

Here the column space is a plane. That plane includes the zero point $(0,0,0)$ which is produced when $x_1 = x_2 = 0$. The plane includes $(5,6,10) = \boldsymbol{a}_1 + \boldsymbol{a}_2$ and $(-1,-2,-4) = \boldsymbol{a}_1 - \boldsymbol{a}_2$. Every combination $x_1\boldsymbol{a}_1 + x_2\boldsymbol{a}_2$ is in this column space. With probability 1 it does **not** include the random point **rand**$(3,1)$! Which points are in the plane ?

$\boldsymbol{b} = (b_1, b_2, b_3)$ is in the column space of $\boldsymbol{A}$ **exactly when** $\boldsymbol{Ax} = \boldsymbol{b}$ has a solution $(\boldsymbol{x_1}, \boldsymbol{x_2})$

When you see that truth, you understand the column space $\mathbf{C}(A)$: The solution $\boldsymbol{x}$ shows how to express the right side $\boldsymbol{b}$ as a combination $x_1\boldsymbol{a}_1 + x_2\boldsymbol{a}_2$ of the columns. For some $\boldsymbol{b}$ this is impossible—they are not in the column space.

Example 2 $\boldsymbol{b} = \begin{bmatrix} 1 \\ 1 \\ 1 \end{bmatrix}$ is not in $\mathbf{C}(A)$. $A\boldsymbol{x} = \begin{bmatrix} 2x_1 + 3x_2 \\ 2x_1 + 4x_2 \\ 3x_1 + 7x_2 \end{bmatrix} = \begin{bmatrix} 1 \\ 1 \\ 1 \end{bmatrix}$ is unsolvable.

The first two equations force $x_1 = \frac{1}{2}$ and $x_2 = 0$. Then equation 3 fails : $3\left(\frac{1}{2}\right) + 7(0) = \mathbf{1.5}$ (**not** 1). This means that $\boldsymbol{b} = (1,1,1)$ is not in the column space—the plane of $\boldsymbol{a}_1$ and $\boldsymbol{a}_2$.

Example 3 What are the column spaces of $A_2 = \begin{bmatrix} 2 & 3 & 5 \\ 2 & 4 & 6 \\ 3 & 7 & 10 \end{bmatrix}$ and $A_3 = \begin{bmatrix} 2 & 3 & 1 \\ 2 & 4 & 1 \\ 3 & 7 & 1 \end{bmatrix}$?

Solution. The column space of A_2 is the same plane as before. The new column $(5,6,10)$ is the sum of column 1 + column 2. So $\boldsymbol{a}_3$ = column 3 is already in the plane and adds nothing new. By including this "*dependent*" column we don't go beyond the original plane.

The column space of A_3 is the whole 3D space $\mathbf{R}^3$. Example 2 showed us that the new third column $(1,1,1)$ is not in the plane $\mathbf{C}(A)$. Our column space $\mathbf{C}(A_3)$ has grown bigger. But there is nowhere to stop between a plane and the full 3D space. Visualize the $x - y$ plane and a third vector (x_3, y_3, z_3) out of the plane (meaning that $z_3 \neq 0$). They combine to give **every vector in** $\mathbf{R}^3$.

Here is a total list of all possible column spaces inside $\mathbf{R}^3$. Dimensions $0, 1, 2, 3$:

Subspaces of $\mathbf{R}^3$ The **zero vector** $(0,0,0)$ by itself
A **line** of all vectors $x_1\boldsymbol{a}_1$
A **plane** of all vectors $x_1\boldsymbol{a}_1 + x_2\boldsymbol{a}_2$
The **whole $\mathbf{R}^3$** with all vectors $x_1\boldsymbol{a}_1 + x_2\boldsymbol{a}_2 + x_3\boldsymbol{a}_3$

In that list we need the vectors $\boldsymbol{a}_1, \boldsymbol{a}_2, \boldsymbol{a}_3$ to be "**independent**". The only combination that gives the zero vector is $0\boldsymbol{a}_1 + 0\boldsymbol{a}_2 + 0\boldsymbol{a}_3$. So $\boldsymbol{a}_1$ by itself gives a line, $\boldsymbol{a}_1$ and $\boldsymbol{a}_2$ give a plane, $\boldsymbol{a}_1$ and $\boldsymbol{a}_2$ and $\boldsymbol{a}_3$ give every vector $\boldsymbol{b}$ in $\mathbf{R}^3$. The zero vector is in every subspace ! In linear algebra language :

- Three independent columns in $\mathbf{R}^3$ produce an **invertible matrix :** $AA^{-1} = A^{-1}A = I$.
- $A\boldsymbol{x} = \boldsymbol{0}$ requires $\boldsymbol{x} = (0,0,0)$. Then $A\boldsymbol{x} = \boldsymbol{b}$ has exactly one solution $\boldsymbol{x} = \boldsymbol{A^{-1}b}$.

You see the picture for the columns of an n by n invertible matrix. Their combinations fill its column space : **all of $\mathbf{R}^n$**. We needed those ideas and that language to go further.

Independent Columns and the Rank of A

After writing those words, I thought this short section was complete. *Wrong*. With just a small effort, we can find a **basis** for the column space of A, we can **factor** A into C times R, and we can prove the **first great theorem** in linear algebra. You will see the rank of a matrix and the dimension of a subspace.

All this comes with an understanding of **independence**. The goal is to create a matrix C whose columns come directly from A—but not to include any column that is a combination of previous columns. The columns of C (as many as possible) will be "independent". Here is a natural construction of C from the n columns of A:

If column 1 of A is not all zero, put it into the matrix C.

If column 2 of A is not a multiple of column 1, put it into C.

If column 3 of A is not a combination of columns 1 and 2, put it into C. *Continue*.

At the end C will have r columns $(r \leq n)$.

They will be a "basis" for the column space of A.

The left out columns are combinations of those basic columns in C.

A **basis** for a subspace is a full set of independent vectors: **All vectors in the space are combinations of the basis vectors**. Examples will make the point.

Example 4 If $A = \begin{bmatrix} 1 & 3 & 8 \\ 1 & 2 & 6 \\ 0 & 1 & 2 \end{bmatrix}$ then $C = \begin{bmatrix} 1 & 3 \\ 1 & 2 \\ 0 & 1 \end{bmatrix}$ $\quad n = 3$ columns in A, $\boldsymbol{r} = \mathbf{2}$ **columns in** C

Column 3 of A is 2 (column 1) + 2 (column 2). Leave it out of the basis in C.

Example 5 If $A = \begin{bmatrix} 1 & 2 & 3 \\ 0 & 4 & 5 \\ 0 & 0 & 6 \end{bmatrix}$ then $C = A$. $\quad n = 3$ columns in A, $\boldsymbol{r} = \mathbf{3}$ **columns in** C

This matrix A is invertible. Its column space is all of $\mathbf{R}^3$. Keep all 3 columns.

Example 6 If $A = \begin{bmatrix} 1 & 2 & 5 \\ 1 & 2 & 5 \\ 1 & 2 & 5 \end{bmatrix}$ then $C = \begin{bmatrix} 1 \\ 1 \\ 1 \end{bmatrix}$ $\quad n = 3$ columns in A, $\boldsymbol{r} = \mathbf{1}$ **column in** C

The number r is the **"rank"** of A. It is also the rank of C. **It counts independent columns.** Admittedly we could have moved from right to left in A, starting with its *last* column. This would not change the final count r. *Different basis, but always the same number of vectors*. That number r is the **"dimension"** of the column space of A and C (same space).

The rank of a matrix is the dimension of its column space.

The matrix C connects to A by a third matrix R: $\boldsymbol{A} = \boldsymbol{CR}$. Their shapes are $(m \text{ by } n) = (m \text{ by } r)(r \text{ by } n)$. I can show this "factorization of A" in Example 4 above:

$$\boldsymbol{A} = \begin{bmatrix} 1 & 3 & 8 \\ 1 & 2 & 6 \\ 0 & 1 & 2 \end{bmatrix} = \begin{bmatrix} 1 & 3 \\ 1 & 2 \\ 0 & 1 \end{bmatrix} \begin{bmatrix} 1 & 0 & 2 \\ 0 & 1 & 2 \end{bmatrix} = \boldsymbol{CR} \tag{2}$$

When C multiplies the first column $\begin{bmatrix} 1 \\ 0 \end{bmatrix}$ of R, this produces column 1 of C and A.

When C multiplies the second column $\begin{bmatrix} 0 \\ 1 \end{bmatrix}$ of R, we get column 2 of C and A.

When C multiplies the third column $\begin{bmatrix} 2 \\ 2 \end{bmatrix}$ of R, we get 2(column 1) + 2(column 2).

This matches column 3 of A. All we are doing is to put the right numbers in R. Combinations of the columns of C produce the columns of A. Then $A = CR$ stores this information as a matrix multiplication. Actually R is a famous matrix in linear algebra:

$\boldsymbol{R} =$ **rref**(A) = **row-reduced echelon form of $\boldsymbol{A}$** (without zero rows).

Example 5 has $C = A$ and then $R = I$ (identity matrix). Example 6 has only *one* column in C, so it has one row in R:

$$A = \begin{bmatrix} 1 & 2 & 5 \\ 1 & 2 & 5 \\ 1 & 2 & 5 \end{bmatrix} = \begin{bmatrix} 1 \\ 1 \\ 1 \end{bmatrix} \begin{bmatrix} 1 & 2 & 5 \end{bmatrix} = CR$$

All three matrices have rank $r = 1$
Column Rank = Row Rank

The number of *independent columns* equals the number of *independent rows*

This rank theorem is true for every matrix. Always columns and rows in linear algebra! The m rows contain the same numbers a_{ij} as the n columns. But different vectors.

The theorem is proved by $A = CR$. Look at that differently—by rows instead of columns. The matrix R has r rows. **Multiplying by $\boldsymbol{C}$ takes combinations of those rows.** Since $A = CR$, we get every row of A from the r rows of R. And those r rows are independent, so they are a **basis for the row space of $\boldsymbol{A}$**. The column space and row space of A both have dimension r, with r basis vectors—columns of C and rows of R.

One minute: Why does R have independent rows? Look again at Example 4.

$$A = \begin{bmatrix} 1 & 3 & 8 \\ 1 & 2 & 6 \\ 0 & 1 & 2 \end{bmatrix} = \begin{bmatrix} 1 & 3 \\ 1 & 2 \\ 0 & 1 \end{bmatrix} \begin{bmatrix} 1 & 0 & 2 \\ 0 & 1 & 2 \end{bmatrix} \begin{matrix} \leftarrow \text{independent} \\ \leftarrow \text{rows of } R \end{matrix}$$

$\uparrow \ \uparrow$ ones and zeros

It is those ones and zeros in R that tell me: No row is a combination of the other rows.

The big factorization for data science is the "SVD" of A—when the first factor C has r *orthogonal* columns and the second factor R has r *orthogonal* rows.

Problem Set I.1

1 Give an example where a combination of three nonzero vectors in $\mathbf{R}^4$ is the zero vector. Then write your example in the form $A\boldsymbol{x} = \mathbf{0}$. What are the shapes of A and $\boldsymbol{x}$ and $\mathbf{0}$?

2 Suppose a combination of the columns of A equals a different combination of those columns. Write that as $A\boldsymbol{x} = A\boldsymbol{y}$. Find two combinations of the columns of A that equal the zero vector (in matrix language, find two solutions to $A\boldsymbol{z} = \mathbf{0}$).

3 (Practice with subscripts) The vectors $\boldsymbol{a}_1, \boldsymbol{a}_2, \ldots, \boldsymbol{a}_n$ are in m-dimensional space $\mathbf{R}^m$, and a combination $c_1\boldsymbol{a}_1 + \cdots + c_n\boldsymbol{a}_n$ is the zero vector. That statement is at the vector level.

(1) Write that statement at the matrix level. Use the matrix A with the $\boldsymbol{a}$'s in its columns and use the column vector $\boldsymbol{c} = (c_1, \ldots, c_n)$.

(2) Write that statement at the scalar level. Use subscripts and sigma notation to add up numbers. The column vector $\boldsymbol{a}_j$ has components $a_{1j}, a_{2j}, \ldots, a_{mj}$.

4 Suppose A is the 3 by 3 matrix **ones**$(3, 3)$ of all ones. Find two independent vectors $\boldsymbol{x}$ and $\boldsymbol{y}$ that solve $A\boldsymbol{x} = \mathbf{0}$ and $A\boldsymbol{y} = \mathbf{0}$. Write that first equation $A\boldsymbol{x} = \mathbf{0}$ (with numbers) as a combination of the columns of A. Why don't I ask for a third independent vector with $A\boldsymbol{z} = \mathbf{0}$?

5 The linear combinations of $\boldsymbol{v} = (1, 1, 0)$ and $\boldsymbol{w} = (0, 1, 1)$ fill a plane in $\mathbf{R}^3$.

(a) Find a vector $\boldsymbol{z}$ that is perpendicular to $\boldsymbol{v}$ and $\boldsymbol{w}$. Then $\boldsymbol{z}$ is perpendicular to every vector $c\boldsymbol{v} + d\boldsymbol{w}$ on the plane: $(c\boldsymbol{v} + d\boldsymbol{w})^{\mathrm{T}}\boldsymbol{z} = c\boldsymbol{v}^{\mathrm{T}}\boldsymbol{z} + d\boldsymbol{w}^{\mathrm{T}}\boldsymbol{z} = 0 + 0$.

(b) Find a vector $\boldsymbol{u}$ that is not on the plane. Check that $\boldsymbol{u}^{\mathrm{T}}\boldsymbol{z} \neq 0$.

6 If three corners of a parallelogram are $(1, 1)$, $(4, 2)$, and $(1, 3)$, what are all three of the possible fourth corners? Draw two of them.

7 Describe the column space of $A = [\boldsymbol{v} \;\; \boldsymbol{w} \;\; \boldsymbol{v} + 2\boldsymbol{w}]$. Describe the nullspace of A: all vectors $\boldsymbol{x} = (x_1, x_2, x_3)$ that solve $A\boldsymbol{x} = \mathbf{0}$. Add the "dimensions" of that plane (the column space of A) and that line (the nullspace of A):

dimension of column space + dimension of nullspace = number of columns

8 $\boldsymbol{A} = \boldsymbol{CR}$ is a representation of the columns of A in the basis formed by the columns of C with coefficients in R. If $A_{ij} = j^2$ is 3 by 3, write down A and C and R.

9 Suppose the column space of an m by n matrix is all of $\mathbf{R}^3$. What can you say about m ? What can you say about n ? What can you say about the rank r ?

10 Find the matrices C_1 and C_2 containing independent columns of A_1 and A_2:

$$A_1 = \begin{bmatrix} 1 & 3 & -2 \\ 3 & 9 & -6 \\ 2 & 6 & -4 \end{bmatrix} \qquad A_2 = \begin{bmatrix} 1 & 2 & 3 \\ 4 & 5 & 6 \\ 7 & 8 & 9 \end{bmatrix}$$

11 Factor each of those matrices into $A = CR$. The matrix R will contain the numbers that multiply columns of C to recover columns of A.

This is one way to look at matrix multiplication: **C times each column of R.**

12 Produce a basis for the column spaces of A_1 and A_2. What are the *dimensions* of those column spaces—the number of independent vectors? What are the *ranks* of A_1 and A_2? How many independent rows in A_1 and A_2?

13 Create a 4 by 4 matrix A of rank 2. What shapes are C and R?

14 Suppose two matrices A and B have the same column space.

(a) Show that their row spaces can be different.

(b) Show that the matrices C (basic columns) can be different.

(c) What number will be the same for A and B?

15 If $A = CR$, the first row of A is a combination of the rows of R. Which part of which matrix holds the coefficients in that combination—the numbers that multiply the rows of R to produce row 1 of A?

16 The rows of R are a basis for the row space of A. *What does that sentence mean*?

17 For these matrices with square blocks, find $A = CR$. What ranks?

$$A_1 = \begin{bmatrix} \text{zeros} & \text{ones} \\ \text{ones} & \text{ones} \end{bmatrix}_{4 \times 4} \qquad A_2 = \begin{bmatrix} A_1 \\ A_1 \end{bmatrix}_{8 \times 4} \qquad A_3 = \begin{bmatrix} A_1 & A_1 \\ A_1 & A_1 \end{bmatrix}_{8 \times 8}$$

18 If $A = CR$, what are the CR factors of the matrix $\begin{bmatrix} 0 & A \\ 0 & A \end{bmatrix}$?

19 "Elimination" subtracts a number ℓ_{ij} times row j from row i: a "row operation." Show how those steps can reduce the matrix A in Example 4 to R (except that this row echelon form R has a row of zeros). The rank won't change!

$$A = \begin{bmatrix} 1 & 3 & 8 \\ 1 & 2 & 6 \\ 0 & 1 & 2 \end{bmatrix} \quad \to \quad \to \quad R = \begin{bmatrix} 1 & 0 & 2 \\ 0 & 1 & 2 \\ 0 & 0 & 0 \end{bmatrix} = \textbf{rref}\,(A).$$

This page is about the factorization $A = CR$ and its close relative $A = \boldsymbol{CMR}$. As before, $\boldsymbol{C}$ has r independent columns taken from A. *The new matrix* $\boldsymbol{R}$ *has* r *independent rows, also taken directly from* A. The r by r **"mixing matrix"** is $\boldsymbol{M}$. This invertible matrix makes $A = \boldsymbol{CMR}$ a true equation.

The rows of R (not bold) were chosen to produce $A = CR$, but those rows of *R did not come directly from A.* We will see that R has the form $\boldsymbol{MR}$ (bold $\boldsymbol{R}$).

$$\textbf{Rank-1 example} \quad A = CR = \boldsymbol{CMR} \qquad \begin{bmatrix} 2 & 4 \\ 3 & 6 \end{bmatrix} = \begin{bmatrix} 2 \\ 3 \end{bmatrix} \begin{bmatrix} 1 & 2 \end{bmatrix} = \begin{bmatrix} \mathbf{2} \\ \mathbf{3} \end{bmatrix} \begin{bmatrix} \frac{1}{2} \end{bmatrix} \begin{bmatrix} \mathbf{2} & \mathbf{4} \end{bmatrix}$$

In this case $\boldsymbol{M}$ is just 1 by 1. How do we find $\boldsymbol{M}$ in other examples of $A = \boldsymbol{CMR}$? $\boldsymbol{C}$ and $\boldsymbol{R}$ are not square. They have *one-sided* inverses. *We invert* $\boldsymbol{C}^{\mathrm{T}}\boldsymbol{C}$ and $\boldsymbol{RR}^{\mathrm{T}}$.

$$\boxed{A = \boldsymbol{CMR}} \quad \boldsymbol{C}^{\mathrm{T}}A\boldsymbol{R}^{\mathrm{T}} = \boldsymbol{C}^{\mathrm{T}}\boldsymbol{C}\,\boldsymbol{M}\,\boldsymbol{RR}^{\mathrm{T}} \quad \boxed{\boldsymbol{M} = (\boldsymbol{C}^{\mathrm{T}}\boldsymbol{C})^{-1}(\boldsymbol{C}^{\mathrm{T}}A\boldsymbol{R}^{\mathrm{T}})(\boldsymbol{RR}^{\mathrm{T}})^{-1}} \quad (*)$$

Here are extra problems to give practice with all these rectangular matrices of rank r. $\boldsymbol{C}^{\mathrm{T}}\boldsymbol{C}$ and $\boldsymbol{RR}^{\mathrm{T}}$ have rank r so they are invertible (see the last page of Section I.3).

20 Show that equation $(*)$ produces $\boldsymbol{M} = \left[\, \frac{1}{2} \,\right]$ in the small example above.

21 The rank-2 example in the text produced $A = CR$ in equation (2):

$$\boldsymbol{A} = \begin{bmatrix} 1 & 3 & 8 \\ 1 & 2 & 6 \\ 0 & 1 & 2 \end{bmatrix} = \begin{bmatrix} 1 & 3 \\ 1 & 2 \\ 0 & 1 \end{bmatrix} \begin{bmatrix} 1 & 0 & 2 \\ 0 & 1 & 2 \end{bmatrix} = \boldsymbol{C}R$$

Choose rows 1 and 2 directly from A to go into $\boldsymbol{R}$. Then from equation $(*)$, find the 2 by 2 matrix $\boldsymbol{M}$ that produces $A = \boldsymbol{CMR}$. Fractions enter the inverse of matrices:

$$\textbf{Inverse of a 2 by 2 matrix} \qquad \begin{bmatrix} a & b \\ c & d \end{bmatrix}^{-1} = \frac{1}{ad - bc} \begin{bmatrix} d & -b \\ -c & a \end{bmatrix} \qquad (**)$$

22 Show that this formula $(**)$ breaks down if $\begin{bmatrix} b \\ d \end{bmatrix} = m \begin{bmatrix} a \\ c \end{bmatrix}$: *dependent columns.*

23 Create a 3 by 2 matrix A with rank 1. Factor A into $A = CR$ and $A = \boldsymbol{CMR}$.

24 Create a 3 by 2 matrix A with rank 2. Factor A into $A = \boldsymbol{CMR}$.

The reason for this page is that the factorizations $A = CR$ and $A = \boldsymbol{CMR}$ have jumped forward in importance for large matrices. When $\boldsymbol{C}$ takes columns directly from A, and $\boldsymbol{R}$ takes rows directly from A, those matrices preserve properties that are lost in the more famous QR and SVD factorizations. Where $A = QR$ and $A = U\Sigma V^{\mathrm{T}}$ involve orthogonalizing the vectors, $\boldsymbol{C}$ and $\boldsymbol{R}$ keep the original data:

If A is nonnegative, so are $\boldsymbol{C}$ and $\boldsymbol{R}$. If A is sparse, so are $\boldsymbol{C}$ and $\boldsymbol{R}$.

I.2 Matrix-Matrix Multiplication AB

Inner products (*rows times columns*) produce each of the numbers in $AB = C$:

$$\begin{array}{c}\textbf{row 2 of } A\\ \textbf{column 3 of } B\\ \textbf{give } c_{23} \textbf{ in } C\end{array} \quad \begin{bmatrix} \cdot & \cdot & \cdot \\ a_{21} & a_{22} & a_{23} \\ \cdot & \cdot & \cdot \end{bmatrix} \begin{bmatrix} \cdot & \cdot & b_{13} \\ \cdot & \cdot & b_{23} \\ \cdot & \cdot & b_{33} \end{bmatrix} = \begin{bmatrix} \cdot & \cdot & \cdot \\ \cdot & \cdot & c_{23} \\ \cdot & \cdot & \cdot \end{bmatrix} \quad (1)$$

That dot product $c_{23} =$ (row 2 of A) $\cdot$ (column 3 of B) is a sum of a's times b's:

$$c_{23} = a_{21}\, b_{13} + a_{22}\, b_{23} + a_{23}\, b_{33} = \sum_{k=1}^{3} a_{2k}\, b_{k3} \quad \text{and} \quad c_{ij} = \sum_{k=1}^{n} a_{ik}\, b_{kj}. \quad (2)$$

This is how we usually compute each number in $AB = C$. But there is another way.

The other way to multiply AB is **columns of A times rows of B**. We need to see this! I start with numbers to make two key points: *one column u* times *one row v^{T}* produces a *matrix*. Concentrate first on that piece of AB. This matrix uv^{T} is especially simple:

$$\begin{array}{c}\textbf{“Outer}\\ \textbf{product”}\end{array} \quad uv^{\mathrm{T}} = \begin{bmatrix} 2 \\ 2 \\ 1 \end{bmatrix} \begin{bmatrix} 3 & 4 & 6 \end{bmatrix} = \begin{bmatrix} 6 & 8 & 12 \\ 6 & 8 & 12 \\ 3 & 4 & 6 \end{bmatrix} = \begin{array}{c}\textbf{“rank one}\\ \textbf{matrix”}\end{array}$$

An m by 1 matrix (a column u) times a 1 by p matrix (a row v^{T}) gives an m by p matrix. Notice what is special about the rank one matrix uv^{T}:

$$\text{All columns of } uv^{\mathrm{T}} \text{ are multiples of } u = \begin{bmatrix} 2 \\ 2 \\ 1 \end{bmatrix} \quad \text{All rows are multiples of } v^{\mathrm{T}} = \begin{bmatrix} 3 & 4 & 6 \end{bmatrix}$$

The column space of uv^{T} is one-dimensional: *the line in the direction of u*. The dimension of the column space (the number of independent columns) is the **rank of the matrix**—a key number. **All nonzero matrices uv^{T} have rank one.** They are the perfect building blocks for every matrix.

Notice also: **The row space of uv^{T} is the line through v.** By definition, the row space of any matrix A is the column space $\mathbf{C}(A^{\mathrm{T}})$ of its transpose A^{T}. That way we stay with column vectors. In the example, we transpose uv^{T} (**exchange rows with columns**) to get the matrix vu^{T}:

$$(uv^{\mathrm{T}})^{\mathrm{T}} = \begin{bmatrix} \mathbf{6} & 8 & 12 \\ \mathbf{6} & 8 & 12 \\ \mathbf{3} & 4 & 6 \end{bmatrix}^{\mathrm{T}} = \begin{bmatrix} \mathbf{6} & \mathbf{6} & \mathbf{3} \\ 8 & 8 & 4 \\ 12 & 12 & 6 \end{bmatrix} = \begin{bmatrix} 3 \\ 4 \\ 6 \end{bmatrix} \begin{bmatrix} 2 & 2 & 1 \end{bmatrix} = vu^{\mathrm{T}}.$$

We are seeing the clearest possible example of the first great theorem in linear algebra:

Row rank = Column rank	r **independent columns** $\Leftrightarrow$ r **independent rows**

A nonzero matrix $\boldsymbol{u}\boldsymbol{v}^{\mathrm{T}}$ has one independent column and one independent row. All columns are multiples of $\boldsymbol{u}$ and all rows are multiples of $\boldsymbol{v}^{\mathrm{T}}$. The rank is $r = 1$ for this matrix.

AB = Sum of Rank One Matrices

We turn to the full product AB, using columns of A times rows of B. Let $\boldsymbol{a}_1, \boldsymbol{a}_2, \ldots, \boldsymbol{a}_n$ be the n columns of A. Then B must have n rows $\boldsymbol{b}_1^*, \boldsymbol{b}_2^*, \ldots, \boldsymbol{b}_n^*$. The matrix A can multiply the matrix B. **Their product AB is the sum of columns $\boldsymbol{a}_k$ times rows $\boldsymbol{b}_k^*$:**

Column-row multiplication of matrices

$$AB = \begin{bmatrix} | & & | \\ \boldsymbol{a}_1 & \cdots & \boldsymbol{a}_n \\ | & & | \end{bmatrix} \begin{bmatrix} - & \boldsymbol{b}_1^* & - \\ & \vdots & \\ - & \boldsymbol{b}_n^* & - \end{bmatrix} = \underbrace{\boldsymbol{a}_1 \boldsymbol{b}_1^* + \boldsymbol{a}_2 \boldsymbol{b}_2^* + \cdots + \boldsymbol{a}_n \boldsymbol{b}_n^*}_{\textbf{sum of rank 1 matrices}}. \quad (3)$$

Here is a 2 by 2 example to show the $n = 2$ pieces (column times row) and their sum AB:

$$\begin{bmatrix} 1 & 0 \\ 3 & 1 \end{bmatrix} \begin{bmatrix} 2 & 4 \\ 0 & 5 \end{bmatrix} = \begin{bmatrix} \mathbf{1} \\ \mathbf{3} \end{bmatrix} \begin{bmatrix} \mathbf{2} & \mathbf{4} \end{bmatrix} + \begin{bmatrix} \mathbf{0} \\ \mathbf{1} \end{bmatrix} \begin{bmatrix} \mathbf{0} & \mathbf{5} \end{bmatrix} = \begin{bmatrix} 2 & 4 \\ 6 & 12 \end{bmatrix} + \begin{bmatrix} 0 & 0 \\ 0 & 5 \end{bmatrix} = \begin{bmatrix} 2 & 4 \\ 6 & 17 \end{bmatrix} \quad (4)$$

We can count the multiplications of number times number. Four multiplications to get $2, 4, 6, 12$. Four more to get $0, 0, 0, 5$. A total of $2^3 = 8$ multiplications. Always there are $\boldsymbol{n^3}$ multiplications when A and B are n by n. And $\boldsymbol{mnp}$ multiplications when $AB =$ (m by n) times (n by p): n rank one matrices, each of those matrices is m by p.

The count is the same for the usual inner product way. Row of A times column of B needs n multiplications. We do this for every number in AB: mp dot products when AB is m by p. The total count is again $\boldsymbol{mnp}$ when we multiply (m by n) times (n by p).

rows times columns	$\boldsymbol{mp}$ **inner products,**	$\boldsymbol{n}$ **multiplications each**	$\boldsymbol{mnp}$
columns times rows	$\boldsymbol{n}$ **outer products,**	$\boldsymbol{mp}$ **multiplications each**	$\boldsymbol{mnp}$

When you look closely, they are exactly the same multiplications $a_{ik}\, b_{kj}$ in different orders. Here is the algebra proof that each number c_{ij} in $C = AB$ is the same by outer products in (3) as by inner products in (2):

$$\text{The } i, j \text{ entry of } \boldsymbol{a}_k \boldsymbol{b}_k^* \text{ is } \boldsymbol{a_{ik} b_{kj}}. \text{ Add to find } \boldsymbol{c_{ij}} = \sum_{k=1}^{n} a_{ik}\, b_{kj} = \textbf{row } \boldsymbol{i} \cdot \textbf{column } \boldsymbol{j}.$$

Insight from Column times Row

Why is the outer product approach essential in data science? The short answer is: *We are looking for the important part of a matrix A.* We don't usually want the biggest number in A (though that could be important). What we want more is the largest piece of A. **And those pieces are rank one matrices uv^{T}.** A dominant theme in applied linear algebra is:

Factor A into CR and look at the pieces $c_k r_k^*$ of $A = CR$.

Factoring A into CR is the reverse of multiplying $CR = A$. Factoring takes longer, especially if the pieces involve *eigenvalues* or *singular values*. But those numbers have inside information about the matrix A. That information is not visible until you factor.

Here are five important factorizations, with the standard choice of letters (usually A) for the original product matrix and then for its factors. This book will explain all five.

$$A = LU \qquad A = QR \qquad S = Q\Lambda Q^{\mathrm{T}} \qquad A = X\Lambda X^{-1} \qquad A = U\Sigma V^{\mathrm{T}}$$

At this point we simply list key words and properties for each of these factorizations.

1 $A = LU$ comes from **elimination**. Combinations of rows take A to U and U back to A. The matrix L is lower triangular and U is upper triangular as in equation (4).

2 $A = QR$ comes from **orthogonalizing** the columns $\boldsymbol{a}_1$ to $\boldsymbol{a}_n$ as in "Gram-Schmidt". Q has orthonormal columns ($Q^{\mathrm{T}}Q = I$) and R is upper triangular.

3 $S = Q\Lambda Q^{\mathrm{T}}$ comes from the **eigenvalues** $\lambda_1, \ldots, \lambda_n$ of a symmetric matrix $S = S^{\mathrm{T}}$. Eigenvalues on the diagonal of Λ. **Orthonormal eigenvectors** in the columns of Q.

4 $A = X\Lambda X^{-1}$ is **diagonalization** when A is n by n with n independent eigenvectors. *Eigenvalues* of A on the diagonal of Λ. *Eigenvectors* of A in the columns of X.

5 $A = U\Sigma V^{\mathrm{T}}$ is the **Singular Value Decomposition** of any matrix A (square or not). **Singular values** $\sigma_1, \ldots, \sigma_r$ in Σ. Orthonormal **singular vectors** in U and V.

Let me pick out a favorite (number **3**) to illustrate the idea. This special factorization $Q\Lambda Q^{\mathrm{T}}$ starts with a symmetric matrix S. That matrix has orthogonal unit eigenvectors $\boldsymbol{q}_1, \ldots, \boldsymbol{q}_n$. Those perpendicular eigenvectors (dot products $= 0$) go into the columns of Q. S and Q are the kings and queens of linear algebra:

Symmetric matrix S	$S^{\mathrm{T}} = S$	All $s_{ij} = s_{ji}$
Orthogonal matrix Q	$Q^{\mathrm{T}} = Q^{-1}$	All $\boldsymbol{q}_i \cdot \boldsymbol{q}_j = \begin{cases} 0 & \text{for } i \neq j \\ 1 & \text{for } i = j \end{cases}$

The diagonal matrix Λ contains real eigenvalues λ_1 to λ_n. Every real symmetric matrix S has n orthonormal eigenvectors $\boldsymbol{q}_1$ to $\boldsymbol{q}_n$. When multiplied by S, the eigenvectors keep the same direction. They are just rescaled by the number λ :

$$\boxed{\textbf{Eigenvector } q \textbf{ and eigenvalue } \boldsymbol{\lambda} \qquad Sq = \lambda q} \tag{5}$$

Finding λ and $\boldsymbol{q}$ is not easy for a big matrix. But n pairs always exist when S is symmetric. Our purpose here is to see how $\boldsymbol{SQ} = \boldsymbol{Q\Lambda}$ comes column by column from $S\boldsymbol{q} = \lambda\boldsymbol{q}$:

$$\boldsymbol{SQ} = S\begin{bmatrix} \boldsymbol{q}_1 & \cdots & \boldsymbol{q}_n \end{bmatrix} = \begin{bmatrix} \lambda_1\boldsymbol{q}_1 & \cdots & \lambda_n\boldsymbol{q}_n \end{bmatrix} = \begin{bmatrix} \boldsymbol{q}_1 & \cdots & \boldsymbol{q}_n \end{bmatrix}\begin{bmatrix} \lambda_1 & & \\ & \ddots & \\ & & \lambda_n \end{bmatrix} = \boldsymbol{Q\Lambda} \tag{6}$$

Multiply $SQ = Q\Lambda$ by $Q^{-1} = Q^{\mathrm{T}}$ to get $\boldsymbol{S = Q\Lambda Q^{\mathrm{T}}}$ = a symmetric matrix. Each eigenvalue λ_k and each eigenvector $\boldsymbol{q}_k$ contribute a rank one piece $\lambda_k\boldsymbol{q}_k\boldsymbol{q}_k^{\mathrm{T}}$ to S.

$$\textbf{Rank one pieces} \qquad S = (Q\Lambda)Q^{\mathrm{T}} = (\lambda_1\boldsymbol{q}_1)\boldsymbol{q}_1^{\mathrm{T}} + (\lambda_2\boldsymbol{q}_2)\boldsymbol{q}_2^{\mathrm{T}} + \cdots + (\lambda_n\boldsymbol{q}_n)\boldsymbol{q}_n^{\mathrm{T}} \tag{7}$$

$$\textbf{All symmetric} \qquad \text{The transpose of } \boldsymbol{q}_i\boldsymbol{q}_i^{\mathrm{T}} \text{ is } \boldsymbol{q}_i\boldsymbol{q}_i^{\mathrm{T}} \tag{8}$$

Please notice that the columns of $Q\Lambda$ are $\lambda_1\boldsymbol{q}_1$ to $\lambda_n\boldsymbol{q}_n$. When you multiply a matrix on the right by the diagonal matrix Λ, you multiply its *columns* by the λ's.

We close with a comment on the proof of this **Spectral Theorem** $\boldsymbol{S = Q\Lambda Q^{\mathrm{T}}}$: Every symmetric S has n real eigenvalues and n orthonormal eigenvectors. Section 1.6 will construct the eigenvalues as the roots of the nth degree polynomial $P_n(\lambda)$ = determinant of $S - \lambda I$. They are real numbers when $S = S^{\mathrm{T}}$. The delicate part of the proof comes when an eigenvalue λ_i is *repeated*— it is a double root or an Mth root from a factor $(\lambda - \lambda_j)^M$. In this case we need to produce M independent eigenvectors. The rank of $S - \lambda_j I$ must be $n - M$. This is true when $S = S^{\mathrm{T}}$. But it requires a proof.

Similarly the Singular Value Decomposition $A = U\Sigma V^{\mathrm{T}}$ requires extra patience when a singular value σ is repeated M times in the diagonal matrix Σ. Again there are M pairs of singular vectors $\boldsymbol{v}$ and $\boldsymbol{u}$ with $A\boldsymbol{v} = \sigma\boldsymbol{u}$. Again this true statement requires proof.

Notation for rows We introduced the symbols $\boldsymbol{b}_1^*, \ldots, \boldsymbol{b}_n^*$ for the rows of the second matrix in AB. You might have expected $\boldsymbol{b}_1^{\mathrm{T}}, \ldots, \boldsymbol{b}_n^{\mathrm{T}}$ and that was our original choice. But this notation is not entirely clear—it seems to mean the transposes of the columns of B. Since that right hand factor could be U or R or Q^{T} or X^{-1} or V^{T}, it is safer to say definitely: *we want the rows of that matrix.*

G. Strang, *Multiplying and factoring matrices*, Amer. Math. Monthly **125** (2018) 223-230.

G. Strang, *Introduction to Linear Algebra*, 5th ed., Wellesley-Cambridge Press (2016).

Problem Set I.2

1 Suppose $Ax = \mathbf{0}$ and $Ay = \mathbf{0}$ (where x and y and $\mathbf{0}$ are vectors). Put those two statements together into one matrix equation $AB = C$. What are those matrices B and C? If the matrix A is m by n, what are the shapes of B and C?

2 Suppose a and b are column vectors with components $a_1, \ldots, a_m$ and $b_1, \ldots, b_p$. Can you multiply a times b^{T} (yes or no)? What is the shape of the answer ab^{T}? What number is in row i, column j of ab^{T}? What can you say about aa^{T}?

3 (Extension of Problem **2**: Practice with subscripts) Instead of that one vector a, suppose you have n vectors a_1 to a_n in the columns of A. Suppose you have n vectors $b_1^{\mathrm{T}}, \ldots, b_n^{\mathrm{T}}$ in the rows of B.

(a) Give a "sum of rank one" formula for the matrix-matrix product AB.

(b) Give a formula for the i, j entry of that matrix-matrix product AB. Use sigma notation to add the i, j entries of each matrix $a_k b_k^{\mathrm{T}}$, found in Problem 2.

4 Suppose B has only one column ($p = 1$). So each row of B just has one number. A has columns a_1 to a_n as usual. Write down the column times row formula for AB. In words, the m by 1 column vector AB is a combination of the ____ .

5 Start with a matrix B. If we want to take combinations of its rows, we premultiply by A to get AB. If we want to take combinations of its columns, we postmultiply by C to get BC. For this question we will do both.

Row operations then column operations First AB then $(AB)C$

Column operations then row operations First BC then $A(BC)$

The **associative law** says that we get the same final result both ways.

Verify $(AB)C = A(BC)$ for $A = \begin{bmatrix} 1 & a \\ 0 & 1 \end{bmatrix}$ $B = \begin{bmatrix} b_1 & b_2 \\ b_3 & b_4 \end{bmatrix}$ $C = \begin{bmatrix} 1 & 0 \\ c & 1 \end{bmatrix}$.

6 If A has columns a_1, a_2, a_3 and $B = I$ is the identity matrix, what are the rank one matrices $a_1 b_1^*$ and $a_2 b_2^*$ and $a_3 b_3^*$? They should add to $AI = A$.

7 *Fact*: The columns of AB are combinations of the columns of A. Then the column *space* of AB is *contained in* the column space of A. Give an example of A and B for which AB has a smaller column space than A.

8 To compute $C = AB = (m \text{ by } n)\,(n \text{ by } p)$, what order of the same three commands leads to columns times rows (outer products)?

Rows times columns	**Columns times rows**
For $i = 1$ to m	For...
For $j = 1$ to p	For...
For $k = 1$ to n	For...
$C(i,j) = C(i,j) + A(i,k) * B(k,j)$	$C =$

I.3 The Four Fundamental Subspaces

This section will explain the "big picture" of linear algebra. That picture shows how every m by n matrix A leads to four subspaces—two subspaces of $\mathbf{R}^m$ and two more of $\mathbf{R}^n$. The first example will be a rank one matrix $\boldsymbol{u}\boldsymbol{v}^{\mathrm{T}}$, where the column space is the line through $\boldsymbol{u}$ and the row space is the line through $\boldsymbol{v}$. The second example moves to 2 by 3.

The third example (a 5 by 4 matrix A) will be *the incidence matrix of a graph.* Graphs have become the most important models in discrete mathematics—this example is worth understanding. All four subspaces have meaning on the graph.

Example 1 $A = \begin{bmatrix} 1 & 2 \\ 3 & 6 \end{bmatrix} = \boldsymbol{u}\boldsymbol{v}^{\mathrm{T}}$ has $m = 2$ and $n = 2$. We have subspaces of $\mathbf{R}^2$.

1 The column space $\mathbf{C}(A)$ is the line through $\boldsymbol{u} = \begin{bmatrix} 1 \\ 3 \end{bmatrix}$. Column 2 is on that line.

2 The row space $\mathbf{C}(A^{\mathrm{T}})$ is the line through $\boldsymbol{v} = \begin{bmatrix} 1 \\ 2 \end{bmatrix}$. Row 2 of A is on that line.

3 The nullspace $\mathbf{N}(A)$ is the line through $\boldsymbol{x} = \begin{bmatrix} 2 \\ -1 \end{bmatrix}$. Then $A\boldsymbol{x} = \mathbf{0}$.

4 The left nullspace $\mathbf{N}(A^{\mathrm{T}})$ is the line through $\boldsymbol{y} = \begin{bmatrix} 3 \\ -1 \end{bmatrix}$. Then $\boldsymbol{A}^{\mathrm{T}}\boldsymbol{y} = \mathbf{0}$.

I constructed those four subspaces in Figure I.1 from their definitions:

The **column space** $\mathbf{C}(A)$ contains all combinations of the columns of A
The **row space** $\mathbf{C}(A^{\mathrm{T}})$ contains all combinations of the columns of A^{T}
The **nullspace** $\mathbf{N}(A)$ contains all solutions $\boldsymbol{x}$ to $A\boldsymbol{x} = \mathbf{0}$
The **left nullspace** $\mathbf{N}(A^{\mathrm{T}})$ contains all solutions $\boldsymbol{y}$ to $A^{\mathrm{T}}\boldsymbol{y} = \mathbf{0}$

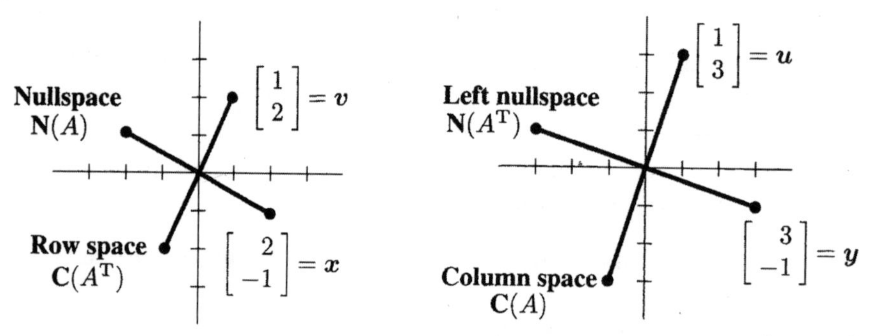

Figure I.1: The four fundamental subspaces (**4 infinite lines**) for $A = \begin{bmatrix} 1 & 2 \\ 3 & 6 \end{bmatrix}$.

That example had exactly one $\boldsymbol{u}$ and $\boldsymbol{v}$ and $\boldsymbol{x}$ and $\boldsymbol{y}$. All four subspaces were 1-dimensional (just lines). Always the $\boldsymbol{u}$'s and $\boldsymbol{v}$'s and $\boldsymbol{x}$'s and $\boldsymbol{y}$'s will be independent vectors—they give a **"basis"** for each of the subspaces. A larger matrix will need more than one basis vector per subspace. The choice of basis vectors is a crucial step in scientific computing.

Example 2 $B = \begin{bmatrix} 1 & -2 & -2 \\ 3 & -6 & -6 \end{bmatrix}$ has $m = 2$ and $n = 3$. Subspaces in $\mathbf{R}^3$ and $\mathbf{R}^2$.

Going from A to B, two subspaces change and two subspaces don't change. The column space of B is still in $\mathbf{R}^2$. It has the same basis vector. But now there are $n = 3$ numbers in the rows of B and the left half of Figure I.2 is in $\mathbf{R}^3$. There is still only one $\boldsymbol{v}$ in the row space ! The rank is still $r = 1$ because both rows of this B go in the same direction.

With $n = 3$ unknowns and only $r = 1$ independent equation, $B\boldsymbol{x} = \mathbf{0}$ will have $3 - 1 = 2$ independent solutions $\boldsymbol{x}_1$ and $\boldsymbol{x}_2$. All solutions go into the nullspace.

$$B\boldsymbol{x} = \begin{bmatrix} 1 & -2 & -2 \\ 3 & -6 & -6 \end{bmatrix} \begin{bmatrix} a \\ b \\ c \end{bmatrix} = \begin{bmatrix} 0 \\ 0 \end{bmatrix} \textbf{ has solutions } \boldsymbol{x}_1 = \begin{bmatrix} 2 \\ 1 \\ 0 \end{bmatrix} \textbf{ and } \boldsymbol{x}_2 = \begin{bmatrix} 2 \\ 0 \\ 1 \end{bmatrix}.$$

In the textbook *Introduction to Linear Algebra*, those vectors $\boldsymbol{x}_1$ and $\boldsymbol{x}_2$ are called "special solutions". They come from the steps of elimination—and you quickly see that $B\boldsymbol{x}_1 = \mathbf{0}$ and $B\boldsymbol{x}_2 = \mathbf{0}$. But those are not perfect choices in the nullspace of B because the vectors $\boldsymbol{x}_1$ and $\boldsymbol{x}_2$ **are not perpendicular**.

This book will give strong preference to perpendicular basis vectors. Section II.2 shows how to produce perpendicular vectors from independent vectors, by "Gram-Schmidt".

Our nullspace $\mathbf{N}(B)$ is a plane in $\mathbf{R}^3$. We can see an **orthonormal basis** $\boldsymbol{v}_2$ and $\boldsymbol{v}_3$ in that plane. The $\boldsymbol{v}_2$ and $\boldsymbol{v}_3$ axes make a 90° angle with each other and with $\boldsymbol{v}_1$.

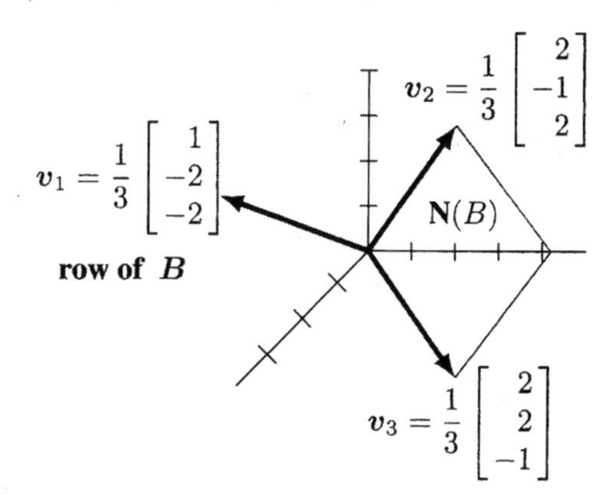

Row space = infinite line through $\boldsymbol{v}_1$
Nullspace = infinite plane of $\boldsymbol{v}_2$ and $\boldsymbol{v}_3$
$n = 3$ columns of B
$r = 1$ independent column

$$\begin{bmatrix} \boldsymbol{v}_1 & \boldsymbol{v}_2 & \boldsymbol{v}_3 \end{bmatrix} = \text{orthonormal basis for } \mathbf{R}^3$$

Figure I.2: Row space and nullspace of $B = \begin{bmatrix} \mathbf{1} & \mathbf{-2} & \mathbf{-2} \\ \mathbf{3} & \mathbf{-6} & \mathbf{-6} \end{bmatrix}$: Line perpendicular to plane !

Counting Law: r independent equations $Ax = 0$ have $n - r$ independent solutions

Example 3 from a graph Here is an example that has five equations (one for every edge in the graph). The equations have four unknowns (one for every node in the graph). The matrix in $Ax = b$ is the **5 by 4 incidence matrix** of the graph.

A has 1 and -1 on every row, to show the end node and the start node for each edge.

Differences $Ax = b$
across edges $1, 2, 3, 4, 5$
between nodes $1, 2, 3, 4$

$$\begin{array}{rrrrl} -x_1 & +x_2 & & & = b_1 \\ -x_1 & & +x_3 & & = b_2 \\ & -x_2 & +x_3 & & = b_3 \\ & -x_2 & & +x_4 & = b_4 \\ & & -x_3 & +x_4 & = b_5 \end{array}$$

When you understand the four fundamental subspaces for this incidence matrix (*the column spaces and the nullspaces for A and A^{T}*) you have captured a central idea of linear algebra.

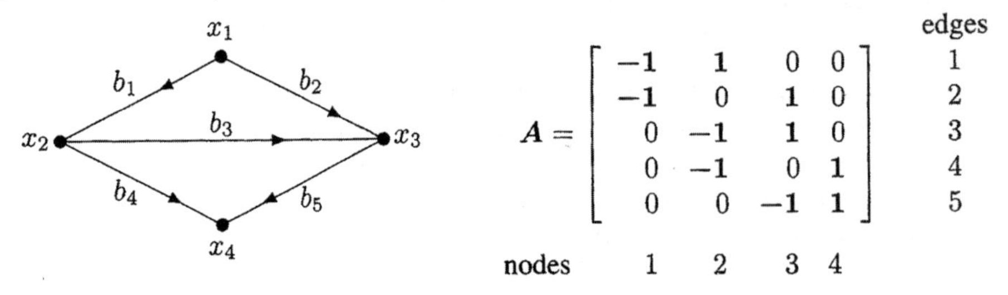

$$A = \begin{bmatrix} -1 & 1 & 0 & 0 \\ -1 & 0 & 1 & 0 \\ 0 & -1 & 1 & 0 \\ 0 & -1 & 0 & 1 \\ 0 & 0 & -1 & 1 \end{bmatrix} \begin{matrix} 1 \\ 2 \\ 3 \\ 4 \\ 5 \end{matrix}$$

nodes 1 2 3 4

This "graph" has 5 edges and 4 nodes. A is its 5 by 4 incidence matrix.

The nullspace $\mathbf{N}(A)$ To find the nullspace we set $b = 0$ in the 5 equations above. Then the first equation says $x_1 = x_2$. The second equation is $x_3 = x_1$. Equation 4 is $x_2 = x_4$. *All four unknowns x_1, x_2, x_3, x_4 have the same value c.* **The vector $x = (1, 1, 1, 1)$ and all vectors $x = (c, c, c, c)$ are the solutions to $Ax = 0$.**

That nullspace is a line in $\mathbf{R}^4$. The special solution $x = (1, 1, 1, 1)$ is a basis for $\mathbf{N}(A)$. The dimension of $\mathbf{N}(A)$ is 1 (one vector in the basis, a line has dimension 1). *The rank of A must be* 3, *since $n - r = 4 - 3 = 1$*. From the rank $r = 3$, we now know the dimensions of all four subspaces.

dimension of row space $= r = 3$ **dimension of column space $= r = 3$**
dimension of nullspace $= n - r = 1$ **dimension of nullspace of $A^{\mathrm{T}} = m - r = 2$**

The column space $\mathbf{C}(A)$ There must be $r = 4 - 1 = 3$ independent columns. The fast way is to look at the first 3 columns. They give a basis for the column space of A:

Columns 1, 2, 3 of this A are independent

$$\begin{matrix} -1 & 1 & 0 \\ -1 & 0 & 1 \\ 0 & -1 & 1 \\ 0 & -1 & 0 \\ 0 & 0 & -1 \end{matrix}$$

Column 4 is a combination of those three basic columns

"Independent" means that the only solution to $Ax = \mathbf{0}$ is $(x_1, x_2, x_3) = (0, 0, 0)$. We know $x_3 = 0$ from the fifth equation $0x_1 + 0x_2 - x_3 = 0$. We know $x_2 = 0$ from the fourth equation $0x_1 - x_2 + 0x_3 = 0$. Then we know $x_1 = 0$ from the first equation.

Column 4 of the incidence matrix A is the sum of those three columns, times -1.

The row space $\mathbf{C}(A^{\mathrm{T}})$ The dimension must again be $r = 3$, the same as for columns. But the first 3 rows of A are *not independent*: row 3 = row 2 − row 1. *The first three independent rows are rows* 1, 2, 4. Those rows are a basis (one possible basis) for the row space.

Edges 1, 2, 3 form a **loop** in the graph: Dependent rows 1, 2, 3.
Edges 1, 2, 4 form a **tree** in the graph: Independent rows 1, 2, 4.

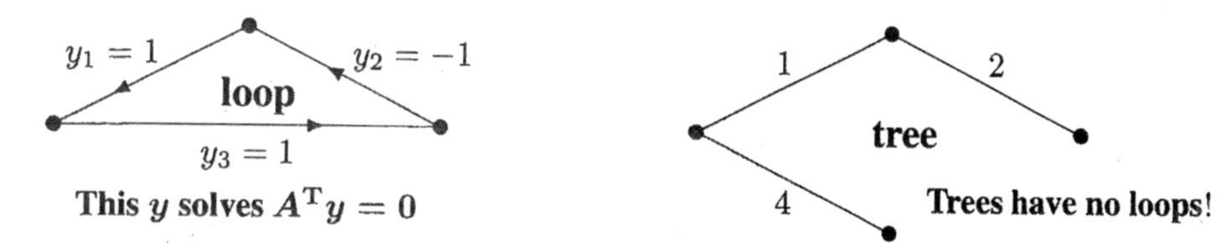

The left nullspace $\mathbf{N}(A^{\mathrm{T}})$ Now we solve $A^{\mathrm{T}}\boldsymbol{y} = \mathbf{0}$. Combinations of the rows give zero. We already noticed that row 3 = row 2 − row 1, so one solution is $\boldsymbol{y} = (1, -1, 1, 0, 0)$. I would say: this $\boldsymbol{y}$ comes from following the upper loop in the graph: forward on edges 1 and 3 and backward on edge 2.

Another $\boldsymbol{y}$ comes from going around the *lower loop* in the graph: forward on 4, back on 5 and 3. This $\boldsymbol{y} = (0, 0, -1, 1, -1)$ is an independent solution of $A^{\mathrm{T}}\boldsymbol{y} = \mathbf{0}$. The dimension of the left nullspace $\mathbf{N}(A^{\mathrm{T}})$ is $\boldsymbol{m} - \boldsymbol{r} = 5 - 3 = \mathbf{2}$. So those two $\boldsymbol{y}$'s are a basis for the left nullspace.

You may ask how "loops" and "trees" got into this problem. That didn't have to happen. We could have used elimination to solve $A^{\mathrm{T}}\boldsymbol{y} = \mathbf{0}$. The 4 by 5 matrix A^{T} would have three pivots. The nullspace of A^{T} has dimension two: $m - r = 5 - 3 = 2$. But *loops* and *trees* identify *dependent rows* and *independent rows* in a beautiful way.

The equations $A^{\mathrm{T}}\boldsymbol{y} = \mathbf{0}$ give "currents" y_1, y_2, y_3, y_4, y_5 on the five edges of the graph. Flows around loops obey **Kirchhoff's Current Law: in = out**. Those words apply to an electrical network. But the ideas behind the words apply all over engineering and science and economics and business. Balancing forces and flows and the budget.

Graphs are *the most important model in discrete applied mathematics*. You see graphs everywhere: roads, pipelines, blood flow, the brain, the Web, the economy of a country or the world. We can understand their incidence matrices A and A^{T}. In Section III.6, the matrix $A^{\mathrm{T}}A$ will be the "*graph Laplacian*". And Ohm's Law will lead to $A^{\mathrm{T}}CA$.

Four subspaces for a connected graph with m edges and n nodes : incidence matrix A

- $\mathbf{N}(A)$ — The constant vectors $(c, c, \ldots, c)$ make up the 1-dimensional nullspace of A.
- $\mathbf{C}(A^{\mathrm{T}})$ — The r edges of a tree give r independent rows of A : rank $= r = n - 1$.
- $\mathbf{C}(A)$ — ***Voltage Law***: The components of $A\boldsymbol{x}$ add to zero around all loops.
- $\mathbf{N}(A^{\mathrm{T}})$ — ***Current Law***: $A^{\mathrm{T}}\boldsymbol{y} =$ **(flow in)** $-$ **(flow out)** $= \mathbf{0}$ is solved by loop currents.
  There are $m - r = m - n + 1$ independent small loops in the graph.

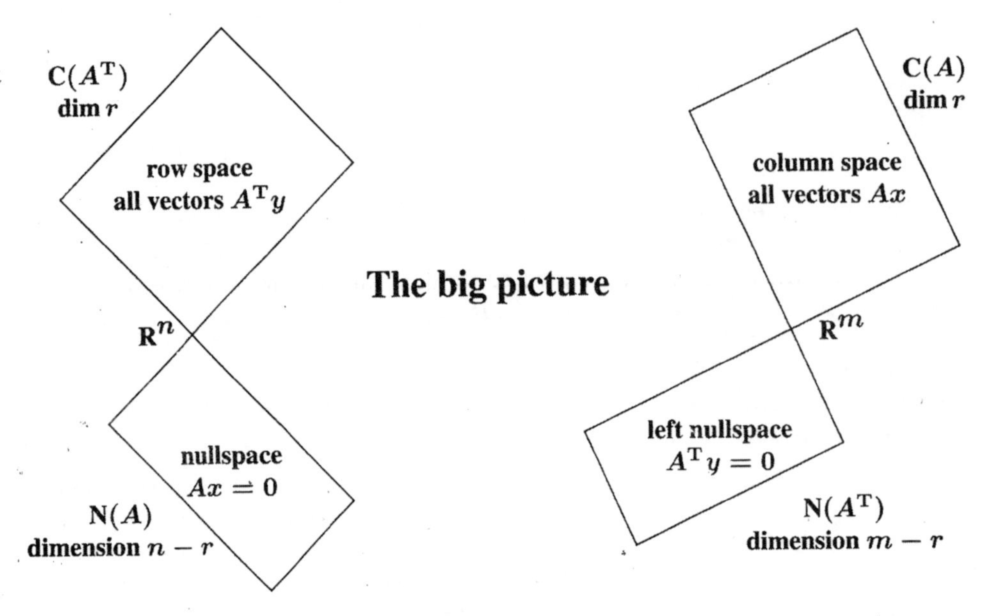

Figure I.3: The Four Fundamental Subspaces : Their dimensions add to n and m.

The Ranks of AB and $A + B$

This page establishes key facts about ranks: **When we multiply matrices, the rank cannot increase.** You will see this by looking at column spaces and row spaces. And there is one special situation when the rank cannot decrease. Then you know the rank of AB. Statement 4 will be important when data science factors a matrix into UV or CR.

Here are five key facts in one place: inequalities and equalities for the rank.

1 **Rank of $AB \leq$ rank of A** **Rank of $AB \leq$ rank of B**
2 **Rank of $A + B \leq$ (rank of A) + (rank of B)**
3 **Rank of $A^{\mathrm{T}}A$ = rank of AA^{T} = rank of A = rank of A^{T}**
4 **If A is m by r and B is r by n**—both with rank r—**then AB also has rank r**

Statement 1 involves the column space and row space of AB:

$$\mathbf{C}(AB) \text{ is contained in } \mathbf{C}(A) \qquad \mathbf{C}((AB)^{\mathrm{T}}) \text{ is contained in } \mathbf{C}(B^{\mathrm{T}})$$

Every column of AB is a combination of the columns of A (*matrix multiplication*)
Every row of AB is a combination of the rows of B (*matrix multiplication*)

Remember from Section I.1 that **row rank = column rank**. We can use rows or columns. *The rank cannot grow when we multiply AB.* Statement 1 in the box is frequently used.

Statement 2 Each column of $A + B$ is the sum of (column of A) + (column of B).

rank $(A + B) \leq$ rank (A) + rank (B) is always true. It combines bases for $\mathbf{C}(A)$ and $\mathbf{C}(B)$

rank $(A + B) =$ rank (A) + rank (B) is not always true. It is certainly false if $A = B = I$.

Statement 3 A and $A^{\mathrm{T}}A$ both have n columns. **They also have the same nullspace.** (This is Problem 6.) So $n - r$ is the same for both, and *the rank r is the same for both.* Then rank$(A^{\mathrm{T}}) \leq$ rank$(A^{\mathrm{T}}A) =$ rank(A). Exchange A and A^{T} to show their equal ranks.

Statement 4 We are told that A and B have rank r. By statement **3**, $A^{\mathrm{T}}A$ and BB^{T} have rank r. Those are r by r matrices so they are invertible. So is their product $A^{\mathrm{T}}ABB^{\mathrm{T}}$. Then

$$r = \textbf{rank of } (A^{\mathrm{T}}ABB^{\mathrm{T}}) \leq \textbf{ rank of } (AB) \text{ by Statement 1 : } A^{\mathrm{T}}, B^{\mathrm{T}} \text{can't increase rank}$$

We also know rank $(AB) \leq$ rank $A = r$. So we have proved that **AB has rank exactly r.**

Note This does not mean that every product of rank r matrices will have rank r. Statement 4 assumes that A has exactly r columns and B has r rows. BA can easily fail.

$$A = \begin{bmatrix} 1 \\ 1 \\ 1 \end{bmatrix} \qquad B = \begin{bmatrix} 1 & 2 & -3 \end{bmatrix} \qquad \textbf{AB has rank 1} \qquad \textbf{But BA is zero!}$$

Problem Set I.3

1 Show that the nullspace of AB contains the nullspace of B. **If $B\boldsymbol{x} = \mathbf{0}$ then...**

2 Find a square matrix with rank $(A^2) <$ rank (A). Confirm that rank $(A^{\mathrm{T}}A) =$ rank (A).

3 How is the nullspace of C related to the nullspaces of A and B, if $C = \begin{bmatrix} A \\ B \end{bmatrix}$?

4 If row space of $A =$ column space of A, and also $\mathbf{N}(A) = \mathbf{N}(A^{\mathrm{T}})$, is A symmetric?

5 Four possibilities for the rank r and size m, n match four possibilities for $A\boldsymbol{x} = \boldsymbol{b}$. Find four matrices A_1 to A_4 that show those possibilities:

$r = m = n$	$A_1\boldsymbol{x} = \boldsymbol{b}$ has **1 solution for every $\boldsymbol{b}$**
$r = m < n$	$A_2\boldsymbol{x} = \boldsymbol{b}$ has **1 or ∞ solutions**
$r = n < m$	$A_3\boldsymbol{x} = \boldsymbol{b}$ has **0 or 1 solution**
$r < m,\ r < n$	$A_4\boldsymbol{x} = \boldsymbol{b}$ has **0 or ∞ solutions**

6 (*Important*) Show that **$A^{\mathrm{T}}A$ has the same nullspace as A**. Here is one approach: First, if $A\boldsymbol{x}$ equals zero then $A^{\mathrm{T}}A\boldsymbol{x}$ equals _____. This proves $\mathbf{N}(A) \subset \mathbf{N}(A^{\mathrm{T}}A)$. Second, if $A^{\mathrm{T}}A\boldsymbol{x} = \mathbf{0}$ then $\boldsymbol{x}^{\mathrm{T}}A^{\mathrm{T}}A\boldsymbol{x} = ||A\boldsymbol{x}||^2 = 0$. Deduce $\mathbf{N}(A^{\mathrm{T}}A) = \mathbf{N}(A)$.

7 Do A^2 and A always have the same nullspace? A is a square matrix.

8 Find the column space $\mathbf{C}(A)$ and the nullspace $\mathbf{N}(A)$ of $A = \begin{bmatrix} 0 & 1 \\ 0 & 0 \end{bmatrix}$. Remember that those are vector spaces, not just single vectors. This is an unusual example with $\mathbf{C}(A) = \mathbf{N}(A)$. It could not happen that $\mathbf{C}(A) = \mathbf{N}(A^{\mathrm{T}})$ because those two subspaces are orthogonal.

9 Draw a square and connect its corners to the center point: 5 nodes and 8 edges. Find the 8 by 5 incidence matrix A of this graph (rank $r = 5 - 1 = 4$). Find a vector $\boldsymbol{x}$ in $\mathbf{N}(A)$ and $8 - 4$ independent vectors $\boldsymbol{y}$ in $\mathbf{N}(A^{\mathrm{T}})$.

10 If $\mathbf{N}(A)$ is the zero vector, what vectors are in the nullspace of $B = [A\ \ A\ \ A]$?

11 For subspaces $\mathbf{S}$ and $\mathbf{T}$ of $\mathbf{R}^{10}$ with dimensions 2 and 7, what are all the possible dimensions of

(i) $\mathbf{S} \cap \mathbf{T} =$ {all vectors that are in both subspaces}

(ii) $\mathbf{S} + \mathbf{T} =$ {all sums $s + t$ with s in $\mathbf{S}$ and t in $\mathbf{T}$}

(iii) $\mathbf{S}^{\perp} =$ {all vectors in $\mathbf{R}^{10}$ that are perpendicular to every vector in $\mathbf{S}$}.

I.4 Elimination and $A = LU$

The first and most fundamental problem of linear algebra is to solve $A\boldsymbol{x} = \boldsymbol{b}$. We are given the n by n matrix A and the n by 1 column vector $\boldsymbol{b}$. We look for the solution vector $\boldsymbol{x}$. Its components $x_1, x_2, \ldots, x_n$ are the n unknowns and we have n equations. Usually a square matrix A means only one solution to $A\boldsymbol{x} = \boldsymbol{b}$ (but not always). We can find $\boldsymbol{x}$ by geometry or by algebra.

This section begins with the row and column pictures of $A\boldsymbol{x} = \boldsymbol{b}$. Then we solve the equations by simplifying them—eliminate x_1 from $n-1$ equations to get a smaller system $A_2\boldsymbol{x}_2 = \boldsymbol{b}_2$ of size $n-1$. Eventually we reach the 1 by 1 system $A_n x_n = b_n$ and we know $x_n = b_n/A_n$. Working backwards produces x_{n-1} and eventually we know x_2 and x_1.

The point of this section is to see those elimination steps in terms of rank 1 matrices. **Every step (from A to A_2 and eventually to A_n) removes a matrix $\boldsymbol{\ell u}^*$.** Then the original A is the sum of those rank one matrices. This sum is exactly the great factorization $A = LU$ into lower and upper triangular matrices L and U—as we will see.

$A = L$ times U is the matrix description of elimination without row exchanges. That will be the algebra. Start with geometry for this 2 by 2 example.

$$\begin{array}{l}\textbf{2 equations and 2 unknowns}\\ \textbf{2 by 2 matrix in } A\boldsymbol{x}=\boldsymbol{b}\end{array} \qquad \begin{bmatrix} 1 & -2 \\ 2 & 3 \end{bmatrix}\begin{bmatrix} x \\ y \end{bmatrix} = \begin{bmatrix} 1 \\ 9 \end{bmatrix} \qquad \begin{array}{r} x - 2y = 1 \\ 2x + 3y = 9 \end{array} \tag{1}$$

Notice! I multiplied $A\boldsymbol{x}$ using inner products (dot products). Each row of the matrix A multiplied the vector $\boldsymbol{x}$. That produced the two equations for x and y, and the two straight lines in Figure I.4. They meet at the solution $x = 3, y = 1$. Here is the **row picture**.

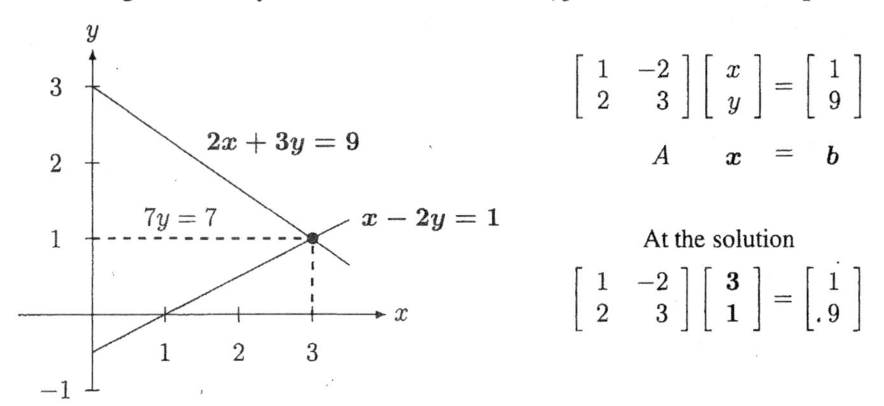

Figure I.4: **The row picture of $A\boldsymbol{x} = \boldsymbol{b}$**: Two lines meet at the solution $x = 3, y = 1$.

Figure I.4 also includes the horizontal line $7y = 7$. I subtracted 2 (equation 1) from (equation 2). The unknown x has been eliminated from $7y = 7$. This is the algebra:

$$\begin{bmatrix} 1 & -2 \\ 2 & 3 \end{bmatrix}\begin{bmatrix} x \\ y \end{bmatrix} = \begin{bmatrix} 1 \\ 9 \end{bmatrix} \quad \text{becomes} \quad \begin{bmatrix} 1 & -2 \\ 0 & 7 \end{bmatrix}\begin{bmatrix} x \\ y \end{bmatrix} = \begin{bmatrix} 1 \\ 7 \end{bmatrix} \qquad \begin{array}{l} x = 3 \\ y = 1 \end{array}$$

Column picture **One vector equation** instead of two scalar equations. We are looking for a combination of the columns of A to match $\boldsymbol{b}$. Figure I.5 shows that the right combination (the solution $\boldsymbol{x}$) has the same $x = 3$ and $y = 1$ that we found in the row picture.

$A\boldsymbol{x}$ is a combination of columns
The columns combine to give b

$$\begin{bmatrix} 1 & -2 \\ 2 & 3 \end{bmatrix} \begin{bmatrix} x \\ y \end{bmatrix} = x \begin{bmatrix} 1 \\ 2 \end{bmatrix} + y \begin{bmatrix} -2 \\ 3 \end{bmatrix} = \begin{bmatrix} 1 \\ 9 \end{bmatrix} \quad (2)$$

Adding **3 (column 1)** to **1 (column 2)** gives $\boldsymbol{b}$ as a combination of the columns.

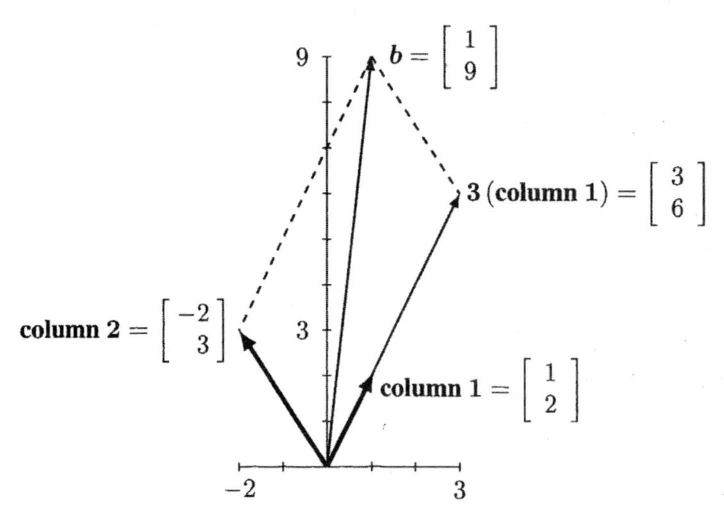

Figure I.5: **The column picture**: 3 times (column 1) + 1 times (column 2) gives $\boldsymbol{b}$.

For $n = 2$, the row picture looked easy. But for $n \geq 3$, the column picture wins. Better to draw three column vectors than three planes! Three equations for $\boldsymbol{x} = (x, y, z)$.

Row picture in 3D	Three planes meet at one point. A plane for each equation.
Column picture in 3D	Three column vectors combine to give the vector $\boldsymbol{b}$.

Solving $A\boldsymbol{x} = \boldsymbol{b}$ by Elimination

To visualize three planes meeting in $\mathbf{R}^3$ is not easy. And n "hyperplanes" meeting at a point in $\mathbf{R}^n$ is truly mind-bending. A combination of the column vectors is simpler: The matrix A must have 3 (or n) **independent** columns. *The columns must not all lie in the same plane in* $\mathbf{R}^3$ (or the same hyperplane in $\mathbf{R}^n$). This translates to a statement in algebra:

Independent columns **The only solution to $A\boldsymbol{x} = \mathbf{0}$ is the zero vector $\boldsymbol{x} = \mathbf{0}$.**

In words, **independence** means that the only combination that adds to the zero vector has zero times every column. Then the only solution to $A\boldsymbol{x} = \mathbf{0}$ is $\boldsymbol{x} = \mathbf{0}$. When that is true, elimination will solve $A\boldsymbol{x} = \boldsymbol{b}$ to find the only combination of columns that produces $\boldsymbol{b}$.

Here is the whole idea, column by column, when elimination succeeds in the usual order:

Column 1. ***Use equation 1 to create zeros below the first pivot. Pivots can't be zero!***

Column 2. ***Use the new equation 2 to create zeros below the second pivot.***

Columns 3 to n. ***Keep going to find the upper triangular U : n pivots on its diagonal.***

$$\textbf{Step 1}\begin{bmatrix} x & x & x & x \\ 0 & x & x & x \\ 0 & x & x & x \\ 0 & x & x & x \end{bmatrix}. \quad \textbf{Step 2}\begin{bmatrix} x & x & x & x \\ 0 & x & x & x \\ 0 & 0 & x & x \\ 0 & 0 & x & x \end{bmatrix}. \quad \textbf{Final } U = \begin{bmatrix} x & x & x & x \\ & x & x & x \\ & & x & x \\ & & & x \end{bmatrix}.$$

Row 1 is the first pivot row—it doesn't change. I multiplied that row by numbers $\ell_{21}, \ell_{31}, \ell_{41}$ and subtracted from rows $2, 3, 4$ of A. The numbers to get zeros in the first column were

$$\textbf{Multipliers} \qquad \ell_{21} = \frac{a_{21}}{a_{11}} \qquad \ell_{31} = \frac{a_{31}}{a_{11}} \qquad \ell_{41} = \frac{a_{41}}{a_{11}}$$

If the corner entry is $a_{11} = 3 =$ first pivot, and a_{21} below it is 12, then $\ell_{21} = 12/3 = 4$.

Step 2 uses the new row 2 (the second pivot row). Multiply that row by ℓ_{32} and ℓ_{42}. Subtract from rows 3 and 4 to get zeros in the second column. Continue all the way to U.

So far we have worked on the matrix A (not on $\boldsymbol{b}$). Elimination on A needs $\frac{1}{3}n^3$ separate multiplications and additions—far more than the n^2 steps for each right hand side $\boldsymbol{b}$. We need a record of that work, and the perfect format is a product $\boldsymbol{A = LU}$ of triangular matrices: **lower triangular $\boldsymbol{L}$ times upper triangular $\boldsymbol{U}$.**

The Factorization $A = LU$

How is the original A related to the final matrix U? The multipliers ℓ_{ij} got us there in three steps. The first step reduced the 4 by 4 problem to a 3 by 3 problem, *by removing multiples of row* 1:

$$\begin{matrix}\textbf{Key idea: Step 1} \\ \textbf{removes } \boldsymbol{\ell_1 u_1^*}\end{matrix} \qquad A = \begin{bmatrix} 1 \text{ times row } 1 \\ \ell_{21} \text{ times row } 1 \\ \ell_{31} \text{ times row } 1 \\ \ell_{41} \text{ times row } 1 \end{bmatrix} + \begin{bmatrix} 0 & 0 \quad 0 \quad 0 \\ 0 & \\ 0 & \boxed{\quad A_2 \quad} \\ 0 & \end{bmatrix}. \qquad (3)$$

What have we done? The first matrix on the right was removed from A. That removed matrix is a column vector $1, \ell_{21}, \ell_{31}, \ell_{41}$ times row 1. **It is the rank 1 matrix $\boldsymbol{\ell_1 u_1^*}$!**

$$\begin{matrix}\textbf{3 by 3 example} \\ \textbf{Remove rank 1 matrix} \\ \textbf{Column / row to zero}\end{matrix} \quad \begin{bmatrix} 1 & 2 & 3 \\ 2 & 5 & 7 \\ 2 & 7 & 8 \end{bmatrix} - \begin{bmatrix} 1 & 2 & 3 \\ 2 & 4 & 6 \\ 2 & 4 & 6 \end{bmatrix} = \begin{bmatrix} 0 & 0 & 0 \\ 0 & \mathbf{1} & \mathbf{1} \\ 0 & \mathbf{3} & \mathbf{2} \end{bmatrix} = \begin{bmatrix} 0 & 0 \quad 0 \\ 0 & \\ 0 & A_2 \end{bmatrix}$$

The next step deals with column 2 of the remaining matrix A_2. The new row 2 is $\boldsymbol{u}_2^* =$ second pivot row. We multiply it by $\ell_{12} = 0$ and $\ell_{22} = 1$ and ℓ_{32} and ℓ_{42}. **Then subtract $\boldsymbol{\ell}_2\boldsymbol{u}_2^*$ from the four rows.** Now row 2 is also zero and A_2 shrinks down to A_3.

$$\textbf{Step 2} \qquad A = \boldsymbol{\ell}_1\boldsymbol{u}_1^* + \begin{bmatrix} \mathbf{0} \text{ times pivot row 2} \\ \mathbf{1} \text{ times pivot row 2} \\ \boldsymbol{\ell_{32}} \text{ times pivot row 2} \\ \boldsymbol{\ell_{42}} \text{ times pivot row 2} \end{bmatrix} + \begin{bmatrix} 0 & 0 & 0 & 0 \\ 0 & 0 & 0 & 0 \\ 0 & 0 & \multicolumn{2}{c}{\boxed{\boldsymbol{A_3}}} \\ 0 & 0 & & \end{bmatrix}. \tag{4}$$

That step was a rank one removal of $\boldsymbol{\ell}_2\boldsymbol{u}_2^*$ with $\boldsymbol{\ell}_2 = (0, 1, \ell_{32}, \ell_{42})$ and $\boldsymbol{u}_2^* =$ pivot row 2. Step 3 will reduce the 2 by 2 matrix A_3 to a single number A_4 (1 by 1). At this point the pivot row $\boldsymbol{u}_3^* =$ row 1 of A_3 has only two nonzeros. And the column $\boldsymbol{\ell}_3$ is $(0, 0, 1, \ell_{43})$.

This way of looking at elimination, a column at a time, directly produces $\boldsymbol{A = LU}$. That matrix multiplication LU is always a sum of columns of L times rows of U:

$$\boldsymbol{A} = \boldsymbol{\ell}_1\boldsymbol{u}_1^* + \boldsymbol{\ell}_2\boldsymbol{u}_2^* + \boldsymbol{\ell}_3\boldsymbol{u}_3^* + \boldsymbol{\ell}_4\boldsymbol{u}_4^* = \begin{bmatrix} 1 & 0 & 0 & 0 \\ \ell_{21} & 1 & 0 & 0 \\ \ell_{31} & \ell_{32} & 1 & 0 \\ \ell_{41} & \ell_{42} & \ell_{43} & 1 \end{bmatrix} \begin{bmatrix} \text{pivot row 1} \\ \text{pivot row 2} \\ \text{pivot row 3} \\ \text{pivot row 4} \end{bmatrix} = \boldsymbol{LU}. \tag{5}$$

Elimination factored $A = LU$ into a lower triangular L times an upper triangular U

Notes on the LU factorization We developed $A = LU$ from the key idea of elimination: Reduce the problem size from n to $n - 1$ by eliminating x_1 from the last $n - 1$ equations. We subtracted multiples of row 1 (the pivot row). So the matrix we removed had *rank one*. After n steps, the whole matrix A is a sum of n rank one matrices. That sum—by the column times row rule for matrix multiplication—is exactly L times U.

This proof is not in my textbook *Introduction to Linear Algebra*. The idea there was to look at rows of U instead of working with columns of A. Row 3 came from subtracting multiples of pivot rows 1 and 2 from row 3 of A:

$$\text{Row 3 of } U = (\text{row 3 of } A) - \ell_{31}\,(\text{row 1 of } U) - \ell_{32}\,(\text{row 2 of } U). \tag{6}$$

Rewrite this equation to see that the row $[\ell_{31} \;\; \ell_{32} \;\; 1]$ of L is multiplying the matrix U:

$$\text{Row 3 of } A = \ell_{31}\,(\text{row 1 of } U) + \ell_{32}\,(\text{row 2 of } U) + 1\,(\text{row 3 of } U). \tag{7}$$

This is row 3 of $A = LU$. The key is that the subtracted rows were pivot rows, and already in U. **With no row exchanges, we have again found $A = LU$.**

The Solution to $Ax = b$

We must apply the same operations to the right side of an equation and to the left side. The direct way is to include b as an additional column—we work with the matrix $[A \ \ b]$. Now our elimination steps on A (they multiplied A by L^{-1} to give U) act also on b:

$$\textbf{Start from} \begin{bmatrix} A & b \end{bmatrix} = \begin{bmatrix} LU & b \end{bmatrix} \quad \textbf{Elimination produces} \begin{bmatrix} U & L^{-1}b \end{bmatrix} = \begin{bmatrix} U & c \end{bmatrix}.$$

The steps from A to U (upper triangular) will change the right side b to c. **Elimination on $Ax = b$ produces the equations $Ux = c$ that are ready for back substitution.**

$$\begin{matrix} 2x + 3y = \ 8 \\ 4x + 7y = 18 \end{matrix} \rightarrow \begin{bmatrix} 2 & 3 & 8 \\ 4 & 7 & 18 \end{bmatrix} \rightarrow \begin{bmatrix} \mathbf{2} & \mathbf{3} & \mathbf{8} \\ 0 & \mathbf{1} & \mathbf{2} \end{bmatrix} = \begin{bmatrix} U & c \end{bmatrix}. \quad (8)$$

L subtracted 2 times row 1 from row 2. Then the triangular system $Ux = c$ is solved upwards—**back substitution**—from bottom to top:

$$\begin{matrix} 2x + 3y = 8 \\ 1y = 2 \end{matrix} \quad \text{gives} \quad y = \mathbf{2} \quad \text{and then} \quad x = \mathbf{1}. \quad Ux = c \quad \text{gives} \quad x = U^{-1}c.$$

Looking closely, the square system $Ax = b$ became two triangular systems:

$Ax = b$ **split into** $Lc = b$ **and** $Ux = c$. **Elimination gave** c **and back substitution gave** x.

The final result is $x = U^{-1}c = U^{-1}L^{-1}b = A^{-1}b$. The correct solution has been found.

Please notice Those steps required nonzero pivots. We divided by those numbers. The first pivot was a_{11}. The second pivot was in the corner of A_2, and the nth pivot was in the 1 by 1 matrix A_n. These numbers ended up on the main diagonal of U.

What do we do if $a_{11} = 0$? Zero cannot be the first pivot. If there is a nonzero number lower down in column 1, its row can be the pivot row. **Good codes will choose the largest number to be the pivot.** They do this to reduce errors, even if a_{11} is not zero.

We look next at the effect of those row exchanges on $A = LU$. A matrix P will enter.

Row Exchanges (Permutations)

Here the largest number in column 1 is found in row 3: $a_{31} = 2$. **Row 3 will be the first pivot row u_1^*.** That row is multiplied by $\ell_{21} = \frac{1}{2}$ and subtracted from row 2.

$$\begin{matrix} u_1^* = \textbf{ row 3 of } A \\ = \textbf{first pivot row} \end{matrix} \qquad A = \begin{bmatrix} 0 & 1 & 1 \\ 1 & 3 & 7 \\ \mathbf{2} & \mathbf{4} & \mathbf{8} \end{bmatrix} \rightarrow \begin{bmatrix} 0 & 1 & 1 \\ 0 & 1 & 3 \\ \mathbf{2} & \mathbf{4} & \mathbf{8} \end{bmatrix} \quad (9)$$

Again that elimination step removed a rank one matrix $\boldsymbol{\ell}_1\boldsymbol{u}_1^*$. But A_2 is in a new place.

$$\begin{bmatrix} 0 & 1 & 1 \\ 1 & 3 & 7 \\ 2 & 4 & 8 \end{bmatrix} = \begin{bmatrix} 0 \\ 1/2 \\ 1 \end{bmatrix} \begin{bmatrix} 2 & 4 & 8 \end{bmatrix} + \begin{bmatrix} 0 & \boxed{\begin{matrix} 1 & 1 \\ 1 & 3 \end{matrix}} \\ 0 & \\ 0 & 0 \quad 0 \end{bmatrix} \leftarrow \boldsymbol{A_2} \qquad (10)$$

Elimination on A_2 produces two more rank one pieces. Then $A = LU$ has three pieces:

$$\boldsymbol{\ell}_1\boldsymbol{u}_1^* + \begin{bmatrix} 1 \\ 1 \\ 0 \end{bmatrix} \begin{bmatrix} 0 & 1 & 1 \end{bmatrix} + \begin{bmatrix} 0 \\ 1 \\ 0 \end{bmatrix} \begin{bmatrix} 0 & 0 & 2 \end{bmatrix} = \begin{bmatrix} 0 & 1 & 0 \\ 1/2 & 1 & 1 \\ 1 & 0 & 0 \end{bmatrix} \begin{bmatrix} 2 & 4 & 8 \\ 0 & 1 & 1 \\ 0 & 0 & 2 \end{bmatrix}. \qquad (11)$$

That last matrix U is triangular but the L matrix is not! The pivot order for this A was **3,1,2**. If we want the pivot rows to be $\mathbf{1, 2, 3}$ we must move row 3 of A to the top:

Row exchange by a permutation P

$$\boldsymbol{PA} = \begin{bmatrix} 0 & 0 & \mathbf{1} \\ \mathbf{1} & 0 & 0 \\ 0 & \mathbf{1} & 0 \end{bmatrix} \begin{bmatrix} 0 & 1 & 1 \\ 1 & 3 & 7 \\ \mathbf{2} & \mathbf{4} & \mathbf{8} \end{bmatrix} = \begin{bmatrix} \mathbf{2} & \mathbf{4} & \mathbf{8} \\ 0 & 1 & 1 \\ 1 & 3 & 7 \end{bmatrix}$$

When both sides of $A\boldsymbol{x} = \boldsymbol{b}$ are multiplied by P, **order is restored and $\boldsymbol{PA = LU}$:**

$$\boldsymbol{PA} = \begin{bmatrix} 2 & 4 & 8 \\ 0 & 1 & 1 \\ 1 & 3 & 7 \end{bmatrix} = \begin{bmatrix} 1 & 0 & 0 \\ 0 & 1 & 0 \\ 1/2 & 1 & 1 \end{bmatrix} \begin{bmatrix} 2 & 4 & 8 \\ 0 & 1 & 1 \\ 0 & 0 & 2 \end{bmatrix} = \boldsymbol{LU}. \qquad (12)$$

Every invertible n by n matrix A leads to $PA = LU$: P = permutation.

There are six 3 by 3 permutations: Six ways to order the rows of the identity matrix.

1 exchange (**odd P**)

$$\boldsymbol{P_{213}} = \begin{bmatrix} 0 & 1 & 0 \\ 1 & 0 & 0 \\ 0 & 0 & 1 \end{bmatrix} \quad \boldsymbol{P_{321}} = \begin{bmatrix} 0 & 0 & 1 \\ 0 & 1 & 0 \\ 1 & 0 & 0 \end{bmatrix} \quad \boldsymbol{P_{132}} = \begin{bmatrix} 1 & 0 & 0 \\ 0 & 0 & 1 \\ 0 & 1 & 0 \end{bmatrix}$$

0 or 2 exchanges (**even P**)

$$\boldsymbol{P_{123}} = \begin{bmatrix} 1 & & \\ & 1 & \\ & & 1 \end{bmatrix} \quad \boldsymbol{P_{312}} = \begin{bmatrix} 0 & 0 & 1 \\ 1 & 0 & 0 \\ 0 & 1 & 0 \end{bmatrix} \quad \boldsymbol{P_{231}} = \begin{bmatrix} 0 & 1 & 0 \\ 0 & 0 & 1 \\ 1 & 0 & 0 \end{bmatrix}$$

The inverse of every permutation matrix P is its transpose P^{T}. The row exchanges will also apply to the right hand side $\boldsymbol{b}$ if we are solving $A\boldsymbol{x} = \boldsymbol{b}$. The computer just remembers the exchanges without actually moving the rows.

There are $\boldsymbol{n!}$ (n factorial) permutation matrices of size n: $3! = (3)(2)(1) = 6$. When A has dependent rows (no inverse) elimination leads to a zero row and stops short.

Problem Set I.4

1 Factor these matrices into $A = LU$:

$$A = \begin{bmatrix} 2 & 1 \\ 6 & 7 \end{bmatrix} \qquad A = \begin{bmatrix} 1 & 1 & 1 \\ 1 & 1 & 1 \\ 1 & 1 & 1 \end{bmatrix} \qquad A = \begin{bmatrix} 2 & -1 & 0 \\ -1 & 2 & -1 \\ 0 & -1 & 2 \end{bmatrix}$$

2 If $a_{11}, \ldots, a_{1n}$ is the first row of a rank-1 matrix A and $a_{11}, \ldots, a_{m1}$ is the first column, *find a formula for* a_{ij}. Good to check when $a_{11} = 2, a_{12} = 3, a_{21} = 4$. When will your formula break down? Then rank 1 is impossible or not unique.

3 What lower triangular matrix E puts A into upper triangular form $EA = U$? Multiply by $E^{-1} = L$ to factor A into LU:

$$A = \begin{bmatrix} 2 & 1 & 0 \\ 0 & 4 & 2 \\ 6 & 3 & 5 \end{bmatrix}$$

4 **This problem shows how the one-step inverses multiply to give L.** You see this best when $A = L$ is already lower triangular with 1's on the diagonal. **Then $U = I$:**

$$\text{Multiply } A = \begin{bmatrix} 1 & 0 & 0 \\ a & 1 & 0 \\ b & c & 1 \end{bmatrix} \text{ by } E_1 = \begin{bmatrix} 1 & & \\ -a & 1 & \\ -b & 0 & 1 \end{bmatrix} \text{ and then } E_2 = \begin{bmatrix} 1 & 0 & 0 \\ 0 & 1 & 0 \\ 0 & -c & 1 \end{bmatrix}.$$

(a) Multiply $E_2 E_1$ to find the single matrix E that produces $EA = I$.

(b) Multiply $E_1^{-1} E_2^{-1}$ to find the matrix $A = L$.

The multipliers a, b, c are mixed up in $E = L^{-1}$ but they are perfect in L.

5 *When zero appears in a pivot position, $A = LU$ is not possible!* (We are requiring nonzero pivots in U.) Show directly why these LU equations are both impossible:

$$\begin{bmatrix} 0 & 1 \\ 2 & 3 \end{bmatrix} = \begin{bmatrix} 1 & 0 \\ \ell & 1 \end{bmatrix} \begin{bmatrix} d & e \\ 0 & f \end{bmatrix} \qquad \begin{bmatrix} 1 & 1 & 0 \\ 1 & 1 & 2 \\ 1 & 2 & 1 \end{bmatrix} = \begin{bmatrix} 1 & & \\ \ell & 1 & \\ m & n & 1 \end{bmatrix} \begin{bmatrix} d & e & g \\ & f & h \\ & & i \end{bmatrix}.$$

These matrices need a row exchange by a *permutation matrix* P.

6 Which number c leads to zero in the second pivot position? A row exchange is needed and $A = LU$ will not be possible. Which c produces zero in the third pivot position? Then a row exchange can't help and elimination fails:

$$A = \begin{bmatrix} 1 & c & 0 \\ 2 & 4 & 1 \\ 3 & 5 & 1 \end{bmatrix}.$$

7 (*Recommended*) Compute L and U for this symmetric matrix A:

$$A = \begin{bmatrix} a & a & a & a \\ a & b & b & b \\ a & b & c & c \\ a & b & c & d \end{bmatrix}.$$

Find four conditions on a, b, c, d to get $A = LU$ with four nonzero pivots.

8 *Tridiagonal matrices* have zero entries except on the main diagonal and the two adjacent diagonals. Factor these into $A = LU$. Symmetry further produces $A = LDL^{\mathrm{T}}$:

$$A = \begin{bmatrix} 1 & 1 & 0 \\ 1 & 2 & 1 \\ 0 & 1 & 2 \end{bmatrix} \quad \text{and} \quad A = \begin{bmatrix} a & a & 0 \\ a & a+b & b \\ 0 & b & b+c \end{bmatrix}.$$

9 *Easy but important.* If A has pivots $5, 9, 3$ with no row exchanges, what are the pivots for the upper left 2 by 2 submatrix A_2 (without row 3 and column 3)?

10 Which invertible matrices allow $A = LU$ (elimination without row exchanges)? *Good question* ! Look at each of the square upper left submatrices $A_1, A_2, \ldots, A_n$.

All upper left submatrices A_k must be invertible : sizes 1 by 1, 2 by $2, \ldots, n$ by n.

Explain that answer: A_k factors into _____ because $LU = \begin{bmatrix} L_k & 0 \\ * & * \end{bmatrix} \begin{bmatrix} U_k & * \\ 0 & * \end{bmatrix}$.

11 In some data science applications, the first pivot is the *largest number* $|a_{ij}|$ *in* A. Then row i becomes the first pivot row $\boldsymbol{u}_1^*$. Column j is the first pivot column. Divide that column by a_{ij} so $\boldsymbol{\ell}_1$ has 1 in row i. Then remove that $\boldsymbol{\ell}_1 \boldsymbol{u}_1^*$ from A.

This example finds $a_{22} = 4$ as the first pivot $(i = j = 2)$. Dividing by 4 gives $\boldsymbol{\ell}_1$:

$$\begin{bmatrix} 1 & 2 \\ 3 & 4 \end{bmatrix} = \begin{bmatrix} 1/2 \\ 1 \end{bmatrix} \begin{bmatrix} 3 & 4 \end{bmatrix} + \begin{bmatrix} -1/2 & 0 \\ 0 & 0 \end{bmatrix} = \boldsymbol{\ell}_1 \boldsymbol{u}_1^* + \boldsymbol{\ell}_2 \boldsymbol{u}_2^* = \begin{bmatrix} 1/2 & 1 \\ 1 & 0 \end{bmatrix} \begin{bmatrix} 3 & 4 \\ -1/2 & 0 \end{bmatrix}$$

For this A, both L and U involve permutations. P_1 exchanges the rows to give L. P_2 exchanges the columns to give an upper triangular U. Then $\boldsymbol{P_1 A P_2 = LU}$.

$$\textbf{Permuted in advance} \qquad \boldsymbol{P_1 A P_2} = \begin{bmatrix} 1 & 0 \\ 1/2 & 1 \end{bmatrix} \begin{bmatrix} 4 & 3 \\ 0 & -1/2 \end{bmatrix} = \begin{bmatrix} \mathbf{4} & \mathbf{3} \\ \mathbf{2} & \mathbf{1} \end{bmatrix}$$

Question for $A = \begin{bmatrix} \mathbf{1} & \mathbf{3} \\ \mathbf{2} & \mathbf{4} \end{bmatrix}$: Apply complete pivoting to produce $P_1 A P_2 = LU$.

12 If the short wide matrix A has $m < n$, how does elimination show that there are nonzero solutions to $A\boldsymbol{x} = \mathbf{0}$? What do we know about the dimension of that "nullspace of A" containing all solution vectors $\boldsymbol{x}$? The nullspace dimension is at least _____ .

Suggestion: First create a specific 2 by 3 matrix A and ask those questions about A.

I.5 Orthogonal Matrices and Subspaces

The word **orthogonal** appears everywhere in linear algebra. It means *perpendicular.* Its use extends far beyond the angle between two vectors. Here are important extensions of that key idea:

1. **Orthogonal vectors x and y.** The test is $x^{\mathrm{T}}y = x_1y_1 + \cdots + x_ny_n = 0$.
   If x and y have complex components, change to $\overline{x}^{\mathrm{T}}y = \overline{x}_1y_1 + \cdots + \overline{x}_ny_n = 0$.

2. **Orthogonal basis for a subspace:** Every pair of basis vectors has $v_i^{\mathrm{T}}v_j = 0$.
   Orthonormal basis: Orthogonal basis of **unit vectors**: every $v_i^{\mathrm{T}}v_i = 1$ (length 1).
   From orthogonal to orthonormal, just divide every basis vector v_i by its length $||v_i||$.

3. **Orthogonal subspaces R and N.** Every vector in the space **R** is orthogonal to every vector in **N**. Notice again! The row space and nullspace are orthogonal:

$$\begin{matrix} Ax = 0 \text{ means} \\ \textbf{each row} \cdot x = 0 \end{matrix} \qquad \begin{bmatrix} \textbf{row 1 of } A \\ \vdots \\ \textbf{row } m \textbf{ of } A \end{bmatrix} \begin{bmatrix} \\ x \\ \\ \end{bmatrix} = \begin{bmatrix} 0 \\ \vdots \\ 0 \end{bmatrix}. \qquad (1)$$

   Every row (and every combination of rows) is orthogonal to all x in the nullspace.

4. **Tall thin matrices Q with orthonormal columns:** $Q^{\mathrm{T}}Q = I$.

$$Q^{\mathrm{T}}Q = \begin{bmatrix} \text{—} & q_1^{\mathrm{T}} & \text{—} \\ & \vdots & \\ \text{—} & q_n^{\mathrm{T}} & \text{—} \end{bmatrix} \begin{bmatrix} \\ q_1 \cdots q_n \\ \\ \end{bmatrix} = \begin{bmatrix} 1 & 0 & 0 \\ 0 & 1 & 0 \\ 0 & 0 & 1 \end{bmatrix} = I \qquad (2)$$

   If this Q multiplies any vector x, the length of the vector does not change:

$$||Qx|| = ||x|| \text{ because } (Qx)^{\mathrm{T}}(Qx) = x^{\mathrm{T}}Q^{\mathrm{T}}Q\,x = x^{\mathrm{T}}x \qquad (3)$$

   If $m > n$ the m rows cannot be orthogonal in $\mathbf{R}^n$. Tall thin matrices have $Q\,Q^{\mathrm{T}} \neq I$.

5. **"Orthogonal matrices" are square with orthonormal columns:** $Q^{\mathrm{T}} = Q^{-1}$.
   For square matrices $Q^{\mathrm{T}}Q = I$ leads to $QQ^{\mathrm{T}} = I$
   For square matrices Q, the left inverse Q^{T} is also a right inverse of Q.
   The columns of this orthogonal n by n matrix are an orthonormal basis for $\mathbf{R}^n$.
   The rows of Q are a (probably different) orthonormal basis for $\mathbf{R}^n$.
   The name "orthogonal matrix" should really be "orthonormal matrix".

The next pages give examples of orthogonal **vectors, bases, subspaces** and **matrices.**

1. Orthogonal vectors. The test $\boldsymbol{x}^{\mathrm{T}}\boldsymbol{y} = 0$ connects to right triangles by $c^2 = a^2 + b^2$:

$$\textbf{Pythagoras Law for right triangles} \quad ||\boldsymbol{x} - \boldsymbol{y}||^2 = ||\boldsymbol{x}||^2 + ||\boldsymbol{y}||^2. \tag{4}$$

The left side is $(\boldsymbol{x} - \boldsymbol{y})^{\mathrm{T}}(\boldsymbol{x} - \boldsymbol{y})$. This expands to $\boldsymbol{x}^{\mathrm{T}}\boldsymbol{x} + \boldsymbol{y}^{\mathrm{T}}\boldsymbol{y} - \boldsymbol{x}^{\mathrm{T}}\boldsymbol{y} - \boldsymbol{y}^{\mathrm{T}}\boldsymbol{x}$. When the last two terms are zero, we have equation (4): $\boldsymbol{x} = (1, 2, 2)$ and $\boldsymbol{y} = (2, 1, -2)$ have $\boldsymbol{x}^{\mathrm{T}}\boldsymbol{y} = 0$. The hypotenuse is $\boldsymbol{x} - \boldsymbol{y} = (-1, 1, 4)$. Then Pythagoras is $\mathbf{18} = \mathbf{9} + \mathbf{9}$.

Dot products $\boldsymbol{x}^{\mathrm{T}}\boldsymbol{y}$ and $\boldsymbol{y}^{\mathrm{T}}\boldsymbol{x}$ always equal $||\boldsymbol{x}||\,||\boldsymbol{y}||\cos\theta$, where θ is the angle between $\boldsymbol{x}$ and $\boldsymbol{y}$. So in all cases we have the Law of Cosines $c^2 = a^2 + b^2 - 2ab\cos\theta$:

$$\textbf{Law of Cosines} \quad ||\boldsymbol{x} - \boldsymbol{y}||^2 = ||\boldsymbol{x}||^2 + ||\boldsymbol{y}||^2 - 2\,||\boldsymbol{x}||\,||\boldsymbol{y}||\cos\theta. \tag{5}$$

Orthogonal vectors have $\cos\theta = 0$ and that last term disappears.

2. Orthogonal basis. The "standard basis" is orthogonal (even orthonormal) in $\mathbf{R}^n$:

$$\text{Standard basis } \boldsymbol{i}, \boldsymbol{j}, \boldsymbol{k} \text{ in } \mathbf{R}^3 \qquad \boldsymbol{i} = \begin{bmatrix} 1 \\ 0 \\ 0 \end{bmatrix} \quad \boldsymbol{j} = \begin{bmatrix} 0 \\ 1 \\ 0 \end{bmatrix} \quad \boldsymbol{k} = \begin{bmatrix} 0 \\ 0 \\ 1 \end{bmatrix}.$$

Here are three Hadamard matrices H_2, H_4, H_8 containing orthogonal bases of $\mathbf{R}^2, \mathbf{R}^4, \mathbf{R}^8$.

Hadamard matrices
Orthogonal columns
sizes 2, 4 and 8

$$\begin{bmatrix} 1 & 1 \\ 1 & -1 \end{bmatrix} \qquad \begin{bmatrix} 1 & 1 & 1 & 1 \\ 1 & -1 & 1 & -1 \\ 1 & 1 & -1 & -1 \\ 1 & -1 & -1 & 1 \end{bmatrix} \qquad \begin{bmatrix} H_4 & H_4 \\ H_4 & -H_4 \end{bmatrix}$$

Are those orthogonal matrices? *No.* The columns have lengths $\sqrt{2}, \sqrt{4}, \sqrt{8}$. If we divide by those lengths, we have the beginning of an infinite list: orthonormal bases in $2, 4, 8, 16, 32, \ldots$ dimensions.

The **Hadamard conjecture** proposes that there is a ± 1 matrix with orthogonal columns whenever 4 divides n. Wikipedia says that $n = 668$ is the smallest of those sizes without a known Hadamard matrix. The construction for $n = 16, 32, \ldots$ follows the pattern above.

Here is a key fact: **Every subspace of $\mathbf{R}^n$ has an orthogonal basis.** Think of a plane in three-dimensional space $\mathbf{R}^3$. The plane has two independent vectors $\boldsymbol{a}$ and $\boldsymbol{b}$. For an orthogonal basis, subtract away from $\boldsymbol{b}$ its component in the direction of $\boldsymbol{a}$:

$$\textbf{Orthogonal basis} \quad \boldsymbol{a} \text{ and } \boldsymbol{c} \qquad \boldsymbol{c} = \boldsymbol{b} - \frac{\boldsymbol{a}^{\mathrm{T}}\boldsymbol{b}}{\boldsymbol{a}^{\mathrm{T}}\boldsymbol{a}}\,\boldsymbol{a}. \tag{6}$$

The inner product $\boldsymbol{a}^{\mathrm{T}}\boldsymbol{c}$ is $\boldsymbol{a}^{\mathrm{T}}\boldsymbol{b} - \boldsymbol{a}^{\mathrm{T}}\boldsymbol{b} = 0$. This idea of "orthogonalizing" applies to any number of basis vectors: *a basis becomes an orthogonal basis.* That is the Gram-Schmidt idea in Section II.2.

3. Orthogonal subspaces. Equation (1) looked at $A\boldsymbol{x} = \mathbf{0}$. Every row of A is multiplying that nullspace vector $\boldsymbol{x}$. So each row (and all combinations of the rows) will be orthogonal to $\boldsymbol{x}$ in $\mathbf{N}(A)$. **The row space of A is orthogonal to the nullspace of A.**

$$A\boldsymbol{x} = \begin{bmatrix} \text{row } 1 \\ \vdots \\ \text{row } m \end{bmatrix} \begin{bmatrix} \\ \boldsymbol{x} \\ \\ \end{bmatrix} = \begin{bmatrix} 0 \\ \vdots \\ 0 \end{bmatrix} \qquad A^{\mathrm{T}}\boldsymbol{y} = \begin{bmatrix} (\text{column } 1)^{\mathrm{T}} \\ \vdots \\ (\text{column } n)^{\mathrm{T}} \end{bmatrix} \begin{bmatrix} \\ \boldsymbol{y} \\ \\ \end{bmatrix} = \begin{bmatrix} 0 \\ \vdots \\ 0 \end{bmatrix} \quad (7)$$

From $A^{\mathrm{T}}\boldsymbol{y} = \mathbf{0}$, the columns of A are all orthogonal to $\boldsymbol{y}$. Their combinations (the whole column space) will also be orthogonal to $\boldsymbol{y}$. **The column space of A is orthogonal to the nullspace of A^{T}.** This produces the "Big Picture of Linear Algebra" in Figure I.6.

Notice the dimensions r and $n - r$ adding to n. The whole space $\mathbf{R}^n$ is accounted for. Every vector $\boldsymbol{v}$ in $\mathbf{R}^n$ has a row space component $\boldsymbol{v}_r$ and a nullspace component $\boldsymbol{v}_n$ with $\boldsymbol{v} = \boldsymbol{v}_r + \boldsymbol{v}_n$. A row space basis ($r$ vectors) together with a nullspace basis ($n - r$ vectors) produces a basis for all of $\mathbf{R}^n$ (n vectors).

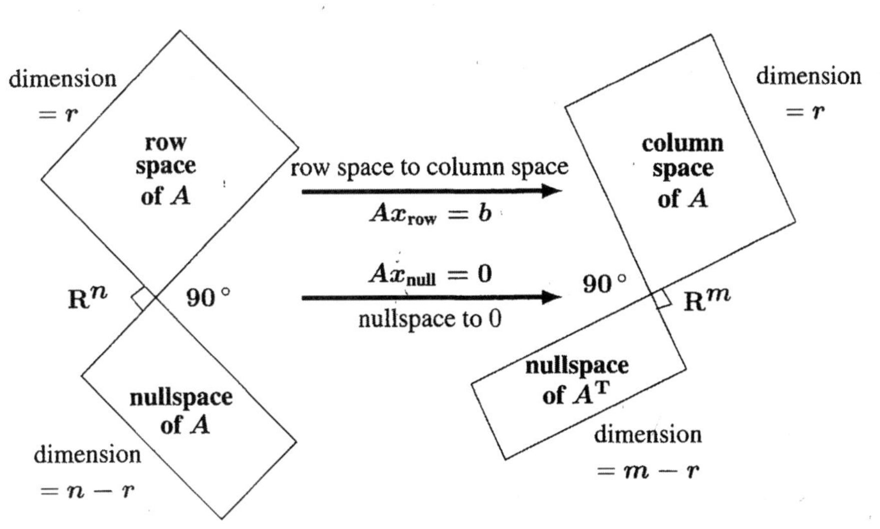

Figure I.6: Two pairs of orthogonal subspaces. The dimensions add to n and add to m. **This is the Big Picture**—two subspaces in $\mathbf{R}^n$ and two subspaces in $\mathbf{R}^m$.

I will mention a big improvement. It comes from the Singular Value Decomposition. The SVD is the most important theorem in data science. It finds orthonormal bases $\boldsymbol{v}_1, \ldots, \boldsymbol{v}_r$ for the row space of A and $\boldsymbol{u}_1, \ldots, \boldsymbol{u}_r$ for the column space of A. Well, Gram-Schmidt can do that. The special bases from the SVD have the extra property that each pair ($\boldsymbol{v}$ and $\boldsymbol{u}$) is connected by A:

$$\boxed{\textbf{Singular vectors} \qquad A\boldsymbol{v}_1 = \sigma_1\boldsymbol{u}_1 \qquad A\boldsymbol{v}_2 = \sigma_2\boldsymbol{u}_2 \qquad \cdots \qquad A\boldsymbol{v}_r = \sigma_r\boldsymbol{u}_r.} \quad (8)$$

In Figure I.6, imagine the $\boldsymbol{v}$'s on the left and the $\boldsymbol{u}$'s on the right. For the bases from the SVD, multiplying by A takes an orthogonal basis of $\boldsymbol{v}$'s to an orthogonal basis of $\boldsymbol{u}$'s.

4. Tall thin Q with orthonormal columns : $Q^{\mathrm{T}}Q = I$.
Here are three possible Q's, growing from (3 by 1) to (3 by 2) to an orthogonal matrix Q_3.

$$Q_1 = \frac{1}{3}\begin{bmatrix} 2 \\ 2 \\ -1 \end{bmatrix} \qquad Q_2 = \frac{1}{3}\begin{bmatrix} 2 & 2 \\ 2 & -1 \\ -1 & 2 \end{bmatrix} \qquad Q_3 = \frac{1}{3}\begin{bmatrix} 2 & 2 & -1 \\ 2 & -1 & 2 \\ -1 & 2 & 2 \end{bmatrix}. \quad (9)$$

Each one of those matrices has $Q^{\mathrm{T}}Q = I$. So Q^{T} is a left inverse of Q. Only the last matrix has $Q_3Q_3^{\mathrm{T}} = I$. Then Q_3^{T} is also a right inverse. Q_3 happens to be symmetric as well as orthogonal. It is a king and also a queen, truly a royal matrix.

Notice that all the matrices $P = QQ^{\mathrm{T}}$ have $P^2 = P$:

$$P^2 = (QQ^{\mathrm{T}})\,(QQ^{\mathrm{T}}) = Q\,(Q^{\mathrm{T}}Q)\,Q^{\mathrm{T}} = QQ^{\mathrm{T}} = P. \quad (10)$$

In the middle we removed $Q^{\mathrm{T}}Q = I$. The equation $P^2 = P$ signals a **projection matrix**.

If $P^2 = P = P^{\mathrm{T}}$ then Pb is the orthogonal projection of b onto the column space of P.

Example 1 To project $b = (3,3,3)$ on the Q_1 line, multiply by $P_1 = Q_1Q_1^{\mathrm{T}}$.

$$P_1 b = \frac{1}{9}\begin{bmatrix} 2 \\ 2 \\ -1 \end{bmatrix}\begin{bmatrix} 2 & 2 & -1 \end{bmatrix}\begin{bmatrix} 3 \\ 3 \\ 3 \end{bmatrix} = \frac{1}{9}\begin{bmatrix} 2 \\ 2 \\ -1 \end{bmatrix} 9 = \begin{bmatrix} \mathbf{2} \\ \mathbf{2} \\ \mathbf{-1} \end{bmatrix} = \begin{matrix}\text{projection} \\ \text{on a line}\end{matrix}$$

That matrix splits b in two perpendicular parts : projection $P_1 b$ and error $e = (I - P_1)\,b$.

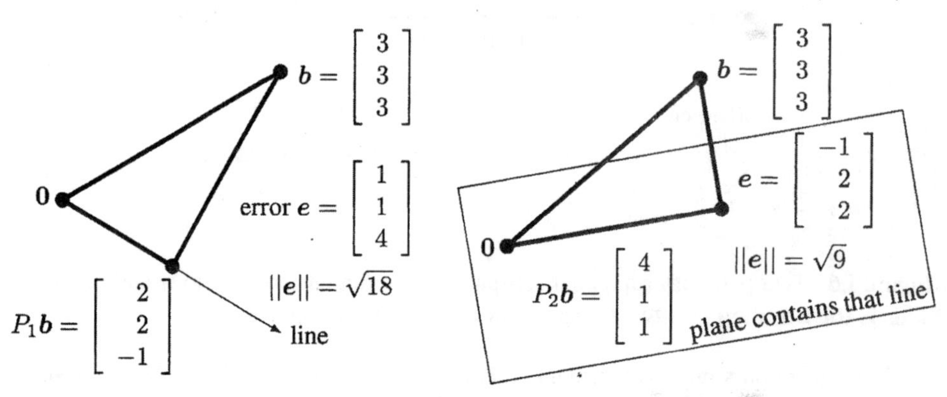

Figure I.7: Projection of b onto a line by $P_1 = Q_1Q_1^{\mathrm{T}}$ and onto a plane by $P_2 = Q_2Q_2^{\mathrm{T}}$.

Now project the same $b = (3,3,3)$ on the column space of Q_2 (a plane). The error vector $b - P_2\,b$ is shorter than $b - P_1 b$ because the plane contains the line.

$$P_2\,b = \frac{1}{9}\begin{bmatrix} 2 & 2 \\ 2 & -1 \\ -1 & 2 \end{bmatrix}\begin{bmatrix} 2 & 2 & -1 \\ 2 & -1 & 2 \end{bmatrix}\begin{bmatrix} 3 \\ 3 \\ 3 \end{bmatrix} = \frac{1}{9}\begin{bmatrix} 2 & 2 \\ 2 & -1 \\ -1 & 2 \end{bmatrix}\begin{bmatrix} 9 \\ 9 \end{bmatrix} = \begin{bmatrix} \mathbf{4} \\ \mathbf{1} \\ \mathbf{1} \end{bmatrix}$$

Question: What is $P_3\,\boldsymbol{b} = Q_3 Q_3^{\mathrm{T}}\,\boldsymbol{b}$? Now you are projecting $\boldsymbol{b}$ onto the whole space $\mathbf{R}^3$.
Answer: $P_3\,\boldsymbol{b} = \boldsymbol{b}$. In fact $P_3 = Q_3 Q_3^{\mathrm{T}}$ = identity matrix ! **The error e is now zero.**
Projections lie at the heart of "least squares" in Section II.2.

5. Orthogonal matrices: Now Q is *square*: $Q^{\mathrm{T}}Q = I$ and $QQ^{\mathrm{T}} = I$. So $\boldsymbol{Q^{-1} = Q^{\mathrm{T}}}$.
These Q's are truly important. For 2 by 2, they are **rotations** of the plane or **reflections**.

When the whole plane rotates around $(0, 0)$, lengths don't change. Angles between vectors don't change. The columns of Q are orthogonal unit vectors, with $\cos^2\theta + \sin^2\theta = 1$:

$$Q_{\text{rotate}} = \begin{bmatrix} \cos\theta & -\sin\theta \\ \sin\theta & \cos\theta \end{bmatrix} = \textbf{rotation} \text{ through an angle } \theta. \tag{11}$$

And if I multiply a column by -1, the two columns are still orthogonal of length 1.

$$Q_{\text{reflect}} = \begin{bmatrix} \cos\theta & \sin\theta \\ \sin\theta & -\cos\theta \end{bmatrix} = \textbf{reflection} \text{ across the } \frac{\theta}{2} - \text{ line.} \tag{12}$$

Now Q reflects every vector in a mirror. It is a reflection with determinant -1 instead of a rotation with determinant $+1$. The x-y plane rotates or the x-y plane flips over.

It is important that multiplying orthogonal matrices produces an orthogonal matrix.

$$\boldsymbol{Q_1 Q_2} \textbf{ is orthogonal} \qquad (Q_1 Q_2)^{\mathrm{T}}(Q_1 Q_2) = Q_2^{\mathrm{T}} Q_1^{\mathrm{T}} Q_1 Q_2 = Q_2^{\mathrm{T}} Q_2 = I.$$

Rotation times rotation = rotation. Reflection times reflection = rotation. Rotation times reflection = *reflection*. All still true in $\mathbf{R}^n$.

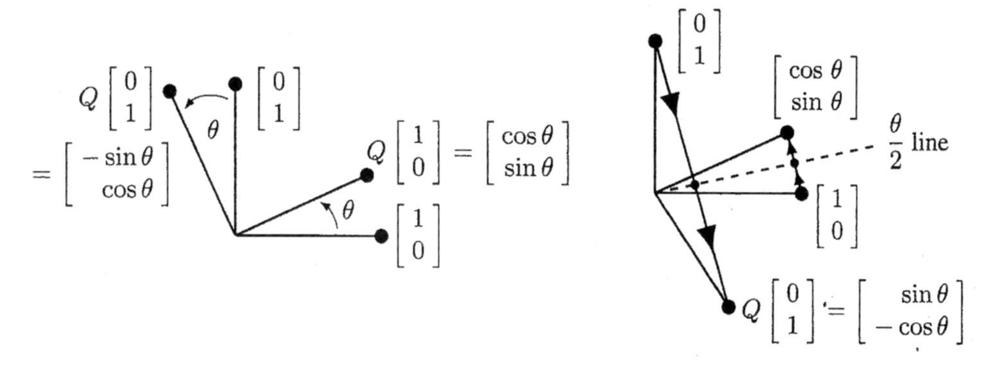

Figure I.8: Rotate the whole plane by θ. Reflect every vector across the line at angle $\theta/2$.

The figure shows how the columns of Q come from $Q\begin{bmatrix}1\\0\end{bmatrix}$ and $Q\begin{bmatrix}0\\1\end{bmatrix}$.

Orthogonal Basis = Orthogonal Axes in $\mathbf{R}^n$

Suppose the n by n orthogonal matrix Q has columns $\boldsymbol{q}_1, \ldots, \boldsymbol{q}_n$. Those unit vectors are a basis for n-dimensional space $\mathbf{R}^n$. Every vector $\boldsymbol{v}$ can be written as a combination of the basis vectors (the $\boldsymbol{q}'s$) :

$$\boldsymbol{v} = c_1\boldsymbol{q}_1 + \cdots + c_n\boldsymbol{q}_n \tag{13}$$

Those $c_1\boldsymbol{q}_1$ and $c_2\boldsymbol{q}_2$ and $c_n\boldsymbol{q}_n$ are the components of $\boldsymbol{v}$ along the axes. They are the projections of $\boldsymbol{v}$ onto the axes ! There is a simple formula for each number c_1 to c_n :

Coefficients in an orthonormal basis

$$\boxed{c_1 = \boldsymbol{q}_1^{\mathrm{T}}\boldsymbol{v} \qquad c_2 = \boldsymbol{q}_2^{\mathrm{T}}\boldsymbol{v} \quad \cdots \quad c_n = \boldsymbol{q}_n^{\mathrm{T}}\boldsymbol{v}} \tag{14}$$

I will give a vector proof and a matrix proof. Take dot products with $\boldsymbol{q}_1$ in equation (13) :

$$\boldsymbol{q}_1^{\mathrm{T}}\boldsymbol{v} = c_1\boldsymbol{q}_1^{\mathrm{T}}\boldsymbol{q}_1 + \cdots + c_n\boldsymbol{q}_1^{\mathrm{T}}\boldsymbol{q}_n = c_1 \tag{15}$$

All terms are zero except $c_1\boldsymbol{q}_1^{\mathrm{T}}\boldsymbol{q}_1 = c_1$. So $\boldsymbol{q}_1^{\mathrm{T}}\boldsymbol{v} = c_1$ and every $\boldsymbol{q}_k^{\mathrm{T}}\boldsymbol{v} = c_k$.

If we write (13) as a matrix equation $\boldsymbol{v} = Q\boldsymbol{c}$, **multiply by Q^{T} to see (14) :**

$$Q^{\mathrm{T}}\boldsymbol{v} = Q^{\mathrm{T}}Q\boldsymbol{c} = \boldsymbol{c} \quad \text{gives all the coefficients } c_k = \boldsymbol{q}_k^{\mathrm{T}}\boldsymbol{v} \text{ at once.}$$

This is the key application of orthogonal bases (for example the basis for Fourier series). When basis vectors are orthonormal, each coefficient c_1 to c_n can be found separately !

Householder Reflections

Here are neat examples of **reflection matrices** $Q = \boldsymbol{H}_n$. Start with the identity matrix. Choose a unit vector $\boldsymbol{u}$. Subtract the rank one symmetric matrix $2\boldsymbol{u}\boldsymbol{u}^{\mathrm{T}}$. Then $I - 2\boldsymbol{u}\boldsymbol{u}^{\mathrm{T}}$ is a "Householder matrix". For example, choose $\boldsymbol{u} = (1, 1, \ldots, 1)/\sqrt{n}$.

Householder example

$$\boxed{\boldsymbol{H}_n = \boldsymbol{I} - 2\boldsymbol{u}\boldsymbol{u}^{\mathrm{T}} = \boldsymbol{I} - \frac{2}{n}\,\textbf{ones}\,(\boldsymbol{n}, \boldsymbol{n}).} \tag{16}$$

With $\boldsymbol{u}\boldsymbol{u}^{\mathrm{T}}$, H_n is surely symmetric. Two reflections give $H^2 = I$ because $\boldsymbol{u}^{\mathrm{T}}\boldsymbol{u} = 1$:

$$\boldsymbol{H}^{\mathrm{T}}\boldsymbol{H} = \boldsymbol{H}^2 = (I - 2\boldsymbol{u}\boldsymbol{u}^{\mathrm{T}})\,(I - 2\boldsymbol{u}\boldsymbol{u}^{\mathrm{T}}) = I - 4\boldsymbol{u}\boldsymbol{u}^{\mathrm{T}} + 4\boldsymbol{u}\boldsymbol{u}^{\mathrm{T}}\boldsymbol{u}\boldsymbol{u}^{\mathrm{T}} = \boldsymbol{I}. \tag{17}$$

The 3 by 3 and 4 by 4 examples are easy to remember, and H_4 is like a Hadamard matrix :

$$\boldsymbol{H}_3 = I - \frac{2}{3}\,\textbf{ones} = \frac{1}{3}\begin{bmatrix} 1 & -2 & -2 \\ -2 & 1 & -2 \\ -2 & -2 & 1 \end{bmatrix} \qquad \boldsymbol{H}_4 = I - \frac{2}{4}\,\textbf{ones} = \frac{1}{2}\begin{bmatrix} \mathbf{1} & -1 & -1 & -1 \\ -1 & \mathbf{1} & -1 & -1 \\ -1 & -1 & \mathbf{1} & -1 \\ -1 & -1 & -1 & \mathbf{1} \end{bmatrix}$$

Householder's n by n reflection matrix has $\boldsymbol{H}_n\boldsymbol{u} = (I - 2\boldsymbol{u}\boldsymbol{u}^{\mathrm{T}})\boldsymbol{u} = \boldsymbol{u} - 2\boldsymbol{u} = -\boldsymbol{u}$. And $\boldsymbol{H}_n\boldsymbol{w} = +\boldsymbol{w}$ whenever $\boldsymbol{w}$ is perpendicular to $\boldsymbol{u}$. The "eigenvalues" of H are -1 (once) and $+1$ ($n - 1$ times). All reflection matrices have eigenvalues -1 and 1.

Problem Set I.5

1 If $\boldsymbol{u}$ and $\boldsymbol{v}$ are orthogonal unit vectors, show that $\boldsymbol{u}+\boldsymbol{v}$ is orthogonal to $\boldsymbol{u}-\boldsymbol{v}$. What are the lengths of those vectors?

2 Draw unit vectors $\boldsymbol{u}$ and $\boldsymbol{v}$ that are *not* orthogonal. Show that $\boldsymbol{w}=\boldsymbol{v}-\boldsymbol{u}(\boldsymbol{u}^{\mathrm{T}}\boldsymbol{v})$ is orthogonal to $\boldsymbol{u}$ (and add $\boldsymbol{w}$ to your picture).

3 Draw any two vectors $\boldsymbol{u}$ and $\boldsymbol{v}$ out from the origin $(0,0)$. Complete two more sides to make a parallelogram with diagonals $\boldsymbol{w}=\boldsymbol{u}+\boldsymbol{v}$ and $\boldsymbol{z}=\boldsymbol{u}-\boldsymbol{v}$. Show that $\boldsymbol{w}^{\mathrm{T}}\boldsymbol{w}+\boldsymbol{z}^{\mathrm{T}}\boldsymbol{z}$ is equal to $2\boldsymbol{u}^{\mathrm{T}}\boldsymbol{u}+2\boldsymbol{v}^{\mathrm{T}}\boldsymbol{v}$.

4 Key property of every orthogonal matrix: $||Q\boldsymbol{x}||^2=||\boldsymbol{x}||^2$ for every vector $\boldsymbol{x}$. More than this, show that $(Q\boldsymbol{x})^{\mathrm{T}}(Q\boldsymbol{y})=\boldsymbol{x}^{\mathrm{T}}\boldsymbol{y}$ for every vector $\boldsymbol{x}$ and $\boldsymbol{y}$. So *lengths and angles are not changed by* Q. **Computations with Q never overflow!**

5 If Q is orthogonal, how do you know that Q is invertible and Q^{-1} is also orthogonal? If $Q_1^{\mathrm{T}}=Q_1^{-1}$ and $Q_2^{\mathrm{T}}=Q_2^{-1}$, show that Q_1Q_2 is also an orthogonal matrix.

6 A **permutation matrix** has the same columns as the identity matrix (in some order). *Explain why this permutation matrix and every permutation matrix is orthogonal*:

$$P=\begin{bmatrix}0&1&0&0\\0&0&1&0\\0&0&0&1\\1&0&0&0\end{bmatrix}\quad \text{has orthonormal columns so } P^{\mathrm{T}}P=\_\_\_\_ \text{ and } P^{-1}=\_\_\_\_.$$

When a matrix is symmetric or orthogonal, **it will have orthogonal eigenvectors.** This is the most important source of orthogonal vectors in applied mathematics.

7 Four eigenvectors of that matrix P are $\boldsymbol{x}_1=(1,1,1,1)$, $\boldsymbol{x}_2=(1,i,i^2,i^3)$, $\boldsymbol{x}_3=(1,i^2,i^4,i^6)$, and $\boldsymbol{x}_4=(1,i^3,i^6,i^9)$. Multiply P times each vector to find $\lambda_1,\lambda_2,\lambda_3,\lambda_4$. The eigenvectors are the columns of the 4 by 4 **Fourier matrix F**.

$$\text{Show that } Q=\frac{F}{2}=\frac{1}{2}\begin{bmatrix}1&1&1&1\\1&i&-1&-i\\1&i^2&1&-1\\1&i^3&-1&i\end{bmatrix}\quad \text{has orthonormal columns}:\ \overline{\boldsymbol{Q}}^{\mathrm{T}}\boldsymbol{Q}=\boldsymbol{I}$$

8 **Haar wavelets are orthogonal vectors (columns of W) using only 1, -1, and 0.**

$n=4$

$$W=\begin{bmatrix}1&1&1&0\\1&1&-1&0\\1&-1&0&1\\1&-1&0&-1\end{bmatrix}$$

Find $W^{\mathrm{T}}W$ and W^{-1} and the eight Haar wavelets for $n=8$.

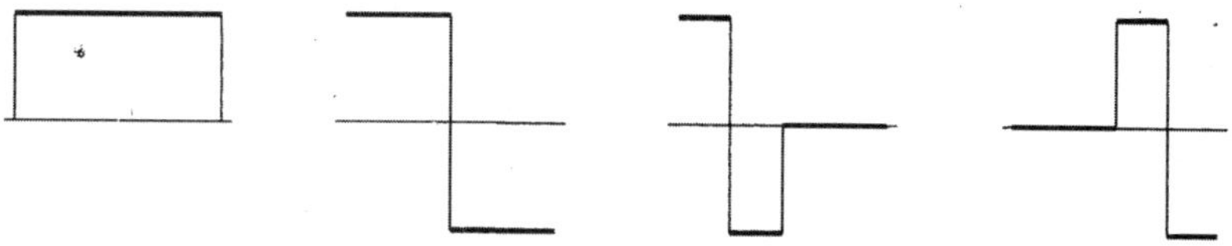

I.6 Eigenvalues and Eigenvectors

The eigenvectors of A don't change direction when you multiply them by A. The output $A\boldsymbol{x}$ is on the same line as the input vector $\boldsymbol{x}$.

$$\boxed{\begin{array}{ll} \boldsymbol{x} = \textbf{eigenvector of } \boldsymbol{A} & \\ \lambda = \textbf{eigenvalue of } \boldsymbol{A} & \quad A\boldsymbol{x} = \lambda \boldsymbol{x} \end{array}} \tag{1}$$

The eigenvector $\boldsymbol{x}$ is just multiplied by its eigenvalue λ. Multiply again by A, to see that **$\boldsymbol{x}$ is also an eigenvector of A^2**: $A^2\boldsymbol{x} = \lambda^2\boldsymbol{x}$.

$$\begin{array}{ll} \boldsymbol{x} \;\; = \text{same eigenvector} & \\ \lambda^2 = \text{squared eigenvalue} & \quad A(A\boldsymbol{x}) = A(\lambda\boldsymbol{x}) = \lambda(A\boldsymbol{x}) = \lambda^2\boldsymbol{x}\ . \end{array} \tag{2}$$

Certainly $A^k\boldsymbol{x} = \lambda^k\boldsymbol{x}$ for all $k = 1, 2, 3, \ldots$ And $A^{-1}\boldsymbol{x} = \frac{1}{\lambda}\boldsymbol{x}$ provided $\lambda \neq 0$.

These eigenvectors are special vectors that depend on A. Most n by n matrices have n independent eigenvectors $\boldsymbol{x}_1$ to $\boldsymbol{x}_n$ with n different eigenvalues λ_1 to λ_n. In that case every n-dimensional vector $\boldsymbol{v}$ will be a combination of the eigenvectors:

$$\boxed{\begin{array}{ll} \textbf{Every } \boldsymbol{v} & \boldsymbol{v} = c_1\boldsymbol{x}_1 + \cdots + c_n\boldsymbol{x}_n \\ \textbf{Multiply by } \boldsymbol{A} & A\boldsymbol{v} = c_1\lambda_1\boldsymbol{x}_1 + \cdots + c_n\lambda_n\boldsymbol{x}_n \\ \textbf{Multiply by } \boldsymbol{A^k} & A^k\boldsymbol{v} = c_1\lambda_1^k\boldsymbol{x}_1 + \cdots + c_n\lambda_n^k\boldsymbol{x}_n \end{array}} \tag{3}$$

Here you see how eigenvalues and eigenvectors are useful. They look into the heart of a matrix. If $|\lambda_1| > 1$ then the component $c_1\lambda_1^n\boldsymbol{x}_1$ will grow as n increases. If $|\lambda_2| < 1$ then that component $c_2\lambda_2^n\boldsymbol{x}_2$ will steadily disappear. **Follow each eigenvector separately!**

Example 1 $S = \begin{bmatrix} 2 & 1 \\ 1 & 2 \end{bmatrix}$ has eigenvectors $S\begin{bmatrix} 1 \\ 1 \end{bmatrix} = 3\begin{bmatrix} 1 \\ 1 \end{bmatrix}$ and $S\begin{bmatrix} 1 \\ -1 \end{bmatrix} = \begin{bmatrix} 1 \\ -1 \end{bmatrix}$

Then $\lambda_1 = 3$ and $\lambda_2 = 1$. **The powers S^k will grow like 3^k.** Those eigenvalues and eigenvectors have four properties to notice:

(**Trace of S**) The sum $\lambda_1 + \lambda_2 = 3 + 1$ equals the diagonal sum $2 + 2 = 4$

(**Determinant**) The product $\lambda_1\lambda_2 = (3)(1) = 3$ equals the determinant $4 - 1$

(**Real eigenvalues**) Symmetric matrices $S = S^{\mathrm{T}}$ always have real eigenvalues

(**Orthogonal eigenvectors**) If $\lambda_1 \neq \lambda_2$ then $\boldsymbol{x}_1 \cdot \boldsymbol{x}_2 = 0$. Here $(1, 1) \cdot (1, -1) = 0$.

Symmetric matrices S are somehow like real numbers (every λ is real). Orthogonal matrices Q are like complex numbers $e^{i\theta} = \cos\theta + i\sin\theta$ of magnitude 1 (every $|\lambda| = 1$). The powers of Q don't grow or decay because $Q^2, Q^3, \ldots$ are orthogonal matrices too.

Example 2 The rotation $Q = \begin{bmatrix} 0 & -1 \\ 1 & 0 \end{bmatrix}$ has imaginary eigenvalues i and $-i$:

$$Q\begin{bmatrix} 1 \\ -i \end{bmatrix} = \begin{bmatrix} 0 & -1 \\ 1 & 0 \end{bmatrix}\begin{bmatrix} 1 \\ -i \end{bmatrix} = (i)\begin{bmatrix} 1 \\ -i \end{bmatrix} \text{ and } Q\begin{bmatrix} 1 \\ i \end{bmatrix} = \begin{bmatrix} 0 & -1 \\ 1 & 0 \end{bmatrix}\begin{bmatrix} 1 \\ i \end{bmatrix} = (-i)\begin{bmatrix} 1 \\ i \end{bmatrix}$$

Certainly $\lambda_1 + \lambda_2 = i - i$ agrees with the trace $0 + 0$ from the main diagonal of Q. And $(\lambda_1)(\lambda_2) = (i)(-i)$ agrees with the determinant of $Q = 1$. The eigenvectors of Q are still orthogonal when we move (as we should) to the dot product of complex vectors. *Change every i in $\boldsymbol{x}_1$ to $-i$.* This produces its conjugate $\overline{\boldsymbol{x}}_1$.

$$\overline{\boldsymbol{x}}_1^{\mathrm{T}}\boldsymbol{x}_2 = \begin{bmatrix} 1 & i \end{bmatrix}\begin{bmatrix} 1 \\ i \end{bmatrix} = 1 + i^2 = 0 \text{ : } \textbf{orthogonal eigenvectors}.$$

Warnings about eigenvalues and eigenvectors

The eigenvalues of $A + B$ are not usually $\lambda(A)$ plus $\lambda(B)$.

The eigenvalues of AB are not usually $\lambda(A)$ times $\lambda(B)$.

A double eigenvalue $\lambda_1 = \lambda_2$ *might or might not* have two independent eigenvectors.

The eigenvectors of a real matrix A are orthogonal if and only if $A^{\mathrm{T}}A = AA^{\mathrm{T}}$.

The matrix A also controls a system of linear differential equations $d\boldsymbol{u}/dt = A\boldsymbol{u}$. The system starts at an initial vector $\boldsymbol{u}(0)$ when $t = 0$. Every eigenvector grows or decays or oscillates according to its own eigenvalue λ. Powers λ^n are changed to exponentials $e^{\lambda t}$:

$$\begin{array}{ll} \textbf{Starting vector} & \boldsymbol{u}(0) = c_1\boldsymbol{x}_1 + \cdots + c_n\boldsymbol{x}_n \\ \textbf{Solution vector} & \boldsymbol{u}(t) = c_1 e^{\lambda_1 t}\boldsymbol{x}_1 + \cdots + c_n e^{\lambda_n t}\boldsymbol{x}_n \end{array} \qquad (4)$$

The difference between growth and decay is now decided by $\operatorname{Re}\lambda > 0$ or $\operatorname{Re}\lambda < 0$, instead of $|\lambda| > 1$ or $|\lambda| < 1$. The real part of $\lambda = a + ib$ is $\operatorname{Re}\lambda = a$. The absolute value of $e^{\lambda t}$ is e^{at}. The other factor $e^{ibt} = \cos bt + i \sin bt$ has $\cos^2 bt + \sin^2 bt = 1$. That part oscillates while e^{at} grows or decays.

Computing the Eigenvalues (by hand)

Notice that $A\boldsymbol{x} = \lambda\boldsymbol{x}$ is the same as $(A - \lambda I)\boldsymbol{x} = 0$. Then $A - \lambda I$ is not invertible: that matrix is *singular*. **The determinant of $A - \lambda I$ must be zero**. This gives an nth degree equation for λ, and this equation $\det(A - \lambda I) = 0$ has n roots. Here $n = 2$ and $A = \begin{bmatrix} a & b \\ c & d \end{bmatrix}$ has two eigenvalues:

$$\textbf{Determinant of } A - \lambda I = \begin{vmatrix} a - \lambda & b \\ c & d - \lambda \end{vmatrix} = \lambda^2 - (a + d)\lambda + (ad - bc) = 0$$

This quadratic equation might factor easily into $(\lambda - \lambda_1)(\lambda - \lambda_2)$. The "quadratic formula" will always give the two roots λ_1 and λ_2 of our equation, from the $+$ sign and the $-$ sign.

$$\lambda = \frac{1}{2}\left[a + d \pm \sqrt{(a+d)^2 - 4\,(ad - bc)}\,\right] = \frac{1}{2}\left[\boldsymbol{a + d \pm \sqrt{(a-d)^2 + 4bc}}\,\right].$$

You see that $\lambda_1 + \lambda_2$ equals $a + d$ (the trace of the matrix). The $\pm$ square roots cancel out.

Notice also that the eigenvalues are real when A is symmetric ($b = c$). Then we are not taking the square root of a negative number to find λ. When bc is very negative, the eigenvalues and eigenvectors go complex!

Example 3 Find the eigenvalues and eigenvectors of $A = \begin{bmatrix} \mathbf{8} & \mathbf{3} \\ \mathbf{2} & \mathbf{7} \end{bmatrix}$: **not symmetric.**

The determinant of $A - \lambda I$ is $\begin{vmatrix} 8-\lambda & 3 \\ 2 & 7-\lambda \end{vmatrix} = \lambda^2 - 15\lambda + 50 = \boldsymbol{(\lambda - 10)\,(\lambda - 5)}$.

$$\boldsymbol{\lambda_1 = 10} \text{ has } \begin{bmatrix} 8-10 & 3 \\ 2 & 7-10 \end{bmatrix}\begin{bmatrix} x_1 \\ x_2 \end{bmatrix} = \begin{bmatrix} 0 \\ 0 \end{bmatrix} \quad \textbf{Eigenvector } \boldsymbol{x}_1 = \begin{bmatrix} x_1 \\ x_2 \end{bmatrix} = \begin{bmatrix} 3 \\ 2 \end{bmatrix}$$

$$\boldsymbol{\lambda_2 = 5} \text{ has } \begin{bmatrix} 8-5 & 3 \\ 2 & 7-5 \end{bmatrix}\begin{bmatrix} x_1 \\ x_2 \end{bmatrix} = \begin{bmatrix} 0 \\ 0 \end{bmatrix} \quad \textbf{Eigenvector } \boldsymbol{x}_2 = \begin{bmatrix} x_1 \\ x_2 \end{bmatrix} = \begin{bmatrix} 1 \\ -1 \end{bmatrix}$$

$\mathbf{10 + 5 = 8 + 7}$. These eigenvectors are not orthogonal. Increase 3 to 30 for complex λ's.

Question: If A is shifted to $A + sI$, what happens to the $\boldsymbol{x}$'s and λ's?
Answer: The eigenvectors $\boldsymbol{x}$ stay the same. Every eigenvalue λ shifts by the number s:

$$\textbf{Shift in } \boldsymbol{A} \Rightarrow \textbf{shift in every } \boldsymbol{\lambda} \qquad \boldsymbol{(A + sI)\,x = \lambda x + s x = (\lambda + s)\,x} \tag{5}$$

Similar Matrices

For every invertible matrix B, the eigenvalues of BAB^{-1} are the same as the eigenvalues of A. The eigenvectors $\boldsymbol{x}$ of A are multiplied by B to give eigenvectors $B\boldsymbol{x}$ of BAB^{-1}:

$$\text{If } A\boldsymbol{x} = \lambda\boldsymbol{x} \text{ then } (BAB^{-1})\,(B\boldsymbol{x}) = BA\boldsymbol{x} = B\lambda\boldsymbol{x} = \lambda(B\boldsymbol{x}). \tag{6}$$

The matrices BAB^{-1} (for every invertible B) are **"similar" to $\boldsymbol{A}$: same eigenvalues.**

We use this idea to compute eigenvalues of large matrices (when the determinant of $A - \lambda I$ would be completely hopeless). The idea is to make BAB^{-1} gradually into a triangular matrix. The eigenvalues are not changing and they gradually show up on the main diagonal of BAB^{-1}:

$$\textbf{The eigenvalues of any triangular matrix } \begin{bmatrix} a & b \\ 0 & d \end{bmatrix} \textbf{ are } \boldsymbol{\lambda_1 = a} \textbf{ and } \boldsymbol{\lambda_2 = d}. \tag{7}$$

You can see that $A - aI$ and $A - dI$ will have determinant zero. So a and d are the eigenvalues of this triangular matrix.

Diagonalizing a Matrix

Suppose A has a full set of n independent eigenvectors. (Most matrices do, but not all matrices.) Put those eigenvectors $\boldsymbol{x}_1, \ldots, \boldsymbol{x}_n$ into an invertible matrix X. Then multiply AX column by column to get the columns $\lambda_1 \boldsymbol{x}_1$ to $\lambda_n \boldsymbol{x}_n$. Important! That matrix splits into X times Λ.

$$A\begin{bmatrix} & & \\ \boldsymbol{x}_1 & \cdot\cdot & \boldsymbol{x}_n \\ & & \end{bmatrix} = \begin{bmatrix} & & \\ A\boldsymbol{x}_1 & \cdot\cdot & A\boldsymbol{x}_n \\ & & \end{bmatrix} = \begin{bmatrix} & & \\ \lambda_1\boldsymbol{x}_1 & \cdot\cdot & \lambda_n\boldsymbol{x}_n \\ & & \end{bmatrix} = \begin{bmatrix} & & \\ \boldsymbol{x}_1 & \cdot\cdot & \boldsymbol{x}_n \\ & & \end{bmatrix}\begin{bmatrix} \lambda_1 & & \\ & \ddots & \\ & & \lambda_n \end{bmatrix}. \tag{8}$$

The eigenvalue matrix Λ goes on the right of X, because the λ's in Λ multiply the columns of X. That equation $AX = X\Lambda$ tells us that $\boldsymbol{A = X\Lambda X^{-1}}$. If we know the eigenvalues and eigenvectors, we know the matrix A. And we can easily compute powers of A:

$\boldsymbol{\Lambda}$ = diagonal eigenvalue matrix	$A = X\Lambda X^{-1}$
$\boldsymbol{X}$ = invertible eigenvector matrix	$A^2 = (X\Lambda X^{-1})(X\Lambda X^{-1}) = X\Lambda^2 X^{-1}$

$\boldsymbol{A = X\Lambda X^{-1}}$ **in Example 3** $\begin{bmatrix} 8 & 3 \\ 2 & 7 \end{bmatrix} = \begin{bmatrix} 3 & 1 \\ 2 & -1 \end{bmatrix}\begin{bmatrix} 10 & \\ & 5 \end{bmatrix}\frac{1}{5}\begin{bmatrix} 1 & 1 \\ 2 & -3 \end{bmatrix} =$ (**eigenvectors**) times (**$\boldsymbol{\lambda}$'s**) times (**left eigenvectors**)

The equation $A^k = X\Lambda^k X^{-1}$ is telling us what we already knew. The eigenvalues of A^k are $\lambda_1^k, \ldots, \lambda_n^k$. The eigenvectors of A^k are the same as the eigenvectors of A. Three steps compute $A^k \boldsymbol{v}$.

Step 1: $\boldsymbol{X^{-1}v}$ — This gives the c's in $\boldsymbol{v} = c_1\boldsymbol{x}_1 + \cdots + c_n\boldsymbol{x}_n$

Step 2: $\boldsymbol{\Lambda^k X^{-1} v}$ — This gives the λ's in $c_1\lambda_1^k\boldsymbol{x}_1 + \cdots + c_n\lambda_n^k\boldsymbol{x}_n$

Step 3: $\boldsymbol{X\Lambda^k X^{-1} v}$ — This adds those pieces in $A^k\boldsymbol{v} = c_1\lambda_1^k\boldsymbol{x}_1 + \cdots + c_n\lambda_n^k\boldsymbol{x}_n$

Example 4 If we divide Example 3 by 10, all eigenvalues are divided by 10. Then $\lambda_1 = 1$ and $\lambda_2 = \frac{1}{2}$. In this case A is a **Markov matrix, with positive columns adding to 1.**

$$A = \begin{bmatrix} 0.8 & 0.3 \\ 0.2 & 0.7 \end{bmatrix} \qquad A^k\boldsymbol{v} = c_1(1)^k\boldsymbol{x}_1 + c_2\left(\tfrac{1}{2}\right)^k\boldsymbol{x}_2$$

As $\boldsymbol{k}$ increases, $\boldsymbol{A^k v}$ approaches $\boldsymbol{c_1 x_1}$ = steady state

We can follow each eigenvector separately. Its growth or decay depends on the eigenvalue λ. The action of the whole matrix A is broken into simple actions (just multiply by λ) on each eigenvector. **To solve a differential equation $\boldsymbol{du/dt = Au}$ we would multiply each eigenvector by $\boldsymbol{e^{\lambda t}}$.**

Nondiagonalizable Matrices (Optional)

Suppose λ is an eigenvalue of A. We discover that fact in two ways:

1. **Eigenvectors** (geometric) There are nonzero solutions to $A\boldsymbol{x} = \lambda\boldsymbol{x}$.

2. **Eigenvalues** (algebraic) The determinant of $A - \lambda I$ is zero.

The number λ may be a simple eigenvalue or a multiple eigenvalue, and we want to know its ***multiplicity***. Most eigenvalues have multiplicity $M = 1$ (simple eigenvalues). Then there is a single line of eigenvectors, and $\det(A - \lambda I)$ does not have a double factor.

For exceptional matrices, an eigenvalue can be ***repeated***. Then there are two different ways to count its multiplicity. Always GM $\leq$ AM for each λ:

1. **(Geometric Multiplicity = GM)** Count the **independent eigenvectors** for λ. Look at the dimension of the nullspace of $A - \lambda I$.

2. **(Algebraic Multiplicity = AM)** Count the **repetitions of λ** among the eigenvalues. Look at the roots of $\det(A - \lambda I) = 0$.

If A has $\lambda = 4, 4, 4$, then that eigenvalue has AM $= 3$ and GM $= 1$ or 2 or 3.

The following matrix A is the standard example of trouble. Its eigenvalue $\lambda = 0$ is repeated. It is a double eigenvalue (AM $= 2$) with only one eigenvector (GM $= 1$).

AM = 2
GM = 1

$$A = \begin{bmatrix} 0 & 1 \\ 0 & 0 \end{bmatrix} \text{ has } \det(A - \lambda I) = \begin{vmatrix} -\lambda & 1 \\ 0 & -\lambda \end{vmatrix} = \lambda^2.$$

$\lambda = 0, 0$ but
1 eigenvector

There "should" be two eigenvectors, because $\lambda^2 = 0$ has a double root. The double factor λ^2 makes AM $= 2$. But there is only one eigenvector $\boldsymbol{x} = (1, 0)$. So GM $= 1$. ***This shortage of eigenvectors when* GM $<$ AM *means that A is not diagonalizable.*** There is no invertible eigenvector matrix. The formula $A = X\Lambda X^{-1}$ fails.

These three matrices all have the same shortage of eigenvectors. Their repeated eigenvalue is $\lambda = 5$. Traces are 10 and determinants are 25:

$$A = \begin{bmatrix} 5 & 1 \\ 0 & 5 \end{bmatrix} \quad \text{and} \quad A = \begin{bmatrix} 6 & -1 \\ 1 & 4 \end{bmatrix} \quad \text{and} \quad A = \begin{bmatrix} 7 & 2 \\ -2 & 3 \end{bmatrix}.$$

Those all have $\det(A - \lambda I) = (\lambda - 5)^2$. The algebraic multiplicity is AM $= 2$. But each $A - 5I$ has rank $r = 1$. The geometric multiplicity is GM $= 1$. There is only one line of eigenvectors for $\lambda = 5$, and these matrices are not diagonalizable.

Problem Set I.6

1 The rotation $Q = \begin{bmatrix} \cos\theta & -\sin\theta \\ \sin\theta & \cos\theta \end{bmatrix}$ has complex eigenvalues $\lambda = \cos\theta \pm i\sin\theta$:

$$Q\begin{bmatrix} 1 \\ -i \end{bmatrix} = (\cos\theta + i\sin\theta)\begin{bmatrix} 1 \\ -i \end{bmatrix} \text{ and } Q\begin{bmatrix} 1 \\ i \end{bmatrix} = (\cos\theta - i\sin\theta)\begin{bmatrix} 1 \\ i \end{bmatrix}.$$

Check that $\lambda_1 + \lambda_2$ equals the trace of Q (sum $Q_{11} + Q_{22}$ down the diagonal). Check that $(\lambda_1)(\lambda_2)$ equals the determinant. Check that those complex eigenvectors are orthogonal, using the complex dot product $\overline{\boldsymbol{x}}_1 \cdot \boldsymbol{x}_2$ (not just $\boldsymbol{x}_1 \cdot \boldsymbol{x}_2$!).

What is Q^{-1} and what are its eigenvalues?

2 Compute the eigenvalues and eigenvectors of A and A^{-1}. Check the trace!

$$A = \begin{bmatrix} 0 & 2 \\ 1 & 1 \end{bmatrix} \quad \text{and} \quad A^{-1} = \begin{bmatrix} -1/2 & 1 \\ 1/2 & 0 \end{bmatrix}.$$

A^{-1} has the _____ eigenvectors as A. When A has eigenvalues λ_1 and λ_2, its inverse has eigenvalues _____.

3 Find the eigenvalues of A and B (easy for triangular matrices) and $A + B$:

$$A = \begin{bmatrix} 3 & 0 \\ 1 & 1 \end{bmatrix} \quad \text{and} \quad B = \begin{bmatrix} 1 & 1 \\ 0 & 3 \end{bmatrix} \quad \text{and} \quad A + B = \begin{bmatrix} 4 & 1 \\ 1 & 4 \end{bmatrix}.$$

Eigenvalues of $A + B$ (*are equal to*)(*are not equal to*) eigenvalues of A plus eigenvalues of B.

4 Find the eigenvalues of A and B and AB and BA:

$$A = \begin{bmatrix} 1 & 0 \\ 1 & 1 \end{bmatrix} \quad \text{and} \quad B = \begin{bmatrix} 1 & 2 \\ 0 & 1 \end{bmatrix} \quad \text{and} \quad AB = \begin{bmatrix} 1 & 2 \\ 1 & 3 \end{bmatrix} \quad \text{and} \quad BA = \begin{bmatrix} 3 & 2 \\ 1 & 1 \end{bmatrix}.$$

(a) Are the eigenvalues of AB equal to eigenvalues of A times eigenvalues of B?

(b) Are the eigenvalues of AB equal to the eigenvalues of BA?

5 (a) If you know that $\boldsymbol{x}$ is an eigenvector, the way to find λ is to _____.

(b) If you know that λ is an eigenvalue, the way to find $\boldsymbol{x}$ is to _____.

6 Find the eigenvalues and eigenvectors for both of these Markov matrices A and A^∞. Explain from those answers why A^{100} is close to A^∞:

$$A = \begin{bmatrix} .6 & .2 \\ .4 & .8 \end{bmatrix} \quad \text{and} \quad A^\infty = \begin{bmatrix} 1/3 & 1/3 \\ 2/3 & 2/3 \end{bmatrix}.$$

7 **The determinant of A equals the product $\lambda_1\lambda_2\cdots\lambda_n$.** Start with the polynomial $\det(A-\lambda I)$ separated into its n factors (always possible). Then set $\lambda=0$:

$$\det(A-\lambda I)=(\lambda_1-\lambda)(\lambda_2-\lambda)\cdots(\lambda_n-\lambda) \quad \text{so} \quad \det A = \_\_\_\_\_ .$$

Check this rule in Example 1 where the Markov matrix has $\lambda=1$ and $\frac{1}{2}$.

8 The sum of the diagonal entries (the *trace*) equals the sum of the eigenvalues:

$$A=\begin{bmatrix} a & b \\ c & d \end{bmatrix} \quad \text{has} \quad \det(A-\lambda I)=\lambda^2-(a+d)\lambda+ad-bc=0.$$

The quadratic formula gives the eigenvalues $\lambda=(a+d+\sqrt{})/2$ and $\lambda=$ _____. Their sum is _____. If A has $\lambda_1=3$ and $\lambda_2=4$ then $\det(A-\lambda I)=$ _____.

9 If A has $\lambda_1=4$ and $\lambda_2=5$ then $\det(A-\lambda I)=(\lambda-4)(\lambda-5)=\lambda^2-9\lambda+20$. Find three matrices that have trace $a+d=9$ and determinant 20 and $\lambda=4,5$.

10 Choose the last rows of A and C to give eigenvalues $4,7$ and $1,2,3$:

Companion matrices $$A=\begin{bmatrix} 0 & 1 \\ * & * \end{bmatrix} \quad C=\begin{bmatrix} 0 & 1 & 0 \\ 0 & 0 & 1 \\ * & * & * \end{bmatrix}.$$

11 ***The eigenvalues of A equal the eigenvalues of A^{T}.*** This is because $\det(A-\lambda I)$ equals $\det(A^{\mathrm{T}}-\lambda I)$. That is true because _____. Show by an example that the eigenvectors of A and A^{T} are *not* the same.

12 This matrix is singular with rank one. Find three λ's and three eigenvectors:

$$A=\begin{bmatrix} 1 \\ 2 \\ 1 \end{bmatrix}\begin{bmatrix} 2 & 1 & 2 \end{bmatrix}=\begin{bmatrix} 2 & 1 & 2 \\ 4 & 2 & 4 \\ 2 & 1 & 2 \end{bmatrix}.$$

13 Suppose A and B have the same eigenvalues $\lambda_1,\ldots,\lambda_n$ with the same independent eigenvectors $\boldsymbol{x}_1,\ldots,\boldsymbol{x}_n$. Then $A=B$. *Reason*: Any vector $\boldsymbol{x}$ is a combination $c_1\boldsymbol{x}_1+\cdots+c_n\boldsymbol{x}_n$. What is $A\boldsymbol{x}$? What is $B\boldsymbol{x}$?

14 Suppose A has eigenvalues $0,3,5$ with independent eigenvectors $\boldsymbol{u},\boldsymbol{v},\boldsymbol{w}$.

(a) Give a basis for the nullspace and a basis for the column space.

(b) Find a particular solution to $A\boldsymbol{x}=\boldsymbol{v}+\boldsymbol{w}$. Find all solutions.

(c) $A\boldsymbol{x}=\boldsymbol{u}$ has no solution. If it did then _____ would be in the column space.

15 (a) Factor these two matrices into $A=X\Lambda X^{-1}$:

$$A=\begin{bmatrix} 1 & 2 \\ 0 & 3 \end{bmatrix} \quad \text{and} \quad A=\begin{bmatrix} 1 & 1 \\ 3 & 3 \end{bmatrix}.$$

(b) If $A=X\Lambda X^{-1}$ then $A^3=(\quad)(\quad)(\quad)$ and $A^{-1}=(\quad)(\quad)(\quad)$.

16 Suppose $A = X\Lambda X^{-1}$. What is the eigenvalue matrix for $A + 2I$? What is the eigenvector matrix? Check that $A + 2I = (\quad)(\quad)(\quad)^{-1}$.

17 True or false: If the columns of X (eigenvectors of A) are linearly independent, then

(a) A is invertible (b) A is diagonalizable

(c) X is invertible (d) X is diagonalizable.

18 Write down the most general matrix that has eigenvectors $\begin{bmatrix}1\\1\end{bmatrix}$ and $\begin{bmatrix}1\\-1\end{bmatrix}$.

19 True or false: If the eigenvalues of A are $2, 2, 5$ then the matrix is certainly

(a) invertible (b) diagonalizable (c) not diagonalizable.

20 True or false: If the only eigenvectors of A are multiples of $(1, 4)$ then A has

(a) no inverse (b) a repeated eigenvalue (c) no diagonalization $X\Lambda X^{-1}$.

21 $A^k = X\Lambda^k X^{-1}$ approaches the zero matrix as $k \to \infty$ if and only if every λ has absolute value less than ______. Which of these matrices has $A^k \to 0$?

$$A_1 = \begin{bmatrix} .6 & .9 \\ .4 & .1 \end{bmatrix} \quad \text{and} \quad A_2 = \begin{bmatrix} .6 & .9 \\ .1 & .6 \end{bmatrix}.$$

22 Diagonalize A and compute $X\Lambda^k X^{-1}$ to prove this formula for A^k:

$$A = \begin{bmatrix} 2 & -1 \\ -1 & 2 \end{bmatrix} \quad \text{has} \quad A^k = \frac{1}{2}\begin{bmatrix} 1+3^k & 1-3^k \\ 1-3^k & 1+3^k \end{bmatrix}.$$

23 The eigenvalues of A are 1 and 9, and the eigenvalues of B are -1 and 9:

$$A = \begin{bmatrix} 5 & 4 \\ 4 & 5 \end{bmatrix} \quad \text{and} \quad B = \begin{bmatrix} 4 & 5 \\ 5 & 4 \end{bmatrix}.$$

Find a matrix square root of A from $R = X\sqrt{\Lambda}\,X^{-1}$. Why is there no real matrix square root of B?

24 Suppose the same X diagonalizes both A and B. They have the *same eigenvectors* in $A = X\Lambda_1 X^{-1}$ and $B = X\Lambda_2 X^{-1}$. Prove that $AB = BA$.

25 The transpose of $A = X\Lambda X^{-1}$ is $A^{\mathrm{T}} = (X^{-1})^{\mathrm{T}}\Lambda X^{\mathrm{T}}$. The eigenvectors in $A^{\mathrm{T}}\boldsymbol{y} = \lambda\boldsymbol{y}$ are the columns of that matrix $(X^{-1})^{\mathrm{T}}$. They are often called ***left eigenvectors of A***, because $\boldsymbol{y}^{\mathrm{T}}A = \lambda\boldsymbol{y}^{\mathrm{T}}$. How do you multiply matrices to find this formula for A?

Sum of rank-1 matrices $A = X\Lambda X^{-1} = \lambda_1\boldsymbol{x}_1\boldsymbol{y}_1^{\mathrm{T}} + \cdots + \lambda_n\boldsymbol{x}_n\boldsymbol{y}_n^{\mathrm{T}}$.

26 **When is a matrix A similar to its eigenvalue matrix Λ?**

A and Λ always have the same eigenvalues. But similarity requires a matrix B with $A = B\Lambda B^{-1}$. Then B is the ______ matrix and A must have n independent ______.

I.7 Symmetric Positive Definite Matrices

Symmetric matrices $S = S^{\mathrm{T}}$ deserve all the attention they get. Looking at their eigenvalues and eigenvectors, you see why they are special:

1 **All n eigenvalues λ of a symmetric matrix S are real numbers.**

2 **The n eigenvectors q can be chosen orthogonal (perpendicular to each other).**

The identity matrix $S = I$ is an extreme case. All its eigenvalues are $\lambda = 1$. Every nonzero vector x is an eigenvector: $Ix = 1x$. This shows why we wrote "can be chosen" in Property 2 above. With repeated eigenvalues like $\lambda_1 = \lambda_2 = 1$, we have a choice of eigenvectors. We can choose them to be orthogonal. And we can rescale them to be unit vectors (length 1). Then those eigenvectors $q_1, \ldots q_n$ are not just orthogonal, they are **orthonormal. The eigenvector matrix for S has $Q^{\mathrm{T}}Q = I$**: orthonormal columns in Q.

$$q_i{}^{\mathrm{T}} q_j = \begin{cases} 0 & i \neq j \\ 1 & i = j \end{cases} \quad \text{leads to} \quad \begin{bmatrix} & q_1^{\mathrm{T}} & \\ & \vdots & \\ & q_n^{\mathrm{T}} & \end{bmatrix} \begin{bmatrix} & & \\ q_1 & \cdots & q_n \\ & & \end{bmatrix} = \begin{bmatrix} 1 & 0 & \cdot & 0 \\ 0 & 1 & 0 & \cdot \\ \cdot & 0 & 1 & 0 \\ 0 & \cdot & 0 & 1 \end{bmatrix}.$$

We write Q instead of X for the eigenvector matrix of S, to emphasize that these eigenvectors are orthonormal: $Q^{\mathrm{T}}Q = I$ and $Q^{\mathrm{T}} = Q^{-1}$. This eigenvector matrix is an **orthogonal matrix**. The usual $A = X\Lambda X^{-1}$ becomes $S = Q\Lambda Q^{\mathrm{T}}$:

Spectral Theorem	**Every real symmetric matrix has the form $S = Q\Lambda Q^{\mathrm{T}}$.**

Every matrix of that form is symmetric: Transpose $Q\Lambda Q^{\mathrm{T}}$ to get $Q^{\mathrm{TT}}\Lambda^{\mathrm{T}}Q^{\mathrm{T}} = Q\Lambda Q^{\mathrm{T}}$.

Quick Proofs: Orthogonal Eigenvectors and Real Eigenvalues

Suppose first that $Sx = \lambda x$ and $Sy = 0y$. The symmetric matrix S has a nonzero eigenvalue λ and a zero eigenvalue. Then y is in the nullspace of S and x is in the column space of S ($x = Sx/\lambda$ is a combination of the columns of S). *But S is symmetric: column space = row space!* Since the row space and nullspace are always orthogonal, we have proved that **x is orthogonal to y**.

When that second eigenvalue is not zero, we have $Sy = \alpha y$. In this case we look at the matrix $S - \alpha I$. Then $(S - \alpha I)y = 0y$ and $(S - \alpha I)x = (\lambda - \alpha)x$ with $\lambda - \alpha \neq 0$. Now y is in the nullspace and x is in the column space (= row space!) of $S - \alpha I$. So $y^{\mathrm{T}}x = 0$: *Orthogonal eigenvectors whenever the eigenvalues $\lambda \neq \alpha$ are different.*

Those paragraphs assumed real eigenvalues and real eigenvectors. To prove this, multiply $Sx = \lambda x$ by the complex conjugate vector $\overline{x}^{\mathrm{T}}$ (every i changes to $-i$). Then $\overline{x}^{\mathrm{T}}Sx = \lambda\overline{x}^{\mathrm{T}}x$. When we show that $\overline{x}^{\mathrm{T}}x$ and $\overline{x}^{\mathrm{T}}Sx$ are real, we know that **λ is real.**

$$\overline{x}^{\mathrm{T}}x = \overline{x}_1 x_1 + \cdots + \overline{x}_n x_n \quad \text{and every } \overline{x}_k x_k \text{ is } (a - ib)(a + ib) = a^2 + b^2 \ (\textbf{real})$$

Also real : $\overline{\boldsymbol{x}}^{\mathrm{T}} S\boldsymbol{x} = S_{11}\overline{x}_1 x_1 + S_{12}(\overline{x}_1 x_2 + x_1\overline{x}_2) + \cdots$ and again $\overline{x}_1 x_1$ is real

$$\overline{x}_1 x_2 + x_1\overline{x}_2 = (a - ib)(c + id) + (a + ib)(c - id) = 2ac + 2bd = \textbf{real}$$

Since $\overline{\boldsymbol{x}}^{\mathrm{T}}\boldsymbol{x} > 0$, **the ratio λ is real.** And $(S - \lambda I)\boldsymbol{x} = \mathbf{0}$ gives a real eigenvector.

Complex comment: Transposing $S\boldsymbol{x} = \lambda\boldsymbol{x}$ and taking complex conjugates gives $\overline{\boldsymbol{x}}^{\mathrm{T}}\,\overline{S}^{\mathrm{T}} = \overline{\lambda}\,\overline{\boldsymbol{x}}^{\mathrm{T}}$. For our real symmetric matrices, $\overline{S}^{\mathrm{T}}$ is exactly S. It is this double step, *transpose and conjugate*, that we depend on to give back S. Since the proof only needs $\overline{S}^{\mathrm{T}} = S$, it allows for complex matrices too : **When $\overline{S}^{\mathrm{T}} = S$ all eigenvalues of S are real.**

Complex example $\quad S = \begin{bmatrix} 2 & 3-3i \\ 3+3i & 5 \end{bmatrix} = \overline{S}^{\mathrm{T}}$ has real eigenvalues 8 and -1.

The key is $\overline{3+3i} = 3 - 3i$. The determinant is $(2)\,(5) - (3+3i)\,(3-3i) = 10 - 18 = -8$. The eigenvectors of this matrix are $\boldsymbol{x}_1 = (1, 1+i)$ and $\boldsymbol{x}_2 = (1-i, -1)$. Those vectors are orthogonal when we adjust complex inner products to $\overline{\boldsymbol{x}}_1^{\mathrm{T}}\boldsymbol{x}_2$. This is the correct inner product for complex vectors, and it produces $\overline{\boldsymbol{x}}_1^{\mathrm{T}}\boldsymbol{x}_2 = 0$:

$$\text{Change } \boldsymbol{x}_1^{\mathrm{T}}\boldsymbol{x}_2 = \begin{bmatrix} 1 & 1+i \end{bmatrix}\begin{bmatrix} 1-i \\ -1 \end{bmatrix} = -2i \text{ to } \overline{\boldsymbol{x}}_1^{\mathrm{T}}\boldsymbol{x}_2 = \begin{bmatrix} 1 & 1-i \end{bmatrix}\begin{bmatrix} 1-i \\ -1 \end{bmatrix} = 0.$$

Systems like MATLAB and **Julia** get the message : The vector $\boldsymbol{x}'$ and the matrix A' are automatically conjugated when they are transposed. *Every i changes to $-i$.* Then $\boldsymbol{x}'$ is $\overline{\boldsymbol{x}}^{\mathrm{T}}$ and A' is $\overline{A}^{\mathrm{T}}$. Another frequently used symbol for $\overline{\boldsymbol{x}}^{\mathrm{T}}$ and $\overline{A}^{\mathrm{T}}$ is a star : $\boldsymbol{x}^*$ and A^*.

Positive Definite Matrices

We are working with real symmetric matrices $S = S^{\mathrm{T}}$. All their eigenvalues are real. Some of those symmetric matrices (*not all*) have a further powerful property that puts them at the center of applied mathematics. Here is that important property :

Test 1 $\quad$ **A positive definite matrix has all positive eigenvalues.**

We would like to check for positive eigenvalues without computing those numbers λ. You will see four more tests for positive definite matrices, after these examples.

1 $S = \begin{bmatrix} 2 & 0 \\ 0 & 6 \end{bmatrix}$ is positive definite. Its eigenvalues 2 and 6 are both positive

2 $S = Q\begin{bmatrix} 2 & 0 \\ 0 & 6 \end{bmatrix}Q^{\mathrm{T}}$ is positive definite if $Q^{\mathrm{T}} = Q^{-1}$: same $\lambda = 2$ and 6

3 $S = C\begin{bmatrix} 2 & 0 \\ 0 & 6 \end{bmatrix}C^{\mathrm{T}}$ is positive definite if C is invertible (not obvious)

4 $S = \begin{bmatrix} a & b \\ b & c \end{bmatrix}$ is positive definite exactly when $a > 0$ and $ac > b^2$

5 $S = \begin{bmatrix} 2 & 0 \\ 0 & 0 \end{bmatrix}$ is only **positive semidefinite** : it has $\boldsymbol{\lambda \geq 0}$ but not $\lambda > 0$

The Energy-based Definition

May I bring forward the most important idea about positive definite matrices? This new approach doesn't directly involve eigenvalues, but it turns out to be a perfect test for $\lambda > 0$. This is a good definition of positive definite matrices: **the energy test.**

S is positive definite if the energy $\boldsymbol{x}^{\mathrm{T}}S\boldsymbol{x}$ is positive for all vectors $\boldsymbol{x} \neq \mathbf{0}$

(1)

Of course $S = I$ is positive definite: All $\lambda_i = 1$. The energy is $\boldsymbol{x}^{\mathrm{T}}I\boldsymbol{x} = \boldsymbol{x}^{\mathrm{T}}\boldsymbol{x}$, positive if $\boldsymbol{x} \neq 0$. Let me show you the energy in a 2 by 2 matrix. It depends on $\boldsymbol{x} = (x_1, x_2)$.

Energy $$\boldsymbol{x}^{\mathrm{T}}S\boldsymbol{x} = \begin{bmatrix} x_1 & x_2 \end{bmatrix}\begin{bmatrix} 2 & 4 \\ 4 & 9 \end{bmatrix}\begin{bmatrix} x_1 \\ x_2 \end{bmatrix} = 2\,x_1^2 + 8\,x_1x_2 + 9\,x_2^2$$

Is this positive for every x_1 and x_2 except $(x_1, x_2) = (0, 0)$? *Yes, it is a sum of squares*:

$$\boldsymbol{x}^{\mathrm{T}}S\boldsymbol{x} = 2x_1^2 + 8x_1x_2 + 9x_2^2 = 2\,(x_1 + 2x_2)^2 + x_2^2 = \text{ positive energy.}$$

We must connect positive energy $\boldsymbol{x}^{\mathrm{T}}S\boldsymbol{x} > 0$ to positive eigenvalues $\lambda > 0$:

If $S\boldsymbol{x} = \lambda\boldsymbol{x}$ then $\boldsymbol{x}^{\mathrm{T}}S\boldsymbol{x} = \lambda\boldsymbol{x}^{\mathrm{T}}\boldsymbol{x}$. So $\lambda > 0$ leads to $\boldsymbol{x}^{\mathrm{T}}S\boldsymbol{x} > 0$.

That line only tested the energy in each separate eigenvector $\boldsymbol{x}$. But the theory says that if every eigenvector has positive energy, **then all nonzero vectors $\boldsymbol{x}$ have positive energy**:

If $\boldsymbol{x}^{\mathrm{T}}S\boldsymbol{x} > 0$ for the eigenvectors of S, then $\boldsymbol{x}^{\mathrm{T}}S\boldsymbol{x} > 0$ for every nonzero vector $\boldsymbol{x}$.

Here is the reason. Every $\boldsymbol{x}$ is a combination $c_1\boldsymbol{x}_1 + \cdots + c_n\boldsymbol{x}_n$ of the eigenvectors. Those eigenvectors can be chosen orthogonal because S is symmetric. We will now show: $\boldsymbol{x}^{\mathrm{T}}S\boldsymbol{x}$ is a positive combination of the energies $\lambda_k\boldsymbol{x}_k^{\mathrm{T}}\boldsymbol{x}_k > 0$ in the separate eigenvectors.

$$\begin{aligned} \boldsymbol{x}^{\mathrm{T}} S \boldsymbol{x} &= (c_1 \boldsymbol{x}_1^{\mathrm{T}} + \cdots + c_n \boldsymbol{x}_n^{\mathrm{T}})\, S\, (c_1 \boldsymbol{x}_1 + \cdots + c_n \boldsymbol{x}_n) \\ &= (c_1 \boldsymbol{x}_1^{\mathrm{T}} + \cdots + c_n \boldsymbol{x}_n^{\mathrm{T}})\, (c_1 \lambda_1 \boldsymbol{x}_1 + \cdots + c_n \lambda_n \boldsymbol{x}_n) \\ &= c_1^2 \lambda_1 \boldsymbol{x}_1^{\mathrm{T}} \boldsymbol{x}_1 + \cdots + c_n^2 \lambda_n \boldsymbol{x}_n^{\mathrm{T}} \boldsymbol{x}_n > \mathbf{0} \textbf{ if every } \boldsymbol{\lambda_i} > \mathbf{0}. \end{aligned}$$

From line 2 to line 3 we used the orthogonality of the eigenvectors of S: $\boldsymbol{x}_i^{\mathrm{T}} \boldsymbol{x}_j = 0$. Here is a typical use for the energy test, without knowing any eigenvalues or eigenvectors.

If S_1 and S_2 are symmetric positive definite, so is $S_1 + S_2$

Proof by adding energies : $\quad \boldsymbol{x}^{\mathrm{T}}(S_1 + S_2)\,\boldsymbol{x} = \boldsymbol{x}^{\mathrm{T}} S_1\, \boldsymbol{x} + \boldsymbol{x}^{\mathrm{T}} S_2\, \boldsymbol{x} > 0 + 0$

The eigenvalues and eigenvectors of $S_1 + S_2$ are not easy to find. Energies just add.

Three More Equivalent Tests

So far we have tests **1** and **2**: positive eigenvalues and positive energy. That energy test quickly produces three more useful tests (and probably others, but we stop with three):

Test 3	$\boldsymbol{S = A^{\mathrm{T}} A}$ **for a matrix $\boldsymbol{A}$ with independent columns**
Test 4	**All the leading determinants $\boldsymbol{D_1, D_2, \ldots, D_n}$ of $\boldsymbol{S}$ are positive**
Test 5	**All the pivots of $\boldsymbol{S}$ are positive** (in elimination)

Test 3 applies to $S = A^{\mathrm{T}} A$. Why must columns of A be independent in this test? Watch these parentheses:

$$\boldsymbol{S = A^{\mathrm{T}} A} \qquad \textbf{Energy} = \boldsymbol{x}^{\mathrm{T}} S \boldsymbol{x} = \boldsymbol{x}^{\mathrm{T}} A^{\mathrm{T}} A \boldsymbol{x} = (A\boldsymbol{x})^{\mathrm{T}} (A\boldsymbol{x}) = ||A\boldsymbol{x}||^2. \tag{2}$$

Those parentheses are the key. The energy is the **length squared** of the vector $A\boldsymbol{x}$. This energy is positive provided $A\boldsymbol{x}$ is not the zero vector. To assure $A\boldsymbol{x} \neq \mathbf{0}$ when $\boldsymbol{x} \neq \mathbf{0}$, the columns of A must be independent. In this 2 by 3 example, A has *dependent columns*:

$$S = A^{\mathrm{T}} A = \begin{bmatrix} 1 & 1 \\ 1 & 2 \\ 1 & 3 \end{bmatrix} \begin{bmatrix} 1 & 1 & 1 \\ 1 & 2 & 3 \end{bmatrix} = \begin{bmatrix} 2 & 3 & 4 \\ 3 & 5 & 7 \\ 4 & 7 & 10 \end{bmatrix} \text{ is } \textbf{not} \text{ positive definite.}$$

This A has column 1 + column 3 = 2 (column 2). Then $\boldsymbol{x} = (1, -2, 1)$ has zero energy. It is an eigenvector of $A^{\mathrm{T}} A$ with $\lambda = 0$. Then $S = A^{\mathrm{T}} A$ is only positive semidefinite.

Equation (2) says that $A^{\mathrm{T}} A$ is at least *semidefinite*, because $\boldsymbol{x}^{\mathrm{T}} S \boldsymbol{x} = ||A\boldsymbol{x}||^2$ is never negative. ***Semidefinite allows energy / eigenvalues / determinants / pivots of S to be zero.***

Determinant Test and Pivot Test

The determinant test is the quickest for a small matrix. I will mark the four "leading determinants" D_1, D_2, D_3, D_4 in this 4 by 4 symmetric second difference matrix.

$$S = \begin{bmatrix} 2 & -1 & & \\ -1 & 2 & -1 & \\ & -1 & 2 & -1 \\ & & -1 & 2 \end{bmatrix} \quad \text{has} \quad \begin{array}{l} \text{1st determinant } D_1 = \mathbf{2} \\ \text{2nd determinant } D_2 = \mathbf{3} \\ \text{3rd determinant } D_3 = \mathbf{4} \\ \text{4th determinant } D_4 = \mathbf{5} \end{array}$$

The determinant test is here passed! The energy $\boldsymbol{x}^{\mathrm{T}} S \boldsymbol{x}$ must be positive too.

Leading determinants are closely related to pivots (the numbers on the diagonal after elimination). Here the first pivot is **2**. The second pivot $\frac{\mathbf{3}}{\mathbf{2}}$ appears when $\frac{1}{2}$(row 1) is added to row 2. The third pivot $\frac{\mathbf{4}}{\mathbf{3}}$ appears when $\frac{2}{3}$(new row 2) is added to row 3. Those fractions $\frac{2}{1}, \frac{3}{2}, \frac{4}{3}$ are ratios of determinants! The last pivot is $\frac{\mathbf{5}}{\mathbf{4}}$.

The kth pivot equals the ratio $\dfrac{D_k}{D_{k-1}}$ of the leading determinants (sizes k and $k-1$)

So the pivots are all positive when the leading determinants are all positive.

I can quickly connect these two tests (**4** and **5**) to the third test $S = A^{\mathrm{T}}A$. In fact elimination on S produces an important choice of A. Remember that *elimination* = *triangular factorization* ($\boldsymbol{S} = \boldsymbol{LU}$). Up to now L has had 1's on the diagonal and U contained the pivots. But with symmetric matrices we can balance S as $\boldsymbol{LDL}^{\mathbf{T}}$:

$$\begin{bmatrix} 2 & -1 & 0 \\ -1 & 2 & -1 \\ 0 & -1 & 2 \end{bmatrix} = \begin{bmatrix} 1 & & \\ -\frac{1}{2} & 1 & \\ 0 & -\frac{2}{3} & 1 \end{bmatrix} \begin{bmatrix} 2 & -1 & 0 \\ & \frac{3}{2} & -1 \\ & & \frac{4}{3} \end{bmatrix} \qquad \boldsymbol{S} = \boldsymbol{LU} \quad (3)$$

$$\begin{array}{c}\text{pull out}\\ \text{the pivots}\\ \text{in } D\end{array} = \begin{bmatrix} 1 & & \\ -\frac{1}{2} & 1 & \\ 0 & -\frac{2}{3} & 1 \end{bmatrix} \begin{bmatrix} \mathbf{2} & & \\ & \frac{\mathbf{3}}{\mathbf{2}} & \\ & & \frac{\mathbf{4}}{\mathbf{3}} \end{bmatrix} \begin{bmatrix} 1 & -\frac{1}{2} & 0 \\ & 1 & -\frac{2}{3} \\ & & 1 \end{bmatrix} = \boldsymbol{LDL}^{\mathbf{T}} \quad (4)$$

$$\begin{array}{c}\textbf{share those pivots}\\ \textbf{between } \boldsymbol{A}^{\mathbf{T}} \textbf{ and } \boldsymbol{A}\end{array} = \begin{bmatrix} \sqrt{2} & & \\ -\sqrt{\frac{1}{2}} & \sqrt{\frac{3}{2}} & \\ 0 & -\sqrt{\frac{2}{3}} & \sqrt{\frac{4}{3}} \end{bmatrix} \begin{bmatrix} \sqrt{2} & -\sqrt{\frac{1}{2}} & 0 \\ & \sqrt{\frac{3}{2}} & -\sqrt{\frac{2}{3}} \\ & & \sqrt{\frac{4}{3}} \end{bmatrix} = \boldsymbol{A}^{\mathbf{T}}\boldsymbol{A} \quad (5)$$

I am sorry about those square roots—but the pattern $S = A^{\mathrm{T}}A$ is beautiful: $\boldsymbol{A} = \sqrt{\boldsymbol{D}}\boldsymbol{L}^{\mathbf{T}}$.

Elimination factors every positive definite S into $A^{\mathrm{T}}A$ (A is upper triangular)

This is the Cholesky factorization $S = A^{\mathrm{T}}A$ with $\sqrt{\text{pivots}}$ on the main diagonal of A.

The Test $S = A^{\mathrm{T}}A$: Two Special Choices for A

To apply the $S = A^{\mathrm{T}}A$ test when S is positive definite, we must find at least one possible A. There are many choices for A, including (1) **symmetric** and (2) **triangular**.

1 If $S = Q\Lambda Q^{\mathrm{T}}$, take square roots of those eigenvalues. Then $A = Q\sqrt{\Lambda}Q^{\mathrm{T}} = A^{\mathrm{T}}$.

2 If $S = LU = LDL^{\mathrm{T}}$ with positive pivots in D, then $S = (L\sqrt{D})\,(\sqrt{D}L^{\mathrm{T}})$.

Summary The five tests for positive definiteness of S involve different parts of linear algebra—pivots from elimination, determinants, eigenvalues, and $S = A^{\mathrm{T}}A$. Each test gives a complete answer by itself: positive definite or semidefinite or neither.

Positive energy $x^{\mathrm{T}}Sx > 0$ is the best definition: it connects them all.

Positive Definite Matrices and Minimum Problems

Suppose S is a symmetric positive definite 2 by 2 matrix. Apply four of the tests:

$$S = \begin{bmatrix} a & b \\ b & c \end{bmatrix} \qquad \begin{array}{ll} \textbf{determinants} & a > 0, ac - b^2 > 0 \\ \textbf{eigenvalues} & \lambda_1 > 0,\ \lambda_2 > 0 \end{array} \qquad \begin{array}{ll} \textbf{pivots} & a > 0, (ac - b^2)/a > 0 \\ \textbf{energy} & ax^2 + 2bxy + cy^2 > 0 \end{array}$$

I will choose an example with $a = c = 5$ and $b = 4$. This matrix S has $\lambda = 9$ and $\lambda = 1$.

$$\textbf{Energy } \boldsymbol{E = x^{\mathrm{T}}Sx} \qquad \begin{bmatrix} x & y \end{bmatrix}\begin{bmatrix} 5 & 4 \\ 4 & 5 \end{bmatrix}\begin{bmatrix} x \\ y \end{bmatrix} = \boldsymbol{5\,x^2 + 8\,xy + 5\,y^2} > 0$$

The graph of that energy function $E(x, y)$ is a **bowl opening upwards**. The bottom point of the bowl has energy $E = 0$ when $x = y = 0$. This connects minimum problems in calculus with positive definite matrices in linear algebra.

Part VI of this book describes numerical minimization. For the best problems, the function is **strictly convex**—like a parabola that opens upward. Here is a perfect test: *The matrix of second derivatives is positive definite at all points.* We are in high dimensions, but linear algebra identifies the crucial properties of the second derivative matrix.

For an ordinary function $f(x)$ of one variable x, the test for a minimum is famous:

$$\text{Minimum if } \textbf{first} \text{ derivative } \frac{df}{dx} = 0 \text{ and } \textbf{second} \text{ derivative } \frac{d^2f}{dx^2} > 0 \text{ at } x = x_0$$

For $f(x, y)$ with two variables, the second derivatives go into a matrix: positive definite!

$$\begin{array}{c}\textbf{Minimum} \\ \textbf{at } \boldsymbol{x_0, y_0}\end{array} \quad \frac{\partial f}{\partial x} = 0 \text{ and } \frac{\partial f}{\partial y} = 0 \text{ and } \begin{bmatrix} \partial^2 f/\partial x^2 & \partial^2 f/\partial x\partial y \\ \partial^2 f/\partial x\partial y & \partial^2 f/\partial y^2 \end{bmatrix} \begin{array}{c}\textbf{is positive definite} \\ \textbf{at } \boldsymbol{x_0, y_0}\end{array}$$

The graph of $z = f(x, y)$ is flat at that point x_0, y_0 because $\partial f/\partial x = \partial f/\partial y = 0$. The graph goes upwards whenever the second derivative matrix is positive definite. So we have a minimum point of the function $f(x, y)$.

Second derivatives $S = \begin{bmatrix} a & b \\ b & c \end{bmatrix}$

$a > 0$ and $ac > b^2$

$f = \frac{1}{2}\boldsymbol{x}^{\mathrm{T}} S \boldsymbol{x} > 0$

x y

The graph of $2f = ax^2 + 2bxy + cy^2$ is a bowl when S is positive definite.

If S has a negative eigenvalue $\lambda < 0$, the graph goes below zero. There is a *maximum* if S is negative definite (all $\lambda < 0$, upside down bowl). Or a *saddle point* when S has both positive and negative eigenvalues. A saddle point matrix is "*indefinite*".

Optimization and Machine Learning

Part VI of this book will describe **gradient descent**. Each step takes the steepest direction, toward the bottom point $\boldsymbol{x}^*$ of the bowl. But that steepest direction changes as we descend. This is where calculus meets linear algebra, at the minimum point $\boldsymbol{x}^*$.

Calculus	The partial derivatives of f are all zero at $\boldsymbol{x}^*$: $\dfrac{\partial f}{\partial x_i} = 0$
Linear algebra	The matrix S of second derivatives $\dfrac{\partial^2 f}{\partial x_i \, \partial x_j}$ is positive definite

If S is positive definite (or semidefinite) at all points $\boldsymbol{x} = (x_1, \ldots, x_n)$, then **the function $f(\boldsymbol{x})$ is convex**. If the eigenvalues of S stay above some positive number δ, then **the function $f(\boldsymbol{x})$ is strictly convex**. These are the best functions to optimize. They have only one minimum, and gradient descent will find it.

Machine learning produces "loss functions" with hundreds of thousands of variables. They measure the error—which we minimize. But computing all the second derivatives is completely impossible. We use first derivatives to tell us a direction to move—the error drops fastest in the steepest direction. Then we take another descent step in a new direction. **This is the central computation in least squares and neural nets and deep learning.**

The Ellipse $ax^2 + 2bxy + cy^2 = 1$

Stay with a positive definite matrix S. The graph of its energy $E = \boldsymbol{x}^{\mathrm{T}} S \boldsymbol{x}$ is a bowl opening upwards. Cut through that bowl at height $\boldsymbol{x}^{\mathrm{T}} S \boldsymbol{x} = 1$. Then the curve that goes around the cut is an **ellipse**.

$S = \begin{bmatrix} 5 & 4 \\ 4 & 5 \end{bmatrix}$ has $\lambda = \mathbf{9}$ and $\mathbf{1}$ Energy ellipse $\mathbf{5x^2 + 8xy + 5y^2 = 1}$ in Figure I.9

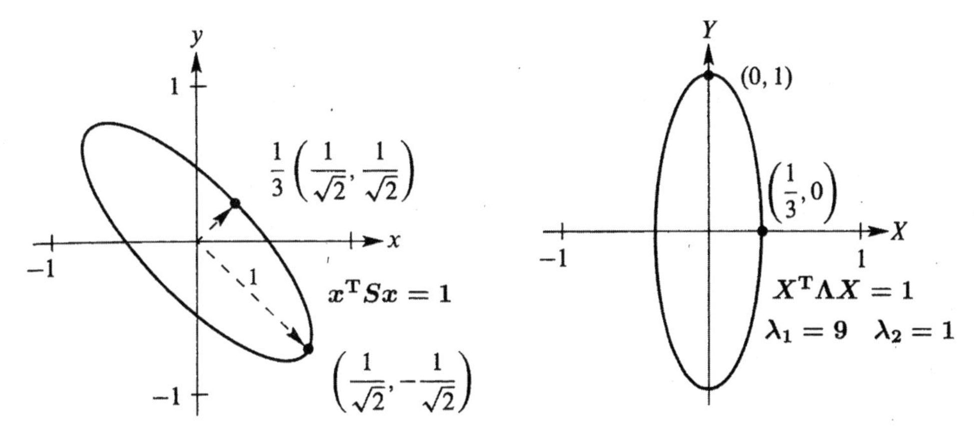

Figure I.9: The tilted ellipse $5x^2 + 8xy + 5y^2 = 1$. Lined up it is $9X^2 + Y^2 = 1$.

The eigenvectors are $\boldsymbol{q}_1 = (1, 1)$ and $\boldsymbol{q}_2 = (1, -1)$. Divide by $\sqrt{2}$ to get unit vectors. Then $S = Q\Lambda Q^T$. Now multiply by $\boldsymbol{x}^T = \begin{bmatrix} x & y \end{bmatrix}$ on the left and $\boldsymbol{x}$ on the right to get the energy $\boldsymbol{x}^T S\boldsymbol{x} = (\boldsymbol{x}^T Q)\Lambda(Q^T\boldsymbol{x})$. The eigenvalues of S are 9 and 1.

$$\boldsymbol{x}^T S\boldsymbol{x} = \textbf{sum of squares} \qquad 5x^2+8xy+5y^2 = 9\left(\frac{x+y}{\sqrt{2}}\right)^2 + 1\left(\frac{x-y}{\sqrt{2}}\right)^2. \tag{6}$$

9 and 1 come from Λ. Inside the squares you see $\boldsymbol{q}_1 = (1, 1)/\sqrt{2}$ and $\boldsymbol{q}_2 = (1, -1)/\sqrt{2}$.

The axes of the tilted ellipse point along those eigenvectors of S. This explains why $S = Q\Lambda Q^T$ is the "principal axis theorem"—it displays the axes. Not only directions (from the eigenvectors) but also the axis lengths (from the λ's): **Length** $= \mathbf{1/\sqrt{\lambda}}$. To see it all, use capital letters for the new coordinates X, Y that line up the ellipse:

$$\textbf{Lined up} \qquad \frac{x+y}{\sqrt{2}} = X \qquad \text{and} \qquad \frac{x-y}{\sqrt{2}} = Y \qquad \text{and} \qquad \mathbf{9X^2 + Y^2 = 1}.$$

The largest value of X^2 is $1/9$. The endpoint of the shorter axis has $X = 1/3$ and $Y = 0$. Notice: The *bigger* eigenvalue $\lambda_1 = 9$ gives the *shorter* axis, of half-length $1/\sqrt{\lambda_1} = 1/3$. The smaller eigenvalue $\lambda_2 = 1$ gives the greater length $1/\sqrt{\lambda_2} = 1$: Y axis in Figure I.9.

In the xy system, the axes are along the eigenvectors of S. In the XY system, the **axes are along the eigenvectors of Λ—the coordinate axes.** All from $S = Q\Lambda Q^T$.

$\boldsymbol{S} = \boldsymbol{Q\Lambda Q}^T$ is positive definite when all $\lambda_i > 0$. The graph of $\boldsymbol{x}^T S\boldsymbol{x} = 1$ is an ellipse, with its axes pointing along the eigenvectors of S.

$$\textbf{Ellipse} \qquad \begin{bmatrix} x & y \end{bmatrix} Q\Lambda Q^T \begin{bmatrix} x \\ y \end{bmatrix} = \begin{bmatrix} X & Y \end{bmatrix} \Lambda \begin{bmatrix} X \\ Y \end{bmatrix} = \boldsymbol{\lambda_1 X^2 + \lambda_2 Y^2 = 1}. \tag{7}$$

Problem Set I.7

1 Suppose $S^{\mathrm{T}} = S$ and $S\boldsymbol{x} = \lambda\boldsymbol{x}$ and $S\boldsymbol{y} = \alpha\boldsymbol{y}$ are all real. Show that

$$\boldsymbol{y}^{\mathrm{T}}S\boldsymbol{x} = \lambda\boldsymbol{y}^{\mathrm{T}}\boldsymbol{x} \quad \text{and} \quad \boldsymbol{x}^{\mathrm{T}}S\boldsymbol{y} = \alpha\boldsymbol{x}^{\mathrm{T}}\boldsymbol{y} \quad \text{and} \quad \boldsymbol{y}^{\mathrm{T}}S\boldsymbol{x} = \boldsymbol{x}^{\mathrm{T}}S\boldsymbol{y}.$$

Show that $\boldsymbol{y}^{\mathrm{T}}\boldsymbol{x}$ must be zero if $\lambda \neq \alpha$: **orthogonal eigenvectors.**

2 Which of S_1, S_2, S_3, S_4 has two positive eigenvalues? Use a test, don't compute the λ's. Also find an $\boldsymbol{x}$ so that $\boldsymbol{x}^{\mathrm{T}}S_1\boldsymbol{x} < 0$, so S_1 is not positive definite.

$$S_1 = \begin{bmatrix} 5 & 6 \\ 6 & 7 \end{bmatrix} \quad S_2 = \begin{bmatrix} -1 & -2 \\ -2 & -5 \end{bmatrix} \quad S_3 = \begin{bmatrix} 1 & 10 \\ 10 & 100 \end{bmatrix} \quad S_4 = \begin{bmatrix} 1 & 10 \\ 10 & 101 \end{bmatrix}.$$

3 For which numbers b and c are these matrices positive definite?

$$S = \begin{bmatrix} 1 & b \\ b & 9 \end{bmatrix} \qquad S = \begin{bmatrix} 2 & 4 \\ 4 & c \end{bmatrix} \qquad S = \begin{bmatrix} c & b \\ b & c \end{bmatrix}.$$

With the pivots in D and multiplier in L, factor each A into LDL^{T}.

4 Here is a quick "proof" that the eigenvalues of every real matrix A are real:

$$\textbf{False proof} \quad A\boldsymbol{x} = \lambda\boldsymbol{x} \quad \text{gives} \quad \boldsymbol{x}^{\mathrm{T}}A\boldsymbol{x} = \lambda\boldsymbol{x}^{\mathrm{T}}\boldsymbol{x} \quad \text{so} \quad \lambda = \frac{\boldsymbol{x}^{\mathrm{T}}A\boldsymbol{x}}{\boldsymbol{x}^{\mathrm{T}}\boldsymbol{x}} = \frac{\text{real}}{\text{real}}.$$

Find the flaw in this reasoning—a hidden assumption that is not justified. You could test those steps on the 90° rotation matrix $[\,0 \;\; -1;\;\; 1 \;\; 0\,]$ with $\lambda = i$ and $\boldsymbol{x} = (i, 1)$.

5 Write S and B in the form $\lambda_1\boldsymbol{x}_1\boldsymbol{x}_1^{\mathrm{T}} + \lambda_2\boldsymbol{x}_2\boldsymbol{x}_2^{\mathrm{T}}$ of the spectral theorem $Q\Lambda Q^T$:

$$S = \begin{bmatrix} 3 & 1 \\ 1 & 3 \end{bmatrix} \qquad B = \begin{bmatrix} 9 & 12 \\ 12 & 16 \end{bmatrix} \qquad (\text{keep } \|\boldsymbol{x}_1\| = \|\boldsymbol{x}_2\| = 1).$$

6 (Recommended) This matrix M is antisymmetric and also _____. Then all its eigenvalues are pure imaginary and they also have $|\lambda| = 1$. ($\|M\boldsymbol{x}\| = \|\boldsymbol{x}\|$ for every $\boldsymbol{x}$ so $\|\lambda\boldsymbol{x}\| = \|\boldsymbol{x}\|$ for eigenvectors.) Find all four eigenvalues from the trace of M:

$$M = \frac{1}{\sqrt{3}}\begin{bmatrix} 0 & 1 & 1 & 1 \\ -1 & 0 & -1 & 1 \\ -1 & 1 & 0 & -1 \\ -1 & -1 & 1 & 0 \end{bmatrix} \quad \text{can only have eigenvalues } i \text{ or } -i.$$

7 Show that this A (**symmetric but complex**) has only one line of eigenvectors:

$$A = \begin{bmatrix} i & 1 \\ 1 & -i \end{bmatrix} \quad \text{is not even diagonalizable: eigenvalues } \lambda = 0 \text{ and } 0.$$

$A^{\mathrm{T}} = A$ is not such a special property for complex matrices. The good property is $\overline{\boldsymbol{A}}^{\mathbf{T}} = \boldsymbol{A}$. Then all eigenvalues are real and A has n orthogonal eigenvectors.

8 This A is nearly symmetric. But its eigenvectors are far from orthogonal:

$$A = \begin{bmatrix} 1 & 10^{-15} \\ 0 & 1+10^{-15} \end{bmatrix} \quad \text{has eigenvectors} \quad \begin{bmatrix} 1 \\ 0 \end{bmatrix} \quad \text{and} \quad [?]$$

What is the angle between the eigenvectors?

9 *Which symmetric matrices S are also orthogonal*? Then $S^{\mathrm{T}} = S$ and $S^{\mathrm{T}} = S^{-1}$.

(a) Show how symmetry and orthogonality lead to $S^2 = I$.

(b) What are the possible eigenvalues of S? Describe all possible Λ.

Then $S = Q\Lambda Q^{\mathrm{T}}$ for one of those eigenvalue matrices Λ and an orthogonal Q.

10 If S is symmetric, show that $A^{\mathrm{T}}SA$ is also symmetric (take the transpose of $A^{\mathrm{T}}SA$). Here A is m by n and S is m by m. Are eigenvalues of S = eigenvalues of $A^{\mathrm{T}}SA$?

In case A is square and invertible, $A^{\mathrm{T}}SA$ is called *congruent* to S. They have the same number of positive, negative, and zero eigenvalues: ***Law of Inertia***.

11 Here is a way to show that a is *in between* the eigenvalues λ_1 and λ_2 of S:

$$S = \begin{bmatrix} a & b \\ b & c \end{bmatrix} \qquad \begin{array}{l} \det(S - \lambda I) = \lambda^2 - a\lambda - c\lambda + ac - b^2 \\ \text{is a parabola opening upwards (because of } \lambda^2) \end{array}$$

Show that $\det(S - \lambda I)$ is negative at $\lambda = a$. So the parabola crosses the axis left and right of $\lambda = a$. It crosses at the two eigenvalues of S so they must enclose a.

The $n-1$ eigenvalues of A always fall between the n eigenvalues of $S = \begin{bmatrix} A & \boldsymbol{b} \\ \boldsymbol{b}^{\mathrm{T}} & c \end{bmatrix}$.

Section III.2 will explain this interlacing of eigenvalues.

12 The energy $\boldsymbol{x}^{\mathrm{T}}S\boldsymbol{x} = 2x_1x_2$ certainly has a saddle point and not a minimum at $(0,0)$. What symmetric matrix S produces this energy? What are its eigenvalues?

13 Test to see if $A^{\mathrm{T}}A$ is positive definite in each case: A needs independent columns.

$$A = \begin{bmatrix} 1 & 2 \\ 0 & 3 \end{bmatrix} \quad \text{and} \quad A = \begin{bmatrix} 1 & 1 \\ 1 & 2 \\ 2 & 1 \end{bmatrix} \quad \text{and} \quad A = \begin{bmatrix} 1 & 1 & 2 \\ 1 & 2 & 1 \end{bmatrix}.$$

14 Find the 3 by 3 matrix S and its pivots, rank, eigenvalues, and determinant:

$$\begin{bmatrix} x_1 & x_2 & x_3 \end{bmatrix} \begin{bmatrix} \\ & S & \\ \\ \end{bmatrix} \begin{bmatrix} x_1 \\ x_2 \\ x_3 \end{bmatrix} = 4(x_1 - x_2 + 2x_3)^2.$$

15 Compute the three upper left determinants of S to establish positive definiteness. Verify that their ratios give the second and third pivots.

$$\textbf{Pivots = ratios of determinants} \qquad S = \begin{bmatrix} 2 & 2 & 0 \\ 2 & 5 & 3 \\ 0 & 3 & 8 \end{bmatrix}.$$

16 For what numbers c and d are S and T positive definite? Test their 3 determinants:

$$S = \begin{bmatrix} c & 1 & 1 \\ 1 & c & 1 \\ 1 & 1 & c \end{bmatrix} \quad \text{and} \quad T = \begin{bmatrix} 1 & 2 & 3 \\ 2 & d & 4 \\ 3 & 4 & 5 \end{bmatrix}.$$

17 Find a matrix with $a > 0$ and $c > 0$ and $a + c > 2b$ that has a negative eigenvalue.

18 A positive definite matrix cannot have a zero (or even worse, a negative number) on its main diagonal. Show that this matrix fails to have $\boldsymbol{x}^{\mathrm{T}} S \boldsymbol{x} > 0$:

$$\begin{bmatrix} x_1 & x_2 & x_3 \end{bmatrix} \begin{bmatrix} 4 & 1 & 1 \\ 1 & 0 & 2 \\ 1 & 2 & 5 \end{bmatrix} \begin{bmatrix} x_1 \\ x_2 \\ x_3 \end{bmatrix} \text{ is not positive when } (x_1, x_2, x_3) = (\ \ ,\ \ ,\ \ ).$$

19 A diagonal entry s_{jj} of a symmetric matrix cannot be smaller than all the λ's. If it were, then $S - s_{jj} I$ would have _____ eigenvalues and would be positive definite. But $S - s_{jj} I$ has a _____ on the main diagonal, impossible by Problem 18.

20 From $S = Q \Lambda Q^{\mathrm{T}}$ compute the positive definite symmetric square root $Q \sqrt{\Lambda} Q^{\mathrm{T}}$ of each matrix. Check that this square root gives $A^{\mathrm{T}} A = S$:

$$S = \begin{bmatrix} 5 & 4 \\ 4 & 5 \end{bmatrix} \quad \text{and} \quad S = \begin{bmatrix} 10 & 6 \\ 6 & 10 \end{bmatrix}.$$

21 Draw the tilted ellipse $x^2 + xy + y^2 = 1$ and find the half-lengths of its axes from the eigenvalues of the corresponding matrix S.

22 In the Cholesky factorization $S = A^{\mathrm{T}} A$, with $A = \sqrt{D} L^{\mathrm{T}}$, the square roots of the pivots are on the diagonal of A. Find A (upper triangular) for

$$S = \begin{bmatrix} 9 & 0 & 0 \\ 0 & 1 & 2 \\ 0 & 2 & 8 \end{bmatrix} \quad \text{and} \quad S = \begin{bmatrix} 1 & 1 & 1 \\ 1 & 2 & 2 \\ 1 & 2 & 7 \end{bmatrix}.$$

23 Suppose C is positive definite (so $\boldsymbol{y}^{\mathrm{T}} C \boldsymbol{y} > 0$ whenever $\boldsymbol{y} \neq \mathbf{0}$) and A has independent columns (so $A\boldsymbol{x} \neq \mathbf{0}$ whenever $\boldsymbol{x} \neq \mathbf{0}$). Apply the energy test to $\boldsymbol{x}^{\mathrm{T}} A^{\mathrm{T}} C A \boldsymbol{x}$ to show that $\boldsymbol{S = A^{\mathrm{T}} C A}$ ***is positive definite: the crucial matrix in engineering.***

The Minimum of a Function $F(x, y, z)$

What tests would you expect for a minimum point? First come zero slopes:

First derivatives are zero $\dfrac{\partial F}{\partial x} = \dfrac{\partial F}{\partial y} = \dfrac{\partial F}{\partial z} = \mathbf{0}$ at the minimum point.

Next comes the linear algebra version of the usual calculus test $d^2 f/dx^2 > 0$:

Second derivative matrix H is positive definite $H = \begin{bmatrix} F_{xx} & F_{xy} & F_{xz} \\ F_{yx} & F_{yy} & F_{yz} \\ F_{zx} & F_{zy} & F_{zz} \end{bmatrix}$

Here $F_{xy} = \dfrac{\partial}{\partial x}\left(\dfrac{\partial F}{\partial y}\right) = \dfrac{\partial}{\partial y}\left(\dfrac{\partial F}{\partial x}\right) = F_{yx}$ is a 'mixed" second derivative.

24 For $F_1(x,y) = \frac{1}{4}x^4 + x^2y + y^2$ and $F_2(x,y) = x^3 + xy - x$ find the second derivative matrices H_1 and H_2 (the **Hessian matrices**):

Test for minimum $H = \begin{bmatrix} \partial^2 F/\partial x^2 & \partial^2 F/\partial x \partial y \\ \partial^2 F/\partial y \partial x & \partial^2 F/\partial y^2 \end{bmatrix}$ is positive definite

H_1 is positive definite so F_1 is concave up (= convex). Find the minimum point of F_1. *Find the saddle point of* F_2 (look only where first derivatives are zero).

25 Which values of c give a bowl and which c give a saddle point for the graph of $z = 4x^2 + 12xy + cy^2$? Describe this graph at the borderline value of c.

26 Without multiplying $S = \begin{bmatrix} \cos\theta & -\sin\theta \\ \sin\theta & \cos\theta \end{bmatrix} \begin{bmatrix} 2 & 0 \\ 0 & 5 \end{bmatrix} \begin{bmatrix} \cos\theta & \sin\theta \\ -\sin\theta & \cos\theta \end{bmatrix}$, find

(a) the determinant of S
(b) the eigenvalues of S
(c) the eigenvectors of S
(d) a reason why S is symmetric positive definite.

27 For which a and c is this matrix positive definite? For which a and c is it positive semidefinite (this includes definite)?

$$S = \begin{bmatrix} a & a & a \\ a & a+c & a-c \\ a & a-c & a+c \end{bmatrix}$$

All 5 tests are possible.
The energy $x^{\mathrm{T}} S x$ equals
$a(x_1 + x_2 + x_3)^2 + c(x_2 - x_3)^2$.

28 **Important**! Suppose S is positive definite with eigenvalues $\lambda_1 \geq \lambda_2 \geq \ldots \geq \lambda_n$.

(a) What are the eigenvalues of the matrix $\lambda_1 I - S$? Is it positive semidefinite?

(b) How does it follow that $\lambda_1 x^{\mathrm{T}} x \geq x^{\mathrm{T}} S x$ for every x?

(c) Draw this conclusion: **The maximum value of $x^{\mathrm{T}} S x / x^{\mathrm{T}} x$ is λ.**

Note Another way to **28** (c): **Maximize $x^{\mathrm{T}} S x$** subject to the condition $x^{\mathrm{T}} x = 1$.

This leads to $\dfrac{\partial}{\partial x}\left[x^{\mathrm{T}} S x - \lambda\left(x^{\mathrm{T}} x - 1\right)\right] = \mathbf{0}$ and then $Sx = \lambda x$ and $\lambda = \lambda_1$.

I.8 Singular Values and Singular Vectors in the SVD

The best matrices (real symmetric matrices S) have real eigenvalues and orthogonal eigenvectors. But for other matrices, the eigenvalues are complex or the eigenvectors are not orthogonal. If A is not square then $A\boldsymbol{x} = \lambda\boldsymbol{x}$ is impossible and eigenvectors fail (left side in $\mathbf{R}^m$, right side in $\mathbf{R}^n$). We need an idea that succeeds for every matrix.

The Singular Value Decomposition fills this gap in a perfect way. In our applications, A is often a matrix of data. The rows could tell us the age and height of 1000 children. Then A is 2 by 1000: definitely rectangular. Unless height is exactly proportional to age, the rank is $r = 2$ and that matrix A has two positive singular values σ_1 and σ_2.

The key point is that we need **two sets of singular vectors**, the $\boldsymbol{u}$'s and the $\boldsymbol{v}$'s. For a real m by n matrix, the n right singular vectors $\boldsymbol{v}_1, \ldots, \boldsymbol{v}_n$ are orthogonal in $\mathbf{R}^n$. The m left singular vectors $\boldsymbol{u}_1, \ldots, \boldsymbol{u}_m$ are perpendicular to each other in $\mathbf{R}^m$. The connection between n $\boldsymbol{v}$'s and m $\boldsymbol{u}$'s is not $A\boldsymbol{x} = \lambda\boldsymbol{x}$. That is for eigenvectors. **For singular vectors, each $A\boldsymbol{v}$ equals $\sigma\boldsymbol{u}$:**

$$\boxed{A\boldsymbol{v}_1 = \sigma_1\boldsymbol{u}_1 \quad \cdots \quad A\boldsymbol{v}_r = \sigma_r\boldsymbol{u}_r} \qquad \boxed{A\boldsymbol{v}_{r+1} = \mathbf{0} \quad \cdots \quad A\boldsymbol{v}_n = \mathbf{0}} \tag{1}$$

I have separated the first r $\boldsymbol{v}$'s and $\boldsymbol{u}$'s from the rest. That number r is the *rank of* A, the number of independent columns (and rows). Then r is the dimension of the column space and the row space. **We will have r positive singular values in descending order $\sigma_1 \geq \sigma_2 \geq \ldots \geq \sigma_r > 0$.** The last $n - r$ $\boldsymbol{v}$'s are in the nullspace of A, and the last $m - r$ $\boldsymbol{u}$'s are in the nullspace of A^{T}.

Our first step is to write equation (1) in matrix form. All of the right singular vectors $\boldsymbol{v}_1$ to $\boldsymbol{v}_n$ go in the columns of V. The left singular vectors $\boldsymbol{u}_1$ to $\boldsymbol{u}_m$ go in the columns of U. Those are **square orthogonal matrices** ($V^{\mathrm{T}} = V^{-1}$ and $U^{\mathrm{T}} = U^{-1}$) because their columns are orthogonal unit vectors. Then equation (1) becomes the full SVD, with square matrices V and U:

$$\boxed{\boldsymbol{AV = U\Sigma} \qquad A\begin{bmatrix} \boldsymbol{v}_1 \ldots \boldsymbol{v}_r \ldots \boldsymbol{v}_n \end{bmatrix} = \begin{bmatrix} \boldsymbol{u}_1 \ldots \boldsymbol{u}_r \ldots \boldsymbol{u}_m \end{bmatrix} \left[\begin{array}{ccc|c} \sigma_1 & & & \\ & \ddots & & \mathbf{0} \\ & & \sigma_r & \\ \hline & \mathbf{0} & & \mathbf{0} \end{array}\right]} \tag{2}$$

You see $A\boldsymbol{v}_k = \sigma_k\boldsymbol{u}_k$ in the first r columns above. That is the important part of the SVD. It shows the basis of $\boldsymbol{v}$'s for the row space of A and then $\boldsymbol{u}$'s for the column space. After the positive numbers $\sigma_1, \ldots, \sigma_r$ on the main diagonal of Σ, the rest of that matrix is all zero from the nullspaces of A and A^{T}.

The eigenvectors give $AX = X\Lambda$. But $AV = U\Sigma$ needs **two sets of singular vectors.**

Example 1 $AV = U\Sigma$

$$\begin{bmatrix} 3 & 0 \\ 4 & 5 \end{bmatrix} \frac{1}{\sqrt{2}} \begin{bmatrix} 1 & -1 \\ 1 & 1 \end{bmatrix} = \frac{1}{\sqrt{10}} \begin{bmatrix} 1 & -3 \\ 3 & 1 \end{bmatrix} \begin{bmatrix} 3\sqrt{5} & \\ & \sqrt{5} \end{bmatrix}$$

The matrix A is not symmetric, so V is different from U. The rank is 2, so there are two singular values $\sigma_1 = 3\sqrt{5}$ and $\sigma_2 = \sqrt{5}$. Their product $3 \cdot 5 = 15$ is the determinant of A (in this respect singular values are like eigenvalues). The columns of V are orthogonal and the columns of U are orthogonal. Those columns are unit vectors after the divisions by $\sqrt{2}$ and $\sqrt{10}$, so **V and U are orthogonal matrices**: $V^{\mathrm{T}} = V^{-1}$ and $U^{\mathrm{T}} = U^{-1}$.

That orthogonality allows us to go from $AV = U\Sigma$ to the usual and famous expression of the SVD: Multiply both sides of $AV = U\Sigma$ by $V^{-1} = V^{\mathrm{T}}$.

$$\boxed{\textbf{The Singular Value Decomposition of } A \textbf{ is} \quad A = U\Sigma V^{\mathrm{T}}.} \tag{3}$$

Then column-row multiplication of $U\Sigma$ times V^{T} separates A into r pieces of rank 1:

Pieces of the SVD
$$\boxed{A = U\Sigma V^{\mathrm{T}} = \sigma_1 u_1 v_1^{\mathrm{T}} + \cdots + \sigma_r u_r v_r^{\mathrm{T}}.} \tag{4}$$

In the 2 by 2 example, the first piece is more important than the second piece because $\sigma_1 = 3\sqrt{5}$ is greater than $\sigma_2 = \sqrt{5}$. To recover A, add the pieces $\sigma_1 u_1 v_1^{\mathrm{T}} + \sigma_2 u_2 v_2^{\mathrm{T}}$:

$$\frac{3\sqrt{5}}{\sqrt{10}\sqrt{2}} \begin{bmatrix} 1 \\ 3 \end{bmatrix} \begin{bmatrix} 1 & 1 \end{bmatrix} + \frac{\sqrt{5}}{\sqrt{10}\sqrt{2}} \begin{bmatrix} -3 \\ 1 \end{bmatrix} \begin{bmatrix} -1 & 1 \end{bmatrix} = \frac{3}{2} \begin{bmatrix} 1 & 1 \\ 3 & 3 \end{bmatrix} + \frac{1}{2} \begin{bmatrix} 3 & -3 \\ -1 & 1 \end{bmatrix} = \begin{bmatrix} 3 & 0 \\ 4 & 5 \end{bmatrix}$$

This simplified because $\sqrt{5}/\sqrt{10}\sqrt{2}$ equals $1/2$. Notice that the right singular vectors $(1,1)$ and $(-1,1)$ in V are transposed to rows $v_1^{\mathrm{T}}, v_2^{\mathrm{T}}$ of V^{T}. We have not yet explained how V and U and Σ were computed!

The Reduced Form of the SVD

The full form $AV = U\Sigma$ in equation (2) can have a lot of zeros in Σ when the rank of A is small and its nullspace is large. Those zeros contribute nothing to matrix multiplication. The heart of the SVD is in the first r v's and u's and σ's. We can reduce $AV = U\Sigma$ to $AV_r = U_r\Sigma_r$ by removing the parts that are sure to produce zeros. This leaves the **reduced SVD where Σ_r is now square:**

$$\boxed{AV_r = U_r\Sigma_r \qquad A \begin{bmatrix} v_1 & \cdots & v_r \\ & \text{row space} & \end{bmatrix} = \begin{bmatrix} u_1 & \cdots & u_r \\ & \text{column space} & \end{bmatrix} \begin{bmatrix} \sigma_1 & & \\ & \ddots & \\ & & \sigma_r \end{bmatrix}} \tag{5}$$

We still have $V_r^{\mathrm{T}} V_r = I_r$ and $U_r^{\mathrm{T}} U_r = I_r$ from those orthogonal unit vectors $\boldsymbol{v}$'s and $\boldsymbol{u}$'s. But when V_r and U_r are not square, we can no longer have two-sided inverses: $V_r V_r^{\mathrm{T}} \neq I$ and $U_r U_r^{\mathrm{T}} \neq I$.

Example $\quad V_r = \begin{bmatrix} 1/3 \\ 2/3 \\ 2/3 \end{bmatrix}$ and $V_r^{\mathrm{T}} V_r = \begin{bmatrix} 1 \end{bmatrix}$ but $V_r V_r^{\mathrm{T}} = \dfrac{1}{9}\begin{bmatrix} 1 & 2 & 2 \\ 2 & 4 & 4 \\ 2 & 4 & 4 \end{bmatrix} =$ rank 1.

Problem 21 shows that **we still have $\boldsymbol{A = U_r \Sigma_r V_r^{\mathrm{T}}}$**. The rest of $U\Sigma V^{\mathrm{T}}$ contributes nothing to A, because of those blocks of zeros in Σ. The key formula is still $A = \sigma_1 \boldsymbol{u}_1 \boldsymbol{v}_1^{\mathrm{T}} + \cdots + \sigma_r \boldsymbol{u}_r \boldsymbol{v}_r^{\mathrm{T}}$. The SVD sees only the r nonzeros in the diagonal matrix Σ.

The Important Fact for Data Science

Why is the SVD so important for this subject and this book? Like the other factorizations $A = LU$ and $A = QR$ and $S = Q\Lambda Q^{\mathrm{T}}$, it separates the matrix into rank one pieces. A special property of the SVD is that **those pieces come in order of importance**. The first piece $\sigma_1 \boldsymbol{u}_1 \boldsymbol{v}_1^{\mathrm{T}}$ is the closest rank one matrix to A. More than that is true: *The sum of the first k pieces is best possible for rank k.*

$\boldsymbol{A_k = \sigma_1 u_1 v_1^{\mathrm{T}} + \cdots + \sigma_k u_k v_k^{\mathrm{T}}}$ **is the best rank $\boldsymbol{k}$ approximation to $\boldsymbol{A}$:**

Eckart-Young $\quad$ **If $\boldsymbol{B}$ has rank $\boldsymbol{k}$ then $\boldsymbol{\|A - A_k\| \leq \|A - B\|}$.** $\quad$ (6)

To interpret that statement you need to know the meaning of the symbol $\|A - B\|$. This is the "**norm**" of the matrix $A - B$, a measure of its size (like the absolute value of a number). The Eckart-Young theorem is proved in Section I.9.

Our first job is to find the $\boldsymbol{v}$'s and $\boldsymbol{u}$'s for equation (1), to reach the SVD.

First Proof of the SVD

Our goal is $A = U\Sigma V^{\mathrm{T}}$. We want to identify the two sets of singular vectors, the $\boldsymbol{u}$'s and the $\boldsymbol{v}$'s. One way to find those vectors is to form the symmetric matrices $A^{\mathrm{T}}A$ and AA^{T}:

$$\boldsymbol{A^{\mathrm{T}}A} = (V\Sigma^{\mathrm{T}}U^{\mathrm{T}})\ (U\Sigma V^{\mathrm{T}}) = \boldsymbol{V\Sigma^{\mathrm{T}}\Sigma V^{\mathrm{T}}} \tag{7}$$

$$\boldsymbol{AA^{\mathrm{T}}} = (U\Sigma V^{\mathrm{T}})\ (V\Sigma^{\mathrm{T}}U^{\mathrm{T}}) = \boldsymbol{U\Sigma\Sigma^{\mathrm{T}}U^{\mathrm{T}}} \tag{8}$$

Both (7) and (8) produced symmetric matrices. Usually $A^{\mathrm{T}}A$ and AA^{T} are different. Both right hand sides have the special form $Q\Lambda Q^{\mathrm{T}}$. Eigenvalues are in $\Lambda = \Sigma^{\mathrm{T}}\Sigma$ or $\Sigma\Sigma^{\mathrm{T}}$. **Eigenvectors are in $\boldsymbol{Q = V}$ or $\boldsymbol{Q = U}$.** So we know from (7) and (8) how V and U and Σ connect to the symmetric matrices $A^{\mathrm{T}}A$ and AA^{T}.

> **V contains orthonormal eigenvectors of $A^{\mathrm{T}}A$**
>
> **U contains orthonormal eigenvectors of AA^{T}**
>
> **σ_1^2 to σ_r^2 are the nonzero eigenvalues of both $A^{\mathrm{T}}A$ and AA^{T}**

We are not quite finished, for this reason. **The SVD requires that $A\boldsymbol{v}_k = \sigma_k \boldsymbol{u}_k$.** It connects each right singular vector $\boldsymbol{v}_k$ to a left singular vector $\boldsymbol{u}_k$, for $k = 1, \ldots, r$. When I choose the $\boldsymbol{v}$'s, that choice will decide the signs of the $\boldsymbol{u}$'s. If $S\boldsymbol{u} = \lambda\boldsymbol{u}$ then also $S(-\boldsymbol{u}) = \lambda(-\boldsymbol{u})$ and I have to know the correct sign. More than that, there is a whole plane of eigenvectors when λ is a double eigenvalue. When I choose two $\boldsymbol{v}$'s in that plane, then $A\boldsymbol{v} = \sigma\boldsymbol{u}$ will tell me both $\boldsymbol{u}$'s. This is in equation (9).

The plan is to start with the $\boldsymbol{v}$'s. **Choose orthonormal eigenvectors $\boldsymbol{v}_1, \ldots, \boldsymbol{v}_r$** of $A^{\mathrm{T}}A$. **Then choose $\sigma_k = \sqrt{\lambda_k}$. To determine the $\boldsymbol{u}$'s we require $A\boldsymbol{v} = \sigma\boldsymbol{u}$:**

$$\boldsymbol{v}\text{'s then } \boldsymbol{u}\text{'s} \qquad A^{\mathrm{T}}A\boldsymbol{v}_k = \sigma_k^2 \boldsymbol{v}_k \quad \text{and then} \quad \boldsymbol{u}_k = \frac{A\boldsymbol{v}_k}{\sigma_k} \quad \text{for} \quad k = 1, \ldots, r \tag{9}$$

This is the proof of the SVD ! Let me check that those $\boldsymbol{u}$'s are eigenvectors of AA^{T}:

$$AA^{\mathrm{T}}\boldsymbol{u}_k = AA^{\mathrm{T}}\left(\frac{A\boldsymbol{v}_k}{\sigma_k}\right) = A\left(\frac{A^{\mathrm{T}}A\boldsymbol{v}_k}{\sigma_k}\right) = A\,\frac{\sigma_k^2\boldsymbol{v}_k}{\sigma_k} = \sigma_k^2\boldsymbol{u}_k \tag{10}$$

The $\boldsymbol{v}$'s were chosen to be orthonormal. I must check that the $\boldsymbol{u}$'s are also orthonormal:

$$\boldsymbol{u}_j^{\mathrm{T}}\boldsymbol{u}_k = \left(\frac{A\boldsymbol{v}_j}{\sigma_j}\right)^{\mathrm{T}}\left(\frac{A\boldsymbol{v}_k}{\sigma_k}\right) = \frac{\boldsymbol{v}_j^{\mathrm{T}}(A^{\mathrm{T}}A\boldsymbol{v}_k)}{\sigma_j\,\sigma_k} = \frac{\sigma_k}{\sigma_j}\,\boldsymbol{v}_j^{\mathrm{T}}\,\boldsymbol{v}_k = \begin{cases} 1 & \text{if } j = k \\ 0 & \text{if } j \neq k \end{cases} \tag{11}$$

Notice that $(AA^{\mathrm{T}})A = A(A^{\mathrm{T}}A)$ was the key to equation (10). The law $(AB)C = A(BC)$ is the key to a great many proofs in linear algebra. Moving the parentheses is a powerful idea. This is the *associative law*.

Finally we have to choose the last $n - r$ vectors $\boldsymbol{v}_{r+1}$ to $\boldsymbol{v}_n$ and the last $m - r$ vectors $\boldsymbol{u}_{r+1}$ to $\boldsymbol{u}_m$. This is easy. **These $\boldsymbol{v}$'s and $\boldsymbol{u}$'s are in the nullspaces of A and A^{T}.** We can choose any orthonormal bases for those nullspaces. They will automatically be orthogonal to the first $\boldsymbol{v}$'s in the row space of A and the first $\boldsymbol{u}$'s in the column space. This is because the whole spaces are orthogonal: $\mathbf{N}(A) \perp \mathbf{C}(A^{\mathrm{T}})$ and $\mathbf{N}(A^{\mathrm{T}}) \perp \mathbf{C}(A)$. *The proof of the SVD is complete.*

Now we have U and V and Σ in the full size SVD of equation (1). You may have noticed that the eigenvalues of $A^{\mathrm{T}}A$ are in $\Sigma^{\mathrm{T}}\Sigma$, and *the same numbers σ_1^2 to σ_r^2 are also eigenvalues of AA^{T} in $\Sigma\Sigma^{\mathrm{T}}$*. An amazing fact: **$BA$ always has the same nonzero eigenvalues as AB : 5 pages ahead.**

Example 1 (completed) Find the matrices U, Σ, V for $A = \begin{bmatrix} 3 & 0 \\ 4 & 5 \end{bmatrix}$.

With rank 2, this A has two positive singular values σ_1 and σ_2. We will see that σ_1 is larger than $\lambda_{\max} = 5$, and σ_2 is smaller than $\lambda_{\min} = 3$. Begin with $A^{\mathrm{T}}A$ and AA^{T}:

$$A^{\mathrm{T}}A = \begin{bmatrix} 25 & 20 \\ 20 & 25 \end{bmatrix} \qquad AA^{\mathrm{T}} = \begin{bmatrix} 9 & 12 \\ 12 & 41 \end{bmatrix}$$

Those have the same trace (50) and the same eigenvalues $\sigma_1^2 = 45$ and $\sigma_2^2 = 5$. The square roots are $\sigma_1 = \sqrt{45}$ and $\sigma_2 = \sqrt{5}$. Then $\sigma_1\sigma_2 = 15$ and this is the determinant of A.

A key step is to find the eigenvectors of $A^{\mathrm{T}}A$ (with eigenvalues 45 and 5):

$$\begin{bmatrix} 25 & 20 \\ 20 & 25 \end{bmatrix}\begin{bmatrix} 1 \\ 1 \end{bmatrix} = \mathbf{45}\begin{bmatrix} \mathbf{1} \\ \mathbf{1} \end{bmatrix} \qquad \begin{bmatrix} 25 & 20 \\ 20 & 25 \end{bmatrix}\begin{bmatrix} -1 \\ 1 \end{bmatrix} = \mathbf{5}\begin{bmatrix} \mathbf{-1} \\ \mathbf{1} \end{bmatrix}$$

Then $\boldsymbol{v}_1$ and $\boldsymbol{v}_2$ are those orthogonal eigenvectors rescaled to length 1. Divide by $\sqrt{2}$.

$$\textbf{Right singular vectors } \boldsymbol{v}_1 = \frac{1}{\sqrt{2}}\begin{bmatrix} 1 \\ 1 \end{bmatrix} \quad \boldsymbol{v}_2 = \frac{1}{\sqrt{2}}\begin{bmatrix} -1 \\ 1 \end{bmatrix} \quad \textbf{Left singular vectors } \boldsymbol{u}_i = \frac{A\boldsymbol{v}_i}{\sigma_i}$$

Now compute $A\boldsymbol{v}_1$ and $A\boldsymbol{v}_2$ which will be $\sigma_1\boldsymbol{u}_1 = \sqrt{45}\,\boldsymbol{u}_1$ and $\sigma_2\boldsymbol{u}_2 = \sqrt{5}\,\boldsymbol{u}_2$:

$$A\boldsymbol{v}_1 = \frac{3}{\sqrt{2}}\begin{bmatrix} 1 \\ 3 \end{bmatrix} = \sqrt{45}\frac{\mathbf{1}}{\sqrt{\mathbf{10}}}\begin{bmatrix} \mathbf{1} \\ \mathbf{3} \end{bmatrix} = \sigma_1\boldsymbol{u}_1$$

$$A\boldsymbol{v}_2 = \frac{1}{\sqrt{2}}\begin{bmatrix} -3 \\ 1 \end{bmatrix} = \sqrt{5}\frac{\mathbf{1}}{\sqrt{\mathbf{10}}}\begin{bmatrix} \mathbf{-3} \\ \mathbf{1} \end{bmatrix} = \sigma_2\boldsymbol{u}_2$$

The division by $\sqrt{10}$ makes $\boldsymbol{u}_1$ and $\boldsymbol{u}_2$ orthonormal. Then $\sigma_1 = \sqrt{45}$ and $\sigma_2 = \sqrt{5}$ as expected. The Singular Value Decomposition of A is U times Σ times V^{T}.

$$\boldsymbol{U} = \frac{1}{\sqrt{10}}\begin{bmatrix} 1 & -3 \\ 3 & 1 \end{bmatrix} \qquad \boldsymbol{\Sigma} = \begin{bmatrix} \sqrt{45} & \\ & \sqrt{5} \end{bmatrix} \qquad \boldsymbol{V} = \frac{1}{\sqrt{2}}\begin{bmatrix} 1 & -1 \\ 1 & 1 \end{bmatrix} \tag{12}$$

U and V contain orthonormal bases for the column space and the row space of A (both spaces are just $\mathbf{R}^2$). The real achievement is that those two bases diagonalize A: AV equals $U\Sigma$. The matrix $A = U\Sigma V^{\mathrm{T}}$ splits into two rank-one matrices, columns times rows, with $\sqrt{2}\,\sqrt{10} = \sqrt{20}$.

$$\sigma_1\boldsymbol{u}_1\boldsymbol{v}_1^{\mathrm{T}} + \sigma_2\boldsymbol{u}_2\boldsymbol{v}_2^{\mathrm{T}} = \frac{\sqrt{\mathbf{45}}}{\sqrt{\mathbf{20}}}\begin{bmatrix} \mathbf{1} & \mathbf{1} \\ \mathbf{3} & \mathbf{3} \end{bmatrix} + \frac{\sqrt{\mathbf{5}}}{\sqrt{\mathbf{20}}}\begin{bmatrix} \mathbf{3} & \mathbf{-3} \\ \mathbf{-1} & \mathbf{1} \end{bmatrix} = \begin{bmatrix} \mathbf{3} & \mathbf{0} \\ \mathbf{4} & \mathbf{5} \end{bmatrix} = \boldsymbol{A}.$$

Every matrix is a sum of rank one matrices with orthogonal $\boldsymbol{u}$'s and orthogonal $\boldsymbol{v}$'s.

Question : If $S = Q\Lambda Q^{\mathrm{T}}$ is symmetric positive definite, what is its SVD ?

Answer: The SVD is exactly $U\Sigma V^{\mathrm{T}} = Q\Lambda Q^{\mathrm{T}}$. The matrix $U = V = Q$ is orthogonal. And the eigenvalue matrix Λ becomes the singular value matrix Σ.

Question : If $S = Q\Lambda Q^{\mathrm{T}}$ has a negative eigenvalue ($S\boldsymbol{x} = -\alpha\,\boldsymbol{x}$), what is the singular value and what are the vectors $\boldsymbol{v}$ and $\boldsymbol{u}$?

Answer: The singular value will be $\boldsymbol{\sigma} = +\boldsymbol{\alpha}$ (positive). One singular vector (either $\boldsymbol{u}$ or $\boldsymbol{v}$) must be $-\boldsymbol{x}$ (reverse the sign). Then $S\boldsymbol{x} = -\alpha\,\boldsymbol{x}$ is the same as $S\boldsymbol{v} = \sigma\boldsymbol{u}$. The two sign changes cancel.

Question : If $A = Q$ is an orthogonal matrix, why does every singular value equal 1 ?

Answer: All singular values are $\sigma = 1$ because $A^{\mathrm{T}}A = Q^{\mathrm{T}}Q = I$. Then $\Sigma = I$. But $U = Q$ and $V = I$ is only one choice for the singular vectors $\boldsymbol{u}$ and $\boldsymbol{v}$:

$$Q = U\Sigma V^{\mathrm{T}} \;\text{ can be }\; Q = QII^{\mathrm{T}} \;\text{ or any }\; Q = (QQ_1)IQ_1^{\mathrm{T}}.$$

Question : Why are all eigenvalues of a square matrix A less than or equal to σ_1 ?

Answer: Multiplying by orthogonal matrices U and V^{T} does not change vector lengths:

$$||A\boldsymbol{x}|| = ||U\Sigma V^{\mathrm{T}}\boldsymbol{x}|| = ||\Sigma V^{\mathrm{T}}\boldsymbol{x}|| \leq \sigma_1||V^{\mathrm{T}}\boldsymbol{x}|| = \boldsymbol{\sigma_1}||\boldsymbol{x}|| \;\text{ for all } \boldsymbol{x}. \tag{13}$$

An eigenvector has $||A\boldsymbol{x}|| = |\lambda|\,||\boldsymbol{x}||$. Then (13) gives $|\lambda|\,||\boldsymbol{x}|| \leq \sigma_1\,||\boldsymbol{x}||$ and $|\boldsymbol{\lambda}| \leq \boldsymbol{\sigma_1}$.

Question : If $A = \boldsymbol{x}\boldsymbol{y}^{\mathrm{T}}$ has rank 1, what are $\boldsymbol{u}_1$ and $\boldsymbol{v}_1$ and σ_1 ? **Check that $|\lambda_1| \leq \sigma_1$.**

Answer: The singular vectors $\boldsymbol{u}_1 = \boldsymbol{x}/||\boldsymbol{x}||$ and $\boldsymbol{v}_1 = \boldsymbol{y}/||\boldsymbol{y}||$ have length 1. Then $\boldsymbol{\sigma}_1 = ||\boldsymbol{x}||\,||\boldsymbol{y}||$ is the only nonzero number in the singular value matrix Σ. Here is the SVD:

$$\textbf{Rank 1 matrix} \qquad \boldsymbol{x}\boldsymbol{y}^{\mathrm{T}} = \frac{\boldsymbol{x}}{||\boldsymbol{x}||}\left(||\boldsymbol{x}||\,||\boldsymbol{y}||\right)\frac{\boldsymbol{y}^{\mathrm{T}}}{||\boldsymbol{y}||} = \boldsymbol{u}_1\sigma_1\boldsymbol{v}_1^{\mathrm{T}}.$$

Observation The only nonzero eigenvalue of $A = \boldsymbol{x}\boldsymbol{y}^{\mathrm{T}}$ is $\lambda = \boldsymbol{y}^{\mathrm{T}}\boldsymbol{x}$. The eigenvector is $\boldsymbol{x}$ because $(\boldsymbol{x}\boldsymbol{y}^{\mathrm{T}})\,\boldsymbol{x} = \boldsymbol{x}\,(\boldsymbol{y}^{\mathrm{T}}\boldsymbol{x}) = \lambda\boldsymbol{x}$. Then $|\lambda_1| = |\boldsymbol{y}^{\mathrm{T}}\boldsymbol{x}| \leq \sigma_1 = ||\boldsymbol{y}||\,||\boldsymbol{x}||$.
The key inequality $|\lambda_1| \leq \sigma_1$ becomes exactly the Schwarz inequality.

Question : What is the Karhunen-Loève transform and its connection to the SVD ?

Answer: KL begins with a covariance matrix V of a zero-mean random process. V is symmetric and positive definite or semidefinite. In general V could be an infinite matrix or a covariance function. Then the KL expansion will be an infinite series.

The eigenvectors of V, in order of decreasing eigenvalues $\sigma_1^2 \geq \sigma_2^2 \geq \ldots \geq 0$, are the basis functions $\boldsymbol{u}_i$ for the KL transform. The expansion of any vector $\boldsymbol{v}$ in an orthonormal basis $\boldsymbol{u}_1, \boldsymbol{u}_2, \ldots$ is $\boldsymbol{v} = \sum(\boldsymbol{u}_i^{\mathrm{T}}\boldsymbol{v})\boldsymbol{u}_i$.

In this stochastic case, that transform decorrelates the random process: the $\boldsymbol{u}_i$ are independent. More than that, the ordering of the eigenvalues means that the first k terms, stopping at $(\boldsymbol{u}_k^{\mathrm{T}}\boldsymbol{v})\boldsymbol{u}_k$, minimize the expected square error. This fact corresponds to the Eckart-Young Theorem in the next section I.9.

The KL transform is a stochastic (random) form of Principal Component Analysis.

The Geometry of the SVD

The SVD separates a matrix into $A = U\Sigma V^{\mathrm{T}}$: **(orthogonal)** $\times$ **(diagonal)** $\times$ **(orthogonal)**. In two dimensions we can draw those steps. The orthogonal matrices U and V rotate the plane. The diagonal matrix Σ stretches it along the axes. Figure I.11 shows **rotation** times **stretching** times **rotation.** **Vectors $\boldsymbol{x}$ on the unit circle go to $A\boldsymbol{x}$ on an ellipse.**

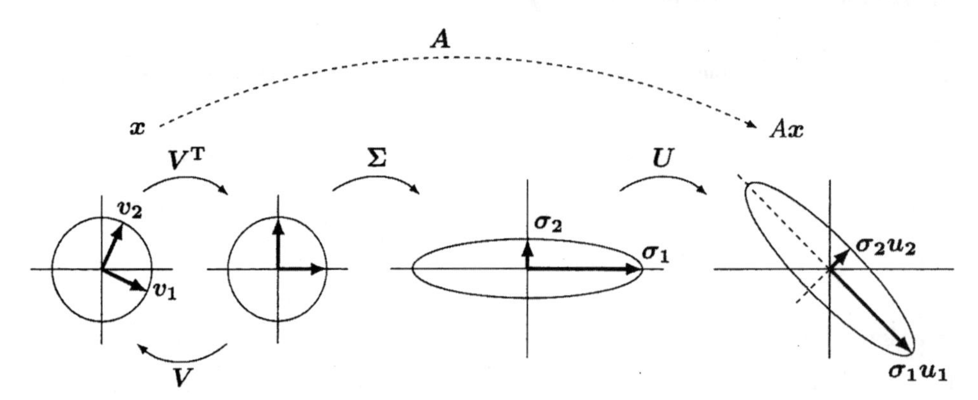

Figure I.10: U and V are rotations and possible reflections. Σ stretches circle to ellipse.

This picture applies to a 2 by 2 invertible matrix (because $\sigma_1 > 0$ and $\sigma_2 > 0$). First is a rotation of any $\boldsymbol{x}$ to $V^{\mathrm{T}}\boldsymbol{x}$. Then Σ stretches that vector to $\Sigma V^{\mathrm{T}}\boldsymbol{x}$. Then U rotates to $A\boldsymbol{x} = U\Sigma V^{\mathrm{T}}\boldsymbol{x}$. We kept all determinants positive to avoid reflections. The four numbers a, b, c, d in the matrix connect to *two angles* θ and ϕ and *two numbers* σ_1 and σ_2.

$$\begin{bmatrix} a & b \\ c & d \end{bmatrix} = \begin{bmatrix} \cos\theta & -\sin\theta \\ \sin\theta & \cos\theta \end{bmatrix} \begin{bmatrix} \sigma_1 & \\ & \sigma_2 \end{bmatrix} \begin{bmatrix} \cos\phi & \sin\phi \\ -\sin\phi & \cos\phi \end{bmatrix}. \tag{14}$$

Question. If the matrix is symmetric then $b = c$ and A has only 3 (not 4) parameters. *How do the* 4 *numbers* θ, ϕ, σ_1, σ_2 *reduce to* 3 *numbers for a symmetric matrix* S?

The First Singular Vector $\boldsymbol{v}_1$

The next page will establish a new way to look at $\boldsymbol{v}_1$. The previous pages chose the $\boldsymbol{v}$'s as eigenvectors of $A^{\mathrm{T}}A$. Certainly that remains true. But there is a valuable way to understand these singular vectors **one at a time instead of all at once**. We start with $\boldsymbol{v}_1$ and the singular value σ_1.

$$\textbf{Maximize the ratio } \frac{||A\boldsymbol{x}||}{||\boldsymbol{x}||}. \textbf{ The maximum is } \sigma_1 \textbf{ at the vector } \boldsymbol{x} = \boldsymbol{v}_1. \tag{15}$$

The ellipse in Figure I.10 showed why the maximizing $\boldsymbol{x}$ is $\boldsymbol{v}_1$. When you follow $\boldsymbol{v}_1$ across the page, it ends at $A\boldsymbol{v}_1 = \sigma_1\boldsymbol{u}_1$ (the longest axis of the ellipse). Its length started at $||\boldsymbol{v}_1|| = 1$ and ended at $||A\boldsymbol{v}_1|| = \sigma_1$.

But we aim for an independent approach to the SVD! We are not assuming that we already know U or Σ or V. How do we recognize that the ratio $||A\boldsymbol{x}||/||\boldsymbol{x}||$ is a maximum when $\boldsymbol{x} = \boldsymbol{v}_1$? Calculus tells us that the first derivatives must be zero. The derivatives will be easier if we square our function:

$$\textbf{Problem: Find the maximum value } \lambda \textbf{ of } \frac{||A\boldsymbol{x}||^2}{||\boldsymbol{x}||^2} = \frac{\boldsymbol{x}^{\mathrm{T}}A^{\mathrm{T}}A\boldsymbol{x}}{\boldsymbol{x}^{\mathrm{T}}\boldsymbol{x}} = \frac{\boldsymbol{x}^{\mathrm{T}}S\boldsymbol{x}}{\boldsymbol{x}^{\mathrm{T}}\boldsymbol{x}}. \tag{16}$$

This "Rayleigh quotient" depends on $x_1, \ldots, x_n$. Calculus uses the quotient rule, so we need

$$\frac{\partial}{\partial x_i}\left(\boldsymbol{x}^{\mathrm{T}}\boldsymbol{x}\right) = \frac{\partial}{\partial x_i}\left(x_1^2 + \cdots + x_i^2 + \cdots + x_n^2\right) = 2(\boldsymbol{x})_i \tag{17}$$

$$\frac{\partial}{\partial x_i}\left(\boldsymbol{x}^{\mathrm{T}}S\boldsymbol{x}\right) = \frac{\partial}{\partial x_i}\left(\sum_i\sum_j S_{ij}x_ix_j\right) = 2\sum_j S_{ij}x_j = 2\left(S\boldsymbol{x}\right)_i \tag{18}$$

The quotient rule finds $\partial/\partial x_i\,(\boldsymbol{x}^{\mathrm{T}}S\boldsymbol{x}/\boldsymbol{x}^{\mathrm{T}}\boldsymbol{x})$. Set those n partial derivatives of (16) to zero:

$$\left(\boldsymbol{x}^{\mathrm{T}}\boldsymbol{x}\right)2\left(S\boldsymbol{x}\right)_i - \left(\boldsymbol{x}^{\mathrm{T}}S\boldsymbol{x}\right)2\left(\boldsymbol{x}\right)_i = \mathbf{0} \text{ for } i = 1, \ldots, n \tag{19}$$

Equation (19) says that the best $\boldsymbol{x}$ is an eigenvector of $S = A^{\mathrm{T}}A$!

$2S\boldsymbol{x} = 2\lambda\boldsymbol{x}$ **and the maximum value of** $\dfrac{\boldsymbol{x}^{\mathrm{T}}S\boldsymbol{x}}{\boldsymbol{x}^{\mathrm{T}}\boldsymbol{x}} = \dfrac{\lvert\lvert A\boldsymbol{x}\rvert\rvert^2}{\lvert\lvert\boldsymbol{x}\rvert\rvert^2}$ **is an eigenvalue λ of S.**

The search is narrowed to eigenvectors of $S = A^{\mathrm{T}}A$. The eigenvector that maximizes is $\boldsymbol{x} = \boldsymbol{v}_1$. The eigenvalue is $\lambda_1 = \sigma_1^2$. Calculus has confirmed the solution (15) of the maximum problem—the first piece of the SVD.

For the full SVD, we need *all* the singular vectors and singular values. To find $\boldsymbol{v}_2$ and σ_2, we adjust the maximum problem so it looks only at vectors $\boldsymbol{x}$ orthogonal to $\boldsymbol{v}_1$.

Maximize $\dfrac{\lvert\lvert A\boldsymbol{x}\rvert\rvert}{\lvert\lvert\boldsymbol{x}\rvert\rvert}$ **under the condition $\boldsymbol{v}_1^{\mathrm{T}}\boldsymbol{x} = 0$. The maximum is σ_2 at $\boldsymbol{x} = \boldsymbol{v}_2$.**

"Lagrange multipliers" were invented to deal with constraints on $\boldsymbol{x}$ like $\boldsymbol{v}_1^{\mathrm{T}}\boldsymbol{x} = 0$. And Problem 3 gives a simple direct way to work with this condition $\boldsymbol{v}_1^{\mathrm{T}}\boldsymbol{x} = 0$.

In the same way, every singular vector $\boldsymbol{v}_{k+1}$ gives the maximum ratio over all vectors $\boldsymbol{x}$ that are perpendicular to the first $\boldsymbol{v}_1, \ldots, \boldsymbol{v}_k$. The left singular vectors would come from maximizing $||A^{\mathrm{T}}\boldsymbol{y}||/||\boldsymbol{y}||$. We are always finding the axes of an ellipsoid and the eigenvectors of symmetric matrices $A^{\mathrm{T}}A$ or AA^{T}.

The Singular Vectors of A^{T}

The SVD connects $\boldsymbol{v}$'s in the row space to $\boldsymbol{u}$'s in the column space. When we transpose $A = U\Sigma V^{\mathrm{T}}$, we see that $\boldsymbol{A^{\mathrm{T}} = V\Sigma^{\mathrm{T}}U^{\mathrm{T}}}$ goes the opposite way, from $\boldsymbol{u}$'s to $\boldsymbol{v}$'s:

$$\boldsymbol{A^{\mathrm{T}}u_k = \sigma_k v_k} \textbf{ for } \boldsymbol{k = 1, \ldots, r} \qquad \boldsymbol{A^{\mathrm{T}}u_k = 0} \textbf{ for } \boldsymbol{k = r+1, \ldots, m} \tag{20}$$

Multiply $A\boldsymbol{v}_k = \sigma_k \boldsymbol{u}_k$ by A^{T}. Remember $A^{\mathrm{T}}A\boldsymbol{v}_k = \sigma_k^2 \boldsymbol{v}_k$ in equation (9). Divide by σ_k.

A Different Symmetric Matrix Also Produces the SVD

We created the SVD from two symmetric matrices $A^{\mathrm{T}}A$ and AA^{T}. Another good way uses one symmetric block matrix S. *This matrix has r pairs of plus and minus eigenvalues.* The nonzero eigenvalues of this matrix S are σ_k and $-\sigma_k$, and its size is $m+n$:

$$\boldsymbol{S} = \begin{bmatrix} \mathbf{0} & \boldsymbol{A} \\ \boldsymbol{A^{\mathrm{T}}} & \mathbf{0} \end{bmatrix} \quad \text{has eigenvectors} \quad \begin{bmatrix} \boldsymbol{u}_k \\ \boldsymbol{v}_k \end{bmatrix} \text{ and } \begin{bmatrix} -\boldsymbol{u}_k \\ \boldsymbol{v}_k \end{bmatrix}.$$

We can check those eigenvectors directly, remembering $A\boldsymbol{v}_k = \sigma_k \boldsymbol{u}_k$ and $A^{\mathrm{T}}\boldsymbol{u}_k = \sigma_k \boldsymbol{v}_k$:

$$\begin{bmatrix} 0 & A \\ A^{\mathrm{T}} & 0 \end{bmatrix} \begin{bmatrix} \pm\boldsymbol{u}_k \\ \boldsymbol{v}_k \end{bmatrix} = \begin{bmatrix} A\boldsymbol{v}_k \\ \pm A^{\mathrm{T}}\boldsymbol{u}_k \end{bmatrix} = \sigma_k \begin{bmatrix} \boldsymbol{u}_k \\ \boldsymbol{v}_k \end{bmatrix} \text{ and } -\sigma_k \begin{bmatrix} -\boldsymbol{u}_k \\ \boldsymbol{v}_k \end{bmatrix}. \tag{21}$$

That gives $2r$ eigenvalues. The eigenvectors are orthogonal: $-\boldsymbol{u}_k^{\mathrm{T}}\boldsymbol{u}_k + \boldsymbol{v}_k^{\mathrm{T}}\boldsymbol{v}_k = -1+1$. Can you see the other $(m-r)+(n-r)$ eigenvectors with $\lambda = 0$ for that block matrix? They must involve the remaining $\boldsymbol{u}$'s and $\boldsymbol{v}$'s in the nullspaces of A^{T} and A.

AB and BA: Equal Nonzero Eigenvalues

If A is m by n and B is n by m, then AB and BA have the same nonzero eigenvalues.

Start with $AB\boldsymbol{x} = \lambda\boldsymbol{x}$ and $\lambda \neq 0$. Multiply both sides by B, to get $BAB\boldsymbol{x} = \lambda B\boldsymbol{x}$. This says that $B\boldsymbol{x}$ is an eigenvector of BA with the same eigenvalue λ—exactly what we wanted. We needed $\lambda \neq 0$ to be sure that this eigenvector $B\boldsymbol{x}$ is not zero.

Notice that if B is square and invertible, then $B^{-1}(BA)B = AB$. This says that BA is *similar* to AB: *same eigenvalues.* But our first proof allows A and B to be m by n and n by m. This covers the important example of the SVD when $B = A^{\mathrm{T}}$. In that case $A^{\mathrm{T}}A$ and AA^{T} both lead to the singular values of A.

If $m > n$, then AB has $m-n$ extra zero eigenvalues compared to BA.

Submatrices Have Smaller Singular Values

The approach to $||A|| = \sigma_1$ by maximizing $||Ax||/||x||$ makes it easy to prove this useful fact. The norm of a submatrix cannot be larger than the norm of the whole matrix: $\boldsymbol{\sigma_1(B) \le \sigma_1(A)}$.

$$\textbf{If } B \textbf{ keeps } M \le m \textbf{ rows and } N \le n \textbf{ columns of } A, \textbf{ then } ||B|| \le ||A||. \quad (22)$$

Proof Look at vectors $\boldsymbol{y}$ with nonzeros only in the N positions that correspond to columns in B. Certainly maximum of $||B\boldsymbol{y}||/||\boldsymbol{y}|| \le$ maximum of $||A\boldsymbol{x}||/||\boldsymbol{x}||$.

Reduce $||B\boldsymbol{y}||$ further by looking only at the M components that correspond to rows of B. So removing columns and rows cannot increase the norm σ_1, and $||B|| \le ||A||$.

The SVD for Derivatives and Integrals

This may be the clearest example of the SVD. It does not start with a matrix (but we will go there). Historically, the first SVD was not for vectors but for *functions*. Then A is not a matrix but an *operator*. One example is the operator that integrates every function. Another example is the (unbounded) operator D that takes the derivative:

Operators on functions
Integral and derivative

$$A\boldsymbol{x}(s) = \int_0^s \boldsymbol{x}(t)\,dt \quad \text{and} \quad D\boldsymbol{x}(t) = \frac{d\boldsymbol{x}}{dt}. \quad (23)$$

Those operators are linear (or calculus would be a lot more difficult than it is). In some way D is the inverse of A, by the Fundamental Theorem of Calculus. More exactly D is a left inverse with $\boldsymbol{DA = I}$: derivative of integral equals original function.

But $\boldsymbol{AD \ne I}$ because the derivative of a constant function is zero. Then D has a nullspace, like a matrix with dependent columns. **$\boldsymbol{D}$ is the pseudoinverse of $\boldsymbol{A}$!** Sines and cosines are the $\boldsymbol{u}$'s and $\boldsymbol{v}$'s for $A =$ integral and $D =$ derivative:

$$\boxed{A\boldsymbol{v} = \sigma\boldsymbol{u} \text{ is } \boldsymbol{A(\cos kt) = \frac{1}{k}(\sin kt)} \qquad \text{Then } \boldsymbol{D(\sin kt) = k\,(\cos kt)}.} \quad (24)$$

The simplicity of those equations is our reason for including them in the book. We are working with *periodic functions*: $x(t + 2\pi) = x(t)$. The input space to A contains the *even functions* like $\cos t = \cos(-t)$. The outputs from A (and the inputs to D) are the *odd functions* like $\sin t = -\sin(-t)$. Those input and output spaces are like $\mathbf{R}^n$ and $\mathbf{R}^m$ for an m by n matrix.

The special property of the SVD is that the $\boldsymbol{v}$'s are orthogonal, and so are the $\boldsymbol{u}$'s. Here those singular vectors have become very nice functions—*the cosines are orthogonal to each other and so are the sines. Their inner products are integrals equal to zero*:

$$\boldsymbol{v}_k^{\mathrm{T}}\boldsymbol{v}_j = \int_0^{2\pi} (\cos kt)(\cos jt)\,dt = \mathbf{0} \quad \text{and} \quad \boldsymbol{u}_k^{\mathrm{T}}\boldsymbol{u}_j = \int_0^{2\pi} (\sin kt)(\sin jt)\,dt = \mathbf{0}.$$

Notice that the inner product of functions $\boldsymbol{x}_1$ and $\boldsymbol{x}_2$ is the integral of $\boldsymbol{x}_1(t)\,\boldsymbol{x}_2(t)$. This copies into function space (*Hilbert space*) the dot product that adds $\boldsymbol{y}\cdot\boldsymbol{z} = \Sigma y_i z_i$. In fact the symbol $\int$ was somehow created from Σ (and integrals are the limits of sums).

Finite Differences

The discrete form of a derivative is a **finite difference**. The discrete form of an integral is a **sum**. Here we choose a 4 by 3 matrix D that corresponds to the backward difference $f(x) - f(x - \Delta x)$:

$$\boldsymbol{D} = \begin{bmatrix} 1 & & \\ -1 & 1 & \\ & -1 & 1 \\ & & -1 \end{bmatrix} \quad \text{with} \quad \boldsymbol{D}^{\mathrm{T}} = \begin{bmatrix} 1 & -1 & & \\ & 1 & -1 & \\ & & 1 & -1 \end{bmatrix}. \tag{25}$$

To find singular values and singular vectors, compute $D^{\mathrm{T}}D$ (3 by 3) and DD^{T} (4 by 4):

$$\boldsymbol{D}^{\mathrm{T}}\boldsymbol{D} = \begin{bmatrix} 2 & -1 & 0 \\ -1 & 2 & -1 \\ 0 & -1 & 2 \end{bmatrix} \quad \text{and} \quad \boldsymbol{D}\boldsymbol{D}^{\mathrm{T}} = \begin{bmatrix} 1 & -1 & 0 & 0 \\ -1 & 2 & -1 & 0 \\ 0 & -1 & 2 & -1 \\ 0 & 0 & -1 & 1 \end{bmatrix}. \tag{26}$$

Their nonzero eigenvalues are always the same! DD^{T} also has a zero eigenvalue with eigenvector $\boldsymbol{u}_4 = \left(\frac{1}{2}, \frac{1}{2}, \frac{1}{2}, \frac{1}{2}\right)$. This is the discrete equivalent of the function $f(x) = \frac{1}{2}$ with $df/dx = 0$.

The nonzero eigenvalues of both symmetric matrices $D^{\mathrm{T}}D$ and DD^{T} are

$$\lambda_1 = \sigma_1^2\,(D) = 2 + \sqrt{2} \qquad \lambda_2 = \sigma_2^2\,(D) = 2 \qquad \lambda_3 = \sigma_3^2\,(D) = 2 - \sqrt{2} \tag{27}$$

The eigenvectors $\boldsymbol{v}$ of $D^{\mathrm{T}}D$ are the right singular vectors of D. They are **discrete sines.** The eigenvectors $\boldsymbol{u}$ of DD^{T} are the left singular vectors of D. They are **discrete cosines**:

$$\sqrt{2}\,\boldsymbol{V} = \begin{bmatrix} \sin\frac{\pi}{4} & \sin\frac{2\pi}{4} & \sin\frac{3\pi}{4} \\ \sin\frac{2\pi}{4} & \sin\frac{4\pi}{4} & \sin\frac{6\pi}{4} \\ \sin\frac{3\pi}{4} & \sin\frac{6\pi}{4} & \sin\frac{9\pi}{4} \end{bmatrix} \qquad \sqrt{2}\,\boldsymbol{U} = \begin{bmatrix} \cos\frac{1}{2}\frac{\pi}{4} & \cos\frac{1}{2}\frac{2\pi}{4} & \cos\frac{1}{2}\frac{3\pi}{4} & 1 \\ \cos\frac{3}{2}\frac{\pi}{4} & \cos\frac{3}{2}\frac{2\pi}{4} & \cos\frac{3}{2}\frac{3\pi}{4} & 1 \\ \cos\frac{5}{2}\frac{\pi}{4} & \cos\frac{5}{2}\frac{2\pi}{4} & \cos\frac{5}{2}\frac{3\pi}{4} & 1 \\ \cos\frac{7}{2}\frac{\pi}{4} & \cos\frac{7}{2}\frac{2\pi}{4} & \cos\frac{7}{2}\frac{3\pi}{4} & 1 \end{bmatrix}.$$

These are the famous DST and DCT matrices—**Discrete Sine Transform** and **Discrete Cosine Transform**. The DCT matrix has been the backbone of JPEG image compression. Actually JPEG increases U to 8 by 8, which reduces the "blockiness" of the image. 8 by 8 blocks of pixels are transformed by a two-dimensional DCT—then compressed and transmitted. Orthogonality of these matrices is the key in Section IV.4.

Our goal was to show the discrete form of the beautiful Singular Value Decomposition $D(\sin kt) = k(\cos kt)$. You could correctly say that this is only one example. But Fourier is always present for linear equations with constant coefficients—and always important.

In signal processing the key letters are L T I : **Linear Time Invariance.**

The Polar Decomposition $A = QS$

Every complex number $x + iy$ has the polar form $re^{i\theta}$. A number $r \geq 0$ multiplies a number $e^{i\theta}$ on the unit circle. We have $x + iy = r\cos\theta + ir\sin\theta = re^{i\theta}$. Think of these numbers as 1 by 1 matrices. Then $e^{i\theta}$ is an *orthogonal matrix* Q and $r \geq 0$ is a *positive semidefinite matrix* (call it S). The ***polar decomposition*** extends the same idea to n by n matrices: orthogonal times positive semidefinite, $\boldsymbol{A = QS}$.

Every real square matrix can be factored into $\boldsymbol{A = QS}$, where Q is ***orthogonal*** and S is ***symmetric positive semidefinite***. If A is invertible, S is positive definite.

$$\textbf{Polar decomposition} \qquad A = U\Sigma V^{\mathrm{T}} = (UV^{\mathrm{T}})(V\Sigma V^{\mathrm{T}}) = (Q)(S). \tag{28}$$

The first factor UV^{T} is Q. The product of orthogonal matrices is orthogonal. The second factor $V\Sigma V^{\mathrm{T}}$ is S. It is positive semidefinite because its eigenvalues are in Σ.

If A is invertible then Σ and S are also invertible. ***S is the symmetric positive definite square root of $A^{\mathrm{T}}A$,*** because $S^2 = V\Sigma^2V^{\mathrm{T}} = A^{\mathrm{T}}A$. So the eigenvalues of S are the singular values of A. The eigenvectors of S are the singular vectors $\boldsymbol{v}$ of A.

There is also a polar decomposition $A = KQ$ in the reverse order. Q is the same but now $K = U\Sigma U^{\mathrm{T}}$. Then K is the symmetric positive definite square root of AA^{T}.

Example Find Q and S (rotation and stretch) in the polar decomposition of $A = \begin{bmatrix} 3 & 0 \\ 4 & 5 \end{bmatrix}$.

Solution The matrices U and Σ and V were found above equation (3):

$$Q = UV^{\mathrm{T}} = \frac{1}{\sqrt{20}}\begin{bmatrix} 1 & -3 \\ 3 & 1 \end{bmatrix}\begin{bmatrix} 1 & -1 \\ -1 & 1 \end{bmatrix} = \frac{1}{\sqrt{20}}\begin{bmatrix} 4 & -2 \\ 2 & 4 \end{bmatrix} = \frac{1}{\sqrt{5}}\begin{bmatrix} \mathbf{2} & \mathbf{-1} \\ \mathbf{1} & \mathbf{2} \end{bmatrix}$$

$$S = V\Sigma V^{\mathrm{T}} = \frac{\sqrt{5}}{2}\begin{bmatrix} 1 & -1 \\ 1 & 1 \end{bmatrix}\begin{bmatrix} 3 & \\ & 1 \end{bmatrix}\begin{bmatrix} 1 & 1 \\ -1 & 1 \end{bmatrix} = \sqrt{5}\begin{bmatrix} \mathbf{2} & \mathbf{1} \\ \mathbf{1} & \mathbf{2} \end{bmatrix}. \text{ Then } \boldsymbol{A = QS}.$$

In mechanics, the polar decomposition separates the ***rotation*** (in Q) from the ***stretching***. The eigenvalues of S give the stretching factors in Figure I.10. The eigenvectors of S give the stretching directions (the principal axes of the ellipse). Section IV.9 on the orthogonal Procrustes problem says that **$\boldsymbol{Q}$ is the nearest orthogonal matrix to $\boldsymbol{A}$.**

Problem Set I.8

1 A symmetric matrix $S = S^{\mathrm{T}}$ has orthonormal eigenvectors $\boldsymbol{v}_1$ to $\boldsymbol{v}_n$. Then any vector $\boldsymbol{x}$ can be written as a combination $\boldsymbol{x} = c_1\boldsymbol{v}_1 + \cdots + c_n\boldsymbol{v}_n$. Explain these two formulas:

$$\boldsymbol{x}^{\mathrm{T}}\boldsymbol{x} = c_1^2 + \cdots + c_n^2 \qquad \boldsymbol{x}^{\mathrm{T}}S\boldsymbol{x} = \lambda_1 c_1^2 + \cdots + \lambda_n c_n^2.$$

2 Problem 1 gives a neat form for the Rayleigh quotient $\boldsymbol{x}^{\mathrm{T}}S\boldsymbol{x}/\boldsymbol{x}^{\mathrm{T}}\boldsymbol{x}$:

$$\boxed{R(\boldsymbol{x}) = \frac{\boldsymbol{x}^{\mathrm{T}}S\boldsymbol{x}}{\boldsymbol{x}^{\mathrm{T}}\boldsymbol{x}} = \frac{\lambda_1 c_1^2 + \cdots + \lambda_n c_n^2}{c_1^2 + \cdots + c_n^2}.}$$

Why is the maximum value of that ratio equal to the largest eigenvalue λ_1? This may be the simplest way to understand the "second construction" of the SVD in equation (15). You can see why the ratio $R(\boldsymbol{x})$ is a maximum when $c_1 = 1$ and $c_2 = c_3 = \cdots = c_n = 0$.

3 Next comes λ_2 when $\boldsymbol{x} = \boldsymbol{v}_2$. We maximize $R(\boldsymbol{x}) = \boldsymbol{x}^{\mathrm{T}}S\boldsymbol{x}/\boldsymbol{x}^{\mathrm{T}}\boldsymbol{x}$ under the condition that $\boldsymbol{x}^{\mathrm{T}}\boldsymbol{v}_1 = 0$. *What does this condition mean for c_1?* Why is the ratio in Problem 2 now maximized when $c_2 = 1$ and $c_1 = c_3 = \cdots = c_n = 0$?

4 Following Problem 3, what maximum problem is solved by $\boldsymbol{x} = \boldsymbol{v}_3$? The best $\boldsymbol{c}$'s are $c_3 = 1$ and $c_1 = c_2 = c_4 \cdots = 0$.

The maximum of $R(\boldsymbol{x}) = \dfrac{\boldsymbol{x}^{\mathrm{T}}S\boldsymbol{x}}{\boldsymbol{x}^{\mathrm{T}}\boldsymbol{x}}$ is $\boldsymbol{\lambda_3}$ subject to what two conditions on $\boldsymbol{x}$?

5 Show that A^{T} has the same (nonzero) singular values as A. Then $||A|| = ||A^{\mathrm{T}}||$ for all matrices. But it's not true that $||A\boldsymbol{x}|| = ||A^{\mathrm{T}}\boldsymbol{x}||$ for all vectors. That needs $A^{\mathrm{T}}A = AA^{\mathrm{T}}$.

6 Find the σ's and $\boldsymbol{v}$'s and $\boldsymbol{u}$'s in the SVD for $A = \begin{bmatrix} 3 & 4 \\ 0 & 5 \end{bmatrix}$. Use equation (12).

7 What is the norm $||A - \sigma_1\,\boldsymbol{u}_1\boldsymbol{v}_1^{\mathrm{T}}||$ when that largest rank one piece of A is removed? What are all the singular values of this reduced matrix, and its rank?

8 Find the σ's and $\boldsymbol{v}$'s and $\boldsymbol{u}$'s, and verify that $A = \begin{bmatrix} 0 & 2 & 0 \\ 0 & 0 & 3 \\ 0 & 0 & 0 \end{bmatrix} = U\Sigma V^{\mathrm{T}}$. For this matrix, the orthogonal matrices U and V are permutation matrices.

9 To maximize $\frac{1}{2}\boldsymbol{x}^{\mathrm{T}}S\boldsymbol{x}$ with $\boldsymbol{x}^{\mathrm{T}}\boldsymbol{x} = 1$, Lagrange would work with $L = \frac{1}{2}\boldsymbol{x}^{\mathrm{T}}S\boldsymbol{x} + \lambda(\boldsymbol{x}^{\mathrm{T}}\boldsymbol{x} - 1)$. Show that $\boldsymbol{\nabla L} = (\partial L/\partial x_1, \ldots, \partial L/\partial x_n) = 0$ is exactly $S\boldsymbol{x} = \lambda\boldsymbol{x}$. Once again max $\mathrm{R}(\boldsymbol{x}) = \lambda_1$.

10 Prove $||B|| \le ||A||$ in equation (22) by a slightly different approach. Remove first the $N - n$ columns of A. The new matrix has $||C|| \le ||A||$. (Why ?). Then transpose C: no change in norm. Finally remove $M - m$ columns of C^{T} to produce B^{T} with no increase in norm. Altogether $||B|| = ||B^{\mathrm{T}}|| \le ||C^{\mathrm{T}}|| = ||C|| \le ||A||$.

11 Check that the trace of $S = \begin{bmatrix} 0 & A \\ A^{\mathrm{T}} & 0 \end{bmatrix}$ from adding up its diagonal entries agrees with the sum of its eigenvalues in equation (21). If A is a square diagonal matrix with entries $1, 2, \ldots, n$, what are the $2n$ eigenvalues and eigenvectors of S ?

12 Find the SVD of the rank 1 matrix $A = \begin{bmatrix} 2 & 4 \\ 1 & 2 \end{bmatrix}$. Factor $A^{\mathrm{T}}A$ into $Q\Lambda Q^{\mathrm{T}}$.

13 Here is my homemade proof of the SVD. Step **2** uses the factorizations $A^{\mathrm{T}}A = V\Lambda V^{\mathrm{T}}$ and $AA^{\mathrm{T}} = U\Lambda U^{\mathrm{T}}$ (same eigenvalues in Λ).

1	$A(A^{\mathrm{T}}A)$	$= (AA^{\mathrm{T}})A$
2	$AV\Lambda V^{\mathrm{T}}$	$= U\Lambda U^{\mathrm{T}}A$
3	$(U^{\mathrm{T}}AV)\Lambda$	$= \Lambda(U^{\mathrm{T}}AV)$
4	$U^{\mathrm{T}}AV$ must be diagonal	

Step 3 multiplied Step 2 on the left by _____ and on the right by _____. Then the matrix $U^{\mathrm{T}}AV$ commutes with the diagonal matrix Λ in Step 3. How does this force the matrix $U^{\mathrm{T}}AV = \Sigma$ to be also a diagonal matrix? Try 3 by 3.

$$\boldsymbol{\Sigma\Lambda} = \begin{bmatrix} \sigma_{11} & \sigma_{12} & \sigma_{13} \\ \sigma_{21} & \sigma_{22} & \sigma_{23} \\ \sigma_{31} & \sigma_{32} & \sigma_{33} \end{bmatrix} \begin{bmatrix} \lambda_1 & & \\ & \lambda_2 & \\ & & \lambda_3 \end{bmatrix} = \begin{bmatrix} \lambda_1 & & \\ & \lambda_2 & \\ & & \lambda_3 \end{bmatrix} \begin{bmatrix} \sigma_{11} & \sigma_{12} & \sigma_{13} \\ \sigma_{21} & \sigma_{22} & \sigma_{23} \\ \sigma_{31} & \sigma_{32} & \sigma_{33} \end{bmatrix} = \boldsymbol{\Lambda\Sigma}$$

Compare the first rows. When can you conclude that $\sigma_{12} = 0$ and $\sigma_{13} = 0$? This shows the limitation on my proof: *It needs the eigenvalues of $A^{\mathrm{T}}A$ to be* _____.

The same bug appears in simple proofs of the spectral theorem $S = Q\Lambda Q^{\mathrm{T}}$. This is easy when S has no repeated λ's. The SVD is easy when A has no repeated σ's.

Both $S = Q\Lambda Q^{\mathrm{T}}$ and $A = U\Sigma V^{\mathrm{T}}$ remain true when λ's or σ's happen to be repeated. The problem is that this produces a whole plane of eigenvectors or singular vectors. You have to choose the singular vectors $\boldsymbol{u}$ specifically as $A\boldsymbol{v}/\sigma$— which is the real proof in equation (9).

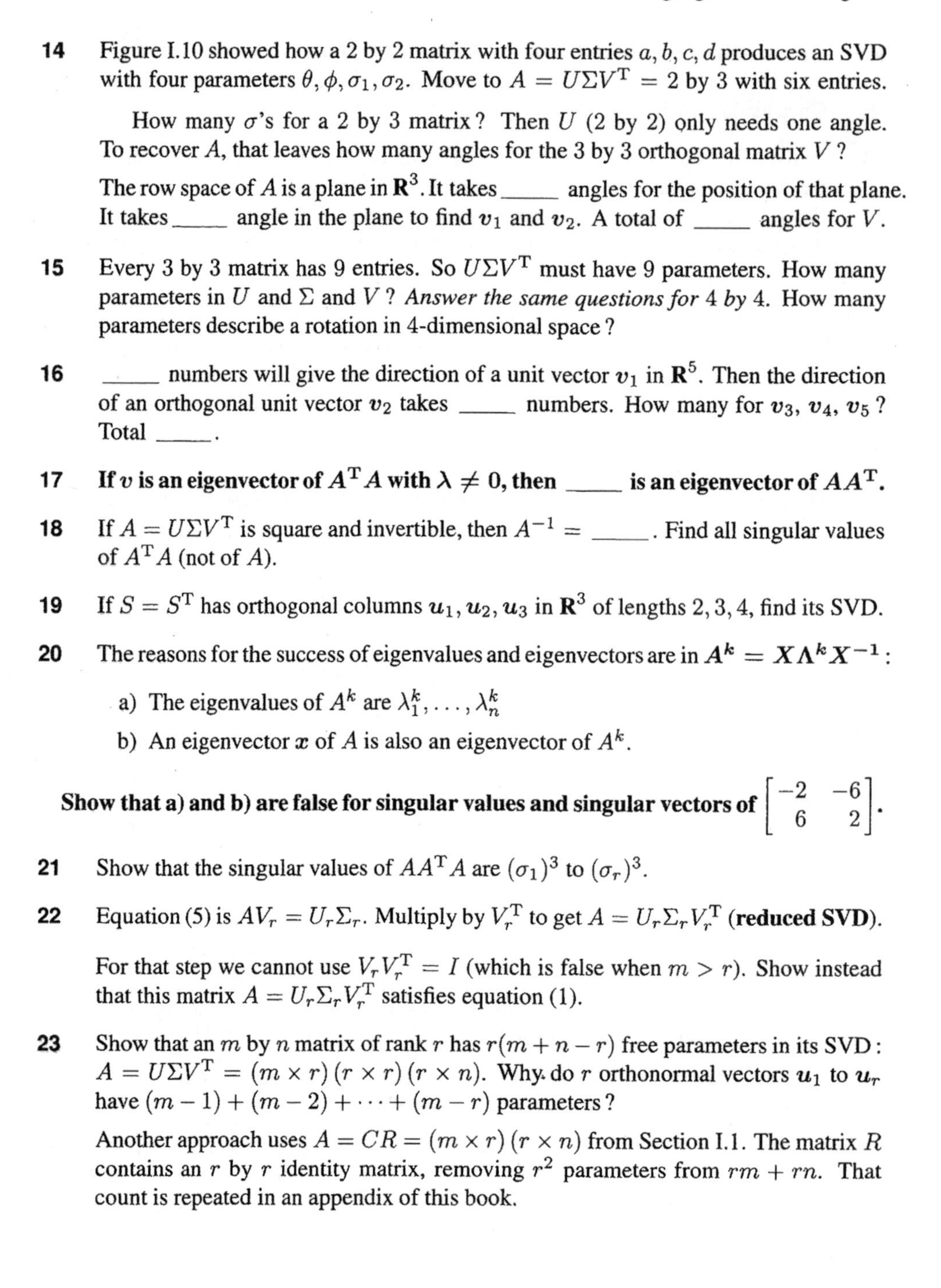

14 Figure I.10 showed how a 2 by 2 matrix with four entries a, b, c, d produces an SVD with four parameters $\theta, \phi, \sigma_1, \sigma_2$. Move to $A = U\Sigma V^{\mathrm{T}} = 2$ by 3 with six entries.

How many σ's for a 2 by 3 matrix? Then U (2 by 2) only needs one angle. To recover A, that leaves how many angles for the 3 by 3 orthogonal matrix V?

The row space of A is a plane in $\mathbf{R}^3$. It takes _____ angles for the position of that plane. It takes _____ angle in the plane to find $\boldsymbol{v}_1$ and $\boldsymbol{v}_2$. A total of _____ angles for V.

15 Every 3 by 3 matrix has 9 entries. So $U\Sigma V^{\mathrm{T}}$ must have 9 parameters. How many parameters in U and Σ and V? *Answer the same questions for* 4 *by* 4. How many parameters describe a rotation in 4-dimensional space?

16 _____ numbers will give the direction of a unit vector $\boldsymbol{v}_1$ in $\mathbf{R}^5$. Then the direction of an orthogonal unit vector $\boldsymbol{v}_2$ takes _____ numbers. How many for $\boldsymbol{v}_3$, $\boldsymbol{v}_4$, $\boldsymbol{v}_5$? Total _____.

17 **If $\boldsymbol{v}$ is an eigenvector of $A^{\mathrm{T}}A$ with $\lambda \neq 0$, then _____ is an eigenvector of AA^{T}.**

18 If $A = U\Sigma V^{\mathrm{T}}$ is square and invertible, then $A^{-1} =$ _____. Find all singular values of $A^{\mathrm{T}}A$ (not of A).

19 If $S = S^{\mathrm{T}}$ has orthogonal columns $\boldsymbol{u}_1, \boldsymbol{u}_2, \boldsymbol{u}_3$ in $\mathbf{R}^3$ of lengths 2, 3, 4, find its SVD.

20 The reasons for the success of eigenvalues and eigenvectors are in $\boldsymbol{A^k = X\Lambda^k X^{-1}}$:

a) The eigenvalues of A^k are $\lambda_1^k, \ldots, \lambda_n^k$

b) An eigenvector $\boldsymbol{x}$ of A is also an eigenvector of A^k.

Show that a) and b) are false for singular values and singular vectors of $\begin{bmatrix} -2 & -6 \\ 6 & 2 \end{bmatrix}$.

21 Show that the singular values of $AA^{\mathrm{T}}A$ are $(\sigma_1)^3$ to $(\sigma_r)^3$.

22 Equation (5) is $AV_r = U_r\Sigma_r$. Multiply by V_r^{T} to get $A = U_r\Sigma_r V_r^{\mathrm{T}}$ (**reduced SVD**).

For that step we cannot use $V_r V_r^{\mathrm{T}} = I$ (which is false when $m > r$). Show instead that this matrix $A = U_r\Sigma_r V_r^{\mathrm{T}}$ satisfies equation (1).

23 Show that an m by n matrix of rank r has $r(m + n - r)$ free parameters in its SVD: $A = U\Sigma V^{\mathrm{T}} = (m \times r)\,(r \times r)\,(r \times n)$. Why do r orthonormal vectors $\boldsymbol{u}_1$ to $\boldsymbol{u}_r$ have $(m-1) + (m-2) + \cdots + (m-r)$ parameters?

Another approach uses $A = CR = (m \times r)\,(r \times n)$ from Section I.1. The matrix R contains an r by r identity matrix, removing r^2 parameters from $rm + rn$. That count is repeated in an appendix of this book.

I.9 Principal Components and the Best Low Rank Matrix

The principal components of A are its singular vectors, the columns $\boldsymbol{u}_j$ and $\boldsymbol{v}_j$ of the orthogonal matrices U and V. **Principal Component Analysis (PCA)** uses the largest σ's connected to the first $\boldsymbol{u}$'s and $\boldsymbol{v}$'s to understand the information in a matrix of data. We are given a matrix A, and we extract its most important part A_k (**largest σ's**):

$$\boldsymbol{A_k = \sigma_1 u_1 v_1^{\mathrm{T}} + \cdots + \sigma_k u_k v_k^{\mathrm{T}}} \quad \textbf{with rank } \boldsymbol{(A_k) = k}.$$

A_k solves a matrix optimization problem—and we start there. **The closest rank k matrix to A is A_k.** In statistics we are identifying the pieces of A with largest variance. This puts the SVD at the center of data science.

In that world, PCA is "unsupervised" learning. Our only instructor is linear algebra—the SVD tells us to choose A_k. When the learning is supervised, we have a big set of training data. Deep Learning (Section VII.1) constructs a (nonlinear!) function F that correctly classifies most of that training data. Then we apply F to new data, as you will see.

Principal Component Analysis is based on matrix approximation by A_k. The proof that A_k is the best choice was begun by Schmidt (1907). His theorem was written for operators A in function space, and it extends directly to matrices in vector spaces. Eckart and Young gave a new proof in 1936 (using the Frobenius norm for matrices). Then Mirsky found a more general proof in 1955 that allows any norm $||A||$ that depends only on the singular values—as in the definitions (2), (3), and (4) below.

Here is that key property of the special rank k matrix $A_k = \sigma_1 \boldsymbol{u}_1 \boldsymbol{v}_1^{\mathrm{T}} + \cdots + \sigma_k \boldsymbol{u}_k \boldsymbol{v}_k^{\mathrm{T}}$:

$$\textbf{Eckart-Young} \qquad \textbf{If } \boldsymbol{B} \textbf{ has rank } \boldsymbol{k} \textbf{ then } \boldsymbol{||A-B|| \geq ||A-A_k||}. \tag{1}$$

Three choices for the matrix norm $||A||$ have special importance and their own names:

$$\textbf{Spectral norm} \qquad \boldsymbol{||A||_2} = \max \frac{||A\boldsymbol{x}||}{||\boldsymbol{x}||} = \boldsymbol{\sigma_1} \quad \text{(often called the } \boldsymbol{\ell^2} \textbf{ norm)} \tag{2}$$

$$\textbf{Frobenius norm} \qquad \boldsymbol{||A||_F} = \sqrt{\sigma_1^2 + \cdots + \sigma_r^2} \quad \text{(12) and (13) also define } \boldsymbol{||A||_F} \tag{3}$$

$$\textbf{Nuclear norm} \qquad \boldsymbol{||A||_N = \sigma_1 + \sigma_2 + \cdots + \sigma_r} \quad \text{(the trace norm)}. \tag{4}$$

These norms have different values already for the n by n identity matrix:

$$||I||_2 = 1 \qquad ||I||_F = \sqrt{n} \qquad ||I||_N = n. \tag{5}$$

Replace I by any orthogonal matrix Q and the norms stay the same (because all $\sigma_i = 1$):

$$||Q||_2 = 1 \qquad ||Q||_F = \sqrt{n} \qquad ||Q||_N = n. \tag{6}$$

More than this, the spectral and Frobenius and nuclear norms of any matrix stay the same when A is multiplied (on either side) by an orthogonal matrix.

The singular values don't change when U and V change to Q_1U and Q_2V. For complex matrices the word *unitary* replaces *orthogonal*. Then $\overline{Q}^{\mathrm{T}}Q = I$. These three norms are **unitarily invariant**: $||Q_1A\overline{Q}_2^{\mathrm{T}}|| = ||A||$. Mirsky's proof of the Eckart-Young theorem in equation (1) applies to all unitarily invariant norms: $||A||$ is computable from Σ.

$$\textbf{All three norms have} \quad ||Q_1AQ_2^{\mathrm{T}}|| = ||A|| \quad \textbf{for orthogonal } Q_1 \textbf{ and } Q_2 \qquad (7)$$

We now give simpler proofs of (1) for the L^2 norm and the Frobenius norm.

Eckart-Young Theorem : Best Approximation by A_k

It helps to see what the theorem tells us, before tackling its proof. In this example, A is diagonal and $k = 2$:

$$\text{The rank two matrix closest to } A = \begin{bmatrix} \mathbf{4} & 0 & 0 & 0 \\ 0 & \mathbf{3} & 0 & 0 \\ 0 & 0 & \mathbf{2} & 0 \\ 0 & 0 & 0 & \mathbf{1} \end{bmatrix} \quad \text{is} \quad A_2 = \begin{bmatrix} \mathbf{4} & 0 & 0 & 0 \\ 0 & \mathbf{3} & 0 & 0 \\ 0 & 0 & \mathbf{0} & 0 \\ 0 & 0 & 0 & \mathbf{0} \end{bmatrix}.$$

This must be true! You might say that this diagonal matrix is too simple, and not typical. But the L^2 norm and Frobenius norm are not changed when the matrix A becomes Q_1AQ_2 (for any orthogonal Q_1 and Q_2). So this example includes *any* 4 by 4 matrix with singular values $4, 3, 2, 1$. The Eckart-Young Theorem tells us to keep 4 and 3 because they are largest. The error in L^2 is $||A - A_2|| = 2$. The Frobenius norm has $||\boldsymbol{A} - \boldsymbol{A}_2||_F = \sqrt{\mathbf{5}}$.

The awkward part of the problem is "rank two matrices". *That set is not convex.* The average of A_2 and B_2 (both rank 2) can easily have rank 4. Is it possible that this B_2 could be closer to A than A_2?

$$\text{Could this } B_2 \text{ be a better rank 2 approximation to } A? \quad \begin{bmatrix} 3.5 & 3.5 & & \\ 3.5 & 3.5 & & \\ & & 1.5 & 1.5 \\ & & 1.5 & 1.5 \end{bmatrix}.$$

The errors $A - B_2$ are only 0.5 on the main diagonal where $A - A_2$ has errors 2 and 1. Of course the errors 3.5 and 1.5 off the diagonal will be too big. But maybe there is another choice that is better than A_2?

No, A_2 is best with rank $k = 2$. We prove this for the L^2 norm and then for Frobenius.

$$\textbf{Eckart-Young in } L^2 \qquad \text{If rank}\,(B) \leq k \text{ then } ||\boldsymbol{A} - \boldsymbol{B}|| = \max \frac{||(A-B)\,\boldsymbol{x}||}{||\boldsymbol{x}||} \geq \sigma_{k+1}. \qquad (8)$$

We know that $||A - A_k|| = \sigma_{k+1}$. The whole proof of $||A - B|| \geq \sigma_{k+1}$ depends on a good choice of the vector $\boldsymbol{x}$ in computing the norm $||A - B||$:

$$\text{Choose } \boldsymbol{x} \neq \mathbf{0} \text{ so that } B\boldsymbol{x} = \mathbf{0} \text{ and } \boldsymbol{x} = \sum_{1}^{k+1} c_i \boldsymbol{v}_i \tag{9}$$

First, the nullspace of B has dimension $\geq n - k$, because B has rank $\leq k$. Second, the combinations of $\boldsymbol{v}_1$ to $\boldsymbol{v}_{k+1}$ produce a subspace of dimension $k + 1$. Those two subspaces must intersect! When dimensions add to $(n - k) + (k + 1) = n + 1$, the subspaces must share a line (at least). Think of two planes through $(0, 0, 0)$ in $\mathbf{R}^3$—they share a line since $2 + 2 > 3$. Choose a nonzero vector $\boldsymbol{x}$ on this line.

Use that $\boldsymbol{x}$ to estimate the norm of $A - B$ in (8). Remember $B\boldsymbol{x} = \mathbf{0}$ and $A\boldsymbol{v}_i = \sigma_i \boldsymbol{u}_i$:

$$||(A - B)\,\boldsymbol{x}||^2 = ||A\boldsymbol{x}||^2 = ||\sum c_i \sigma_i \boldsymbol{u}_i||^2 = \sum_{1}^{k+1} c_i^2 \sigma_i^2. \tag{10}$$

That sum is at least as large as $(\sum c_i^2)\, \sigma_{k+1}^2$, which is exactly $||\boldsymbol{x}||^2 \sigma_{k+1}^2$. Equation (10) proves that $||(A-B)\boldsymbol{x}|| \geq \sigma_{k+1} ||\boldsymbol{x}||$. This $\boldsymbol{x}$ gives the lower bound we want for $||A-B||$:

$$\boxed{\frac{||(A - B)\,\boldsymbol{x}||}{||\boldsymbol{x}||} \geq \sigma_{k+1} \text{ means that } ||\boldsymbol{A} - \boldsymbol{B}|| \geq \boldsymbol{\sigma}_{k+1} = ||\boldsymbol{A} - \boldsymbol{A}_k||. \textbf{ Proved!}} \tag{11}$$

The Frobenius Norm

Eckart-YoungFrobenius norm Now we go to the Frobenius norm, to show that A_k is the best approximation there too.
It is useful to see three different formulas for this norm. The first formula treats A as a long vector, and takes the usual ℓ^2 norm of that vector. The second formula notices that the main diagonal of $A^{\mathrm{T}}A$ contains the ℓ^2 norm (squared) of each column of A.

For example, the 1, 1 entry of $A^{\mathrm{T}}A$ is $|a_{11}|^2 + \cdots + |a_{m1}|^2$ from column 1. So (12) is the same as (13), we are just taking the numbers $|a_{ij}|^2$ a column at a time.

Then formula (14) for the Frobenius norm uses the eigenvalues σ_i^2 of $A^{\mathrm{T}}A$. (The trace is always the sum of the eigenvalues.) Formula (14) also comes directly from the SVD—the Frobenius norm of $A = U\Sigma V^{\mathrm{T}}$ is not affected by U and V, so $||\boldsymbol{A}||_F^2 = ||\boldsymbol{\Sigma}||_F^2$. This is $\boldsymbol{\sigma}_1^2 + \cdots + \boldsymbol{\sigma}_r^2$.

$$||\boldsymbol{A}||_F^2 = |a_{11}|^2 + |a_{12}|^2 + \cdots + |a_{mn}|^2 \quad (\text{every } a_{ij}^2) \tag{12}$$

$$||\boldsymbol{A}||_F^2 = \text{trace of } A^{\mathrm{T}}A = (A^{\mathrm{T}}A)_{11} + \cdots + (A^{\mathrm{T}}A)_{nn} \tag{13}$$

$$||\boldsymbol{A}||_F^2 = \boldsymbol{\sigma}_1^2 + \boldsymbol{\sigma}_2^2 + \cdots + \boldsymbol{\sigma}_r^2 \tag{14}$$

Eckart-Young in the Frobenius Norm

For the norm $||A-B||_F$, Pete Stewart has found and generously shared this neat proof.

Suppose the matrix B of rank $\leq k$ is closest to A. We want to prove that $B = A_k$. The surprise is that we start with the singular value decomposition of B (*not* A):

$$\boldsymbol{B} = U \begin{bmatrix} D & 0 \\ 0 & 0 \end{bmatrix} V^{\mathrm{T}} \quad \text{where the diagonal matrix } D \text{ is } k \text{ by } k. \tag{15}$$

Those orthogonal matrices U and V from B will not necessarily diagonalize A:

$$\boldsymbol{A} = U \begin{bmatrix} L+E+R & F \\ G & H \end{bmatrix} V^{\mathrm{T}} \tag{16}$$

Here L is strictly lower triangular in the first k rows, E is diagonal, and R is strictly upper triangular. Step 1 will show that L, R, and F are all zero by comparing A and B with this matrix C that clearly has rank $\leq k$:

$$\boldsymbol{C} = U \begin{bmatrix} L+D+R & F \\ 0 & 0 \end{bmatrix} V^{\mathrm{T}} \tag{17}$$

This is Stewart's key idea, to construct C with zero rows to show its rank. Those orthogonal matrices U and V^{T} leave the Frobenius norm unchanged. Square all matrix entries and add, noticing that $A - C$ has zeros where $A - B$ has the matrices L, R, F:

$$||\boldsymbol{A} - \boldsymbol{B}||_F^2 = ||\boldsymbol{A} - \boldsymbol{C}||_F^2 + ||\boldsymbol{L}||_F^2 + ||\boldsymbol{R}||_F^2 + ||\boldsymbol{F}||_F^2. \tag{18}$$

Since $||A - B||_F^2$ was as small as possible we learn that L, R, F are zero! Similarly we find $G = 0$. At this point we know that $U^{\mathrm{T}}AV$ has two blocks and E is diagonal (like D):

$$\boldsymbol{U}^{\mathrm{T}}\boldsymbol{A}\boldsymbol{V} = \begin{bmatrix} \boldsymbol{E} & \boldsymbol{0} \\ \boldsymbol{0} & \boldsymbol{H} \end{bmatrix} \quad \text{and} \quad \boldsymbol{U}^{\mathrm{T}}\boldsymbol{B}\boldsymbol{V} = \begin{bmatrix} \boldsymbol{D} & \boldsymbol{0} \\ \boldsymbol{0} & \boldsymbol{0} \end{bmatrix}.$$

If B is closest to A then $U^{\mathrm{T}}BV$ is closest to $U^{\mathrm{T}}AV$. And now we see the truth.

The matrix D must be the same as $E = \operatorname{diag}(\sigma_1, \ldots, \sigma_k)$.

The singular values of H must be the smallest $n - k$ singular values of A.

The smallest error $||A-B||_F$ must be $||H||_F = \sqrt{\sigma_{k+1}^2 + \cdots + \sigma_r^2}$ = **Eckart-Young**.

In the 4 by 4 example starting this section, A_2 is best possible: $||A - A_2||_F = \sqrt{5}$. It is exceptional to have this explicit solution A_k for a non-convex optimization.

Minimizing the Frobenius Distance $||A - B||_F^2$

Here is a different and more direct approach to prove Eckart-Young: *Set derivatives of* $||A - B||_F^2$ *to zero.* Every rank k matrix factors into $\boldsymbol{B} = \boldsymbol{CR} = (m \times k)\,(k \times n)$. By the SVD, we can require r orthogonal columns in C (so $C^{\mathrm{T}}C =$ diagonal matrix D) and r orthonormal rows in R (so $RR^{\mathrm{T}} = I$). We are aiming for $C = U_k\Sigma_k$ and $R = V_k^{\mathrm{T}}$.

Take derivatives of $\boldsymbol{E} = ||\boldsymbol{A} - \boldsymbol{CR}||_F^2$ to find the matrices $\boldsymbol{C}$ and $\boldsymbol{R}$ that minimize $\boldsymbol{E}$:

$$\frac{\partial E}{\partial C} = 2(CR - A)R^{\mathrm{T}} = 0 \qquad\qquad \frac{\partial E}{\partial R} = 2(R^{\mathrm{T}}C^{\mathrm{T}} - A^{\mathrm{T}})\,C = 0 \qquad (19)$$

The first gives $AR^{\mathrm{T}} = CRR^{\mathrm{T}} = C$. Then the second gives $\boldsymbol{R}^{\mathrm{T}}\boldsymbol{D} = A^{\mathrm{T}}C = \boldsymbol{A}^{\mathrm{T}}\boldsymbol{A}\boldsymbol{R}^{\mathrm{T}}$. Since D is diagonal, this means:

The columns of R^{T} are eigenvectors of $A^{\mathrm{T}}A$. They are right singular vectors $\boldsymbol{v}_j$ of A. Similarly the columns of C are eigenvectors of AA^{T}: $\boldsymbol{AA}^{\mathrm{T}}\boldsymbol{C} = AR^{\mathrm{T}}D = \boldsymbol{CD}$. Then C contains left singular vectors $\boldsymbol{u}_j$. Which singular vectors actually minimize the error E?

E is a sum of all the σ^2 that were *not involved in C and R.* To minimize, those should be the smallest singular values of A. That leaves the largest singular values to produce the best $B = CR = A_k$, with $||A - CR||_F^2 = \sigma_{k+1}^2 + \cdots + \sigma_r^2$. This neat proof is in Nathan Srebro's MIT doctoral thesis: **ttic.uchicago.edu/~nati/Publications/thesis.pdf**

Principal Component Analysis

Now we start using the SVD. *The matrix A is full of data.* We have n samples. For each sample we measure m variables (like height and weight). The data matrix A_0 has n columns and m rows. In many applications it is a very large matrix.

The first step is to find the average (the sample mean) along each row of A_0. Subtract that mean from all m entries in the row. Now each row of the centered matrix A has *mean zero.* The columns of A are n points in $\mathbf{R}^m$. Because of centering, the sum of the n column vectors is zero. So the average column is the zero vector.

Often those n points are clustered near a line or a plane or another low-dimensional subspace of $\mathbf{R}^m$. Figure I.11 shows a typical set of data points clustered along a line in $\mathbf{R}^2$ (after centering A_0 to shift the points left-right and up-down for mean $(0,0)$ in A).

How will linear algebra find that closest line through $(0,0)$? **It is in the direction of the first singular vector $\boldsymbol{u}_1$ of $\boldsymbol{A}$.** This is the key point of PCA!

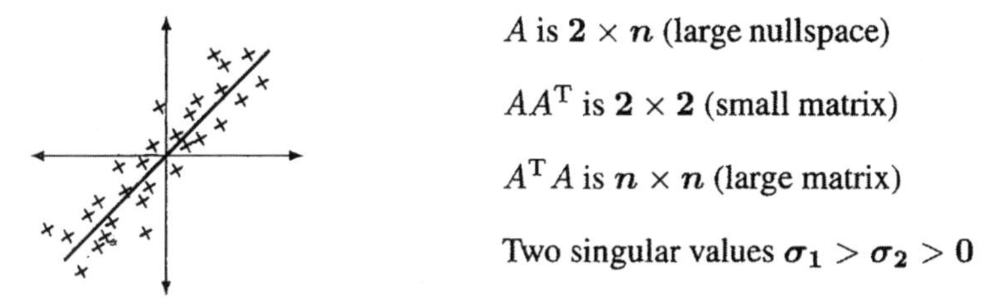

Figure I.11: Data points (columns of A) are often close to a line in $\mathbf{R}^2$ or a subspace in $\mathbf{R}^m$.

Let me express the problem (which the SVD solves) first in terms of statistics and then in terms of geometry. After that come the linear algebra and the examples.

The Statistics Behind PCA

The key numbers in probability and statistics are the **mean** and **variance**. The "mean" is an average of the data (in each row of A_0). Subtracting those means from each row of A_0 produced the centered A. The crucial quantities are the "variances" and "covariances". The variances are sums of squares of distances from the mean—along each row of A.

The variances are the diagonal entries of the matrix AA^{T}.

Suppose the columns of A correspond to a child's age on the x-axis and its height on the y-axis. (Those ages and heights are measured from the average age and height.) We are looking for the straight line that stays closest to the data points in the figure. And we have to account for the *joint age-height distribution* of the data.

The covariances are the off-diagonal entries of the matrix AA^{T}.

Those are dot products (row i of A) $\cdot$ (row j of A). High covariance means that increased height goes with increased age. (Negative covariance means that one variable increases when the other decreases.) Our example has only two rows from age and height: the symmetric matrix AA^{T} is 2 by 2. As the number n of sample children increases, we divide by $n-1$ to give AA^{T} its statistically correct scale.

The sample covariance matrix is defined by $S = \dfrac{AA^{\mathrm{T}}}{n-1}$.

The factor is $n-1$ because one degree of freedom has already been used for mean $= 0$. Here is an example with six ages and heights already centered to make each row add to zero:

$$A = \begin{bmatrix} 3 & -4 & 7 & 1 & -4 & -3 \\ 7 & -6 & 8 & -1 & -1 & 7 \end{bmatrix}$$

For this data, the sample covariance matrix S is easily computed. It is positive definite.

Variances and covariances $$S = \frac{1}{6-1}AA^{\mathrm{T}} = \begin{bmatrix} 20 & 25 \\ 25 & 40 \end{bmatrix}.$$

The two orthogonal eigenvectors of S are $\boldsymbol{u}_1$ and $\boldsymbol{u}_2$. Those are the left singular vectors (*principal components*) of A. The Eckart-Young theorem says that **the vector $\boldsymbol{u}_1$ points along the closest line in Figure I.11. Eigenvectors of S are singular vectors of A.**

The second singular vector $\boldsymbol{u}_2$ will be perpendicular to that closest line.

Important note PCA can be described using the symmetric $S = AA^{\mathrm{T}}/(n-1)$ or the rectangular A. No doubt S is the nicer matrix. But given the data in A, computing S would be a computational mistake. For large matrices, a direct SVD of A is faster and more accurate.

In the example, S has eigenvalues near 57 and 3. Their sum is $20 + 40 = 60$, the trace of S. The first rank one piece $\sqrt{57}\boldsymbol{u}_1\boldsymbol{v}_1^{\mathrm{T}}$ is much larger than the second piece $\sqrt{3}\boldsymbol{u}_2\boldsymbol{v}_2^{\mathrm{T}}$. **The leading eigenvector $\boldsymbol{u}_1 \approx (0.6, 0.8)$ tells us that the closest line in the scatter plot has slope near $8/6$.** The direction in the graph nearly produces a $6-8-10$ right triangle.

I will move now from the algebra of PCA to the geometry. In what sense will the line in the direction of $\boldsymbol{u}_1$ be the *closest line* to the centered data ?

The Geometry Behind PCA

The best line in Figure I.11 solves a problem in **perpendicular least squares**. This is also called *orthogonal regression*. It is different from the standard least squares fit to n data points, or the least squares solution to a linear system $A\boldsymbol{x} = \boldsymbol{b}$. That classical problem in Section II.2 minimizes $||A\boldsymbol{x} - \boldsymbol{b}||^2$. It measures distances up and down to the best line. Our problem minimizes *perpendicular* distances. The older problem leads to a linear system $A^{\mathrm{T}}A\widehat{\boldsymbol{x}} = A^{\mathrm{T}}\boldsymbol{b}$. Our problem leads to *eigenvalues* σ^2 and singular vectors $\boldsymbol{u}_i$ (eigenvectors of S). Those are the two sides of linear algebra : not the same side.

The sum of squared distances from the data points to the $\boldsymbol{u}_1$ line is a minimum.

To see this, separate each column $\boldsymbol{a}_j$ of A into its components along $\boldsymbol{u}_1$ and $\boldsymbol{u}_2$:

$$\sum_1^n ||\boldsymbol{a}_j||^2 = \sum_1^n |\boldsymbol{a}_j^{\mathrm{T}}\boldsymbol{u}_1|^2 + \sum_1^n |\boldsymbol{a}_j^{\mathrm{T}}\boldsymbol{u}_2|^2. \tag{20}$$

The sum on the left is fixed by the data. The first sum on the right has terms $\boldsymbol{u}_1^{\mathrm{T}}\boldsymbol{a}_j\boldsymbol{a}_j^{\mathrm{T}}\boldsymbol{u}_1$. It adds to $\boldsymbol{u}_1^{\mathrm{T}}(AA^{\mathrm{T}})\boldsymbol{u}_1$. So when we maximize that sum in PCA by choosing the top eigenvector $\boldsymbol{u}_1$ of AA^{T}, we minimize the second sum. That second sum of squared distances from data points to the best line (or best subspace) is the smallest possible.

The Linear Algebra Behind PCA

Principal Component Analysis is a way to understand n sample points $\boldsymbol{a}_1, \ldots, \boldsymbol{a}_n$ in m-dimensional space—the data. That data plot is centered : all rows of A add to zero ($A\mathbf{1} = \mathbf{0}$). The crucial connection to linear algebra is in the singular values σ_i and the singular vectors $\boldsymbol{u}_i$ of A. Those come from the eigenvalues $\lambda_i = \sigma_i^2$ and the eigenvectors of the sample covariance matrix $S = AA^{\mathrm{T}}/(n-1)$.

The total variance in the data comes from the Frobenius norm (squared) of A :

$$\textbf{Total variance } T = ||A||_F^2/(n-1) = (||\boldsymbol{a}_1||^2 + \cdots + ||\boldsymbol{a}_n||^2)/(n-1). \tag{21}$$

This is the **trace of S**—the sum down the diagonal. Linear algebra tells us that the trace equals the **sum of the eigenvalues $\sigma_i^2/(n-1)$ of the sample covariance matrix S.**

The trace of S connects the total variance to the sum of variances of the principal components $\boldsymbol{u}_1, \ldots, \boldsymbol{u}_r$:

$$\textbf{Total variance} \qquad T = (\sigma_1^2 + \cdots + \sigma_r^2)/(n-1). \tag{22}$$

Exactly as in equation (20), the first principal component $\boldsymbol{u}_1$ accounts for (or "*explains*") a fraction σ_1^2/T of the total variance. The next singular vector $\boldsymbol{u}_2$ of A explains the next largest fraction σ_2^2/T. Each singular vector is doing its best to capture the meaning in a matrix—and together they succeed.

The point of the Eckart-Young Theorem is that k singular vectors (acting together) explain *more of the data than any other set of k vectors*. So we are justified in choosing $\boldsymbol{u}_1$ to $\boldsymbol{u}_k$ as a basis for the k-dimensional subspace closest to the n data points.

The reader understands that our Figure I.11 showed a cluster of data points around a straight line ($k = 1$) in dimension $m = 2$. Real problems often have $k > 1$ and $m > 2$.

The "effective rank" of A and S is the number of singular values above the point where noise drowns the true signal in the data. Often this point is visible on a "**scree plot**" showing the dropoff in the singular values σ_i (or their squares σ_i^2). Figure I.12 shows the "elbow" in the scree plot where signal ends and noise takes over.

In this example the noise comes from roundoff error in computing singular values of the badly conditioned **Hilbert matrix**. The dropoff in the true singular values remains very steep. In practice the noise is in the data matrix itself—errors in the measurements of A_0. Section III.3 of this book studies matrices like H with rapidly decaying σ's.

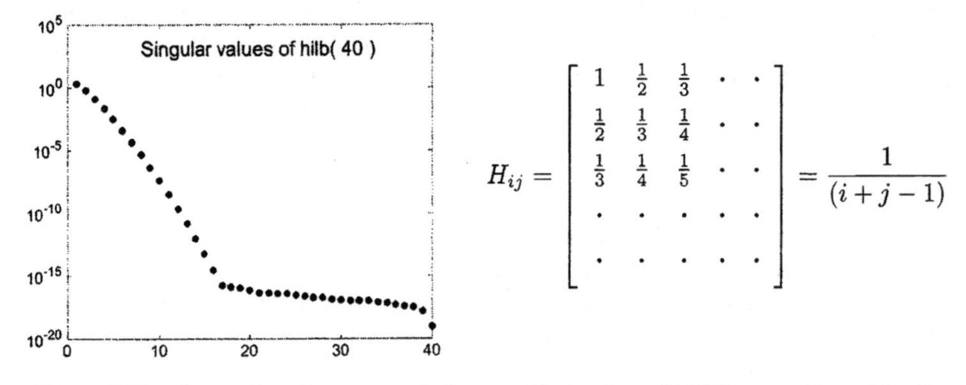

$$H_{ij} = \begin{bmatrix} 1 & \frac{1}{2} & \frac{1}{3} & \cdot & \cdot \\ \frac{1}{2} & \frac{1}{3} & \frac{1}{4} & \cdot & \cdot \\ \frac{1}{3} & \frac{1}{4} & \frac{1}{5} & \cdot & \cdot \\ \cdot & \cdot & \cdot & \cdot & \cdot \\ \cdot & \cdot & \cdot & \cdot & \cdot \end{bmatrix} = \frac{1}{(i+j-1)}$$

Figure I.12: Scree plot of $\sigma_1, \ldots, \sigma_{39}$ ($\sigma_{40} = 0$) for the evil Hilbert matrix, with elbow at the effective rank: $r \approx 17$ and $\sigma_r \approx 10^{-16}$.

One-Zero Matrices and Their Properties

Alex Townsend and the author began a study of matrices with 1's inside a circle and 0's outside. As the matrices get larger, their rank goes up. The graph of singular values approaches a limit—which we can't yet predict. But we understand the rank.

Three shapes are drawn in Figure I.13: **square, triangle, quarter circle.** Any square of 1's will have rank 1. The triangle has all eigenvalues $\boldsymbol{\lambda} = \mathbf{1}$, and its singular values are more interesting. The rank of the quarter circle matrix was our first puzzle, solved below.

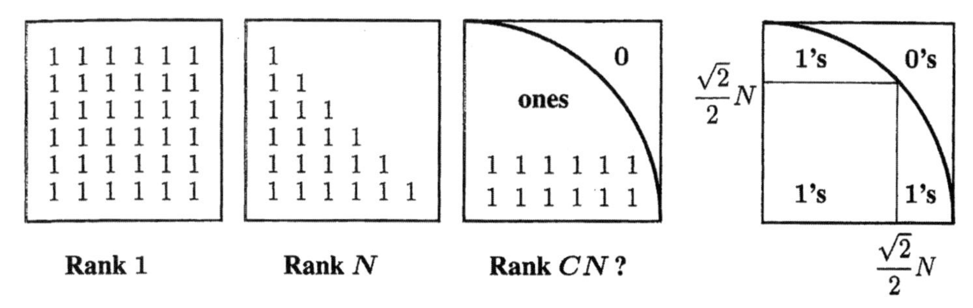

Figure I.13: Square and triangle and quarter circle of 1's in matrices with $N = 6$.

Reflection of these figures in the x-axis will produce a rectangle and larger triangle and semicircle with side $2N$. The ranks will not change because the new rows are copies of the old rows. Then reflection in the y-axis will produce a square and a diamond and a full circle. This time the new columns are copies of the old columns: again the same rank.

From the square and triangle we learn that low rank goes with horizontal-vertical alignment. Diagonals bring high rank, and $45°$ diagonals bring the highest.

What is the "*asymptotic rank*" of the quarter circle as the radius $N = 6$ increases? We are looking for the leading term CN in the rank.

The fourth figure shows a way to compute C. Draw a square of maximum size in the quarter circle. That square submatrix (all 1's) has rank 1. The shape above the square has $N - \frac{\sqrt{2}}{2}N$ rows (about $0.3N$) and the shape beside it has $N - \frac{\sqrt{2}}{2}N$ columns. Those rows and those columns are independent. Adding those two numbers produces the leading term in the rank—and it is numerically confirmed:

$$\textbf{Rank of quarter circle matrix} \approx (\mathbf{2} - \sqrt{\mathbf{2}})\, \boldsymbol{N} \text{ as } N \to \infty.$$

We turn to the (nonzero) singular values of these matrices—trivial for the square, known for the triangle, computable for the quarter circle. For these shapes and others, we have always seen a **"singular gap"**. The singular values don't approach zero. All the σ's stay above some limit L—and we don't know why.

The graphs show σ's for the quarter circle (computed values) and the triangle (exact values). For the triangle of 1's, the inverse matrix just has a diagonal of 1's above a diagonal of -1's. Then $\boldsymbol{\sigma_i} = \frac{\mathbf{1}}{\mathbf{2}} \sin \boldsymbol{\theta}$ for N equally spaced angles $\theta_i = (2i-1)\pi/(4N+2)$. Therefore the gap with no singular values reaches up to $\sigma_{\text{min}} \approx \frac{1}{2} \sin \frac{\pi}{2} = \frac{1}{2}$. The quarter circle also has $\sigma_{\text{min}} \approx \frac{1}{2}$. See the student project on **math.mit.edu/learningfromdata.**

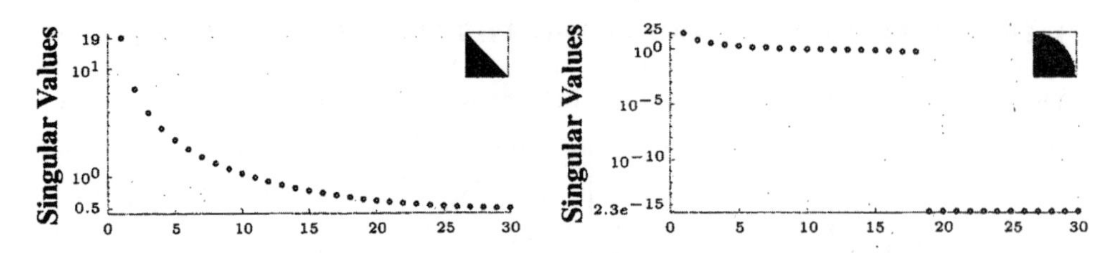

Figure I.14: The (nonzero) singular values for the triangle and quarter circle matrices.

Problem Set I.9

1 What are the singular values (in descending order) of $\boldsymbol{A} - \boldsymbol{A}_k$? Omit any zeros.

2 Find a closest rank-1 approximation to these matrices (L^2 or Frobenius norm):

$$A = \begin{bmatrix} 3 & 0 & 0 \\ 0 & 2 & 0 \\ 0 & 0 & 1 \end{bmatrix} \qquad A = \begin{bmatrix} 0 & 3 \\ 2 & 0 \end{bmatrix} \qquad A = \begin{bmatrix} 2 & 1 \\ 1 & 2 \end{bmatrix}$$

3 Find a closest rank-1 approximation in the L^2 norm to $A = \begin{bmatrix} \cos\theta & -\sin\theta \\ \sin\theta & \cos\theta \end{bmatrix}$

4 The Eckart-Young theorem is false in the matrix norm $||A||_\infty = $ max row sum:

$$A = \begin{bmatrix} a & b \\ c & d \end{bmatrix} \text{ has } ||A||_\infty = \max \frac{||\boldsymbol{Ax}||_\infty}{||\boldsymbol{x}||_\infty} = \max(|a| + |b|, |c| + |d|)$$

Find a rank-1 matrix closer to $A = \begin{bmatrix} 3 & 0 \\ 4 & 5 \end{bmatrix}$ than $A_1 = \dfrac{3}{2}\begin{bmatrix} 1 & 1 \\ 3 & 3 \end{bmatrix}$.

5 Show that this norm $||A||_\infty = \max(|a| + |b|, |c| + |d|)$ is not orthogonally invariant:

Find a diagonal matrix A where $||QA||_\infty \neq ||A||_\infty$ for $Q = \begin{bmatrix} \cos\theta & -\sin\theta \\ \sin\theta & \cos\theta \end{bmatrix}$.

6 If $S = Q\Lambda Q^{\mathrm{T}}$ is a symmetric positive definite matrix, explain from Eckart-Young why $\boldsymbol{q}_1\lambda_1\boldsymbol{q}_1^{\mathrm{T}}$ is the closest rank-1 approximation in the L^2 matrix norm $||S||_2$.

7 Explain the derivatives $\partial E/\partial C$ and $\partial E/\partial R$ in equation (19) for size $n = 2$.

8 Which rank-3 matrices have $||A - A_1||_2 = ||A - A_2||_2$? A_2 is $\sigma_1\boldsymbol{u}_1\boldsymbol{v}_1^{\mathrm{T}} + \sigma_2\boldsymbol{u}_2\boldsymbol{v}_2^{\mathrm{T}}$.

9 Replace the quarter circle in Figure I.13 by the parabola $\boldsymbol{y} = \boldsymbol{1} - \boldsymbol{x^2}$. Estimate the rank CN with all 1's under the parabola (N 1's along the axes). First remove a rectangle of 1's, touching the parabola where slope $= -1$.

10 If A is a 2 by 2 matrix with $\sigma_1 \geq \sigma_2 > 0$, find $||A^{-1}||_2$ and $||A^{-1}||_F^2$.

I.10 Rayleigh Quotients and Generalized Eigenvalues

This section picks up and extends a theme from Section I.8. There we connected the eigenvalues and eigenvectors of a symmetric matrix S to the **Rayleigh quotient** $R(\boldsymbol{x})$:

$$R(\boldsymbol{x}) = \frac{\boldsymbol{x}^{\mathrm{T}} S \boldsymbol{x}}{\boldsymbol{x}^{\mathrm{T}} \boldsymbol{x}}. \tag{1}$$

The maximum value of $R(\boldsymbol{x})$ is the largest eigenvalue λ_1 of S. That maximum is achieved at the eigenvector $\boldsymbol{x} = \boldsymbol{q}_1$ where $S\boldsymbol{q}_1 = \lambda \boldsymbol{q}_1$:

$$\textbf{Maximum} \qquad R(\boldsymbol{q}_1) = \frac{\boldsymbol{q}_1^{\mathrm{T}} S \boldsymbol{q}_1}{\boldsymbol{q}_1^{\mathrm{T}} \boldsymbol{q}_1} = \frac{\boldsymbol{q}_1^{\mathrm{T}} \lambda_1 \boldsymbol{q}_1}{\boldsymbol{q}_1^{\mathrm{T}} \boldsymbol{q}_1} = \lambda_1. \tag{2}$$

Similarly the minimum value of $R(\boldsymbol{x})$ equals the smallest eigenvalue λ_n of S. That minimum is attained at the "bottom eigenvector" $\boldsymbol{x} = \boldsymbol{q}_n$. More than that, all the eigenvectors $\boldsymbol{x} = \boldsymbol{q}_k$ of S for eigenvalues *between* λ_n and λ_1 are saddle points of $R(\boldsymbol{x})$. Saddles have first derivatives = zero but they are not maxima or minima.

$$\textbf{Saddle point} \qquad \text{All } \frac{\partial R}{\partial x_i} = 0 \text{ at } \boldsymbol{x} = \boldsymbol{q}_k \text{ Then } R(\boldsymbol{q}_k) = \frac{\boldsymbol{q}_k^{\mathrm{T}} \lambda_k \boldsymbol{q}_k}{\boldsymbol{q}_k^{\mathrm{T}} \boldsymbol{q}_k} = \lambda_k. \tag{3}$$

These facts connected to the Singular Value Decomposition of A. The connection was through $S = A^{\mathrm{T}} A$. For that positive definite (or semidefinite) matrix S, the Rayleigh quotient led to the norm (squared) of A. And the largest eigenvalue of S is $\sigma_1^2(A)$:

$$\boxed{||A||^2 = \max \frac{||A\boldsymbol{x}||^2}{||\boldsymbol{x}||^2} = \max \frac{\boldsymbol{x}^{\mathrm{T}} A^{\mathrm{T}} A \boldsymbol{x}}{\boldsymbol{x}^{\mathrm{T}} \boldsymbol{x}} = \max \frac{\boldsymbol{x}^{\mathrm{T}} S \boldsymbol{x}}{\boldsymbol{x}^{\mathrm{T}} \boldsymbol{x}} = \lambda_1(S) = \sigma_1^2(A).} \tag{4}$$

In this way a symmetric eigenvalue problem is also an optimization: **Maximize $R(\boldsymbol{x})$.**

Generalized Eigenvalues and Eigenvectors

Applications in statistics and data science lead us to the next step. Applications in engineering and mechanics point the same way. *A second symmetric matrix M enters* the denominator of $R(\boldsymbol{x})$:

$$\textbf{Generalized Rayleigh quotient} \qquad R(\boldsymbol{x}) = \frac{\boldsymbol{x}^{\mathrm{T}} S \boldsymbol{x}}{\boldsymbol{x}^{\mathrm{T}} M \boldsymbol{x}} \tag{5}$$

In dynamical problems M is often the "mass matrix" or the "inertia matrix". In statistics M is generally the **covariance matrix**. The construction of covariance matrices and their application to classifying data will come in the chapter on probability and statistics.

Here our goal is to see how the eigenvalue problem $Sx = \lambda x$ changes to $Sx = \lambda Mx$, when $R(x)$ becomes $x^{\mathrm{T}}Sx/x^{\mathrm{T}}Mx$. This is the *generalized* symmetric eigenvalue problem.

If M is positive definite, the maximum of $R(x)$ is the largest eigenvalue of $M^{-1}S$.

We will reduce this generalized problem $Sx = \lambda Mx$ to an ordinary eigenvalue problem $Hy = \lambda y$. But you have to see that the choice $H = M^{-1}S$ is not really perfect. The reason is simple: *$M^{-1}S$ is not usually symmetric*! Even a diagonal matrix M will make this point clear. The square root $M^{1/2}$ of that same diagonal matrix will suggest the right way to hold on to symmetry.

$$M^{-1}S = \begin{bmatrix} m_1 & 0 \\ 0 & m_2 \end{bmatrix}^{-1} \begin{bmatrix} a & b \\ b & c \end{bmatrix} = \begin{bmatrix} a/m_1 & b/m_1 \\ b/m_2 & c/m_2 \end{bmatrix} \quad \textbf{is not symmetric}$$

$$H = M^{-1/2}\,S\,M^{-1/2} = \begin{bmatrix} a/m_1 & b/\sqrt{m_1m_2} \\ b/\sqrt{m_1m_2} & c/m_2 \end{bmatrix} \quad \textbf{is symmetric.}$$

Those matrices $M^{-1}S$ and $H = M^{-1/2}SM^{-1/2}$ have the same eigenvalues. This H looks awkward, but symmetry is saved when we choose the symmetric square root of M and M^{-1}. **Every positive definite M has a positive definite square root.**

The diagonal example above had $M^{1/2} = \operatorname{diag}(\sqrt{m_1}, \sqrt{m_2})$. Its inverse is $M^{-1/2}$. In all cases, we just diagonalize M and take the square root of each eigenvalue:

$$\textbf{If } M = Q\Lambda Q^{\mathrm{T}} \textbf{ has } \Lambda > 0 \textbf{ then } M^{1/2} = Q\Lambda^{1/2}Q^{\mathrm{T}} \textbf{ has } \Lambda^{1/2} > 0. \quad (6)$$

Squaring $M^{1/2}$ recovers $Q\Lambda^{1/2}Q^{\mathrm{T}}Q\Lambda^{1/2}Q^{\mathrm{T}} = Q\Lambda Q^{\mathrm{T}}$ which is M. We will not use $M^{1/2}$ or $M^{-1/2}$ numerically! The generalized eigenvalue problem $Sx = \lambda Mx$ is solved in MATLAB by the command **eig**(S, M). **Julia** and **Python** and **R** and all full linear algebra systems include this extension to $Sx = \lambda Mx$.

A Rayleigh quotient with $x^{\mathrm{T}}Mx$ is easily converted to a quotient with $y^{\mathrm{T}}y$:

$$\textbf{Set } x = M^{-1/2}y \quad \text{Then} \quad \frac{x^{\mathrm{T}}Sx}{x^{\mathrm{T}}Mx} = \frac{y^{\mathrm{T}}(M^{-1/2})^{\mathrm{T}}SM^{-1/2}y}{y^{\mathrm{T}}y} = \frac{y^{\mathrm{T}}Hy}{y^{\mathrm{T}}y}. \quad (7)$$

This changes the generalized problem $Sx = \lambda Mx$ to an ordinary symmetric problem $Hy = \lambda y$. If S and M are positive definite, so is $H = M^{-1/2}SM^{-1/2}$. The largest Rayleigh quotient still gives the largest eigenvalue λ_1. And we see the top eigenvector y_1 of H and the top eigenvector x_1 of $M^{-1}S$:

$$\max \frac{y^{\mathrm{T}}Hy}{y^{\mathrm{T}}y} = \lambda_1 \text{ when } \boldsymbol{Hy_1 = \lambda_1 y_1}. \text{ Then } \boldsymbol{Sx_1 = \lambda Mx_1} \text{ for } x_1 = M^{-1/2}y_1.$$

Example 1 Solve $Sx = \lambda Mx$ when $S = \begin{bmatrix} 4 & -2 \\ -2 & 4 \end{bmatrix}$ and $M = \begin{bmatrix} 1 & 0 \\ 0 & 2 \end{bmatrix}$.

Solution Our eigenvalue problems are $(S - \lambda M)\boldsymbol{x} = \mathbf{0}$ and $(H - \lambda I)\boldsymbol{y} = \mathbf{0}$. We will find the **same** $\boldsymbol{\lambda}$'s from both determinants: $\det(S - \lambda M) = 0$ and $\det(H - \lambda I) = 0$.

$$\det(S - \lambda M) = \det \begin{bmatrix} 4-\lambda & -2 \\ -2 & 4-2\lambda \end{bmatrix} = 2\lambda^2 - 12\lambda + 12 = 0 \quad \text{gives} \quad \boldsymbol{\lambda = 3 \pm \sqrt{3}}.$$

If you prefer to work with one matrix $H = M^{-1/2} S M^{-1/2}$, we must first compute it:

$$H = \begin{bmatrix} 1 & 0 \\ 0 & 1/\sqrt{2} \end{bmatrix} \begin{bmatrix} 4 & -2 \\ -2 & 4 \end{bmatrix} \begin{bmatrix} 1 & 0 \\ 0 & 1/\sqrt{2} \end{bmatrix} = \begin{bmatrix} 4 & -\sqrt{2} \\ -\sqrt{2} & 2 \end{bmatrix}.$$

Then its eigenvalues come from the determinant of $H - \lambda I$:

$$\det \begin{bmatrix} 4-\lambda & -\sqrt{2} \\ -\sqrt{2} & 2-\lambda \end{bmatrix} = \lambda^2 - 6\lambda + 6 = 0 \text{ also gives } \lambda = 3 \pm \sqrt{3}.$$

This equation is just half of the previous $2\lambda^2 - 12\lambda + 12 = 0$. Same λ's for H and $M^{-1}S$.

In mechanical engineering those λ's would tell us the frequencies $\omega = \sqrt{\lambda}$ for two oscillating masses $m_1 = 1$ and $m_2 = 2$ in a line of springs. S tells us the stiffness in the three springs that connect these two masses to fixed endpoints.

The differential equation is Newton's Law $Md^2\boldsymbol{u}/dt^2 = -S\boldsymbol{u}$.

Generalized Eigenvectors are M-orthogonal

A crucial fact about a symmetric matrix S is that any two eigenvectors are orthogonal (when the eigenvalues are different). Does this extend to $S\boldsymbol{x}_1 = \lambda M\boldsymbol{x}_1$ with two symmetric matrices? The immediate answer is **no**, but the right answer is **yes**. For that answer, we have to assume that M is positive definite, and we have to change from $\boldsymbol{x}_1^{\mathrm{T}}\boldsymbol{x}_2 = 0$ to "M-orthogonality" of $\boldsymbol{x}_1$ and $\boldsymbol{x}_2$. **Two vectors are M-orthogonal if $\boldsymbol{x}_1^{\mathrm{T}}M\boldsymbol{x}_2 = 0$.**

$$\boldsymbol{x}_1^{\mathrm{T}}M\boldsymbol{x}_2 = 0 \textbf{ if } S\boldsymbol{x}_1 = \lambda_1 M\boldsymbol{x}_1 \textbf{ and } S\boldsymbol{x}_2 = \lambda_2 M\boldsymbol{x}_2 \textbf{ and } \lambda_1 \neq \lambda_2. \tag{8}$$

Proof. Multiply one equation by $\boldsymbol{x}_2^{\mathrm{T}}$ and multiply the other equation by $\boldsymbol{x}_1^{\mathrm{T}}$:

$$\boldsymbol{x}_2^{\mathrm{T}}S\boldsymbol{x}_1 = \lambda_1 \boldsymbol{x}_2^{\mathrm{T}}M\boldsymbol{x}_1 \quad \text{and} \quad \boldsymbol{x}_1^{\mathrm{T}}S\boldsymbol{x}_2 = \lambda_2 \boldsymbol{x}_1^{\mathrm{T}}M\boldsymbol{x}_2$$

Because S and M are symmetric, transposing the first equation gives $\boldsymbol{x}_1^{\mathrm{T}}S\boldsymbol{x}_2 = \lambda_1 \boldsymbol{x}_1^{\mathrm{T}}M\boldsymbol{x}_2$. Subtract the second equation:

$$(\lambda_1 - \lambda_2)\,\boldsymbol{x}_1^{\mathrm{T}}M\boldsymbol{x}_2 = 0 \quad \text{and with} \quad \lambda_1 \neq \lambda_2 \quad \text{this requires} \quad \boldsymbol{x}_1^{\mathrm{T}}M\boldsymbol{x}_2 = 0. \tag{9}$$

Then also $\boldsymbol{x}_1^{\mathrm{T}}S\boldsymbol{x}_2 = 0$. We can test this conclusion on the matrices S and M in Example 1.

Example 2 Find the eigenvectors for $\lambda_1 = 3+\sqrt{3}$ and $\lambda_2 = 3-\sqrt{3}$. Test $\boldsymbol{x}^{\mathrm{T}} M \boldsymbol{y} = 0$.

The eigenvectors $\boldsymbol{x}$ and $\boldsymbol{y}$ are in the nullspaces where $(S-\lambda_1 M)\boldsymbol{x}=\mathbf{0}$ and $(S-\lambda_2 M)\boldsymbol{y}=\mathbf{0}$.

$$(S-\lambda_1 M)\boldsymbol{x} = \begin{bmatrix} 4-(3+\sqrt{3}) & -2 \\ -2 & 4-2(3+\sqrt{3}) \end{bmatrix} \begin{bmatrix} x_1 \\ x_2 \end{bmatrix} \quad \text{gives} \quad \boldsymbol{x} = c \begin{bmatrix} \mathbf{2} \\ \mathbf{1}+\sqrt{\mathbf{3}} \end{bmatrix}$$

$$(S-\lambda_2 M)\boldsymbol{y} = \begin{bmatrix} 4-(3-\sqrt{3}) & -2 \\ -2 & 4-2(3-\sqrt{3}) \end{bmatrix} \begin{bmatrix} y_1 \\ y_2 \end{bmatrix} \quad \text{gives} \quad \boldsymbol{y} = c \begin{bmatrix} \mathbf{2} \\ \mathbf{1}-\sqrt{\mathbf{3}} \end{bmatrix}$$

Those eigenvectors $\boldsymbol{x}$ and $\boldsymbol{y}$ are not orthogonal. But they are M-orthogonal because

$$\boldsymbol{x}^{\mathrm{T}} M \boldsymbol{y} = \begin{bmatrix} 2 & 1+\sqrt{3} \end{bmatrix} \begin{bmatrix} 1 & 0 \\ 0 & 2 \end{bmatrix} \begin{bmatrix} 2 \\ 1-\sqrt{3} \end{bmatrix} = 0.$$

Positive Semidefinite M : Not Invertible

There are important applications in which the matrix M is only **positive semidefinite. Then $\boldsymbol{x}^{\mathrm{T}} M \boldsymbol{x}$ can be zero!** The matrix M will not be invertible. The quotient $\boldsymbol{x}^{\mathrm{T}} S \boldsymbol{x} / \boldsymbol{x}^{\mathrm{T}} M \boldsymbol{x}$ can be infinite. The matrices $M^{-1/2}$ and H do not even exist. The eigenvalue problem $S\boldsymbol{x} = \lambda M \boldsymbol{x}$ is still to be solved, but an infinite eigenvalue $\lambda = \infty$ is now very possible.

In statistics M is often a covariance matrix. Its diagonal entries tell us the separate variances of two or more measurements. Its off-diagonal entries tell us the "covariances between the measurements". If we are foolishly repeating exactly the same observations—or if one experiment is completely determined by another—then the covariance matrix M is singular. Its determinant is zero and it is not invertible. The Rayleigh quotient (which divides by $\boldsymbol{x}^{\mathrm{T}} M \boldsymbol{x}$) may become infinite.

One way to look at this mathematically is to write $S\boldsymbol{x} = \lambda M \boldsymbol{x}$ in a form with α and β.

$$\boldsymbol{\alpha} S\boldsymbol{x} = \boldsymbol{\beta} M \boldsymbol{x} \text{ with } \boldsymbol{\alpha} \geq \mathbf{0} \text{ and } \boldsymbol{\beta} \geq \mathbf{0} \text{ and } \textbf{eigenvalues } \lambda = \frac{\beta}{\alpha}. \tag{10}$$

λ will be an ordinary positive eigenvalue if $\alpha > 0$ and $\beta > 0$. We can even normalize those two numbers by $\alpha^2 + \beta^2 = 1$. But now we see three other possibilities in equation (10):

$\alpha > 0$ and $\beta = 0$ Then $\boldsymbol{\lambda} = \mathbf{0}$ and $S\boldsymbol{x} = 0\boldsymbol{x}$: a normal zero eigenvalue of S

$\alpha = 0$ and $\beta > 0$ Then $\boldsymbol{\lambda} = \infty$ and $M\boldsymbol{x} = 0$: M is not invertible

$\alpha = 0$ and $\beta = 0$ Then $\boldsymbol{\lambda} = \frac{\mathbf{0}}{\mathbf{0}}$ is undetermined: $M\boldsymbol{x} = \mathbf{0}$ and also $S\boldsymbol{x} = \mathbf{0}$.

$\alpha = 0$ can occur when we have clusters of data, if the number of samples in a cluster is smaller than the number of features we measure. This is the problem of **small sample size.** It happens.

You will understand that the mathematics becomes more delicate. The SVD approach (when you factor a data matrix into $A = U\Sigma V^{\mathrm{T}}$ with singular vectors $\boldsymbol{v}$ coming from eigenvectors of $S = A^{\mathrm{T}} A$) is not sufficient. **We need to generalize the SVD. We need to allow for a second matrix M. This led to the GSVD.**

The Generalized SVD (Simplified)

In its full generality, this factorization is complicated. It allows for two matrices S and M and it allows them to be singular. In this book it makes sense to stay with the usual and best case, when these symmetric matrices are positive definite. Then we can see the primary purpose of the GSVD, to factor two matrices at the same time.

Remember that the classical SVD factors a rectangular matrix A into $U\Sigma V^{\mathrm{T}}$. It begins with A and not with $S = A^{\mathrm{T}}A$. Similarly here, **we begin with two matrices A and B.** Our simplification is to assume that both are tall thin matrices of rank n. Their sizes are m_A by n and m_B by n. Then $\boldsymbol{S = A^{\mathrm{T}}A}$ and $\boldsymbol{M = B^{\mathrm{T}}B}$ are n by n and positive definite.

Generalized Singular Value Decomposition

A and B can be factored into $A = U_A\Sigma_A Z$ and $B = U_B\Sigma_B Z$ (same Z)

U_A and U_B are orthogonal matrices (sizes m_A and m_B)

Σ_A and Σ_B are positive diagonal matrices (with $\Sigma_A^{\mathrm{T}}\Sigma_A + \Sigma_B^{\mathrm{T}}\Sigma_B = I_{n\times n}$)

Z is an invertible matrix (size n)

Notice that Z is probably not an orthogonal matrix. That would be asking too much. The remarkable property of Z is to **simultaneously diagonalize** $S = A^{\mathrm{T}}A$ and $M = B^{\mathrm{T}}B$:

$$A^{\mathrm{T}}A = Z^{\mathrm{T}}\Sigma_A^{\mathrm{T}}U_A^{\mathrm{T}}U_A\Sigma_A Z = \boldsymbol{Z^{\mathrm{T}}\left(\Sigma_A^{\mathrm{T}}\Sigma_A\right)Z} \quad \text{and} \quad B^{\mathrm{T}}B = \boldsymbol{Z^{\mathrm{T}}\left(\Sigma_B^{\mathrm{T}}\Sigma_B\right)Z}. \quad (11)$$

So this is a fact of linear algebra: Any two positive definite matrices can be diagonalized by the same matrix Z. By equation (9), its columns can be $\boldsymbol{x}_1, \ldots, \boldsymbol{x}_n$! That was known before the GSVD was invented. And because orthogonality is not required, we can scale Z so that $\Sigma_A^{\mathrm{T}}\Sigma_A + \Sigma_B^{\mathrm{T}}\Sigma_B = I$. We can also order its columns $\boldsymbol{x}_k$ to put the n positive numbers σ_A in decreasing order (in Σ_A).

Please also notice the meaning of "diagonalize". Equation (11) does not contain Z^{-1} and Z, it contains Z^{T} and Z. With Z^{-1} we have a **similarity** transformation, preserving eigenvalues. With Z^{T} we have a **congruence** transformation $Z^{\mathrm{T}}SZ$, preserving symmetry. (Then the eigenvalues of S and $Z^{\mathrm{T}}SZ$ have the **same signs**. This is Sylvester's Law of Inertia in PSet III.2. Here the signs are all positive.) The symmetry of $S = A^{\mathrm{T}}A$ and the positive definiteness of $M = B^{\mathrm{T}}B$ allows one Z to diagonalize both matrices.

The Problem Set (Problem 5) will guide you to a proof of this simplified GSVD.

Fisher's Linear Discriminant Analysis (LDA)

Here is a nice application in statistics and machine learning. We are given samples from two different populations and they are mixed together. We know the basic facts about each population—its average value m and its average spread σ around that mean m. So we have a mean m_1 and variance σ_1 for the first population and m_2, σ_2 for the second population. If all the samples are mixed together and we pick one, *how do we tell if it probably came from population* 1 *or population* 2 ?

Fisher's "linear discriminant" answers that question.

Actually the problem is one step more complicated. Each sample has several features, like a child's age and height and weight. This is normal for machine learning, to have a "feature vector" like $\boldsymbol{f} =$ (age, height, weight) for each sample. If a sample has feature vector $\boldsymbol{f}$, which population did it probably come from? We start with vectors not scalars.

We have an average age $\boldsymbol{m_a}$ and average height $\boldsymbol{m_h}$ and average weight $\boldsymbol{m_w}$ for each population. **The mean (average) of population 1 is a vector $\boldsymbol{m_1} = (\boldsymbol{m_{a1}}, \boldsymbol{m_{h1}}, \boldsymbol{m_{w1}})$.** Population 2 also has a vector $\boldsymbol{m_2}$ of average age, average height, average weight. And the variance σ for each population, to measure its spread around its mean, becomes a **3 × 3 matrix** Σ. This "covariance matrix" will be a key to Chapter V on statistics. For now, we have $\boldsymbol{m_1}, \boldsymbol{m_2}, \Sigma_1, \Sigma_2$ and we want a rule to discriminate between the two populations.

Fisher's test has a simple form: *He finds a separation vector* $\boldsymbol{v}$. If the sample has $\boldsymbol{v}^{\mathrm{T}}\boldsymbol{f} > c$ then our best guess is population 1. If $\boldsymbol{v}^{\mathrm{T}}\boldsymbol{f} < c$ then the sample probably came from population 2. The vector $\boldsymbol{v}$ is trying to separate the two populations (as much as possible). It maximizes the separation ratio R:

$$\textbf{Separation ratio} \qquad R = \frac{(\boldsymbol{x}^{\mathrm{T}}\boldsymbol{m_1} - \boldsymbol{x}^{\mathrm{T}}\boldsymbol{m_2})^2}{\boldsymbol{x}^{\mathrm{T}}\Sigma_1\boldsymbol{x} + \boldsymbol{x}^{\mathrm{T}}\Sigma_2\boldsymbol{x}} \tag{12}$$

That ratio R has the form $\boldsymbol{x}^{\mathrm{T}}S\boldsymbol{x}/\boldsymbol{x}^{\mathrm{T}}M\boldsymbol{x}$. The matrix S is $(\boldsymbol{m_1} - \boldsymbol{m_2})(\boldsymbol{m_1} - \boldsymbol{m_2})^{\mathrm{T}}$. The matrix M is $\Sigma_1 + \Sigma_2$. **We know the rule $S\boldsymbol{v} = \lambda M\boldsymbol{v}$ for the vector $\boldsymbol{x} = \boldsymbol{v}$ that maximizes the separation ratio R.**

Fisher could actually find that eigenvector $\boldsymbol{v}$ of $M^{-1}S$. So can we, because the matrix $S = (\boldsymbol{m_1} - \boldsymbol{m_2})(\boldsymbol{m_1} - \boldsymbol{m_2})^{\mathrm{T}}$ *has rank one*. So $S\boldsymbol{v}$ is always in the direction $\boldsymbol{m_1} - \boldsymbol{m_2}$. Then $M\boldsymbol{v}$ must be in that direction, to have $S\boldsymbol{v} = \lambda M\boldsymbol{v}$. *So* $\boldsymbol{v} = M^{-1}(\boldsymbol{m_1} - \boldsymbol{m_2})$.

This was a nice problem because we found the eigenvector $\boldsymbol{v}$. It makes sense that when the unknown sample has feature vector $\boldsymbol{f} =$ (age, height, weight), we would look at the numbers $\boldsymbol{m}_1^{\mathrm{T}}\boldsymbol{f}$ and $\boldsymbol{m}_2^{\mathrm{T}}\boldsymbol{f}$. If we were ready for a full statistical discussion (*which we are not*), then we could see how the weighting matrix $M = \Sigma_0 + \Sigma_1$ enters into the final test on $\boldsymbol{v}^{\mathrm{T}}\boldsymbol{f}$. Here it is enough to say: The feature vectors $\boldsymbol{f}$ from the two populations are separated as well as possible by a plane that is perpendicular to $\boldsymbol{v}$.

Summary. We have two clouds of points in 3-dimensional feature space. We try to separate them by a plane—not always possible. Fisher proposed one reasonable plane.

Neural networks will succeed by allowing separating surfaces that are *not just planes*.

Problem Set I.10

1 Solve $(S - \lambda M)\boldsymbol{x} = \mathbf{0}$ and $(H - \lambda I)\boldsymbol{y} = \mathbf{0}$ after computing the matrix $H = M^{-1/2}SM^{-1/2}$:

$$S = \begin{bmatrix} 5 & 4 \\ 4 & 5 \end{bmatrix} \qquad M = \begin{bmatrix} 1 & 0 \\ 0 & 4 \end{bmatrix}$$

Step 1 is to find λ_1 and λ_2 from $\det(S - \lambda M) = 0$. The equation $\det(H - \lambda I) = 0$ should produce the same λ_1 and λ_2. Those eigenvalues will produce two eigenvectors $\boldsymbol{x}_1$ and $\boldsymbol{x}_2$ of $S - \lambda M$ and two eigenvectors $\boldsymbol{y}_1$ and $\boldsymbol{y}_2$ of $H - \lambda I$.

Verify that $\boldsymbol{x}_1^{\mathrm{T}}\boldsymbol{x}_2$ is not zero but $\boldsymbol{x}_1^{\mathrm{T}}M\boldsymbol{x}_2 = 0$. H is symmetric so $\boldsymbol{y}_1^{\mathrm{T}}\boldsymbol{y}_2 = 0$.

2 (a) For $\boldsymbol{x} = (a, b)$ and $\boldsymbol{y} = (c, d)$ write the Rayleigh quotients in Problem 1 as

$$R^*(\boldsymbol{x}) = \frac{\boldsymbol{x}^{\mathrm{T}}S\boldsymbol{x}}{\boldsymbol{x}^{\mathrm{T}}M\boldsymbol{x}} = \frac{(5a^2 + 8ab + 5b^2)}{(\cdots + \cdots + \cdots)} \text{ and } R(\boldsymbol{y}) = \frac{\boldsymbol{y}^{\mathrm{T}}H\boldsymbol{y}}{\boldsymbol{y}^{\mathrm{T}}\boldsymbol{y}} = \frac{5c^2 + 16cd + 20d^2}{(\cdots + \cdots)}$$

(b) Take the c and d derivatives of $R(\boldsymbol{y})$ to find its maximum and minimum.

(c) Take the a and b derivatives of $R^*(\boldsymbol{x})$ to find its maximum and minimum.

(d) Verify that those maxima occur at eigenvectors from $(S - \lambda M)\boldsymbol{x} = \mathbf{0}$ and $(H - \lambda I)\boldsymbol{y} = \mathbf{0}$.

3 How are the eigenvectors $\boldsymbol{x}_1$ and $\boldsymbol{x}_2$ related to the eigenvectors $\boldsymbol{y}_1$ and $\boldsymbol{y}_2$?

4 Change M to $\begin{bmatrix} 1 & 0 \\ 0 & 0 \end{bmatrix}$ and solve $S\boldsymbol{x} = \lambda M\boldsymbol{x}$. Now M is singular and one of the eigenvalues λ is infinite. But its eigenvector $\boldsymbol{x}_2$ is still M-orthogonal to the other eigenvector $\boldsymbol{x}_1$.

5 Start with symmetric positive definite matrices S and M. Eigenvectors of S fill an orthogonal matrix Q so that $Q^{\mathrm{T}}SQ = \Lambda$ is diagonal. What is the diagonal matrix D so that $D^{\mathrm{T}}\Lambda D = I$? Now we have $D^{\mathrm{T}}Q^{\mathrm{T}}SQD = I$ and we look at $D^{\mathrm{T}}Q^{\mathrm{T}}MQD$. Its eigenvector matrix Q_2 gives $Q_2^{\mathrm{T}}IQ_2 = I$ and $Q_2^{\mathrm{T}}D^{\mathrm{T}}Q^{\mathrm{T}}MQDQ_2 = \Lambda_2$. *Show that $Z = QDQ_2$ diagonalizes both congruences $Z^{\mathrm{T}}SZ$ and $Z^{\mathrm{T}}MZ$ in the GSVD.*

6 (a) Why does every congruence $Z^{\mathrm{T}}SZ$ preserve the symmetry of S?

(b) Why is $Z^{\mathrm{T}}SZ$ positive definite when S is positive definite and Z is square and invertible? Apply the energy test to $Z^{\mathrm{T}}SZ$. Be sure to explain why $Z\boldsymbol{x}$ is not the zero vector.

7 Which matrices $Z^{\mathrm{T}}IZ$ are congruent to the identity matrix for invertible Z?

8 Solve this matrix problem basic to Fisher's Linear Discriminant Analysis:

$$\text{If } R(\boldsymbol{x}) = \frac{\boldsymbol{x}^{\mathrm{T}}S\boldsymbol{x}}{\boldsymbol{x}^{\mathrm{T}}M\boldsymbol{x}} \text{ and } S = \boldsymbol{u}\boldsymbol{u}^{\mathrm{T}} \text{ what vector } \boldsymbol{x} \text{ minimizes } R(\boldsymbol{x})\text{?}$$

I.11 Norms of Vectors and Functions and Matrices

The norm of a nonzero vector $\boldsymbol{v}$ is a positive number $||\boldsymbol{v}||$. That number measures the "length" of the vector. There are many useful measures of length (many different norms). Every norm for vectors or functions or matrices must share these two properties of the absolute value $|c|$ of a number:

All norms

Multiply $\boldsymbol{v}$ by c (**Rescaling**) $\qquad ||c\,\boldsymbol{v}|| = |c|\,||\boldsymbol{v}|| \qquad (1)$

Add $\boldsymbol{v}$ to $\boldsymbol{w}$ (**Triangle inequality**) $\qquad ||\boldsymbol{v}+\boldsymbol{w}|| \leq ||\boldsymbol{v}|| + ||\boldsymbol{w}|| \qquad (2)$

We start with three special norms—by far the most important. They are the $\boldsymbol{\ell}^2$ norm and $\boldsymbol{\ell}^1$ norm and $\boldsymbol{\ell}^\infty$ norm of the vector $\boldsymbol{v} = (v_1, \ldots, v_n)$. The vector $\boldsymbol{v}$ is in $\mathbf{R}^n$ (real v_i) or in $\mathbf{C}^n$ (complex v_i):

$\boldsymbol{\ell}^2$ **norm**	= **Euclidean norm**	$\|\|\boldsymbol{v}\|\|_2$	$= \sqrt{\|v_1\|^2 + \cdots + \|v_n\|^2}$
$\boldsymbol{\ell}^1$ **norm**	= **1–norm**	$\|\|\boldsymbol{v}\|\|_1$	$= \|v_1\| + \|v_2\| + \cdots + \|v_n\|$
$\boldsymbol{\ell}^\infty$ **norm**	= **max norm**	$\|\|\boldsymbol{v}\|\|_\infty$	= **maximum of** $\|v_1\|, \ldots, \|v_n\|$

The all-ones vector $\boldsymbol{v} = (1, 1, \ldots, 1)$ has norms $||\boldsymbol{v}||_2 = \sqrt{n}$ and $||\boldsymbol{v}||_1 = n$ and $||\boldsymbol{v}||_\infty = 1$.

These three norms are the particular cases $p = 2$ and $p = 1$ and $p = \infty$ of the ℓ^p norm $||\boldsymbol{v}||_p = (|v_1|^p + \cdots + |v_n|^p)^{1/p}$. This figure shows vectors with **norm 1** : $p = \frac{1}{2}$ is illegal.

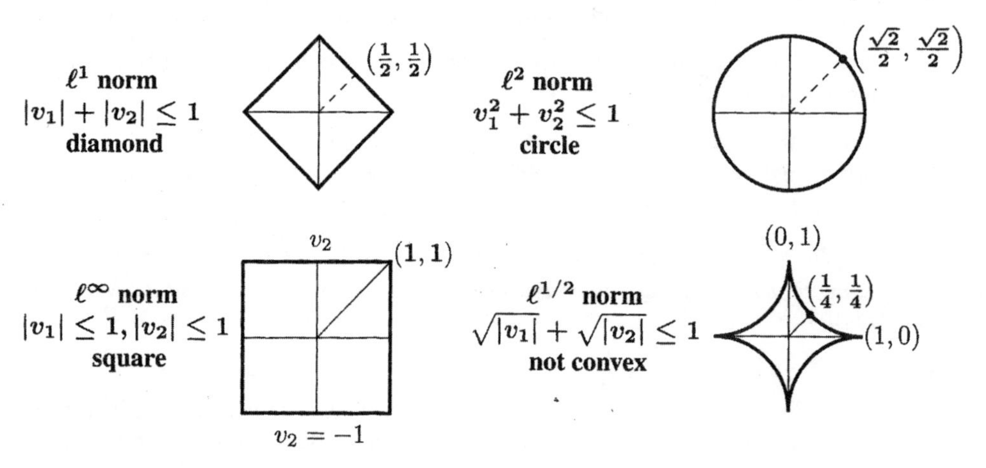

Figure I.15: The important vector norms $||\boldsymbol{v}||_1, ||\boldsymbol{v}||_2, ||\boldsymbol{v}||_\infty$ and a failure ($p = 0$ fails too).

The failure for $p = \frac{1}{2}$ is in the triangle inequality: $(1, 0)$ and $(0, 1)$ have norm 1, but their sum $(1, 1)$ has norm $2^{1/p} = 4$. Only $1 \leq p \leq \infty$ produce an acceptable norm $||\boldsymbol{v}||_p$.

The Minimum of $||v||_p$ on the line $a_1v_1 + a_2v_2 = 1$

Which point on a diagonal line like $3v_1 + 4v_2 = 1$ is closest to $(0,0)$? The answer (and the meaning of "closest") will depend on the norm. This is another way to see important differences between ℓ^1 and ℓ^2 and ℓ^∞. We will see a first example of a very special feature: **Minimization in ℓ^1 produces sparse solutions**

To see the closest point to $(0,0)$, expand the ℓ^1 diamond and ℓ^2 circle and ℓ^∞ square **until they touch the diagonal line**. For each p, that touching point $\boldsymbol{v}^*$ will solve our optimization problem:

Minimize $\|\|v\|\|_p$ among vectors (v_1, v_2) on the line $3v_1 + 4v_2 = 1$

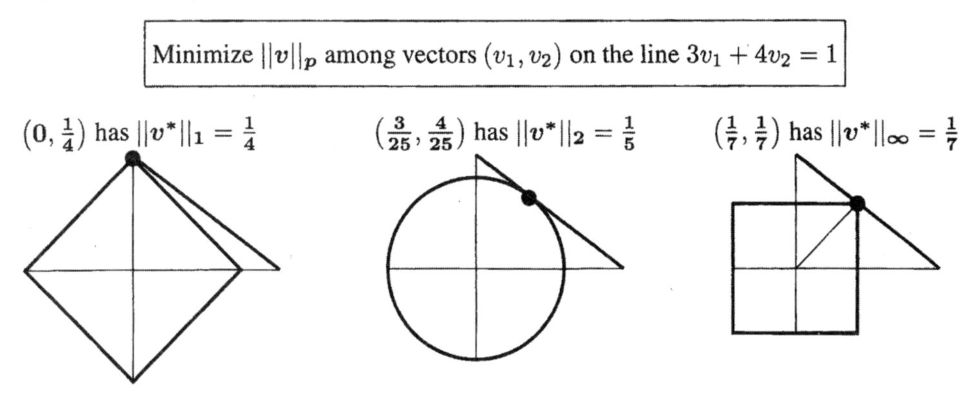

Figure I.16: The solutions $\boldsymbol{v}^*$ to the ℓ^1 and ℓ^2 and ℓ^∞ minimizations. The first is **sparse**.

The first figure displays a highly important property of the minimizing solution to the ℓ^1 problem: **That solution $\boldsymbol{v}^*$ has zero components.** The vector $\boldsymbol{v}^*$ is "sparse". This is because **a diamond touches a line at a sharp point.** The line (or hyperplane in high dimensions) contains the vectors that solve the m constraints $A\boldsymbol{v} = \boldsymbol{b}$. The surface of the diamond contains vectors with the same ℓ^1 norm. The diamond expands to meet the line at a corner of the diamond! The Problem Set and also Section III.4 will return to this "**basis pursuit**" problem and closely related ℓ^1 problems.

The essential point is that **the solutions to those problems are sparse.** They have few nonzero components, and those components have meaning. By contrast the least squares solution (using ℓ^2) has many small and non-interesting components. By squaring, those components become very small and hardly affect the ℓ^2 distance.

One final observation: **The "ℓ^0 norm"** of a vector $\boldsymbol{v}$ counts the number of nonzero components. But this is not a true norm. The points with $||v||_0 = 1$ lie on the x axis or y axis—one nonzero component only. The figure for $p = \frac{1}{2}$ on the previous page becomes even more extreme—just a cross or a skeleton along the two axes.

Of course this skeleton is not at all convex. The "zero norm" violates the fundamental requirement that $||2\boldsymbol{v}|| = 2\,||\boldsymbol{v}||$. In fact $||2\boldsymbol{v}||_0 = ||\boldsymbol{v}||_0 =$ number of nonzeros in $\boldsymbol{v}$.

The wonderful observation is that we can find the sparsest solution to $A\boldsymbol{v} = \boldsymbol{b}$ *by using the ℓ^1 norm.* We have "convexified" that ℓ^0 skeleton along the two axes. We filled in the skeleton, and the result is the ℓ^1 diamond.

Inner Products and Angles

The ℓ^2 norm has a special place. When we write $||v||$ with no subscript, this is the norm we mean. It connects to the ordinary geometry of inner products $(v, w) = v^{\mathrm{T}} w$ and angles θ between vectors:

$$\textbf{Inner product} = \textbf{length squared} \qquad v \cdot v = v^{\mathrm{T}} v = ||v||^2 \tag{3}$$

$$\textbf{Angle } \theta \textbf{ between vectors } v \textbf{ and } w \qquad v^{\mathrm{T}} w = ||v|| \, ||w|| \cos \theta \tag{4}$$

Then v is orthogonal to w when $\theta = 90°$ and $\cos \theta = 0$ and $v^{\mathrm{T}} w = 0$.

Those connections (3) and (4) lead to the most important inequalities in mathematics:

$$\textbf{Cauchy-Schwarz } |v^{\mathrm{T}} w| \leq ||v|| \, ||w|| \qquad \textbf{Triangle Inequality } ||v + w|| \leq ||v|| + ||w||$$

The Problem Set includes a direct proof of Cauchy-Schwarz. Here in the text we connect it to equation (4) for the cosine: $|\cos \theta| \leq 1$ means that $|v^{\mathrm{T}} w| \leq ||v|| \, ||w||$. And this in turn leads to the triangle inequality in equation (2)—connecting the sides v, w, and $v + w$ of an ordinary triangle in n dimensions:

$$\text{Equality} \qquad ||v + w||^2 = (v + w, v + w) = v^{\mathrm{T}} v + v^{\mathrm{T}} w + w^{\mathrm{T}} v + w^{\mathrm{T}} w$$

$$\text{Inequality} \qquad ||v + w||^2 \leq ||v||^2 + 2 \, ||v|| \, ||w|| + ||w||^2 = (||v|| + ||w||)^2 \tag{5}$$

This confirms our intuition: Any side length in a triangle is less than the sum of the other two side lengths: $||v + w|| \leq ||v|| + ||w||$. Equality in the ℓ^2 norm is only possible when the triangle is totally flat and all angles have $|\cos \theta| = 1$.

Inner Products and S-Norms

A final question about vector norms. Is ℓ^2 the only norm connected to inner products (dot products) and to angles? There are no dot products for ℓ^1 and ℓ^∞. But we can find other inner products that match other norms:

Choose any symmetric positive definite matrix S

$$||v||_S^2 = v^{\mathrm{T}} S v \text{ gives a norm for } v \text{ in } \mathbf{R}^n \text{ (called the } S\text{-norm)} \tag{6}$$

$$(v, w)_S = v^{\mathrm{T}} S w \text{ gives the } S\text{-inner product for } v, w \text{ in } \mathbf{R}^n \tag{7}$$

The inner product $(v, v)_S$ agrees with $||v||_S^2$. We have angles from (4). We have inequalities from (5). The proof is in (5) when every norm includes the matrix S.

We know that every positive definite matrix S can be factored into $A^{\mathrm{T}} A$. Then the S-norm and S-inner product for v and w are exactly the standard ℓ^2 norm and the standard inner product for Av and Aw.

$$(\boldsymbol{v}, \boldsymbol{w})_S = \boldsymbol{v}^{\mathrm{T}} S \boldsymbol{w} = (A\boldsymbol{v})^{\mathrm{T}}(A\boldsymbol{w}) \text{ because } S = A^{\mathrm{T}}A \tag{8}$$

This is not an impressive idea but it is convenient. The matrices S and A are "weighting" the vectors and their lengths. Then weighted least squares is just ordinary least squares in this weighted norm.

Einstein needed a new definition of length and distance in 4-dimensional space-time. Lorentz proposed this one, which Einstein accepted (c = speed of light):

$$\boldsymbol{v} = (x, y, z, t) \qquad ||\boldsymbol{v}||^2 = x^2 + y^2 + z^2 - c^2t^2 \qquad \textit{Is this a true norm in } \mathbf{R}^4\,?$$

Norms and Inner Products of Functions

A function $f(x)$ is a **"vector in function space"**. That simple idea gives linear algebra an importance that goes beyond n-dimensional space $\mathbf{R}^n$. All the intuition associated with linearity carries us from finite dimensions onward to infinite dimensions. The fundamental requirement for a vector space is to allow *linear combinations* $c\boldsymbol{v} + d\boldsymbol{w}$ of vectors $\boldsymbol{v}$ and $\boldsymbol{w}$. This idea extends directly to linear combinations $cf + dg$ of functions f and g.

It is exactly with norms that new questions arise in infinite dimensions. Think about the particular vectors $\boldsymbol{v}_n = \left(1, \frac{1}{2}, \ldots, \left(\frac{1}{2}\right)^n, 0, 0, \ldots\right)$ in the usual ℓ_2 norm. Those vectors come closer together since $||\boldsymbol{v}_n - \boldsymbol{v}_N|| \to 0$ as $n \to \infty$ and $N \to \infty$. **For a vector space to be "complete", every converging sequence $\boldsymbol{v}_n$ must have a limit $\boldsymbol{v}_\infty$ in the space**: $||\boldsymbol{v}_n - \boldsymbol{v}_\infty|| \to 0$.

1. The space of infinite vectors $\boldsymbol{v} = (v_1, \ldots, v_N, 0, 0, \ldots)$ ending in all zeros is *not complete*.

2. The space of vectors with $||\boldsymbol{v}||^2 = |v_1|^2 + |v_2|^2 + \cdots < \infty$ *is complete*. A vector like $\boldsymbol{v}_\infty = \left(1, \frac{1}{2}, \frac{1}{4}, \frac{1}{8}, \ldots\right)$ is included in this space but not in 1. It doesn't end in zeros.

Two famous names are associated with complete infinite-dimensional vector spaces:

A **Banach space** is a complete vector space with a norm $||\boldsymbol{v}||$ satisfying rules (1) and (2)
A **Hilbert space** is a Banach space that also has an inner product with $(\boldsymbol{v}, \boldsymbol{v})$ equal to $||\boldsymbol{v}||^2$

Those spaces are infinite-dimensional when the vectors have infinitely many components:

$\boldsymbol{\ell}^1$ is a Banach space with norm $||\boldsymbol{v}||_1 = |v_1| + |v_2| + \cdots$

$\boldsymbol{\ell}^2$ is a Hilbert space because it has an inner product $(\boldsymbol{v}, \boldsymbol{w}) = v_1w_1 + v_2w_2 + \cdots$

$\boldsymbol{\ell}^\infty$ is a Banach space with norm $||\boldsymbol{v}||_\infty$ = supremum of the numbers $|v_1|, |v_2|, \ldots$

Our special interest is in function spaces. The vectors can be functions $f(x)$ for $0 \le x \le 1$.

$\boldsymbol{L}^1[0,1]$ is a Banach space with $||\boldsymbol{f}||_1 = \int_0^1 |f(x)|dx$

$\boldsymbol{L}^2[0,1]$ is a Hilbert space with $(\boldsymbol{f}, \boldsymbol{g}) = \int_0^1 f(x)\, g(x)\, dx$ and $||\boldsymbol{f}||_2^2 = \int_0^1 |f(x)|^2 dx$

$\boldsymbol{L}^\infty[0,1]$ is a Banach space with $||\boldsymbol{f}||_\infty$ = supremum of $|f(x)|$.

Notice the parallel between sums of components in ℓ^1 and integrals of functions in L^1. Similarly for sums of squares in ℓ^2 and integrals of $|f(x)|^2$ in L^2. *Add or integrate.*

Smoothness of Functions

There are many types of function spaces. This part of mathematics is "*functional analysis*". Often a function space brings together all functions with a specific level of smoothness. An outstanding example is the space $C[0,1]$ containing all **continuous functions**:

$$f \text{ belongs to } C[0,1] \text{ and } ||f||_C = \max|f(x)| \quad \text{if } f(x) \text{ is continuous for all } 0 \le x \le 1.$$

The max norm in the function space $\mathbf{C}$ is like the ℓ^∞ norm for vectors. We can increase the level of smoothness to $C^1[0,1]$ or $C^2[0,1]$. Then the first derivative or second derivative must also be continuous. These are Banach spaces but not Hilbert spaces. Their norms do not come from inner products—compare (9) and (10):

$$||f||_{C^1} = ||f||_C + \left|\left|\frac{df}{dx}\right|\right|_C \qquad ||f||_{C^2} = ||f||_C + \left|\left|\frac{d^2f}{dx^2}\right|\right|_C \tag{9}$$

If we want a Hilbert space H^1, then we build on the usual L^2 space (which is H^0):

$$||f||^2_{H^1} = ||f||^2 + \left|\left|\frac{df}{dx}\right|\right|^2 \quad \text{and} \quad (f,g)_{H^1} = \int_0^1 f(x)\,g(x)\,dx + \int_0^1 \frac{df}{dx}\frac{dg}{dx}\,dx \tag{10}$$

We bring this wild excursion in function space to an end with three examples.

1. The infinite vector $v = \left(1, \frac{1}{2}, \frac{1}{3}, \frac{1}{4}, \ldots\right)$ is in ℓ^2 and ℓ^∞. But it is not in ℓ^1. The sum of its components is infinite.

2. A step function is in L^1 and L^2 and L^∞, but not in C. The function has a jump.

3. The ramp function $\max(0, x)$ is in C and H^1 but not in C^1. The slope has a jump.

Norms of Matrices: The Frobenius Norm

A space of matrices follows all the rules for a vector space. So a matrix norm $||A||$ must follow the three rules for a vector norm, and also a new rule when A multiplies B:

$$||A|| > 0 \text{ if } A \text{ is not the zero matrix} \tag{11}$$

$$||cA|| = |c|\,||A|| \text{ and } ||A+B|| \le ||A|| + ||B|| \tag{12}$$

- **New rule for a matrix norm** $\quad ||AB|| \le ||A||\,||B|| \quad (13)$

We need to verify (13) for the Frobenius norm, which treats matrices as long vectors.

$$||A||_F^2 = (\textbf{Frobenius norm of } A)^2 = |a_{11}|^2 + \cdots + |a_{1n}|^2 + \cdots + |a_{mn}|^2. \quad (14)$$

Frobenius is the ℓ^2 norm (Euclidean norm) for a vector with mn components. So (11) and (12) are sure to be true. And when AB is a column vector $\boldsymbol{a}$ times a row vector $\boldsymbol{b}^{\mathrm{T}}$— in other words AB is a rank one matrix $\boldsymbol{ab}^{\mathrm{T}}$—the norm inequality (13) is an exact equality $||\boldsymbol{ab}^{\mathrm{T}}||_F = ||\boldsymbol{a}||_F ||\boldsymbol{b}^{\mathrm{T}}||_F$:

$$\boldsymbol{ab}^{\mathrm{T}} = \begin{bmatrix} a_1 \\ \vdots \\ a_m \end{bmatrix} \begin{bmatrix} b_1 & \cdots & b_p \end{bmatrix} \text{ has } ||\boldsymbol{ab}^{\mathrm{T}}||_F^2 = \begin{matrix} |a_1|^2 \left(|b_1|^2 + \cdots + |b_p|^2\right) \\ + \cdots + \\ |a_m|^2 \left(|b_1|^2 + \cdots + |b_p|^2\right) \end{matrix} = ||\boldsymbol{a}||_F^2 \, ||\boldsymbol{b}||_F^2. \quad (15)$$

This leads to a first proof that $||\boldsymbol{AB}||_F \leq ||\boldsymbol{A}||_F \, ||\boldsymbol{B}||_F$. AB is a sum of rank-1 matrices.

$$\begin{aligned} ||\boldsymbol{AB}||_F &= ||\boldsymbol{a}_1\boldsymbol{b}_1^{\mathrm{T}} + \cdots + \boldsymbol{a}_n\boldsymbol{b}_n^{\mathrm{T}}||_F \text{ by column-row multiplication} \\ &\leq ||\boldsymbol{a}_1\boldsymbol{b}_1^{\mathrm{T}}||_F + \cdots + ||\boldsymbol{a}_n\boldsymbol{b}_n^{\mathrm{T}}||_F \text{ by the triangle inequality (12)} \\ &= ||\boldsymbol{a}_1||_F||\boldsymbol{b}_1||_F + \cdots + ||\boldsymbol{a}_n||_F||\boldsymbol{b}_n||_F \text{ by equation (15)} \\ &\leq \left(||\boldsymbol{a}_1||_F^2 + \cdots + ||\boldsymbol{a}_n||_F^2\right)^{1/2} \left(||\boldsymbol{b}_1||_F^2 + \cdots + ||\boldsymbol{b}_n||_F^2\right)^{1/2} \text{ by Cauchy-Schwarz} \\ &= ||\boldsymbol{A}||_F \, ||\boldsymbol{B}||_F \text{ by the definition (14) of the Frobenius norm} \end{aligned}$$

The Problem Set finds a different and quicker proof, multiplying AB by rows times columns.

When Q is an orthogonal matrix, we know that $Q\boldsymbol{x}$ has the same ℓ^2 length as $\boldsymbol{x}$:

Orthogonal Q $\quad ||Q\boldsymbol{x}||_2 = ||\boldsymbol{x}||_2 \qquad$ **Q multiplies columns of B** $\quad ||QB||_F = ||B||_F$

This connects the Frobenius norm of $A = U\Sigma V^{\mathrm{T}}$ to its singular values in Σ:

$$||\boldsymbol{A}||_F = ||\boldsymbol{U\Sigma V}^{\mathrm{T}}||_F = ||\boldsymbol{\Sigma V}^{\mathrm{T}}||_F = ||\boldsymbol{\Sigma}||_F = \sqrt{\boldsymbol{\sigma}_1^2 + \cdots + \boldsymbol{\sigma}_r^2} \quad (16)$$

Here is another way to reach that good formula for the Frobenius norm.

Multiplying $A^{\mathrm{T}}A$ brings all the numbers $|a_{ij}|^2$ onto the main diagonal

$$||\boldsymbol{A}||_F^2 = \textbf{trace of } \boldsymbol{A}^{\mathrm{T}}\boldsymbol{A} = \textbf{sum of eigenvalues } = \boldsymbol{\sigma}_1^2 + \cdots + \boldsymbol{\sigma}_r^2 \quad (17)$$

The Frobenius norm (squared) of A is easy to compute: Just square the entries and add.

The inequality $||AB|| \leq ||A|| \, ||B||$ will be built in for the matrix norms that come next.

Matrix Norms $||A||$ from Vector Norms $||v||$

Start with any norm $||v||$ for the vectors in $\mathbf{R}^n$. When we compare $||Av||$ to $||v||$, this measures the growth factor—the increase or decrease in size produced when we multiply by A. If we choose the vector v with the *largest growth factor*, that gives an important matrix norm $||A||$.

Vector norm leads to a matrix norm

$$||A|| = \max_{v \neq 0} \frac{||Av||}{||v||} = \textbf{largest growth factor} \tag{18}$$

The largest ratio $||A||$ automatically satisfies all the conditions for a matrix norm—because $||v||$ satisfied all the conditions for a vector norm. The identity matrix will have $||I|| = 1$ because its growth factor is always $||Iv||/||v|| = 1$. The key point of (18) is that $||\boldsymbol{Av}|| \leq ||\boldsymbol{A}||\,||\boldsymbol{v}||$ because $||\boldsymbol{A}||$ is the largest value reached by $||\boldsymbol{Av}||/||\boldsymbol{v}||$. Then $||\boldsymbol{ABv}|| \leq ||\boldsymbol{A}||\,||\boldsymbol{Bv}|| \leq ||\boldsymbol{A}||\,||\boldsymbol{B}||\,||\boldsymbol{v}||$. Therefore $||\boldsymbol{AB}|| \leq ||\boldsymbol{A}||\,||\boldsymbol{B}||$.

We think of $||v||_2$ and $||v||_1$ and $||v||_\infty$ as the important vector norms. Then (19), (20), (21) produce the three matrix norms $||A||_2$ and $||A||_1$ and $||A||_\infty$. They all have $||AB|| \leq ||A||\,||B||$. For a given matrix A, how do we compute those three norms—how do we maximize the ratio $||Av||/||v||$?

ℓ^2 norm	$\|\|A\|\|_2$ = **largest singular value σ_1** of A	(19)
ℓ^1 norm	$\|\|A\|\|_1$ = **largest ℓ_1 norm of the columns** of A	(20)
ℓ^∞ norm	$\|\|A\|\|_\infty$ = **largest ℓ_1 norm of the rows** of A	(21)

This book has emphasized $||A||_2$ = largest ratio $||Av||_2/||v||_2 = \sigma_1$. That comes from $A = U\Sigma V^{\mathrm{T}}$, because orthogonal matrices U and V^{T} have no effect on ℓ^2 norms. This leaves the diagonal matrix Σ which has ℓ^2 norm $= \sigma_1$. Notice that $A^{\mathrm{T}}A$ has norm σ_1^2 (again no effect from U and V).

The three matrix norms have two especially beautiful connections:

$$||\boldsymbol{A}||_\infty = ||\boldsymbol{A}^{\mathrm{T}}||_1 \qquad ||\boldsymbol{A}||_2^2 \leq ||\boldsymbol{A}||_1\,||\boldsymbol{A}||_\infty. \tag{22}$$

The rows of A are the columns of A^{T}. So $||A||_\infty = ||A^{\mathrm{T}}||_1$ comes directly from (20)-(21).

For the inequality (22) that involves all three norms, look at the first singular vector v of A. That vector has $A^{\mathrm{T}}Av = \sigma_1^2 v$. Take the ℓ_1 norm of this particular v and use $||A||_\infty = ||A^{\mathrm{T}}||_1$:

$$\sigma_1^2\,||v||_1 = ||A^{\mathrm{T}}Av||_1 \leq ||A^{\mathrm{T}}||_1\,||Av||_1 \leq ||A||_\infty\,||A||_1\,||v||_1.$$

Since $\sigma_1 = ||A||_2$, this tells us that $||A||_2^2 \leq ||A||_\infty\,||A||_1$.

The Nuclear Norm

$||A||_{\mathbf{nuclear}}$ comes directly from the singular values of A. It is also called the **trace norm**. Along with ℓ^2 and Frobenius, $||A||_N$ starts from $A = U\Sigma V^T$ (the SVD). Those three norms are not affected by U and V ("unitary invariance"). We just take the ℓ^1 and ℓ^2 and ℓ^∞ norms of the vector $\sigma = (\sigma_1, \sigma_2, \ldots, \sigma_r)$ on the main diagonal of Σ:

$$\boxed{||A||_{\text{nuclear}} = \sigma_1 + \cdots + \sigma_r \qquad ||A||_F^2 = \sigma_1^2 + \cdots + \sigma_r^2 \qquad ||A||_2 = \sigma_1}$$

In III.4, the nuclear norm is the key to matrix completion, when data is missing. A remarkable fact: $||A||_N$ *is the minimum value of* $||U||_F\,||V||_F$ *subject to* $UV = A$. And a related but much easier fact: $||A^TA||_N = ||A||_F^2$.

Notice that $||A||_\infty = \max||A\boldsymbol{v}||_\infty/||\boldsymbol{v}||_\infty$ is entirely different from $|||A|||_\infty = \max|a_{ij}|$. We call $|||A|||_\infty$ the *medical norm* because $|||A - B|||_\infty$ is small only when every entry (every pixel) a_{ij} in A is close to b_{ij} in B. Then $||A - B||_\infty$ is also small.

Example 1 $A = \begin{bmatrix} 1 & 2 \\ 3 & 6 \end{bmatrix}$ has $||A||_2 = \sqrt{50}$ because $A^TA = \begin{bmatrix} 5 & 15 \\ 15 & 45 \end{bmatrix}$ has $\lambda_1 = 50$.

The ℓ^1 and ℓ^∞ norms are $||\boldsymbol{A}||_1 = \mathbf{8}$ (column sum) and $||\boldsymbol{A}||_\infty = \mathbf{9}$ (row sum).

$A = \begin{bmatrix} 1 & 2 \\ 1 & 2 \end{bmatrix}$ has $||\boldsymbol{A}||_2 = \sqrt{\mathbf{10}}$ and $||\boldsymbol{A}||_1 = \mathbf{4}$ (column 2) and $||\boldsymbol{A}||_\infty = \mathbf{3}$ (rows).

The ℓ^2 norm is $\sqrt{10}$ because $A^TA = \begin{bmatrix} 5 & 5 \\ 5 & 5 \end{bmatrix}$ has eigenvalues 0 and 10. And $10 < (4)(3)$.

Important The largest eigenvalue $|\lambda|_{\max}$ of A is not on our list of matrix norms!

The Spectral Radius

That number $|\lambda|_{\max} = \max|\lambda_i|$ fails on all of the three main requirements for a norm. A and B can have all zero eigenvalues when they are *not* zero matrices (A and B below). The tests on $A + B$ and AB (triangle inequality and $||AB|| \le ||A||\,||B||$) also fail for the largest eigenvalue. *Every norm has* $||A|| \ge |\lambda|_{\max}$.

$$A = \begin{bmatrix} 0 & 1 \\ 0 & 0 \end{bmatrix} \quad B = \begin{bmatrix} 0 & 0 \\ 1 & 0 \end{bmatrix} \quad \begin{aligned} \lambda_{\max}(A+B) &= 1 > \lambda_{\max}(A) + \lambda_{\max}(B) = (0) + (0) \\ \lambda_{\max}(AB) &= 1 > \lambda_{\max}(A) \times \lambda_{\max}(B) = (0) \times (0) \end{aligned}$$

This number $|\lambda|_{\max}$ is the "**spectral radius**". It is not a norm but it is important for this reason: $||\boldsymbol{A}^n|| \to \mathbf{0}$ **exactly when** $|\boldsymbol{\lambda}|_{\max} < 1$.

When we multiply again and again by a matrix A (as we will do for Markov chains in Section V.6) the largest eigenvalue $|\lambda|_{\max}$ of A begins to dominate. This is the basis of the "power method" to compute $|\lambda|_{\max}$.

Problem Set I.11

1. Show directly this fact about ℓ^1 and ℓ^2 and ℓ^∞ vector norms: $||v||_2^2 \leq ||v||_1 \, ||v||_\infty$.

2. Prove the Cauchy-Schwarz inequality $|v^{\mathrm{T}}w| \leq ||v||_2 \, ||w||_2$ for vectors in $\mathbf{R}^n$. You could verify and use this identity for a length squared:

$$0 \leq \left(v - \frac{v^{\mathrm{T}}w}{w^{\mathrm{T}}w}w, v - \frac{v^{\mathrm{T}}w}{w^{\mathrm{T}}w}w\right) = v^{\mathrm{T}}v - \frac{|v^{\mathrm{T}}w|^2}{w^{\mathrm{T}}w}.$$

3. Show that always $||v||_2 \leq \sqrt{n}\, ||v||_\infty$. Also prove $||v||_1 \leq \sqrt{n}\, ||v||_2$ by choosing a suitable vector w and applying the Cauchy-Schwarz inequality.

4. The ℓ^p and ℓ^q norms are "*dual*" if $p^{-1} + q^{-1} = 1$. The ℓ^1 and ℓ^∞ vector norms are dual (and ℓ^2 is self-dual). Hölder's inequality extends Cauchy-Schwarz to all those dual pairs. What does it say for $p = 1$ and $q = \infty$?

   Hölder's inequality $\quad |v^{\mathrm{T}}w| \leq ||v||_p \, ||w||_q$ when $p^{-1} + q^{-1} = 1$.

5. What rules should be satisfied by any "*inner product*" of vectors v and w?

6. The first page of I.11 shows *unit balls* for the ℓ^1 and ℓ^2 and ℓ^∞ norms. Those are the three sets of vectors $v = (v_1, v_2)$ with $||v||_1 \leq 1, ||v||_2 \leq 1, ||v||_\infty \leq 1$. *Unit balls are always convex because of the triangle inequality for vector norms*:

   If $||v|| \leq 1$ and $||w|| \leq 1$ show that $||\frac{v}{2} + \frac{w}{2}|| \leq 1$.

7. A short proof of $||AB||_F \leq ||A||_F \, ||B||_F$ starts from multiplying rows times columns:

   $|(AB)_{ij}|^2 \leq ||\text{row } i \text{ of } A||^2 \, ||\text{column } j \text{ of } B||^2$ is the Cauchy-Schwarz inequality

   Add up both sides over all i and j to show that $||AB||_F^2 \leq ||A||_F^2 \, ||B||_F^2$.

8. Test $||AB||_F \leq ||A||_F \, ||B||_F$ for $A = B = I$ and $A = B =$ "all ones matrix".

9. (Conjecture) The only matrices with $||AB||_F = ||A||_F \, ||B||_F$ and no zero entries have the rank one form $A = uv^{\mathrm{T}}$ and $B = vw^{\mathrm{T}}$ with a shared vector v.

10. The space of m by n matrices with the Frobenius norm is actually a Hilbert space— **it has an inner product $(A, B) = \text{trace}(A^{\mathrm{T}}B)$**. Show that $||A||_F^2 = (A, A)$.

11. Why is (21) a true formula for $||A||_\infty$? Which v with ± 1's has $||Av||_\infty = ||A||_\infty$?

12. Suppose A, B, and AB are m by n, n by p, and m by p. The "medical norm" of A is its largest entry: $|||A|||_\infty = \max|a_{ij}|$.

    Show that $|||AB|||_\infty \leq n \, |||A|||_\infty \, |||B|||_\infty$ (this is false without the factor n).

    Rewrite that in the form $(\sqrt{mp}\, |||AB|||_\infty) \leq (\sqrt{mn}\, |||A|||_\infty)(\sqrt{np}\, |||B|||_\infty)$. The rescaling by those square roots gives a true matrix norm.

I.12 Factoring Matrices and Tensors : Positive and Sparse

This section opens a wider view of factorizations for data matrices (and extends to tensors). Up to now we have given full attention to the SVD. $\boldsymbol{A} = \boldsymbol{U\Sigma V}^{\mathrm{T}}$ gave the perfect factors for Principal Component Analysis—perfect until issues of sparseness or nonnegativity or tensor data enter the problem. For many applications these issues are important. Then we must see the SVD as a first step but *not the last step*.

Here are factorizations of A and T with new and important properties:

Nonnegative Matrices	$\min \lVert \boldsymbol{A} - \boldsymbol{UV} \rVert_F^2$ with $U \geq 0$ and $V \geq 0$
Sparse and Nonnegative	$\min \lVert \boldsymbol{A} - \boldsymbol{UV} \rVert_F^2 + \lambda \lVert \boldsymbol{UV} \rVert_N$ with $U \geq 0$ and $V \geq 0$
CP Tensor Decomposition	$\min \lVert \boldsymbol{T} - \sum_{i=1}^{R} \boldsymbol{a}_i \circ \boldsymbol{b}_i \circ \boldsymbol{c}_i \rVert$

We will work with matrices A and then tensors T. A matrix is just a two-way tensor.

To compute a factorization $A = UV$, we introduce a simple **alternating iteration. Update U with V fixed, then update V with U fixed**. Each half step is quick because it is effectively linear (the other factor being fixed). This idea applies to the ordinary SVD, if we include the diagonal matrix Σ with U. The algorithm is simple and often effective. Section III.4 will do even better.

This UV idea also fits the famous **k-means algorithm** in Section IV.7 on graphs. The problem is to put n vectors $\boldsymbol{a}_1, \ldots, \boldsymbol{a}_n$ into r clusters. If $\boldsymbol{a}_k$ is in the cluster around $\boldsymbol{u}_j$, *this fact* $\boldsymbol{a}_k \approx \boldsymbol{u}_j$ *is expressed by column k of* $\boldsymbol{A} \approx \boldsymbol{UV}$. Then column k of V is column j of the r by r identity matrix.

Nonnegative Matrix Factorization (NMF)

The goal of NMF is to approximate a nonnegative matrix $A \geq 0$ by a lower rank product UV of two nonnegative matrices $U \geq 0$ and $V \geq 0$. The purpose of lower rank is simplicity. The purpose of nonnegativity (no negative entries) is to produce numbers that *have a meaning*. Features are recognizable with no plus-minus cancellation. A negative weight or volume or count or probability is wrong from the start.

But nonnegativity can be difficult. When $A \geq 0$ is symmetric positive definite, we hope for a matrix $B \geq 0$ that satisfies $B^{\mathrm{T}}B = A$. Very often no such matrix exists. (The matrix $A = A^{\mathrm{T}}$ with constant diagonals $1 + \sqrt{5}, 2, 0, 0, 2$ is a 5×5 example.) We are forced to accept the matrix $B^{\mathrm{T}}B$ (with $B \geq 0$) that is closest to A (when $A \geq 0$). The question is how to find B. The unsymmetric case, probably not square, looks for U and V.

Lee and Seung focused attention on NMF in a letter to *Nature* **401** (1999) 788–791.

The basic problem is clear. Sparsity and nonnegativity are very valuable properties. For a sparse vector or matrix, the few nonzeros will have meaning—when 1000 or $100,000$ individual numbers cannot be separately understood. And it often happens that numbers are naturally nonnegative. But singular vectors in the SVD almost always have many small components of mixed signs. In practical problems, we must be willing to give up the orthogonality of U and V. Those are beautiful properties, but the Lee-Seung essay urged the value of sparse PCA and no negative numbers.

These new objectives require new factorizations of A.

NMF	Find **nonnegative matrices U and V** so that $A \approx UV$	(1)
SPCA	Find **sparse low rank matrices B and C** so that $A \approx BC$.	(2)

First we recall the meaning and purpose of a factorization. $A = BC$ expresses every column of A as a combination of columns of B. The coefficients in that combination are in a column of C. So each column $\boldsymbol{a}_j$ of A is the approximation $c_{1j}\boldsymbol{b}_1 + \cdots + c_{nj}\boldsymbol{b}_n$. A good choice of BC means that this sum is nearly exact.

If C has fewer columns than A, this is *linear dimensionality reduction*. It is fundamental to compression and feature selection and visualization. In many of those problems it can be assumed that the noise is Gaussian. Then the Frobenius norm $||A - BC||_F$ is a natural measure of the approximation error. Here is an excellent essay describing two important applications, and a recent paper with algorithms and references.

N. Gillis, *The Why and How of Nonnegative Matrix Factorization*, arXiv: 1401.5226.

L. Xu, B. Yu, and Y. Zhang, *An alternating direction and projection algorithm for structure-enforced matrix factorization*, Computational Optimization Appl. **68** (2017) 333-362.

Facial Feature Extraction

Each column vector of the data matrix A will represent a face. Its components are the intensities of the pixels in that image, so $A \geq 0$. The goal is to find a few "basic faces" in B, so that their combinations come close to the many faces in A. We may hope that a few variations in the geometry of the eyes and nose and mouth will allow a close reconstruction of most faces. The development of *eigenfaces* by Turk and Pentland finds a set of basic faces, and matrix factorization $A \approx BC$ is another good way.

Text Mining and Document Classification

Now each column of A represents a document. Each row of A represents a word. A simple construction (not in general the best: it ignores the ordering of words) is a sparse nonnegative matrix. To classify the documents in A, we look for sparse nonnegative factors:

$$\text{Document} \quad \boldsymbol{a}_j \approx \sum (\text{importance } c_{ij})\,(\text{topic } \boldsymbol{b}_i) \tag{3}$$

Since $B \geq 0$, each topic vector $\boldsymbol{b}_i$ can be seen as a document. Since $C \geq 0$, we are combining but *not subtracting* those topic documents. Thus NMF identifies topics and classifies the whole set of documents with respect to those topics. Related methods are "latest semantic analysis" and indexing.

Note that NMF is an NP-hard problem, unlike the SVD. Even exact solutions $A = BC$ are not always unique. More than that, the number of topics (columns of A) is unknown.

Optimality Conditions for Nonnegative U and V

Given $A \geq 0$, here are the conditions for $U \geq 0$ and $V \geq 0$ to minimize $||A - UV||_F^2$:

$$\begin{aligned} Y &= UVV^{\mathrm{T}} - AV^{\mathrm{T}} \geq 0 \quad \text{with } Y_{ij} \text{ or } U_{ij} = 0 \text{ for all } i, j \\ Z &= U^{\mathrm{T}}UV - U^{\mathrm{T}}A \geq 0 \quad \text{with } Z_{ij} \text{ or } V_{ij} = 0 \text{ for all } i, j \end{aligned} \tag{4}$$

Those last conditions already suggest that U and V may turn out to be sparse.

Computing the Factors : Basic Methods

Many algorithms have been suggested to compute U and V and B and C. A central idea is **alternating factorization**: Hold one factor fixed, and optimize the other factor. Hold that one fixed, optimize the first factor, and repeat. Using the Frobenius norm, each step is a form of least squares. This is a natural approach and it generally gives a good result. *But convergence to the best factors is not sure.* We may expect further developments in the theory of optimization. And there is a well-established improvement of this method to be presented in Section III.4 : **Alternating Direction Method of Multipliers**.

This ADMM algorithm uses a penalty term and duality to promote convergence.

Sparse Principal Components

Many applications do allow both negative and positive numbers. We are not counting or building actual objects. In finance, we may buy or sell. In other applications the zero point has no intrinsic meaning. Zero temperature is a matter of opinion, between Centigrade and Fahrenheit. Maybe water votes for Centigrade, and super-cold physics resets $0°$.

The number of nonzero components is often important. That is the difficulty with the singular vectors $\boldsymbol{u}$ and $\boldsymbol{v}$ in the SVD. They are full of nonzeros, as in least squares. We cannot buy miniature amounts of a giant asset, because of transaction costs. If we learn 500 genes that affect a patient's outcome, we cannot deal with them individually. To be understood and acted on, the number of nonzero decision variables must be under control.

One possibility is to remove the very small components of the $\boldsymbol{u}$'s and $\boldsymbol{v}$'s. But if we want real control, we are better with a direct construction of sparse vectors. A good number of algorithms have been proposed.

H. Zou, T. Hastie, R. Tibshirani, *Sparse principal component analysis*, J. Computational and Graphical Statistics **15** (2006) 265–286. See https://en.wikipedia.org/wiki/Sparse_PCA

Sparse PCA starts with a data matrix A or a positive (semi)definite sample covariance matrix S. Given S, a natural idea is to include Card $(\boldsymbol{x})$ = number of nonzero components in a penalty term or a constraint on $\boldsymbol{x}$:

$$\underset{||\boldsymbol{x}||=1}{\text{Maximize}} \ \boldsymbol{x}^{\mathrm{T}} S \boldsymbol{x} - \rho \,\text{Card}\,(\boldsymbol{x}) \quad \text{or} \quad \underset{||\boldsymbol{x}||=1}{\text{Maximize}} \ \boldsymbol{x}^{\mathrm{T}} S \boldsymbol{x} \ \text{ subject to } \ \text{Card}\,(\boldsymbol{x}) \leq k. \quad (5)$$

But the cardinality of $\boldsymbol{x}$ is not the best quantity for optimization algorithms.

Another direction is *semidefinite programming*, discussed briefly in Section VI.3. The unknown vector $\boldsymbol{x}$ becomes an unknown symmetric matrix X. Inequalities like $\boldsymbol{x} \geq 0$ (meaning that every $x_i \geq 0$) are replaced by $X \geq 0$ (X must be positive semidefinite). Sparsity is achieved by including an ℓ^1 penalty on the unknown matrix X. Looking ahead to IV.5, that penalty uses the *nuclear norm* $||X||_N$ *: the sum of singular values* σ_i.

The connection between ℓ^1 and sparsity was in the figures at the start of Section I.11. **The ℓ^1 minimization had the sparse solution $\boldsymbol{x} = (0, \frac{1}{4})$.** That zero may have looked accidental or insignificant, for this short vector in $\mathbf{R}^2$. On the contrary, that zero is the important fact. For matrices, replace the ℓ^1 norm by the nuclear norm $||X||_N$.

A penalty on $||\boldsymbol{x}||_1$ or $||X||_N$ produces sparse vectors $\boldsymbol{x}$ and sparse matrices X.

In the end, for sparse vectors $\boldsymbol{x}$, our algorithm must *select* the important variables. This is the great property of ℓ^1 optimization. It is the key to the LASSO:

$$\textbf{LASSO} \qquad \text{Minimize } ||A\boldsymbol{x} - \boldsymbol{b}||^2 + \lambda \sum_{1}^{n} |x_k| \qquad (6)$$

Finding that minimum efficiently is a triumph of nonlinear optimization. The ADMM and Bregman algorithms are presented and discussed in Section III.4.

One note about LASSO: The optimal $\boldsymbol{x}^*$ will not have more nonzero components than the number of samples. Adding an ℓ^2 penalty produces an "elastic net" without this disadvantage. This ℓ^1+ ridge regression can be solved as quickly as least squares.

$$\textbf{Elastic net} \qquad \text{Minimize } ||A\boldsymbol{x} - \boldsymbol{b}||_2^2 + \lambda||\boldsymbol{x}||_1 + \beta||\boldsymbol{x}||_2^2 \qquad (7)$$

Section III.4 will present the **ADMM** algorithm that splits ℓ^1 from ℓ^2. And it adds a penalty using Lagrange multipliers and duality. That combination is powerful.

1. R. Tibshirani, Regression shrinkage and selection via the Lasso, *Journal of the Royal Statistical Society*, Series B **58** (1996) 267–288.

2. H. Zou and T. Hastie, Regularization and variable selection via the elastic net, *Journal of the Royal Statistical Society*, Series B **67** (2005) 301–320.

Tensors

A column vector is a 1-way tensor. A matrix is a 2-way tensor. Then a 3-way tensor T has **elements T_{ijk} with three indices** : row number, column number, and "tube number". **Slices of T** are two-dimensional sections, so the 3-way tensor below has three horizontal slices and four lateral slices and two frontal slices. The rows and columns and tubes are the **fibers of T**, with only one varying index.

We can stack m by n matrices (p of them) into a 3-way tensor. And we can stack m by n by p tensors into a 4-way tensor = 4-way array.

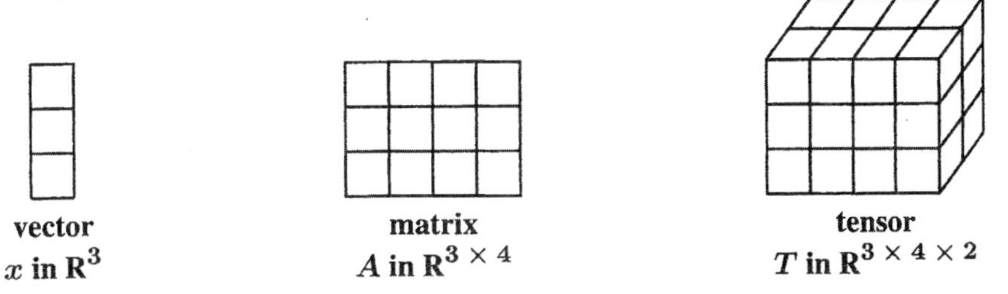

Example 1 : A Color Image is a Tensor with 3 Slices

A black and white image is just a matrix of pixels. The numbers in the matrix are the grayscales of each pixel. Normally those numbers are between zero (black) and 255 (white). Every entry in A has $2^8 = 256$ possible grayscales.

A color image is a tensor. It has 3 slices corresponding to red-green-blue. Each slice shows the density of one of the colors RGB. Processing this tensor T (for example in deep learning : Section VII.2) is not more difficult than a black-white image.

Example 2 : The Derivative $\partial w / \partial A$ of $w = Av$

This is a tensor that we didn't see coming. The $m \times n$ matrix A contains "weights" to be optimized in deep learning. That matrix multiplies a vector v to produce $w = Av$. Then the algorithm to optimize A (Sections VI.4 to VII.3) involves the derivative of each output w_i with respect to each weight A_{jk}. So we have three indices i, j, k.

In matrix multiplication, we know that row j of A has no effect on row i of $w = Av$. So the derivative formula includes the symbol δ_{ij}, which is 1 if $i = j$ and 0 otherwise. In proper tensor notation that symbol becomes δ^i_j (our authority on tensors is Pavel Grinfeld). The derivatives of the linear function $w = Av$ with respect to the weights A_{jk} are in T :

$$\boxed{T_{ijk} = \frac{\partial w_i}{\partial A_{jk}} = v_k \, \delta_{ij} \qquad \text{as in } T_{111} = v_1 \text{ and } T_{122} = 0} \tag{8}$$

Section VII.3 has a $2 \times 2 \times 2$ example. This tensor $T_{ijk} = v_k \delta_{ij}$ is of particular interest:

1. The slices k = constant are multiples v_k of the identity matrix.
2. The key function of deep learning connects each layer of a neural net to the next layer. If one layer contains a vector $\boldsymbol{v}$, the next layer contains the vector $\boldsymbol{w} = (A\boldsymbol{v} + \boldsymbol{b})_+$. *$A$ is a matrix of "weights". We optimize those weights to match the training data.* So the derivatives of the loss function L will be zero for the optimal weights.

Using the chain rule from calculus, the derivatives of L are found by multiplying the derivatives $\partial \boldsymbol{w}/\partial A$ from each layer to the next layer. That is a linear step to $A\boldsymbol{v} + \boldsymbol{b}$, followed by the nonlinear ReLU function that sets all negative components to zero. The derivative of the linear step is our 3-way tensor $v_k \delta_{ij}$.

All this will appear again in Section VII.3. But we won't explicitly use tensor calculus. The idea of **backpropagation** is to compute all the derivatives of L "automatically". For every step in computing L, the derivative of that step enters the derivative of L. Our simple formula (8) for an interesting tensor will get buried in backpropagation.

Example 3 : The Joint Probability Tensor

Suppose we measure age a in years and height h in inches and weight w in pounds. We put N children into I age groups and J height groups and K weight groups. So a typical child is in an age group i and a height group j and a weight group k— where the numbers i, j, k are between $1, 1, 1$ and I, J, K.

Pick a random child. Suppose the I age groups contain $a_1, a_2, \ldots, a_I$ children (adding to N children). Then a random child is in age group i with probability $\boldsymbol{a_i/N}$. Similarly the J height groups contain $h_i, h_2, \ldots, h_J$ children and the K weight groups contain $w_1, w_2, \ldots, w_K$ children. For that random child,

Probability of height group j is $\dfrac{h_j}{N}$ Probability of weight group k is $\dfrac{w_k}{N}$

Now comes our real goal: **joint probabilities $\boldsymbol{p_{ijk}}$**. For each combination i, j, k we count only the children who are in age group i and also height group j and also weight group k. Each child has *I times J times K possibilities.* (Possibly $\boldsymbol{p_{I11}}$ is zero— no oldest children with the lowest height and weight.) Suppose N_{ijk} children are found in the *intersection* of age group i and height group j and weight group k:

$$\textbf{The joint probability of this age-height-weight combination is } \boldsymbol{p_{ijk}} = \frac{\boldsymbol{N_{ijk}}}{\boldsymbol{N}}. \qquad (9)$$

We have I times J times K numbers p_{ijk}. All those numbers are between 0 and 1. They fit into a 3D tensor T of joint probabilities. This tensor T has I rows and J columns and K "tubes". The sum of all the entries N_{ijk}/N is 1.

To appreciate this I by J by K tensor, suppose you add all the numbers $\boldsymbol{p_{2jk}}$. You are accounting for all children in age group 2:

$$\sum_{j=1}^{J} \sum_{k=1}^{K} \boldsymbol{p_{2jk}} = \boldsymbol{p_2^a} = \textbf{probability that a child is in age group 2}. \qquad (10)$$

We are seeing a 2D **slice** of the tensor T. Certainly the sum $p_1^a + p_2^a + \cdots + p_I^a$ equals 1.

Similarly you could add all the numbers $\boldsymbol{p_{2j5}}$. Now you are accounting for all children in age group 2 and weight group 5 :

$$\sum_{j=1}^{J} \boldsymbol{p_{2j5}} = \boldsymbol{p_{2\,5}^{aw}} = \textbf{probability of age group 2 and weight group 5.} \tag{11}$$

These numbers are in a *column* of T. We could combine columns to make a slice of T. We could combine slices to produce the whole tensor T :

$$\sum_{i=1}^{I} \sum_{k=1}^{K} \boldsymbol{p_{ik}^{aw}} = \sum_{i=1}^{I} \boldsymbol{p_i^a} = \mathbf{1}$$

By measuring three properties we were led to this 3-way tensor T with entries T_{ijk}.

The Norm and Rank of a Tensor

In general a tensor is a d-way array. In the same way that a matrix entry needs two numbers i, j to identify its position, a d-way tensor needs d numbers. We will concentrate here on $d = 3$ and 3-*way tensors* (also called tensors of order 3). After vectors and matrices, $d = 3$ is the most common and the easiest to understand. The norm of T is like the Frobenius norm of a matrix : **Add all T_{ijk}^2 to find $||T||^2$.**

The theory of tensors is still part of linear algebra (or perhaps multilinear algebra). Just like a matrix, a tensor can have two different roles in science and engineering :

1 A tensor can multiply vectors, matrices, or tensors. Then it is a **linear operator**.

2 A tensor can **contain data**. Its entries could give the brightness of pixels in an image. A color image is 3-way, stacking RGB. A color video will be a 4-way tensor.

The operator tensor could multiply the data tensor—in the same way that a permutation matrix or a reflection matrix or any orthogonal matrix operates on a data matrix.

The analogies are clear but tensors need more indices and they look more complicated. *They are*. We could succeed with tensor multiplication (as for matrices, the operations can come in different orders). *We will not succeed so well for tensor factorization.* This has been and still is an intense research direction—to capture as much as possible of the matrix factorizations that are so central to linear algebra : $LU, QR, Q\Lambda Q^{\mathrm{T}}, U\Sigma V^{\mathrm{T}}$.

Even the definition and computation of the **"rank of a tensor"** is not so simple or successful as the rank of a matrix. But rank one tensors = outer products are still the simplest and clearest : They are created from three vectors $\boldsymbol{a}, \boldsymbol{b}, \boldsymbol{c}$.

$$\boxed{\textbf{3-way tensor } \boldsymbol{T = a \circ b \circ c} \textbf{ of rank one} \qquad T_{ijk} = a_i b_j c_k} \tag{12}$$

This outer product $\boldsymbol{a} \circ \boldsymbol{b} \circ \boldsymbol{c}$ is defined by the $m + n + p$ numbers in those three vectors. *The rank of a tensor is the smallest number of rank-*1 *tensors that add to* T.

If we add several of these outer products, we have a convenient low rank tensor—even if we don't always know its exact rank. Here is an example to show why.

$$T = \boldsymbol{u} \circ \boldsymbol{u} \circ \boldsymbol{v} + \boldsymbol{u} \circ \boldsymbol{v} \circ \boldsymbol{u} + \boldsymbol{v} \circ \boldsymbol{u} \circ \boldsymbol{u} \quad \textbf{(three rank-1 tensors with } ||\boldsymbol{u}|| = ||\boldsymbol{v}|| = \mathbf{1})$$

T seems to have rank 3. But it is the limit of these rank-**2** tensors T_n when $n \to \infty$:

$$T_n = n\left(\boldsymbol{u} + \frac{1}{n}\boldsymbol{v}\right) \circ \left(\boldsymbol{u} + \frac{1}{n}\boldsymbol{v}\right) \circ \left(\boldsymbol{u} + \frac{1}{n}\boldsymbol{v}\right) - n\,\boldsymbol{u} \circ \boldsymbol{u} \circ \boldsymbol{u} \tag{13}$$

Why could this never happen for matrices? Because the closest approximation to A by a matrix of rank k is fixed by the Eckart-Young theorem. That best approximation is A_k from the k leading singular vectors in the SVD. The distance from rank 3 to rank 2 is fixed.

Unfortunately there seems to be no SVD for general 3-way tensors. But the next paragraphs show that we still try—because for computations we want a good low rank approximation to T. Two options are CP and Tucker.

The CP Decomposition of a Tensor

A fundamental problem in tensor analysis and fast tensor computations is to approximate a given tensor T by a **sum of rank one tensors**: an approximate factorization.

$$\textbf{CP Decomposition} \qquad T \approx \boldsymbol{a}_1 \circ \boldsymbol{b}_1 \circ \boldsymbol{c}_1 + \cdots + \boldsymbol{a}_R \circ \boldsymbol{b}_R \circ \boldsymbol{c}_R \tag{14}$$

This decomposition has several discoverers: Hitchcock, Carroll, Chang, and Harshman. It also has unfortunate names like CANDECOMP and PARAFAC. Eventually it became a **CP Decomposition of T**.

This looks like an extension to tensors of the SVD. But there are important differences. *The vectors $\boldsymbol{a}_1, \ldots, \boldsymbol{a}_R$ are not orthogonal* (the same for $\boldsymbol{b}$'s and $\boldsymbol{c}$'s). We don't have orthogonal invariance (which gave $Q_1 A Q_2^{\mathrm{T}}$ the same singular values as A). And the Eckart-Young theorem is not true—we often don't know the rank R tensor closest to T. There are other approaches to tensor decomposition—but so far **CP** has been the most useful. Kruskal proved that closest rank-one tensors are unique (if they exist). If we change R, the best $\boldsymbol{a}, \boldsymbol{b}, \boldsymbol{c}$ will change.

So we are faced with an entirely new problem. From the viewpoint of computability, the problem is NP-hard (unsolvable in polynomial time unless it turns out that $P = NP$, which would surprise almost everyone). Lim and Hillar have proved that many simpler-sounding problems for tensors are also NP-hard. The route of exact computations is closed.

C. Hillar and L.-H. Lim, *Most tensor problems are NP-hard*, J. ACM **60** (2013) Article 45.

We look for an algorithm that computes the $\boldsymbol{a}, \boldsymbol{b}, \boldsymbol{c}$ vectors in a reasonably efficient way. A major step in tensor computations is to come close to the best CP decomposition. A simple idea (*alternating least squares*) works reasonably well for now. The overall problem is not convex, but the subproblems cycle in improving A then B then C—and each subproblem (for A and B and C) is convex least squares.

$$\begin{array}{l|ll|}
\textbf{Alternate} & \text{(a) Fix } B, C \text{ and vary } \boldsymbol{A} & \textbf{Minimize } \|T_1 - A(C \circ B)^{\mathrm{T}}\|_F^2 \\
\boldsymbol{A}, \boldsymbol{B}, \textbf{ and } \boldsymbol{C} & \text{(b) Fix } A, C \text{ and vary } \boldsymbol{B} & \text{(c) Fix } A, B \text{ and vary } \boldsymbol{C}
\end{array} \qquad (15)$$

This alternating algorithm is using the three matricized forms T_1, T_2, T_3, described next. $C \circ B$ is the "Khatri-Rao product" coming in equation (17). **Then (15) is all matrices.**

Matricized Form of a Tensor T

Suppose A, B, C are the matrices whose columns are the $\boldsymbol{a}$'s and $\boldsymbol{b}$'s and $\boldsymbol{c}$'s in (14). Each matrix has R columns. If the third order tensor T has dimensions I, J, K then the three matrices are I by R, J by R, and K by R. It is good to "matricize" the tensor T, so as to compute the CP decomposition.

Start by separating off the $\boldsymbol{a}$'s in the I by R matrix A. Then we look for an R by JK matrix M_1 so that AM_1 expresses our sum (14) of rank-one tensors. M_1 must come from the $\boldsymbol{b}$'s and $\boldsymbol{c}$'s. But in what way will a matrix product AM_1 express a 3-way tensor? The answer is that we have to **unfold the tensor T into a matrix T_1**. After that we can compare T_1 with AM_1.

An example of Kolda and Bader shows how tensors unfold into matrices. We have $I \times J \times K = 3 \times 4 \times 2 = 24$ numbers in T. The matrix unfolding of T can be 3×8 or 4×6 or 2×12. We have three unfoldings T_1, T_2, T_3, slicing T three ways:

First way
Front and back slices
$I \times JK = 3 \times 8$

$$T_1 = \left[\begin{array}{cccc|cccc} 1 & 4 & 7 & 10 & 13 & 16 & 19 & 22 \\ 2 & 5 & 8 & 11 & 14 & 17 & 20 & 23 \\ 3 & 6 & 9 & 12 & 15 & 18 & 21 & 24 \end{array}\right] \qquad (16)$$

Second way
$J \times IK = 4 \times 6$
Same 24 numbers

$$T_2 = \begin{bmatrix} 1 & 2 & 3 & 13 & 14 & 15 \\ 4 & 5 & 6 & 16 & 17 & 18 \\ 7 & 8 & 9 & 19 & 20 & 21 \\ 10 & 11 & 12 & 22 & 23 & 24 \end{bmatrix}$$

Third way
$K \times IJ = 2 \times 12$
Same 24 numbers

$$T_3 = \begin{bmatrix} 1 & 2 & 3 & 4 & 5 & 6 & 7 & 8 & 9 & 10 & 11 & 12 \\ 13 & 14 & 15 & 16 & 17 & 18 & 19 & 20 & 21 & 22 & 23 & 24 \end{bmatrix}$$

The Khatri-Rao Product $A \odot B$

Section IV.3 will introduce the **Kronecker product $K = A \otimes B$** of matrices A and B. It contains all the products a_{ij} times b_{kl} of entries in A and B (so it can be a big matrix). If A and B are just *column vectors* (J by 1 and K by 1 matrices) then **$A \otimes B$ is a long column**: JK by 1. First comes a_{11} times each entry in B, then a_{21} times those same K entries, and finally a_{J1} times those K entries. **$A \odot B$ has R of these long columns.**

Khatri-Rao multiplies all $a_{ij}b_{Ij}$ to find column j. A and B and $A \odot B$ have R columns:

$$\textbf{Khatri-Rao} \quad \textbf{Column } j \textbf{ of } A \odot B = (\textbf{column } j \textbf{ of } A) \otimes (\textbf{column } j \textbf{ of } B) \tag{17}$$

Thus C and B (K by R and J by R) produce $C \odot B$ with R long columns (JK by R).

Summary T is approximated by $\sum \boldsymbol{a}_i \circ \boldsymbol{b}_i \circ \boldsymbol{c}_i$. We are slicing T in three directions and placing the slices next to each other in the three matrices T_1, T_2, T_3. Then we look for three matrices M_1, M_2, M_3 that give us nearly correct equations by ordinary matrix multiplication:

$$T_1 \approx AM_1 \quad \text{and} \quad T_2 \approx BM_2 \quad \text{and} \quad T_3 \approx CM_3. \tag{18}$$

Recall that A is I by R, with columns $\boldsymbol{a}_1$ to $\boldsymbol{a}_R$. T_1 is I by JK. So the correct M_1 must be R by JK. This M_1 comes from the matrices B and C, with columns $\boldsymbol{b}_j$ and $\boldsymbol{c}_k$.

M_1 is the transpose of the Khatri-Rao product $C \odot B$ which is JK by R.

The ith column of $C \odot B$ comes from the ith columns of C and B (for $i = 1$ to R). That ith column of $C \odot B$ contains all of the JK numbers c_{ki} and b_{ji}, for $1 \le k \le K$ and $1 \le j \le J$. $C \odot B$ contains all JKR products of c_{ki} and b_{ji}, coming from the ith columns of C and B for $i = 1$ to R.

The three matricized forms T_1, T_2, T_3 of T are now approximate matrix products. The Khatri-Rao definition was invented to make equation (19) true.

$$T_1 \approx A(C \odot B)^{\mathrm{T}} \qquad T_2 \approx B(C \odot A)^{\mathrm{T}} \qquad T_3 \approx C(B \odot A)^{\mathrm{T}} \tag{19}$$

Computing the CP Decomposition of T

We aim to compute the $\boldsymbol{a}$'s and $\boldsymbol{b}$'s and $\boldsymbol{c}$'s in the approximation (14) to the tensor T. The plan is to use **alternating minimizations**. With the $\boldsymbol{b}$'s and $\boldsymbol{c}$'s in the matrices B and C fixed, we solve a linear least squares problem for the $\boldsymbol{a}$'s in A.

Put T in its matricized form T_1. That matrix is I by JK. By equation (19), we are aiming for $T_1 \approx A\,(C \odot B)^{\mathrm{T}} = (I \times R)(R \times JK)$. Fix B and C for now.

$$\textbf{Choose the best } A \textbf{ in } \|T_1 - A\,(C \odot B)^{\mathrm{T}}\|_F^2 = \|T_1^{\mathrm{T}} - (C \odot B)\,A^{\mathrm{T}}\|_F^2. \tag{20}$$

Here $C \odot B$ is the $JK \times R$ coefficient matrix. It multiplies each of the I columns of A^{T}. With the Frobenius norm, we have I ordinary least squares problems to solve for A: one for every column of A^{T} (= row of A).

Please note: A is not in its usual position $A\boldsymbol{x} = \boldsymbol{b}$ for least squares. The rows of A are the unknowns! **The coefficient matrix is $C \odot B$** (not A). We expect that matrix to be tall and thin, with $JK \ge R$. Similarly we will want $IK \ge R$ and $IJ \ge R$ when the unknowns are alternated to become B and C.

The solution to a least squares problem $A\boldsymbol{x} = \boldsymbol{b}$ is given by the pseudoinverse $\widehat{\boldsymbol{x}} = A^+\boldsymbol{b}$.

If that matrix A has independent columns (as least squares often assumes), A^+ is a left inverse $(A^{\mathrm{T}}A)^{-1}A^{\mathrm{T}}$ of the matrix A. There you see the coefficient matrix $A^{\mathrm{T}}A$ in the usual normal equations for the best $\widehat{\boldsymbol{x}}$. In our case that coefficient matrix is not A but the Khatri-Rao product $C \odot B$. By good fortune the pseudoinverse of our coefficient matrix $C \odot B$ can be expressed as

$$(C \odot B)^+ = [(C^{\mathrm{T}}C) \,.\!*\, (B^{\mathrm{T}}B)]^+(C \odot B)^{\mathrm{T}}. \tag{21}$$

This formula is borrowed from equation (2.2) of Kolda and Bader. It allows us to form $C^{\mathrm{T}}C$ and $B^{\mathrm{T}}B$ in advance (R by R matrices). The symbol $.*$ (or frequency $\circ$) represents the element-by-element product (the **Hadamard product**). Transposing, the least squares problem (20) with fixed B and C is solved by the matrix

$$\boxed{A = T_1\,(C \odot B)\,(C^{\mathrm{T}}C \,.\!*\, B^{\mathrm{T}}B)^+} \tag{22}$$

Next we use this A together with C to find the best B. One cycle of the alternating algorithm ends by finding the best C, with A and B fixed at their new values.

The Tucker Decomposition

The SVD separates a matrix (a 2-tensor) into $U\Sigma V^{\mathrm{T}}$. The columns of U and V are orthonormal. That decomposition is generally impossible for higher-order tensors. This is the reason for the CP approximation and now the Tucker approximation.

Tucker allows P column vectors $\boldsymbol{a}_p$ and Q column vectors $\boldsymbol{b}_q$ and R column vectors $\boldsymbol{c}_r$. *Then all rank-one combinations* $\boldsymbol{a}_p \circ \boldsymbol{b}_q \circ \boldsymbol{c}_r$ *are allowed.* A core tensor G with dimensions P, Q, R determines the coefficients in those combinations :

$$\textbf{Tucker decomposition of } T \qquad T \approx \sum_1^P \sum_1^Q \sum_1^R g_{pqr}\, \boldsymbol{a}_p \circ \boldsymbol{b}_q \circ \boldsymbol{c}_r \tag{23}$$

With this extra freedom in G—which was just a diagonal tensor in the CP decomposition—we can ask that the $\boldsymbol{a}$'s and $\boldsymbol{b}$'s and $\boldsymbol{c}$'s be *three sets of orthonormal columns.* So Tucker is a combination of PQR rank-one tensors, instead of only R.

Remember that (23) is an approximation and not an equality. It generalizes to d-way tensors. The 3-way case has a matricized form, where T_1, T_2, T_3 and similarly G_1, G_2, G_3 are the unfolding matrices in equation (16). The CP matrices in (14) change to Tucker matrices, and now we have Kronecker products instead of Khatri-Rao :

$$\textbf{Tucker} \quad T_1 \approx A\,G_1\,(C \otimes B)^{\mathrm{T}} \quad T_2 \approx B\,G_2\,(C \otimes A)^{\mathrm{T}} \quad T_3 \approx C\,G_3\,(B \otimes A)^{\mathrm{T}} \tag{24}$$

The *higher-order SVD (HOSVD)* is a particular Tucker decomposition. Its properties and its computation are best explained in the work of De Lathauwer.

Decomposition and Randomization for Large Tensors

This section has described basic steps in computing with tensors. *Data is arriving in tensor form.* We end with two newer constructions (since CP), and with references.

1. Tensor train decomposition (Oseledets and Tyrtyshnikov) The problem is to handle d-way tensors. The full CP decomposition would approximate T by a sum of rank-one tensors. For large dimensions d, CP becomes unworkable. A better idea is to reduce T to a train of 3-way tensors. Then linear algebra and CP can operate in this tensor train format.

2. CURT decompositions (Song, Woodruff, and Zhong) This is low rank approximation for tensors. This paper aims to compute a rank k tensor within ϵ of the closest to T. For matrices, this is achieved (with $\epsilon = 0$) by the SVD and the Eckart-Young Theorem. For tensors we have no SVD. Computations are based instead on a column-row CUR approximation (Section III.3) with a *mixing tensor* U.

The algorithm comes near the goal of nnz steps: equal to the number of nonzeros in T. It uses **randomized factorizations**—the powerful tool in Section II.4 for very large computations. Tensors are at the frontier of numerical linear algebra.

1. T. Kolda and B. Bader, *Tensor decompositions and applications*, SIAM Review **52** (2009) 455–500.
2. M. Mahoney, M. Maggioni, and P. Drineas, *Tensor-CUR decompositions for tensor-based data*, SIAM J. Matrix Analysis Appl. **30** (2008) 957–987.
3. B. Bader and T. Kolda, MATLAB Tensor Toolbox, version 2.2 (2007).
4. C. Andersson and R. Bro, The N-Way Toolbox for MATLAB (2000).
5. R. A. Harshman, http://www.psychology.uwo.ca/faculty/harshman
6. P. Paatero and U. Tapper, *Positive matrix factorization*, Environmetrics **5** (1994) 111-126.
7. D. D. Lee and H. S. Seung, *Learning the parts of objects by non-negative matrix factorization*, Nature **401** (1999) 788–791.
8. L. De Lathauwer, B. de Moor, and J. Vandewalle, SIAM J. Matrix Anal. Appl. **21** (2000) 1253–1278 and 1324–1342 (*and more recent papers on tensors*).
9. S. Ragnarsson and C. Van Loan, *Block tensor unfoldings*, SIAM J. Matrix Anal. Appl. **33** (2013) 149–169, arXiv: 1101.2005, 2 Oct 2011.
10. C. Van Loan, **www.alm.unibo.it/~simoncin/CIME/vanloan1.pdf—vanloan4.pdf**
11. I. Oseledets, *Tensor-train decomposition*, SIAM J. Sci. Comp. **33** (2011) 2295-2317.
12. Z. Song, D. P. Woodruff, and P. Zhong, *Relative error tensor low rank approximation*, arXiv: 1704.08246, 29 Mar 2018.
13. P. Grinfeld, *Introduction to Tensor Calculus and the Calculus of Moving Surfaces*, Springer (2013).

Problem Set I.12

The first 5 questions are about minimizing $||A - UV||_F^2$ when UV has rank 1.

Problem Minimize $\left\| \begin{bmatrix} a & c \\ b & d \end{bmatrix} - \begin{bmatrix} u_1 \\ u_2 \end{bmatrix} \begin{bmatrix} v_1 & v_2 \end{bmatrix} \right\|_F^2$

1 Look at the first column of $A - UV$ with A and U fixed:

$$\text{Minimize} \quad \left\| \begin{bmatrix} a - v_1 u_1 \\ b - v_1 u_2 \end{bmatrix} \right\|^2 = (a - v_1 u_1)^2 + (b - v_1 u_2)^2$$

Show by calculus that the minimizing number v_1 has $(u_1^2 + u_2^2)\, v_1 = u_1 a + u_2 b$. In vector notation $\boldsymbol{u}^{\mathrm{T}}\boldsymbol{u}\, v_1 = \boldsymbol{u}^{\mathrm{T}}\boldsymbol{a}_1$ where $\boldsymbol{a}_1$ is column 1 of A.

2 Which point $v_1 \boldsymbol{u}$ in this picture minimizes $||\boldsymbol{a}_1 - v_1\boldsymbol{u}||^2$?

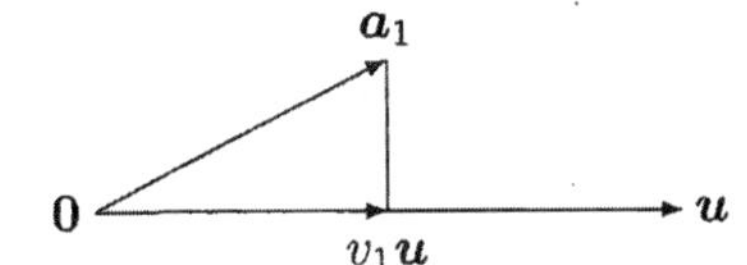

The error vector $\boldsymbol{a}_1 - v_1\boldsymbol{u}$ is _____ to $\boldsymbol{u}$.

From that fact, find again the number v_1.

3 The second column of $A - UV$ is $\begin{bmatrix} c \\ d \end{bmatrix} - \begin{bmatrix} u_1 \\ u_2 \end{bmatrix} v_2 = \boldsymbol{a}_2 - v_2\boldsymbol{u}$. Which number v_2 minimizes $||\boldsymbol{a}_2 - v_2\boldsymbol{u}||^2$?

Vector form The best $\boldsymbol{v} = \begin{bmatrix} v_1 & v_2 \end{bmatrix}$ solves $(\boldsymbol{u}^{\mathrm{T}}\boldsymbol{u})\,\boldsymbol{v} = \boldsymbol{u}^{\mathrm{T}}A$

4 Problems 1 to 3 minimized $||A - UV||_F^2$ with fixed U, when UV has rank 1.

With fixed $V = \begin{bmatrix} v_1 & v_2 \end{bmatrix}$, which $U = \begin{bmatrix} u_1 \\ u_2 \end{bmatrix}$ gives the minimum of $||A - UV||_F^2$?

5 (Computer) Starting from any U_0, does this alternating minimization converge to the closest rank 1 matrix $A_1 = \sigma_1 \boldsymbol{u}_1 \boldsymbol{v}_1^{\mathrm{T}}$ from the SVD ?

$$\underset{V}{\text{Minimize}} \quad ||A - U_n V||_F^2 \quad \text{at } V = V_n \qquad A \text{ is } 3 \times 3$$

$$U \text{ is } 3 \times 1$$

$$\underset{U}{\text{Minimize}} \quad ||A - U V_n||_F^2 \quad \text{at } U = U_{n+1} \qquad V \text{ is } 1 \times 3$$

Note : These questions are also an introduction to least squares (Section II.2). For fixed V or fixed U, each minimization is a least squares problem—even when the rank of UV increases beyond 1. But requiring nonnegativity or sparsity of U and V makes each minimization more difficult and new methods are needed.

Problems 6 to 11 are about tensors. We are not doing calculus (with derivatives) or tensor algebra (with spaces of tensors). Our focus is on a single tensor $\boldsymbol{T}$ that contains multidimensional data. If we have samples 1 to n and each sample is an image (a matrix) then we have a tensor of order 3. *How can we approximate this tensor T by a combination of simple tensors?* This is the data science question.

First we must decide: Which tensors are simple? Our answer: $\boldsymbol{a} \otimes \boldsymbol{b} \otimes \boldsymbol{c}$ is a simple (rank 1) tensor. Its i, j, k entry is the number a_i times b_j times c_k—just as a rank-1 matrix $\boldsymbol{a}\boldsymbol{b}^{\mathrm{T}}$ has entries $a_i b_j$. A sum of simple tensors approximates $\boldsymbol{T}$.

6 Given an m by n matrix A, how do you decide if A has rank 1?

7 Given an m by n by p tensor T, how do you decide if T has rank 1?

8 The largest possible rank of a 2 by 2 by 2 tensor is 3. Can you find an example?

9 (a) Suppose you know the row sums r_1 to r_m and the column sums c_1 to c_n of an m by n matrix A. What condition must be satisfied by those numbers?

(b) For an m by n by p tensor, the slices are n by p matrices and m by p matrices and m by n matrices. Suppose you add up the entries in each of those m slices and n slices and p slices. What conditions would be guaranteed to connect those m numbers and n numbers and p numbers?

10 Suppose all entries are 1 in a $2 \times 2 \times 2$ tensor T, except the first entry is $T_{111} = 0$. Write T as a sum of two rank-1 tensors. What is the closest rank-1 tensor to T (in the usual Frobenius norm)?

11 A 2 by 2 by 2 tensor $\boldsymbol{T}$ times a vector $\boldsymbol{v}$ in $\mathbf{R}^2$ should produce a matrix A in $\mathbf{R}^{2\times 2}$. How could you define that output $A = \boldsymbol{T}\boldsymbol{v}$?

Part II

Computations with Large Matrices

II.1 **Numerical Linear Algebra**

II.2 **Least Squares : Four Ways**

II.3 **Three Bases for the Column Space**

II.4 **Randomized Linear Algebra**

II : Computations with Large Matrices

This part of the book discusses $Ax = b$ in its many variations. Ordinary elimination might compute an accurate x—or maybe not. There might be too many equations ($m > n$) and no solution. A square matrix might be singular. The solution might be impossible to compute (A is extremely ill-conditioned, or simply too large). In deep learning we have *too many solutions*—and we want one that will generalize well to unseen test data.

These two pages will try to separate those sources of difficulty. We are like doctors performing triage— identify the problem and suggest a course of action for each one. The "pseudoinverse" of A proposes an inverse for every matrix—but this might not help.

0. Every matrix $A = U\Sigma V^{\mathrm{T}}$ has a **pseudoinverse** $\boldsymbol{A^+ = V\Sigma^+U^{\mathrm{T}}}$. For the diagonal matrix Σ the pseudoinverse Σ^+ contains $1/\sigma_k$ for each nonzero singular value. *But the pseudoinverse of* 0 *is* 0. To know when a number is exactly zero is an extremely rigid requirement—impossible in many computations.

The pseudoinverse in Section II.2 is one way to solve $Ax = b$. Here are other ways.

1. Suppose A is square and invertible, its size is reasonable, and its condition number σ_1/σ_n is not large. Then **elimination** will succeed (possibly with row exchanges). We have $PA = LU$ or $A = LU$ (row exchanges or no row exchanges) as in Section I.4.

The backslash command $A\backslash b$ is engineered to make A block diagonal when possible.

2. Suppose $m > n = r$: There are too many equations $Ax = b$ to expect a solution. If the columns of A are independent and not too ill-conditioned, then we solve the **normal equations** $\boldsymbol{A^{\mathrm{T}}A\widehat{x} = A^{\mathrm{T}}b}$ to find the least squares solution $\widehat{x}$.

The vector b is probably not in the column space of A, and $Ax = b$ is probably impossible. $A\widehat{x}$ is the **projection of b** onto that column space in Section II.2.

Those are two good problems to have—an invertible A or an invertible $A^{\mathrm{T}}A$, well conditioned and not too large. The next page describes four computations (still linear equations) that are more difficult.

3. Suppose $m < n$. Now the equation $Ax = b$ has many solutions, if it has one. A has a nonzero nullspace. **The solution x is underdetermined**. We want to choose the best x for our purpose. Two possible choices are x^+ and x_1 :

$x = x^+ = A^+b$. The pseudoinverse A^+ gives the **minimum ℓ^2 norm solution** with nullspace component = zero.
$x = x_1$ = **minimum ℓ^1 norm solution**. This solution is often sparse (many zero components) and very desirable. It comes from "basis pursuit" in Section III.4.

4. The columns of A may be in **bad condition**. Now the ratio σ_1/σ_r is too large. Then x is not well determined (as in high-order interpolation: Section III.3). The usual remedy is to **orthogonalize the columns** by a Gram-Schmidt or Householder algorithm. Create orthonormal vectors $q_1, \ldots, q_n$ from the columns $a_1, \ldots, a_n$.

Section II.2 explains two important forms of Gram-Schmidt. The standard way is to orthogonalize each column a_{k+1} against the known directions q_1 to q_k. The safer way orthogonalizes all the $n - k$ remaining columns against q_k as soon as that vector is found. Then a small column $k + 1$ can be exchanged for a later column when necessary. The very safest way picks the largest available column at each step.

5. A may be nearly singular (as in 4). In this case $A^{\mathrm{T}}A$ will have a very large inverse. Gram-Schmidt may fail. A different approach is to add a **penalty term**:

$$\text{Minimize } ||Ax - b||^2 + \delta^2||x||^2 \qquad \text{Solve } (A^{\mathrm{T}}A + \delta^2 I)x_\delta = A^{\mathrm{T}}b$$

As the penalty δ^2 approaches zero, $(A^{\mathrm{T}}A + \delta^2 I)^{-1}A^{\mathrm{T}}$ approaches the pseudoinverse A^+ (Section II.2). We achieve invertibility by adding $\delta^2 I$ to make $A^{\mathrm{T}}A$ more positive. This connects to ridge regression in statistics.

Penalty terms like $\delta^2||x||^2$ are common in **inverse problems**, which aim to reconstruct a system from knowledge of its outputs. Usually we know the system (like an electrical network) and we find its outputs (currents and voltages). The inverse problem starts with outputs (like CT or MRI scans). Reconstructing the system is ill-conditioned.

6. Suppose A is **way too big**. It spills outside fast memory. We can ask to see a few columns but we cannot go forward with elimination. Multiplying A^{T} times A would be impossible, even on the petascale computer that is expected at Oak Ridge and the exascale machine planned for Argonne (*New York Times*, 28 February 2018). What to do for such a large matrix ?

The best solution is **random sampling** of the columns (Section II.4). If A is oversize but reasonably coherent, each Ax will be a useful sample of the column space. The results from random sampling are never certain, but the probability of going wrong is low. *Randomized numerical linear algebra* has led to algorithms with a secure statistical base.

This is a necessary evolution or revolution based on deep results in probability.

II.1 Numerical Linear Algebra

This section summarizes the central ideas of (classical) numerical linear algebra. They won't be explained at length, because so many books do this well. Their aim is to solve $Ax = b$ or $Ax = \lambda x$ or $Av = \sigma u$. They are the foundation on which new computational methods are built.

It is those new methods—aiming to extract information from the data in a matrix or a tensor—that are the real goals of this part of the book. When a matrix or a tensor is really large ("big data") we will often have to **sample the matrix**. It may seem impossible that random sampling would give a reliable answer. But in fact this is highly probable.

The bible for numerical linear algebra is *Matrix Computations*. Its authors are Gene Golub and Charles Van Loan. The fourth edition was published in 2013 by Johns Hopkins University Press. Gene lectured at Johns Hopkins more than 30 years earlier and that led to the first edition—I don't think the publisher had any idea of a 2013 book with 750 pages.

Many other books are listed on **math.mit.edu/learningfromdata**. Here we choose one outstanding textbook: *Numerical Linear Algebra* by Trefethen and Bau. Its chapter titles provide a good outline of the central ideas and algorithms:

I. **Fundamentals** (reaching the SVD and Eckart-Young)

II. $\boldsymbol{QR}$ **Factorization and Least Squares** (all 3 ways: A^+ and $(A^{\mathrm{T}}A)^{-1}A^{\mathrm{T}}$ and QR)

III. **Conditioning and Stability** (condition numbers, backward stability, perturbations)

IV. **Systems of Equations** (direct elimination: $PA = LU$ and Cholesky's $S = A^{\mathrm{T}}A$)

V. **Eigenvalues** (reduction to tridiagonal-Hessenberg-bidiagonal; QR with shifts)

VI. **Iterative Methods** (Arnoldi, Lanczos, GMRES, conjugate gradients, Krylov).

Our plan in this section is to outline those important iterative algorithms from parts **VI** and **V**. All these are included in major codes to solve $Ax = b$ and $Sq = \lambda q$ and $Ax = \lambda x$ and $Av = \sigma u$. That word "iterative" indicates that we repeat a simple and fast step, aiming to approach the solution to a larger and more difficult problem.

A model for iteration (but not a specially fast algorithm!) is to split A into $A = S - T$:

$$\textbf{Prepare for iteration} \qquad \textbf{Rewrite } Ax = b \textbf{ as } Sx = Tx + b. \tag{1}$$

Start with any x_0 and solve $Sx_1 = Tx_0 + b$. Continue to $Sx_2 = Tx_1 + b$. A hundred iterations are very common. If S is well chosen, each step $\boldsymbol{Sx_{k+1} = Tx_k + b}$ is fast.

Subtracting the iteration $Sx_{k+1} = Tx_k + b$ from the exact $Sx = Tx + b$, the error $x - x_k$ obeys the error equation (and b cancels out):

$$\textbf{Error equation} \qquad S(x - x_{k+1}) = T(x - x_k) \tag{2}$$

Every step multiplies the error by $S^{-1}T$. When $||S^{-1}T|| << 1$ the convergence is fast. But in practice $S^{-1}T$ often has an eigenvalue near 1. Then a better idea will be needed—like the conjugate gradient method.

Let me add this note first. A textbook might find eigenvalues by solving $\det(A - \lambda I) = 0$. It might find singular values by working with the matrix $A^{\mathrm{T}}A$. In reality, those determinants are unthinkable and a large $A^{\mathrm{T}}A$ can be numerically very unwise. $Ax = \lambda x$ and $Av = \sigma u$ are serious problems. We solve them in this section for matrices of size 100 or more. For $n = 10^4$, read onward to Section II.4.

Krylov Subspaces and Arnoldi Iteration

Key idea: **Matrix-vector multiplication Ab is fast, especially if A is sparse.** If we start with A and b, we can quickly compute each of the vectors $b, Ab, \ldots, A^{n-1}b$. (Never compute A^2 or A^3. Only compute vectors.) The combinations of those n vectors make up the nth **Krylov subspace**. We look inside this subspace K_n for a close approximation to the desired solution x.

The first problem is to find a much better basis than those vectors $b, Ab, \ldots, A^{n-1}b$. An orthogonal basis $q_1, \ldots, q_n$ is usually best! The Gram-Schmidt idea of subtracting off the projections of $v = Aq_k$ onto all the earlier vectors $q_1, \ldots, q_k$ is so natural. **This is Arnoldi's method to find q_{k+1}.**

Arnoldi Iteration	$q_1 = b/\|\|b\|\|,\ q_2, \ldots, q_k$ are known
$v = Aq_k$	**Start with new v**
for $j = 1$ to k	For each known q
$\quad h_{jk} = q_j^{\mathrm{T}} v$	Compute inner product
$\quad\quad v = v - h_{jk} q_j$	Subtract projection
$h_{k+1,k} = \|\|v\|\|$	Compute norm
$q_{k+1} = v / h_{k+1,k}$	**New basis vector with norm 1**

You have to see this in matrix language. The last column is Aq_k = combination of q_1 to q_{k+1}:

$$\begin{bmatrix} \\ & A & \\ \\ \end{bmatrix} \begin{bmatrix} \\ q_1 & \cdots & q_k \\ \\ \end{bmatrix} = \begin{bmatrix} \\ q_1 & \cdots & q_{k+1} \\ \\ \end{bmatrix} \begin{bmatrix} h_{11} & \cdots & h_{1k} \\ h_{21} & \cdots & \cdot \\ & \ddots & \vdots \\ & & h_{k+1,k} \end{bmatrix} \tag{3}$$

This is $AQ_k = Q_{k+1}H_{k+1,k}$. Multiply both sides by Q_k^{T}. The result is important.

$$\boxed{Q_k^{\mathrm{T}} A Q_k = Q_k^{\mathrm{T}} Q_{k+1} H_{k+1,k} = \begin{bmatrix} I_{k\times k} & 0_{k\times 1} \end{bmatrix} \begin{bmatrix} H_k \\ \text{row } k+1 \end{bmatrix} = H_k.} \tag{4}$$

The square matrix H_k has lost the last row that was in equation (3). This leaves an upper triangular matrix plus *one subdiagonal containing h_{21} to $h_{k,k-1}$*. Matrices with only one nonzero subdiagonal are called **Hessenberg matrices**. This H_k has a neat interpretation:

$H_k = Q_k^{\mathrm{T}} A Q_k$ is the projection of A onto the Krylov space, using the basis of q's.

The Arnoldi process to find H_k is one of the great algorithms of numerical linear algebra. It is numerically stable and the q's are orthonormal.

Eigenvalues from Arnoldi

The numbers in $H_k = Q_k^{\mathrm{T}} A Q_k$ are computed during the Arnoldi process. If we go all the way to $k =$ size of A, then we have a Hessenberg matrix $H = Q^{-1} A Q$ that is **similar to A**: same eigenvalues. We compute those eigenvalues by the shifted QR algorithm below, applied to H.

In reality we don't go all the way with Arnoldi. We stop at a decent value of k. Then the k eigenvalues of H are (usually) good approximations to k extreme eigenvalues of A. Trefethen and Bau emphasize for *non-symmetric* A that we may not want eigenvalues of A in the first place! When they are badly conditioned, this led Trefethen and Embree to the theory of pseudospectra.

Linear Systems by Arnoldi and GMRES

Arnoldi has given us a great basis (orthonormal q's) for the growing Krylov subspaces spanned by $b, Ab, \ldots, A^{k-1}b$. So Arnoldi is the first step. In that subspace, the GMRES idea for $Ax = b$ is to **find the vector x_k that minimizes $||b - Ax_k||$**: the Generalized Minimum RESidual. With an orthonormal basis, we can compute accurately and safely:

> **GMRES with Arnoldi's basis $q_1, \ldots, q_k$**
>
> Find y_k to minimize the length of $H_{k+1,k}\, y - (||b||, 0, \ldots, 0)^{\mathrm{T}}$.
>
> Then $x_k = Q_k y_k$

Finding y_k is a least squares problem with a $k + 1$ by k Hessenberg matrix. The zeros below the first subdiagonal of $H_{k+1,k}$ make GMRES especially fast.

Symmetric Matrices : Arnoldi Becomes Lanczos

Suppose our matrix is *symmetric* : $A = S$. In this important case, two extra facts are true.

1. $\boldsymbol{H_k = Q_k^{\mathrm{T}} S Q_k}$ **is also symmetric**. Its transpose is clearly H_k.
2. $\boldsymbol{H_k}$ **is tridiagonal** : only one diagonal above because only one diagonal below.

A tridiagonal matrix H gives a major saving in cost—the Arnoldi iteration needs only one orthogonalization step. The other orthogonalities are built in because H is **symmetric Hessenberg** (so it is tridiagonal).

Here is Lanczos with simpler letters a_1 to a_k (on the main diagonal) and b_1 to b_{k-1} (on the diagonals above and below). The a's and b's in T replace the h's of Arnoldi's matrix H.

Lanczos iteration for $\boldsymbol{Sx = \lambda x}$ (symmetric Arnoldi)

$\boldsymbol{q}_0 = \boldsymbol{0},\ \boldsymbol{q}_1 = \boldsymbol{b}/\lVert\boldsymbol{b}\rVert$	Orthogonalize $\boldsymbol{b}, S\boldsymbol{b}, S\boldsymbol{b}^2, \ldots$
For $k = 1, 2, 3, \ldots$	
$\boldsymbol{v} = S\boldsymbol{q}_k$	Start with new $\boldsymbol{v}$
$a_k = \boldsymbol{q}_k^{\mathrm{T}}\boldsymbol{v}$	Diagonal entry in T is a_k
$\boldsymbol{v} = \boldsymbol{v} - b_{k-1}\boldsymbol{q}_{k-1} - a_k\,\boldsymbol{q}_k$	Orthogonal to earlier $\boldsymbol{q}$'s
$b_k = \lVert\boldsymbol{v}\rVert$	Off-diagonal entry in T is b_k
$\boldsymbol{q}_{k+1} = \boldsymbol{v}/b_k$	Next basis vector

Writing T for tridiagonal instead of H for Hessenberg, here are the key facts for Lanczos. They are simply copied from Arnoldi :

$$\text{Equations (3) and (4)} \qquad \boxed{\boldsymbol{T_k = Q_k^{\mathrm{T}} S Q_k} \quad \text{and} \quad \boldsymbol{S Q_k = Q_{k+1}\, T_{k+1,k}}} \qquad (5)$$

"The eigenvalues of T_k (fast to compute) approximate the eigenvalues of S." If only that were exactly and always true ! Trefethen and Bau create a diagonal matrix S with 201 equally spaced eigenvalues from 0 to 2, and also two larger eigenvalues 2.5 and 3.0. Starting from a random vector $\boldsymbol{b}$, Lanczos at step $k = 9$ approximates $\lambda = 2.5$ and 3.0 exponentially well. The other 7 eigenvalues of $T_9 = Q_9^{\mathrm{T}} S Q_9$ bunch near 0 and 2. But they don't capture individual eigenvalues in that group of 201 λ's.

The problem comes from non-orthogonal $\boldsymbol{q}$'s when exact Lanczos iterations would guarantee orthogonal $\boldsymbol{q}$'s. **Lanczos is valuable**. But special care is needed to keep all the $\boldsymbol{q}$'s orthogonal in practice—which was true also of Gram-Schmidt.

Eigenvalues of Tridiagonal T by the QR Iteration

How to compute the eigenvalues of a symmetric tridiagonal matrix T ? This is the key question for the symmetric eigenvalue problem. The original matrix S has been simplified by Lanczos to the tridiagonal $T = Q^{\mathrm{T}}SQ = Q^{-1}SQ$ (no change in eigenvalues because T is similar to S). Or those zeros could also come from 2 by 2 "Givens rotations".

At this point we have a tridiagonal symmetric matrix $T = T_0$. To find its eigenvalues, an amazing idea appeared almost from nowhere:

1. **By Gram-Schmidt or Householder, factor T_0 into QR. Notice $R = Q^{-1}T_0$.**
2. **Reverse those factors Q and R to produce $T_1 = RQ = Q^{-1}T_0Q$.**
3. Repeat... Repeat... Repeat...

The new $T_1 = Q^{-1}T_0Q$ is similar to T: *same eigenvalues.* More than that, the new T_1 is *still tridiagonal* (Problem 1). So the next step and all the later steps are still fast. And best of all, **the similar matrices $T, T_1, T_2, \ldots$ approach a diagonal matrix Λ.** That diagonal matrix reveals the (unchanged) eigenvalues of the original matrix T. The first eigenvalue to appear is in the last entry T_{nn}.

This is the "**QR algorithm**" to compute eigenvalues. As it gradually became known, it caused a sensation in numerical linear algebra. But numerical analysts are serious people. If you give them a good algorithm, they immediately start to make it better. In this case they succeeded triumphantly, because the improvements came at virtually no cost (and they really worked).

The improved algorithm is **shifted QR**, or **QR with shifts**. The "shift" subtracts a multiple s_kI of the identity matrix before the QR step, and adds it back after the RQ step:

QR algorithm with shifts to find eigenvalues	**Choose a shift s_k at step k** **Factor $T_k - s_kI = Q_kR_k$** **Reverse factors and shift back: $T_{k+1} = R_kQ_k + s_kI$**

The T's all have the same eigenvalues because they are similar matrices. Each new T_{k+1} is $Q_k^{-1}T_kQ_k$. It is still symmetric because $Q_k^{-1} = Q_k^{\mathrm{T}}$:

$$R_k = Q_k^{-1}(T_k - s_kI) \text{ and then } T_{k+1} = Q_k^{-1}(T_k - s_kI)Q_k + s_kI = Q_k^{-1}T_kQ_k. \quad (6)$$

Well-chosen shifts s_k will greatly speed up the approach of the T's to a diagonal matrix Λ. A good shift is $s = T_{nn}$. A shift suggested by Wilkinson is based on the last 2 by 2 submatrix of T_k:

$$\textbf{Wilkinson shift} \qquad s_k = \text{ the eigenvalue of } \begin{bmatrix} a_{n-1} & b_{n-1} \\ b_{n-1} & a_n \end{bmatrix} \text{ closest to } a_n.$$

Shifted QR achieves cubic convergence (very rare). In the example that follows, the off-diagonal goes from $\sin\theta$ to $-(\sin\theta)^3$. The n eigenvalues of a typical tridiagonal T take only $O(n^3/\epsilon)$ flops for accuracy ϵ.

$$\boldsymbol{T_0} = \begin{bmatrix} \cos\theta & \sin\theta \\ \boldsymbol{\sin\theta} & 0 \end{bmatrix} \quad \text{Shift } = 0 \quad \boldsymbol{Q_0} = \begin{bmatrix} \cos\theta & -\sin\theta \\ \sin\theta & \cos\theta \end{bmatrix} \quad \boldsymbol{R_0} = \begin{bmatrix} 1 & \sin\theta\cos\theta \\ 0 & -\sin^2\theta \end{bmatrix}$$

$$\boldsymbol{T_1} = \boldsymbol{R_0}\,\boldsymbol{Q_0} = \begin{bmatrix} \cos\theta(1+\sin^2\theta) & -\sin^3\theta \\ \boldsymbol{-\sin^3\theta} & -\sin^2\theta\cos\theta \end{bmatrix} \textbf{ has cubed the error in one step.}$$

Computing the SVD

What is the main difference between the symmetric eigenvalue problem $Sx = \lambda x$ and $A = U\Sigma V^{\mathrm{T}}$? How much can we simplify S and A before computing λ's and σ's ?

Eigenvalues are the same for S and $Q^{-1}SQ = Q^{\mathrm{T}}SQ$ because Q is orthogonal.

So we have limited freedom to create zeros in $Q^{-1}SQ$ (which stays symmetric). If we try for too many zeros in $Q^{-1}S$, the final Q can destroy them. The good $Q^{-1}SQ$ will be **tridiagonal**: only three diagonals.

Singular values are the same for A and $Q_1AQ_2^{\mathrm{T}}$ even if Q_1 is different from Q_2.

We have more freedom to create zeros in $Q_1AQ_2^{\mathrm{T}}$. With the right Q's, this will be **bidiagonal** (two diagonals). We can quickly find Q and Q_1 and Q_2 so that

$$\boldsymbol{Q^{-1}SQ} = \begin{bmatrix} a_1 & b_1 & & \\ b_1 & a_2 & b_2 & \\ & b_2 & \cdot & \cdot \\ & & \cdot & a_n \end{bmatrix} \begin{matrix} \leftarrow \text{for } \lambda\text{'s} \\ \boldsymbol{Q_1AQ_2^{\mathrm{T}}} = \\ \text{for } \sigma\text{'s} \rightarrow \end{matrix} \begin{bmatrix} c_1 & d_1 & & \\ 0 & c_2 & d_2 & \\ & 0 & \cdot & \cdot \\ & & 0 & c_n \end{bmatrix} \qquad (7)$$

The reader will know that the singular values of A are the square roots of the eigenvalues of $S = A^{\mathrm{T}}A$. And the unchanged singular values of $Q_1AQ_2^{\mathrm{T}}$ are the square roots of the unchanged eigenvalues of $(Q_1AQ_2^{\mathrm{T}})^{\mathrm{T}}(Q_1AQ_2^{\mathrm{T}}) = Q_2A^{\mathrm{T}}AQ_2^{\mathrm{T}}$. **Multiply (bidiagonal)^{T}(bidiagonal) to see tridiagonal.**

This offers an option that we should not take. Don't multiply $A^{\mathrm{T}}A$ and find its eigenvalues. This is unnecessary work and the condition of the problem will be unnecessarily squared. The Golub-Kahan algorithm for the SVD works directly on A, in two steps:

1. Find Q_1 and Q_2 so that $Q_1AQ_2^{\mathrm{T}}$ is bidiagonal as in (8).
2. Adjust the shifted QR algorithm to preserve singular values of this bidiagonal matrix.

Step 1 requires $O(mn^2)$ multiplications to put an m by n matrix A into bidiagonal form. Then later steps will work only with bidiagonal matrices. Normally it then takes $O(n^2)$ multiplications to find singular values (correct to nearly machine precision). The full algorithm is described on pages 489 to 492 of Golub-Van Loan (4th edition).

Those operation counts are very acceptable for many applications—an SVD is computable. Other algorithms are proposed and successful. But the cost is not trivial (you can't just do SVD's by the thousands). When A is truly large, the next sections of this book will introduce methods which including "*random sampling*" of the original matrix A—this approach can handle big matrices. With very high probability the results are accurate. Most gamblers would say that a good outcome from careful random sampling is certain.

Conjugate Gradients for $Sx = b$

The conjugate gradient algorithm applies to symmetric positive definite matrices S. It solves $Sx = b$. Theoretically it gives the exact solution in n steps (but those steps are slower than elimination). Practically it gives excellent results for large matrices much sooner than the nth step (this discovery revived the whole idea and now **CG** is one of the very best algorithms). Such is the history of the most celebrated of all Krylov methods.

Key point: Because S is symmetric, the Hessenberg matrix H in Arnoldi becomes the tridiagonal matrix T in Lanczos. Three nonzeros in the rows and columns of T makes the symmetric case especially fast.

And now S is not only symmetric. It is also positive definite. In that case $||x||_S^2 = x^{\mathrm{T}}Sx$ gives a very appropriate norm (the S-norm) to measure the error after n steps. In fact the kth conjugate gradient iterate x_k has a remarkable property:

x_k minimizes the error $||x - x_k||_S$ over the kth Krylov subspace
x_k is the best combination of $b, Sb, \ldots, S^{k-1}b$

Here are the steps of the conjugate gradient iteration to solve $Sx = b$:

Conjugate Gradient Iteration for Positive Definite S

$x_0 = \mathbf{0}, r_0 = b, d_0 = r_0$

for $k = 1$ **to** N

$\alpha_k = (r_{k-1}^{\mathrm{T}} r_{k-1})/(d_{k-1}^{\mathrm{T}} S d_{k-1})$	step length x_{k-1} to x_k
$x_k = x_{k-1} + \alpha_k d_{k-1}$	approximate solution
$r_k = r_{k-1} - \alpha_k S d_{k-1}$	new residual $b - Sx_k$
$\beta_k = (r_k^{\mathrm{T}} r_k)/(r_{k-1}^{\mathrm{T}} r_{k-1})$	improvement this step
$d_k = r_k + \beta_k d_{k-1}$	next search direction

% Notice: only 1 matrix-vector multiplication Sd in each step

Here are the two great facts that follow (with patience) from those steps. Zigzags are gone!

1. **The error residuals $r_k = b - Sx_k$ are orthogonal: $r_k^{\mathrm{T}} r_j = 0$**
2. **The search directions d_k are S-orthogonal: $d_k^{\mathrm{T}} S d_j = 0$**

Notice Solving $Sx - b = \mathbf{0}$ is the same as minimizing the quadratic $\frac{1}{2}x^{\mathrm{T}}Sx - x^{\mathrm{T}}b$. One is the gradient of the other. So conjugate gradients is also a minimization algorithm. It can be generalized to nonlinear equations and nonquadratic cost functions. It could be considered for deep learning in Part VII of this book—but the matrices there are simply too large for conjugate gradients.

We close with the cleanest estimate for the error after k conjugate gradient steps. The success is greatest when the eigenvalues λ of S are well spaced.

$$\textbf{CG Method} \qquad ||x - x_k||_S \leq 2||x - x_0||_S \left(\frac{\sqrt{\lambda_{\max}} - \sqrt{\lambda_{\min}}}{\sqrt{\lambda_{\max}} + \sqrt{\lambda_{\min}}}\right)^k. \qquad (8)$$

Preconditioning for $Ax = b$

The idea of preconditioning is to find a "nearby" problem that can be solved quickly. Explaining the idea is fairly easy. *Choosing a good preconditioner is a serious problem.* For a given matrix A, the idea is to choose a simpler matrix P that is close to A. Those matrices may be close in the sense that $A - P$ has small norm or it has low rank. Working with $P^{-1}A$ is faster:

$$\textbf{Preconditioned} \qquad P^{-1}A\boldsymbol{x} = P^{-1}\boldsymbol{b} \text{ instead of } A\boldsymbol{x} = \boldsymbol{b}. \tag{9}$$

The convergence test (for whichever algorithm is used) applies to $P^{-1}A$ in place of A.

If the algorithm is conjugate gradients (which works on symmetric positive definite matrices) we likely change from A to $P^{-1/2}AP^{-1/2}$.

Here are frequent choices of the preconditioner P:

1 $P =$ diagonal matrix (copying the main diagonal of A): *Jacobi iteration*

2 $P =$ triangular matrix (copying that part of A): *Gauss-Seidel method*

3 $P = L_0U_0$ omits fill-in from $A = LU$ (elimination) to preserve sparsity: *incomplete LU*

4 $P =$ same difference matrix as A but on a coarser grid: *multigrid method*

Multigrid is a powerful and highly developed solution method. It uses a whole sequence of grids or meshes. The given problem on the finest grid has the most meshpoints (large matrix A). Successive problems on coarser grids have fewer meshpoints (smaller matrices). Those can be solved quickly and the results can be interpolated back to the fine mesh. Highly efficient with fast convergence.

Kaczmarz Iteration

Equation (10) is fast to execute but not easy to analyze. Now we know that convergence is exponentially fast with high probability when each step solves one random equation of $A\boldsymbol{x} = \boldsymbol{b}$. Step k of Kaczmarz gets the ith equation right:

$$\boldsymbol{x}_{k+1} \textbf{ satisfies } \boldsymbol{a}_i^{\mathrm{T}}\boldsymbol{x} = b_i \qquad \boldsymbol{x}_{k+1} = \boldsymbol{x}_k + \frac{b_i - \boldsymbol{a}_i^{\mathrm{T}}\boldsymbol{x}_k}{||\boldsymbol{a}_i||^2}\,\boldsymbol{a}_i \tag{10}$$

Each step projects the previous $\boldsymbol{x}_k$ onto the plane $\boldsymbol{a}_i^{\mathrm{T}}\boldsymbol{x} = b_i$. Cycling through the m equations in order is classical Kaczmarz. The randomized algorithm chooses row i with probability proportional to $||\boldsymbol{a}_i||^2$ (**norm-squared sampling in II.4**).

Kaczmarz iteration is an important example of *stochastic gradient descent* (stochastic because equation i is a random choice at step k). We return to this algorithm in Section VI.5 on optimizing the weights in deep learning.

T. Strohmer and R. Vershynin, *A randomized Kaczmarz algorithm with exponential convergence*, J. Fourier Anal. Appl. **15** (2009) 262-278; arXiv: math/0702226.

Problem Set II.1

These problems start with a bidiagonal n by n backward difference matrix $D = I - S$. Two tridiagonal second difference matrices are DD^{T} and $A = -S + 2I - S^{\mathrm{T}}$. The shift S has one nonzero subdiagonal $S_{i,i-1} = 1$ for $i = 2, \ldots, n$. A has diagonals $-1, 2, -1$.

1 Show that DD^{T} equals A except that $1 \neq 2$ in their $(1,1)$ entries. Similarly $D^{\mathrm{T}}D = A$ except that $1 \neq 2$ in their (n, n) entries.

Note $A\boldsymbol{u}$ corresponds to $-d^2u/dx^2$ for $0 \leq x \leq 1$ with fixed boundaries $u(0) = 0$ and $u(1) = 0$. DD^{T} changes the first condition to $du/dx(0) = 0$ (*free*). $D^{\mathrm{T}}D$ changes the second condition to $du/dx(1) = 0$ (*free*). Highly useful matrices.

2 Show that the inverse of $D = I - S$ is D^{-1} = lower triangular "*sum matrix*" of 1's. $DD^{-1} = I$ is like the Fundamental Theorem of Calculus: *derivative of integral of f equals f*. Multiply $(D^{-1})^{\mathrm{T}}$ times D^{-1} to find $(DD^{\mathrm{T}})^{-1}$ for $n = 4$.

3 Problem 1 says that $A = DD^{\mathrm{T}} + \boldsymbol{e}\boldsymbol{e}^{\mathrm{T}}$ where $\boldsymbol{e} = (1, 0, \ldots, 0)$. Section III.1 will show that $A^{-1} = (DD^{\mathrm{T}})^{-1} - \boldsymbol{z}\boldsymbol{z}^{\mathrm{T}}$. For $n = 3$, can you discover the vector $\boldsymbol{z}$? Rank-one change in DD^{T} produces rank-one change in its inverse.

4 Suppose you split A into $-S + 2I$ (lower triangular) and $-S^{\mathrm{T}}$ (upper triangular). The Jacobi iteration to solve $A\boldsymbol{x} = \boldsymbol{b}$ will be $(-S + 2I)\,\boldsymbol{x}_{k+1} = S^{\mathrm{T}}\boldsymbol{x}_k + \boldsymbol{b}$.

This iteration converges provided all eigenvalues of $(-S + 2I)^{-1}\,S^{\mathrm{T}}$ have $|\lambda| < 1$. Find those eigenvalues for sizes $n = 2$ and $n = 3$.

5 For $\boldsymbol{b} = (1, 0, 0)$ and $n = 3$, the vectors $\boldsymbol{b}, A\boldsymbol{b}, A^2\boldsymbol{b}$ are a non-orthogonal basis for $\mathbf{R}^3$. Use the Arnoldi iteration with A to produce an orthonormal basis $\boldsymbol{q}_1, \boldsymbol{q}_2, \boldsymbol{q}_3$. Find the matrix H that gives $AQ_2 = Q_3H$ as in equation (3).

6 In Problem 5, verify that $Q_2^{\mathrm{T}}AQ_2$ is a tridiagonal matrix.

7 Apply one step of the *QR algorithm* to the 3 by 3 second difference matrix A. The actual eigenvalues of A are $\lambda = 2 - \sqrt{2}, 2, 2 + \sqrt{2}$.

8 Try one step of the QR algorithm with the recommended shift $s = A_{33} = 2$.

9 Solve $A\boldsymbol{x} = (1, 0, 0)$ by hand. Then by computer using the conjugate gradient method.

II.2 Least Squares : Four Ways

Many applications lead to **unsolvable linear equations** $Ax = b$. It is ironic that this is such an important problem in linear algebra. We can't throw equations away, we need to produce a best solution $\widehat{x}$. **The least squares method chooses** $\widehat{x}$ **to make** $||b - A\widehat{x}||^2$ **as small as possible**. Minimizing that error means that its derivatives are zero : those are the *normal equations* $\boldsymbol{A}^{\mathrm{T}}\boldsymbol{A}\widehat{\boldsymbol{x}} = \boldsymbol{A}^{\mathrm{T}}\boldsymbol{b}$. Their geometry will be in Figure II.2.

This section explains four ways to solve those important (and solvable !) equations :

1 The SVD of A leads to its **pseudoinverse** $\boldsymbol{A}^{+}$. Then $\widehat{x} = \boldsymbol{A}^{+}\boldsymbol{b}$: One short formula.

2 $\boldsymbol{A}^{\mathrm{T}}\boldsymbol{A}\widehat{\boldsymbol{x}} = \boldsymbol{A}^{\mathrm{T}}\boldsymbol{b}$ can be solved directly when A has **independent columns.**

3 The Gram-Schmidt idea produces **orthogonal columns** in Q. Then $\boldsymbol{A} = \boldsymbol{QR}$.

4 **Minimize** $||b - Ax||^2 + \delta^2||x||^2$. That penalty changes the normal equations to $(\boldsymbol{A}^{\mathrm{T}}\boldsymbol{A} + \delta^2 \boldsymbol{I})\boldsymbol{x}_\delta = \boldsymbol{A}^{\mathrm{T}}\boldsymbol{b}$. Now the matrix is invertible and $\boldsymbol{x}_\delta$ goes to $\widehat{x}$ as $\delta \to 0$.

$A^{\mathrm{T}}A$ has an attractive symmetry. But its size may be a problem. And its condition number—measuring the danger of unacceptable roundoff error—is the *square* of the condition number of A. In well-posed problems of moderate size we go ahead to solve the least squares equation $A^{\mathrm{T}}A\widehat{x} = A^{\mathrm{T}}\boldsymbol{b}$, but in large or ill-posed problems we find another way.

We could orthogonalize the columns of A. We could use its SVD. For really large problems we sample the column space of A by simply multiplying Av for **random vectors** $\boldsymbol{v}$. This seems to be the future for very big computations : a high probability of success.

First of all, please allow us to emphasize the importance of $A^{\mathrm{T}}A$ and $A^{\mathrm{T}}CA$. That matrix C is often a positive diagonal matrix. It gives stiffnesses or conductances or edge capacities or inverse variances $1/\sigma^2$—the constants from science or engineering or statistics that define our particular problem : the "weights" in weighted least squares.

Here is a sample of the appearances of $A^{\mathrm{T}}A$ and $A^{\mathrm{T}}CA$ in applied mathematics :

In mechanical engineering, $A^{\mathrm{T}}A$ (or $A^{\mathrm{T}}CA$) is the **stiffness matrix**

In circuit theory, $A^{\mathrm{T}}A$ (or $A^{\mathrm{T}}CA$) is the **conductance matrix**

In graph theory, $A^{\mathrm{T}}A$ (or $A^{\mathrm{T}}CA$) is the (weighted) **graph Laplacian**

In mathematics, $A^{\mathrm{T}}A$ is the **Gram matrix** : inner products of columns of A

In large problems, $A^{\mathrm{T}}A$ is expensive and often dangerous to compute. We avoid it if we can ! The Gram-Schmidt way replaces A by QR (orthogonal Q, triangular R). Then $A^{\mathrm{T}}A$ is the same as $R^{\mathrm{T}}Q^{\mathrm{T}}QR = R^{\mathrm{T}}R$. And the fundamental equation $A^{\mathrm{T}}A\widehat{x} = A^{\mathrm{T}}\boldsymbol{b}$ becomes $R^{\mathrm{T}}R\widehat{x} = R^{\mathrm{T}}Q^{\mathrm{T}}\boldsymbol{b}$. Finally this is $\boldsymbol{R}\widehat{\boldsymbol{x}} = \boldsymbol{Q}^{\mathrm{T}}\boldsymbol{b}$, safe to solve and fast too.

Thus $A^{\mathrm{T}}A$ and $A^{\mathrm{T}}CA$ are crucial matrices—but paradoxically, we try not to compute them. Orthogonal matrices and triangular matrices : Those are the good ones.

A^+ is the Pseudoinverse of A

I will first describe the pseudoinverse A^+ in words. If A is invertible then A^+ is A^{-1}. If A is m by n then A^+ is n by m. When A multiplies a vector x in its row space, this produces Ax in the column space. Those two spaces have equal dimension r (the rank). Restricted to these spaces A is always invertible—and A^+ inverts A. **Thus $A^+Ax = x$ exactly when x is in the row space.** And $AA^+b = b$ when b is in the column space.

The nullspace of A^+ is the nullspace of A^T. It contains the vectors y in $\mathbf{R}^m$ with $A^Ty = 0$. Those vectors y are perpendicular to every Ax in the column space. For these y, we accept $x^+ = A^+y = 0$ as the best solution to the unsolvable equation $Ax = y$. Altogether **A^+ inverts A where that is possible :**

$$\textbf{The pseudoinverse of } A = \begin{bmatrix} 2 & 0 \\ 0 & \mathbf{0} \end{bmatrix} \textbf{ is } A^+ = \begin{bmatrix} 1/2 & 0 \\ 0 & \mathbf{0} \end{bmatrix}.$$

The whole point is to produce a suitable "pseudoinverse" when A has no inverse.

Rule 1 If A has independent columns, then $A^+ = (A^TA)^{-1}A^T$ and so $A^+A = I$.
Rule 2 If A has independent rows, then $A^+ = A^T(AA^T)^{-1}$ and so $AA^+ = I$.
Rule 3 A diagonal matrix Σ is inverted where possible—otherwise Σ^+ has zeros :

$$\Sigma = \begin{bmatrix} \sigma_1 & 0 & 0 & 0 \\ 0 & \sigma_2 & 0 & 0 \\ 0 & 0 & \mathbf{0} & \mathbf{0} \end{bmatrix} \qquad \Sigma^+ = \begin{bmatrix} 1/\sigma_1 & 0 & 0 \\ 0 & 1/\sigma_2 & 0 \\ 0 & 0 & \mathbf{0} \\ 0 & 0 & 0 \end{bmatrix}$$

On the four subspaces
$\Sigma^+\Sigma = I \quad \Sigma\Sigma^+ = I$
$\Sigma^+\Sigma = 0 \quad \Sigma\Sigma^+ = 0$

All matrices $\quad$ **The pseudoinverse of $A = U\Sigma V^T$ is $A^+ = V\Sigma^+U^T$.** $\quad$ (1)

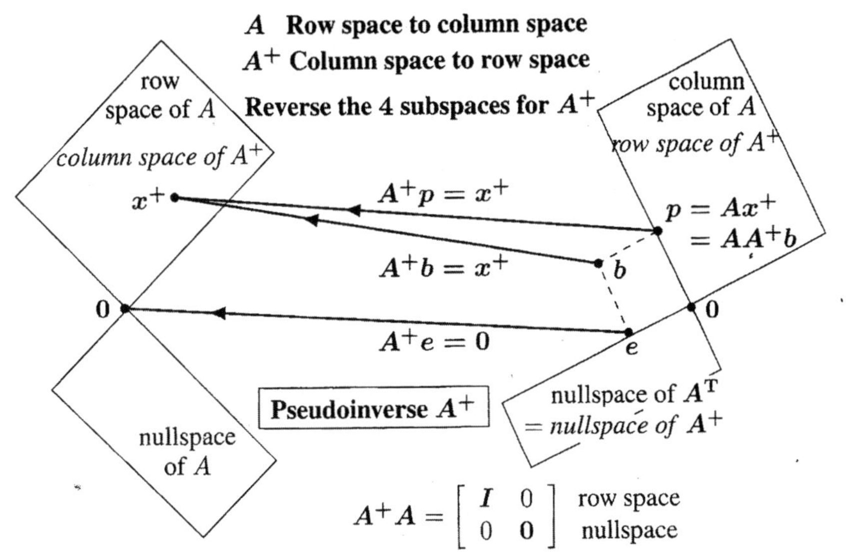

Figure II.1: Vectors $p = Ax^+$ in the column space of A go back to x^+ in the row space.

This pseudoinverse A^+ (*sometimes written* $A^\dagger$ with a dagger instead of a plus sign) solves the least squares equation $A^\mathrm{T}A\widehat{x} = A^\mathrm{T}b$ in one step. This page verifies that $x^+ = A^+b = V\Sigma^+U^\mathrm{T}b$ is best possible. At the end of this section, we look in more detail at A^+.

Question: The formula $A^+ = V\Sigma^+U^\mathrm{T}$ uses the SVD. Is the SVD essential to find A^+?
Answer: No, A^+ could also be computed directly from A by modifying the elimination steps that usually produce A^{-1}. However each step of arithmetic would have to be exact! You need to distinguish exact zeros from small nonzeros. That is the hard part of A^+.

The Least Squares Solution to $Ax = b$ is $x^+ = A^+b$

I have written x^+ instead of $\widehat{x}$ because the vector x^+ has two properties:

1 $x = x^+ = A^+b$ makes $\|\|b - Ax\|\|^2$ as small as possible.	**Least squares solution**
2 If another $\widehat{x}$ achieves that minimum then $\|\|x^+\|\| < \|\|\widehat{x}\|\|$.	**Minimum norm solution**

$x^+ = A^+b$ **is the minimum norm least squares solution.** When A has independent columns and rank $r = n$, this is the only least squares solution. But if there are nonzero vectors x in the nullspace of A (so $r < n$), they can be added to x^+. The error $b - A(x^+ + x)$ is not affected when $Ax = 0$. But the length $||x^+ + x||^2$ will grow to $||x^+||^2 + ||x||^2$. Those pieces are orthogonal: Row space $\perp$ nullspace.

So the minimum norm (shortest) solution of $A^\mathrm{T}A\widehat{x} = A^\mathrm{T}b$ is $x^+ = A^+b$, x^+ **has a zero component in the nullspace of** A.

Example 1 The shortest least squares solution to $\begin{bmatrix} 3 & 0 \\ 0 & 0 \end{bmatrix}\begin{bmatrix} x_1 \\ x_2 \end{bmatrix} = \begin{bmatrix} 6 \\ 8 \end{bmatrix}$ is x^+

$x^+ = A^+b = \begin{bmatrix} 1/3 & 0 \\ 0 & 0 \end{bmatrix}\begin{bmatrix} 6 \\ 8 \end{bmatrix} = \begin{bmatrix} \mathbf{2} \\ \mathbf{0} \end{bmatrix}$. All vectors $\begin{bmatrix} \mathbf{0} \\ \boldsymbol{x_2} \end{bmatrix}$ are in the nullspace of A.

All the vectors $\widehat{x} = \begin{bmatrix} \mathbf{2} \\ \boldsymbol{x_2} \end{bmatrix}$ minimize $||b - A\widehat{x}||^2 = 64$. But $x^+ = \begin{bmatrix} 2 \\ 0 \end{bmatrix}$ is shortest.

That example shows the least squares solutions when A is a diagonal matrix like Σ. To allow every matrix $U\Sigma V^\mathrm{T}$, we have to account for the orthogonal matrices U and V. We can freely multiply by U^T without changing any lengths, because $U^\mathrm{T}U = I$:

$$\textbf{Squared error} \qquad ||b - Ax||^2 = ||b - U\Sigma V^\mathrm{T}x||^2 = ||U^\mathrm{T}b - \Sigma V^\mathrm{T}x||^2. \tag{2}$$

Set $w = V^\mathrm{T}x$ to get $||U^\mathrm{T}b - \Sigma w||^2$. **The best w is $\Sigma^+U^\mathrm{T}b$.** And finally x^+ is A^+b:

$$w = V^\mathrm{T}x^+ = \Sigma^+U^\mathrm{T}b \text{ and } V^\mathrm{T} = V^{-1} \text{ lead to } x^+ = \boldsymbol{V\Sigma^+U^\mathrm{T}b = A^+b}. \tag{3}$$

The SVD solved the least squares problem in one step A^+b. The only question is the computational cost. Singular values and singular vectors cost more than elimination. The next two proposed solutions work directly with the linear equations $A^\mathrm{T}A\widehat{x} = A^\mathrm{T}b$. This succeeds when $A^\mathrm{T}A$ is invertible—and then $\widehat{x}$ is the same as x^+.

When is $A^{\mathrm{T}}A$ Invertible ?

The invertibility (or not) of the matrix $A^{\mathrm{T}}A$ is an important question with a nice answer:
$\boldsymbol{A^{\mathrm{T}}A}$ **is invertible exactly when** $\boldsymbol{A}$ **has independent columns. If** $\boldsymbol{Ax = 0}$ **then** $\boldsymbol{x = 0}$

Always A and $A^{\mathrm{T}}A$ have the same nullspace ! This is because $A^{\mathrm{T}}A\boldsymbol{x} = \mathbf{0}$ always leads to $\boldsymbol{x}^{\mathrm{T}}A^{\mathrm{T}}A\boldsymbol{x} = 0$. This is $||A\boldsymbol{x}||^2 = 0$. Then $A\boldsymbol{x} = \mathbf{0}$ and $\boldsymbol{x}$ is in $\mathbf{N}(A)$. For all matrices :

$$\mathbf{N}(A^{\mathrm{T}}A) = \mathbf{N}(A) \text{ and } \mathbf{C}(AA^{\mathrm{T}}) = \mathbf{C}(A) \text{ and } \mathbf{rank}\,(A^{\mathrm{T}}A) = \mathbf{rank}\,(AA^{\mathrm{T}}) = \mathbf{rank}\,(A)$$

We now go forward when $A^{\mathrm{T}}A$ is invertible to solve the normal equations $A^{\mathrm{T}}A\widehat{\boldsymbol{x}} = A^{\mathrm{T}}\boldsymbol{b}$.

The Normal Equations $A^{\mathrm{T}}A\widehat{\boldsymbol{x}} = A^{\mathrm{T}}\boldsymbol{b}$

Figure II.2 shows a picture of the least squares problem and its solution. The problem is that $\boldsymbol{b}$ is not in the column space of A, so $A\boldsymbol{x} = \boldsymbol{b}$ has no solution. The best vector $\boldsymbol{p} = A\widehat{\boldsymbol{x}}$ is a projection. **We project $\boldsymbol{b}$ onto the column space of $\boldsymbol{A}$.** The vectors $\widehat{\boldsymbol{x}}$ and $\boldsymbol{p} = A\widehat{\boldsymbol{x}}$ come from solving a famous system of linear equations : $A^{\mathrm{T}}A\widehat{\boldsymbol{x}} = A^{\mathrm{T}}\boldsymbol{b}$. To invert $A^{\mathrm{T}}A$, we need to know that A has independent columns.

The picture shows the all-important right triangle with sides $\boldsymbol{b}, \boldsymbol{p}$, and $\boldsymbol{e}$.

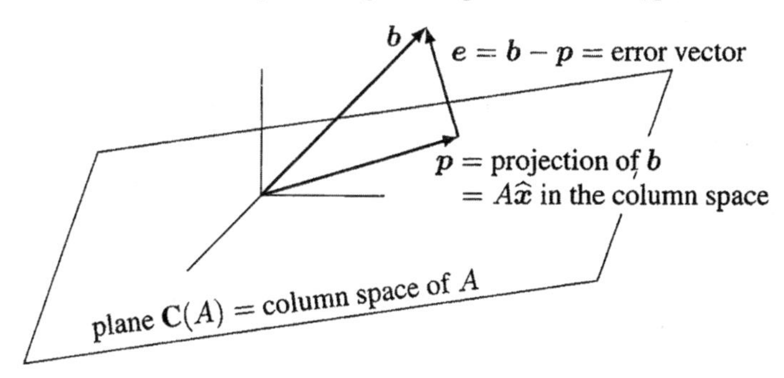

Figure II.2: The projection $\boldsymbol{p} = A\widehat{\boldsymbol{x}}$ is the point in the column space that is closest to $\boldsymbol{b}$.

Everybody understands that $\boldsymbol{e}$ is perpendicular to the plane (the column space of A). This says that $\boldsymbol{b} - \boldsymbol{p} = \boldsymbol{b} - A\widehat{\boldsymbol{x}}$ is perpendicular to all vectors $A\boldsymbol{x}$ in the column space :

$$(A\boldsymbol{x})^{\mathrm{T}}(\boldsymbol{b} - A\widehat{\boldsymbol{x}}) = \boldsymbol{x}^{\mathrm{T}}A^{\mathrm{T}}(\boldsymbol{b} - A\widehat{\boldsymbol{x}}) = 0 \text{ for all } \boldsymbol{x} \text{ forces } A^{\mathrm{T}}(\boldsymbol{b} - A\widehat{\boldsymbol{x}}) = \mathbf{0}. \quad (4)$$

Everything comes from that last equation, when we write it as $\boldsymbol{A^{\mathrm{T}}A\widehat{x} = A^{\mathrm{T}}b}$.

Normal equation for $\widehat{\boldsymbol{x}}$	$\boldsymbol{A^{\mathrm{T}}A\widehat{x} = A^{\mathrm{T}}b}$	(5)
Least squares solution to $A\boldsymbol{x} = \boldsymbol{b}$	$\boldsymbol{\widehat{x} = (A^{\mathrm{T}}A)^{-1}A^{\mathrm{T}}b}$	(6)
Projection of $\boldsymbol{b}$ onto the column space of A	$\boldsymbol{p = A\widehat{x} = A(A^{\mathrm{T}}A)^{-1}A^{\mathrm{T}}b}$	(7)
Projection matrix that multiplies $\boldsymbol{b}$ to give $\boldsymbol{p}$	$\boldsymbol{P = A(A^{\mathrm{T}}A)^{-1}A^{\mathrm{T}}}$	(8)

A now has independent columns: $r = n$. That makes $A^{\mathrm{T}}A$ positive definite and invertible. We could check that our $\widehat{x}$ is the same vector $x^+ = A^+b$ that came from the pseudoinverse. There are no other $\widehat{x}$'s because the rank is assumed to be $r = n$. The nullspace of A only contains the zero vector.

Projection matrices have the special property $P^2 = P$. When we project a second time, the projection p stays exactly the same. Use equation (8) for P:

$$P^2 = A\,(A^{\mathrm{T}}A)^{-1}A^{\mathrm{T}}A\,(A^{\mathrm{T}}A)^{-1}A^{\mathrm{T}} = A\,(A^{\mathrm{T}}A)^{-1}A^{\mathrm{T}} = P. \tag{9}$$

The Third Way to Compute $\widehat{x}$: Gram-Schmidt

The columns of A are still assumed to be independent: $r = n$. But they are not assumed to be orthogonal! Then $A^{\mathrm{T}}A$ is not a diagonal matrix and solving $A^{\mathrm{T}}A\widehat{x} = A^{\mathrm{T}}b$ needs work. Our third approach **orthogonalizes the columns of A**, and then $\widehat{x}$ is easy to find.

You could say: The work is now in producing orthogonal (even orthonormal) columns. Exactly true. The operation count is actually doubled compared to $A^{\mathrm{T}}A\widehat{x} = A^{\mathrm{T}}b$, but orthogonal vectors provide numerical stability. Stability becomes important when $A^{\mathrm{T}}A$ is nearly singular. The **condition number** of $A^{\mathrm{T}}A$ is its norm $||A^{\mathrm{T}}A||$ times $||(A^{\mathrm{T}}A)^{-1}||$. When this number σ_1^2/σ_n^2 is large, it is wise to orthogonalize the columns of A in advance. Then work with an orthogonal matrix Q.

The condition number of Q is $||Q||$ times $||Q^{-1}||$. *Those norms equal* 1 *: best possible.*

Here is the famous Gram-Schmidt idea, starting with A and ending with Q. Independent columns $a_1, \ldots, a_n$ lead to orthonormal $q_1, \ldots, q_n$. This is a fundamental computation in linear algebra. The first step is $q_1 = a_1/||a_1||$. That is a unit vector: $||q_1|| = 1$. Then subtract from a_2 its component in the q_1 direction:

Gram-Schmidt step	Orthogonalize	$A_2 = a_2 - (a_2^{\mathrm{T}} q_1)\, q_1$	(10)
	Normalize	$q_2 = A_2/||A_2||$	(11)

Subtracting that component $(a_2^{\mathrm{T}} q_1)\, q_1$ produced the vector A_2 orthogonal to q_1:

$$(a_2 - (a_2^{\mathrm{T}} q_1)\, q_1)^{\mathrm{T}} q_1 = a_2^{\mathrm{T}} q_1 - a_2^{\mathrm{T}} q_1 = 0 \;\text{ since }\; q_1^{\mathrm{T}} q_1 = 1.$$

The algorithm goes onward to a_3 and A_3 and q_3, normalizing each time to make $||q|| = 1$. Subtracting the components of a_3 along q_1 and q_2 leaves A_3:

$$\textbf{Orthogonalize}\quad A_3 = a_3 - (a_3^{\mathrm{T}} q_1)\, q_1 - (a_3^{\mathrm{T}} q_2)\, q_2 \quad \textbf{Normalize}\quad q_3 = \frac{A_3}{||A_3||} \tag{12}$$

$$A_3^{\mathrm{T}} q_1 = A_3^{\mathrm{T}} q_2 = 0 \quad\text{and}\quad ||q_3|| = 1$$

Each q_k is a combination of a_1 to a_k. Then each a_k is a combination of q_1 to q_k.

$$\textbf{a's from q's}\qquad \begin{aligned} a_1 &= ||a_1||\, q_1 \\ a_2 &= (a_2^{\mathrm{T}} q_1)\, q_1 + ||A_2||\, q_2 \\ a_3 &= (a_3^{\mathrm{T}} q_1)\, q_1 + (a_3^{\mathrm{T}} q_2)\, q_2 + ||A_3||\, q_3 \end{aligned} \tag{13}$$

Those equations tell us that **the matrix $R = Q^{\mathrm{T}}A$ with $r_{ij} = q_i^{\mathrm{T}}a_j$ is upper triangular :**

$$\begin{bmatrix} a_1 & a_2 & a_3 \end{bmatrix} = \begin{bmatrix} q_1 & q_2 & q_3 \end{bmatrix} \begin{bmatrix} r_{11} & r_{12} & r_{13} \\ 0 & r_{22} & r_{23} \\ 0 & 0 & r_{33} \end{bmatrix} \quad \text{is } \boldsymbol{A = QR}. \tag{14}$$

Gram-Schmidt produces orthonormal q's from independent a's. Then $A = QR$.

If $A = QR$ then $R = Q^{\mathrm{T}}A$ = inner products of q's with a's! Later a's are not involved in earlier q's, so R is triangular. And certainly $A^{\mathrm{T}}A = R^{\mathrm{T}}Q^{\mathrm{T}}QR = R^{\mathrm{T}}R$:

$$\textbf{The least squares solution to} \quad Ax = b \ \textbf{ is } \ \widehat{x} = R^{-1}Q^{\mathrm{T}}b.$$

The MATLAB command is $[Q,\ R] = qr(A)$. Every $r_{ij} = q_i^{\mathrm{T}}a_j$ because $R = Q^{\mathrm{T}}A$. The vector $\widehat{x} = (A^{\mathrm{T}}A)^{-1}A^{\mathrm{T}}b$ is $(R^{\mathrm{T}}R)^{-1}R^{\mathrm{T}}Q^{\mathrm{T}}b$. This is exactly $\widehat{x} = R^{-1}Q^{\mathrm{T}}b$.

Gram-Schmidt with Column Pivoting

That straightforward description of Gram-Schmidt worked with the columns of A in their original order $a_1, a_2, a_3, \ldots$ This could be dangerous ! We could never live with a code for elimination that didn't allow row exchanges. Then roundoff error could wipe us out.

Similarly, each step of Gram-Schmidt should begin with a new column that is as independent as possible of the columns already processed. **We need column exchanges to pick the largest remaining column. Change the order of columns as we go.**

To choose correctly from the remaining columns of A, we make a simple change in Gram-Schmidt :

Old Accept column a_j as next. Subtract its components in the directions q_1 to q_{j-1}

New When q_{j-1} is found, subtract the q_{j-1} component from **all remaining columns**

This might look like more work, but it's not. Sooner or later we had to remove $(a_i^{\mathrm{T}}q_{j-1})q_{j-1}$ from each remaining column a_i. Now we do it sooner—as soon as we know q_{j-1}. Then we have a free choice of the next column to work with, and we choose the largest.

Elimination Row exchanges on A left us with $PA = LU$ (permutation matrix P)
Gram-Schmidt Column exchanges leave us with $AP = QR$ (permutation matrix P)

Here is the situation after $j-1$ Gram-Schmidt steps with column pivoting. We have $j-1$ orthogonal unit vectors q_1 to q_{j-1} in the columns of a matrix Q_{j-1}. We have the square matrix R_{j-1} that combines those columns of Q_{j-1} to produce $j-1$ columns of A. *They might not be the* first $j-1$ *columns of* A—we are optimizing the column order. All the remaining columns of A have been *orthogonalized* against the vectors q_1 to q_{j-1}.

Step j. Choose the largest of the remaining columns of A. Normalize it to length 1.

This is q_j. Then from each of the $n-j$ vectors still waiting to be chosen, subtract the component in the direction of this latest q_j. Ready now for step $j+1$.

We will follow Gunnar Martinsson's 2016 course notes for APPM 5720, to express step j in pseudocode. The original A is A_0 and the matrices Q_0 and R_0 are empty.

Step j is the following loop, which starts with A_{j-1} and ends at A_j. The code stops after j reaches $\min(m,n)$. **Here is column pivoting in Gram-Schmidt:**

i	$= \text{argmax}\,\|\|A_{j-1}(:,\ \ell)\|\|$ finds the largest column not yet chosen for the basis
$\boldsymbol{q}_j$	$= A_{j-1}(:,\ i)/\|\|A_{j-1}(:,\ i)\|\|$ normalizes that column to give the new unit vector $\boldsymbol{q}_j$
Q_j	$= \begin{bmatrix} Q_{j-1} & \boldsymbol{q}_j \end{bmatrix}$ updates Q_{j-1} with the new orthogonal unit vector $\boldsymbol{q}_j$
$\boldsymbol{r}_j$	$= \boldsymbol{q}_j^{\mathrm{T}} A_{j-1}$ finds the row of inner products of $\boldsymbol{q}_j$ with remaining columns of A
R_j	$= \begin{bmatrix} R_{j-1} \\ \boldsymbol{r}_j \end{bmatrix}$ updates R_{j-1} with the new row of inner products
A_j	$= A_{j-1} - \boldsymbol{q}_j \boldsymbol{r}_j$ subtracts the new rank-one piece from each column to give A_j

When this loop ends, we have Q and R and $A = QR$. This R is a *permutation* of an upper triangular matrix. (It will be upper triangular if the largest columns in Step 1 come first, so each $i = j$.) The actual output can be an upper triangular matrix plus a vector with the numbers $1, \ldots, n$ in the permutation order we need to know, to construct R.

In practice, this QR algorithm with pivoting is made safer by *reorthonormalizing*:

$$\boldsymbol{q}_j = \boldsymbol{q}_j - Q_{j-1}(Q_{j-1}^{\mathrm{T}} \boldsymbol{q}_j)$$
$$\boldsymbol{q}_j = \boldsymbol{q}_j \,/\, \|\boldsymbol{q}_j\| \quad \textbf{(to make sure!)}$$

There is a similar reordering for "QR with Householder matrices" to reduce roundoff error. You have seen the essential point of pivoting: **good columns come first.**

Question: Both Q from Gram-Schmidt and U from the SVD contain an orthonormal basis for the column space **C**(A). *Are they likely to be the same basis?*

Answer: No, they are not the same. The columns of U are eigenvectors of AA^{T}. You cannot find eigenvectors (or eigenvalues) in a finite number of exact "arithmetic" steps for matrices of size $n > 4$. The equation $\det(A - \lambda I) = 0$ will be 5th degree or higher: *No formula can exist for the roots λ of a 5th degree equation (Abel).* Gram-Schmidt just requires inner products and square roots so Q must be different from U.

In the past, computing a nearly accurate eigenvalue took a much larger multiple of n^3 floating point operations than elimination or Gram-Schmidt. That is not true now.

Another Way to Q : Householder Reflections

Without column exchanges, Gram-Schmidt subtracts from each vector $\boldsymbol{a}_j$ its components in the directions $\boldsymbol{q}_1$ to $\boldsymbol{q}_{j-1}$ that are already set. For numerical stability those subtractions must be done *one at a time*. Here is $\boldsymbol{q}_3$ with *two separate subtractions from* $\boldsymbol{a}_3$:

$$\text{Compute } \boldsymbol{a}_3 - (\boldsymbol{a}_3^{\mathrm{T}}\boldsymbol{q}_1)\,\boldsymbol{q}_1 = \boldsymbol{a}_3' \text{ and } \boldsymbol{a}_3' - (\boldsymbol{a}_3'^{\mathrm{T}}\boldsymbol{q}_2)\,\boldsymbol{q}_2 = \boldsymbol{A}_3 \text{ and } \boldsymbol{q}_3 = \boldsymbol{A}_3/||\boldsymbol{A}_3||.$$

But still the computed $\boldsymbol{q}_3$ won't be *exactly* orthogonal to $\boldsymbol{q}_1$ and $\boldsymbol{q}_2$. That is just hard. The good way to create an exactly orthogonal Q is to build it that way, as Householder did :

$$\textbf{Householder reflection matrix} \qquad H = I - 2\frac{\boldsymbol{v}\boldsymbol{v}^{\mathrm{T}}}{||\boldsymbol{v}||^2} = I - 2\boldsymbol{u}\boldsymbol{u}^{\mathrm{T}}. \tag{15}$$

$\boldsymbol{u}$ is the unit vector $\boldsymbol{v}/||\boldsymbol{v}||$. Then $\boldsymbol{u}\boldsymbol{u}^{\mathrm{T}} = \boldsymbol{v}\boldsymbol{v}^{\mathrm{T}}/||\boldsymbol{v}||^2$. **$H$ is symmetric and orthogonal.**

$$H^{\mathrm{T}}H = (I - 2\boldsymbol{u}\boldsymbol{u}^{\mathrm{T}})^2 = I - 4\boldsymbol{u}\boldsymbol{u}^{\mathrm{T}} + 4\boldsymbol{u}(\boldsymbol{u}^{\mathrm{T}}\boldsymbol{u})\,\boldsymbol{u}^{\mathrm{T}} = I. \tag{16}$$

Key point: If $\boldsymbol{v} = \boldsymbol{a} - \boldsymbol{r}$ and $||\boldsymbol{a}|| = ||\boldsymbol{r}||$ then $H\boldsymbol{a} = \boldsymbol{r}$ (see Problem 6). To produce zeros in column k below the main diagonal, use this H_k with $\boldsymbol{v} = (\boldsymbol{a}_{\text{lower}} - \boldsymbol{r}_{\text{lower}})$ and $\boldsymbol{u} = \boldsymbol{v}/||\boldsymbol{v}||$. **We are creating zeros in $H\boldsymbol{A}$.** And we have a choice of signs in $\boldsymbol{r}$:

$$\boldsymbol{H}_k[\,\textbf{column } \boldsymbol{k}\,] = \begin{bmatrix} I & \\ & I - 2\boldsymbol{u}\boldsymbol{u}^{\mathrm{T}} \end{bmatrix}\begin{bmatrix} \boldsymbol{a}_{\text{upper}} \\ \boldsymbol{a}_{\text{lower}} \end{bmatrix} = \begin{bmatrix} \boldsymbol{a}_{\text{upper}} \\ \pm||\boldsymbol{a}_{\textbf{lower}}|| \\ \boldsymbol{n} - \boldsymbol{k} \textbf{ zeros} \end{bmatrix} = \boldsymbol{r}_k \tag{17}$$

A_1 is the original matrix A. The first step produces H_1A and the next step finds H_2H_1A. By marching across the columns, the reflections H_1 to H_{n-1} multiply to give Q. And in opposite order, those reflections create zeros in A and produce a triangular matrix R:

$$H_{n-1}\ldots H_2H_1A = \begin{bmatrix} & & & \\ r_1 & r_2 & \cdot\ \cdot & r_n \\ & & & \end{bmatrix} \quad \text{becomes } Q^{\mathrm{T}}A = R. \tag{18}$$

The key is to keep a record of the H_j by storing only the vectors $\boldsymbol{v}_j = \boldsymbol{a}_j - \boldsymbol{r}_j$, not the matrix. Then every $H_j = I - 2\boldsymbol{v}_j\boldsymbol{v}_j^{\mathrm{T}}/||\boldsymbol{v}_j||^2$ is exactly orthogonal. To solve $A\boldsymbol{x} = \boldsymbol{b}$ by least squares, you can start with the matrix $[A \quad \boldsymbol{b}]$ as in elimination. Multiply by all the H's to reach $[R \quad Q^{\mathrm{T}}\boldsymbol{b}]$. Then solve the triangular system $R\widehat{\boldsymbol{x}} = Q^{\mathrm{T}}\boldsymbol{b}$ by ordinary back substitution. **You have found the least squares solution $\widehat{\boldsymbol{x}} = \boldsymbol{R}^{-1}\boldsymbol{Q}^{\mathrm{T}}\boldsymbol{b}$.**

Example $A = \begin{bmatrix} 4 & x \\ 3 & x \end{bmatrix}$ has $\boldsymbol{a} = \begin{bmatrix} 4 \\ 3 \end{bmatrix}$ and $\boldsymbol{r} = \begin{bmatrix} 5 \\ 0 \end{bmatrix}$ and $||\boldsymbol{a}|| = ||\boldsymbol{r}||$

$$\text{Choose } \boldsymbol{v} = \boldsymbol{a} - \boldsymbol{r} = \begin{bmatrix} -1 \\ 3 \end{bmatrix} \text{ and } \boldsymbol{u} = \frac{\boldsymbol{v}}{||\boldsymbol{v}||} = \frac{1}{\sqrt{10}}\begin{bmatrix} -1 \\ 3 \end{bmatrix}$$

$$\text{Then } \boldsymbol{H} = I - 2\boldsymbol{u}\boldsymbol{u}^{\mathrm{T}} = \frac{1}{5}\begin{bmatrix} 4 & 3 \\ 3 & -4 \end{bmatrix} = \boldsymbol{Q}^{\mathrm{T}} \text{ and } \boldsymbol{H}\boldsymbol{A} = \begin{bmatrix} 5 & x \\ 0 & x \end{bmatrix} = \boldsymbol{R}$$

Least Squares with a Penalty Term

If A has dependent columns and $Ax = \mathbf{0}$ has nonzero solutions, then $A^{\mathrm{T}}A$ cannot be invertible. This is where we need A^+. A gentle approach will "regularize" least squares:

Penalty term Minimize $\|\|Ax-b\|\|^2+\delta^2\,\|\|x\|\|^2$ Solve $(\boldsymbol{A^{\mathrm{T}}A+\delta^2 I})\,\widehat{x}=A^{\mathrm{T}}\boldsymbol{b}$	(19)

This fourth approach to least squares is called *ridge regression.* We will show that $\widehat{x}$ approaches the shortest solution $x^+ = A^+ b$ as the penalty disappears (δ goes to zero).

Section III.4 describes a different penalty: **Add the ℓ^1 norm $\lambda\|\|x\|\|_1$.** That has a beautiful result: Instead of x^+ with minimum norm, *the ℓ^1 norm leads to sparse solutions.*

The Pseudoinverse A^+ is the Limit of $(A^{\mathrm{T}}A+\delta^2 I)^{-1}A^{\mathrm{T}}$

Equation (19) creates the pseudoinverse A^+ from the positive definite matrices $A^{\mathrm{T}}A+\delta^2 I$. Those are invertible matrices up to the very last minute when $\delta = 0$. At that moment we have a sudden split in A^+. You see it best if A is a 1 by 1 matrix (*just a single number* σ):

$$\textbf{For } \delta > 0 \qquad (A^{\mathrm{T}}A+\delta^2 I)^{-1}A^{\mathrm{T}} = \left[\frac{\sigma}{\sigma^2+\delta^2}\right] \text{ is 1 by 1} \qquad \textbf{Now let } \delta \to 0$$

The limit is zero if $\sigma = 0$. The limit is $\dfrac{1}{\sigma}$ if $\sigma \neq 0$. This is exactly A^+ = **zero or** $\dfrac{1}{\sigma}$.

Now allow any diagonal matrix Σ. This is easy because all matrices stay diagonal. We are seeing the 1 by 1 case at every position along the main diagonal. Σ has positive entries σ_1 to σ_r and otherwise all zeros. The penalty makes the whole diagonal positive:

$$(\Sigma^{\mathrm{T}}\Sigma+\delta^2 I)^{-1}\Sigma^{\mathrm{T}} \text{ has positive diagonal entries } \frac{\sigma_i}{\sigma_i^2+\delta^2} \text{ and otherwise all zeros.}$$

Positive numbers approach $\frac{1}{\sigma_i}$. Zeros stay zero. When $\delta \to 0$ the limit is again Σ^+.

To prove that the limit is A^+ for *every matrix* A, bring in the SVD: $A = U\Sigma V^{\mathrm{T}}$. Substitute this matrix A into $(A^{\mathrm{T}}A+\delta^2 I)^{-1}A^{\mathrm{T}}$. The orthogonal matrices U and V move out of the way because $U^{\mathrm{T}} = U^{-1}$ and $V^{\mathrm{T}} = V^{-1}$:

$$A^{\mathrm{T}}A+\delta^2 I = V\Sigma^{\mathrm{T}}U^{\mathrm{T}}U\Sigma V^{\mathrm{T}}+\delta^2 I = V(\Sigma^{\mathrm{T}}\Sigma+\delta^2 I)V^{\mathrm{T}}$$

$$(A^{\mathrm{T}}A+\delta^2 I)^{-1}A^{\mathrm{T}} = V(\Sigma^{\mathrm{T}}\Sigma+\delta^2 I)^{-1}V^{\mathrm{T}}V\Sigma^{\mathrm{T}}U^{\mathrm{T}} = \boldsymbol{V}\left[(\Sigma^{\mathrm{T}}\Sigma+\delta^2 I)^{-1}\Sigma^{\mathrm{T}}\right]\boldsymbol{U^{\mathrm{T}}}$$

Now is the moment for $\delta \to 0$. The matrices V and U^{T} stay in their places. **The diagonal matrix in brackets approaches Σ^+** (this is exactly the diagonal case established above). **The limit of $(A^{\mathrm{T}}A+\delta^2 I)^{-1}A^{\mathrm{T}}$ is our pseudoinverse A^+.**

$$\boxed{\lim_{\delta\to 0}\ \boldsymbol{V}\left[(\boldsymbol{\Sigma^{\mathrm{T}}\Sigma+\delta^2 I})^{-1}\boldsymbol{\Sigma^{\mathrm{T}}}\right]\boldsymbol{U^{\mathrm{T}}} = \boldsymbol{V\Sigma^+U^{\mathrm{T}}} = \boldsymbol{A^+}.} \qquad (20)$$

The difficulty of computing A^+ is to know if a singular value is **zero or very small.** The diagonal entry in Σ^+ is zero or extremely large! Σ^+ and A^+ are far from being continuous functions of Σ and A. The value of A^+ is to warn us when λ or σ is very close to zero— then we often treat it as zero without knowing for sure.

Here are small matrices and their pseudoinverses, to bring home the point that A^+ does not follow all the rules for A^{-1}. It is *discontinuous* at the moment when singular values touch zero.

$$\textbf{From 0 to } \mathbf{2^{10}} \quad \begin{bmatrix} 2 & 0 \\ 0 & \mathbf{0} \end{bmatrix}^+ = \begin{bmatrix} 1/2 & 0 \\ 0 & \mathbf{0} \end{bmatrix} \quad \text{but} \quad \begin{bmatrix} 2 & 0 \\ 0 & \mathbf{2^{-10}} \end{bmatrix}^+ = \begin{bmatrix} 1/2 & 0 \\ 0 & \mathbf{2^{10}} \end{bmatrix}$$

It is not true that $(AB)^+ = B^+A^+$. Pseudoinverses do not obey all the rules of inverse matrices! They do obey $(A^{\mathrm{T}})^+ = (A^+)^{\mathrm{T}}$ and $(A^{\mathrm{T}}A)^+ = A^+(A^{\mathrm{T}})^+$.

If $A = \begin{bmatrix} 1 & 0 \end{bmatrix}$ and $B = \begin{bmatrix} 1 \\ 1 \end{bmatrix}$ then $(AB)^+$ is not equal to B^+A^+:

$$AB = \begin{bmatrix} 1 \end{bmatrix} \text{ and } (AB)^+ = \begin{bmatrix} 1 \end{bmatrix} \text{ but } B^+ = \begin{bmatrix} \frac{1}{2} & \frac{1}{2} \end{bmatrix} \text{ and } A^+ = \begin{bmatrix} 1 \\ 0 \end{bmatrix} \text{ and } B^+A^+ = \begin{bmatrix} \frac{1}{2} \end{bmatrix}.$$

If C has full column rank and R has full row rank then $(CR)^+ = R^+C^+$ is true.

This is a surprisingly useful fact. It means that the pseudoinverse of any matrix can be computed without knowing its SVD (and without computing any eigenvalues). Here is the reasoning. The first step came in Section I.1, page 4 of the book.

Every m by n matrix A of rank r can be factored into $A = CR = (m \times r)(r \times n)$.

The matrix C gives a column space basis. R gives a row space basis.

$C^+ = (C^{\mathrm{T}}C)^{-1}C^{\mathrm{T}}$ = left inverse of C and $R^+ = R^{\mathrm{T}}(RR^{\mathrm{T}})^{-1}$ = right inverse of R.

Then $A = CR$ has $A^+ = R^+C^+$ computed without eigenvalues or singular values.

The catch is that we need exact computations (no roundoff) to know the exact rank r. The pseudoinverse is discontinuous when the rank suddenly drops. The large number $1/\sigma$ suddenly becomes zero. Always $\mathbf{0}^+ = \mathbf{0}$.

In the example above with $(AB)^+ \neq B^+A^+$, you could verify that if we put those matrices in reverse order then $(BA)^+ = A^+B^+$ is true.

The pseudoinverse is also called the Moore-Penrose inverse. In MATLAB it is pinv(A).

Weighted Least Squares

By choosing to minimize the error $||\boldsymbol{b} - A\boldsymbol{x}||^2$, we are implicitly assuming that all the observations $b_1, \ldots, b_m$ are *equally reliable*. The errors in b_i have mean = average value = 0, and the variances are equal. That assumption could be false. Some b_i may have less noise and more accuracy. The variances $\sigma_1^2, \ldots, \sigma_m^2$ in those m measurements might not be equal. In this case we should assign greatest weight to the most reliable data (the b's with the smallest variance).

The natural choice is to *divide equation k by σ_k*. Then b_k/σ_k has variance 1. All observations are normalized to the same unit variance. Note that σ is the accepted notation for the square root of the variance (see Section V.1 for variances and V.4 for covariances). Here σ is a variance and *not* a singular value of A.

When the observations b_k are independent, all covariances are zero. The only nonzeros in the variance-covariance matrix C are $\sigma_1^2, \ldots \sigma_m^2$ on the diagonal. So our weights $1/\sigma_k$ have effectively multiplied $A\boldsymbol{x} = \boldsymbol{b}$ by the matrix $C^{-1/2}$.

Multiplying by $C^{-1/2}$ is still the right choice when C has nonzeros off the diagonal.

We are "whitening" the data. The quantity to minimize is not $||\boldsymbol{b} - A\boldsymbol{x}||^2$. That error should be weighted by C^{-1}.

Weighted least squares minimizes $||C^{-1/2}(\boldsymbol{b} - A\boldsymbol{x})||^2 = (\boldsymbol{b} - A\boldsymbol{x})^{\mathrm{T}} C^{-1}(\boldsymbol{b} - A\boldsymbol{x})$.

Now the normal equation $A^{\mathrm{T}}A\widehat{\boldsymbol{x}} = A^{\mathrm{T}}\boldsymbol{b}$ for the best $\widehat{\boldsymbol{x}}$ includes C^{-1} =inverse covariances:

$$\textbf{Weighted normal equation} \qquad A^{\mathrm{T}} C^{-1} A\widehat{\boldsymbol{x}} = A^{\mathrm{T}} C^{-1}\boldsymbol{b}. \tag{21}$$

Example 2 Suppose $x = b_1$ and $x = b_2$ are independent noisy measurements of the number x. We multiply those equations by their weights $1/\sigma$.

Solve $Ax = \begin{bmatrix} 1 \\ 1 \end{bmatrix} x = \begin{bmatrix} b_1 \\ b_2 \end{bmatrix}$ by weighted least squares. The weights are $1/\sigma_1$ and $1/\sigma_2$.

The equations become $x/\sigma_1 = b_1/\sigma_1$ and $x/\sigma_2 = b_2/\sigma_2$: $\begin{bmatrix} 1/\sigma_1 \\ 1/\sigma_2 \end{bmatrix} x = \begin{bmatrix} b_1/\sigma_1 \\ b_2/\sigma_2 \end{bmatrix}$

$$\textbf{Weighted normal equation} \qquad \left[\frac{1}{\sigma_1^2} + \frac{1}{\sigma_2^2} \right] \widehat{\boldsymbol{x}} = \frac{b_1}{\sigma_1^2} + \frac{b_2}{\sigma_2^2}. \tag{22}$$

The statistically best estimate of x is a weighted average of b_1 and b_2 :

$$\widehat{\boldsymbol{x}} = \left(\frac{\sigma_1^2 \sigma_2^2}{\sigma_1^2 + \sigma_2^2} \right) \left(\frac{b_1}{\sigma_1^2} + \frac{b_2}{\sigma_2^2} \right) = \frac{\sigma_2^2 b_1 + \sigma_1^2 b_2}{\sigma_1^2 + \sigma_2^2}$$

Problem Set II.2

1 (A new proof that $\mathbf{N}(A^{\mathrm{T}}A) = \mathbf{N}(A)$) Suppose $A^{\mathrm{T}}A\boldsymbol{x} = \mathbf{0}$. Then $A\boldsymbol{x}$ is in the nullspace of A^{T}. But always $A\boldsymbol{x}$ is in the column space of A. Those two subspaces are orthogonal, so *if* $A^{\mathrm{T}}A\boldsymbol{x} = \mathbf{0}$ *then* $A\boldsymbol{x} = \mathbf{0}$.

Prove the opposite statement to reach $\mathbf{N}(A^{\mathrm{T}}A) = \mathbf{N}(A)$.

2 Why do A and A^+ have the same rank? If A is square, do A and A^+ have the same eigenvectors? What are the eigenvalues of A^+?

3 From A and A^+ show that A^+A is correct and $(A^+A)^2 = A^+A =$ projection.

$$A = \sum \sigma_i \boldsymbol{u}_i \boldsymbol{v}_i^{\mathrm{T}} \qquad A^+ = \sum \frac{\boldsymbol{v}_i \boldsymbol{u}_i^{\mathrm{T}}}{\sigma_i} \qquad A^+A = \sum \boldsymbol{v}_i \boldsymbol{v}_i^{\mathrm{T}} \qquad AA^+ = \sum \boldsymbol{u}_i \boldsymbol{u}_i^{\mathrm{T}}$$

4 Which matrices have $A^+ = A$? Why are they square? Look at A^+A.

5 Suppose A has independent columns (rank $r = n$; nullspace = zero vector).

(a) Describe the m by n matrix Σ in $A = U\Sigma V^{\mathrm{T}}$. How many nonzeros in Σ?

(b) Show that $\Sigma^{\mathrm{T}}\Sigma$ is invertible by finding its inverse.

(c) Write down the n by m matrix $(\Sigma^{\mathrm{T}}\Sigma)^{-1}\Sigma^{\mathrm{T}}$ and identify it as Σ^+.

(d) Substitute $A = U\Sigma V^{\mathrm{T}}$ into $(A^{\mathrm{T}}A)^{-1}A^{\mathrm{T}}$ and identify that matrix as A^+.

$\boldsymbol{A^{\mathrm{T}}A\widehat{x} = A^{\mathrm{T}}b}$ **leads to** $\boldsymbol{A^+ = (A^{\mathrm{T}}A)^{-1}A^{\mathrm{T}}}$**, but only if** $\boldsymbol{A}$ **has rank** $\boldsymbol{n}$**.**

6 The Householder matrix H in equation (17) chooses $\boldsymbol{v} = \boldsymbol{a} - \boldsymbol{r}$ with $||\boldsymbol{a}||^2 = ||\boldsymbol{r}||^2$. Check that this choice of the vector $\boldsymbol{v}$ always gives $H\boldsymbol{a} = \boldsymbol{r}$:

$$\text{Verify that } H\boldsymbol{a} = \boldsymbol{a} - 2\,\frac{(\boldsymbol{a}-\boldsymbol{r})\,(\boldsymbol{a}-\boldsymbol{r})^{\mathrm{T}}}{(\boldsymbol{a}-\boldsymbol{r})^{\mathrm{T}}(\boldsymbol{a}-\boldsymbol{r})}\boldsymbol{a} \text{ reduces to } \boldsymbol{r}.$$

7 According to Problem 6, which n by n Householder matrix H gives $H\boldsymbol{a} = \begin{bmatrix} ||\boldsymbol{a}|| \\ \textbf{zeros} \end{bmatrix}$?

8 What multiple of $\boldsymbol{a} = \begin{bmatrix} 1 \\ 1 \end{bmatrix}$ should be subtracted from $\boldsymbol{b} = \begin{bmatrix} 4 \\ 0 \end{bmatrix}$ to make the result $\boldsymbol{A_2}$ orthogonal to $\boldsymbol{a}$? Sketch a figure to show $\boldsymbol{a}$, $\boldsymbol{b}$, and $\boldsymbol{A_2}$.

9 Complete the Gram-Schmidt process in Problem 8 by computing $\boldsymbol{q}_1 = \boldsymbol{a}/||\boldsymbol{a}||$ and $\boldsymbol{A}_2 = \boldsymbol{b} - (\boldsymbol{a}^{\mathrm{T}}\boldsymbol{q}_1)\boldsymbol{q}_1$ and $\boldsymbol{q}_2 = \boldsymbol{A_2}/||\boldsymbol{A_2}||$ and factoring into QR:

$$\begin{bmatrix} 1 & 4 \\ 1 & 0 \end{bmatrix} = \begin{bmatrix} \boldsymbol{q}_1 & \boldsymbol{q}_2 \end{bmatrix} \begin{bmatrix} ||\boldsymbol{a}|| & ? \\ 0 & ||\boldsymbol{A_2}|| \end{bmatrix}.$$

10 If $A = QR$ then $A^{\mathrm{T}}A = R^{\mathrm{T}}R =$ _____ triangular times _____ triangular. *Gram-Schmidt on* A *corresponds to elimination on* $A^{\mathrm{T}}A$.

11 *If* $Q^{\mathrm{T}}Q = I$ *show that* $Q^{\mathrm{T}} = Q^+$. *If* $A = QR$ *for invertible* R, *show that* $QQ^+ = AA^+$. *On the last page* 155 *of Part* II, *this will be the key to computing an SVD.*

This page is devoted to the simplest and most important application of least squares: **Fitting a straight line to data**. A line $b = C + Dt$ has $n = 2$ parameters C and D. We are given $m > 2$ measurements b_i at m different times t_i. The equations $Ax = b$ (unsolvable) and $A^{\mathrm{T}}A\widehat{x} = A^{\mathrm{T}}b$ (solvable) are

$$Ax = \begin{bmatrix} 1 & t_1 \\ 1 & t_2 \\ : & : \\ 1 & t_m \end{bmatrix} \begin{bmatrix} C \\ D \end{bmatrix} = \begin{bmatrix} b_1 \\ b_2 \\ : \\ b_m \end{bmatrix} \qquad A^{\mathrm{T}}A\widehat{x} = \begin{bmatrix} m & \sum t_i \\ \sum t_i & \sum t_i^2 \end{bmatrix} \begin{bmatrix} \widehat{C} \\ \widehat{D} \end{bmatrix} = \begin{bmatrix} \sum b_i \\ \sum b_i t_i \end{bmatrix}.$$

The column space $\mathbf{C}(A)$ is a 2-dimensional plane in $\mathbf{R}^m$. The vector $\boldsymbol{b}$ is in this column space if and only if the m points (t_i, b_i) actually lie on a straight line. In that case only, $Ax = \boldsymbol{b}$ is solvable: the line is $C + Dt$. **Always $\boldsymbol{b}$ is projected to the closest $\boldsymbol{p}$ in $\mathbf{C}(A)$.**

The best line (the least squares fit) passes through the points (t_i, p_i). The error vector $e = A\widehat{x} - b$ has components $b_i - p_i$. And e is perpendicular to p.

There are two important ways to draw this least squares regression problem. One way shows the best line $b = \widehat{C} + \widehat{D}t$ and the errors e_i (vertical distances to the line). The second way is in $\mathbf{R}^m = m$-dimensional space. There we see the data vector $\boldsymbol{b}$, its projection $\boldsymbol{p}$ onto $\mathbf{C}(A)$, and the error vector e. This is a right triangle with $||\boldsymbol{p}||^2 + ||e||^2 = ||\boldsymbol{b}||^2$.

Problems 12 to 22 use four data points $\boldsymbol{b} = (0, 8, 8, 20)$ to bring out the key ideas.

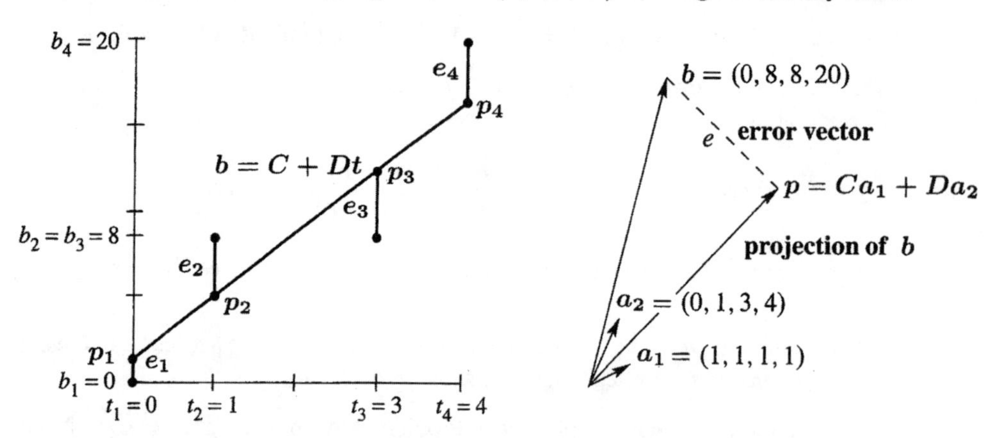

Figure II.3: The closest line $C + Dt$ in the $t - b$ plane matches $Ca_1 + Da_2$ in $\mathbf{R}^4$.

12 With $b = 0, 8, 8, 20$ at $t = 0, 1, 3, 4$, set up and solve the normal equations $A^{\mathrm{T}}A\widehat{x} = A^{\mathrm{T}}b$. For the best straight line in Figure II.3a, find its four heights p_i and four errors e_i. What is the minimum squared error $E = e_1^2 + e_2^2 + e_3^2 + e_4^2$?

13 (Line $C + Dt$ does go through p's) With $b = 0, 8, 8, 20$ at times $t = 0, 1, 3, 4$, write down the four equations $A\boldsymbol{x} = \boldsymbol{b}$ (unsolvable). Change the measurements to $p = 1, 5, 13, 17$ and find an exact solution to $A\widehat{\boldsymbol{x}} = \boldsymbol{p}$.

14 Check that $\boldsymbol{e} = \boldsymbol{b} - \boldsymbol{p} = (-1, 3, -5, 3)$ is perpendicular to both columns of the same matrix A. What is the shortest distance $\|\boldsymbol{e}\|$ from $\boldsymbol{b}$ to the column space of A?

15 (By calculus) Write down $E = \|A\boldsymbol{x} - \boldsymbol{b}\|^2$ as a sum of four squares—the last one is $(C + 4D - 20)^2$. Find the derivative equations $\partial E/\partial C = 0$ and $\partial E/\partial D = 0$. Divide by 2 to obtain the normal equations $A^{\mathrm{T}} A\widehat{\boldsymbol{x}} = A^{\mathrm{T}}\boldsymbol{b}$.

16 Find the height C of the best *horizontal line* to fit $\boldsymbol{b} = (0, 8, 8, 20)$. An exact fit would solve the unsolvable equations $C = 0$, $C = 8$, $C = 8$, $C = 20$. Find the 4 by 1 matrix A in these equations and solve $A^{\mathrm{T}} A\widehat{x} = A^{\mathrm{T}}\boldsymbol{b}$. Draw the horizontal line at height $\widehat{x} = C$ and the four errors in $\boldsymbol{e}$.

17 Project $\boldsymbol{b} = (0, 8, 8, 20)$ onto the line through $\boldsymbol{a} = (1, 1, 1, 1)$. Find $\widehat{x} = \boldsymbol{a}^{\mathrm{T}}\boldsymbol{b}/\boldsymbol{a}^{\mathrm{T}}\boldsymbol{a}$ and the projection $\boldsymbol{p} = \widehat{x}\boldsymbol{a}$. Check that $\boldsymbol{e} = \boldsymbol{b} - \boldsymbol{p}$ is perpendicular to $\boldsymbol{a}$, and find the shortest distance $\|\boldsymbol{e}\|$ from $\boldsymbol{b}$ to the line through $\boldsymbol{a}$.

18 Find the closest line $b = Dt$, *through the origin*, to the same four points. An exact fit would solve $D \cdot 0 = 0, D \cdot 1 = 8, D \cdot 3 = 8, D \cdot 4 = 20$. Find the 4 by 1 matrix and solve $A^{\mathrm{T}} A\widehat{x} = A^{\mathrm{T}}\boldsymbol{b}$. Redraw Figure II.3a showing the best line $b = Dt$.

19 Project $\boldsymbol{b} = (0, 8, 8, 20)$ onto the line through $\boldsymbol{a} = (0, 1, 3, 4)$. Find $\widehat{x} = D$ and $\boldsymbol{p} = \widehat{x}\boldsymbol{a}$. The best C in Problem 16 and the best D in Problem 18 *do not agree* with the best $(\widehat{C}, \widehat{D})$ in Problems 11–14. That is because the two columns $(1, 1, 1, 1)$ and $(0, 1, 3, 4)$ are _____ perpendicular.

20 For the closest parabola $b = C + Dt + Et^2$ to the same four points, write down the unsolvable equations $A\boldsymbol{x} = \boldsymbol{b}$ in three unknowns $\boldsymbol{x} = (C, D, E)$. Set up the three normal equations $A^{\mathrm{T}} A\widehat{\boldsymbol{x}} = A^{\mathrm{T}}\boldsymbol{b}$ (solution not required). In Figure II.3a you are now fitting a parabola to 4 points—what is happening in Figure II.3b?

21 For the closest cubic $b = C + Dt + Et^2 + Ft^3$ to the same four points, write down the four equations $A\boldsymbol{x} = \boldsymbol{b}$. Solve them by elimination. In Figure II.3a this cubic now goes exactly through the points. What are $\boldsymbol{p}$ and $\boldsymbol{e}$?

22 The averages of the t_i and b_i are $\bar{t} = 2$ and $\bar{b} = 9$. Verify that $C + D\bar{t} = \bar{b}$. Explain !

(a) Verify that the best line goes through the center point $(\bar{t}, \bar{b}) = (2, 9)$.

(b) Explain why $C + D\bar{t} = \bar{b}$ comes from the first equation in $A^{\mathrm{T}} A\widehat{\boldsymbol{x}} = A^{\mathrm{T}}\boldsymbol{b}$.

II.3 Three Bases for the Column Space

This section touches on serious computational questions. The matrices get large. Their rank is also large if there is random noise. But the effective rank, when the noise is removed, may be considerably smaller than m and n. Modern linear algebra has developed fast algorithms to solve $A\boldsymbol{x} = \boldsymbol{b}$ and $A\boldsymbol{x} = \lambda\boldsymbol{x}$ and $A\boldsymbol{v} = \sigma\boldsymbol{u}$ for large matrices.

Mostly we will leave special algorithms to the professionals. Numerical linear algebra has developed quickly and well—we are safe with the experts. But you and I can recognize some basic rules of common computational sense, including these two :

1. Don't use $A^{\mathrm{T}}A$ and AA^{T} when you can operate directly on A.

2. Don't assume that the original order of the rows and columns is necessarily best.

The first warning would apply to least squares and the SVD (and computational statistics). By forming $A^{\mathrm{T}}A$ we are squaring the condition number σ_1/σ_n. This measures the sensitivity and vulnerability of A. And for really large matrices the cost of computing and storing $A^{\mathrm{T}}A$ is just unthinkable. It is true that the stiffness matrix of mechanics and the conductance matrix of electronics and the graph Laplacian matrix for networks have the form $A^{\mathrm{T}}A$ or $A^{\mathrm{T}}CA$ (with physical constants in C). But for data matrices we want "square root algorithms" that work directly with the basic matrix A.

What do we really need from A ? Often the answer goes to the heart of pure and applied algebra : **We need a good basis for the column space**. From that starting point we can do anything ! For $A\boldsymbol{x} = \boldsymbol{b}$, we can find a combination of basic columns that comes near $\boldsymbol{b}$. For $A\boldsymbol{v} = \sigma\boldsymbol{u}$, we can compute accurate singular vectors.

Again, what we cannot do is to accept the first r independent columns of A as automatically a good basis for computations in the column space $\mathbf{C}(A)$.

Three Good Bases

Let me reveal immediately the three bases that we propose to study. You may be tempted to put them in the order gold, silver, bronze.

In part those three prizes represent the guaranteed quality of the three bases. But they also suggest that the SVD has highest cost. All three are winners in the construction of a basis for the column space.

1. **Singular vectors** $\boldsymbol{u}_1, \ldots, \boldsymbol{u}_r$ from the SVD, with $\sigma_1 \geq \sigma_2 \geq \ldots \geq \sigma_r$.
2. **Orthonormal vectors** $\boldsymbol{q}_1, \ldots, \boldsymbol{q}_r$ from Gram-Schmidt. Use column pivoting !
3. **Independent columns** $\boldsymbol{c}_1, \ldots, \boldsymbol{c}_r$ taken directly from A after column exchanges.

Each choice of basis for $\mathbf{C}(A)$ gives the column matrix in a "rank-revealing factorization"

$$A = \text{column matrix times row matrix}: (\boldsymbol{m} \textbf{ by } \boldsymbol{n}) \textbf{ equals } (\boldsymbol{m} \textbf{ by } \boldsymbol{r})(\boldsymbol{r} \textbf{ by } \boldsymbol{n}).$$

1. The factors of A in the reduced SVD are $\boldsymbol{U}_r$ times $\boldsymbol{\Sigma}_r \boldsymbol{V}_r^{\mathrm{T}}$
2. The factors of A in Gram-Schmidt are $\boldsymbol{Q}_{m \times r}$ times $\boldsymbol{R}_{r \times n}$
3. The factors of A in pivoted elimination are $\boldsymbol{C}_{m \times r}$ times $\boldsymbol{Z}_{r \times n}$

Let me comment right away on the properties of these three important factorizations

1. The column basis $\boldsymbol{u}_1, \ldots, \boldsymbol{u}_r$ in U is *orthonormal.* It has the extra property that also the rows $\sigma_k \boldsymbol{v}_k^{\mathrm{T}}$ of the second factor $\Sigma_r V_r^{\mathrm{T}}$ are orthogonal.

2. The column basis $\boldsymbol{q}_1, \ldots, \boldsymbol{q}_r$ in Q is *orthonormal.* The rows of the second factor R are not orthogonal but they are well-conditioned (safely independent).

3. The column basis $\boldsymbol{c}_1, \ldots, \boldsymbol{c}_r$ in C is *not orthogonal.* But both C and the second factor Z can be well conditioned. This is achieved by allowing column exchanges. (Not just allowing but insisting.) C contains r "good columns" from A.

The third factorization $A = CZ$ can be called an Interpolative Decomposition (**ID** rather than **SVD** or **QR**). Since this is the new idea for this section, I will focus now on its properties. Notes and articles by Gunnar Martinsson and distinguished coauthors are the basis for this exposition of ID.

The columns of C come directly from A, but the rows of Z do not. That is asking too much. Later we will have $\boldsymbol{CMR}$ with columns of A in C and rows of A in R—and an invertible **mixing matrix** $\boldsymbol{M}$ to make the product close to A.

Interpolative Decomposition = Column / Row Factorization

Here are four important advantages of CZ compared to QR and $U\Sigma V^{\mathrm{T}}$. Remember that the columns of C are **actual columns of** $\boldsymbol{A}$—thoughtfully chosen. That gives serious advantages to $A = CZ$.

- $A = CZ$ takes **less computing time** and **less storage** than $A = U\Sigma V^{\mathrm{T}}$ and $A = QR$.
- When A is **sparse or nonnegative or both**, so is C. C comes directly from A.
- When A comes from discretizing a differential or integral equation, the columns in C **have more meaning** than the orthonormal bases in U and Q.
- When A is a matrix of data, the columns kept in C have **simple interpretations**.

Those last three points can make a major difference in our understanding of the output, after the computations are complete. This has been a criticism of the SVD: the singular vectors are algebraically and geometrically perfect but they are "humanly" hard to know.

When the numbers in A are intrinsically positive, the goal may be a Nonnegative Matrix Factorization (NMF). *Then both factors in* $A \approx MN$ *have entries* ≥ 0. This is important for moderate sizes but it is asking a lot in the world of big data.

$A = CZ$ is right for large matrices when Z has small entries (say $|z_{ij}| \leq 2$).

Factoring A into CZ

When C contains a basis for the column space of A, the factor Z is completely determined. Every column of A is a unique combination of those basic columns in C:

$$(\text{column } j \text{ of } A) = (\text{matrix } C)\ (\text{column vector } \boldsymbol{z}_j). \tag{1}$$

This is exactly the statement $A = CZ$, column by column. This idea was in Section I.1 (with R instead of Z). It gave a neat proof of row rank = column rank. But the columns for that C were possibly not very independent. This section will soon become serious about choosing C quickly and well (but not yet).

First we can take a step beyond $A = CZ$, by looking at the row space of C. Somewhere C has r independent rows. If we put those rows $\boldsymbol{y}_1^{\mathrm{T}}, \ldots, \boldsymbol{y}_r^{\mathrm{T}}$ into an r by r matrix B, then B is invertible. Every row of C is a unique combination of those basic rows in B:

$$(\text{row } i \text{ of } C) = (\text{row vector } \boldsymbol{y}_i^{\mathrm{T}})\ (\text{invertible matrix } B). \tag{2}$$

This is exactly the statement $C = YB$, row by row. Combine it with $A = CZ$:

$$\boxed{\boldsymbol{A}_{m\times n} = \boldsymbol{C}_{m\times r}\ \boldsymbol{Z}_{r\times n} = \boldsymbol{Y}_{m\times r}\ \boldsymbol{B}_{r\times r}\ \boldsymbol{Z}_{r\times n}} \tag{3}$$

All those matrices have rank r.

Now use the fact that the columns of C came directly from A and the rows of B came directly from C. For those columns and rows, the vectors $\boldsymbol{z}_j$ in (1) and $\boldsymbol{y}_i^{\mathrm{T}}$ in (2) came from the identity matrix I_r! If you allow me to suppose that $\boldsymbol{z}_j$ and $\boldsymbol{y}_i^{\mathrm{T}}$ are the **first $\boldsymbol{r}$ columns and rows in $\boldsymbol{A}$ and $\boldsymbol{C}$**, then $\boldsymbol{A} = \boldsymbol{YBZ}$ has a special form:

$$\boxed{\boldsymbol{Y} = \begin{bmatrix} \boldsymbol{I}_r \\ \boldsymbol{C}_{m-r}\boldsymbol{B}^{-1} \end{bmatrix} \qquad \boldsymbol{B} = \text{submatrix of } \boldsymbol{A} \qquad \boldsymbol{Z} = \begin{bmatrix} \boldsymbol{I}_r & \boldsymbol{B}^{-1}\boldsymbol{Z}_{n-r} \end{bmatrix}} \tag{4}$$

We are just supposing that the upper left r by r corner B of A is invertible. Every matrix A of rank r has an invertible r by r submatrix B somewhere! (Maybe many such B's.) When B is in that upper left corner, elimination finds Y and Z.

Example 1 This 3 by 4 matrix of rank 2 begins with an invertible 2 by 2 matrix B:

$$A = \begin{bmatrix} 1 & 2 & 4 & 2 \\ 0 & 1 & 2 & 1 \\ 1 & 3 & 6 & 3 \end{bmatrix} = \begin{bmatrix} 1 & 0 \\ 0 & 1 \\ 1 & 1 \end{bmatrix} \begin{bmatrix} 1 & 2 \\ 0 & 1 \end{bmatrix} \begin{bmatrix} 1 & 0 & 0 & 0 \\ 0 & 1 & 2 & 1 \end{bmatrix} = YBZ \tag{5}$$

$$(3 \times 2)(2 \times 2)(2 \times 4)$$

Please understand. If C_{m-r} or Z_{n-r} or both happen to contain large numbers, the factors in (5) are a bad choice. We are happiest if those other entries of Y and Z are small. The best is $|y_{ij}| \le 1$ and $|z_{ij}| \le 1$. And we can find a **submatrix B that makes this true!**

Choose B as the r by r submatrix of A with the largest determinant.

Then all entries of Y and Z have $|\, y_{ij} \,| \le 1$ and $|\, z_{ij} \,| \le 1$.

Proof. Those y_{ij} and z_{ij} are the numbers in $C_{m-r}B^{-1}$ and in $B^{-1}Z_{n-r}$. We start with the z_{ij}. Since $B(B^{-1}Z_{n-r}) = Z_{n-r}$, we know that every column $\boldsymbol{z}_j$ of $B^{-1}Z_{n-r}$ solves this system of linear equations:

$$\boldsymbol{B z_j} = \textbf{column } \boldsymbol{j} \textbf{ of } \boldsymbol{Z_{n-r}}.$$

By Cramer's Rule, the numbers z_{ij} in the solution $\boldsymbol{z}_j$ are *ratios of determinants*:

$$z_{ij} = \frac{\det(B \text{ with its } i\text{th column replaced by that column } j)}{\text{determinant of } B}$$

Remember that B is the r by r submatrix of A with largest determinant. The matrix in the numerator is one of the many submatrices that we did *not* choose. Its determinant is smaller than $\det B$. So $|z_{ij}| \le 1$.

Similarly the rows of Y come from $C_{m-r}B^{-1}$. Since $(C_{m-r}B^{-1})B = C_{m-r}$, those rows solve the linear equations $\boldsymbol{y}_i^{\mathrm{T}} B = \text{row } i \text{ of } C_{m-r}$. When we transpose and use Cramer's Rule, the components y_{ij} of $\boldsymbol{y}_i^{\mathrm{T}}$ are again ratios of determinants. And $\det B$ is again the denominator! Since $\det B$ is as large as possible, we have $|y_{ij}| \le 1$.

We admit to one big problem. We said that "we can find B so that $|y_{ij}| \le 1$ and $|z_{ij}| \le 1$." **This is not true**. In reality, we can't find the submatrix B with maximum determinant. Not without a quantum computer (which doesn't yet exist). B is somewhere inside A, but we have no idea which submatrix it is.

The amazing thing is that by randomization we can identify a good submatrix B. Then there is a **very high probability** (not a certainty) that all $|y_{ij}| \le 2$ and $|z_{ij}| \le 2$. So the next section will complete this presentation of Interpolative Decomposition (when columns or rows come directly from A). The selection of columns or rows will be **random with carefully chosen probabilities.**

Example 2 The matrix in Example 1 has this $B_{\max}$ with maximum determinant $= 2$. Then $|y_{ij}| \le 1$ and $|z_{ij}| \le 1$:

$$A = \begin{bmatrix} \mathbf{1} & 2 & \mathbf{4} & 2 \\ \mathbf{0} & 1 & \mathbf{2} & 1 \\ 1 & 3 & 6 & 3 \end{bmatrix} = \begin{bmatrix} \mathbf{1} & \mathbf{0} \\ \mathbf{0} & \mathbf{1} \\ 1 & 1 \end{bmatrix} \begin{bmatrix} 1 & 4 \\ 0 & 2 \end{bmatrix} \begin{bmatrix} \mathbf{1} & 0 & \mathbf{0} & 0 \\ \mathbf{0} & -.5 & \mathbf{1} & -.5 \end{bmatrix} = Y B_{\max} Z \quad (6)$$

CMR Factorization : Selecting C and R

Now we describe a recommended choice of columns and rows of A to go directly into C and R. A **mixing matrix** M is always needed to achieve $A \approx CMR$. This can be an equality $A = CMR$ if A actually has low rank. For large matrices that are *approximately low rank* (this is what we assume) we can start with approximate singular vectors in $A \approx U\Sigma V^{\mathrm{T}}$. We will use U and V to achieve a rank r factorization CMR that is close to A.

Our guide is D. C. Sorensen and M. Embree, *A DEIM induced CUR factorization*, arXiv : 1407.5516v2, 18 Sep 2015; SIAM J. Scientific Computing **38** (2016) 1454-1482.

The Discrete Empirical Interpolation Method chooses C and R. We write M in place of U.

It seems a little strange that CMR (which uses columns and rows directly from A) begins with $U\Sigma V^{\mathrm{T}}$ (which finds orthonormal combinations of those columns and rows). And the approximate computation of $U\Sigma V^{\mathrm{T}}$ often begins with QR, to get one orthogonal matrix Q in a fast way. So the bases C, U, Q for the column space of A (*approximate* !) are all linked by fast algorithms.

The accuracy of the final approximation is controlled by the next singular value σ_{r+1} in $A = U\Sigma V^{\mathrm{T}}$—and the r columns of U and V corresponding to our choices for C and R:

$$(\boldsymbol{m} \times \boldsymbol{r})\,(\boldsymbol{r} \times \boldsymbol{r})\,(\boldsymbol{r} \times \boldsymbol{m}) \qquad ||\boldsymbol{A} - \boldsymbol{CMR}|| \leq (||\boldsymbol{U}_r^{-1}|| + ||\boldsymbol{V}_r^{-1}||)\,\sigma_{r+1}. \qquad (7)$$

Selection of Columns from A to Enter C

Start with the r columns of $U_{m\times r}$ (the approximate left singular vectors of A). Suppose s columns of $E_{m\times r}$ come directly from $I_{m\times r}$. If $E^{\mathrm{T}}U$ is an invertible matrix, then

1 $P = U(E^{\mathrm{T}}U)^{-1}E^{\mathrm{T}}$ has $P^2 = P$ = projection matrix

2 $P\boldsymbol{x}$ equals $\boldsymbol{x}$ in those s chosen positions, so P is an **interpolatory projection**

The crucial property is **2** (see Sorensen and Embree for all proofs). For $s = 1$, the DEIM algorithm chooses the largest entry in the first singular vector $\boldsymbol{u}_1$. That leads to P_1. The next choice (leading to P_2) is decided by the largest entry in $\boldsymbol{u}_2 - P_1\boldsymbol{u}_2$. Every later P_j is decided by the largest entry in $\boldsymbol{u}_j - P_{j-1}\boldsymbol{u}_j$. This corresponds to maximizing each pivot in ordinary elimination. A simple pseudocode is in [Sorensen-Embree].

The rows of A to enter R are selected in the same way. A big advantage comes from this sequential processing, compared to norm-squared sampling based on all the row norms of U and V. The next step is to estimate the error in interpolatory projection :

$$||A - C(C^{\mathrm{T}}C)^{-1}C^{\mathrm{T}}A|| \leq \eta_C\sigma_{r+1} \qquad \text{and} \qquad ||A - AR^{\mathrm{T}}(RR^{\mathrm{T}})^{-1}R|| \leq \eta_R\sigma_{r+1} \quad (8)$$

In practice, those constants η_C and η_R are modest (of order less than 100). Again this is similar to partial pivoting : fast growth is possible in theory but never seen in practice.

The Mixing Matrix M

The final decision is the mixing matrix in $A \approx CMR$. Without that matrix M, the product CR is normally not close to A. The natural choice for M is

$$\boxed{M = (C^{\mathrm{T}}C)^{-1}C^{\mathrm{T}}AR^{\mathrm{T}}(RR^{\mathrm{T}})^{-1} = [\text{left inverse of } C]\, A\, [\text{right inverse of } R].} \quad (9)$$

On the last page of Section I.1, this is the choice that produced equality in $A = CMR$. On that page the rank of A was exactly r (we were innocent then). Now r is only the approximate rank of a large matrix A.

For the randomized algorithms of the next section II.4 (where A is too large for the DEIM algorithm) you will see that Halko-Martinsson-Tropp and Mahoney-Drineas also made this choice. The early analysis of Stewart (*Numerische Math.* **83** (1999) 313-323) pointed the way to the error estimate in (8).

Starting from $A = QR$ with Column Pivoting

The QR factorization is a favorite starting point for numerical linear algebra with large matrices. At reasonable cost, it produces an orthonormal basis in Q for the column space of A. From Q we can go quickly and accurately to the other two fundamental bases:

Columns of C (coming directly from A: Interpolatory Decomposition $\boldsymbol{A = CMR}$)

Columns of U (orthonormal vectors: Singular Value Decomposition $\boldsymbol{A = U\Sigma V^{\mathrm{T}}}$)

Please understand that good choices for C and U will depend on a good choice for Q. Since that matrix is orthogonal (thus perfectly conditioned) it is the other factor R that decides the quality of $A = QR$.

The ordinary Gram-Schmidt process in Section II.2 kept the columns of A in their original order. *Pivoted* QR chooses the largest remaining column at the start of each new step (*column pivoting*). This produces a permutation Π so that the first k columns of $A\Pi$ (and Q) are the important columns:

$$\boldsymbol{A\Pi} = QR = \boldsymbol{Q}_{m\times m}\begin{bmatrix} \boldsymbol{A} & \boldsymbol{B} \\ \boldsymbol{0} & \boldsymbol{C} \end{bmatrix} \quad \text{with triangular } A_{k\times k} \quad (10)$$

A "strong rank-revealing factorization" has this form with extra conditions on the blocks: *$\sigma_i(A)$ is not small, $\sigma_j(C)$ is not large, $A^{-1}B$ is not large.* Those properties are valuable in finding a basis for the (computational) nullspace of A. Allow us to assume that these properties hold, and go forward to use the first k columns of Q in the CMR and $U\Sigma V^{\mathrm{T}}$ factorizations of A.

Low Rank Approximation by a Partial QR

The previous pages assumed that the rank r of the matrix A is relatively small. Often this assumption is false but effectively true. We mean that

$$A = (\textbf{matrix } A_r \textbf{ of low rank } r) + (\textbf{matrix } E \textbf{ of small norm}).$$

We want to find and compute with the low rank matrix A_r. We can estimate the error in A_r that comes from ignoring the matrix E (large matrix, small norm).

Martinsson proposes that the low rank approximation A_r can be computed using QR. The first r steps of that algorithm (with column pivoting) produce $Q_r R_r$:

$$A = Q_r R_r + E = (r \text{ columns } q_j)(r \text{ rows } r_j^{\mathrm{T}}) + (n-r \text{ columns orthogonal to those } q_j).$$

$||A_{n-r}||$ is small! Since this algorithm includes column exchanges, those columns in A_{n-r} might not be the *last* $n-r$ columns of A. But an n by n column permutation P will move those columns to the back of AP. The r important columns are $Q_r R_r$ at the front:

$$\boxed{AP = \left[\begin{array}{cc} Q_r R_r & A_{n-r} \end{array}\right] \quad \textbf{and} \quad A = \left[\begin{array}{cc} Q_r R_r P^{\mathrm{T}} & A_{n-r} P^{\mathrm{T}} \end{array}\right].} \tag{11}$$

$Q_r R_r P^{\mathrm{T}}$ is the good rank r approximation to A. Fortunately the QR algorithm with pivoting computes the column norms, so we know when $||A_{n-r}||_F$ is below our preset bound ϵ. Then we stop.

This is an effective algorithm. But it surrendered on the goal of choosing columns directly from A. It succeeded on the goal of orthonormal columns (in Q).

We come back to low rank approximation in II.4 and in Part III of this book.

An Approximate SVD from the Partial QR

To complete the circle we aim now for a good approximation to the SVD. This comes from the close approximation in equation (11) to $A = QR$. The error matrix E has $||\boldsymbol{E}||_F < \epsilon$.

$$\textbf{Small error } \boldsymbol{E} \qquad \underset{m \times n}{(\boldsymbol{A})} = \underset{m \times r}{(\boldsymbol{Q}_r)} \; \underset{r \times n}{(\boldsymbol{R}_r \boldsymbol{P}^{\mathrm{T}})} + \underset{m \times n}{(\boldsymbol{E})} \tag{12}$$

Two quick steps are enough to produce an SVD very close to A, with the same error E:

First, find the SVD of the matrix $R_r P^{\mathrm{T}}$ with only r rows: $\boldsymbol{R}_r \boldsymbol{P}^{\mathrm{T}} = \boldsymbol{U}_r \boldsymbol{\Sigma} \boldsymbol{V}^{\mathrm{T}}$

Second, multiply Q_r times U_r to find $U = Q_r U_r$ = orthogonal times orthogonal:

$$\begin{array}{c}\textbf{Approximate SVD}\\ \textbf{with error } \boldsymbol{E}\end{array} \qquad \boxed{\boldsymbol{A} = \boldsymbol{Q}_r \boldsymbol{U}_r \boldsymbol{\Sigma} \boldsymbol{V}^{\mathrm{T}} + \boldsymbol{E} = \boldsymbol{U} \boldsymbol{\Sigma} \boldsymbol{V}^{\mathrm{T}} + \boldsymbol{E}} \tag{13}$$

So a small SVD and a large QR give a large (approximate) SVD.

Problem Set II.3

1 The usual measure of danger from roundoff error is the *condition number* $||A||\,||A^{-1}||$. Show why that number is squared if we work with $A^{\mathrm{T}}A$ instead of A.

2 Write down a 2 by 2 matrix A with condition number > 1000. What is A^{-1}? Why does A^{-1} also have condition number > 1000?

3 The reason that $||A||$ and $||A^{-1}||$ both appear is that we work with *relative error*. If $A\boldsymbol{x} = \boldsymbol{b}$ and $A(\boldsymbol{x} + \Delta\boldsymbol{x}) = \boldsymbol{b} + \Delta\boldsymbol{b}$ then $A\,\Delta\boldsymbol{x} = \Delta\boldsymbol{b}$. Show that

$$\frac{||\Delta\boldsymbol{x}||}{||\boldsymbol{x}||} \leq ||A||\,||A^{-1}||\,\frac{||\Delta\boldsymbol{b}||}{||\boldsymbol{b}||}.$$

4 Why does $\lambda_{\max}/\lambda_{\min}$ equal the condition number for positive definite A?

5 *Important* What is the condition number of an orthogonal matrix Q?

6 Suppose the columns of C contains an orthonormal basis for the column space of A and the rows of R contains an orthonormal basis for the row space. Will those bases contain the singular vectors $\boldsymbol{v}$ and $\boldsymbol{u}$ in the SVD?

7 If C and R contain bases for the column space and row space of A, why does $A = CMR$ for some square invertible matrix M?

8 Here is a matrix whose *numerical rank* is 2. The number ϵ = machine epsilon is 2^{-16}. What orthonormal vectors $\boldsymbol{q}_1$ and $\boldsymbol{q}_2$ will give a good basis for the column space—a basis that pivoted QR will probably choose?

$$A = \begin{bmatrix} 1 & 1 & 1 \\ 1 & 1+\epsilon & 0 \\ 1 & 1 & 0 \end{bmatrix}$$

9 Approximate that matrix A as $QRP^{\mathrm{T}}+$ (order ϵ) for a permutation matrix P.

10 Which 2 by 2 submatrix $B_{\max}$ of A (rank 2) has the largest determinant?

$$A = \begin{bmatrix} 2 & 5 & 1 \\ 1 & 3 & 5 \\ 3 & 1 & 1 \end{bmatrix} \qquad \text{Factor } A = YB_{\max}Z \text{ as in equation (6).}$$

II.4 Randomized Linear Algebra

This section on randomization will be incomplete. The first reason is that it is not written by an expert. This book cannot be a detailed guide to computing the SVD or QR. Yet it still seems possible—and very worthwhile—to report on key ideas and algorithms that have made those computations possible for large matrices.

Among those ideas are important new approaches that begin with random vectors $\boldsymbol{x}$. **Then the products $A\boldsymbol{x}$ are random samples from the column space of A.** With r of those vectors (or $r + 10$ to be on the safe side, protecting against random accidents) we have a potentially thin matrix to compute. The speedup is impressive. (This is also the starting point in Section III.5 for "*compressed sensing*" that speeds up acquisition and processing of digital signals.)

This section will introduce and describe the basic steps of randomized computations. That idea has brought a revolution in numerical linear algebra for big matrices.

The first example is **matrix multiplication**. If A and B are m by n and n by p, then $C = AB$ normally needs mnp individual multiplications: n multiplications for each of the mp inner products in AB, or mp multiplications for each of the n outer products (columns times rows). *Multiplying very large matrices is expensive.*

Suppose we **just sample A and B** instead of using the complete matrices. A few entries a_{ij} and b_{jk} don't tell us much. But s columns $\boldsymbol{a}_k$ from A and s corresponding rows $\boldsymbol{b}_k^{\mathrm{T}}$ from B will give us s rank one matrices $\boldsymbol{a}_k\boldsymbol{b}_k^{\mathrm{T}}$. *If those are "typical" outer products, we can multiply their sum by n/s*—to estimate the true AB = sum of n products. Notice that this uses column-row products (highly recommended) and not row-column inner products (low level).

There is more to this idea. Random sampling uses some basic statistics. Large products $\boldsymbol{a}_k\boldsymbol{b}_k^{\mathrm{T}}$ obviously make greater contributions to $C = AB$. We can and will increase the chances of those larger samples by changing from uniform probability to "**norm-squared sampling**". We have to compensate in our formulas, which aim to have the correct expected value and the lowest variance. You will see the value of statistical ideas !

Our presentation will mostly follow Michael Mahoney's lecture notes for his course at UC Berkeley. They are well organized and well written—a generous and important contribution. The 2013 course notes were posted in 2016 and they begin with this quick overview of random matrix multiplication:

A **sampling matrix S** will act on the columns of A and rows of B to produce C and R:

$$\boldsymbol{C} = \boldsymbol{AS} \quad \text{and} \quad \boldsymbol{R} = \boldsymbol{S}^{\mathrm{T}}\boldsymbol{B} \quad \text{and} \quad \boldsymbol{CR} = \boldsymbol{ASS}^{\mathrm{T}}\boldsymbol{B} \approx \boldsymbol{AB}. \tag{1}$$

We multiply C and R instead of the full and correct matrices A and B. It will not be true that SS^{T} is close to I. But it will be true that *the expected value of SS^{T} is I*. There you see the key to randomization.

1 **Column-row sampling** Each column of S has one nonzero entry (call it s_k). Then AS picks out individual columns $\boldsymbol{a}_k$ directly from A, and $S^{\mathrm{T}}B$ picks out the corresponding rows $\boldsymbol{b}_k^{\mathrm{T}}$ of B. Multiplying AS times $S^{\mathrm{T}}B$ gives a sum of **column-row products $\boldsymbol{a}_k\boldsymbol{b}_k^{\mathrm{T}}$ weighted by the numbers s_k^2.**

One product $s_k^2\boldsymbol{a}_k\boldsymbol{b}_k^{\mathrm{T}}$ comes from one column of AS and one row of $S^{\mathrm{T}}B$:

$$\begin{bmatrix} \cdot & \boldsymbol{a}_k & \cdot \end{bmatrix}\begin{bmatrix} 0 \\ s_k \\ 0 \end{bmatrix} = s_k\boldsymbol{a}_k \qquad \begin{bmatrix} 0 & s_k & 0 \end{bmatrix}\begin{bmatrix} \cdot \\ \boldsymbol{b}_k^{\mathrm{T}} \\ \cdot \end{bmatrix} = s_k\boldsymbol{b}_k^{\mathrm{T}} \quad (2)$$

Altogether AB is approximated by the weighted sum $CR = \sum s_k^2\boldsymbol{a}_k\boldsymbol{b}_k^{\mathrm{T}}$ of a *random selection* of s rank-one matrices. The selection is random, but we choose the weights.

2 **Sampling by random projections S** The matrix S is still thin, so AS has many fewer columns than A. The columns of S now contain multiple nonzeros, so AS mixes columns as it projects $\mathbf{C}(A)$ into lower dimensions. Then $ASS^{\mathrm{T}}B$ is a more uniform approximation to AB.

We start with random sampling. Later we describe random projections. Those can produce fast and useful **preconditioners** for the original matrix AB—and they may reduce the computational cost of graph clustering.

Practice with Computing Mean and Variance

Here is a greatly simplified sampling problem. Instead of a matrix, we start with a vector $\boldsymbol{v} = (a, b)$. We will sample it twice (2 independent trials). Then we compute the mean m and the variance σ^2. Section V.1 will describe many more examples of m and σ^2.

First sample: With probabilities $\frac{1}{2}$ and $\frac{1}{2}$, choose $(a, 0)$ or $(0, b)$
Second sample: Repeat exactly. Then add the two samples to get (x_1, x_2)

Computing the mean $m = \mathrm{E}[(x_1, x_2)] =$ **expected value** $=$ **average output** (x_1, x_2)
First way: The average value of sample 1 is $\frac{1}{2}(a, 0) + \frac{1}{2}(0, b) = \frac{1}{2}(a, b)$
We have two independent identical trials (two samples). Add their means:

$$\textbf{Overall mean } m = \mathrm{E}[(x_1, x_2)] = \frac{1}{2}(a, b) + \frac{1}{2}(a, b) = (\boldsymbol{a}, \boldsymbol{b})$$

Our two-sample experiment was *unbiased*. The desired mean was achieved.
Second way: The experiment had the following 4 outcomes each with probability $\frac{1}{4}$

$$(a, 0)+(a, 0) = (\boldsymbol{2a}, \boldsymbol{0}) \quad (a, 0)+(0, b) = (\boldsymbol{a}, \boldsymbol{b}) = (0, b)+(a, 0) \quad (0, b)+(0, b) = (\boldsymbol{0}, \boldsymbol{2b})$$

The mean is the sum of all four outputs weighted by their probabilities $\left(\text{all } \frac{1}{4}\right)$:

$$\textbf{Overall mean } m = \frac{1}{4}(2a, 0) + \frac{1}{4}(a, b) + \frac{1}{4}(a, b) + \frac{1}{4}(0, 2b) = (\boldsymbol{a}, \boldsymbol{b}) \text{ as before}$$

The variance σ^2 = weighted average of the squared distances from outputs to mean. We will use two equivalent ways to compute $\sigma^2 = \mathrm{E}\,[(x - \text{mean})]^2 = \mathrm{E}\,[x^2] - (\text{mean})^2$. *First way to find variance*: Add all **(outputs − mean)²** weighted by their probabilities $\frac{1}{4}$

$$\frac{1}{4}\Big[(2a,0)-(a,b)\Big]^2+\frac{1}{4}\Big[(a,b)-(a,b)\Big]^2+\frac{1}{4}\Big[(a,b)-(a,b)\Big]^2+\frac{1}{4}\Big[(0,2b)-(a,b)\Big]^2=$$

$$\frac{1}{4}(a^2,b^2)+\frac{1}{4}(0,0)+\frac{1}{4}(0,0)+\frac{1}{4}(a^2,b^2)=\frac{\mathbf{1}}{\mathbf{2}}(\boldsymbol{a^2},\boldsymbol{b^2})$$

Second way: Add all **(outputs)²** weighted by their probabilities and subtract **(mean)²**

$$\boldsymbol{\sigma^2}=\frac{1}{4}(2a,0)^2+\frac{1}{4}(a,b)^2+\frac{1}{4}(a,b)^2+\frac{1}{4}(0,2b)^2-(a,b)^2$$

$$=\left(a^2+\frac{a^2}{4}+\frac{a^2}{4}+0-a^2,0+\frac{b^2}{4}+\frac{b^2}{4}+b^2-b^2\right)=\frac{\mathbf{1}}{\mathbf{2}}\left(\boldsymbol{a^2},\boldsymbol{b^2}\right) \tag{3}$$

Observation: If b is larger than a, we could keep the correct mean (a, b) and reduce the variance σ^2 by giving *greater probability* to choosing the larger samples $(0, b)$. Matrix sampling will do this (see Problem 7 at the end of this Section II.4).

This page used $s = 2$ trials for $n = 2$ numbers a, b. *Not useful, no time was saved.* The next pages use $s << n$ trials for a matrix with n columns. The mean of AB stays correct.

Random Matrix Multiplication with the Correct Mean AB

The n by s sampling matrix S will contain s columns. Each column of S has *one nonzero*. For column j of S, the position of that nonzero is random! If the random choice is row $k = k(j)$, the nonzero in row k, column j of S is s_{kj}. The sampled matrix is AS:

Columns 1 to s of AS are numbers s_{kj} times columns $k(1)$ to $k(s)$ of A.

Here is an example with $s = 2$ trials. It samples columns $k(1) = 1$ and $k(2) = 3$ from A:

$$AS=\begin{bmatrix} & & \\ \boldsymbol{a}_1 & \boldsymbol{a}_2 & \boldsymbol{a}_3 \\ & & \end{bmatrix}\begin{bmatrix} s_{11} & 0 \\ 0 & 0 \\ 0 & s_{32}\end{bmatrix}=\begin{bmatrix} & \\ s_{11}\boldsymbol{a}_1 & s_{32}\boldsymbol{a}_3 \\ & \end{bmatrix}$$

The key question is: **How do we choose those numbers s_{kj} ?** The answer is: **They come from probabilities!** We intend to do random sampling. So we must choose those s columns of A in a random way (allowing all columns a chance) in random multiplication:

Assign **probabilities p_j** to all of the n columns of A, with $p_1 + \cdots + p_n = 1$

Choose **s columns with replacement** (so columns can be chosen more than once)

If column k of A is chosen (with row k of B), **multiply both of those by $1/\sqrt{sp_k}$**

Then (column k of A) (row k of B)/sp_k goes into our random product AB.

Conclusion to be verified: **The expected value of the n by n matrix SS^{T} is I**

Same conclusion in other words: **The expected value of $ASS^{\mathrm{T}}B$ is AB**

Thus random sampling computes the matrix product with the **correct mean AB**

Proof. There are s identical trials. Each trial chooses a column-row pair of A and B with probabilities p_1 to p_n (column from A times row from B divided by $\sqrt{sp_j}^{\,2} = sp_j$). Then the expected value $=$ *mean* $=$ average outcome from each trial is

$$p_1 \frac{(\text{column 1 of } A)\ (\text{row 1 of } B)}{sp_1} + \cdots + p_n \frac{(\text{column } n \text{ of } A)\ (\text{row } n \text{ of } B)}{sp_n}. \tag{4}$$

The p's cancel. **This is exactly AB/s.** And since there are s trials, **the expected value for the randomized multiplication $(AS)(S^{\mathrm{T}}B)$ is AB.**

Conclusion All well so far—but we have to choose the probabilities p_1 to p_n. Any choice (adding to 1) gives the correct expected value AB (the mean). But the choice of the p's can strongly affect the variance!

Uniform sampling would choose equal probabilities $p = 1/n$. This is reasonable if the columns of A (and also the rows of B) have similar lengths. But suppose that (column 1 of A) (row 1 of B) makes up most of AB—it dominates the other column-row outer products. Then we don't want to randomly miss it.

Our approach here is to use *unequal probabilities p_j*. We now state and compute the best p's. And we mention an entirely different option: Introduce a mixing matrix M and work with AM and $M^{-1}B$. Then use equal probabilities p_j for the randomly mixed columns of A and rows of B.

Norm-squared Sampling Minimizes the Variance

Norm-squared sampling chooses the probabilities p_j proportional to the numbers $||\text{column } j \text{ of } A||\ ||\text{row } j \text{ of } B||$. **In the important case $B = A^{\mathrm{T}}$**, the p_j are proportional to $||\text{column } j \text{ of } A||^2$. The name "*norm-squared*" or "*length-squared*" is then natural. We still have to scale all the probabilities p_j by a suitable constant C so they add to 1:

$$\boxed{p_j = \frac{1}{C}\, ||\textbf{column } j \textbf{ of } A||\ ||\textbf{row } j \textbf{ of } B|| = \frac{||\boldsymbol{a}_j||\ ||\boldsymbol{b}_j^{\mathrm{T}}||}{C} \text{ with } C = \textstyle\sum_{j=1}^{n} ||\boldsymbol{a}_j||\ ||\boldsymbol{b}_j^{\mathrm{T}}||.} \tag{5}$$

Now we will painstakingly compute the variance for randomized matrix multiplication using any probabilities p_j—and we will verify that *the choice of p's in equation* (5) *minimizes the variance.* Best to choose large columns and rows more often. The line after equation (4) showed that all choices give the correct mean $\mathrm{E}\,[ASS^{\mathrm{T}}B] = AB$.

Each of the s trials produces a matrix $X_j = \boldsymbol{a}_j \boldsymbol{b}_j^{\mathrm{T}}/sp_j$ with probability p_j. Its i,k entry is $(X_j)_{ik} = a_{ij} b_{jk}/sp_j$. In each trial we compute the mean: the p_j cancel.

$$\textbf{Mean} \qquad \mathrm{E}[X] = \sum_{j=1}^{n} p_j X_j = \frac{1}{s} \sum_{1}^{n} \boldsymbol{a}_j \boldsymbol{b}_j^{\mathrm{T}} = \frac{1}{s} AB \text{ as in (4)}$$

The variance for one trial is by definition $\mathbf{E}[X^2] - (\mathbf{E}[X])^2$. Adding the results for s independent trials multiplies the one-trial mean and also the one-trial variance by s. The mean becomes AB as we know. The variance will be computed in the Frobenius norm (sum of squares of matrix entries). Compare the correct AB with the random CR:

$$\begin{aligned} \textbf{Variance} \quad \mathrm{E}\left[||AB - CR||_F^2 \right] &= \sum_{i,k} \sum_{j=1}^{n} p_j \frac{a_{ij}^2 \, b_{jk}^2}{sp_j^2} - \frac{1}{s} ||AB||_F^2 \\ \text{(sum first over } i \text{ and } k) &= \sum_{j=1}^{n} \frac{||\boldsymbol{a}_j||^2 \, ||\boldsymbol{b}_j^{\mathrm{T}}||^2}{sp_j} - \frac{1}{s} ||AB||_F^2 \end{aligned} \qquad (6)$$

Finally we choose probabilities $p_1, \ldots, p_n$ to minimize this variance. Equation (5) reveals the minimizing choice, proved below. For that choice $p_j = ||\boldsymbol{a}_j|| \, ||\boldsymbol{b}_j^{\mathrm{T}}||/C$ from (5), the terms $||\boldsymbol{a}_j||^2 \, ||\boldsymbol{b}_j^{\mathrm{T}}||^2/p_j$ in equation (6) become $C||\boldsymbol{a}_j|| \, ||\boldsymbol{b}_j^{\mathrm{T}}||$. **Their sum is $\boldsymbol{C^2}$.** The **smallest variance** (using those optimal p_j) is our final result:

$$\boxed{\mathrm{E}\left[||AB - CR||_F^2 \right] = \frac{1}{s} \left[\sum_{j=1}^{n} ||\boldsymbol{a}_j|| \, ||\boldsymbol{b}_j^{\mathrm{T}}|| \right]^2 - \frac{1}{s} ||AB||_F^2 = \frac{1}{s} \left(C^2 - ||AB||_F^2 \right).} \qquad (7)$$

Here is the proof that (5) gives the probabilities p_j that minimize the variance in (6). Multiply the constraint $p_1 + \cdots + p_n = 1$ by a Lagrange multiplier λ. Add it to the function in (6). This is the key to Lagrange multipliers:

$$L(p_1, \ldots, p_n, \lambda) = \sum_{j=1}^{n} \frac{||\boldsymbol{a}_j||^2 \, ||\boldsymbol{b}_j^{\mathrm{T}}||^2}{sp_j} - \frac{1}{s} ||AB||_F^2 + \lambda \left(\sum_{1}^{n} p_j - 1 \right)$$

Take the partial derivatives $\partial L / \partial p_j$ to find the minimizing p_j (the optimal probabilities):

$$\boxed{\frac{\partial L}{\partial p_j} = 0 \text{ becomes } \frac{1}{sp_j^2} ||\boldsymbol{a}_j||^2 \, ||\boldsymbol{b}_j^{\mathrm{T}}||^2 = \lambda} \qquad (8)$$

This says that $p_j = ||\boldsymbol{a}_j|| \, ||\boldsymbol{b}_j^{\mathrm{T}}||/\sqrt{s\lambda}$. Choose the Lagrange multiplier λ so that $\sum p_j = 1$.

$$\sum_{1}^{n} p_j = \sum_{1}^{n} \frac{||\boldsymbol{a}_j|| \, ||\boldsymbol{b}_j^{\mathrm{T}}||}{\sqrt{s\lambda}} = 1 \text{ gives } \sqrt{s\lambda} = C \text{ and } p_j = \frac{||\boldsymbol{a}_j|| \, ||\boldsymbol{b}_j^{\mathrm{T}}||}{C} \text{ as predicted in (5).}$$

Norm-squared sampling uses the optimal probabilities p_j for minimum variance.

For very large matrices, stored outside of Random Access Memory (RAM), norm-squared sampling may ask to read the matrix twice. The first pass computes the squared length of each column of A and row of B. Then (inside RAM) the probabilities p_j are found, and s columns and rows are chosen for sampling. The second pass puts the sampling approximation CR into fast memory.

Applications of Randomized Matrix Multiplication

Norm-squared sampling = length-squared sampling can help to solve these central problems of numerical linear algebra:

1	**Interpolative approximation $A \approx CMR$: C and R use columns and rows of A**
2	**Approximation of A by a low rank matrix**
3	**Approximation of the SVD of A**

CMR aims to produce an accurate "sketch" of A from k of its columns and its rows. The columns will go into C and the rows will go into R. Then a *mixing matrix* M connects C with R to produce $CMR \approx A$. The dimensions are $(m \times k)\,(k \times k)\,(k \times n) = (m \times n)$. If A is sparse then C and R will be sparse, because they come directly from A.

Notice the fast multiplication $(C(M(Rv)))$. We never explicitly multiply CMR.

Understand first that $A \approx CR$ is probably not true. The column space of A will be accurately captured by C (we hope). The row space of A will be captured by R. "*The spaces are right but not the matrix.*" Every matrix of the form CMR has the same good column and row spaces (for invertible M). We want to choose M so that CMR is close to A. We still avoid the notation CUR and reserve the letter U for the SVD.

To start, I will look for the mixing matrix M that is theoretically best. Good choices of M have been developed and tested in the sampling literature. Here are six important references to randomized linear algebra:

N. Halko, P. -G. Martinsson, and J. A. Tropp, Finding structure with randomness: probabilistic algorithms for constructing approximate matrix decompositions. *SIAM Review* **53** (2011) 217-288.

R. Kannan and S. Vempala, Randomized algorithms in numerical linear algebra, *Acta Numerica* **26** (2017) 95-135.

E. Liberty, F. Woolfe, P. -G. Martinsson, V. Rokhlin, M. Tygert, Randomized algorithms for the low-rank approximation of matrices, *PNAS* **104** (2007) no. 51, 20167-20172.

M. W. Mahoney, Lecture Notes on Randomized Linear Algebra. arXiv:1608.04481.

P.-G. Martinsson, Compressing rank-structured matrices via randomized sampling, arXiv:1503.07152.

D. P. Woodruff, Sketching as a tool for numerical linear algebra. *Foundations and Trends in Theoretical Computer Science* **10** (2014) 1-157.

Best M in $A \approx CMR$: Frobenius Norm and L^2 Norm

We are given A, C, R. Suppose Q_C contains an orthonormal basis for the column space of C. Then $Q_C Q_C^{\mathrm{T}}$ is the projection matrix onto that subspace. Similarly Q_R contains an orthonormal basis for $\mathbf{C}(R^{\mathrm{T}})$ and $Q_R Q_R^{\mathrm{T}}$ is the projection matrix onto that row space. $Q_R^{\perp}$ contains an orthonormal basis for the nullspace $\mathbf{N}(R)$, and $Q_C^{\perp}$ for $\mathbf{N}(C^{\mathrm{T}})$.

By definition, the projection of A into row/column spaces is $\widehat{A} = Q_C^{\mathrm{T}} A Q_R$. The purpose of these projections is to separate the subspaces where we choose M from the subspaces that we cannot change. Yuji Nakatsukasa clarified and solved this problem (conversations in Oxford). In the Frobenius norm, that solution is especially neat. The orthogonal matrices $\left[\, Q_C \;\; Q_C^{\perp} \,\right]$ and $\left[\, Q_R \;\; Q_R^{\perp} \,\right]$ won't change the Frobenius and L^2 norms of $A - CMR$. But they help us to see the optimal M to minimize this error.

$$\left[\, Q_C \;\; Q_C^{\perp} \,\right]^{\mathrm{T}} (A - CMR) \left[\, Q_R \;\; Q_R^{\perp} \,\right] = \begin{bmatrix} \widehat{A} & X \\ Y & Z \end{bmatrix} - \begin{bmatrix} \widehat{C} M \widehat{R} & 0 \\ 0 & 0 \end{bmatrix}. \quad (9)$$

Should we choose M so that $\widehat{A} = \widehat{C} M \widehat{R}$? In the Frobenius norm, this is optimal!

Frobenius deals separately with every entry in each block of these matrices. The error is smallest when the top corner is exactly right: $\widehat{C} M \widehat{R} = \widehat{A}$. This corner is the (only) problem that we control. We have found the same M as in sections II.3 and I.1.

$$\textbf{Frobenius norm} \qquad \boxed{\min_M \, \|A - CMR\|_F = \left\| \begin{bmatrix} 0 & X \\ Y & Z \end{bmatrix} \right\|_F .} \quad (10)$$

In the L^2 matrix norm, we might expect the same plan to succeed. But a zero block in the corner may not give the smallest L^2 norm!

$$\textbf{Example} \qquad \left\| \begin{bmatrix} -1 & 1 \\ 1 & 1 \end{bmatrix} \right\|_{L^2} \text{ is smaller than } \left\| \begin{bmatrix} 0 & 1 \\ 1 & 1 \end{bmatrix} \right\|_{L^2} \quad (11)$$

On the left, the columns are orthogonal with length $\sqrt{2}$. Both singular values have $\sigma^2 = 2$. On the right, the larger singular value has $\sigma_1^2 = \frac{1}{2}(3 + \sqrt{5}) > 2$. **The zero submatrix produced a larger L^2 norm for the whole matrix.**

The optimal submatrix is a beautiful problem solved by Davis, Kahan, and Weinberger in 1982. The optimal M often does not make $\widehat{C} M \widehat{R} = \widehat{A}$. But it reduces the L^2 norm to the larger of $\|[Y \;\; Z]\|$ and $\|[X^{\mathrm{T}} \;\; Z^{\mathrm{T}}]\|$—a smaller L^2 norm is clearly impossible in (10).
In the example the smallest L^2 norm was achieved by that submatrix -1 (**not zero**).

Randomized Matrix Factorizations

Now we come to a summary of our subject. Orthogonal matrices are the goal in $A = QR$ and $A = U\Sigma V^{\mathrm{T}}$. The matrix A is too large for exact factorizations—maybe too large to read every entry a_{ij}. If we start from an $(m \times k)\,(k \times n)$ approximation $\boldsymbol{A \approx CB}$, then Halko-Martinsson-Tropp find the QR and $U\Sigma V^{\mathrm{T}}$ factorizations quickly and accurately.

Here are fast decompositions starting from a randomized $A \approx CB$.

$(\boldsymbol{A \approx QR})$	Factor C into Q_1R_1. Factor R_1B into Q_2R_2. Then $A \approx (Q_1Q_2)R_2$.
$(\boldsymbol{A \approx U\Sigma V^{\mathrm{T}}})$	From $C = Q_1R_1$ factor R_1B into $U_2\Sigma V^{\mathrm{T}}$. Choose $U = Q_1U_2$.

Projections J and Projection Matrices P

A "*projection*" of m-dimensional space $\mathbf{R}^m$ to k-dimensional space $\mathbf{R}^k$ is a k by m matrix. In case that matrix $\boldsymbol{J}$ has full row rank k, then $k \le m$ and the projections fill all of $\mathbf{R}^k$.

A specially nice case is when the k rows of J are orthonormal. This means that $JJ^{\mathrm{T}} = I$. Then the rank is certainly k and the column space $\mathbf{C}(J)$ is all of $\mathbf{R}^k$.

Notice that a *projection* J is different from a *projection matrix* P. P is square (m by m). Its rank is $k < m$ (except when $P = I$). Its column space is a k-dimensional subspace of $\mathbf{R}^m$ (quite different from the vector space $\mathbf{R}^k$). And the key property of the projection matrix is $\boldsymbol{P^2 = P}$. If we project *perpendicularly* onto the column space, the projection $P\boldsymbol{b}$ is closest to $\boldsymbol{b}$ in the usual norm $||\boldsymbol{b} - P\boldsymbol{b}||$. Then P is also symmetric.

J and P have one nice connection. **In case $\boldsymbol{J}$ has orthonormal rows, then $\boldsymbol{J^{\mathrm{T}}J}$ is a symmetric projection matrix $\boldsymbol{P}$.** You see that $P^2 = J^{\mathrm{T}}(JJ^{\mathrm{T}})J = J^{\mathrm{T}}IJ = P$.

Example 1 **Projection** $\boldsymbol{J} = [\cos\theta \ \ \sin\theta\,]$ with $JJ^{\mathrm{T}} = [\cos^2\theta + \sin^2\theta] = [1]$

$$\textbf{Projection matrix } \boldsymbol{P = J^{\mathrm{T}}J} = \begin{bmatrix} \cos^2\theta & \cos\theta\sin\theta \\ \cos\theta\sin\theta & \sin^2\theta \end{bmatrix} = \boldsymbol{P^2}$$

A symmetric P projects **orthogonally** onto its column space. Here that is $\mathbf{C}(J^{\mathrm{T}})$.

Random Projections

Suppose the entries of a k by m projection J are independent random variables. In the simplest case they are drawn from a normal distribution (a Gaussian) with mean $m = 0$ and variance $\sigma^2 = 1/k$. We will show that **the expected value of $\boldsymbol{P = J^{\mathrm{T}}J}$ is the identity matrix**. In other words, the **expected length of the projection $\boldsymbol{v = Ju}$ equals the length of $\boldsymbol{u}$.**

1. The (i, i) entry on the diagonal of $J^{\mathrm{T}}J$ is the sum $J_{1i}^2 + \cdots + J_{ki}^2$. Those squares are independent samples, each with mean $1/k$. (With zero mean for each entry of J, the expected value of the square is the variance $\sigma^2 = 1/k$.) Then the mean (the expected value) of the sum of k terms is $\boldsymbol{k(1/k) = 1}$.

2. The (i, j) entry off the diagonal of $J^{\mathrm{T}}J$ is the sum $J_{1i}J_{1j} + \cdots + J_{ki}J_{kj}$. Each of those terms is the product of two independent variables with mean zero. So the mean of each term is zero, and the mean of their sum $(J^{\mathrm{T}}J)_{ij}$ is also zero.

Thus $\mathbf{E}[\boldsymbol{J^{\mathrm{T}}J}] = \boldsymbol{I}$ = identity matrix of size m.

Using columns times rows, each matrix (row i of J)^{T} (row i of J) has expectation I/k. The sum has expectation I. Please notice what that means.

"In expectation" the m columns of J are orthonormal.
Therefore in expectation Jx has the same length as x:

$$\mathbf{E}\left[\, ||Jx||^2 \,\right] = \mathbf{E}\left[\, x^{\mathrm{T}} J^{\mathrm{T}} J x \,\right] = \mathbf{E}\left[\, x^{\mathrm{T}} x \,\right] = ||x||^2. \tag{12}$$

This random projection shows why linear algebra and probability sometimes seem to be different worlds within mathematics. In linear algebra, a k by m matrix J with $k < m$ could not have rank m. **Then $J^{\mathrm{T}}J$ could not have rank m.** But now take expected values ! E[J] can be the zero matrix and E[$J^{\mathrm{T}}J$] can be the identity matrix. For 1 by 1 matrices this would be the most ordinary thing in the world : *mean zero and variance one.*

Here is the key point. On average, the squared distance between x and y in $\mathbf{R}^m$ is the same as the squared distance between Jx and Jy in $\mathbf{R}^k$. "The projection J into a lower dimensional space preserves distances in an average sense." **What we really want is to preserve the actual distances within a factor $1 + \epsilon$, for a set of n given points in $\mathbf{R}^m$.**

That property of J is the subject of the famous Johnson-Lindenstrauss Lemma. It seems amazing that points in a high dimension can be transformed linearly to points in a low dimension, with very small change in the distance between every pair of points.

How low can the low dimension be ? This is a crucial question.

Dimension $k = 1$ is Too Low

Suppose we have $n = 3$ points x_1, x_2, x_3 in the plane $\mathbf{R}^2$. Is there a 1 by 2 matrix J that nearly preserves their distances ? We can certainly achieve $||Jx_1 - Jx_2|| = ||x_1 - x_2||$. If the third point $x_3 = \frac{1}{2}(x_1 + x_2)$ is halfway between, then by linearity Jx_3 will be $\frac{1}{2}(Jx_1 + Jx_2)$ and all distances are perfect. But if x_1, x_2, x_3 give an equilateral triangle, then x_3 has a component in the nullspace of J. That component will be lost (projected to zero). The lengths $||Jx_1 - Jx_3||$ and $||Jx_2 - Jx_3||$ will be seriously reduced.

Johnson-Lindenstrauss looked at **random k by m projections of n points in $\mathbf{R}^m$.** They proved that **if the dimension k is large enough**, then one of those projections (in fact most of them) will nearly preserve distances between the n points.

"*In high dimensions, random vectors are orthogonal with probability near* 1."

The Johnson-Lindenstrauss Lemma

Suppose $x_1, \ldots, x_n$ are any n points in $\mathbf{R}^m$, and $k \geq (8 \log n)/\epsilon^2$. Then there is a projection J from $\mathbf{R}^m$ to $\mathbf{R}^k$ so that all distances between the n points are nearly preserved :

$$(1 - \epsilon)\, ||x_i - x_j||^2 \leq ||Jx_i - Jx_j||^2 \leq (1 + \epsilon)\, ||x_i - x_j||^2 \tag{13}$$

A key point is that the dimension k must grow like $(\log n)/\epsilon^2$. An amazing step in the proof shows that a *random k by m projection J* is very likely to keep the n points apart. Then there must be *many specific J*'s that confirm equation (13) in the Lemma.

> If a random choice has a positive probability
> of success then a successful choice must exist.
> Probabilistic hypothesis, deterministic conclusion.

That is the "*probabilistic method*". Multiple proofs of Johnson-Lindenstrauss are easily found, and we liked the one on this website: **cseweb.ucsd.edu/~dasgupta/papers/jl.pdf.**

One of the 18.065 class projects chose an example in which the Lemma requires $k > 2800$. The accuracy of distances $||\boldsymbol{x}_i - \boldsymbol{x}_j||$ did break down for $k = 2700$. Also interesting: Clustering of points survived even when distances went wrong.

Randomized Matrix Approximation

To summarize this topic we will follow Per-Gunnar Martinsson: *Randomized methods for matrix computations*, arXiv: 1607.01649. First we identify the goals.

1 Rank k factorizations $\boldsymbol{A} \approx \boldsymbol{Y}(\boldsymbol{Y}^+\boldsymbol{A})$ and $\boldsymbol{S} \approx \boldsymbol{UDU}^{\mathrm{T}}$ and $\boldsymbol{A} \approx (\boldsymbol{QU})\boldsymbol{DV}^{\mathrm{T}}$.
2 Interpolative decompositions $\boldsymbol{A} \approx \boldsymbol{CMR}$ and $\boldsymbol{A} \approx \boldsymbol{CZ}$ using columns of A in C.

The randomized m by k factor Y is AG, for an n by k Gaussian random matrix G. Always YY^+ is the orthogonal projection onto the column space of Y. So YY^+A is almost surely a very good rank k approximation to A. This is the random part.

How does an approximate SVD follow from $A \approx YY^+A$? For an orthonormal column basis, first apply the QR factorization to Y. QQ^T is the same projection as YY^+. Then find the SVD of the small matrix $Q^{\mathrm{T}}A = UDV^{\mathrm{T}}$. The desired SVD is $\boldsymbol{A} \approx (\boldsymbol{QU})\boldsymbol{DV}^{\mathrm{T}}$.

Notice the two-stage construction of that SVD. First we fix an approximate column space, reducing the problem to "size k". Then any desired factorization is deterministic and fast and essentially exact. Martinsson shows how those stages can combine into a *streaming algorithm* that accesses each entry of A only once. The price is ill-conditioning and the remedy is over-sampling.

For positive semidefinite S, an extra *Nyström step* improves the factors at low cost.

Finally we look at $A \approx CZ$ or CMR, where C contains actual columns of A: preserving sparsity and nonnegativity. The accuracy of a deterministic algorithm is the same as pivoted $AP \approx QR$ (not always as close as Eckart-Young). Randomized algorithms are faster and better, if A has rapidly decaying singular values as in Section III.3.

For a clear picture of randomized matrix algorithms, read Martinsson's paper and Kannan-Vempala: Randomized algorithms in numerical linear algebra, *Acta Numerica* (2017), 95-135. These authors discovered norm-squared sampling.

Problem Set II.4

1 Given positive numbers $a_1, \ldots, a_n$ find positive numbers $p_1 \ldots, p_n$ so that

$$p_1 + \cdots + p_n = 1 \quad \text{and} \quad V = \frac{a_1^2}{p_1} + \cdots + \frac{a_n^2}{p_n} \text{ reaches its minimum } (a_1 + \cdots + a_n)^2.$$

The derivatives of $L(p, \lambda) = V - \lambda(p_1 + \cdots + p_n - 1)$ are zero as in equation (8).

2 (*for functions*) Given $a(x) > 0$ find $p(x) > 0$ by analogy with Problem 1, so that

$$\int_0^1 p(x)\,dx = 1 \quad \text{and} \quad \int_0^1 \frac{(a(x))^2}{p(x)}\,dx \quad \text{is a minimum.}$$

3 Prove that $n(a_1^2 + \cdots + a_n^2) \geq (a_1 + \cdots + a_n)^2$. This is Problem 1 with $p_i = 1/n$. Back in Problem Set I.11 you proved that $||\boldsymbol{a}||_1 \leq \sqrt{n}||\boldsymbol{a}||_2$.

4 If $M = \mathbf{1}\,\mathbf{1}^{\mathrm{T}}$ is the n by n matrix of 1's, prove that $nI - M$ is positive semidefinite. Problem 3 was the energy test. For Problem 4, find the eigenvalues of $nI - M$.

5 In case $B = A^{\mathrm{T}}$ show that the "norm-squared" or "length-squared" probabilities p_j in the text equation (5) are $||\boldsymbol{a}_j||^2/||A||_F^2$. Why is $C = \sum ||\boldsymbol{a}_j||\,||\boldsymbol{b}_j|| = ||A||_F^2$?

6 The variance computed in equation (7) cannot be negative! Show this directly:

$$||AB||_F^2 \leq (\textstyle\sum ||\boldsymbol{a}_j||\,||\boldsymbol{b}_j^{\mathrm{T}}||)^2.$$

Problem 7 returns to the example in the text of sampling (a, b) to get $(a, 0)$ or $(0, b)$. If $b > a$ then the variance will be reduced when b is chosen more often. This is achieved by optimizing the probabilities p and $1 - p$ to minimize σ^2 :

$$\text{Variance} \quad \sigma^2 = p\,\frac{a^2}{p^2} + (1-p)\,\frac{b^2}{(1-p)^2} - (\text{mean})^2$$

7 Show that $p = a/(a + b)$ and $1 - p = b/(a + b)$ minimize that variance. (The mean is the same for all p.) This optimal p agrees with equation (5) when applied to the small matrix multiplication $AB = [1]\,[a \;\; b]$. In this case $C = a + b$ in equation (5).

8 In the randomized construction on the previous page, show why $(QU)DV^{\mathrm{T}}$ (*the approximate* SVD *in italics*) is close to A. Use the steps $A \approx YY^+A$ and $Y \approx QR$ and $Q^{\mathrm{T}}A = UDV^{\mathrm{T}}$. *Problem* II.2.11 *on page* 135 *is a key.*

Part III

Low Rank and Compressed Sensing

III.1 **Changes in A^{-1} from Changes in A**

III.2 **Interlacing Eigenvalues and Low Rank Signals**

III.3 **Rapidly Decaying Singular Values**

III.4 **Split Algorithms for $\ell^2 + \ell^1$**

III.5 **Compressed Sensing and Matrix Completion**

Part III : Low Rank and Compressed Sensing

This part of the book looks at three types of low rank matrices :

1. Matrices that truly have a **small rank** ($\boldsymbol{u}\boldsymbol{v}^{\mathrm{T}}$ is an extreme case with rank $= 1$)
2. Matrices that have exponentially decreasing singular values (**low effective rank**).
3. **Incomplete matrices** (missing entries) that are completed to low rank matrices.

The first type are not invertible (because rank $< n$). The second type are invertible in theory but not in practice. The matrix with entries $(i + j - 1)^{-1}$ is a famous example. How can you recognize that this matrix or another matrix has very low effective rank ?

The third question—matrix completion—is approached in Section III.5. We create a minimization problem that applies to recommender matrices :

Minimize $||A||_N$ over all possible choices of the missing entries.

That "nuclear norm" gives a well-posed problem to replace a nonconvex problem : minimizing rank. *Nuclear norms are conjectured to be important in gradient descent.*

The rank of a matrix corresponds in some deep way to the number of nonzeros in a vector. In that analogy, a low rank matrix is like a **sparse vector**. Again, the number of nonzeros in $\boldsymbol{x}$ is not a norm ! That number is sometimes written as $||\boldsymbol{x}||_0$, but this "**ℓ^0 norm**" violates the rule $||2\boldsymbol{x}|| = 2||\boldsymbol{x}||$. We don't double the number of nonzeros.

It is highly important to find sparse solutions to $A\boldsymbol{x} = \boldsymbol{b}$. By a seeming miracle, sparse solutions come by minimizing the **ℓ^1 norm** $||\boldsymbol{x}||_1 = |x_1| + \cdots + |x_n|$. This fact has led to a new world of **compressed sensing**, with applications throughout engineering and medicine (including changes in the machines for Magnetic Resonance Imaging). Algorithms for ℓ^1 minimization are described and compared in III.4.

Section III.1 opens this chapter with a famous formula for $(I - \boldsymbol{u}\boldsymbol{v}^{\mathrm{T}})^{-1}$ and $(A - \boldsymbol{u}\boldsymbol{v}^{\mathrm{T}})^{-1}$. This is the Sherman-Morrison-Woodbury formula. It shows that the change in the inverse matrix also has rank 1 (if the matrix remains invertible). This formula with its extension to higher rank perturbations $(A - UV^{\mathrm{T}})^{-1}$ is fundamental.

We also compute the derivatives of $A(t)^{-1}$ and $\lambda(t)$ and $\sigma(t)$ when A varies with t.

III.1 Changes in A^{-1} from Changes in A

Suppose we subtract a low rank matrix from A. The next section estimates the change in eigenvalues and the change in singular values. This section finds an exact formula for the change in A^{-1}. The formula is called the **matrix inversion lemma** by some authors. To others it is better known as the **Sherman-Morrison-Woodbury formula**. Those names from engineering and statistics correspond to *updating and downdating* formulas in numerical analysis.

This formula is the key to updating the solution to a linear system $Ax = b$. The change could be in an existing row or column of A, or in adding/removing a row or column. Those would be rank one changes. We start with this simple example, when $A = I$.

$$\textbf{The inverse of } \boldsymbol{M = I - uv^{\mathrm{T}}} \textbf{ is } \boldsymbol{M^{-1} = I + \frac{uv^{\mathrm{T}}}{1 - v^{\mathrm{T}}u}} \tag{1}$$

There are two striking features of this formula. The first is that the correction to M^{-1} is **also rank one**. That is the final term $\boldsymbol{uv}^{\mathrm{T}}/(1-\boldsymbol{v}^{\mathrm{T}}\boldsymbol{u})$ in the formula. The second feature is that this correction term can become infinite. **Then M is not invertible:** no M^{-1}.

This occurs if the number $\boldsymbol{v}^{\mathrm{T}}\boldsymbol{u}$ happens to be 1. Equation (1) ends with a division by zero. In this case the formula fails. $M = I - \boldsymbol{uv}^{\mathrm{T}}$ is not invertible because $M\boldsymbol{u} = \boldsymbol{0}$:

$$M\boldsymbol{u} = (I - \boldsymbol{uv}^{\mathrm{T}})\boldsymbol{u} = \boldsymbol{u} - \boldsymbol{u}(\boldsymbol{v}^{\mathrm{T}}\boldsymbol{u}) = \boldsymbol{0} \text{ if } \boldsymbol{v}^{\mathrm{T}}\boldsymbol{u} = 1. \tag{2}$$

The simplest proof of formula (1) is a direct multiplication of M times M^{-1}:

$$MM^{-1} = (I - \boldsymbol{uv}^{\mathrm{T}})\left(I + \frac{\boldsymbol{uv}^{\mathrm{T}}}{1 - \boldsymbol{v}^{\mathrm{T}}\boldsymbol{u}}\right) = I - \boldsymbol{uv}^{\mathrm{T}} + \frac{(I - \boldsymbol{uv}^{\mathrm{T}})\boldsymbol{uv}^{\mathrm{T}}}{1 - \boldsymbol{v}^{\mathrm{T}}\boldsymbol{u}} = I - \boldsymbol{uv}^{\mathrm{T}} + \boldsymbol{uv}^{\mathrm{T}}. \tag{3}$$

You see how the number $\boldsymbol{v}^{\mathrm{T}}\boldsymbol{u}$ moves outside the matrix $\boldsymbol{uv}^{\mathrm{T}}$ in that key final step.

Now we have shown that formula (1) is correct. But we haven't shown where it came from. One good way is to introduce an "extension" of I to a matrix E with a new row and a new column:

$$\textbf{Extended matrix} \quad E = \begin{bmatrix} I & \boldsymbol{u} \\ \boldsymbol{v}^{\mathrm{T}} & 1 \end{bmatrix} \text{ has determinant } D = 1 - \boldsymbol{v}^{\mathrm{T}}\boldsymbol{u}$$

Elimination gives two ways to find E^{-1}. First, subtract $\boldsymbol{v}^{\mathrm{T}}$ times row 1 of E from row 2:

$$\begin{bmatrix} I & 0 \\ -\boldsymbol{v}^{\mathrm{T}} & 1 \end{bmatrix} E = \begin{bmatrix} I & \boldsymbol{u} \\ 0 & D \end{bmatrix}. \quad \text{Then } E^{-1} = \begin{bmatrix} I & \boldsymbol{u} \\ 0 & D \end{bmatrix}^{-1} \begin{bmatrix} I & 0 \\ -\boldsymbol{v}^{\mathrm{T}} & 1 \end{bmatrix} \tag{4}$$

The second way subtracts u times the second row of E from its first row:

$$\begin{bmatrix} I & -u \\ 0 & 1 \end{bmatrix} E = \begin{bmatrix} I - uv^{\mathrm{T}} & 0 \\ v^{\mathrm{T}} & 1 \end{bmatrix} \quad \text{Then } E^{-1} = \begin{bmatrix} I - uv^{\mathrm{T}} & 0 \\ v^{\mathrm{T}} & 1 \end{bmatrix}^{-1} \begin{bmatrix} I & -u \\ 0 & 1 \end{bmatrix}. \tag{5}$$

Now compare those two formulas for **the same** E^{-1}. Problem 2 does the algebra:

$$\textbf{Two forms of } E^{-1} \quad \begin{bmatrix} I + uD^{-1}v^{\mathrm{T}} & -uD^{-1} \\ -D^{-1}v^{\mathrm{T}} & D^{-1} \end{bmatrix} = \begin{bmatrix} M^{-1} & -M^{-1}u \\ -v^{\mathrm{T}}M^{-1} & 1 + v^{\mathrm{T}}M^{-1}u \end{bmatrix} \tag{6}$$

The $1,1$ blocks say that $M^{-1} = I + uD^{-1}v^{\mathrm{T}}$. *This is formula* (1), with $D = 1 - v^{\mathrm{T}}u$.

The Inverse of $M = I - UV^{\mathrm{T}}$

We can take a big step with no effort. Instead of a perturbation uv^{T} of rank 1, suppose we have a perturbation UV^{T} of rank k. The matrix **U is n by k** and the matrix **V^{T} is k by n.** So we have k columns and k rows exactly as we had one column u and one row v^{T}.

The formula for M^{-1} stays exactly the same! But there are two sizes I_n and I_k:

$$\textbf{The inverse of } M = I_n - UV^{\mathrm{T}} \textbf{ is } M^{-1} = I_n + U(I_k - V^{\mathrm{T}}U)^{-1}V^{\mathrm{T}} \tag{7}$$

This brings out an important point about these inverse formulas. **We are exchanging an inverse of size n for an inverse of size k.** Since $k = 1$ at the beginning of this section, we had an inverse of size 1 which was just an ordinary division by the number $1 - v^{\mathrm{T}}u$. Now $V^{\mathrm{T}}U$ is $(k \times n)\,(n \times k)$. We have a k by k matrix $I_k - V^{\mathrm{T}}U$ to invert, not n by n.

The fast proof of formula (7) is again a direct check that $MM^{-1} = I$:

$$(I_n - UV^{\mathrm{T}})(I_n + U(I_k - V^{\mathrm{T}}U)^{-1}V^{\mathrm{T}}) = I_n - UV^{\mathrm{T}} + (I_n - UV^{\mathrm{T}})U(I_k - V^{\mathrm{T}}U)^{-1}V^{\mathrm{T}}.$$

Replace $(I_n - UV^{\mathrm{T}})U$ in that equation by $U(I_k - V^{\mathrm{T}}U)$. This is a neat identity! The right side reduces to $I_n - UV^{\mathrm{T}} + UV^{\mathrm{T}}$ which is I_n. This proves formula (7).

Again there is an extended matrix E of size $n + k$ that holds the key:

$$E = \begin{bmatrix} I_n & U \\ V^{\mathrm{T}} & I_k \end{bmatrix} \textbf{ has determinant} = \det(I_n - UV^{\mathrm{T}}) = \det(I_k - V^{\mathrm{T}}U). \tag{8}$$

If $k << n$, the right hand side of (7) is probably easier and faster than a direct attack on the left hand side. **The matrix $V^{\mathrm{T}}U$ of size k is smaller than UV^{T} of size n.**

Example 1 What is the inverse of $M = I - \begin{bmatrix} 1 & 1 & 1 \\ 1 & 1 & 1 \\ 1 & 1 & 1 \end{bmatrix}$? In this case $u = v = \begin{bmatrix} 1 \\ 1 \\ 1 \end{bmatrix}$.

Solution Here $v^{\mathrm{T}}u = 3$ and $M^{-1} = I + \dfrac{uv^{\mathrm{T}}}{1-3}$. So M^{-1} equals $I - \dfrac{1}{2}\begin{bmatrix} 1 & 1 & 1 \\ 1 & 1 & 1 \\ 1 & 1 & 1 \end{bmatrix}$.

Example 2 If $M = I - \begin{bmatrix} 0 & 1 & 1 \\ 0 & 0 & 1 \\ 0 & 0 & 0 \end{bmatrix} = \boldsymbol{I - UV^{\mathrm{T}}}$ then $\boldsymbol{M^{-1}} = \begin{bmatrix} 1 & 1 & 0 \\ 0 & 1 & 1 \\ 0 & 0 & 1 \end{bmatrix}$

That came from writing the first displayed matrix as UV^{T} and reversing to $V^{\mathrm{T}}U$:

$$UV^{\mathrm{T}} = \begin{bmatrix} 1 & 0 \\ 0 & 1 \\ 0 & 0 \end{bmatrix} \begin{bmatrix} 0 & 1 & 1 \\ 0 & 0 & 1 \end{bmatrix} \quad \text{and} \quad V^{\mathrm{T}}U = \begin{bmatrix} 0 & 1 & 1 \\ 0 & 0 & 1 \end{bmatrix} \begin{bmatrix} 1 & 0 \\ 0 & 1 \\ 0 & 0 \end{bmatrix} = \begin{bmatrix} 0 & 1 \\ 0 & 0 \end{bmatrix}.$$

Then M^{-1} above is $\boldsymbol{I_3 + U\left[\ I_2 - V^{\mathrm{T}}U\ \right]^{-1} V^{\mathrm{T}}} = I_3 + U \begin{bmatrix} \mathbf{1} & \mathbf{-1} \\ \mathbf{0} & \mathbf{1} \end{bmatrix}^{-1} V^{\mathrm{T}}.$

The whole point is that the 3 by 3 matrix M^{-1} came from inverting that bold 2 by 2.

Perturbing any Invertible Matrix A

Up to now we have started with the identity matrix $I = I_n$. We modified it to $I - \boldsymbol{u}\boldsymbol{v}^{\mathrm{T}}$ and then to $I - UV^{\mathrm{T}}$. Change by rank 1 and then by rank k. To get the benefit of the full Sherman-Morrison-Woodbury idea, we now go further: *Start with A instead of I.*

Perturb any invertible A by a rank k matrix UV^{T}. Now $M = A - UV^{\mathrm{T}}$.

Sherman-Morrison-Woodbury formula

$$\boldsymbol{M^{-1} = (A - UV^{\mathrm{T}})^{-1} = A^{-1} + A^{-1}U(I - V^{\mathrm{T}}A^{-1}U)^{-1}V^{\mathrm{T}}A^{-1}} \tag{9}$$

Up to now A was I_n. The final formula (9) still connects to an extension matrix E.

Suppose A is invertible $\quad E = \begin{bmatrix} A & U \\ V^{\mathrm{T}} & I \end{bmatrix}$ is invertible when $M = A - UV^{\mathrm{T}}$ is invertible.

To find that inverse of E, we can do row operations to replace V^{T} by zeros:

Multiply row 1 by $V^{\mathrm{T}}A^{-1}$ and subtract from row 2 to get $\begin{bmatrix} A & U \\ 0 & I - V^{\mathrm{T}}A^{-1}U \end{bmatrix}$

Or we can do column operations to replace U by zeros:

Multiply column 1 by $A^{-1}U$ and subtract from column 2 to get $\begin{bmatrix} A & 0 \\ V^{\mathrm{T}} & I - V^{\mathrm{T}}A^{-1}U \end{bmatrix}$

As in equation (6), we have two ways to invert E. These two forms of E^{-1} must be equal.

$$\begin{bmatrix} A^{-1} + A^{-1}UC^{-1}V^{\mathrm{T}}A^{-1} & -A^{-1}UC^{-1} \\ -C^{-1}VA^{-1} & C^{-1} \end{bmatrix} = \begin{bmatrix} M^{-1} & -M^{-1}U \\ V^{\mathrm{T}}B^{-1} & I_k + V^{\mathrm{T}}M^{-1}U \end{bmatrix} \tag{10}$$

Here C is $I - V^{\mathrm{T}}A^{-1}U$ and M is $A - UV^{\mathrm{T}}$. The desired matrix is M^{-1} (the inverse when A is perturbed). Comparing the $(1,1)$ blocks in equation (10) produces equation (9).

Summary The n by n inverse of $M = A - UV^{\mathrm{T}}$ comes from the n by n inverse of A and the k by k inverse of $C = I - V^{\mathrm{T}}A^{-1}U$. **For a fast proof, multiply (9) by $\boldsymbol{A - UV^{\mathrm{T}}}$.**

This is a good place to collect four closely related matrix identities. In every case, a matrix B or A^{T} or U on the left reappears on the right, even if it doesn't commute with A or V. As in many proofs, the *associative law* is hiding in plain sight: $B(AB) = (BA)B$.

$$\boxed{\begin{aligned} B(I_m + AB) &= (I_n + BA)B \\ B(I_m + AB)^{-1} &= (I_n + BA)^{-1}B \\ A^{\mathrm{T}}(AA^{\mathrm{T}} + \lambda I_n)^{-1} &= (A^{\mathrm{T}}A + \lambda I_m)^{-1}A^{\mathrm{T}} \\ U(I_k - V^{\mathrm{T}}U) &= (I_n - UV^{\mathrm{T}})U \end{aligned}}$$

A is m by n and B is n by m. The second identity includes the fact that $I + AB$ is invertible exactly when $I + BA$ is invertible. In other words, -1 is not an eigenvalue of AB exactly when -1 is not an eigenvalue of BA. *AB and BA have the same nonzero eigenvalues.*

The key as in Section I.6 is that $(I + AB)\boldsymbol{x} = \mathbf{0}$ leads to $(I + BA)B\boldsymbol{x} = \mathbf{0}$.

The Derivative of A^{-1}

In a moment this section will turn to applications of the inverse formulas. First I turn to matrix calculus! The whole point of derivatives is to find the change in a function $f(x)$ when x is *moved very slightly.* That change Δx produces a change Δf. Then the ratio of Δf to Δx approaches the derivative df/dx.

Here x is a matrix A. The function is $f(A) = A^{-1}$. How does A^{-1} change when A changes? Up to now the change $\boldsymbol{u}\boldsymbol{v}^{\mathrm{T}}$ or UV^{T} was small in rank. Now the desired change in A will be infinitesimally small, of any rank.

I start with the letter $B = A + \Delta A$, and write down this very useful matrix formula:

$$\boxed{\boldsymbol{B^{-1} - A^{-1} = B^{-1}\,(A - B)\,A^{-1}}} \tag{11}$$

You see that this equation is true. On the right side AA^{-1} is I and $B^{-1}B$ is I. In fact (11) could lead to the earlier formulas for $(A - UV^{\mathrm{T}})^{-1}$. It shows instantly that if $A - B$ has rank 1 (or k), then $B^{-1} - A^{-1}$ has rank 1 (or k). The matrices A and B are assumed invertible, so multiplication by B^{-1} or A^{-1} has no effect on the rank.

Now think of $A = A(t)$ as a matrix that changes with the time t. Its derivative at each time t is dA/dt. Of course A^{-1} is also changing with the time t. **We want to find its derivative $d\boldsymbol{A}^{-1}/dt$.** So we divide those changes $\Delta A = B - A$ and $\Delta A^{-1} = B^{-1} - A^{-1}$ by Δt. Now insert $A + \Delta A$ for B in equation (11) and let $\Delta t \to 0$.

$$\boxed{\frac{\Delta A^{-1}}{\Delta t} = -(A + \Delta A)^{-1}\,\frac{\Delta A}{\Delta t}\,A^{-1} \text{ approaches } \frac{\boldsymbol{dA^{-1}}}{\boldsymbol{dt}} = \boldsymbol{-A^{-1}\frac{dA}{dt}A^{-1}}} \tag{12}$$

For a 1 by 1 matrix $A = t$, with $dA/dt = 1$, we recover the derivative of $1/t$ as $-1/t^2$.

Problem 7 points out that the derivative of A^2 **is not** $2A\,dA/dt$!

Updating Least Squares

Section II.2 discussed the least squares equation $A^{\mathrm{T}}A\widehat{\boldsymbol{x}} = A^{\mathrm{T}}\boldsymbol{b}$—the "normal equations" to minimize $||\boldsymbol{b} - A\boldsymbol{x}||^2$. *Suppose that a new equation arrives.* Then A has a new row $\boldsymbol{r}$ (1 by n) and there is a new measurement b_{m+1} and a new $\widehat{\boldsymbol{x}}$:

$$\begin{bmatrix} A^{\mathrm{T}} & \boldsymbol{r}^{\mathrm{T}} \end{bmatrix}\begin{bmatrix} A \\ \boldsymbol{r} \end{bmatrix}\widehat{\boldsymbol{x}} = \begin{bmatrix} A^{\mathrm{T}} & \boldsymbol{r}^{\mathrm{T}} \end{bmatrix}\begin{bmatrix} \boldsymbol{b} \\ b_{m+1} \end{bmatrix} \quad \text{is} \quad \begin{bmatrix} A^{\mathrm{T}}A + \boldsymbol{r}^{\mathrm{T}}\boldsymbol{r} \end{bmatrix}\widehat{\boldsymbol{x}} = A^{\mathrm{T}}\boldsymbol{b} + \boldsymbol{r}^{\mathrm{T}}b_{m+1}. \quad (13)$$

The matrix in the new normal equations is $A^{\mathrm{T}}A + \boldsymbol{r}^{\mathrm{T}}\boldsymbol{r}$. This is a rank one correction to the original $A^{\mathrm{T}}A$. To update $\widehat{\boldsymbol{x}}$, we do *not* want to create and solve a whole new set of normal equations. Instead we use the update formula:

$$\begin{bmatrix} A^{\mathrm{T}}A + \boldsymbol{r}^{\mathrm{T}}\boldsymbol{r} \end{bmatrix}^{-1} = (A^{\mathrm{T}}A)^{-1} - c\,(A^{\mathrm{T}}A)^{-1}\boldsymbol{r}^{\mathrm{T}}\boldsymbol{r}\,(A^{\mathrm{T}}A)^{-1} \text{ with } c = 1/(1+\boldsymbol{r}(A^{\mathrm{T}}A)^{-1}\boldsymbol{r}^{\mathrm{T}}) \quad (14)$$

To find c quickly we only need to solve the old equation $(A^{\mathrm{T}}A)\,\boldsymbol{y} = \boldsymbol{r}^{\mathrm{T}}$.

Problem 4 will produce the least squares solution $\widehat{\boldsymbol{x}}_{\text{new}}$ as an update of $\widehat{\boldsymbol{x}}$. The same idea applies when A has M new rows instead of one. This is **recursive least squares.**

The Kalman Filter

Kalman noticed that this update idea also applies to **dynamic least squares**. That word *dynamic* means that even without new data, the state vector $\boldsymbol{x}$ is changing with time. If $\boldsymbol{x}$ gives the position of a GPS satellite, that position will move by about $\Delta\boldsymbol{x} = \boldsymbol{v}\Delta t$ ($\boldsymbol{v}$ = velocity). This approximation or a better one will be the **state equation** for $\boldsymbol{x}_{n+1}$ at the new time. Then a new measurement b_{m+1} at time $t + \Delta t$ will further update that approximate position to $\widehat{\boldsymbol{x}}_{\text{new}}$. I hope you see that we are now adding **two new equations** (state equation and measurement equation) to the original system $A\boldsymbol{x} \approx \boldsymbol{b}$:

$$\begin{matrix} \textbf{Original} \\ \textbf{State update} \\ \textbf{Measurement update} \end{matrix} \qquad A_{\text{new}} = \begin{bmatrix} A & 0 \\ -I & I \\ 0 & \boldsymbol{r} \end{bmatrix}\begin{bmatrix} \boldsymbol{x}_{\text{old}} \\ \boldsymbol{x}_{\text{new}} \end{bmatrix} = \begin{bmatrix} \boldsymbol{b} \\ \boldsymbol{v}\Delta t \\ b_{m+1} \end{bmatrix}. \quad (15)$$

We want the least squares solution of (15). And there is one more twist that makes the Kalman filter formulas truly impressive (or truly complicated). The state equation and the measurement equation have their own covariance matrices. Those equations are inexact (of course). The variance or covariance V measures their different reliabilities. The normal equations $A^{\mathrm{T}}A\widehat{\boldsymbol{x}} = A^{\mathrm{T}}\boldsymbol{b}$ should properly be weighted by V^{-1} to become $A^{\mathrm{T}}V^{-1}A\widehat{\boldsymbol{x}} = A^{\mathrm{T}}V^{-1}\boldsymbol{b}$. And in truth V itself has to be updated at each step.

Through all this, Kalman pursued the goal of using update formulas. Instead of solving the full normal equations to learn $\widehat{\boldsymbol{x}}_{\text{new}}$, he updated $\widehat{\boldsymbol{x}}_{\text{old}}$ in two steps.

The prediction $\widehat{\boldsymbol{x}}_{\text{state}}$ comes from the state equation. Then comes the correction to $\widehat{\boldsymbol{x}}_{\text{new}}$, using the new measurement b_{m+1}: zero correction.

$$\boldsymbol{K = \textbf{Kalman gain matrix}} \qquad \widehat{\boldsymbol{x}}_{\text{new}} = \widehat{\boldsymbol{x}}_{\text{state}} + K(b_{m+1} - \boldsymbol{r}\,\widehat{\boldsymbol{x}}_{\text{state}}) \quad (16)$$

The *gain matrix* K is created from A and $\boldsymbol{r}$ and the covariance matrices V_{state} and $\boldsymbol{V_b}$. You see that if the new b_{m+1} agrees perfectly with the prediction $\widehat{\boldsymbol{x}}_{\text{state}}$, then there is a zero correction in (16) from $\widehat{\boldsymbol{x}}_{\text{state}}$ to $\widehat{\boldsymbol{x}}_{\text{new}}$.

We also need to update the covariance of the whole system—measuring the reliability of $\widehat{x}_{\text{new}}$. In fact this V is often the most important output. It measures the accuracy of the whole system of sensors that produced $\widehat{x}_{\text{final}}$.

For the GPS application, our text with Kai Borre provides much more detail: *Algorithms for Global Positioning* (Wellesley-Cambridge Press). The goal is to estimate the accuracy of GPS measurements: very high accuracy for the measurement of tectonic plates, lower accuracy for satellites, and much lower accuracy for the position of your car.

Quasi-Newton Update Methods

A completely different update occurs in approximate Newton methods to solve $\boldsymbol{f}(\boldsymbol{x}) = \mathbf{0}$. Those are n equations for n unknowns $x_1, \ldots, x_n$. The classical Newton's method uses the Jacobian matrix $\boldsymbol{J}(\boldsymbol{x})$ containing the first derivatives of each component of $\boldsymbol{f}$:

$$\textbf{Newton} \qquad J_{ik} = \frac{\partial f_i}{\partial x_k} \quad \text{and} \quad \boldsymbol{x}_{\text{new}} = \boldsymbol{x}_{\text{old}} - \boldsymbol{J}(\boldsymbol{x}_{\text{old}})^{-1} \boldsymbol{f}(\boldsymbol{x}_{\text{old}}). \qquad (17)$$

That is based on the fundamental approximation $\boldsymbol{J}\,\Delta\boldsymbol{x} = \Delta\boldsymbol{f}$ of calculus. Here $\Delta\boldsymbol{f} = \boldsymbol{f}(\boldsymbol{x}_{\text{new}}) - \boldsymbol{f}(\boldsymbol{x}_{\text{old}})$ is $-\boldsymbol{f}(\boldsymbol{x}_{\text{old}})$ because our whole plan is to achieve $\boldsymbol{f}(\boldsymbol{x}_{\text{new}}) \approx \mathbf{0}$.

The difficulty is the Jacobian matrix J. For large n, even automatic differentiation (the key to backpropagation in Chapter VII) will be slow. Instead of recomputing $\boldsymbol{J}$ at each iteration, *quasi-Newton methods use an update formula* $\boldsymbol{J}(\boldsymbol{x}_{\text{new}}) = \boldsymbol{J}(\boldsymbol{x}_{\text{old}}) + \Delta\boldsymbol{J}$.

In principle $\Delta\boldsymbol{J}$ involves the derivatives of J and therefore **second derivatives of $\boldsymbol{f}$**. The reward is second order accuracy and fast convergence of Newton's method. But the price of computing all second derivatives when n is large (as in deep learning) may be impossibly high.

Quasi-Newton methods create a low rank update to $J(\boldsymbol{x}_{\text{old}})$ instead of computing an entirely new Jacobian at $\boldsymbol{x}_{\text{new}}$. The update reflects the new information that comes with computing $\boldsymbol{x}_{\text{new}}$ in (17). Because it is J^{-1} that appears in Newton's method, the update formula accounts for its rank one change to J_{new}^{-1}—**without recomputing $\boldsymbol{J}^{-1}$**. Here is the key, and derivatives of $f_1, \ldots, f_n$ are not needed:

$$\textbf{Quasi-Newton condition} \qquad J_{\text{new}}\,(\boldsymbol{x}_{\text{new}} - \boldsymbol{x}_{\text{old}}) = \boldsymbol{f}_{\text{new}} - \boldsymbol{f}_{\text{old}} \qquad (18)$$

This is information $\boldsymbol{J}\Delta\boldsymbol{x} = \Delta\boldsymbol{f}$ in the direction we moved. Since equation (17) uses J^{-1} instead of J, we update J^{-1} to satisfy (18). The Sherman-Morrison formula will do this. Or the "BFGS correction" is a rank-2 matrix discovered by four authors at the same time. Another approach is to update the LU or the LDL^{T} factors of J_{old}.

Frequently the original n equations $\boldsymbol{f}(\boldsymbol{x}) = \mathbf{0}$ come from minimizing a function $F(x_1, \ldots, x_n)$. Then $\boldsymbol{f} = (\partial F/\partial x_1, \ldots, \partial F/\partial x_n)$ is the gradient of this function F, and $\boldsymbol{f} = \mathbf{0}$ at the minimum point. Now the Jacobian matrix J (first derivatives of $\boldsymbol{f}$) becomes a Hessian matrix H (second derivatives of F). Its entries are $H_{jk} = \partial^2 F/\partial x_j\,\partial x_k$.

If all goes well, Newton's method quickly finds the point $\boldsymbol{x}^*$ where F is minimized and its derivatives are $\boldsymbol{f}(\boldsymbol{x}^*) = \mathbf{0}$. The quasi-Newton method that updates J approximately instead of recomputing J is far more affordable for large n. For extremely large n (as in many problems of machine learning) the cost may still be excessive.

Problem Set III.1

1 Another approach to $(I - \boldsymbol{u}\boldsymbol{v}^{\mathrm{T}})^{-1}$ starts with the formula for a geometric series: $(1-x)^{-1} = 1 + x + x^2 + x^3 + \cdots$ Apply that formula when $x = \boldsymbol{u}\boldsymbol{v}^{\mathrm{T}}$ = matrix:

$$\begin{aligned}(I - \boldsymbol{u}\boldsymbol{v}^{\mathrm{T}})^{-1} &= I + \boldsymbol{u}\boldsymbol{v}^{\mathrm{T}} + \boldsymbol{u}\boldsymbol{v}^{\mathrm{T}}\boldsymbol{u}\boldsymbol{v}^{\mathrm{T}} + \boldsymbol{u}\boldsymbol{v}^{\mathrm{T}}\boldsymbol{u}\boldsymbol{v}^{\mathrm{T}}\boldsymbol{u}\boldsymbol{v}^{\mathrm{T}} + \cdots \\ &= I + \boldsymbol{u}\left[1 + \boldsymbol{v}^{\mathrm{T}}\boldsymbol{u} + \boldsymbol{v}^{\mathrm{T}}\boldsymbol{u}\boldsymbol{v}^{\mathrm{T}}\boldsymbol{u} + \cdots\right]\boldsymbol{v}^{\mathrm{T}}.\end{aligned}$$

Take $x = \boldsymbol{v}^{\mathrm{T}}\boldsymbol{u}$ to see $I + \dfrac{\boldsymbol{u}\boldsymbol{v}^{\mathrm{T}}}{1 - \boldsymbol{v}^{\mathrm{T}}\boldsymbol{u}}$. This is exactly equation (1) for $(I - \boldsymbol{u}\boldsymbol{v}^{\mathrm{T}})^{-1}$.

2 Find E^{-1} from equation (4) with $D = 1 - \boldsymbol{v}^{\mathrm{T}}\boldsymbol{u}$ and also from equation (5):

From (4) $\quad E^{-1} = \begin{bmatrix} I & -\boldsymbol{u}D^{-1} \\ 0 & D^{-1} \end{bmatrix}\begin{bmatrix} I & 0 \\ -\boldsymbol{v}^{\mathrm{T}} & 1 \end{bmatrix} = \begin{bmatrix} & \\ & \end{bmatrix}$

From (5) $\quad E^{-1} = \begin{bmatrix} (I - \boldsymbol{u}\boldsymbol{v}^{\mathrm{T}})^{-1} & 0 \\ -\boldsymbol{v}^{\mathrm{T}}(I - \boldsymbol{u}\boldsymbol{v}^{\mathrm{T}})^{-1} & 1 \end{bmatrix}\begin{bmatrix} I & -\boldsymbol{u} \\ 0 & 1 \end{bmatrix} = \begin{bmatrix} & \\ & \end{bmatrix}$

Compare the $1, 1$ blocks to find $M^{-1} = (I - \boldsymbol{u}\boldsymbol{v}^{\mathrm{T}})^{-1}$ in formula (1).

3 The final Sherman-Morrison-Woodbury formula (9) perturbs A by UV^{T} (rank k). Write down that formula in the important case when $k = 1$:

$$M^{-1} = (A - \boldsymbol{u}\boldsymbol{v}^{\mathrm{T}})^{-1} = A^{-1} + \underline{\hspace{3cm}}.$$

Test the formula on this small example:

$$A = \begin{bmatrix} 3 & 0 \\ 0 & 2 \end{bmatrix} \quad \boldsymbol{u} = \begin{bmatrix} 1 \\ 0 \end{bmatrix} \quad \boldsymbol{v} = \begin{bmatrix} 1 \\ 0 \end{bmatrix} \quad A - \boldsymbol{u}\boldsymbol{v}^{\mathrm{T}} = \begin{bmatrix} 2 & 0 \\ 0 & 2 \end{bmatrix}$$

4 Problem 3 found the inverse matrix $M^{-1} = (A - \boldsymbol{u}\boldsymbol{v}^{\mathrm{T}})^{-1}$. In solving the equation $M\boldsymbol{y} = \boldsymbol{b}$, we compute **only the solution** $\boldsymbol{y}$ and not the whole inverse matrix M^{-1}. You can find $\boldsymbol{y}$ in two easy steps:

Step 1 Solve $A\boldsymbol{x} = \boldsymbol{b}$ and $A\boldsymbol{z} = \boldsymbol{u}$. Compute $D = 1 - \boldsymbol{v}^{\mathrm{T}}\boldsymbol{z}$.

Step 2 Then $\boldsymbol{y} = \boldsymbol{x} + \dfrac{\boldsymbol{v}^{\mathrm{T}}\boldsymbol{x}}{D}\boldsymbol{z}$ is the solution to $M\boldsymbol{y} = (A - \boldsymbol{u}\boldsymbol{v}^{\mathrm{T}})\boldsymbol{y} = \boldsymbol{b}$.

Verify $(A - \boldsymbol{u}\boldsymbol{v}^{\mathrm{T}})\boldsymbol{y} = \boldsymbol{b}$. **We solved two equations using A, no equations using M.**

5 **Prove that the final formula (9) is correct! Multiply equation (9) by $A - UV^{\mathrm{T}}$.**

Watch for the moment when $(A - UV^{\mathrm{T}})A^{-1}U$ becomes $U(I - V^{\mathrm{T}}A^{-1}U)$.

6 In the foolish case $U = V = I_n$, equation (9) gives what formula for $(A - I)^{-1}$? Can you prove it directly?

7 Problem 4 extends to a rank k change $M^{-1} = (A - UV^{\mathrm{T}})^{-1}$. To solve the equation $My = b$, we compute only the solution y and not the whole inverse matrix M^{-1}.

Step 1 Solve $\boldsymbol{Ax} = \boldsymbol{b}$ and the k equations $\boldsymbol{AZ} = \boldsymbol{U}$ (U and Z are n by k)

Step 2 Form the matrix $C = I - V^{\mathrm{T}}Z$ and solve $Cw = V^{\mathrm{T}}\boldsymbol{x}$. The desired $y = M^{-1}b$ is $y = x + Zw$.

Use (9) to verify that $(A - UV^{\mathrm{T}})\,y = b$. We solved $k + 1$ equations using A and we multiplied $V^{\mathrm{T}}Z$, but we never used $M = A - UV^{\mathrm{T}}$.

8 What is the derivative of $(A(t))^2$? **The correct derivative is not $2\,A(t)\,\dfrac{dA}{dt}$.** You must compute $(A + \Delta A)^2$ and subtract A^2. Divide by Δt and send Δt to 0.

9 Test formula (12) for the derivative of $A^{-1}(t)$ when

$$A(t) = \begin{bmatrix} 1 & t^2 \\ 0 & 1 \end{bmatrix} \quad \text{and} \quad A^{-1}(t) = \begin{bmatrix} 1 & -t^2 \\ 0 & 1 \end{bmatrix}.$$

10 Suppose you know the average $\widehat{x}_{\text{old}}$ of $b_1, b_2, \ldots, b_{999}$. When b_{1000} arrives, check that the new average is a combination of $\widehat{x}_{\text{old}}$ and the mismatch $b_{1000} - \widehat{x}_{\text{old}}$:

$$\widehat{x}_{\text{new}} = \frac{b_1 + \cdots + b_{1000}}{1000} = \frac{b_1 + \cdots + b_{999}}{999} + \frac{1}{1000}\left(b_{1000} - \frac{b_1 + \cdots + b_{999}}{999}\right).$$

This is a "Kalman filter" $\widehat{x}_{\text{new}} = \widehat{x}_{\text{old}} + \frac{1}{1000}\,(b_{1000} - \widehat{x}_{\text{old}})$ with gain matrix $\frac{1}{1000}$.

11 The Kalman filter includes also a *state equation* $x_{k+1} = Fx_k$ *with its own error variance* s^2. The dynamic least squares problem allows x to "drift" as k increases:

$$\begin{bmatrix} 1 & \\ -F & 1 \\ & 1 \end{bmatrix} \begin{bmatrix} x_0 \\ x_1 \end{bmatrix} = \begin{bmatrix} b_0 \\ 0 \\ b_1 \end{bmatrix} \quad \text{with variances} \quad \begin{bmatrix} \sigma^2 \\ s^2 \\ \sigma^2 \end{bmatrix}.$$

With $F = 1$, divide both sides of those three equations by σ, s, and σ. Find $\widehat{x_0}$ and $\widehat{x_1}$ by least squares, which gives more weight to the recent b_1.

Bill Hager's paper on *Updating the Inverse of a Matrix* was extremely useful in writing this section of the book: SIAM Review **31** (1989).

III.2 Interlacing Eigenvalues and Low Rank Signals

The previous section found the change in A^{-1} produced by a change in A. We could allow infinitesimal changes dA and also finite changes $\Delta A = -UV^{\mathrm{T}}$. The results were an infinitesimal change or a finite change in the inverse matrix:

$$\frac{dA^{-1}}{dt} = -A^{-1}\frac{dA}{dt}A^{-1} \quad \text{and} \quad \Delta A^{-1} = A^{-1}U(I - V^{\mathrm{T}}A^{-1}U)^{-1}V^{\mathrm{T}}A^{-1} \tag{1}$$

This section asks the same questions about the eigenvalues and singular values of A.

How do each λ and each σ change as the matrix A changes?

You will see nice formulas for $d\lambda/dt$ and $d\sigma/dt$. But not much is linear about eigenvalues or singular values. Calculus succeeds for infinitesimal changes $d\lambda$ and $d\sigma$, because the derivative is a linear operator. But we can't expect to know exact values in the jumps to $\lambda(A + \Delta A)$ or $\sigma(A + \Delta A)$. Eigenvalues are more complicated than inverses.

Still there is good news. What can be achieved is remarkable. Here is a taste for a symmetric matrix S. Suppose S changes to $S + uu^{\mathrm{T}}$ (a "positive" change of rank 1). Its eigenvalues change from $\lambda_1 \geq \lambda_2 \geq \ldots$ to $z_1 \geq z_2 \geq \ldots$ We expect increases in eigenvalues since uu^{T} was positive semidefinite. *But how large are the increases?*

Each eigenvalue z_i of $S + uu^{\mathrm{T}}$ is not smaller than λ_i or greater than λ_{i-1}. So the λ's and z's are "interlaced". Each $z_2, \ldots, z_n$ is between two λ's:

$$z_1 \geq \lambda_1 \geq z_2 \geq \lambda_2 \geq \ldots \geq z_n \geq \lambda_n. \tag{2}$$

We have upper bounds on the eigenvalue changes even if we don't have formulas for $\Delta\lambda$. There is one point to notice because it could be misunderstood. Suppose the change uu^{T} in the matrix is $Cq_2q_2^{\mathrm{T}}$ (where q_2 is the second unit eigenvector of S). Then $Sq_2 = \lambda_2 q_2$ will see a jump in that eigenvalue to $\lambda_2 + C$, because $(S + Cq_2q_2^{\mathrm{T}})q_2 = (\lambda_2 + C)q_2$. That jump is large if C is large. *So how could the second eigenvalue of $S + uu^{\mathrm{T}}$ possibly have $z_2 = \lambda_2 + C \leq \lambda_1$?*

Answer: If C is a big number, then $\lambda_2 + C$ is not the second eigenvalue of $S + uu^{\mathrm{T}}$! It becomes z_1, the **largest eigenvalue** of the new matrix $S + Cq_2q_2^{\mathrm{T}}$ (and its eigenvector is q_2). The original top eigenvalue λ_1 of S is now the second eigenvalue z_2 of the new matrix. So the statement (2) that $z_2 \leq \lambda_1 \leq z_1$ is the completely true statement (in this example) that $z_2 = \lambda_1$ is below $z_1 = \lambda_2 + C$.

We will connect this interlacing to the fact that the eigenvectors between $\lambda_1 = \lambda_{\max}$ and $\lambda_n = \lambda_{\min}$ are all **saddle points of the ratio $R(x) = x^{\mathrm{T}}Sx/x^{\mathrm{T}}x$.**

The Derivative of an Eigenvalue

We have a matrix $A(t)$ that is changing with the time t. So its eigenvalues $\lambda(t)$ are also changing. We will suppose that *no eigenvalues of* $A(0)$ *are repeated*—each eigenvalue $\lambda(0)$ of $A(0)$ can be safely followed for at least a short time t, as $\lambda(0)$ changes to an eigenvalue $\lambda(t)$ of $A(t)$. **What is its derivative $d\lambda/dt$?**

The key to $d\lambda/dt$ is to assemble the facts we know. The first is $A(t)\boldsymbol{x}(t) = \lambda(t)\boldsymbol{x}(t)$. The second is that the transpose matrix $A^{\mathrm{T}}(t)$ also has the eigenvalue $\lambda(t)$, because $\det(A^{\mathrm{T}} - \lambda I) = \det(A - \lambda I)$. Probably A^{T} has a different eigenvector $\boldsymbol{y}(t)$. When $\boldsymbol{x}$ is column k of the eigenvector matrix X for A, $\boldsymbol{y}$ is column k of the eigenvector matrix $(X^{-1})^{\mathrm{T}}$ for A^{T}. (The reason is that $A = X^{-1}\Lambda X$ leads to $A^{\mathrm{T}} = X^{\mathrm{T}}\Lambda(X^{-1})^{\mathrm{T}}$). The lengths of $\boldsymbol{x}$ and $\boldsymbol{y}$ are normalized by $X^{-1}X = I$. *This requires* $\boldsymbol{y}^{\mathrm{T}}(t)\boldsymbol{x}(t) = \mathbf{1}$ *for all* t.

Here are these facts on one line with the desired formula for $d\lambda/dt$ on the next line.

$$\textbf{Facts} \quad \boldsymbol{A}(t)\,\boldsymbol{x}(t) = \boldsymbol{\lambda}(t)\,\boldsymbol{x}(t) \quad \boldsymbol{y}^{\mathrm{T}}(t)\boldsymbol{A}(t) = \boldsymbol{\lambda}(t)\boldsymbol{y}^{\mathrm{T}}(t) \quad \boldsymbol{y}^{\mathrm{T}}(t)\,\boldsymbol{x}(t) = 1 \tag{3}$$

$$\textbf{Formulas} \quad \boldsymbol{\lambda}(t) = \boldsymbol{y}^{\mathrm{T}}(t)\boldsymbol{A}(t)\,\boldsymbol{x}(t) \quad \text{and} \quad \frac{d\boldsymbol{\lambda}}{dt} = \boldsymbol{y}^{\mathrm{T}}(t)\frac{d\boldsymbol{A}}{dt}\,\boldsymbol{x}(t) \tag{4}$$

To find that formula $\lambda = \boldsymbol{y}^{\mathrm{T}}A\boldsymbol{x}$, just multiply the first fact $A\boldsymbol{x} = \lambda\boldsymbol{x}$ by $\boldsymbol{y}^{\mathrm{T}}$ and use $\boldsymbol{y}^{\mathrm{T}}\boldsymbol{x} = 1$. Or multiply the second fact $\boldsymbol{y}^{\mathrm{T}}A = \lambda\boldsymbol{y}^{\mathrm{T}}$ on the right side by $\boldsymbol{x}$.

Now take the derivative of $\boldsymbol{\lambda} = \boldsymbol{y}^{\mathrm{T}}A\,\boldsymbol{x}$. The product rule gives three terms in $d\lambda/dt$:

$$\boxed{\frac{d\lambda}{dt}} = \frac{d\boldsymbol{y}^{\mathrm{T}}}{dt}A\,\boldsymbol{x} + \boxed{\boldsymbol{y}^{\mathrm{T}}\frac{d\boldsymbol{A}}{dt}\boldsymbol{x}} + \boldsymbol{y}^{\mathrm{T}}A\frac{d\boldsymbol{x}}{dt} \tag{5}$$

The middle term is the correct derivative $d\lambda/dt$. The first and third terms add to zero:

$$\frac{d\boldsymbol{y}^{\mathrm{T}}}{dt}A\,\boldsymbol{x} + \boldsymbol{y}^{\mathrm{T}}A\frac{d\boldsymbol{x}}{dt} = \lambda\left(\frac{d\boldsymbol{y}^{\mathrm{T}}}{dt}\boldsymbol{x} + \boldsymbol{y}^{\mathrm{T}}\frac{d\boldsymbol{x}}{dt}\right) = \lambda\frac{d}{dt}\left(\boldsymbol{y}^{\mathrm{T}}\boldsymbol{x}\right) = \lambda\frac{d}{dt}(1) = \mathbf{0}. \tag{6}$$

There are also formulas for $d^2\lambda/dt^2$ and $d\boldsymbol{x}/dt$ (but they are more complicated).

Example $A = \begin{bmatrix} 2t & 1 \\ 2t & 2 \end{bmatrix}$ has $\lambda^2 - 2(1+t)\lambda + 2t = 0$ and $\boldsymbol{\lambda} = \mathbf{1} + \boldsymbol{t} \pm \sqrt{\mathbf{1} + \boldsymbol{t}^2}$.

At $t = 0$, $\lambda_1 = \mathbf{2}$ and $\lambda_2 = \mathbf{0}$ and the derivatives of λ_1 and λ_2 are $1 \pm t(1+t^2)^{-1/2} = \mathbf{1}$.

The eigenvectors for $\lambda_1 = 2$ at $t = 0$ are $\boldsymbol{y}_1^{\mathrm{T}} = \begin{bmatrix} 0 & 1 \end{bmatrix}$ and $\boldsymbol{x}_1 = \begin{bmatrix} 1/2 \\ 1 \end{bmatrix}$.

The eigenvectors for $\lambda_2 = 0$ at $t = 0$ are $\boldsymbol{y}_2^{\mathrm{T}} = \begin{bmatrix} 1 & -\frac{1}{2} \end{bmatrix}$ and $\boldsymbol{x}_2 = \begin{bmatrix} 1 \\ 0 \end{bmatrix}$.

Now equation (5) confirms that $\dfrac{d\boldsymbol{\lambda}_1}{dt} = \boldsymbol{y}_1^{\mathrm{T}}\dfrac{d\boldsymbol{A}}{dt}\boldsymbol{x}_1 = \begin{bmatrix} \mathbf{0} & \mathbf{1} \end{bmatrix}\begin{bmatrix} \mathbf{2} & \mathbf{0} \\ \mathbf{2} & \mathbf{0} \end{bmatrix}\begin{bmatrix} \mathbf{1/2} \\ \mathbf{1} \end{bmatrix} = 1.$

The Derivative of a Singular Value

A similar formula for $d\sigma/dt$ (derivative of a non-repeated $\sigma(t)$) comes from $A\boldsymbol{v} = \sigma\boldsymbol{u}$:

$$U^{\mathrm{T}}AV = \Sigma \qquad \boldsymbol{u}^{\mathrm{T}}(t)\,A(t)\,\boldsymbol{v}(t) = \boldsymbol{u}^{\mathrm{T}}(t)\,\sigma(t)\,\boldsymbol{u}(t) = \sigma(t). \tag{7}$$

The derivative of the left side has three terms from the product rule, as in (5). The first and third terms are zero because $A\boldsymbol{v} = \sigma\boldsymbol{u}$ and $A^{\mathrm{T}}\boldsymbol{u} = \sigma\boldsymbol{v}$ and $\boldsymbol{u}^{\mathrm{T}}\boldsymbol{u} = \boldsymbol{v}^{\mathrm{T}}\boldsymbol{v} = 1$. The derivatives of $\boldsymbol{u}^{\mathrm{T}}\boldsymbol{u}$ and $\boldsymbol{v}^{\mathrm{T}}\boldsymbol{v}$ are zero, so

$$\frac{d\boldsymbol{u}^{\mathrm{T}}}{dt}A(t)\,\boldsymbol{v}(t) = \sigma(t)\,\frac{d\boldsymbol{u}^{\mathrm{T}}}{dt}\boldsymbol{u}(t) = \mathbf{0} \quad \text{and} \quad \boldsymbol{u}^{\mathrm{T}}(t)\,A(t)\frac{d\boldsymbol{v}}{dt} = \sigma(t)\,\boldsymbol{v}^{\mathrm{T}}(t)\,\frac{d\boldsymbol{v}}{dt} = \mathbf{0}. \tag{8}$$

The third term from the product rule for $\boldsymbol{u}^{\mathrm{T}}A\boldsymbol{v}$ gives the formula for $d\sigma/dt$:

$$\textbf{Derivative of a Singular Value} \qquad \boxed{\boldsymbol{u}^{\mathrm{T}}(t)\,\frac{d\boldsymbol{A}}{d\boldsymbol{t}}\,\boldsymbol{v}(t) = \frac{d\boldsymbol{\sigma}}{d\boldsymbol{t}}} \tag{9}$$

When $A(t)$ is symmetric positive definite, $\boldsymbol{\sigma}(t) = \boldsymbol{\lambda}(t)$ and $\boldsymbol{u} = \boldsymbol{v} = \boldsymbol{y} = \boldsymbol{x}$ in (4) and (9).

Note First derivatives of *eigenvectors* go with second derivatives of eigenvalues—not so easy. The Davis-Kahan bound on the angle θ between unit eigenvectors of S and $S+T$ is $\sin\theta \leq ||T||/d$ (d is the smallest distance from the eigenvalue of $S+T$ to all other eigenvalues of S). Tighter bounds that use the structure of S and T are highly valuable for applications to stochastic gradient descent (see Eldridge, Belkin, and Wang).

C. Davis and W. M. Kahan, *Some new bounds on perturbation of subspaces*, Bull. Amer. Math. Soc. **75** (1969) 863 – 868.

J. Eldridge, M. Belkin, and Y. Wang, *Unperturbed: Spectral analysis beyond Davis-Kahan*, arXiv: 1706.06516v1, 20 Jun 2017.

A Graphical Explanation of Interlacing

This page owes everything to Professor Raj Rao Nadakuditi of the University of Michigan. In his visits to MIT, he explained the theory and its applications to the 18.065 class. His **OptShrink** software to find low rank signals is described in *IEEE Transactions on Information Theory* **60** (May 2014) 3002 – 3018.

What is the change in the $\boldsymbol{\lambda}$'s, when a low rank matrix $\theta\boldsymbol{u}\boldsymbol{u}^{\mathrm{T}}$ is added to a full rank symmetric matrix S? We are thinking of S as noise and $\theta\boldsymbol{u}\boldsymbol{u}^{\mathrm{T}}$ as the rank one signal. How are the eigenvalues of S affected by adding that signal?

Let me make clear that *all* the eigenvalues of S can be changed by adding $\theta\boldsymbol{u}\boldsymbol{u}^{\mathrm{T}}$, not just one or two. But we will see that only one or two have changes of order θ. This makes them easy to find. If our vectors represent videos, and $\theta\boldsymbol{u}\boldsymbol{u}^{\mathrm{T}}$ represents a light turned on or off during filming (a rank-one signal), we will see the effect on the λ's.

Start with an eigenvalue z and its eigenvector $\boldsymbol{v}$ of the new matrix $S + \theta\boldsymbol{u}\boldsymbol{u}^{\mathrm{T}}$:

$$(S + \theta\boldsymbol{u}\boldsymbol{u}^{\mathrm{T}})\boldsymbol{v} = z\boldsymbol{v}. \tag{10}$$

Rewrite that equation as

$$(zI - S)\boldsymbol{v} = \theta\boldsymbol{u}(\boldsymbol{u}^{\mathrm{T}}\boldsymbol{v}) \quad \text{or} \quad \boldsymbol{v} = (zI - S)^{-1}\theta\boldsymbol{u}(\boldsymbol{u}^{\mathrm{T}}\boldsymbol{v}). \tag{11}$$

Multiply by $\boldsymbol{u}^{\mathrm{T}}$ and cancel the common factor $\boldsymbol{u}^{\mathrm{T}}\boldsymbol{v}$. This removes $\boldsymbol{v}$. Then divide by θ. That connects the new eigenvalue z to the change $\theta\boldsymbol{u}\boldsymbol{u}^{\mathrm{T}}$ in the symmetric matrix S.

$$\boxed{\frac{1}{\theta}=\boldsymbol{u}^{\mathrm{T}}(zI-S)^{-1}\boldsymbol{u}.} \tag{12}$$

To understand this equation, use the eigenvalues and eigenvectors of S. If $S\boldsymbol{q}_k=\lambda_k\boldsymbol{q}_k$ then $(zI-S)\boldsymbol{q}_k=(z-\lambda_k)\boldsymbol{q}_k$ and $(zI-S)^{-1}\boldsymbol{q}_k=\boldsymbol{q}_k/(z-\lambda_k)$:

$$\boldsymbol{u}=\sum c_k\boldsymbol{q}_k \quad \text{leads to} \quad (zI-S)^{-1}\boldsymbol{u}=\sum c_k(zI-S)^{-1}\boldsymbol{q}_k=\sum\frac{c_k\boldsymbol{q}_k}{z-\lambda_k}. \tag{13}$$

Finally equation (12) multiplies $(zI-S)^{-1}\boldsymbol{u}$ by $\boldsymbol{u}^{\mathrm{T}}=\sum c_k\boldsymbol{q}_k^{\mathrm{T}}$. **The result is $1/\theta$.** Remember that the $\boldsymbol{q}$'s are orthogonal unit vectors :

Secular equation
$$\boxed{\frac{1}{\theta}=\boldsymbol{u}^{\mathrm{T}}(zI-S)^{-1}\boldsymbol{u}=\sum_{k=1}^{n}\frac{c_k^2}{z-\lambda_k}} \tag{14}$$

We can graph the left side and right side. The left side is constant, the right side blows up at each eigenvalue $z=\lambda_k$ of S. The two sides are equal at the n points $z_1,\ldots,z_n$ where the flat $1/\theta$ line meets the steep **curves. Those z's are the n eigenvalues of $S+\theta\boldsymbol{u}\boldsymbol{u}^{\mathrm{T}}$. The graph shows that each z_i is above λ_i and below λ_{i-1} : Interlacing.**

The top eigenvalue z_1 is most likely above λ_1. The z's will increase as θ increases, because the $1/\theta$ line moves down.

Of course the z's depend on the vector $\boldsymbol{u}$ in the signal (as well as θ). If $\boldsymbol{u}$ happened to be also an eigenvector of S, then its eigenvalue λ_k would increase by exactly θ. All other eigenvalues would stay the same. It is much more likely that each eigenvalue λ_k moves up a little to z_k. **The point of the graph is that z_k doesn't go beyond λ_{k-1}.**

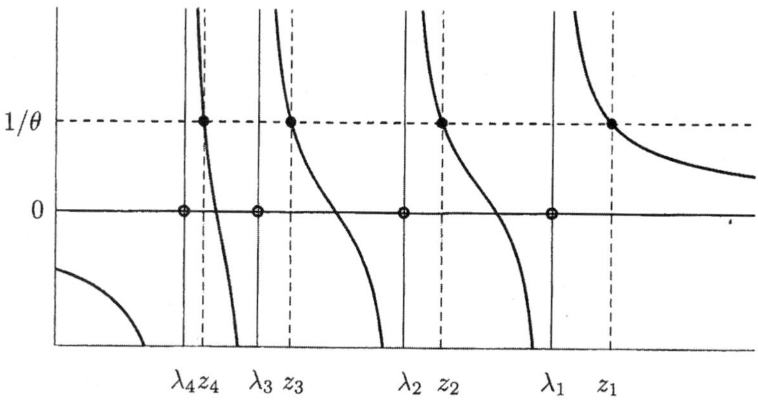

Figure III.1: Eigenvalues z_i of $S+\theta\boldsymbol{u}\boldsymbol{u}^{\mathrm{T}}$ where the $\dfrac{1}{\theta}$ line meets the curves in (14).

Raj Rao Nadakuditi, *When are the most informative components for inference also the principal components*? arXiv : 1302.1232, 5 Feb 2013.

The Largest Eigenvalue of $S + T$

The largest eigenvalue of a symmetric matrix S is the maximum value of $\boldsymbol{x}^{\mathrm{T}} S\boldsymbol{x}/\boldsymbol{x}^{\mathrm{T}}\boldsymbol{x}$. This statement applies also to T (still symmetric). Right away we know about the largest eigenvalue of $S + T$.

$$\boxed{\lambda_{\max}(S+T) \leq \lambda_{\max}(S) + \lambda_{\max}(T)} \tag{15}$$

The left side is the maximum value of $\dfrac{\boldsymbol{x}^{\mathrm{T}}(S+T)\boldsymbol{x}}{\boldsymbol{x}^{\mathrm{T}}\boldsymbol{x}}$. That maximum is reached at an eigenvector $\boldsymbol{v}$ of $S+T$:

$$\lambda_{S+T} = \frac{\boldsymbol{v}^{\mathrm{T}}(S+T)\boldsymbol{v}}{\boldsymbol{v}^{\mathrm{T}}\boldsymbol{v}} = \frac{\boldsymbol{v}^{\mathrm{T}}S\boldsymbol{v}}{\boldsymbol{v}^{\mathrm{T}}\boldsymbol{v}} + \frac{\boldsymbol{v}^{\mathrm{T}}T\boldsymbol{v}}{\boldsymbol{v}^{\mathrm{T}}\boldsymbol{v}} \leq \max\frac{\boldsymbol{x}^{\mathrm{T}}S\boldsymbol{x}}{\boldsymbol{x}^{\mathrm{T}}\boldsymbol{x}} + \max\frac{\boldsymbol{x}^{\mathrm{T}}T\boldsymbol{x}}{\boldsymbol{x}^{\mathrm{T}}\boldsymbol{x}} = \lambda_S + \lambda_T.$$

The eigenvector $\boldsymbol{v}$ of $S+T$ maximizes that first ratio. But it probably doesn't maximize the last two. Therefore $\lambda_S + \lambda_T$ can only increase beyond λ_{S+T}.

This shows that a maximum principle is convenient. So is a minimum principle for the *smallest eigenvalue*. There we expect $\lambda_{\min}(S+T) \geq \lambda_{\min}(S) + \lambda_{\min}(T)$. The same reasoning will prove it—the separate minimum principles for S and T will bring us lower than $\lambda_{\min}(S+T)$. Or we can apply the maximum principle to $-S$ and $-T$.

The difficulties come for the in-between eigenvalues λ_2 to λ_{n-1}. Their eigenvectors are **saddle points** of the function $R(\boldsymbol{x}) = \boldsymbol{x}^{\mathrm{T}}S\boldsymbol{x}/\boldsymbol{x}^{\mathrm{T}}\boldsymbol{x}$. The derivatives of $R(\boldsymbol{x})$ are all zero at the eigenvectors $\boldsymbol{q}_2$ to $\boldsymbol{q}_{n-1}$. But the matrix of second derivatives of R is indefinite (plus and minus eigenvalues) at these saddle points. That makes the eigenvalues hard to estimate and hard to calculate.

We now give attention to the saddle points. One reason is their possible appearance in the algorithms of deep learning. We need some experience with them, in the basic problem of eigenvalues. Saddle points also appear when there are constraints. If possible we want to connect them to *maxima of minima* or to *minima of maxima*.

Those ideas lead to the best possible bound for each eigenvalue of $S+T$:

$$\textbf{Weyl upper bounds} \qquad \lambda_{i+j-1}(S+T) \leq \lambda_i(S) + \lambda_j(T). \tag{16}$$

Saddle Points from Lagrange Multipliers

Compared to saddle points, computing a maximum or minimum of $F(\boldsymbol{x})$ is relatively easy. When we have an approximate solution $\widehat{\boldsymbol{x}}$, we know that $F(\widehat{\boldsymbol{x}})$ is not below the minimum of F and not above the maximum: *by definition*. But we don't know whether $F(\widehat{\boldsymbol{x}})$ is *above or below* a saddle point value. Similarly, the matrix $H(\boldsymbol{x})$ of second derivatives of F is positive definite (or semidefinite) at a minimum and negative definite at a maximum.

The second derivative matrix H at a saddle point is symmetric but indefinite.

H has both positive and negative eigenvalues—this makes saddle points more difficult. The conjugate gradient method is not usually available to find saddle points of $\boldsymbol{x}^{\mathrm{T}}H\boldsymbol{x}$.

Lagrange is responsible for a lot of saddle point problems. We start by minimizing a positive definite energy $\frac{1}{2}\boldsymbol{x}^{\mathrm{T}}S\boldsymbol{x}$, but there are m constraints $A\boldsymbol{x} = \boldsymbol{b}$ on the solution. Those constraints are multiplied by new unknowns $\lambda_1, \ldots, \lambda_m$ (the Lagrange multipliers) and they are built into the Lagrangian function:

$$\textbf{Lagrangian} \qquad L(\boldsymbol{x}, \boldsymbol{\lambda}) = \frac{1}{2}\boldsymbol{x}^{\mathrm{T}}S\boldsymbol{x} + \boldsymbol{\lambda}^{\mathrm{T}}(A\boldsymbol{x} - \boldsymbol{b}).$$

The $m + n$ equations $\partial L/\partial \boldsymbol{x} = \mathbf{0}$ and $\partial L/\partial \boldsymbol{\lambda} = \mathbf{0}$ produce an indefinite block matrix:

$$\begin{bmatrix} \partial L/\partial \boldsymbol{x} \\ \partial L/\partial \boldsymbol{\lambda} \end{bmatrix} = \begin{bmatrix} S\boldsymbol{x} + A^{\mathrm{T}}\boldsymbol{\lambda} \\ A\boldsymbol{x} - \boldsymbol{b} \end{bmatrix} \quad \text{and} \quad H\begin{bmatrix} \boldsymbol{x} \\ \boldsymbol{\lambda} \end{bmatrix} = \begin{bmatrix} S & A^{\mathrm{T}} \\ A & 0 \end{bmatrix}\begin{bmatrix} \boldsymbol{x} \\ \boldsymbol{\lambda} \end{bmatrix} = \begin{bmatrix} \mathbf{0} \\ \boldsymbol{b} \end{bmatrix} \qquad (17)$$

A small example would be $H = \begin{bmatrix} 1 & 1 \\ 1 & 0 \end{bmatrix}$ with negative determinant -1: its eigenvalues have opposite signs. The Problem Set confirms that this **"KKT matrix"** in equation (17) is *indefinite*. The solution $(\boldsymbol{x}, \boldsymbol{\lambda})$ is a saddle point of Lagrange's function L.

Saddle Points from Rayleigh Quotients

The maximum and minimum of the Rayleigh quotient $R(\boldsymbol{x}) = \boldsymbol{x}^{\mathrm{T}}S\boldsymbol{x}/\boldsymbol{x}^{\mathrm{T}}\boldsymbol{x}$ are λ_1 and λ_n:

$$\textbf{Maximum } \frac{\boldsymbol{q}_1^{\mathrm{T}}S\boldsymbol{q}_1}{\boldsymbol{q}_1^{\mathrm{T}}\boldsymbol{q}_1} = \boldsymbol{q}_1^{\mathrm{T}}\lambda_1\boldsymbol{q}_1 = \lambda_1 \qquad \textbf{Minimum } \frac{\boldsymbol{q}_n^{\mathrm{T}}S\boldsymbol{q}_n}{\boldsymbol{q}_n^{\mathrm{T}}\boldsymbol{q}_n} = \boldsymbol{q}_n^{\mathrm{T}}\lambda_n\boldsymbol{q}_n = \lambda_n$$

Our question is about the saddle points—the other points where all derivatives of the quotient $R(\boldsymbol{x})$ are zero. We will confirm that **those saddle points occur at the other eigenvectors $\boldsymbol{q}_2$ to $\boldsymbol{q}_{n-1}$ of S. Our goal is to see $\boldsymbol{\lambda}_2$ to $\boldsymbol{\lambda}_{n-1}$ as maxima of minima.** That max-min insight is the key to interlacing.

Notice that the vectors $\boldsymbol{x}$ and $2\boldsymbol{x}$ and $c\boldsymbol{x}$ $(c \neq 0)$ all produce the same quotient R:

$$R\,(2\boldsymbol{x}) = \frac{(2\boldsymbol{x})^{\mathrm{T}}S\,(2\boldsymbol{x})}{(2\boldsymbol{x})^{\mathrm{T}}(2\boldsymbol{x})} = \frac{4\,\boldsymbol{x}^{\mathrm{T}}S\boldsymbol{x}}{4\,\boldsymbol{x}^{\mathrm{T}}\boldsymbol{x}} = \frac{\boldsymbol{x}^{\mathrm{T}}S\boldsymbol{x}}{\boldsymbol{x}^{\mathrm{T}}\boldsymbol{x}} = R\,(\boldsymbol{x}).$$

So we only need to consider unit vectors with $\boldsymbol{x}^{\mathrm{T}}\boldsymbol{x} = 1$. That can become a constraint:

$$\max \frac{\boldsymbol{x}^{\mathrm{T}}S\boldsymbol{x}}{\boldsymbol{x}^{\mathrm{T}}\boldsymbol{x}} \quad \text{is the same as} \quad \textbf{max } \boldsymbol{x}^{\mathrm{T}}S\boldsymbol{x} \;\; \textbf{subject to } \boldsymbol{x}^{\mathrm{T}}\boldsymbol{x} = \mathbf{1}. \qquad (18)$$

The constraint $\boldsymbol{x}^{\mathrm{T}}\boldsymbol{x} = 1$ can be handled by one Lagrange multiplier!

$$\textbf{Lagrangian} \qquad L(\boldsymbol{x}, \lambda) = \boldsymbol{x}^{\mathrm{T}}S\boldsymbol{x} - \lambda(\boldsymbol{x}^{\mathrm{T}}\boldsymbol{x} - 1). \qquad (19)$$

The max-min-saddle points will have $\partial L/\partial \boldsymbol{x} = 0$ and $\partial L/\partial \lambda = 0$ (as in Section I.9):

$$\frac{\partial L}{\partial \boldsymbol{x}} = 2S\boldsymbol{x} - 2\lambda\boldsymbol{x} = \mathbf{0} \quad \text{and} \quad \frac{\partial L}{\partial \lambda} = 1 - \boldsymbol{x}^{\mathrm{T}}\boldsymbol{x} = 0. \qquad (20)$$

This says that the unit vector $\boldsymbol{x}$ is an eigenvector with $S\boldsymbol{x} = \lambda\boldsymbol{x}$.

Example **Suppose S is the diagonal matrix with entries 5, 3, 1.** Write x as (u, v, w) :

$$R = \frac{x^T Sx}{x^T x} = \frac{5u^2 + 3v^2 + w^2}{u^2 + v^2 + w^2} \text{ has a } \begin{array}{l} \textbf{maximum value 5 at } x = (1,0,0) \\ \textbf{minimum value 1 at } x = (0,0,1) \\ \textbf{saddle point value 3 at } x = (0,1,0) \end{array}$$

By looking at R, you see its maximum of 5 and its minimum of 1. All partial derivatives of $R(u, v, w)$ are zero at those three points $(1, 0, 0)$, $(0, 0, 1)$, $(0, 1, 0)$. These are eigenvectors of the diagonal matrix S. $R(x)$ equals the eigenvalues $5, 1, 3$ at those three points.

Maxima and Minima over Subspaces

All the middle eigenvectors $q_2, \ldots, q_{n-1}$ of S are saddle points of the quotient $x^T Sx/x^T x$. The quotient equals $\lambda_2, \ldots, \lambda_{n-1}$ at those eigenvectors. All the middle singular vectors $v_2, \ldots, v_{n-1}$ are saddle points of the growth ratio $||Ax||/||x||$. The ratio equals $\sigma_2, \ldots, \sigma_{n-1}$ at those singular vectors. Those statements are directly connected by the fact that $x^T Sx = x^T A^T Ax = ||Ax||^2$.

But saddle points are more difficult to study than maxima or minima. A function moves both ways, up and down, as you leave a saddle. At a maximum the only movement is down. At a minimum the only movement is up. So the best way to study saddle points is to capture them by a "max-min" or "min-max" principle.

Max-min for λ_2

$$\boxed{\lambda_2 = \max_{\textbf{all 2D spaces } Y} \; \min_{x \textbf{ is in } Y} \frac{x^T Sx}{x^T x}} \qquad (21)$$

In the $5, 3, 1$ example, one choice of the 2D subspace Y is all vectors $x = (u, v, 0)$. Those vectors are combinations of q_1 and q_2. Inside this Y, the minimum ratio $x^T Sx/x^T x$ will certainly be $\lambda_2 = 3$. That minimum is at $x = q_2 = (0, 1, 0)$ (we understand minima).

Key point: Every 2D space Y must intersect the 2D space of all vectors $(0, v, w)$. Those 2D spaces in $\mathbf{R}^3$ will surely meet because $2 + 2 > 3$. For any $x = (0, v, w)$ we definitely know that $x^T Sx/x^T x \le \lambda_2$. So for each Y the minimum in (21) is $\le \lambda_2$.

Conclusion: The maximum possible minimum is λ_2 in (21) and λ_i in (22).

$$\boxed{\lambda_i(S) = \max_{\dim V = i} \; \min_{x \text{ in } V} \frac{x^T Sx}{x^T x} \qquad \sigma_i(A) = \max_{\dim W = i} \; \min_{x \text{ in } W} \frac{||Ax||}{||x||}} \qquad (22)$$

For $i = 1$, the spaces V and W are one-dimensional lines. The line V through $x = q_1$ (first eigenvector) makes $x^T Sx/x^T x = \lambda_1$ a maximum. The line W through $x = v_1$ (first singular vector) makes $||Ax||/||x|| = \sigma_1$ a maximum.

For $i = 2$, the spaces V and W are two-dimensional planes. The maximizing V contains the eigenvectors q_1, q_2 and the maximizing W contains the singular vectors v_1, v_2. The minimum over that V is λ_2, the minimum over that W is σ_2. This pattern continues for every i. It produces the Courant-Fischer *max-min principles* in equation (22).

Interlacing and the Weyl Inequalities

For any symmetric matrices S and T, Weyl found bounds on the eigenvalues of $S+T$.

Weyl inequalities $\lambda_{i+j-1}(S+T) \le \lambda_i(S) + \lambda_j(T)$ (23)

$$\lambda_k(S) + \lambda_n(T) \le \lambda_k(S+T) \le \lambda_k(S) + \lambda_1(T) \quad (24)$$

The interlacing of the z's that we saw in Figure III.1 is also proved by equation (23). The rank one matrix T is $\theta \boldsymbol{u}\boldsymbol{u}^{\mathrm{T}}$ and its largest eigenvalue is $\lambda_1(T) = \theta$. All of the other eigenvalues $\lambda_j(T)$ are zero. Then for every $j = 2, 3, \ldots$ Weyl's inequality gives $\boldsymbol{\lambda_{i+1}(S+T) \le \lambda_i(S)}$. Each eigenvalue z_{i+1} of $S+T$ cannot go past the next eigenvalue λ_i of S. And for $j = 1$ we have $\boldsymbol{\lambda_1(S+T) \le \lambda_1(S) + \theta}$: an upper bound on the largest eigenvalue of signal plus noise.

Here is a beautiful interlacing theorem for eigenvalues, when the last column and row of a symmetric matrix S are removed. That leaves a matrix $\boldsymbol{S_{n-1}}$ of size $n-1$.

The $n-1$ eigenvalues α_i of the matrix S_{n-1} interlace the n eigenvalues of S.

The idea of the proof is that removing the last row and column is the same as forcing all vectors to be orthogonal to $(0, \ldots, 0, 1)$. Then the minimum in (22) could move below λ_i. But α_i won't move below λ_{i+1}, because λ_{i+1} allows a free choice with $\dim V = i+1$.

Example

$\lambda_i \ge \alpha_i \ge \lambda_{i+1}$

$$\begin{bmatrix} 2 & -1 & -1 \\ -1 & 2 & -1 \\ -1 & -1 & 2 \end{bmatrix} \qquad \begin{bmatrix} 2 & -1 \\ -1 & 2 \end{bmatrix} \qquad \begin{bmatrix} 2 \end{bmatrix}$$

$$\boldsymbol{\lambda = 3,\ 3,\ 0} \qquad \boldsymbol{\alpha = 3,\ 1} \qquad \boldsymbol{3 > 2 > 1}$$

Interlacing of Singular Values

Suppose A is not square and symmetric—so its singular values are involved. Each column of A represents one frame in a video. We want to identify a rank one signal $\beta \boldsymbol{x}\boldsymbol{y}^{\mathrm{T}}$ hidden in those columns. That signal is obscured by random noise. If a light was turned on or off during the video, the goal is to see when that happened.

This leads us to ask: **How much do the singular values change from A to $A+B$?** Changes in eigenvalues of symmetric matrices are now understood. So we can study $A^{\mathrm{T}}A$ or AA^{T} or this symmetric matrix of size $m+n$ with eigenvalues σ_i and $-\sigma_i$:

$$\begin{bmatrix} 0 & A \\ A^{\mathrm{T}} & 0 \end{bmatrix} \begin{bmatrix} \boldsymbol{u}_i \\ \boldsymbol{v}_i \end{bmatrix} = \sigma_i \begin{bmatrix} \boldsymbol{u}_i \\ \boldsymbol{v}_i \end{bmatrix} \quad \text{and} \quad \begin{bmatrix} 0 & A \\ A^{\mathrm{T}} & 0 \end{bmatrix} \begin{bmatrix} -\boldsymbol{u}_i \\ \boldsymbol{v}_i \end{bmatrix} = -\sigma_i \begin{bmatrix} -\boldsymbol{u}_i \\ \boldsymbol{v}_i \end{bmatrix} \quad (25)$$

Instead we recommend the amazing notes by Terry Tao: **https://terrytao.wordpress.com/ 2010/01/12/254a-notes-3a-eigenvalues-and-sums-of- hermitian-matrices/**

Weyl inequalities $\sigma_{i+j-1}(A+B) \le \sigma_i(A) + \sigma_j(B)$ (26)

$i \le m \le n$ $\quad |\sigma_i(A+B) - \sigma_i(A)| \le ||B||$ (27)

Problem Set III.2

1 A unit vector $\boldsymbol{u}(t)$ describes a point moving around on the unit sphere $\boldsymbol{u}^{\mathrm{T}}\boldsymbol{u} = 1$. Show that the velocity vector $d\boldsymbol{u}/dt$ is orthogonal to the position: $\boldsymbol{u}^{\mathrm{T}}(d\boldsymbol{u}/dt) = 0$.

2 Suppose you add a positive semidefinite **rank two** matrix to S. What interlacing inequalities will connect the eigenvalues λ of S and α of $S + \boldsymbol{u}\boldsymbol{u}^{\mathrm{T}} + \boldsymbol{v}\boldsymbol{v}^{\mathrm{T}}$?

3 (a) Find the eigenvalues $\lambda_1(t)$ and $\lambda_2(t)$ of $A = \begin{bmatrix} 2 & 1 \\ 1 & 0 \end{bmatrix} + t\begin{bmatrix} 1 & 1 \\ 1 & 1 \end{bmatrix}$.

(b) At $t = 0$, find the eigenvectors of $A(0)$ and verify $\dfrac{d\lambda}{dt} = \boldsymbol{y}^{\mathrm{T}}\dfrac{dA}{dt}\boldsymbol{x}$.

(c) Check that the change $A(t) - A(0)$ is positive semidefinite for $t > 0$. Then verify the interlacing law $\lambda_1(t) \geq \lambda_1(0) \geq \lambda_2(t) \geq \lambda_2(0)$.

4 S is a symmetric matrix with eigenvalues $\lambda_1 > \lambda_2 > \ldots > \lambda_n$ and eigenvectors $\boldsymbol{q}_1, \boldsymbol{q}_2, \ldots, \boldsymbol{q}_n$. Which i of those eigenvectors are a basis for an i-dimensional subspace Y with this property: The minimum of $\boldsymbol{x}^{\mathrm{T}}S\boldsymbol{x}/\boldsymbol{x}^{\mathrm{T}}\boldsymbol{x}$ for $\boldsymbol{x}$ in Y is λ_i.

5 Find the eigenvalues of A_3 and A_2 and A_1. Show that they are interlacing:

$$A_3 = \begin{bmatrix} 1 & -1 & 0 \\ -1 & 2 & -1 \\ 0 & -1 & 1 \end{bmatrix} \qquad A_2 = \begin{bmatrix} 1 & -1 \\ -1 & 2 \end{bmatrix} \qquad A_1 = \begin{bmatrix} 1 \end{bmatrix}$$

6 Suppose D is the diagonal matrix diag $(1, 2, \ldots, n)$ and S is positive definite.

1) Find the derivatives at $t = 0$ of the eigenvalues $\lambda(t)$ of $D + tS$.

2) For a small $t > 0$ show that the λ's interlace the numbers $1, 2, \ldots, n$.

3) For any $t > 0$, find bounds on $\lambda_{\min}(D + tS)$ and $\lambda_{\max}(D + tS)$.

7 Suppose D is again diag $(1, 2, \ldots, n)$ and A is any n by n matrix.

1) Find the derivatives at $t = 0$ of the singular values $\sigma(t)$ of $D + tA$.

2) What do Weyl's inequalities say about $\sigma_{\max}(D + tA)$ and $\sigma_{\min}(D + tA)$?

8 (a) Show that every i-dimensional subspace $\boldsymbol{V}$ contains a nonzero vector $\boldsymbol{z}$ that is a combination of $\boldsymbol{q}_i, \boldsymbol{q}_{i+1}, \ldots, \boldsymbol{q}_n$. (Those $\boldsymbol{q}$'s span a space $\boldsymbol{Z}$ of dimension $n - i + 1$. Based on the dimensions i and $n - i + 1$, why does $\boldsymbol{Z}$ intersect $\boldsymbol{V}$?)

(b) Why does that vector $\boldsymbol{z}$ have $\boldsymbol{z}^{\mathrm{T}}S\boldsymbol{z}/\boldsymbol{z}^{\mathrm{T}}\boldsymbol{z} \leq \lambda_i$? Then explain -:

$$\boxed{\lambda_i = \max_{\dim \boldsymbol{V} = i} \; \min_{\boldsymbol{z} \text{ in } \boldsymbol{V}} \; \frac{\boldsymbol{z}^{\mathrm{T}}S\boldsymbol{z}}{\boldsymbol{z}^{\mathrm{T}}\boldsymbol{z}}}$$

The Law of Inertia

Definition If S is symmetric and C is invertible, then the matrix $C^{\mathrm{T}}SC$ is "congruent to S". This is not similarity $B^{-1}SB$! Eigenvalues of $C^{\mathrm{T}}SC$ can change from eigenvalues of S, *but they can't change sign.* That is called the **"Law of Inertia"**:

$C^{\mathrm{T}}SC$ has the same number of (positive) (negative) (zero) eigenvalues as S.

My favorite proof starts with $C = QR$ (by Gram-Schmidt). As R changes gradually to I, $C^{\mathrm{T}}SC$ changes gradually to $Q^{\mathrm{T}}SQ = Q^{-1}SQ$. Now we do have similarity ($Q^{-1}SQ$ has the same eigenvalues as S). If R is invertible all the way to I, then no eigenvalues can cross zero on the way. *Their signs are the same for $C^{\mathrm{T}}SC$ and $Q^{-1}SQ$ and S.*

The max-min principles also prove the Law of Inertia.

9 If $S = LDL^{\mathrm{T}}$ has n nonzero pivots in elimination, show that **the signs of the pivots of S (in D) match the signs of the eigenvalues of S.** Apply the Law to S and D.

10 Show that this $2n \times 2n$ KKT matrix H has n positive and n negative eigenvalues:

S positive definite
C invertible

$$H = \begin{bmatrix} S & C \\ C^{\mathrm{T}} & 0 \end{bmatrix}$$

The first n pivots from S are positive. The last n pivots come from $-C^{\mathrm{T}}S^{-1}C$.

11 The KKT matrix H is symmetric and indefinite—this problem counts eigenvalues:

$$H = \begin{bmatrix} S & A^{\mathrm{T}} \\ A & 0 \end{bmatrix} \begin{matrix} n \\ m \end{matrix} \quad \text{as in equation (17)}$$
$$\quad n \quad m$$

H comes from minimizing $\frac{1}{2}\boldsymbol{x}^{\mathrm{T}}S\boldsymbol{x}$ (positive definite) with m constraints $A\boldsymbol{x} = \boldsymbol{b}$. Elimination on H begins with S. We know its n pivots are all positive.

Then elimination multiplies AS^{-1} times $\begin{bmatrix} S & A^{\mathrm{T}} \end{bmatrix}$ and subtracts from $\begin{bmatrix} A & 0 \end{bmatrix}$ to get $\begin{bmatrix} 0 & -AS^{-1}A^{\mathrm{T}} \end{bmatrix}$. That Schur complement $-AS^{-1}A^{\mathrm{T}}$ is negative definite. Why? Then the last m pivots of H (coming from $-AS^{-1}A^{\mathrm{T}}$) are negative.

12 If $\boldsymbol{x}^{\mathrm{T}}S\boldsymbol{x} > 0$ for all $\boldsymbol{x} \neq \mathbf{0}$ and C is invertible, why is $(C\boldsymbol{y})^{\mathrm{T}}S(C\boldsymbol{y})$ also positive? This shows again that if S has all positive eigenvalues, so does $C^{\mathrm{T}}SC$.

III.3 Rapidly Decaying Singular Values

There are important matrices whose singular values have $\sigma_k \leq Ce^{-ak}$. Those numbers decay quickly. Often the matrices are invertible (their inverses are incredibly large). And often we have a *family* of matrices of all sizes—Hilbert and Vandermonde matrices, Hankel and Cauchy and Krylov and spectral difference matrices and more.

These matrices are at the same time easy and also hard to work with. Easy because only a few singular values are significant. Not easy when the inverse has a giant norm, increasing exponentially with the matrix size N. We will focus on two of many examples:

1 The *nonuniform* discrete Fourier transform (**NUDFT**) has $\boldsymbol{U} = \boldsymbol{A} .* \boldsymbol{F}$ in place of F.

2 The *Vandermonde matrix* $\boldsymbol{V}$ fits a polynomial of degree $N-1$ to N data points.

Actually those examples are connected. A standard DFT fits N values f_0 to f_{N-1} at the N points $\omega^k = e^{-2\pi ik/N}$. The Fourier matrix is a Vandermonde matrix! But instead of real points between -1 and 1, the DFT is interpolating at *complex points*—equally spaced around the unit circle $|e^{i\theta}| = 1$. The real case produces terrible Vandermonde matrices (virtually singular). The complex case produces a beautiful Fourier matrix (orthogonal).

We start with that complex Fourier matrix F (equal spacing as usual). Multiplying by F is superfast with the Fast Fourier Transform in Section IV.1: $\frac{1}{2}N\log_2 N$ operations.

For nonuniform spacing $\boldsymbol{x_j \neq j/N}$, the special identities behind the FFT are gone. But Ruiz-Antolin and Townsend showed how you can recover almost all of the speed: Write the nonuniform U as $A_{jk}F_{jk}$ where **$\boldsymbol{A}$ is near a low rank matrix:**

$$\boldsymbol{F_{jk} = e^{-2\pi ikj/N}} \textbf{ and } \boldsymbol{U_{jk} = e^{-2\pi ikx_j} = A_{jk}F_{jk}}$$

When $U = F$ is the DFT with equal spacing, every $A_{jk} = 1$. *A has rank one.* With unequal spacing, a low rank matrix virtually agrees with A, and fast transforms are possible. Here are symbols for element by element multiplication $A_{jk}F_{jk}$ and division U_{jk}/F_{jk}:

$$\textbf{Multiplication} \quad \boldsymbol{U = A .* F = A \bigcirc F} \qquad \textbf{Division} \quad \boldsymbol{A = U \oslash F}. \tag{1}$$

The operation to execute quickly is the NUDFT: $\boldsymbol{U}$ **times** $\boldsymbol{c}$. The Fast Fourier Transform computes F times $\boldsymbol{c}$. The ratios U_{jk}/F_{jk} give a matrix A that is **nearly low rank. So the nonuniform transform $\boldsymbol{U}$ comes from a correction $\boldsymbol{A}$ to the Fourier matrix $\boldsymbol{F}$:**

$$\boxed{\boldsymbol{A \approx y_1 z_1^{\mathrm{T}} + \cdots + y_r z_r^{\mathrm{T}}} \quad \text{and} \quad \boldsymbol{Uc \approx Y_1 F Z_1 c + \cdots + Y_r F Z_r c}.} \tag{2}$$

Y_i and Z_i are diagonal N by N matrices with $\boldsymbol{y}_i$ and $\boldsymbol{z}_i$ along their main diagonals. Equal spacing has $U = F$ and $A =$ **ones** and $\boldsymbol{y}_1 = \boldsymbol{z}_1 = (1,\ldots,1)$ and $Y_1 = Z_1 = I$. For unequal spacing, r is determined by the nonuniformity of the sampling points x_i.

Sample Points x_j near j/N

In this "lightly perturbed" case, we can match the unequally spaced x_j with the equally spaced fractions j/N for $0 \le j < N$. For each entry of U, A_{jk} is a correction to F_{jk}:

$$\boxed{U_{jk} = A_{jk}F_{jk} \text{ is } e^{-2\pi ikx_j} = e^{-2\pi ik(x_j - j/N)}\, e^{-2\pi ikj/N}} \tag{3}$$

Then the key step for a fast algorithm is to find a close approximation to this matrix A (we know F is fast). The Eckart-Young theorem would suggest to use the SVD of A. But the SVD is a more expensive step than the rest of the fast unequally spaced transform.

A is an interesting matrix. All its entries in equation (3) have the form $e^{i\theta}$. If we replace $A_{jk} = e^{-i\theta}$ by its power series $1 - i\theta + \cdots$ then A will begin with the all-ones matrix. The rest of this section finds low rank approximations by working with the Sylvester equation. Here Townsend takes a different route: *Approximate the function* e^{-ixy}.

The key idea is to replace the Taylor series for each of those matrix entries by a **Chebyshev expansion**. This reflects a rule of great importance for numerical analysis:

Fourier series are good for periodic functions on an interval like $|\theta| \le \pi$.
Chebyshev series are good for nonperiodic functions on an interval like $|x| \le 1$.

The connection between Fourier and Chebyshev is $\cos\theta = x$. The Fourier basis function $\cos n\theta$ becomes the Chebyshev polynomial $T_n(x) = \cos(n \arccos x)$. The Chebyshev basis starts with $T_0 = 1$ and $T_1 = x$ and $T_2 = 2x^2 - 1$ because $\cos 2\theta = 2\cos^2\theta - 1$. All those basis functions have $\max|T_n(x)| = \max|\cos n\theta| = 1$ from $x = -1$ to $x = 1$.

An important point is that the n solutions of $T_n(x) = 0$ are *not* equally spaced. The zeros of $\cos n\theta$ are equally spaced, but in the x-variable those points come close together near the boundaries -1 and 1. Interpolation at these $x = \cos(\pi(2k-1)/2n)$ is *far more stable* than equally spaced interpolation.

The highly developed computational system at **chebfun.org** is based on Chebyshev polynomials (for functions in one or more dimensions). It computes a polynomial very close to $f(x)$. Then all operations on functions produce new polynomials as finite Chebyshev series. The degree needed for high accuracy is chosen by the **chebfun** code.

This approach to the matrix A leads Ruiz-Antolin and Townsend to a proof of *low effective rank* (close approximation to A by a low rank matrix). Their paper provides a very efficient code for the nonuniform Fourier transform.

The Sylvester Equation

We turn to the central problem of the subject: **Which families of matrices have low effective rank?** The goal is to find a test that will reveal this property. To the examples of the Hilbert matrix and the NUDFT matrix A we want to add Vandermonde. And we hope for a test that can be applied to much wider classes of matrices—going far beyond the Vandermonde example.

Here is the **"Sylvester test"** = "$\boldsymbol{A, B, C}$ **test**" developed by Beckermann and Townsend. The words *low displacement rank* and *structured matrix* are often applied to X.

> If $\boldsymbol{AX - XB = C}$ **has rank** $\boldsymbol{r}$ for normal matrices A, B with no shared eigenvalues, then the singular values of X decay at an exponential rate decided by A, B, and C.

A matrix is normal if $\overline{A}^{\mathrm{T}}A = A\overline{A}^{\mathrm{T}}$. Then A has orthogonal eigenvectors: $A = Q\Lambda\overline{Q}^{\mathrm{T}}$. Symmetric and orthogonal matrices are normal, because the test gives $S^2 = S^2$ and $I = I$. The Sylvester matrix equation $AX - XB = C$ is important in control theory, and the particular case when $B = -\overline{A}^{\mathrm{T}}$ is called the Lyapunov equation.

This Sylvester test requires us to find A, B, and C! That has been done for highly important families of Toeplitz and Hankel and Cauchy and Krylov matrices (including the Vandermonde matrix V). All those matrices solve Sylvester's equation for simple choices of A, B, C. We take V as our prime example:

$$\textbf{Vandermonde matrix} \qquad V = \begin{bmatrix} 1 & x_1 & x_1^2 & \cdots & x_1^{n-1} \\ 1 & x_2 & x_2^2 & \cdots & x_2^{n-1} \\ \cdot & \cdot & \cdot & & \cdot \\ 1 & x_n & x_n^2 & \cdots & x_n^{n-1} \end{bmatrix} \tag{4}$$

V is the n by n "interpolation matrix". It is invertible as long as the points $x_1, \ldots, x_n$ are all different. We solve $V\boldsymbol{c} = \boldsymbol{f}$ when we want the coefficients of a polynomial $p = c_0 + c_1x + \cdots + c_{n-1}x^{n-1}$. Multiplying V times $\boldsymbol{c}$ gives us the value of p at the points $x = x_1, \ldots, x_n$. Then $V\boldsymbol{c} = \boldsymbol{f}$ says that the interpolating polynomial has the desired values $f_1, \ldots, f_n$ at those n points. **The polynomial exactly fits the data.**

We noted that V becomes the Fourier matrix F when we choose complex points $x_1 = \omega = e^{-2\pi i/N}$ and $x_k = \omega^k$. This F has full rank. In that Fourier case A and B (below) have the same eigenvalues ω^k: *not allowed.* A and B don't satisfy our requirement of *well-separated eigenvalues* and all the singular values have equal size: no decay.

It is Vandermonde matrices with *real numbers* x_1 to x_n that have exponential decay in their singular values. To confirm this, apply the A, B, C test using these matrices:

$$A = \begin{bmatrix} x_1 & & & \\ & x_2 & & \\ & & \cdot & \\ & & & x_n \end{bmatrix} \quad B = \begin{bmatrix} 0 & 0 & \cdot & -1 \\ 1 & 0 & 0 & \cdot \\ 0 & 1 & 0 & 0 \\ 0 & 0 & 1 & 0 \end{bmatrix} \quad C = \begin{bmatrix} 0 & 0 & 0 & x_1^n + 1 \\ 0 & 0 & 0 & x_2^n + 1 \\ 0 & 0 & 0 & \cdot \\ 0 & 0 & 0 & x_n^n + 1 \end{bmatrix} \tag{5}$$

Those matrices A and B are certainly normal. (This requirement could be weakened but here it's not necessary.) The eigenvalues of B are equally spaced around the unit circle. Those λ's are at angles $\pi/n, 3\pi/n, \ldots, (2n-1)\pi/n$, so they are not real numbers provided n is even. Then they don't touch the real eigenvalues $x_1, \ldots, x_n$ of A. And C has rank 1. *The A, B, C Sylvester test is passed.*

The graph of singular values confirms that V is highly ill-conditioned. So is K.

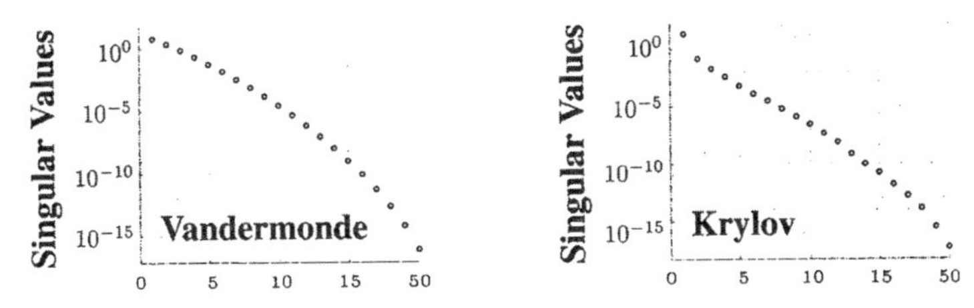

Figure III.2: Vandermonde and Krylov singular values for V and $K = [\boldsymbol{b}\; A\boldsymbol{b} \ldots A^{n-1}\boldsymbol{b}]$. Here $\boldsymbol{b}$ = all ones and A_{ij} = random numbers (standard normal). V equals K when $A = \text{diag}\,(1/n, 2/n, \ldots, 1)$.

An Improved Sylvester Test

The requirement that $AX - XB = C$ has low rank is much more strict than the conclusion that X has rapidly decaying singular values. A perfect theorem would match the hypothesis on the C_n with the conclusion about X_n. Certainly it is not true that Vandermonde or Krylov matrices X_n of increasing size n have bounded rank. So we have to weaken that low rank requirement on C_n, while preserving the rapid singular value decay for X_n.

Townsend has found such a theorem: **"log-rank" for C leads to "log-rank" for X.** And a recent paper with Udell establishes that log-rank is a very widespread property. Here is the definition (and many log-rank examples have $q = 0$ or $q = 1$).

A family of matrices C_n has **log-rank** if $|(C_n - E_n)_{ij}| < \epsilon$ for nearby matrices E_n that have **rank** $(\boldsymbol{E_n}) < c\,(\log n)^q$ (6)

Example 1 The **radial basis function kernel** is often used in support vector machines:

$$\textbf{RBF Kernel} \qquad K(\boldsymbol{x}, \boldsymbol{x}') = \exp\left(-\frac{||\boldsymbol{x} - \boldsymbol{x}'||^2}{2\sigma^2}\right)$$

For a set of feature vectors $\boldsymbol{x}_i$, this produces the entries $0 < K_{ij} < 1$ of a *full matrix* K. Calculating all of them is impossible. With good approximations, we solve nonlinear classification problems by the "kernel trick" in VII.5. That matrix has low effective rank.

1. M. D. Buhmann, *Radial basis functions*, Acta Numerica **9** (2000) 1-38.
2. B. Fornberg and N. Flyer, *A Primer on Radial Basis Functions*, SIAM (2015).
3. T. Hofmann, B. Scholkopf, and A. J. Smola, *Kernel methods in machine learning*, Annals of Statistics **36** (2008) 1171-1220 (with extensive references).

ADI and the Zolotarev Problem

In a short paragraph we can point to two ideas that lead to fast decay for the singular values of X. The first is an ADI iteration to solve Sylvester's equation $AX - XB = C$. The *Alternating Direction Implicit* algorithm gives a computationally efficient solution. The second idea connects the eigenvalues of A and B (they are required not to overlap) to a problem in *rational approximation.*

That "Zolotarev problem" looks for a ratio $r(x) = p(x)/q(x)$ of polynomials that is small at the eigenvalues of A and large at the eigenvalues of B. Approximating by rational functions $r(x)$ can be exponentially better than polynomials—a famous example is Newman's approximation of the absolute value $|x|$. The exponential accuracy of $r(x)$ becomes connected to the exponential decay of singular values of X.

ADI solves $AX - XB = C$

Matrices $X_{j+1/2}$ and X_{j+1}

$$\begin{aligned} X_{j+1/2}(B - p_j I) &= C - (A - p_j I)\, X_j \\ (A - q_j I) X_{j+1} &= C - X_{j+1/2}(B - q_j I) \end{aligned}$$

The good rational function $r(x) = p(x)/q(x)$ has roots p_j in the numerator and q_j in the denominator. It was Zolotarev in 1877 (!) who found the best p's and q's in a model problem. With A and B in the Sylvester test, the bound on $\sigma_{1+kr}(X)$ is the "Z-number" times $\sigma_1(X)$—which means exponential decay of singular values.

Townsend and Fortunato developed this idea into a superfast Poisson solver on a square. When X is the usual 5-point finite difference approximation to $\partial^2 u/\partial x^2 + \partial^2 u/\partial y^2$, fast solvers are already known. Their goal was a spectral method with optimal complexity.

1. B. Beckermann, The condition number of real Vandermonde, Krylov, and positive definite Hankel matrices, *Numerische Mathematik* **85** (2000) 553-577.
2. B. Beckermann and A. Townsend, On the singular values of matrices with displacement structure, *SIAM J. Matrix Analysis*, arXiv : 1609.09494v1, 29 Sep 2016.
3. P. Benner, R.-C. Li, and N. , On the ADI method for Sylvester equations, *J. Comput. Appl. Math.* **233** (2009) 1035-1045.
4. D. Fortunato and A. Townsend, Fast Poisson solvers for spectral methods, arXiv : 1710.11259v1, 30 Oct 2017.
5. D. Ruiz-Antolin and A. Townsend, A nonuniform fast Fourier transform based on low rank approximation, arXiv : 1701.04492, *SIAM J. Sci. Comp.* **40-1** (2018).
6. A. Townsend and H. Wilber, On the singular values of matrices with high displacement rank, arXiv :17120.5864, Linear Alg. Appl. **548** (2018) 19-41.
7. A. Townsend, **www.math.cornell.edu/~ ajt/presentations/LowRankMatrices.pdf**
8. M. Udell and A. Townsend, Nice latent variable models have log-rank, arXiv : 1705.07474v1, *SIAM J. Math. of Data Science*, to appear.

Problem Set III.3

1 Verify that a **Krylov matrix** $K = [\boldsymbol{b}\ A\boldsymbol{b} \ldots A^{n-1}\boldsymbol{b}]$ satisfies a Sylvester equation $AK - KB = C$ with B as in equation (6). Find the matrix C.

2 Show that the evil **Hilbert matrix** H passes the Sylvester test $AH - HB = C$

$$H_{ij} = \frac{1}{i+j-1} \qquad A = \frac{1}{2}\text{diag}\,(1, 3, \ldots, 2n-1) \qquad B = -A \qquad C = \textbf{ones}(n)$$

3 A **Toeplitz matrix** T has constant diagonals (IV.5). Compute $AT - TA^{\mathrm{T}} = C$:

$$T = \begin{bmatrix} t_0 & t_{-1} & \cdot & \cdot \\ t_1 & t_0 & t_{-1} & \cdot \\ \cdot & t_1 & t_0 & \cdot \\ \cdot & \cdot & \cdot & \cdot \end{bmatrix} \qquad A = \begin{bmatrix} 0 & 0 & 0 & \cdot \\ 1 & 0 & 0 & 0 \\ 0 & 1 & 0 & 0 \\ \cdot & 0 & 1 & \cdot \end{bmatrix} \qquad B = A^{\mathrm{T}}$$

4 A **Hankel matrix** H has constant *anti*diagonals, like the Hilbert matrix. Then H_{ij} depends only on $i + j$. When H is symmetric positive definite, Beckermann and Townsend show that $H = \overline{K}^{\mathrm{T}} K$ for a Krylov matrix K (as in Problem 1 above). Then $\sigma_j(H) = |\sigma_j(K)|^2$ (*why*?) and the singular values of H decay quickly.

5 A **Pick matrix** has entries $P_{jk} = (s_j + s_k)/(x_j + x_k)$ where $\boldsymbol{x} = (x_1, \ldots, x_n) > 0$ and $s = (s_1, \ldots, s_n)$ can be complex. Show that $\boldsymbol{AP - P(-A) = s1^{\mathrm{T}} + 1s^{\mathrm{T}}}$ where $\mathbf{1}^{\mathrm{T}} = [1 \ \ 1 \ldots 1]$ and $A = \text{diag}\,(x_1, \ldots, x_n)$. A has positive eigenvalues, $B = -A$ has negative eigenvalues, and the Sylvester test is passed.

6 If an invertible matrix X satisfies the Sylvester equation $AX - XB = C$, find a Sylvester equation for X^{-1}.

7 If $A = Q\Lambda\overline{Q}^{\mathrm{T}}$ has complex orthogonal eigenvectors $\boldsymbol{q}_1, \ldots, \boldsymbol{q}_n$ in the columns of Q, verify that $\overline{A}^{\mathrm{T}} A = A\overline{A}^{\mathrm{T}}$: *then A is normal.* The eigenvalues can be complex.

8 If $S^{\mathrm{T}} = S$ and $Z^{\mathrm{T}} = -Z$ and $SZ = ZS$, verify that $A = S + Z$ is normal. Since S has real eigenvalues and Z has imaginary eigenvalues, $A = S + Z$ probably has complex eigenvalues.

9 Show that equation (3) for $U\boldsymbol{c}$ follows from equation (2) for A.

III.4 Split Algorithms for $\ell^2 + \ell^1$

These topics are truly important. They deserve a whole book. They include **basis pursuit** and **LASSO optimization**—plus **matrix completion** and **compressed sensing** in III.5. What we can do here is to present basic ideas and successful algorithms.

Start with a linear system $Ax = b$. Suppose A has many more columns than rows $(m << n)$. Then $Ax_n = 0$ has many solutions (A has a large nullspace). If $Ax = b$ has one solution, then every $x + x_n$ is another solution. Which one to choose?

To minimize the ℓ^2 norm of x, we remove its nullspace component. That leaves the minimum norm solution $x^+ = A^+b$, coming from the pseudoinverse of A. This is the solution in II.2 that uses the SVD. But x^+ will generally have many small components—it can be very difficult to interpret. In MATLAB, **pinv(A)** $* \; b$ finds x^+. Backslash $A\backslash b$ finds a fairly sparse solution—but probably not the optimal x for basis pursuit.

A solution to $Ax = b$ with **many zero components** (*a sparse solution*, if it exists) comes from minimizing the ℓ^1 norm instead of the ℓ^2 norm:

Basis pursuit	**Minimize** $\|\|x\|\|_1 = \|x_1\| + \cdots + \|x_n\|$ **subject to** $Ax = b$.	(1)

This is a convex optimization problem, because the ℓ^1 norm is a convex function of x. The ℓ^1 norm is *piecewise linear*, unlike ℓ^2 and all other ℓ^p norms except $||x||_\infty = \max|x_i|$. Basis pursuit is not solved by the SVD of A, which connects to the ℓ^2 norm. But ℓ^1 has become famous for giving sparse solutions, and fast algorithms have now been found.

The sparsest solution of $Ax = b$ minimizes $||x||_0$ = *number of nonzero components of* x. But this is not a true norm: $||2x||_0 = ||x||_0$. The vectors with $||x||_0 = 1$ and only one nonzero component lie along the coordinate axes, like $i = (1,0)$ and $j = (0,1)$. So we "relax" or "convexify" the ℓ^0 problem to get a true norm—**and that norm is** ℓ^1. The vectors with $||x||_1 = |x_1| + |x_2| \leq 1$ fill the diamond with corners at $\pm i$ and $\pm j$.

A related problem allows for *noise* in $Ax = b$; an exact solution is not required. Now the ℓ^1 norm enters as a penalty $\lambda||x||_1$ or a constraint $||x||_1 \leq t$:

LASSO (in statistics)	**Minimize** $\frac{1}{2}\|\|Ax - b\|\|_2^2 + \lambda\|\|x\|\|_1$ or **Minimize** $\frac{1}{2}\|\|Ax - b\|\|_2^2$ with $\|\|x\|\|_1 \leq t$	(2)

LASSO was invented by Tibshirani to improve on least squares regression. It has sparser solutions than "ridge regression" which uses ℓ^2 norms in the penalty and the constraint. Geometrically, the difference between $||x||_1 \leq 1$ and $||x||_2 \leq 1$ is in the shape of those two sets: **diamond for ℓ^1 versus sphere for ℓ^2**. Please see both figures in Section I.11.

A convex set $||Ax - b||_2^2 = C$ has a good chance to hit that diamond at one of its sharp points. *The sharpest points are sparse vectors.* But the sphere has no sharp points. The optimal x is almost never sparse in ℓ^2 optimization. Since that round convex set can touch the sphere anywhere, the ℓ^2 solution has many nonzeros.

Split Algorithms for ℓ^1 Optimization

Sparse solutions are a key reason for minimizing in ℓ^1 norms. The small nonzeros that appear in ℓ^2 will disappear in ℓ^1. This was frustrating to know, in the days when ℓ^1 algorithms were very slow. Now the numerical implementation of ℓ^1 optimization has essentially caught up with the theory.

Here is the picture. Many important optimization problems combine **two terms**:

$$\boxed{\textbf{Convex } F_1 \textbf{ and } F_2 \qquad \textbf{Minimize } F_1(x) + F_2(x) \textbf{ for } x \textbf{ in a convex set } K} \tag{3}$$

F_1 involves an ℓ^1-type norm and F_2 involves an ℓ^2-type norm. Their convexity is very valuable. But they don't mix well—the ℓ^2 iterations (ordinarily fast enough) are slowed down waiting for ℓ^1 to learn which components should be nonzero.

The solution is to "split" the algorithm. Alternate between ℓ^2 steps and ℓ^1 steps. In important cases, it becomes possible to solve ℓ^1 problems explicitly by a "shrinkage" operation. The split iterations are much faster and more effective than a mixed ℓ^1-ℓ^2 step. Among the leading algorithms are **ADMM** and **split Bregman** and **split Kaczmarz** (all described below).

One way to start is to see key words that describe the development of the algorithms. Those words describe forward steps that improved on the previous method—and made the final ADMM algorithm a success:

Dual decomposition
Augmented Lagrangian
Method of multipliers
ADMM: Alternating direction method of multipliers

This step-by-step presentation is following the plan established by Boyd, Parikh, Chu, Peleato, and Eckstein in their excellent online book on ADMM: *Distributed Optimization and Statistical Learning via the Alternating Direction Method of Multipliers.* It was published in Foundations and Trends in Machine Learning, **3** (2010) 1-122.

Those authors include a neat history (with references) for each of the four steps to ADMM. Here is a statement of the problem, leaving ample freedom in the convex function $f(x)$.

$$\boxed{\textbf{Minimize } f(x) \textbf{ subject to } Ax = b\text{: } A \text{ is } m \text{ by } n} \tag{4}$$

This is the *primal problem* for $x = (x_1, \ldots, x_n)$. The first step is to combine $Ax = b$ with the cost function $f(x)$ by introducing Lagrange multipliers $y_1, \ldots, y_m$:

$$\boxed{\textbf{Lagrangian} \qquad L(x, y) = f(x) + y^{\mathrm{T}}(Ax - b) = f(x) + y^{\mathrm{T}}Ax - y^{\mathrm{T}}b} \tag{5}$$

The combined solution $\boldsymbol{x}^*, \boldsymbol{y}^*$ is a *saddle point* of L: $\min_{\boldsymbol{x}} \max_{\boldsymbol{y}} L = \max_{\boldsymbol{y}} \min_{\boldsymbol{x}} L$. The equations to solve are $\partial L/\partial \boldsymbol{x} = \mathbf{0}$ and $\partial L/\partial \boldsymbol{y} = \mathbf{0}$. In fact $\partial L/\partial \boldsymbol{y} = \mathbf{0}$ gives us back exactly the constraint $A\boldsymbol{x} = \boldsymbol{b}$. This is a fundamental idea in optimization with constraints. In Chapter VI, the y's are seen as the derivatives $\partial L/\partial b$ at the optimal $\boldsymbol{x}^*$.

Now comes the key step from the *primal problem* for $\boldsymbol{x} = (x_1, \ldots, x_n)$ to the *dual problem* for $\boldsymbol{y} = (y_1, \ldots, y_m)$. **Minimize $L(\boldsymbol{x}, \boldsymbol{y})$ over $\boldsymbol{x}$.** The minimum occurs at a point $\boldsymbol{x}^*$ depending on $\boldsymbol{y}$. Then the dual problem is to maximize $\boldsymbol{m(y)} = \boldsymbol{L(x^*(y), y)}$.

Steepest Increase of $m(y)$

To maximize a function $m(\boldsymbol{y})$, we look for a point $\boldsymbol{y}^*$ where all the partial derivatives are zero. In other words, **the gradient is zero**: $\boldsymbol{\nabla m} = (\partial m/\partial y_1, \ldots, \partial m/\partial y_m) = \mathbf{0}$. The present problem has a neat formula for those $\boldsymbol{y}$-derivatives of m: $\boldsymbol{\nabla m} = A\boldsymbol{x}^* - \boldsymbol{b}$.

How to maximize a function $m(\boldsymbol{y})$ when we know its first derivatives? This will be the central question of Chapter VI, leading to the algorithm that optimizes the weights in deep learning. There we will minimize a loss function; here we are maximizing $m(\boldsymbol{y})$. There we will follow the gradient downhill (*steepest descent*). Here we will follow the gradient uphill (*steepest ascent*). In both cases the gradient tells us the steepest direction.

Steepest increase for max min L = max m

Find $\boldsymbol{x}_{k+1}$ and follow $\boldsymbol{\nabla m} = A\boldsymbol{x}_{k+1} - \boldsymbol{b}$

$$\boldsymbol{x}_{k+1} = \operatorname{argmin} L(\boldsymbol{x}, \boldsymbol{y}_k) \tag{6}$$

$$\boldsymbol{y}_{k+1} = \boldsymbol{y}_k + s_k(A\boldsymbol{x}_{k+1} - \boldsymbol{b}) \tag{7}$$

That number s_k is the **stepsize**. It determines how far we move in the uphill direction $\boldsymbol{\nabla m}$. We do not expect to hit the maximum of m in one step! We just take careful steps upward. It is a common experience that the first steps are successful and later steps give only small increases in m. In financial mathematics that dual variable $\boldsymbol{y}$ often represents "prices".

Note that "**argmin**" in equation (6) is the point $\boldsymbol{x}$ where that function L is minimized. This introduction of duality—minimizing over $\boldsymbol{x}$ and maximizing over $\boldsymbol{y}$, in either order—was not just a wild impulse. The reason comes next.

Dual Decomposition

Suppose that the original function $f(\boldsymbol{x})$ is separable: $f(\boldsymbol{x}) = f_1(\boldsymbol{x}_1) + \cdots + f_N(\boldsymbol{x}_N)$. Those $\boldsymbol{x}_i$ are subvectors of $\boldsymbol{x} = (x_1, \ldots, x_n)$. We partition the columns of A in the same way, so that $A = [A_1 \ldots A_N]$. Then Lagrange's function $L(\boldsymbol{x}, \boldsymbol{y})$ splits into N simpler Lagrangians, L_1 to L_N:

$$f(\boldsymbol{x}) + \boldsymbol{y}^{\mathrm{T}}(A\boldsymbol{x} - \boldsymbol{b}) = \sum_1^N L_i(\boldsymbol{x}_i, \boldsymbol{y}) = \sum_1^N \left[f_i(\boldsymbol{x}_i) + \boldsymbol{y}^{\mathrm{T}} A_i \boldsymbol{x}_i - \frac{1}{N} \boldsymbol{y}^{\mathrm{T}} \boldsymbol{b} \right] \tag{8}$$

Now the $\boldsymbol{x}$-minimization of L splits into N minimizations to be solved in parallel.

Decomposed dual problem	$x_i^{k+1} = \text{argmin}\, L_i\,(x_i, y^k)$	(9)
***N* dual problems in parallel**	$y^{k+1} = y^k + s_k(Ax^{k+1} - b)$	(10)

The N new x_i^{k+1} from (9) are gathered into Ax^{k+1} in (10). Then the resulting y^{k+1} is distributed to the N processors that execute separate minimizations in the next iteration of (9). This can give an enormous saving compared to one large minimization (6) when $f(x)$ is not separable.

Augmented Lagrangians

To make the iterations (9)–(10) more robust—to help them converge if $f(x)$ is not strictly convex—we can augment $f(x)$ by a penalty term with a variable factor ρ:

$$\textbf{Augmented Lagrangian} \quad L_\rho(x, y) = f(x) + y^{\mathrm{T}}(Ax - b) + \frac{1}{2}\,\rho\,||Ax - b||_2^2. \quad (11)$$

This is the Lagrangian for minimizing $f(x) + \frac{1}{2}\,\rho\,||Ax - b||^2$ with constraint $Ax = b$.

The same steps still lead to maximization as in (6)–(7). And the penalty constant ρ becomes an appropriate stepsize s: We can show that each new (x_{k+1}, y_{k+1}) satisfies $\nabla f(x_{k+1}) + A^{\mathrm{T}} y_{k+1} = \mathbf{0}$. But there is a big drawback to adding the penalty term with ρ: *The Lagrangian L_ρ is not separable even if $f(x)$ is separable!*

We need one more step to keep the advantage of separability (leading to N simpler maximizations solved in parallel) together with the greater safety that comes from including the penalty term $\frac{1}{2}\,\rho\,||Ax - b||^2$.

ADMM : Alternating Direction Method of Multipliers

The key new idea is splitting. The original $f(x)$ is broken into two parts (possibly an ℓ^1 part and an ℓ^2 part). We could call these parts f_1 and f_2, but to avoid subscripts they will be f and g. And we allow g to have a new variable z (instead of x), but *we recover the original problem by adding the constraint $x = z$*. This dodgy but legal maneuver allows us to keep a separable problem, with the big advantage of parallel computations—the pieces of x_{k+1} and z_{k+1} are distributed to separate computers.

That new constraint $x = z$ joins the original $Ax = b$ in a total of p linear constraints $Ax + Bz = c$. **We are now maximizing $f(x) + g(z)$**. And as before, we augment the Lagrangian for safer convergence:

$$\boxed{L_\rho(x, z, y) = f(x) + g(z) + y^{\mathrm{T}}(Ax + By - c) + \tfrac{1}{2}\,\rho\,||Ax + By - c||^2.} \quad (12)$$

Now we have an extra equation to update z at each step. As before, the stepsize s can be the regularizing coefficient ρ. The key advantage of ADMM is that x and z are updated sequentially and not jointly. *The two functions $f(x)$ and $g(z)$ are alternated.* A separable f or g will allow the distribution of pieces, for minimization in parallel.

Here are the three steps that reach a new $\boldsymbol{x}, \boldsymbol{z}, \boldsymbol{y}$ closer to $\boldsymbol{x}^*, \boldsymbol{z}^*, \boldsymbol{y}^*$:

ADMM

$$\boldsymbol{x}_{k+1} = \underset{\boldsymbol{x}}{\text{argmin}}\; L_\rho(\boldsymbol{x}, \boldsymbol{z}_k, \boldsymbol{y}_k) \tag{13}$$

$$\boldsymbol{z}_{k+1} = \underset{\boldsymbol{z}}{\text{argmin}}\; L_\rho(\boldsymbol{x}_{k+1}, \boldsymbol{z}, \boldsymbol{y}_k) \tag{14}$$

$$\boldsymbol{y}_{k+1} = \boldsymbol{y}_k + \rho(A\boldsymbol{x}_{k+1} + B\boldsymbol{z}_{k+1} - \boldsymbol{c}) \tag{15}$$

In practice, ADMM can be slow to reach high accuracy—but surprisingly fast to achieve acceptable accuracy. We mention here (looking ahead to Chapter VII on deep learning) that modest accuracy is often sufficient and even desirable—in situations where overfitting the training data leads to unhappy results on test data.

We continue to follow Boyd et al. by rescaling the dual variable $\boldsymbol{y}$. The new variable is $\boldsymbol{u} = \boldsymbol{y}/\rho$ and the linear and quadratic terms are combined in the Lagrangian. This produces a **scaled ADMM** that has advantages in practice.

$$\boldsymbol{x}_{k+1} = \underset{\boldsymbol{x}}{\text{argmin}}\; \left(f(\boldsymbol{x}) + \tfrac{1}{2}\rho\,||A\boldsymbol{x} + B\boldsymbol{z}_k - \boldsymbol{c} + \boldsymbol{u}_k||^2\right) \tag{16}$$

$$\boldsymbol{z}_{k+1} = \underset{\boldsymbol{z}}{\text{argmin}}\; \left(g(\boldsymbol{x}) + \tfrac{1}{2}\rho\,||A\boldsymbol{x}_{k+1} + B\boldsymbol{z} - \boldsymbol{c} + \boldsymbol{u}_k||^2\right) \tag{17}$$

$$\boldsymbol{u}_{k+1} = \boldsymbol{u}_k + A\boldsymbol{x}_{k+1} + B\boldsymbol{z}_{k+1} - \boldsymbol{c} \tag{18}$$

In any optimization, and certainly in this one, we should identify the equations we are solving. Those equations are satisfied (as we wanted) by the optimal $\boldsymbol{x}^*, \boldsymbol{z}^*$ in the primal problem and $\boldsymbol{y}^*$ or $\boldsymbol{u}^*$ in the dual problem:

$$\mathbf{0} = \nabla f(\boldsymbol{x}^*) + A^{\mathrm{T}}\boldsymbol{y}^* \qquad \mathbf{0} \in \partial f(\boldsymbol{x}^*) + A^{\mathrm{T}}\boldsymbol{y}^* \tag{19}$$

$$\mathbf{0} = \nabla g(\boldsymbol{z}^*) + B^{\mathrm{T}}\boldsymbol{y}^* \qquad \mathbf{0} \in \partial g(\boldsymbol{z}^*) + B^{\mathrm{T}}\boldsymbol{y}^* \tag{20}$$

Gradients ∇f and ∇g Subdifferentials ∂f and ∂g

A proper treatment of this problem (and a convergence proof for ADMM) would require more convex analysis than we are prepared for. But the reader will see why each step was taken, in reaching the scaled ADMM in (16)–(18). Convergence holds (see recent papers of Hajinezhad) even if f and g are not strictly convex. (They must be *closed* and *proper*. This allows $f = 0$ on a closed nonempty convex set and $f = +\infty$ otherwise. Subdifferentials = *multivalued derivatives* enter for such a function, at the edge of K.) And the unaugmented Lagrangian must have a saddle point.

The next pages follow Boyd's ADMM book by developing four major examples.

1 D. P. Bertsekas, *Constrained Optimization and Lagrange Multiplier Methods*, Athena Scientific (1986).

2 M. Fortin and R. Glowinski, *Augmented Lagrangian Methods*, North-Holland (1983-5).

3 D. Hajinezhad and Q. Shi, *ADMM for a class of nonconvex bilinear optimization*, Journal of Global Optimization **70** (2018) 261–288.

Example 1 This classical problem of convex optimization will be the starting point of Chapter VI. The function $f(\boldsymbol{x})$ is convex. The set K is closed and convex.

$$\boxed{\text{Minimize } f(\boldsymbol{x}) \text{ for } \boldsymbol{x} \text{ in } K.} \tag{21}$$

ADMM rewrites "$\boldsymbol{x}$ in K" as a minimization of g. It connects $\boldsymbol{x}$ to $\boldsymbol{z}$ by a constraint.

$$\boxed{\text{Minimize } f(\boldsymbol{x}) + g(\boldsymbol{z}) \text{ subject to } \boldsymbol{x} - \boldsymbol{z} = \mathbf{0}.} \tag{22}$$

g is the **indicator function** of the set K: $g(\boldsymbol{z}) = 0$ or $+\infty$ for $\boldsymbol{z}$ in or out of K. So $g = 0$ at the minimum, which forces the minimizer to be in K. The indicator function $g(\boldsymbol{z})$ is closed and convex, because the "cylinder" above its graph (above K) is closed and convex. The scaled augmented Lagrangian includes the penalty term:

$$L(\boldsymbol{x}, \boldsymbol{z}, \boldsymbol{u}) = f(\boldsymbol{x}) + g(\boldsymbol{z}) + \frac{1}{2}\rho\,||\boldsymbol{x} - \boldsymbol{z} + \boldsymbol{u}||^2. \tag{23}$$

Notice how the usual $\boldsymbol{\lambda}^{\mathrm{T}}(\boldsymbol{x} - \boldsymbol{z})$ was folded into equation (23) by the scaling step. Then ADMM splits (22) into *a minimization alternated with a projection*:

$$\textbf{ADMM} \qquad \begin{aligned} \boldsymbol{x}_{k+1} &= \text{argmin}\left[f(\boldsymbol{x}) + \tfrac{1}{2}\rho\,||\boldsymbol{x} - \boldsymbol{z}_k + \boldsymbol{u}_k||^2\right] \\ \boldsymbol{z}_{k+1} &= \text{projection of } \boldsymbol{x}_{k+1} + \boldsymbol{u}_{k+1} \text{ onto } K \\ \boldsymbol{u}_{k+1} &= \boldsymbol{u}_k + \boldsymbol{x}_{k+1} - \boldsymbol{z}_{k+1} \end{aligned}$$

Example 2 **Soft thresholding** An important ℓ^1 problem has an exact solution. $f(\boldsymbol{x}) = \lambda||\boldsymbol{x}||_1 = \lambda|x_1| + \cdots + \lambda|x_n|$ splits into n scalar functions $\lambda\,|x|_i$.
Separation by ADMM leads to the minimization of each special function f_1 to f_n:

$$f_i(x_i) = \lambda|x_i| + \frac{1}{2}\rho\,(x_i - v_i)^2 \text{ with } v_i = z_i - u_i$$

The solution x_i^* is the "soft thresholding" of v_i drawn in Section VI.4:

$$x_i^* = \left(v_i - \frac{\lambda}{\rho}\right)_+ - \left(-v_i - \frac{\lambda}{\rho}\right)_+ = v_i\left(1 - \frac{\lambda}{\rho\,|v_i|}\right)_+ \tag{24}$$

Not only is this thresholding function x_i^* an explicit solution to an important nonlinear problem, it is also a **shrinkage operator: every v_i moves toward zero.**

We will soon see this soft thresholding as a "proximal" operator.

Example 3 **Nonnegative Matrix Factorization** $A \approx CR$ with $C_{ij} \geq 0$ and $R_{ij} \geq 0$

ADMM begins with an *alternating minimization*—our favorite way to factor a matrix.

Find $C \geq 0$	Minimize $\|\|A - CR\|\|_F^2$ with $R \geq 0$ fixed
Find $R \geq 0$	Minimize $\|\|A - CR\|\|_F^2$ with $C \geq 0$ fixed

Boyd et al. point out an equivalent problem with C and R in the constraint $X = CR$:

$$\textbf{NMF} \qquad \text{Minimize } ||A - X||_F^2 + I_+(C) + I_+(R) \text{ with } X = CR$$

ADMM adds a third step that updates the dual variable U. And it introduces a new variable X constrained by $X = CR \geq 0$. The indicator function $I_+(C)$ is zero for $C \geq 0$ and infinite otherwise. Now ADMM splits into a minimization over X and C, alternating with a minimization over R:

$$(X_{k+1}, C_{k+1}) = \text{argmin}\left[||A - X||_F^2 + \frac{1}{2}\rho\,||X - CR_k + U_k||_F^2\right] \text{ with } X \geq 0, C \geq 0$$

$$R_{k+1} = \text{argmin}\,||X_{k+1} - C_{k+1}R + U_k||_F^2 \text{ with } R \geq 0$$

$$U_{k+1} = U_k + X_{k+1} - C_{k+1}R_{k+1}$$

The rows of X_{k+1}, C_{k+1} and then the columns of R_{k+1} can all be found separately. The splitting promotes parallel computation.

Example 4 LASSO aims for a sparse solution to $A\boldsymbol{x} = \boldsymbol{b}$ by including an ℓ^1 penalty:

$$\textbf{LASSO} \qquad \text{Minimize } \frac{1}{2}||A\boldsymbol{x} - \boldsymbol{b}||^2 + \lambda\,||\boldsymbol{x}||_1 \tag{25}$$

Immediately that problem splits into $f(\boldsymbol{x}) + g(\boldsymbol{z})$ with the constraint $\boldsymbol{x} - \boldsymbol{z} = \boldsymbol{0}$. The subproblem for $\boldsymbol{x}$ is least squares so we meet $A^{\mathrm{T}}A$. Augment with $\frac{1}{2}\rho\,||A\boldsymbol{x} - \boldsymbol{b}||^2$.

Scaled ADMM $\quad \boldsymbol{x}_{k+1} = (A^{\mathrm{T}}A + \rho I)^{-1}(A^{\mathrm{T}}\boldsymbol{b} + \rho\,(\boldsymbol{z}_k - \boldsymbol{u}_k))$

Soft thresholding $\quad \boldsymbol{z}_{k+1} = S_{\lambda/\rho}(\boldsymbol{x}_{k+1} + \boldsymbol{u}_k)$

Dual variable $\quad \boldsymbol{u}_{k+1} = \boldsymbol{u}_k + \boldsymbol{x}_{k+1} - \boldsymbol{z}_{k+1}$

By storing the LU factors of $A^{\mathrm{T}}A + \rho I$, the first step reduces to back substitution. Boyd et al remark that replacing $||\boldsymbol{x}||_1 = \sum |x_i|$ by $||F\boldsymbol{x}||_1 = \sum |x_{i+1} - x_i|$ converts this example into "total variation denoising". Their online book offers excellent and convincing examples of ADMM.

Matrix Splitting and Proximal Algorithms

A first connection of $Ax = b$ to ADMM comes from splitting that matrix into $A = B + C$. In classical examples B could be the diagonal part of A or the lower triangular part of A (Jacobi or Gauss-Seidel splitting). Those were improved by Douglas-Rachford and Peaceman-Rachford, who regularized by adding αI and alternated between B and C:

$$\boxed{\begin{aligned}(B + \alpha I)\boldsymbol{x}_{k+1} &= \boldsymbol{b} + (\alpha I - C)\boldsymbol{z}_k \\ (C + \alpha I)\boldsymbol{z}_{k+1} &= \boldsymbol{b} + (\alpha I - B)\boldsymbol{x}_{k+1}\end{aligned}} \tag{26}$$

This idea appeared in 1955, when problems were linear. It led to deeper nonlinear ideas: proximal operators and monotone operators. A special feature is the appearance of exact formulas for several important proximal operators—particularly the ℓ^1 minimization that led to soft thresholding and shrinkage in Example 2. We will define the **Prox** operator and connect it to ADMM.

In optimization, the analog to those matrix equations (26) has $\boldsymbol{b} = \mathbf{0}$ and $B = \boldsymbol{\nabla F_2}$. Here F_2 is the ℓ^2 part and its gradient $\boldsymbol{\nabla F_2}$ (*or its subgradient* $\boldsymbol{\partial F_2}$) is effectively linear. Then $\boldsymbol{\partial F_2} + \alpha I$ corresponds to $B + \alpha I$ and its inverse is a **"proximal operator"**:

$$\boxed{\mathbf{Prox}_F(\boldsymbol{v}) = \operatorname{argmin}\left(F(\boldsymbol{x}) + \tfrac{1}{2}||\boldsymbol{x} - \boldsymbol{v}||_2^2\right)} \tag{27}$$

The key facts to justify this proximal approach (and the splitting into $F_1 + F_2$) are

1 The ℓ^2 problem is well understood and **fast to solve**

2 The ℓ^1 problem often has a **closed form solution (by shrinkage).**

The double-step combination—the analog for optimization of the double-step matrix equations (26)—is the "proximal version" of ADMM with scaling factor α:

ADMM

$$\boldsymbol{x}_{k+1} = (\boldsymbol{\partial F_2} + \boldsymbol{\alpha I})^{-1}(\boldsymbol{\alpha z}_k - \boldsymbol{\alpha u}_k) \tag{28a}$$

$$\boldsymbol{z}_{k+1} = (\boldsymbol{\partial F_1} + \boldsymbol{\alpha I})^{-1}(\boldsymbol{\alpha x}_{k+1} + \boldsymbol{\alpha u}_k) \tag{28b}$$

$$\boldsymbol{u}_{k+1} = \boldsymbol{u}_k + \boldsymbol{x}_{k+1} - \boldsymbol{z}_{k+1} \tag{28c}$$

Overall, proximal algorithms apply to convex optimization. For the right problems they are fast. This includes distributed problems—minimizing a sum of terms in parallel (as for ADMM). Parikh and Boyd remark that **Prox** compromises between minimizing a function and not moving too far. They compare it to gradient descent $\boldsymbol{x} - \alpha\boldsymbol{\nabla F}(\boldsymbol{x})$ in which α plays the role of the stepsize. The fixed points of **Prox** are the minimizers of F.

Here are two excellent references and a website with source code for examples:

P. Combettes and J.- C. Pesquet, *Proximal splitting methods in signal processing, in Fixed-Point Algorithms for Inverse Problems*, Springer (2011). arXiv: 0912.3522

N. Parikh and S. Boyd, *Proximal algorithms*, Foundations and Trends in Optimization **1** (2013) 123-321.

Book and codes: **http://stanford.edu/~boyd/papers/pdf/prox_algs.pdf**

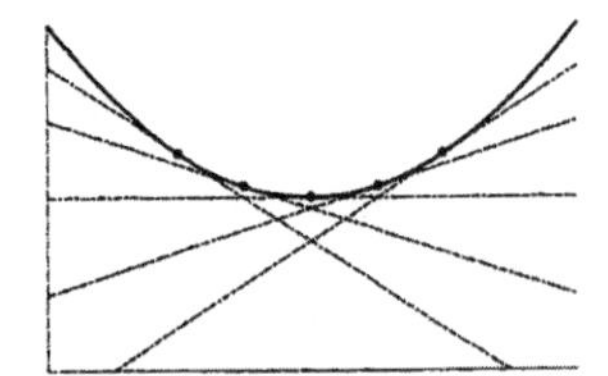
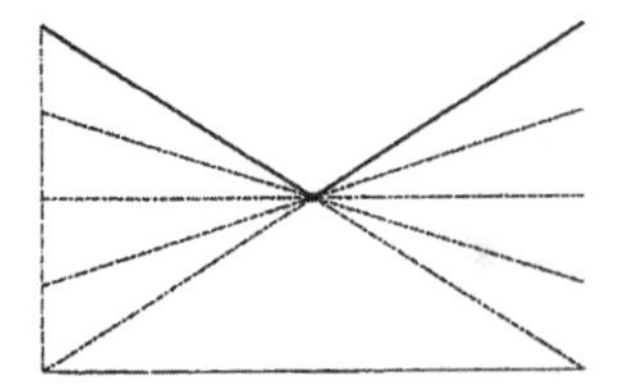

Figure III.3: Smooth function $f(x)$: One tangent at each point with slope ∇f. Pointed function $f(x) = |x|$: Many tangents with slopes ∂f (the subgradient) at the sharp point.

Bregman Distance

The name Bregman is appearing for the first time in this book. But "Bregman distance" is an increasingly important idea. Its unusual feature is the lack of symmetry: $D(u,v) \neq D(v,u)$. And the distance from u to v depends on a function f. We do have $D \geq 0$ and also $D(w,v) \leq D(u,v)$ along a straight line with $u < w < v$:

$$\textbf{Bregman Distance } D \qquad D_f(u,v) = f(u) - f(v) - (\nabla f(v), u - v) \tag{29}$$

Always the gradient ∇f is replaced by a **subgradient** ∂f at points where the graph of $f(x)$ has a corner. Then ∂f can be the slope of any tangent plane that stays below the convex function $f(x)$. The standard examples in one dimension are the absolute value $|x|$ and the ReLU function $\mathbf{max}\,(\mathbf{0}, x) = \frac{1}{2}(x + |x|)$ with corners at $x = 0$. The tangent planes to $|x|$ can have slopes $\partial\,|x|$ between -1 and 1. For ∂(ReLU) the slopes are between 0 and 1. These subgradients make ∂f an effective (but multivalued) replacement for ∇f.

Split Bregman and Split Kaczmarz

For minimization with an ℓ^1 penalty term (like basis pursuit), two iterative algorithms need to be seen. Start from $Ax = b$:

Linearized Bregman iteration with parameter $\lambda \geq 0$

$$y_{k+1} = y_k - s_k A^{\mathrm{T}}(Ax_k - b) \tag{30}$$

$$x_{k+1} = S(y_{k+1}) = \text{sign}\,(y_{k+1})\,\max\,(|y_{k+1}| - \lambda, 0) \tag{31}$$

(30) is a normal adjustable step to reduce $||Ax - b||^2$. Its stepsize is $s > 0$. Then the *soft thresholding function* S applies to each component of the vector y_{k+1}. If $y_{k+1} = (1, -3)$ and $\lambda = 2$, the output x_{k+1} from this nonlinear function S in (31) will be the vector x_{k+1} :

$$x_{k+1} = ((1)\max\,(1-2,0), (-1)\max\,(3-2,0)) = (0, -1).$$

If $Ax = b$ has a solution and if $s\lambda_{\max}(A^{\mathrm{T}}A) < 1$, the Bregman vectors x_k will converge to the optimal vector x^* :

$$x^* \text{ minimizes } \lambda||x||_1 + \frac{1}{2}||x||_2^2 \text{ subject to } Ax = b. \tag{32}$$

Kaczmarz iteration is effective for big data with sparse vectors. The standard steps solve $Ax = b$. They cycle sequentially or randomly through those n equations $\boldsymbol{a}_i^{\mathrm{T}}\boldsymbol{x} = b_i$ (where $\boldsymbol{a}_i^{\mathrm{T}}$ is row i of A). Each step adjusts $\boldsymbol{x}_k$ by a multiple $c\boldsymbol{a}_i$ of one column of A^{T}, to satisfy the ith equation $\boldsymbol{a}_i^{\mathrm{T}}\boldsymbol{x}_{k+1} = b_i$:

$$\boldsymbol{x}_{k+1} = \boldsymbol{x}_k + \frac{b_i - \boldsymbol{a}_i^{\mathrm{T}}\boldsymbol{x}_k}{||\boldsymbol{a}_i||^2}\boldsymbol{a}_i \quad \text{solves} \quad \boldsymbol{a}_i^{\mathrm{T}}\boldsymbol{x}_{k+1} = \boldsymbol{a}_i^{\mathrm{T}}\boldsymbol{x}_k + \boldsymbol{a}_i^{\mathrm{T}}\boldsymbol{a}_i\frac{b_i - \boldsymbol{a}_i^{\mathrm{T}}\boldsymbol{x}_k}{||\boldsymbol{a}_i||^2} = b_i. \quad (33)$$

The sparse Kaczmarz steps combine (33) with the soft threshold in (31):

$$\textbf{Sparse Kaczmarz} \quad \boldsymbol{y}_{k+1} = \boldsymbol{x}_k - s_k\boldsymbol{a}_i \quad \text{and} \quad \boldsymbol{x}_{k+1} = S(\boldsymbol{y}_{k+1}). \quad (34)$$

There is always freedom in the stepsize s_k. The choice $s_k = (\boldsymbol{a}_i^{\mathrm{T}}\boldsymbol{x}_k - b_i)/||\boldsymbol{a}_i||^2$ is consistent with Kaczmarz. Lin and Zhou have proposed an **online learning algorithm** for the same problem. This means that the step $k \to k+1$ is taken as soon as the new observation b_k and the new equation $\boldsymbol{a}_k^{\mathrm{T}}\boldsymbol{x} = b_k$ arrive to join the computation. Then the learning algorithm (34) immediately computes $\boldsymbol{y}_{k+1}$ and $\boldsymbol{x}_{k+1}$.

1 T. Goldstein and S. Osher, *The split Bregman method for L^1 regularized problems*, SIAM Journal of Imaging Sciences **2** (2009) 323 – 343.

2 W. Yin, S. Osher, D. Goldfarb, and J. Darbon , *Bregman iterative algorithms for ℓ^1 minimization with applications to compressed sensing*, SIAM J. Imaging Sciences **1** (2008) 143-168.

3 Y. Lei and D.-X. Zhou, *Learning theory of randomized sparse Kaczmarz method*, SIAM J. Imaging Sciences **11** (2018) 547–574.

Bounded Variation : L^1 Norm for the Gradient

Natural images have *edges*. Across those edges, the defining function $u(x, y)$ can have a jump. Its gradient $\nabla u = (\partial u/\partial x, \partial u/\partial y)$ can be smooth in the edge direction, but ∇u **has a delta function** in the perpendicular direction. The energy norm of u is infinite but its bounded variation norm is finite:

$$||\nabla u||_2^2 = \iint (u_x^2 + u_y^2)\, dx\, dy = \infty \quad \text{but} \quad ||u||_{\mathrm{BV}} = ||\nabla u||_1 = \iint \sqrt{u_x^2 + u_y^2}\, dx\, dy < \infty$$

In one dimension $u(x)$ could be a unit *step function*. Its derivative is a *delta function* $\delta(x)$. The integral of $\delta(x)^2$ is infinite. But the integral of $\delta(x)$ is 1. We can't work with the L^2 norm of the derivative $\delta(x)$, but the L^1 norm of $\delta(x)$ is good.

In applications to image denoising, that BV norm is a very successful penalty term. We fit the data, and we use the BV norm to prevent wild oscillations. The image can be smooth or piecewise smooth, but it can't have random "speckled" noise.

L. Rudin, S. Osher, and E. Fatemi, *Nonlinear total variation based noise removal algorithms*, Physica D **60** (1992) 259-268. (This paper was fundamental in applying **BV**.)

Problem Set III.4

1 What vector $\boldsymbol{x}$ minimizes $f(\boldsymbol{x}) = ||\boldsymbol{x}||^2$ for $\boldsymbol{x}$ on the line $\boldsymbol{x}^{\mathrm{T}}\boldsymbol{v} = 1$?

2 Following Example 1 in this section, write Problem 1 as a minimization of $f(\boldsymbol{x}) + g(\boldsymbol{z})$ with $\boldsymbol{x} = \boldsymbol{z}$. Describe that 0–1 indicator function $g(\boldsymbol{z})$, and take one ADMM step starting from $\boldsymbol{x} = \boldsymbol{0}, \boldsymbol{z} = \boldsymbol{0}, \boldsymbol{u} = \boldsymbol{0}$.

3 Which vector $\boldsymbol{x}$ minimizes $\lambda||\boldsymbol{x}||_1$ on the line $\boldsymbol{x}^{\mathrm{T}}\boldsymbol{w} = 1$ if $\boldsymbol{w} = (2, 3)$?

4 Following Example 2, take one ADMM step in Problem 3 from $\boldsymbol{x} = (1, 1)$.

5 What matrices $C \geq 0$ and $R \geq 0$ minimize $||A - CR||_F^2$ if $A = \begin{bmatrix} 2 & 1 \\ -1 & 2 \end{bmatrix}$?

6 Following Example 3 in this section, take one ADMM step in Problem 5 :

$$\text{From } A = \begin{bmatrix} 2 & 1 \\ -1 & 2 \end{bmatrix} \quad R_0 = \begin{bmatrix} 2 & 0 \\ 0 & 1 \end{bmatrix} \quad U_0 = \begin{bmatrix} 1 & 0 \\ 0 & 1 \end{bmatrix} \quad \text{compute } X_1, C_1, R_1, U_1.$$

7 Find the LASSO vector $\boldsymbol{x}$ that minimizes $\frac{1}{2}||A\boldsymbol{x} - \boldsymbol{b}||_2^2 + \lambda||\boldsymbol{x}||_1$ with

$$A = \begin{bmatrix} 4 & 1 \\ 0 & 1 \end{bmatrix} \qquad \boldsymbol{b} = \begin{bmatrix} 1 \\ 1 \end{bmatrix} \qquad \lambda = 2$$

8 Following Example 4, take one ADMM step in Problem 7 from $\boldsymbol{u}_0 = (1, 0) = \boldsymbol{z}_0$ with $\rho = 2$.

9 Find $\mathbf{Prox}_F(\boldsymbol{v})$ = argmin $\left(\frac{1}{2}||\boldsymbol{x}||^2 + \frac{1}{2}||\boldsymbol{x} - \boldsymbol{v}||^2\right)$ in equation (27). This is the proximal operator for $F(\boldsymbol{x}) = \frac{1}{2}||\boldsymbol{x}||^2$. It is a function of $\boldsymbol{v}$.

Here are three function spaces (each contained in the next) and three examples of $u(x, y)$:

Smooth $u(x, y)$ **Lipschitz** (slope can jump) **Bounded variation** (u can jump)

Bowl x^2+y^2 **Flat base** $\max(x^2+y^2-1, 0)$ **Cylinder base** (add step up along $r=1$)

A neat "*coarea formula*" expresses the BV norm of u as the integral of the lengths of the level sets $L(t)$ where $u(x, y) = t$. $\iint ||\mathbf{grad}\ \boldsymbol{u}||\, \boldsymbol{dx}\, \boldsymbol{dy} = \int (\mathbf{length\ of}\ \boldsymbol{L(t)})\, \boldsymbol{dt}$.

Example Compute both sides of the coarea formula for a bowl : $u(x, y) = x^2 + y^2$.

On the left side, $||\text{grad}\, u|| = ||(2x, 2y)|| = 2r$. The bowl area integrates from 0 to R: $||u||_{\mathrm{BV}} = ||\text{grad}\, u||_1 = \iint (2r)\, r\, dr\, d\theta = 4\pi R^3/3$.

On the right side, the level set where $u = t$ is the circle $x^2 + y^2 = t$ with length $2\pi\sqrt{t}$. The integral of that length $L(t)$ from $t = 0$ to R^2 is $\int 2\pi\sqrt{t}\, dt = 4\pi R^3/3$.

10 What is $||u||_{\mathrm{BV}}$ if $u(x, y) = x+y$ in the triangle with sides $x = 0, y = 0, x+y = 1$? What is $||u||_{\mathrm{BV}}$ if $u = 0$ in a unit square and $u = 1$ outside ?

III.5 Compressed Sensing and Matrix Completion

The basic principle of compressed sensing is that a sparse signal can be exactly recovered from incomplete data. It is a remarkable experience to see a perfect image emerging from too few measurements. The geometry of an ℓ^1 diamond versus an ℓ^2 sphere is one way to explain it. Now mathematical analysis has identified the conditions on A and the minimum number of measurements. We will summarize.

The reader will understand that the Nyquist-Shannon theorem is still in force. To recover a noisy signal exactly, you must sample with at least twice its top frequency. Otherwise part of that signal is missed. But sparse signals are not noisy! If a signal is expressed by a small number of sinusoids or wavelets, then that sparse signal can be recovered (with probability extremely close to 1) by a small number of measurements.

It is important to remember: **Sparsity depends on the basis $\boldsymbol{v}_1, \ldots, \boldsymbol{v}_n$ in which the signal is represented**. It uses only a few $\boldsymbol{v}$'s. And the signal can be sensed in a different basis $\boldsymbol{w}_1, \ldots, \boldsymbol{w}_n$. The $\boldsymbol{v}$'s are the columns of a representing matrix V, and the $\boldsymbol{w}$'s are the columns of an acquiring matrix W. A common example: The $\boldsymbol{v}$'s are a Fourier basis and the $\boldsymbol{w}$'s are a spike basis. Then V is the Fourier matrix F and W is the identity matrix I.

A key requirement for compressed sensing is **"incoherence" of V and W**: the entries of $V^{\mathrm{T}}W$ are small. Those are the inner products $\boldsymbol{v}_i^{\mathrm{T}}\boldsymbol{w}_j$. Luckily we do have low coherence for F and I: all entries of F have equal size. And even luckier: Random matrices with $||\text{columns}|| = 1$ are almost sure to be incoherent with any fixed basis. So randomness and probabilities close to 1 are key ideas in compressed sensing.

The sensing step produces only $m < n$ nonzero coefficients y_k of the unknown signal $\boldsymbol{f}$: $\boldsymbol{y} = W^{\mathrm{T}}\boldsymbol{f}$. To reconstruct $\boldsymbol{f}^* = V\boldsymbol{x}^*$ close to this $\boldsymbol{f}$, we use ℓ^1 optimization to find $\boldsymbol{x}^*$:

$$\text{Minimize } ||\boldsymbol{x}||_1 \text{ subject to } W^{\mathrm{T}}V\boldsymbol{x} = \boldsymbol{y} \tag{1}$$

This is a linear programming problem (it is basis pursuit). That tells us again to expect a sparse solution $\boldsymbol{x}^*$ (at a corner of the set of vectors with $W^{\mathrm{T}}V\boldsymbol{x} = \boldsymbol{y}$). So the simplex method is a potential algorithm for locating $\boldsymbol{x}^*$. And there are faster ways to solve (1).

The basic theorem was established by Candès and Tao and Donoho:

Suppose V and W are incoherent and $\boldsymbol{x}^*$ is sparse ($\leq S$ nonzeros). The probability is overwhelming that if $m > C\,S\log n$, the solution to (1) will reproduce $\boldsymbol{f}$ exactly.

Note One unusual situation arises when the true signal is supersparse in the basis of $\boldsymbol{w}$'s. Maybe it is exactly one of the $\boldsymbol{w}$'s. Then probes might find only zeros and we totally miss that $\boldsymbol{w}$ in $\boldsymbol{x}^*$. This is one reason that probability enters into compressed sensing.

The basic theorem above is not the full story. The system will have noise. So we are recovering a vector x that solves $Ax = b + z$, where z can be stochastic and unknown. The sparse problem is nearby, but it doesn't perfectly match the noisy data. We want **stable recovery**: Solve (1) with noisy data and obtain an x^* that is **near the sparse x^{**} ($\leq S$ nonzeros) that would have come from noiseless data.**

This conclusion is true when A has the **"restricted isometry property"** with $\delta < \sqrt{2} - 1$:

$$\textbf{(RIP)} \qquad (1-\delta)\,||x||_2^2 \leq ||Ax||_2^2 \leq (1+\delta)\,||x||_2^2 \text{ if } x \text{ is } S\text{-sparse.} \qquad (2)$$

Fortunately, the matrix A can have columns chosen randomly on the unit sphere—or chosen randomly from the normal distribution $N(0, 1/m)$—or with independent random entries equal to $\pm 1/\sqrt{m}$—or in other random ways. The RIP requirement is almost surely satisfied if $m > C\,S \log(n/S)$.

This finally connects the number of measurements m with the sparsity S. And the noisy data leads us to replace basis pursuit (where Ax exactly equals b) with LASSO:

$$\boxed{\textbf{LASSO with noise} \qquad \text{Minimize } ||x||_1 \text{ subject to } ||Ax - b||_2 \leq \epsilon} \qquad (3)$$

This exposition of compressed sensing has followed the article by Candès and Wakin: *IEEE Signal Processing Magazine* **21** (March 2008). That article refers to the early work of Candès, Tao, and Donoho—which achieved a highly valuable goal. Instead of acquiring massive accounts of data and then compressing it all (as a camera does), only $m = O(S \log n/S)$ is acquired and used. *A one-pixel camera becomes possible, with no lens*, as Rich Baraniuk has shown. Perhaps the medical applications are the most valuable—we will start and end with those.

The start was an unexpected observation by Candès in 2004. The Logan-Shepp test image (an abstract model of the human head) was corrupted by noise. It looked as if Magnetic Resonance Imaging (MRI) had been stopped too soon. To improve the image, Candès tried an idea using ℓ^1. To his surprise, *the phantom image became perfect* (even though the data was incomplete). It was almost an online equivalent of Roentgen's discovery of X-rays in 1895—an accident that gave birth to an industry.

Incredibly, in the week of writing these words, an essay by David Donoho appeared in the online Notices of the American Math Society (January 2018). It describes how compressed sensing has accelerated MRI. Scan times are reduced from 8 minutes to 70 seconds, with high quality images. Dynamic heart imaging becomes feasible even for children. Michael Lustig was a pioneer in this MRI success, and a host of mathematicians contributed to the theory and the algorithms. The FDA has approved a change that will come gradually in the United States, as thousands of scanners are upgraded.

The purpose of Donoho's essay was to show that funding and collaboration and theory and experiment (and a little luck) have produced something wonderful.

Matrix Completion in the Nuclear Norm

The rank of a matrix is like the number of nonzeros in a vector. In some way the rank measures sparsity. For low rank, the matrix Σ of singular values is literally sparse (r nonzeros). Just as the number of nonzeros is a "0-norm" for vectors, **the rank is a "0-norm" for matrices.** But $||v||_0$ **and** $||A||_0$ **are not true vector and matrix norms,** because $||v||_0 = ||2v||_0$ and $\text{rank}(A) = \text{rank}(2A)$. Multiplying by 2 doesn't change the count of nonzeros or the rank—but it always doubles any true norm.

Nevertheless we often want sparsity for vectors and low rank for matrices. The **matrix completion problem** starts with missing entries in a matrix A_0. We want to complete A_0 to A, keeping the rank as low as possible. We are introducing a minimum of unexplained data. This problem of **missing data** is widespread in all areas of observational science. Some assumption about the completed A is needed to fill in the blanks, and *minimum rank* is a natural choice. Look at these examples:

$$A_0 = \begin{bmatrix} 1 & 2 \\ * & * \end{bmatrix} \qquad B_0 = \begin{bmatrix} 1 & * \\ * & 4 \end{bmatrix} \qquad C_0 = \begin{bmatrix} 1 & 2 \\ 3 & * \end{bmatrix}$$

All can be completed with rank 1. A_0 allows any multiple of $(1, 2)$ in the second row. B_0 allows any numbers b and c with $bc = 4$. C_0 allows only 6 in its last entry.

With vectors, we relaxed the sparsity norm $||v||_0$ to the ℓ^1 norm $||v||_1$. With matrices, **we now relax the rank norm** $||A||_0$ **to the nuclear norm** $||A||_N$. This nuclear norm is the ℓ^1 norm of the diagonal of Σ. **We are minimizing the sum of the singular values**:

$$\textbf{Nuclear norm} \qquad ||A||_N = \sigma_1 + \sigma_2 + \cdots + \sigma_r \tag{4}$$

Now we have a convex norm, but not strictly convex: $||B_1 + B_2||_N = ||B_1||_N + ||B_2||_N$ is possible in the triangle "inequality". In fact the example matrix B_0 could be completed symmetrically by any $b = c$ between -2 and 2. The nuclear norm $||B||_N$ stays at 5 (singular values = eigenvalues and $||B||_N$ = trace for this positive semidefinite B). Rank one matrices are roughly analogous to sharp points in the diamond $||x||_1 = 1$.

A famous example for matrix completion was the *Netflix competition*. The ratings of m films by n viewers went into A_0. But the customers didn't see all movies. Many ratings were missing. Those had to be predicted by a *recommender system*. The nuclear norm gave a good solution that needed to be adjusted for human psychology—Netflix was not a competition in pure mathematics.

Another application is computing the covariance matrix from a lot of sampled data. Finding all covariances σ_{ij} can be expensive (typical case: the covariances of all stocks over 365 days). We may compute some and estimate the rest by matrix completion. A scientific example: All the covariances in measuring the ocean surface at 10^4 positions.

Here is the "convex relaxation" of rank minimization to nuclear norm minimization:

Matrix completion **Minimize** $||A||_N$ **subject to** $A = A_0$ **in the known entries.**

The mathematical question is: By knowing K entries in an n by n matrix of rank r, can we expect a perfect recovery of the whole matrix? The remarkable answer is *yes*—provided K is large enough. Candès and Recht proved that if $K > Cn^{5/4} r \log n$, then with high probability (in a suitable model!) the recovery of A is perfect.

Here are a few comments on their 20-page proof, followed by references.

1. "With high probability" does not mean "certain". We must work with a model for random matrices A. One choice takes U and V as random orthogonal matrices in $A = U\Sigma V^{\mathrm{T}}$.

2. The analysis and proof use matrix versions of the key inequalities of statistics in V.3.

3. Finding A with smallest nuclear norm can be expressed as a **semidefinite program**:

$$\text{Minimize the trace of } \begin{bmatrix} W_1 & X \\ X & W_2 \end{bmatrix} \quad \begin{matrix} X \text{ contains the known entries} \\ W_1, W_2 \text{ are positive semidefinite} \end{matrix}$$

1. D. Donoho, *Compressed sensing*, IEEE Trans. Inform. Th. **52** (2006) 1289-1306.

2. E. Candès, J. Romberg, and T. Tao, *Robust uncertainty principles: Exact signal reconstruction from highly incomplete Fourier information*, IEEE Transactions on Information Theory **52** (2006) 489-509.

3. E. Candès and B. Recht, *Exact matrix completion via convex optimization*, Foundations of Comp. Math. **9** (2009) 717 – 736; arXiv: 0805.4471v1, 29 May 2008.

4. T. Hastie, B. Mazumder, J. Lee, and R. Zadeh, Matrix completion and low-rank SVD via fast alternating least squares, arXiv: 1410.2596, 9 Oct 2014.

Algorithms for Matrix Completion

We need an algorithm that completes the matrix A, with fixed entries A_{known}. Reference 4 above traces the development of three alternating iterations—each new method improving on the previous algorithm. This is how numerical analysis evolves.

1 (*Mazumder et. al*) Soft-threshold the SVD of A_k (see Section VI.5 for S_λ):

$$A_k = U_k \Sigma_k V_k^{\mathrm{T}} \text{ and } B_k = U_k S_\lambda(\Sigma_k) V_k^{\mathrm{T}} \text{ with } S_\lambda(\sigma) = \max(\sigma - \lambda, 0)$$

S_λ sets the smaller singular values of B_k to zero, reducing its rank. Then

$$A_{k+1} \text{ minimizes } \frac{1}{2}||(A - B_k)_{\text{known}}||_F^2 + \lambda ||B_k||_N \tag{5}$$

The drawback is the task of computing the SVD of A_k at each step. In the Netflix competition, A had 8×10^9 entries. The authors succeeded to reduce this heavy cost. And each previous SVD gives a warm start to the next one. And new ideas kept coming.

2 (*Srebro et. al*) These authors write the solution matrix as $CR^{\mathrm{T}} = (m \times r)(r \times n)$:

$$\underset{C \text{ and } R}{\textbf{Minimize}} \quad \frac{1}{2}||(A - CR^{\mathrm{T}})_{\text{known}}||_F^2 + \frac{\lambda}{2}(||C||_F^2 + ||R||_F^2) \tag{6}$$

This is convex in C alone and in R alone (so biconvex). An alternating algorithm is natural: *Update C and then update R.* Each problem is an ℓ^2 ridge regression for each column of C and then each column of R. This produces a "maximum margin" factorization CR.

3 (*Hastie et. al*, see 4 above) The third algorithm came in 2014. It is a variation on the first—but it alternates between C and R as in the second. The minimization now includes all entries of $A - CR^{\mathrm{T}}$, not only those in the known positions. This greatly simplifies the least squares problem, to work with full columns of both known and unknown entries.

Start with this (alternating) least squares problem, when A is *fully known*:

$$\underset{C \text{ and } R}{\textbf{Minimize}} \quad \frac{1}{2}\|A - CR^{\mathrm{T}}\|_F^2 + \frac{\lambda}{2}(\|C\|_F^2 + \|R\|_F^2) \tag{7}$$

An explicit solution is $C = U_r S_\lambda(\Sigma_r)^{1/2}$ and $R = V_r S_\lambda(\Sigma_r)^{1/2}$. All solutions including this one have $CR^{\mathrm{T}} = U_r S_\lambda(\Sigma_r) V_r^{\mathrm{T}}$ = soft SVD of A. The soft thresholding S_λ shows the effect of the penalty term in (7). Amazingly and beautifully, that product $B = CR^{\mathrm{T}}$ solves this rank r nuclear norm problem:

$$\underset{\operatorname{rank}(B) \leq r}{\textbf{Minimize}} \quad \frac{1}{2}\|A - B\|_F^2 + \lambda\|B\|_N \tag{8}$$

This connection to (7) yields the following fast alternating computation of $B = CR^{\mathrm{T}}$.

For the moment assume all entries of A are known: none missing. Start with C = random $m \times r$ and $D = I_r$. Each step updates R, C, and D.

1. Minimize over R $(n \times r)$ $\quad \|A - CR^{\mathrm{T}}\|_F^2 + \lambda\|R\|_F^2$	(9)
2. Compute this SVD $\quad RD = U\Sigma V^{\mathrm{T}} \quad$ Set $D = \sqrt{\Sigma}$ and $R_{\text{new}} = VD$	
3. Minimize over C $(m \times r)$ $\quad \|A - CR^{\mathrm{T}}\|_F^2 + \lambda\|C\|_F^2$	(10)
4. Compute this SVD $\quad CD = U\Sigma V^{\mathrm{T}} \quad$ Set $D = \sqrt{\Sigma}$ and $C_{\text{new}} = UD$	

Repeat those steps until CR^{T} converges to the solution B of (8).

Now Hastie et. al return to the real problem, in which A has missing entries: not known. At each iteration, those are taken from the current matrix CR^{T}. This produces a very efficient sparse plus low rank representation of A.

$$A = A_{\text{known}} + (CR^{\mathrm{T}})_{\text{unknown}} = (A - CR^{\mathrm{T}})_{\text{known}} + CR^{\mathrm{T}}. \tag{11}$$

The final algorithm is a slight modification of (9)-(10), by using A from equation (13). The solutions to (9)-(10) have remarkably simple forms, because those problems are essentially ridge regressions (least squares):

$$R^{\mathrm{T}} = (D^2 + \lambda I)^{-1} D U^{\mathrm{T}} A \text{ in (9)} \qquad C = AVD(D^2 + \lambda I)^{-1} \text{ in (10)}$$

Altogether the algorithm is described as **Soft-impute Alternating Least Squares**. Hastie et al analyze its convergence and numerical stability. They test an implementation on the data for the Netflix competition—with success.

Problem Set III.5

1. For one or more of these examples in the text, can you find the completion that minimizes $||A||_N$?

$$A_0 = \begin{bmatrix} 1 & 2 \\ * & * \end{bmatrix} \qquad B_0 = \begin{bmatrix} 1 & * \\ * & 4 \end{bmatrix} \qquad C_0 = \begin{bmatrix} 1 & 2 \\ 3 & * \end{bmatrix}$$

2. Corresponding to the important Figure I.16 near the start of Section I.11, can you find the matrix with smallest "*sum norm*" $||A||_S = |a| + |b| + |c| + |d|$ so that

$$\begin{bmatrix} a & b \\ c & d \end{bmatrix} \begin{bmatrix} 3 \\ 4 \end{bmatrix} = \begin{bmatrix} 1 \\ 0 \end{bmatrix}?$$

3. For 2 by 2 matrices, how would you describe the unit ball $||A||_S \leq 1$ in that sum norm $|a| + |b| + |c| + |d|$?

4. Can you find inequalities that connect the sum norm to the nuclear norm for n by n matrices ?

$$||A||_N \leq c(n)\, ||A||_S \quad \text{and} \quad ||A||_S \leq d(n)\, ||A||_N$$

5. If only one entry of a matrix A is unknown, how would you complete A to minimize $||A||_S$ or $||A||_N$?

Here are two neat formulas for the nuclear norm (thank you to Yuji Nakatsukasa).

$$\boxed{||A||_N = \min_{UV=A} ||U||_F\, ||V||_F = \min_{UV=A} \frac{1}{2}||U||_F^2 + \frac{1}{2}||V||_F^2} \qquad (*)$$

6. Start from $A = U\Sigma V^{\mathrm{T}} = (U\Sigma^{1/2})\,(\Sigma^{1/2}V^{\mathrm{T}})$. If you rename those two factors U^* and V^*, so that $A = U^*V^*$, show that $||U^*||_F^2 = ||V^*||_F^2 = ||A||_N$: equality in $(*)$.

7. If A is positive semidefinite then $||A||_N$ = trace of A (why ?). Then if $A = UV$, explain how the Cauchy-Schwarz inequality gives

$$||A||_N = \text{trace}\,(UV) = \Sigma\Sigma\, U_{ij}V_{ji} \leq ||U||_F\, ||V||_F.$$

Part IV

Special Matrices

IV.1 Fourier Transforms : Discrete and Continuous

IV.2 Shift Matrices and Circulant Matrices

IV.3 The Kronecker Product $A \otimes B$

IV.4 Sine and Cosine Transforms from Kronecker Sums

IV.5 Toeplitz Matrices and Shift Invariant Filters

IV.6 Graphs and Laplacians and Kirchhoff's Laws

IV.7 Clustering by Spectral Methods and k-means

IV.8 Completing Rank One Matrices

IV.9 The Orthogonal Procrustes Problem

IV.10 Distance Matrices

Part IV : Special Matrices

This chapter develops two large topics : the key matrices for **Discrete Fourier Transforms** and for **graphical models**. Both topics appear in machine learning when the problem has special structure—which is reflected in the architecture of the neural net.

Fourier is for problems with *shift invariance*. The operations on one pixel of an image are the same as the operations on the next pixel. In that case the operation is a *convolution*. Each row of the underlying matrix is a shift of the previous row. Each column is a shift of the previous column. Convolutional nets use the same weights around each pixel.

With that shift-invariant structure, a convolution matrix (a "*filter*" in image processing) has constant diagonals. The N by N matrix in a 1-dimensional problem is fully determined by its first row and column. Very often the matrix is banded—the filter has finite length—the matrix has zeros outside a band of diagonals. For 2-dimensional problems with repeated one-dimensional blocks the saving is enormous.

This makes it possible to use a Convolutional Neural Net (**CNN** or **ConvNet**) when an N^2 by N^2 matrix full of independent weights would be impossible. A major breakthrough in the history of CNN's was this NIPS 2012 paper :

> Alex Krizhevsky, Ilya Sutskever, and Geoffrey Hinton : *ImageNet Classification with Deep Convolutional Neural Networks.* Their neural net had 60 million parameters.

Graphs have a different structure. They consist of n nodes connected by m edges. Those connections are expressed by the incidence matrix A : m rows for the edges and n columns for the nodes. Every graph is associated with four important matrices :

Incidence matrix A	m by n with -1 and 1 in each row
Adjacency matrix M	n by n with $a_{ij} = 1$ when nodes i and j are connected
Degree matrix D	n by n diagonal matrix with row sums of M
Laplacian matrix $L = A^{\mathrm{T}}A = D - M$	positive semidefinite matrix

From the viewpoint of deep learning, Fourier problems are associated with CNN's and graphical models lead to graphical nets. Graph theory has become our most valuable tool to understand discrete problems on networks.

IV.1 Fourier Transforms : Discrete and Continuous

The classical Fourier transforms apply to functions. The discrete Fourier transform (DFT) applies to vectors :

Real Fourier series : real periodic functions $f(x + 2\pi) = f(x)$

Complex Fourier series : complex periodic functions

Fourier integral transforms : complex functions $f(x)$ for $-\infty < x < \infty$

Discrete Fourier series : complex vectors $\boldsymbol{f} = (f_0, f_1, \ldots, f_{N-1})$

Our focus is on the last one—transforming vectors to vectors. That is achieved by an N by N matrix. The inverse transform uses the inverse matrix. And all these transforms share **basis functions** of the same type that Fourier used :

Real Fourier series : cosines $\cos nx$ and sines $\sin nx$

Complex Fourier series : complex exponentials e^{inx} for $n = 0, \pm 1, \pm 2, \ldots$

Fourier integral transforms : complex exponentials e^{ikx} for $-\infty < k < \infty$

Discrete Fourier series : N basis vectors $\boldsymbol{b}_k$ with $(\boldsymbol{b}_k)_j = e^{2\pi ijk/N} = (e^{2\pi i/N})^{jk}$.

Each function and each vector is expressed as a combination of Fourier basis functions. **What is the "transform"?** It is the rule that connects f to its coefficients $a_k, b_k, c_k, \widehat{f}(k)$ in these combinations of basis functions :

Real series	$f(x) = a_0 + a_1 \cos x + b_1 \sin x + a_2 \cos 2x + b_2 \sin 2x + \cdots$
Complex series	$f(x) = c_0 + c_1 e^{ix} + c_{-1} e^{-ix} + c_2 e^{2ix} + c_{-2} e^{-2ix} + \cdots$
Fourier integrals	$f(x) = \int_{-\infty}^{\infty} \widehat{f}(k) e^{ikx}\, dk$
Discrete series	$\boldsymbol{f} = c_0 \boldsymbol{b}_0 + c_1 \boldsymbol{b}_1 + \cdots + c_{N-1} \boldsymbol{b}_{N-1} =$ **Fourier matrix F times c**

Each **Fourier transform** takes f in "x-space" to its coefficients in "frequency space". The **inverse transform** starts with the coefficients and reconstructs the original function.

Computing coefficients is *analysis*. Reconstructing the original $\boldsymbol{f}$ (shown in the box) is *synthesis*. For vectors the commands **fft** and **ifft** produce $\boldsymbol{c}$ from $\boldsymbol{f}$, $\boldsymbol{f}$ from $\boldsymbol{c}$. Those commands are executed by the **Fast Fourier Transform**.

Orthogonality

In all four transforms, the coefficients have nice formulas. That is because **the basis functions are orthogonal.** Orthogonality is the key to all the famous transforms (often the basis contains the eigenfunctions of a symmetric operator). This allows us to find each number c_k separately. For vectors, we just take the (complex !) dot product $(\boldsymbol{b}_k, \boldsymbol{f})$ between $\boldsymbol{f}$ and each of the orthogonal basis vectors $\boldsymbol{b}_k$. Here is that key step :

$$(\boldsymbol{b}_k, \boldsymbol{f}) = (\boldsymbol{b}_k, c_0\boldsymbol{b}_0 + c_1\boldsymbol{b}_1 + \cdots + c_{N-1}\boldsymbol{b}_{N-1}) = c_k(\boldsymbol{b}_k, \boldsymbol{b}_k). \textbf{ Then } \boxed{c_k = \frac{(\boldsymbol{b}_k, \boldsymbol{f})}{(\boldsymbol{b}_k, \boldsymbol{b}_k)}.} \quad (1)$$

All inner products $(\boldsymbol{b}_k, \boldsymbol{b}_j)$ are zero in equation (1), except $(\boldsymbol{b}_k, \boldsymbol{b}_k)$. Those denominators $(\boldsymbol{b}_k, \boldsymbol{b}_k) = ||\boldsymbol{b}_k||^2 = \pi$ or 2π or N are a pleasure to compute for Fourier basis functions :

Real series : $\int (\cos nx)^2\, dx = \pi$ and $\int (\sin nx)^2\, dx = \pi$ and $\int (1)^2 dx = 2\pi$

Complex series : $\int e^{ikx} e^{-ikx} dx = \int 1\, dx = 2\pi$

Discrete series : $\overline{\boldsymbol{b}}_k^{\mathrm{T}} \boldsymbol{b}_k = 1 \cdot 1 + e^{2\pi ik/N} \cdot e^{-2\pi ik/N} + e^{4\pi ik/N} \cdot e^{-4\pi ik/N} + \cdots = N$

If we normalize basis vectors to $||\boldsymbol{b}_k|| = 1$ then Fourier matrices are orthogonal.

The Fourier integral case is the subtle one and we stop after writing down equation (2). Its infinite integrals come from Fourier series when the period 2π increases to $2\pi T$ with $T \to \infty$:

$$\widehat{f}(k) = \int_{x=-\infty}^{\infty} f(x) e^{-ikx}\, dx \text{ and } f(x) = \frac{1}{2\pi} \int_{k=-\infty}^{\infty} \widehat{f}(k) e^{ikx}\, dk. \quad (2)$$

Fourier Matrix F and DFT Matrix Ω

The matrices F_N and Ω_N are N by N. They are both symmetric (but complex). They have the same columns, but the order of columns is different. F_N contains powers of $w = e^{2\pi i/N}$ and Ω_N contains powers of the complex conjugate $\overline{w} = \omega = e^{-2\pi i/N}$.

Roman w and Greek ω. In fact the two matrices are complex conjugates : $\overline{F}_N = \Omega_N$. Here are F_4 and Ω_4, containing powers of $w = e^{2\pi i/4} = i$ and $\omega = e^{-2\pi i/4} = -i$. We count rows and columns starting at zero, so F and Ω have rows $0, 1, 2, 3$:

$$\textbf{Fourier matrix}\quad F_4 = \begin{bmatrix} 1 & 1 & 1 & 1 \\ 1 & i & i^2 & i^3 \\ 1 & i^2 & i^4 & i^6 \\ 1 & i^3 & i^6 & i^9 \end{bmatrix} \qquad \textbf{DFT matrix}\quad \Omega_4 = \begin{bmatrix} 1 & 1 & 1 & 1 \\ 1 & -i & (-i)^2 & (-i)^3 \\ 1 & (-i)^2 & (-i)^4 & (-i)^6 \\ 1 & (-i)^3 & (-i)^6 & (-i)^9 \end{bmatrix} \quad (3)$$

Ω times $\boldsymbol{f}$ produces the discrete Fourier coefficients $\boldsymbol{c}$

F times $\boldsymbol{c}$ brings back the vector $4\boldsymbol{f} = N\boldsymbol{f}$ because $F\Omega = NI$

If you multiply F_N times Ω_N, you discover a fundamental identity for discrete Fourier transforms. *It reveals the inverse transform*:

$$\boldsymbol{F_N \Omega_N = NI} \quad \text{and therefore} \quad \boldsymbol{F_N^{-1} = \frac{1}{N}\Omega_N = \frac{1}{N}\overline{F}_N} \tag{4}$$

To see this, look closely at the N numbers $1, w, \ldots, w^{N-1}$ in the second column of F. Those numbers all have magnitude 1. They are equally spaced around the unit circle in the complex plane (Figure IV.1). They are the N solutions to the Nth degree equation $z^N = 1$.

Their complex conjugates $1, \omega, \ldots, \omega^{N-1}$ are **the same N numbers.** These powers of ω go around the unit circle in the opposite direction. Figure IV.1 shows 8 powers of $w = e^{2\pi i/8}$ and $\omega = e^{-2\pi i/8}$. **Their angles are 45 ° and −45 °.**

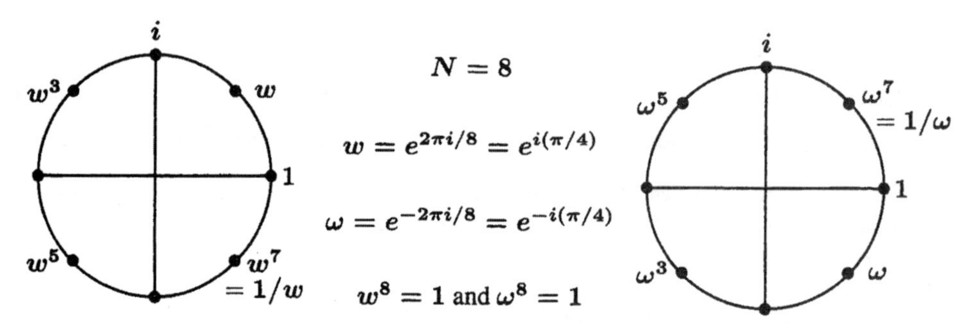

Figure IV.1: The powers of w are also the powers of ω. They are the N solutions to $z^N = 1$.

To prove that $F_N\Omega_N = NI$, here is the property we need. The N points add to zero!

$$\textbf{For every } \boldsymbol{N} \textbf{ the sum } \boldsymbol{S = 1 + w + w^2 + \cdots + w^{N-1}} \textbf{ is zero.} \tag{5}$$

The proof is to multiply S by w. This produces $w + \cdots + w^N$. That is the same as S, because w^N at the end is the same as 1 (at the start). Then $Sw = S$. So S must be zero.

You see right away that the eight numbers around the circle in Figure IV.1 add to zero, because opposite pairs already add to zero. For odd N that pairing doesn't work.

This fact $S = 0$ tells us that every off-diagonal entry in $F_N\Omega_N$ is zero. The diagonal entries (sum of 1's) are certainly N. So $\boldsymbol{F_N \Omega_N = NI}$. If we divide F and Ω by $\sqrt{N}$, then the two matrices are inverses and also complex conjugates: they are *unitary*.

$$\textbf{Unitary matrices} \quad \left(\frac{1}{\sqrt{N}} F_N\right)\left(\frac{1}{\sqrt{N}} \Omega_N\right) = \left(\frac{1}{\sqrt{N}} F_N\right)\left(\frac{1}{\sqrt{N}} \overline{F}_N\right) = I. \tag{6}$$

Unitary matrices are the complex version of orthogonal matrices. Instead of $Q^{\mathrm{T}} = Q^{-1}$ we have $\overline{Q}^{\mathrm{T}} = Q^{-1}$. It is appropriate to take complex conjugates when you transpose a complex matrix (and most linear algebra codes do that automatically). In the same way, the complex version of a real symmetric matrix is a **Hermitian matrix** with $\overline{S}^{\mathrm{T}} = S$.

The DFT Matrix Ω is a Permutation of the Fourier Matrix F

The matrices F and Ω have the same columns. So F and Ω are connected by a permutation matrix. That permutation P leaves the zeroth columns alone: the column of 1's in both matrices. Then P exchanges the next column $(1, w, w^2, \ldots, w^{N-1})$ of F for its last column $(1, \omega, \omega^2, \ldots, \omega^{N-1})$. After the $\mathbf{1}$ in its zeroth row and column, P contains the reverse identity matrix J (with 1's on the antidiagonal):

$$\boldsymbol{P} = \begin{bmatrix} \mathbf{1} & 0 \\ 0 & \boldsymbol{J} \end{bmatrix} = \begin{bmatrix} \mathbf{1} & 0 & 0 & 0 \\ 0 & 0 & 0 & \mathbf{1} \\ 0 & 0 & \mathbf{1} & 0 \\ 0 & \mathbf{1} & 0 & 0 \end{bmatrix} \qquad \boldsymbol{P^2} = \boldsymbol{I} \text{ and } \boldsymbol{\Omega} = \boldsymbol{FP} \text{ and } \boldsymbol{\Omega P} = \boldsymbol{FP^2} = \boldsymbol{F}$$

Here are the full matrices for $\boldsymbol{\Omega} = \boldsymbol{FP}$ when $N = 4$:

$$\boldsymbol{\Omega} = \begin{bmatrix} 1 & \mathbf{1} & 1 & 1 \\ 1 & \boldsymbol{-i} & (-i)^2 & (-i)^3 \\ 1 & \boldsymbol{(-i)^2} & (-i)^4 & (-i)^6 \\ 1 & \boldsymbol{(-i)^3} & (-i)^6 & (-i)^9 \end{bmatrix} = \begin{bmatrix} 1 & 1 & 1 & \mathbf{1} \\ 1 & i & i^2 & \boldsymbol{i^3} \\ 1 & i^2 & i^4 & \boldsymbol{i^6} \\ 1 & i^3 & i^6 & \boldsymbol{i^9} \end{bmatrix} \begin{bmatrix} \mathbf{1} & & & \\ & & & \mathbf{1} \\ & & \mathbf{1} & \\ & \mathbf{1} & & \end{bmatrix} \quad \text{because} \quad \begin{matrix} 1 = 1 \\ -i = i^3 \\ (-i)^2 = i^2 \\ (-i)^3 = i \end{matrix}$$

These matrix identities lead to the remarkable fact that $\boldsymbol{F^4} = \boldsymbol{\Omega^4} = \boldsymbol{N^2 I}$. Four transforms bring back the original vector (times N^2). Just combine $F\Omega = NI$ and $FP = \Omega$ with $P^2 = I$:

$$\boxed{\boldsymbol{F^2 P} = \boldsymbol{F\Omega} = \boldsymbol{NI} \quad \text{so} \quad \boldsymbol{PF^2} = \boldsymbol{NI} \quad \text{and} \quad \boldsymbol{F^4} = \boldsymbol{F^2 PPF^2} = \boldsymbol{N^2 I}.}$$

From $F^4 = N^2 I$, it follows that the Fourier matrix F and the DFT matrix Ω have only **four possible eigenvalues**! They are the numbers $\lambda = \sqrt{N}$ and $i\sqrt{N}$ and $-\sqrt{N}$ and $-i\sqrt{N}$ that solve $\lambda^4 = N^2$. For sizes $N > 4$ there must be and will be repeated λ's. The eigenvectors of F are not so easy to find.

The Discrete Fourier Transform

Start with any N-dimensional vector $\boldsymbol{f} = (f_0, \ldots, f_{N-1})$. The Discrete Fourier Transform expresses $\boldsymbol{f}$ as a combination of $c_0\boldsymbol{b}_0 + c_1\boldsymbol{b}_1 + \cdots + c_{N-1}\boldsymbol{b}_{N-1}$ of the Fourier basis vectors. Those basis vectors are the columns $\boldsymbol{b}$ (containing powers of w) in the Fourier matrix F_N:

$$\boxed{\begin{bmatrix} f_0 \\ \vdots \\ f_{N-1} \end{bmatrix} = \begin{bmatrix} & & \\ \boldsymbol{b}_0 & \cdots & \boldsymbol{b}_{N-1} \\ & & \end{bmatrix} \begin{bmatrix} c_0 \\ \vdots \\ c_{N-1} \end{bmatrix} \qquad \begin{aligned} \boldsymbol{f} &= F_N \boldsymbol{c} \\ \boldsymbol{c} &= F_N^{-1} \boldsymbol{f} \\ \boldsymbol{c} &= \frac{1}{N} \Omega_N \boldsymbol{f} \end{aligned}} \qquad (7)$$

The forward transform $\boldsymbol{c} = \textbf{fft}\,(\boldsymbol{f})$ multiplies $\boldsymbol{f}$ by the DFT matrix Ω (and divides by N). That is the analysis step, to separate $\boldsymbol{f}$ into N orthogonal pieces. The DFT finds the coefficients in a finite Fourier series $\boldsymbol{f} = c_0\boldsymbol{b}_0 + \cdots + c_{N-1}\boldsymbol{b}_{N-1}$.

The synthesis step is the inverse transform $f = \textbf{ifft}(c) = Fc$. It starts with those coefficients $c = (c_0, \ldots, c_{N-1})$. It carries out the matrix-vector multiplication Fc to recover f.

Thus $\textbf{ifft}(\textbf{fft}(f)) = f$. Examples will throw light on the two vectors f and c.

Example 1 The transform of $f = (1, 0, \ldots, 0)$ is $c = \frac{1}{N}(1, 1, \ldots, 1)$.

That vector f with one spike is a discrete delta function. It is concentrated at one point. Its transform c spreads out over all frequencies. Multiplying Ωf picks out the zeroth column of Ω. Therefore c shows the same Fourier coefficients $1/N$ from all frequencies. Here $N = 4$:

$$c = \frac{1}{4}\Omega f = \frac{1}{4}\begin{bmatrix} 1 & \cdot & \cdot & \cdot \\ 1 & \cdot & \cdot & \cdot \\ 1 & \cdot & \cdot & \cdot \\ 1 & \cdot & \cdot & \cdot \end{bmatrix}\begin{bmatrix} 1 \\ 0 \\ 0 \\ 0 \end{bmatrix} = \frac{1}{4}\begin{bmatrix} 1 \\ 1 \\ 1 \\ 1 \end{bmatrix} \qquad f = Fc = \frac{1}{4}\left(\begin{bmatrix} 1 \\ 1 \\ 1 \\ 1 \end{bmatrix} + \begin{bmatrix} 1 \\ i \\ i^2 \\ i^3 \end{bmatrix} + \begin{bmatrix} 1 \\ i^2 \\ i^4 \\ i^6 \end{bmatrix} + \begin{bmatrix} 1 \\ i^3 \\ i^6 \\ i^9 \end{bmatrix}\right) = \begin{bmatrix} 1 \\ 0 \\ 0 \\ 0 \end{bmatrix}$$

This transform c of a "delta vector" f is like the continuous transform of a "delta function". The delta function concentrates everything at $x = 0$. Its Fourier transform spreads out flat:

$$\delta(x) = \sum_{-\infty}^{\infty} c_k e^{ikx} \text{ has } c_k = \frac{1}{2\pi}\int_{-\pi}^{\pi} \delta(x) e^{-ikx}\, dx = \frac{1}{2\pi} \textbf{ for every frequency } k. \tag{8}$$

You see how 2π in the continuous transform matches N in the discrete transform.

Example 2 $f = (1, 1, \ldots, 1)$ will transform back to the delta vector $c = (N, 0, \ldots, 0)$.

Example 3 A shift in the delta vector to $f = (0, 1, 0, \ldots, 0)$ produces a "modulation" in its transform. This shifted f picks out the next column $(1, \omega, \ldots, \omega^{N-1})$ of F:

$$c = \frac{1}{N}\Omega_N f = \frac{1}{N}\begin{bmatrix} 1 \\ \omega \\ \vdots \\ \omega^{N-1} \end{bmatrix} \text{ and } f = Fc = \frac{1}{N}\begin{bmatrix} 1 & 1 & \cdot\cdot & 1 \\ 1 & w & \cdot\cdot & w^{N-1} \\ \cdot & \cdot & \cdot\cdot & \cdot \\ 1 & w^{N-1} & \cdot\cdot & w^{(N-1)^2} \end{bmatrix}\begin{bmatrix} 1 \\ \omega \\ \vdots \\ \omega^{N-1} \end{bmatrix} = \begin{bmatrix} 0 \\ 1 \\ \vdots \\ 0 \end{bmatrix}. \tag{9}$$

Fourier series would have a shift from $f(x)$ to $f(x - s)$. Each coefficient c_k is multiplied by e^{-iks}. This is exactly like the multiplications of $(1, 1, \ldots, 1)$ in Example 1 to produce $(1, \omega, \ldots, \omega^{N-1})$ in Example 3.

Shift rule **A shift in x-space is multiplication of the coefficients in k-space.**

One Step of the Fast Fourier Transform

We want to multiply F times c as quickly as possible. Normally a matrix times a vector takes N^2 separate multiplications—the matrix has N^2 entries. You might think it is impossible to do better. (If the matrix has zeros then multiplications can be skipped. But the Fourier matrix has no zeros!) By using the special patterns ω^{jk} and w^{jk} for their entries, Ω and F can be factored in a way that produces many zeros. This is the **FFT**.

The key idea is to connect F_N with the half-size Fourier matrix $F_{N/2}$. Assume that N is a power of 2 (for example $N = 2^{10} = 1024$). $\boldsymbol{F_{1024}}$ **connects to** $\boldsymbol{F_{512}}$.

When $N = 4$, the key is in the relation between F_4 and two copies of F_2 :

$$\boldsymbol{F_4} = \begin{bmatrix} 1 & 1 & 1 & 1 \\ 1 & i & i^2 & i^3 \\ 1 & i^2 & i^4 & i^6 \\ 1 & i^3 & i^6 & i^9 \end{bmatrix} \quad \text{and} \quad \begin{bmatrix} \boldsymbol{F_2} & \\ & \boldsymbol{F_2} \end{bmatrix} = \begin{bmatrix} 1 & 1 & & \\ 1 & i^2 & & \\ & & 1 & 1 \\ & & 1 & i^2 \end{bmatrix}.$$

On the left is F_4, with no zeros. On the right is a matrix that is half zero. The work is cut in half. But wait, those matrices are not the same. We need two sparse and simple matrices to complete the FFT factorization:

The FFT has three matrices

$$F_4 = \begin{bmatrix} 1 & & 1 & \\ & 1 & & i \\ 1 & & -1 & \\ & 1 & & -i \end{bmatrix} \begin{bmatrix} 1 & 1 & & \\ 1 & i^2 & & \\ & & 1 & 1 \\ & & 1 & i^2 \end{bmatrix} \begin{bmatrix} 1 & & & \\ & & 1 & \\ & 1 & & \\ & & & 1 \end{bmatrix}. \qquad (10)$$

The last matrix is a permutation. It puts the even c's (c_0 and c_2) ahead of the odd c's (c_1 and c_3). The middle matrix performs **half-size transforms** F_2 and F_2 on the even c's and odd c's separately. The matrix at the left combines the two half-size outputs—in a way that produces the correct full-size output $\boldsymbol{y} = F_4\boldsymbol{c}$.

The same idea applies when $N = 1024$ and $M = \frac{1}{2}N = 512$. The number w is $e^{2\pi i/1024}$. It is at the angle $\theta = 2\pi/1024$ on the unit circle. The Fourier matrix F_{1024} is full of powers of w. The first stage of the FFT is the great factorization discovered by Cooley and Tukey (and foreshadowed in 1805 by Gauss):

$$\boxed{\boldsymbol{F_{1024}} = \begin{bmatrix} I_{512} & D_{512} \\ I_{512} & -D_{512} \end{bmatrix} \begin{bmatrix} \boldsymbol{F_{512}} & \\ & \boldsymbol{F_{512}} \end{bmatrix} \begin{bmatrix} \text{even-odd} \\ \text{permutation} \end{bmatrix}.} \qquad (11)$$

I_{512} is the identity matrix. D_{512} is the diagonal matrix with entries $(1, w, \ldots, w^{511})$. The two copies of F_{512} are what we expected. Don't forget that they use the 512th root of unity (which is nothing but w^2!!) The permutation matrix separates the incoming vector $\boldsymbol{c}$ into its even and odd parts $\boldsymbol{c}' = (c_0, c_2, \ldots, c_{1022})$ and $\boldsymbol{c}'' = (c_1, c_3, \ldots, c_{1023})$.

Here are the algebra formulas which say the same thing as that factorization of F_{1024}:

(One step of the FFT) Set $M = \frac{1}{2}N$. The first M and last M components of $\boldsymbol{y} = F_N\boldsymbol{c}$ combine the two half-size transforms $\boldsymbol{y}' = F_M\boldsymbol{c}'$ and $\boldsymbol{y}'' = F_M\boldsymbol{c}''$. Equation (11) shows this step from N to $M = N/2$ as $I\boldsymbol{y}' + D\boldsymbol{y}''$ and $I\boldsymbol{y}' - D\boldsymbol{y}''$:

$$\begin{aligned} y_j &= y'_j + (w_N)^j y''_j, \quad j = 0, \ldots, M-1 \\ y_{j+M} &= y'_j - (w_N)^j y''_j, \quad j = 0, \ldots, M-1. \end{aligned} \tag{12}$$

Split $\boldsymbol{c}$ into $\boldsymbol{c}'$ and $\boldsymbol{c}''$. Transform them by F_M into $\boldsymbol{y}'$ and $\boldsymbol{y}''$. Then (12) reconstructs $\boldsymbol{y}$. Those formulas come from separating $c_0 \ldots, c_{N-1}$ into even c_{2k} and odd c_{2k+1} in (13).

$$y_j = \sum_0^{N-1} w^{jk} c_k = \sum_0^{M-1} w^{2jk} c_{2k} + \sum_0^{M-1} w^{j(2k+1)} c_{2k+1} \text{ with } M = \tfrac{1}{2}N, w = w_N. \tag{13}$$

Even c's go into $\boldsymbol{c}' = (c_0, c_2, \ldots)$ and odd c's go into $\boldsymbol{c}'' = (c_1, c_3, \ldots)$. Then come the transforms $F_M\boldsymbol{c}'$ and $F_M\boldsymbol{c}''$. **The key is** $\boldsymbol{w_N^2 = w_M}$. This gives $w_N^{2jk} = w_M^{jk}$.

Rewrite (13) $$y_j = \sum (w_M)^{jk} c'_k + (w_N)^j \sum (w_M)^{jk} c''_k = \boldsymbol{y}'_j + (w_N)^j \boldsymbol{y}''_j \tag{14}$$

For $j \geq M$, the minus sign in (12) comes from factoring out $(w_N)^M = -1$ from $(w_N)^j$.

MATLAB easily separates even c's from odd c's and multiplies by w_N^j. We use conj(F) or equivalently MATLAB's inverse transform ifft to produce $\boldsymbol{y} = F\boldsymbol{c}$. Remember that fft is based on $\omega = \overline{w} = e^{-2\pi i/N}$ and it produces $\boldsymbol{c}$ from $\boldsymbol{y}$ using Ω.

FFT step from N to $N/2$ in MATLAB		
	Transform even c's	$y' =$ ifft $(c(0:2:N-2)) * N/2;$
	Transform odd c's	$y'' =$ ifft $(c(1:2:N-1)) * N/2;$
	Vector $1, w, \ldots$ is $\boldsymbol{d}$	$d = w.\hat{}(0:N/2-1)';$
	Combine $\boldsymbol{y}'$ and $\boldsymbol{y}''$	$y = [y' + d.*y''\,;\, y' - d.*y''];$

The flow graph on the next page shows $\boldsymbol{c}'$ and $\boldsymbol{c}''$ going through the half-size F_2. Those steps are called "*butterflies*," from their shape. Then the outputs $\boldsymbol{y}'$ and $\boldsymbol{y}''$ are combined (multiplying y'' by $1, i$ from D and also by $-1, -i$ from $-D$) to produce $\boldsymbol{y} = F_4\boldsymbol{c}$.

This reduction from F_N to two F_M's almost cuts the work in half—you see the zeros in the matrix factorization. That 50% reduction is good but not great. The complete **FFT** is much more powerful. It saves much more than half the time. The key idea is **recursion**.

And the factorization of F applies equally to the conjugate matrix $\Omega = \overline{F}$.

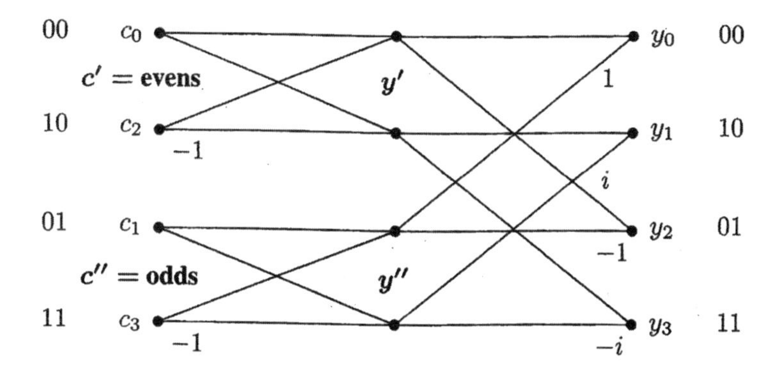

The Full FFT by Recursion

If you have read this far, you probably guessed what comes next. We reduced F_N to $F_{N/2}$. ***Keep on going to*** $\boldsymbol{F_{N/4}}$. Every F_{512} leads to F_{256}. Then 256 leads to 128. *That is recursion.*

Recursion is a basic principle of many fast algorithms. Here is step 2 with four copies of F_{256} and D (256 powers of w_{512}). Evens of evens $c_0, c_4, c_8, \ldots$ come before $c_2, c_6, c_{10}, \ldots$

$$\begin{bmatrix} F_{512} & \\ & F_{512} \end{bmatrix} = \begin{bmatrix} I & D & & \\ I & -D & & \\ & & I & D \\ & & I & -D \end{bmatrix} \begin{bmatrix} F & & & \\ & F & & \\ & & F & \\ & & & F \end{bmatrix} \begin{bmatrix} \text{pick } 0,4,8,\ldots \\ \text{pick } 2,6,10,\ldots \\ \text{pick } 1,5,9,\ldots \\ \text{pick } 3,7,11,\ldots \end{bmatrix}.$$

We will count the individual multiplications, to see how much is saved. Before the **FFT** was invented, the count was the usual $N^2 = (1024)^2$. This is about a million multiplications. I am not saying that they take a long time. The cost becomes large when we have many, many transforms to do—which is typical. Then the saving by the FFT is also large:

The final count for size $\boldsymbol{N = 2^\ell}$ ***is reduced from*** $\boldsymbol{N^2}$ ***to*** $\boldsymbol{\frac{1}{2}N\ell}$.

The number 1024 is 2^{10}, so $\ell = 10$. The original count of $(1024)^2$ is reduced to (5)(1024). The saving is a factor of 200. A million is reduced to five thousand. That is why the FFT has revolutionized signal processing.

Here is the reasoning behind $\frac{1}{2}N\ell$. There are ℓ levels, going from $N = 2^\ell$ down to $N = 1$. Each level has $N/2$ multiplications from the diagonal D's, to reassemble the half-size outputs from the lower level. This yields the final count $\frac{1}{2}N\ell$, which is $\frac{1}{2}N\log_2 N$.

One last note about this remarkable algorithm. There is an amazing rule for the order that the c's enter the FFT, after all the even-odd permutations. Write the numbers 0 to $n-1$ in binary (like $00, 01, 10, 11$ for $n = 4$). Reverse the order of those digits: $00, 10, 01, 11$. That gives the **bit-reversed order** $\mathbf{0, 2, 1, 3}$ with evens before odds.

The complete picture shows the c's in bit-reversed order, the $\ell = \log_2 N$ steps of the recursion, and the final output $y_0, \ldots, y_{N-1}$ which is F_N times $\mathbf{c}$.

The FFT recursion for the DFT matrix $\Omega = \overline{F}$ uses exactly the same ideas.

Problem Set IV.1

1 After $S = 0$ in equation (5), the text says that all off-diagonal entries of $F_n\Omega_N$ are zero. Explain (row i of F) $\cdot$ (column j of Ω) $=$ complex dot product $= 0$?

$$\text{Why is } (1, w^i, w^{2i}, \ldots, w^{(N-1)i}) \cdot (1, \omega^j, \omega^{2j}, \ldots, \omega^{(N-1)j}) = 0 \text{ if } i \neq j?$$

You must use the dot product for complex vectors, not $x_1y_1 + \cdots + x_Ny_N$.

2 If $M = \frac{1}{2}N$ show that $(w_N)^M = -1$. This is used in the FFT equation (12).

3 What are the matrices F_3 and Ω_3 (using $w = e^{2\pi i/3}$ and $\omega = \overline{w}$)? What 3 by 3 permutation matrix P will connect them by $\Omega = FP$ and $F = \Omega P$?

4 Find the Discrete Fourier Transform $\boldsymbol{c}$ of $\boldsymbol{f} = (0, 1, 0, 0)$. Verify that the inverse transform of $\boldsymbol{c}$ is $\boldsymbol{f}$.

5 Connect F_6 to two copies of F_3 by a matrix equation like (10) and (11).

6 For $N = 6$, how do you see that $1 + w + w^2 + w^3 + w^4 + w^5 = 0$?

7 Suppose $f(x) = 1$ for $|x| \leq \pi/2$ and $f(x) = 0$ for $\pi/2 < |x| \leq \pi$. This function is "even" because $f(-x) = f(x)$. Even functions can be expanded in *cosine series*:

$$f(x) = a_0 + a_1 \cos x + a_2 \cos 2x + \cdots$$

Integrate both sides from $x = -\pi$ to π, to find a_0. Multiply both sides by $\cos x$ and integrate from $-\pi$ to π, to find a_1.

8 Every real matrix A with n columns has $AA^{\mathrm{T}}\boldsymbol{x} = \boldsymbol{a}_1(\boldsymbol{a}_1^{\mathrm{T}}\boldsymbol{x}) + \cdots + \boldsymbol{a}_n(\boldsymbol{a}_n^{\mathrm{T}}\boldsymbol{x})$.

If A is an orthogonal matrix Q, what is special about those n pieces?

For the Fourier matrix (complex), what is changed in that formula?

9 What vector $\boldsymbol{x}$ has $F_4\boldsymbol{x} = (1, 0, 1, 0)$? What vector has $F_4\boldsymbol{y} = (0, 0, 0, 1)$?

IV.2 Shift Matrices and Circulant Matrices

When this matrix P multiplies a vector x, the components of x shift upward:

$$\textbf{Upward shift} \quad \textbf{Cyclic permutation} \qquad Px = \begin{bmatrix} 0 & 1 & 0 & 0 \\ 0 & 0 & 1 & 0 \\ 0 & 0 & 0 & 1 \\ 1 & 0 & 0 & 0 \end{bmatrix} \begin{bmatrix} x_1 \\ x_2 \\ x_3 \\ x_4 \end{bmatrix} = \begin{bmatrix} x_2 \\ x_3 \\ x_4 \\ x_1 \end{bmatrix} \tag{1}$$

The words "cyclic" and "circular" apply to P because the first component x_1 moves to the end. If we think of numbers x_1, x_2, x_3, x_4 around a circle, P moves them all by one position. And P^2 turns the circle by two positions:

$$P^2x = \begin{bmatrix} 0 & 1 & 0 & 0 \\ 0 & 0 & 1 & 0 \\ 0 & 0 & 0 & 1 \\ 1 & 0 & 0 & 0 \end{bmatrix} \begin{bmatrix} 0 & 1 & 0 & 0 \\ 0 & 0 & 1 & 0 \\ 0 & 0 & 0 & 1 \\ 1 & 0 & 0 & 0 \end{bmatrix} \begin{bmatrix} x_1 \\ x_2 \\ x_3 \\ x_4 \end{bmatrix} = \begin{bmatrix} 0 & 0 & 1 & 0 \\ 0 & 0 & 0 & 1 \\ 1 & 0 & 0 & 0 \\ 0 & 1 & 0 & 0 \end{bmatrix} \begin{bmatrix} x_1 \\ x_2 \\ x_3 \\ x_4 \end{bmatrix} = \begin{bmatrix} x_3 \\ x_4 \\ x_1 \\ x_2 \end{bmatrix} \tag{2}$$

Every new factor P gives one additional shift. **Then P^4 gives a complete 360° turn: $P^4x = x$ and $P^4 = I$.** The next powers P^5, P^6, P^7, P^8 repeat the pattern and cycle around again. Notice that P^3 is the inverse of P, because $(P^3)(P) = P^4 = I$.

The next matrix is called a **circulant**. It is simply a combination of P, P^2, P^3, and $P^4 = I$. It has **constant diagonals**:

$$\textbf{Circulant matrix} \quad C = c_0I + c_1P + c_2P^2 + c_3P^3 = \begin{bmatrix} c_0 & c_1 & c_2 & c_3 \\ c_3 & c_0 & c_1 & c_2 \\ c_2 & c_3 & c_0 & c_1 \\ c_1 & c_2 & c_3 & c_0 \end{bmatrix} \tag{3}$$

Each diagonal in that matrix cycles around exactly like the 1's in P. The diagonal with c_1, c_1, c_1 is completed by the fourth c_1 at the bottom. Important: **If you multiply two circulant matrices C and D, their product $CD = DC$ is again a circulant matrix.**

When you multiply CD, you are multiplying powers of P to get more powers of P. And DC is doing exactly the same. This example has $N = 3$ and $P^3 = I$:

$$CD = \begin{bmatrix} 1 & 2 & 3 \\ 3 & 1 & 2 \\ 2 & 3 & 1 \end{bmatrix} \begin{bmatrix} 5 & 0 & 4 \\ 4 & 5 & 0 \\ 0 & 4 & 5 \end{bmatrix} = \begin{bmatrix} 13 & 22 & 19 \\ 19 & 13 & 22 \\ 22 & 19 & 13 \end{bmatrix} = \textbf{circulant} \tag{4}$$

$$\begin{matrix} (I + 2P + 3P^2)\,(5I + 4P^2) \\ (5I + 4P^2)\,(I + 2P + 3P^2) \end{matrix} = 5I + 10P + (15+4)P^2 + 8P^3 + 12P^4 = \mathbf{13I + 22P + 19P^2}$$

At that last step, I needed and used the circular facts that $P^3 = I$ and $P^4 = P$. So the vectors $(1, 2, 3)$ for C and $(5, 0, 4)$ for D produce the vector $(13, 22, 19)$ for CD and DC. This operation on vectors is called **cyclic convolution.**

To summarize: When we multiply N by N circulant matrices C and D, we take the cyclic convolution of the vectors $(c_0, c_1, \ldots, c_{N-1})$ and $(d_0, d_1, \ldots, d_{N-1})$. Ordinary convolution finds the coefficients when we multiply $(c_0 I + c_1 P + \cdots + c_{N-1} P^{N-1})$ times $(d_0 I + d_1 P + \cdots + d_{N-1} P^{N-1})$. Then cyclic convolution uses the crucial fact that $P^N = I$.

Convolution	$(1, 2, 3) * (5, 0, 4) = (\mathbf{5, 10, 19, 8, 12})$	(5)
Cyclic convolution	$(1, 2, 3) \circledast (5, 0, 4) = (5{+}8, 10{+}12, 19) = (\mathbf{13, 22, 19})$	(6)

Ordinary convolution is the multiplication you learned in second grade (made easier because there is no "carrying" to the next column):

$$\begin{array}{rrrrr} & & 1 & 2 & 3 \\ & & 5 & 0 & 4 \\ \hline & & 4 & 8 & 12 \\ & 0 & 0 & 0 & \\ 5 & 10 & 15 & & \\ \hline \mathbf{5} & \mathbf{10} & \mathbf{19} & \mathbf{8} & \mathbf{12} \end{array} \quad = \mathbf{c} * \mathbf{d}$$

The cyclic step combines $5 + 8$ because $P^3 = I$. It combines $10 + 12$ because $P^4 = P$. **The result is $(13, 22, 19)$.**

$$\text{Practice} \quad \begin{aligned} (0, 1, 0) \circledast (d_0, d_1, d_2) &= (d_1, d_2, d_0) \\ (1, 1, 1) \circledast (d_0, d_1, d_2) &= (d_0 + d_1 + d_2, d_0 + d_1 + d_2, d_0 + d_1 + d_2) \\ (c_0, c_1, c_2) \circledast (d_0, d_1, d_2) &= (d_0, d_1, d_2) \circledast (c_0, c_1, c_2) \end{aligned}$$

That last line means that $\boldsymbol{CD = DC}$ for circulant matrices and $\mathbf{c} \circledast \mathbf{d} = \mathbf{d} \circledast \mathbf{c}$ for cyclic convolutions. Powers of P in C commute with powers of P in D.

If you add $1 + 2 + 3 = 6$ and $5 + 0 + 4 = 9$, you have a quick check on convolution. Multiply 6 times 9 to get 54. Then 54 should be (and is) equal to $5 + 10 + 19 + 8 + 12 = 54$. And also $13 + 22 + 19 = 54$ for cyclic convolution.

The sum of the c's times the sum of the d's equals the sum of the outputs. That is because every c multiplies every d in $\mathbf{c} * \mathbf{d}$ and in $\mathbf{c} \circledast \mathbf{d}$.

Eigenvalues and Eigenvectors of P

With $N = 4$, the equation $Px = \lambda x$ leads directly to four eigenvalues and eigenvectors:

$$Px = \begin{bmatrix} 0 & 1 & 0 & 0 \\ 0 & 0 & 1 & 0 \\ 0 & 0 & 0 & 1 \\ 1 & 0 & 0 & 0 \end{bmatrix} \begin{bmatrix} x_1 \\ x_2 \\ x_3 \\ x_4 \end{bmatrix} = \begin{bmatrix} x_2 \\ x_3 \\ x_4 \\ x_1 \end{bmatrix} = \lambda \begin{bmatrix} x_1 \\ x_2 \\ x_3 \\ x_4 \end{bmatrix} \quad \text{gives} \quad \begin{matrix} x_2 = \lambda x_1 \\ x_3 = \lambda x_2 \\ x_4 = \lambda x_3 \\ x_1 = \lambda x_4 \end{matrix} . \tag{7}$$

Start with the last equation $x_1 = \lambda x_4$ and work upwards:

$$x_1 = \lambda x_4 = \lambda^2 x_3 = \lambda^3 x_2 = \lambda^4 x_1 \quad \text{leading to} \quad \boldsymbol{\lambda^4 = 1}.$$

The eigenvalues of P are the fourth roots of 1. They are all the powers $i, i^2, i^3, 1$ of $w = i$.

$$\lambda = \boldsymbol{i}, \qquad \lambda = i^2 = \boldsymbol{-1}, \qquad \lambda = i^3 = \boldsymbol{-i}, \qquad \text{and } \lambda = i^4 = \boldsymbol{1}. \tag{8}$$

These are the four solutions to $\det(P - \lambda I) = \lambda^4 - 1 = 0$. The eigenvalues $i, -1, -i, 1$ are equally spaced around the unit circle in the complex plane (Figure IV.2).

When P is N by N, the same reasoning leads from $P^N = I$ to $\lambda^N = 1$. The N eigenvalues are again equally spaced around the circle, and now they are powers of the **complex number w** at $360/N$ degrees $= 2\pi/N$ radians.

$$\boxed{\textbf{The solutions to } z^N = 1 \textbf{ are } \boldsymbol{\lambda = w, w^2, \ldots, w^{N-1}, 1} \textbf{ with } \boldsymbol{w = e^{2\pi i/N}}.} \tag{9}$$

In the complex plane, the first eigenvalue w is $e^{i\theta} = \cos\theta + i \sin\theta$ and the angle θ is $2\pi/N$. The angles for the other eigenvalues are $2\theta, 3\theta, \ldots, N\theta$. Since θ is $2\pi/N$, that last angle is $N\theta = 2\pi$ and that eigenvalue is $\lambda = e^{2\pi i}$ which is $\cos 2\pi + i \sin 2\pi = 1$.

For powers of complex numbers, the polar form with $e^{i\theta}$ is much better than using $\cos\theta + i \sin\theta$.

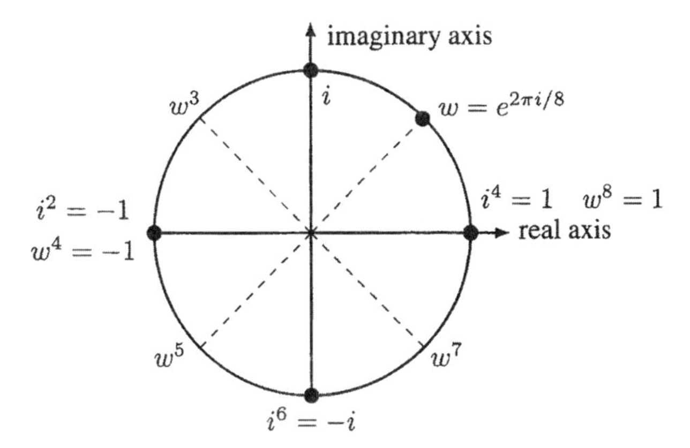

Figure IV.2: Eigenvalues of P_N: The 4 powers of $\lambda = i$ for $N = 4$ will add to zero. The 8 powers of $\lambda = w = e^{2\pi i/8}$ for $N = 8$ must also add to zero.

Knowing the N eigenvalues $\lambda = 1, w, \ldots, w^{N-1}$ of P_N, we quickly find N eigenvectors:

Set the first component of $\boldsymbol{q}$ to 1. The other components of $\boldsymbol{q}$ are λ and λ^2 and λ^3:

Eigenvectors for $\boldsymbol{\lambda = 1, i, i^2, i^3}$

$$q_0 = \begin{bmatrix} 1 \\ 1 \\ 1 \\ 1 \end{bmatrix} \quad q_1 = \begin{bmatrix} 1 \\ i \\ i^2 \\ i^3 \end{bmatrix} \quad q_2 = \begin{bmatrix} 1 \\ i^2 \\ i^4 \\ i^6 \end{bmatrix} \quad q_3 = \begin{bmatrix} 1 \\ i^3 \\ i^6 \\ i^9 \end{bmatrix} \qquad (10)$$

We have started the numbering at zero, as Fourier people always do. The zeroth eigenvector has eigenvalue $\lambda = 1 = w^0$. The eigenvector matrix has columns $0, 1, 2, 3$ containing $\boldsymbol{q}_0, \boldsymbol{q}_1, \boldsymbol{q}_2, \boldsymbol{q}_3$. **That eigenvector matrix for $\boldsymbol{P}$ is the Fourier matrix.**

Eigenvector matrix
$\boldsymbol{N = 4}$
Fourier matrix

$$\begin{bmatrix} 1 & 1 & 1 & 1 \\ 1 & i & i^2 & i^3 \\ 1 & i^2 & i^3 & i^6 \\ 1 & i^3 & i^4 & i^9 \end{bmatrix} \quad \text{has} \quad \overline{F}^{\mathrm{T}} F = 4I. \qquad (11)$$

The pattern stays the same for any size N. The kth eigenvalue is $w^k = (e^{2\pi i/N})^k = e^{2\pi ik/N}$. Again the count starts at zero: $\lambda_0 = w^0 = 1, \lambda_1 = w, \ldots, \lambda_{N-1} = w^{N-1}$.

The $\boldsymbol{k}$th eigenvector contains the powers of $\boldsymbol{w^k}$. The eigenvector matrix contains all N eigenvectors. It is the N by N Fourier matrix with $\overline{F}^{\mathrm{T}} F = NI$.

Fourier matrix
Eigenvectors of $\boldsymbol{P}$

$$F_N = \begin{bmatrix} 1 & 1 & 1 & \cdot & 1 \\ 1 & w & w^2 & \cdot & w^{N-1} \\ 1 & w^2 & w^4 & \cdot & w^{2(N-1)} \\ \cdot & \cdot & \cdot & \cdot & \cdot \\ 1 & w^{N-1} & w^{2(N-1)} & \cdot & w^{(N-1)(N-1)} \end{bmatrix}. \qquad (12)$$

We see again that **the columns of the Fourier matrix are orthogonal.** We must use the complex inner product $(\boldsymbol{x}, \boldsymbol{y}) = \overline{\boldsymbol{x}}^{\mathrm{T}} \boldsymbol{y}$. Here is the new proof of orthogonality:

Orthogonal matrices like $\boldsymbol{P}$ have orthogonal eigenvectors.

Then $P = F \Lambda F^{\mathrm{T}}/N$ with its diagonal eigenvalue matrix $\Lambda = \text{diag}\,(1, w, \ldots, w^{N-1})$.

The next page moves from the special permutation P to any circulant matrix C.

Eigenvalues and Eigenvectors of a Circulant C

The eigenvectors of a circulant matrix C are especially easy. **Those eigenvectors are the same as the eigenvectors of the permutation P.** So they are the columns $\boldsymbol{q}_0, \boldsymbol{q}_1, \ldots, \boldsymbol{q}_{N-1}$ of the same Fourier matrix F. Here is $C\boldsymbol{q}_k = \lambda \boldsymbol{q}_k$ for the kth eigenvector and eigenvalue:

$$\left(c_0 I + c_1 P + \cdots + c_{N-1} P^{N-1}\right) \boldsymbol{q}_k = \left(c_0 + c_1 \lambda_k + \cdots + c_{N-1} \lambda_k^{N-1}\right) \boldsymbol{q}_k. \tag{13}$$

Remember that $\boldsymbol{\lambda}_k = \boldsymbol{w}^k = e^{2\pi i k/N}$ is the kth eigenvalue of P. Those numbers are in the Fourier matrix F. Then the eigenvalues of C in equation (13) have an almost magical formula: **Multiply F times the vector c in the top row of C to find the eigenvalues.**

$$\begin{bmatrix} \lambda_0(C) \\ \lambda_1(C) \\ \lambda_2(C) \\ \cdot \\ \lambda_{N-1}(C) \end{bmatrix} = \begin{bmatrix} c_0 + c_1 + \cdots + c_{N-1} \\ c_0 + c_1 w + \cdots + c_{N-1} w^{N-1} \\ c_0 + c_1 w^2 + \cdots + c_{N-1} w^{2(N-1)} \\ \cdot\ \cdot\ \cdot \\ c_0 + c_1 w^{N-1} + \cdots + c_{N-1} w^{(N-1)(N-1)} \end{bmatrix} = F \begin{bmatrix} c_0 \\ c_1 \\ c_2 \\ \cdot \\ c_{N-1} \end{bmatrix} = \boldsymbol{F c}. \tag{14}$$

The N eigenvalues of C are the components of $F\boldsymbol{c}$ = inverse Fourier transform of $\boldsymbol{c}$.

Example for $N = 2$ with $w = e^{2\pi i/2} = -1$ in the Fourier matrix F

$$P = \begin{bmatrix} 0 & 1 \\ 1 & 0 \end{bmatrix} \quad \text{and} \quad C = \begin{bmatrix} c_0 & c_1 \\ c_1 & c_0 \end{bmatrix} \quad \text{and} \quad \boldsymbol{c} = \begin{bmatrix} c_0 \\ c_1 \end{bmatrix} \quad \text{and} \quad F = \begin{bmatrix} 1 & 1 \\ 1 & -1 \end{bmatrix}$$

The eigenvectors of P and also C are the columns of F. The eigenvalues of P are ± 1. The eigenvalues of $C = c_0 I + c_1 P$ are $c_0 + c_1$ and $c_0 - c_1$. These eigenvalues of C are the components of F times $\boldsymbol{c}$:

$$\textbf{Eigenvalues of } C \qquad F\boldsymbol{c} = \begin{bmatrix} 1 & 1 \\ 1 & -1 \end{bmatrix} \begin{bmatrix} c_0 \\ c_1 \end{bmatrix} = \begin{bmatrix} c_0 + c_1 \\ c_0 - c_1 \end{bmatrix}. \tag{15}$$

For any N, the permutation P is a circulant matrix C with $\boldsymbol{c} = (0, 1, 0, \ldots, 0)$.
The eigenvalues of P are in the column vector $F\boldsymbol{c}$ with this $\boldsymbol{c}$.
That is the column $(1, w, w^2, \ldots, w^{N-1})$ of the Fourier matrix F.
This agrees with the eigenvalues $1, i, i^2, i^3$ of P in equation (8), for $N = 4$.

The Convolution Rule

This rule compares **convolution** with **multiplication**. Please understand that they are quite different. But they are beautifully connected by the Fourier matrix F.

I will start with two circulant matrices C and D. Their top rows are the vectors $\boldsymbol{c}$ and $\boldsymbol{d}$. Equation (4) at the start of this section showed an example of the top row of CD:

$$\textbf{Top row of } \boldsymbol{CD} = \textbf{ cyclic convolution } = \boldsymbol{c} \circledast \boldsymbol{d}. \qquad (16)$$

Then the eigenvalues of CD according to equation (14) are in the vector $\boldsymbol{F}(\boldsymbol{c} \circledast \boldsymbol{d})$.

Now find those eigenvalues of CD in another way. The eigenvalues $\lambda(C)$ are in the vector $F\boldsymbol{c}$. The eigenvalues $\lambda(D)$ are in the vector $F\boldsymbol{d}$. The eigenvectors $\boldsymbol{q}_k$ are the same for C and D! They are the columns of F. **So each eigenvalue $\lambda_k(CD)$ is just $\lambda_k(C)$ times $\lambda_k(D)$. This term by term "Hadamard product" is denoted in MATLAB by .***

$$\begin{bmatrix} \lambda_0(CD) \\ \vdots \\ \lambda_{N-1}(CD) \end{bmatrix} = \begin{bmatrix} \lambda_0(C)\lambda_0(D) \\ \vdots \\ \lambda_{N-1}(C)\lambda_{N-1}(D) \end{bmatrix} = \begin{bmatrix} \lambda_0(C) \\ \vdots \\ \lambda_{N-1}(C) \end{bmatrix} .* \begin{bmatrix} \lambda_0(D) \\ \vdots \\ \lambda_{N-1}(D) \end{bmatrix} = \boldsymbol{Fc} \;.*\; \boldsymbol{Fd}$$

That notation **.*** denotes component-by-component multiplication of two vectors.

The convolution rule compares our two formulas for the eigenvalues of CD:

Convolve vectors
Multiply transforms

$$\boxed{\textbf{Convolution Rule} \qquad F\,(\boldsymbol{c} \circledast \boldsymbol{d}) = (F\boldsymbol{c}).*(F\boldsymbol{d}).} \qquad (17)$$

Left side Convolve $\boldsymbol{c}$ and $\boldsymbol{d}$ first, then transform by F
Right side Transform by F first, then multiply $F\boldsymbol{c}$ times $F\boldsymbol{d}$ component by component.

This is the fundamental identity of signal processing! Transforms are fast by the FFT.

Another way to see the convolution rule is by multiplying the diagonal matrices $\Lambda(C)$ and $\Lambda(D)$ that contain the eigenvalues of C and D. C is diagonalized by $F^{-1}CF = \Lambda(C)$:

$$(F^{-1}CF)\,(F^{-1}DF) = F^{-1}(CD)F \text{ is exactly } \boldsymbol{\Lambda(C)\;\Lambda(D) = \Lambda(CD)}. \qquad (18)$$

This succeeds because all circulant matrices have the same eigenvectors (columns of F).

The convolution rule can be checked directly (Problem 1). Good to see it for $N = 2$:

$$F = \begin{bmatrix} 1 & 1 \\ 1 & -1 \end{bmatrix} \qquad \boldsymbol{c} \circledast \boldsymbol{d} = \begin{bmatrix} c_0d_0 + c_1d_1 \\ c_0d_1 + c_1d_0 \end{bmatrix} \qquad F\boldsymbol{c} = \begin{bmatrix} c_0 + c_1 \\ c_0 - c_1 \end{bmatrix} \qquad F\boldsymbol{d} = \begin{bmatrix} d_0 + d_1 \\ d_0 - d_1 \end{bmatrix}$$

The convolution rule (17) says that $\boldsymbol{F}(\boldsymbol{c} \circledast \boldsymbol{d})$ is the component by component product.

$$F(\boldsymbol{c} \circledast \boldsymbol{d}) = \begin{bmatrix} c_0d_0 + c_1d_1 + c_0d_1 + c_1d_0 \\ c_0d_0 + c_1d_1 - c_0d_1 - c_1d_0 \end{bmatrix} = \begin{bmatrix} (c_0 + c_1)(d_0 + d_1) \\ (c_0 - c_1)(d_0 - d_1) \end{bmatrix} = (F\boldsymbol{c}).*(F\boldsymbol{d})$$

Multiplication and Convolution of Functions

If $f(x) = \Sigma\, c_k e^{ikx}$ and $g(x) = \Sigma\, d_m e^{imx}$, what are the Fourier coefficients of $f(x)\, g(x)$? We are multiplying the 2π-periodic functions f and g.

The multiplication fg is in "x-space". By the convolution rule, we expect to convolve $c * d$ in "frequency space". These are periodic functions f and g $(-\pi \le x \le \pi)$ and their Fourier series are infinite $(k = 0, \pm 1, \pm 2, \ldots)$. So cyclic convolution is gone and we are multiplying two infinite Fourier series :

$$f(x)\, g(x) = \left(\sum_{k=-\infty}^{\infty} c_k e^{ikx} \right) \left(\sum_{m=-\infty}^{\infty} d_m e^{imx} \right) = \sum_{n=-\infty}^{\infty} h_n e^{inx}. \qquad (19)$$

When does e^{ikx} times e^{imx} produce e^{inx} ? The requirement is $\boldsymbol{k + m = n}$.

The coefficient h_n combines all products $c_k d_m$ with $k+m=n$. Then $\boldsymbol{m = n - k}$:

$$\boldsymbol{h_n} = \sum_{k=-\infty}^{\infty} \boldsymbol{c_k d_{n-k}} \text{ is convolution } \boldsymbol{h = c * d} \text{ for infinite vectors.} \qquad (20)$$

Next we multiply coefficients $c_k d_k$ in k-space. So we convolve $f * g$ in x-space !

Convolution of 2π-periodic functions	$\displaystyle (\boldsymbol{f * g})\,(x) = \int_{t=-\pi}^{\pi} \boldsymbol{f(t)\, g(x-t)}\, dt$	(21)
Convolution rule for periodic functions	The Fourier coefficients of $\boldsymbol{f * g}$ are $\boldsymbol{2\pi c_k d_k}$	(22)

To see that $\boldsymbol{f * g = g * f}$ in equation (21), change variables to $T = x - t$ and $t = x - T$.

The delta function $\delta(x)$ is the identity for convolution—like I for multiplication :

$$(\boldsymbol{\delta * g})\,(x) = \int \delta(t)\, g(x-t)\, dt = \boldsymbol{g(x)} \quad \text{and} \quad (1,0,0) \circledast (a,b,c) = (a,b,c).$$

Cross-correlation and Autocorrelation

Cross-correlation is like convolution, but an important difference is indicated here by $**$

$$[\textbf{Not } n-k] \qquad h_n = \sum_k c_k d_{n+k} \text{ is } \textbf{cross-correlation } \boldsymbol{h = c ** d} \text{ for vectors} \qquad (23)$$

$$(\boldsymbol{f ** g})\,(x) = \int f(t)\, g(x+t)\, dt = \boldsymbol{f(x) * g(-x)} \text{ is cross-correlation for functions} \qquad (24)$$

We are shifting the vector $\boldsymbol{d}$ and taking its dot products with $\boldsymbol{c}$. We are sliding the function g along the x-axis and taking its inner products with f. These dot products will be largest when the vectors $\boldsymbol{c}, \boldsymbol{d}$ and the functions f, g are *best aligned*.

It is valuable to have this simple way to find the best alignment.

The special cases $c = d$ and $f = g$ are the most important and certainly the best aligned. In those cases, the cross-correlations $c ** c$ and $f ** f$ are called **autocorrelations.**

Convolution and cross-correlation correspond perfectly to matrix multiplications!

$c * d$ gives the entries in CD (infinite constant-diagonal matrices)

$c \circledast d$ gives the entries in CD (finite circulant matrices)

$c ** d$ gives the entries in $C^{\mathrm{T}}D$ and $a ** a$ gives the entries in $A^{\mathrm{T}}A$

Problem Set IV.2

1 Find $c * d$ and $c \circledast d$ for $c = (2, 1, 3)$ and $d = (3, 1, 2)$.

2 Prove the convolution rule for $N = 3$: The kth component of $F(c \circledast d)$ equals $(Fc)_k$ times $(Fd)_k$. Start from $(c \circledast d)_p = c_0 d_p + c_1 d_{p-1} + c_2 d_{p-2}$.

$$\text{Prove} \quad \sum_{p=0}^{2} w^{kp}(c \circledast d)_p = \left(\sum_{m=0}^{2} w^{km} c_m\right)\left(\sum_{n=0}^{2} w^{kn} d_n\right) \quad \text{with } w^3 = 1$$

3 If $c * d = e$, why is $(\sum c_i)(\sum d_i) = (\sum e_i)$? Why was our check successful? $(1 + 2 + 3)(5 + 0 + 4) = (6)(9) = 54 = 5 + 10 + 19 + 8 + 12$.

4 Any two circulant matrices of the same size commute: $CD = DC$. They have the same eigenvectors q_k (the columns of the Fourier matrix F). Show that the eigenvalues $\lambda_k(CD)$ are equal to $\lambda_k(C)$ times $\lambda_k(D)$.

5 What are the eigenvalues of the 4 by 4 circulant $C = I + P + P^2 + P^3$? Connect those eigenvalues to the discrete transform Fc for $c = (1, 1, 1, 1)$. For which three real or complex numbers z is $1 + z + z^2 + z^3 = 0$?

6 "A circulant matrix C is invertible when the vector Fc has no zeros." Connect that true statement to this test on the frequency response:

$$C(e^{i\theta}) = \sum_{0}^{N-1} c_j e^{ij\theta} \neq 0 \text{ at the } N \text{ points } \theta = 2\pi/N, 4\pi/N, \ldots, 2\pi.$$

7 How would you solve for d in a convolution equation $c * d = e$ or $c \circledast d = e$? With matrices this is $CD = E$ and then $D = C^{-1}E$. But deconvolution is often faster using the convolution rule $(Fc).*(Fd) = (Fe)$. Then $Fd = ??$

8 The nth component of the autocorrelation $c ** c$ is the dot product of the vectors c and $S^n c$ (the vector c shifted by n places). *Why is* $c^{\mathrm{T}} S^n c \leq c^{\mathrm{T}} c$? Then the largest component of $c ** c$ is the zeroth component $c^{\mathrm{T}} c$ (with no shift).

IV.3 The Kronecker Product $A \otimes B$

Section IV.1 described the Discrete Fourier Transform of a 1-dimensional signal $\boldsymbol{f}$. This section will describe the 2-dimensional DFT—which is needed for image processing. When the 1-dimensional transform uses a matrix of size N, the 2-dimensional transform needs a matrix of size N^2 (and video will introduce a third dimension). The 2D matrices will be large. We hope to construct them easily from the 1D Fourier matrices F and Ω.

This construction uses the **Kronecker products $F \otimes F$ and $\Omega \otimes \Omega$**. The earlier word was *tensor product*. The MATLAB command is **kron**(F, F) and **kron**(Ω, Ω).

This operation is extremely convenient for many purposes. So this section develops the key ideas and operations on $K = A \otimes B$: to invert K, to solve $(A \otimes B)\,\boldsymbol{x} = \boldsymbol{y}$, and to find the eigenvalues and eigenvectors and SVD of $A \otimes B$.

The first thing to know about Kronecker products is the size of $A \otimes B = \text{kron}(A, B)$:

1 If A and B are n by n, then **$A \otimes B$ is n^2 by n^2.**

2 If A is m by n and B is M by N, then **$A \otimes B$ has mM rows and nN columns.**

The entries of $A \otimes B$ are (all mn entries of A) times (all MN entries of B).

The next fact is the position of those products in the large matrix. The rule is to **multiply each entry of A times the whole matrix B.** Then $A \otimes B$ is a block matrix. Every block is a multiple of B:

$$\textbf{Kronecker product} \qquad A \otimes B = \begin{bmatrix} a_{11}B & \cdots & a_{1n}B \\ \vdots & & \vdots \\ a_{m1}B & \cdots & a_{mn}B \end{bmatrix}. \tag{1}$$

The simplest case has identity matrices for A and B: (2 by 2) $\circledast$ (3 by 3) = 6 by 6.

$$I_2 \otimes I_3 = I_6 \qquad \begin{bmatrix} 1 & 0 \\ 0 & 1 \end{bmatrix} \otimes \begin{bmatrix} 1 & 0 & 0 \\ 0 & 1 & 0 \\ 0 & 0 & 1 \end{bmatrix} = \begin{bmatrix} 1\,I_3 & 0\,I_3 \\ 0\,I_3 & 1\,I_3 \end{bmatrix} = I_6. \tag{2}$$

A harder case (but not impossible) comes from multiplying two Kronecker products:

$$\begin{array}{llll} A \otimes B & \text{times} & C \otimes D & \text{equals} \quad AC \otimes BD \qquad (3) \\ A \otimes B & \text{times} & A^{-1} \otimes B^{-1} & \text{equals} \quad I \otimes I \qquad (4) \end{array}$$

Equation (3) allows rectangular matrices. Equation (4) is for invertible square matrices. $I \otimes I$ is the identity matrix of size nN. **So the inverse of $A \otimes B$ is $A^{-1} \otimes B^{-1}$.**

The proof of equation (3) comes from block multiplication of Kronecker products:

$$\begin{bmatrix} a_{11}B & a_{12}B \\ a_{21}B & a_{22}B \end{bmatrix}\begin{bmatrix} c_{11}D & c_{12}D \\ c_{21}D & c_{22}D \end{bmatrix} = \begin{bmatrix} (a_{11}c_{11}+a_{12}c_{21})\,BD & (a_{11}c_{12}+a_{12}c_{22})\,BD \\ (a_{21}c_{11}+a_{22}c_{21})\,BD & (a_{21}c_{12}+a_{22}c_{22})\,BD \end{bmatrix}. \quad (5)$$

That is exactly $AC \otimes BD$. A matrix of size N^2 times a matrix of size N^2 is still a matrix of size N^2. The final basic identity produces the transpose of $A \otimes B$:

$$(\boldsymbol{A} \otimes \boldsymbol{B})^{\mathrm{T}} = \boldsymbol{A}^{\mathrm{T}} \otimes \boldsymbol{B}^{\mathrm{T}} \qquad \begin{bmatrix} a_{11}B & a_{12}B \\ a_{21}B & a_{22}B \end{bmatrix}^{\mathrm{T}} = \begin{bmatrix} a_{11}B^{\mathrm{T}} & a_{21}B^{\mathrm{T}} \\ a_{12}B^{\mathrm{T}} & a_{22}B^{\mathrm{T}} \end{bmatrix} \quad (6)$$

Two-dimensional Discrete Fourier Transforms

Start with an N by N image. It could be a photograph with N^2 small pixels, taken by your phone. It could be a painting by Chuck Close (who realized that your eyes merge the small squares into a continuous image—this is how the phone camera works too). We have N^2 numbers.

And we have a choice. We could unfold those numbers into a long vector of length N^2. Or we could keep them in an N by N matrix, so that pixels which are close in the image remain close. The unfolding into a vector (by an operator called *vec*) will be described later in this section. For now we think of a vector $\boldsymbol{f}$ in $\mathbf{R}^{n^2}$ with its n^2 components in a square.

We intend to apply a $2D$ Discrete Fourier Transform to $\boldsymbol{f}$. The result will be a $2D$ vector $\boldsymbol{c}$. You could think of this process in two steps:

Row by row Apply the $1D$ DFT to each row of pixels separately.
Column by column Rearrange the output by columns and transform each column.

The matrix for each step is N^2 by N^2. First think of the N^2 pixels a row at a time and multiply each row in that long vector by the one-dimensional DFT matrix Ω_N:

$$\boldsymbol{\Omega}_{\mathbf{row}}\boldsymbol{f} = \begin{bmatrix} \Omega_N & & & \\ & \Omega_N & & \\ & & \cdot & \\ & & & \Omega_N \end{bmatrix}\begin{bmatrix} \text{row 1} \\ \text{row 2} \\ \text{row 3} \\ \text{row 4} \end{bmatrix} \qquad (\boldsymbol{f} \text{ and } \boldsymbol{\Omega}_{\mathbf{row}}\boldsymbol{f} \text{ have length } N^2) \quad (7)$$

That matrix is $\Omega_{\mathbf{row}} = I_N \otimes \Omega_N$. It is a Kronecker product of size N^2.

Now the output $\boldsymbol{\Omega}_{\mathbf{row}}\boldsymbol{f}$ is (mentally not electronically) rearranged into columns. The second step of the $2D$ transform multiplies each column of that "halfway" image $\boldsymbol{\Omega}_{\mathbf{row}}\boldsymbol{f}$ by Ω_N. Again we are multiplying by a matrix $\boldsymbol{\Omega}_{\mathbf{column}}$ of size N^2. The full 2D transform is $\Omega_N \otimes \Omega_N$.

That matrix $\boldsymbol{\Omega}_{\mathbf{column}}$ is the Kronecker product $\boldsymbol{\Omega}_N \otimes \boldsymbol{I}_N$.

The $2D$ transform puts the row and column steps together into $\boldsymbol{\Omega}_{N\times N}$.

$$\boldsymbol{\Omega}_{N\times N} = \boldsymbol{\Omega}_{\mathbf{column}}\,\boldsymbol{\Omega}_{\mathbf{row}} = (\boldsymbol{\Omega}_N \otimes \boldsymbol{I}_N)(\boldsymbol{I}_N \otimes \boldsymbol{\Omega}_N) = \boldsymbol{\Omega}_N \otimes \boldsymbol{\Omega}_N. \quad (8)$$

Example 1 With $N = 4$ there are $N^2 = 16$ pixels in a square. The block matrix $\Omega_{4\times4}$ for the $2D$ transform is 16 by 16:

$$\boldsymbol{\Omega_{4\times 4} = \Omega_4 \otimes \Omega_4} = \begin{bmatrix} \Omega_4 & \Omega_4 & \Omega_4 & \Omega_4 \\ \Omega_4 & -i\,\Omega_4 & (-i)^2\,\Omega_4 & (-i)^3\,\Omega_4 \\ \Omega_4 & (-i)^2\,\Omega_4 & (-i)^4\,\Omega_4 & (-i)^6\,\Omega_4 \\ \Omega_4 & (-i)^3\,\Omega_4 & (-i)^6\,\Omega_4 & (-i)^9\,\Omega_4 \end{bmatrix}$$

That matrix has to be divided by 4 times $4 = 16$ to correctly match the $1D$ transform $\Omega_4/4$. Then the inverse of this 16 by 16 matrix gives the two-dimensional inverse DFT— which is $F_4 \otimes F_4$. Apply the inverse formula (4) for Kronecker products:

$$\textbf{The } \boldsymbol{2D} \textbf{ inverse is } \left(\frac{1}{16}\Omega_4 \otimes \Omega_4\right)^{-1} = \boldsymbol{F_4 \otimes F_4} = \textbf{Kronecker product of } \boldsymbol{1D} \textbf{ inverses.}$$

The Kronecker Sum $A \oplus B$

The $2D$ Fourier transform is the **product** of two steps: transform by rows and transform the result by columns. Those steps were $I \otimes \Omega$ and $\Omega \otimes I$. In other problems we want the **sum** instead of the product. That produces a different matrix. Its size is still N^2 or MN.

A is M by M **B is N by N**	**The Kronecker sum $A \oplus B = A \otimes I_N + I_M \otimes B$ is MN by MN.**	(9)

This construction is natural for Laplace's equation (or Poisson's equation) in 2D:

$$\textbf{Laplace equation in a square} \quad -\frac{\partial^2 u}{\partial x^2} - \frac{\partial^2 u}{\partial y^2} = F(x,y) \quad \text{for } 0 \le x \le 1, 0 \le y \le 1$$

Divide that unit square into N^2 small squares with sides $h = 1/N$. Put nodes or meshpoints at the corners of those squares. There will be $(N+1)^2$ nodes. Then replace Laplace's second order differential equation by a second order difference equation that connects the values U_{jk} at the nodes:

$$\text{In one dimension } -\frac{\partial^2 u}{\partial x^2} \text{ becomes a second difference } \frac{-u(x+h) + 2u(x) - u(x-h)}{h^2}$$

For a line of $N+1$ nodes those second differences go into a matrix Δ of size $N+1$:

$$\begin{matrix} N = 4 \\ h = 1/4 \end{matrix} \qquad \boldsymbol{\Delta_5} = \frac{1}{(1/4)^2} \begin{bmatrix} \mathbf{1} & -1 & & & \\ -1 & \mathbf{2} & -1 & & \\ & -1 & \mathbf{2} & -1 & \\ & & -1 & \mathbf{2} & -1 \\ & & & -1 & \mathbf{1} \end{bmatrix} \qquad (10)$$

Notice the first and last entries on the diagonal: **1 and not 2**. Those reflect the boundary conditions at the ends of the row. We are choosing the free condition $\partial u/\partial x = 0$ in rows 1 and 5. (Fixed conditions $u = 0$ would lead to 2's on the whole diagonal.)

The second difference matrix Δ_5 is **positive semidefinite**. Its nullspace contains the column vector $\mathbf{1} = (1,1,1,1,1)$.

This matrix Δ_5 replaces $-\partial^2/\partial x^2$ along each row and $-\partial^2/\partial y^2$ down each column. We want the 25×25 matrix $\boldsymbol{\Delta}_{\textbf{row}}$ that finds second differences along *all rows at once*, plus second differences $\boldsymbol{\Delta}_{\textbf{column}}$ down *all columns at once*:

$$\boldsymbol{\Delta}_{\textbf{row}} = \begin{bmatrix} \Delta_5 & & \\ & \ddots & \\ & & \Delta_5 \end{bmatrix} = \boldsymbol{I_5 \otimes \Delta_5}$$

$$\boldsymbol{\Delta}_{\textbf{column}} = \text{(column at a time)} = \boldsymbol{\Delta_5 \otimes I_5}$$

So far this is like the discrete Fourier transform: all-row operation and all-column. The difference is that we *add* these matrices. We are aiming to approximate $-\partial^2/\partial x^2$ *plus* $-\partial^2/\partial y^2$. The overall $2D$ finite difference matrix is the 25×25 **Kronecker sum**:

$$\boldsymbol{\Delta}_{\mathbf{5\times 5}} = I_5 \otimes \Delta_5 + \Delta_5 \otimes I_5 = \boldsymbol{\Delta}_{\textbf{row}} + \boldsymbol{\Delta}_{\textbf{column}}. \tag{11}$$

The same matrix $\Delta_{5\times 5}$ comes from the finite element method (for linear finite elements).

This matrix is also a graph Laplacian! The graph is the 5 by 5 square array of 25 nodes. It has 20 horizontal edges and 20 vertical edges. Its incidence matrix A is 40×25. Then its graph Laplacian $A^{\mathrm{T}}A$ is the same 25×25 matrix $\Delta_{5\times 5}$. This matrix is positive semidefinite but not invertible. We study graph Laplacians in Section IV.6.

The nullspace of $\Delta_{5\times 5}$ contains the vector of 25 ones. That vector is $\mathbf{1} \otimes \mathbf{1}$.

Eigenvectors and Eigenvalues of $A \otimes B$ and $A \oplus B$

Suppose $\boldsymbol{x}$ is an eigenvector of $A : A\boldsymbol{x} = \lambda\boldsymbol{x}$. Suppose $\boldsymbol{y}$ is an eigenvector of $B : B\boldsymbol{y} = \mu\boldsymbol{y}$. Then the Kronecker product of $\boldsymbol{x}$ and $\boldsymbol{y}$ is an eigenvector of $A \otimes B$. **The eigenvalue is $\lambda\mu$:**

$$(A \otimes B)(\boldsymbol{x} \otimes \boldsymbol{y}) = (A\boldsymbol{x}) \otimes (B\boldsymbol{y}) = (\lambda\boldsymbol{x}) \otimes (\mu\boldsymbol{y}) = \lambda\mu(\boldsymbol{x} \otimes \boldsymbol{y}). \tag{12}$$

Here A is n by n and B is N by N. Therefore $\boldsymbol{x}$ is n by 1 and $\boldsymbol{y}$ is N by 1. So $A \otimes B$ is a square matrix of size nN and $\boldsymbol{x} \otimes \boldsymbol{y}$ is a vector of length nN. The same pattern also succeeds for Kronecker sums—with the same eigenvector $\boldsymbol{x} \otimes \boldsymbol{y}$:

$$(A \oplus B)(\boldsymbol{x} \otimes \boldsymbol{y}) = (A \otimes I_N)(\boldsymbol{x} \otimes \boldsymbol{y}) + (I_n \otimes B)(\boldsymbol{x} \otimes \boldsymbol{y}) = (\lambda + \mu)(\boldsymbol{x} \otimes \boldsymbol{y}). \tag{13}$$

The eigenvalue of $A \oplus B$ is $\lambda + \mu$. The vector y is certainly an eigenvector of the identity matrix I_N (with eigenvalue 1). The vector $\boldsymbol{x}$ is certainly an eigenvector of I_n (with eigenvalue 1). So equation (13) for the Kronecker sum just comes from two applications of equation (12) for the Kronecker product. $A \otimes I_N$ has eigenvalue λ times 1, and $I_n \otimes B$ has eigenvalue 1 times μ. The eigenvector in both cases is $\boldsymbol{x} \otimes \boldsymbol{y}$. So we add to see equation (13) with eigenvalue $\lambda + \mu$.

Separation of Variables

The last page was very formal. But the idea is simple and important. It is the matrix equivalent of the most useful trick for eigenfunctions of Laplace's equation—and all similar equations that have x-derivatives added to y-derivatives.

$$\textbf{Eigenvalues } \boldsymbol{\alpha} \textbf{ of the Laplacian} \qquad \frac{\partial^2 u}{\partial x^2} + \frac{\partial^2 u}{\partial y^2} = \alpha\, u(x,y). \tag{14}$$

The trick is to look for u in the separated form $\boldsymbol{u(x,y) = v(x)\, w(y)}$. Substitute vw for u in Laplace's eigenvalue equation (14):

$$\textbf{Separation of variables } \boldsymbol{x} \textbf{ and } \boldsymbol{y} \qquad \left(\frac{d^2 v}{dx^2}\right) w(y) + v(x)\frac{d^2 w}{dy^2} = \alpha\, v(x)\, w(y). \tag{15}$$

Ordinary derivatives instead of partials, v depends only on x and w depends only on y.

Divide equation (15) by $\boldsymbol{v}$ times $\boldsymbol{w}$. The result on the left side is a function only of x plus a function only of y.

$$\frac{d^2v/dx^2}{v(x)} + \frac{d^2w/dy^2}{w(y)} = \alpha = \text{constant}. \tag{16}$$

If this is true for every x and y then each term is a constant (*think about this* !):

$$\boxed{\textbf{Separated equations} \quad \frac{d^2 v}{dx^2} = \lambda v \ \text{ and } \ \frac{d^2 w}{dy^2} = \mu w \ \text{ with } \ \lambda + \mu = \alpha.} \tag{17}$$

So $\lambda + \mu$ is an eigenvalue of Laplace's equation (14). The Laplacian on the left side of (14) is just the **Kronecker sum** of $A = \partial^2/\partial x^2$ and $B = \partial^2/\partial y^2$. The key point is that the eigenfunction $u(x,y)$ is the product of two 1D eigenfunctions $v(x)$ and $w(y)$, This is continuous instead of discrete, with derivatives instead of difference matrices and eigenfunctions instead of eigenvectors.

The partial differential equation is reduced to two ordinary differential equations (17).

The Matrix to Vector Operation vec($\boldsymbol{A}$)

We often want to *vectorize a matrix*. Starting with an m by n matrix A, we stack its n columns to get one column vector **vec**(A) of length mn:

$$\textbf{vec}(A) = \begin{bmatrix} \text{column 1} \\ \vdots \\ \text{column } n \end{bmatrix} \qquad \textbf{vec}\left(\begin{bmatrix} a & b \\ c & d \end{bmatrix}\right) = \begin{bmatrix} a \\ c \\ b \\ d \end{bmatrix}. \tag{18}$$

This vector of length N^2 is multiplied by the $2D$ Fourier matrix (a Kronecker product of size N^2) to produce the vector of N^2 Fourier coefficients. And this vector of length $(N+1)^2$ is multiplied by the graph Laplacian matrix (a Kronecker sum) in the finite difference approximation to Laplace's equation on a square.

So we need to see how this simple **vec** operation interacts with matrix multiplication. Here is the key identity, when a matrix B is multiplied on the left by a matrix A and on the right by a matrix C:

$$\boxed{\textbf{vec}(ABC) = (C^{\mathrm{T}} \otimes A)\,\textbf{vec}(B).} \tag{19}$$

First check dimensions when all three matrices A, B, C are n by n. Then ABC is n by n and **vec** makes it n^2 by 1. The Kronecker product is n^2 by n^2 and it multiplies the n^2 by 1 vector **vec**(B). So the right hand side of (19) also has length n^2. And if ABC is m by p then the matrix $C^{\mathrm{T}} \otimes A$ has mp rows as it should. *Good.*

But notice a big difference in the operation count. In the n by n case, the two multiplications in ABC need $2n^3$ separate multiply-adds. The Kronecker product on the right has n^4 entries! So if we don't notice that it is a Kronecker product, this matrix-vector multiplication needs n^4 multiply-adds. Reshaping is essential to take advantage of the Kronecker structure:

A is $m \times n$ and $mn = MN$ $\quad$ $\boldsymbol{B} =$ **reshape** (A, M, N) is $M \times N$ $\quad$ **vec** $(B) =$ **vec** (A)

If A is 3×2 then $\boldsymbol{B} =$ **reshape** $(\boldsymbol{A}, \mathbf{1}, \mathbf{6})$ produces $B = [a_{11} \;\; a_{21} \;\; a_{31} \;\; a_{12} \;\; a_{22} \;\; a_{32}]$.

We need to understand the **vec** identity (19). Start with the case when $B = [\boldsymbol{x}_1 \;\; \boldsymbol{x}_2]$ has 2 columns and C is 2 by 2. The right side of (19) is simply a matrix-vector multiplication $K\boldsymbol{x}$ where $\boldsymbol{x} =$ **vec** (B) and we have recognized that K is a Kronecker product $C^{\mathrm{T}} \otimes A$. That multiplication produces a vector $\boldsymbol{y}$:

$$\boldsymbol{y} = K\boldsymbol{x} = \left(C^{\mathrm{T}} \otimes A\right) \begin{bmatrix} \boldsymbol{x}_1 \\ \boldsymbol{x}_2 \end{bmatrix} = \begin{bmatrix} c_{11}A & c_{21}A \\ c_{12}A & c_{22}A \end{bmatrix} \begin{bmatrix} \boldsymbol{x}_1 \\ \boldsymbol{x}_2 \end{bmatrix} = \begin{bmatrix} c_{11}A\boldsymbol{x}_1 + c_{21}A\boldsymbol{x}_2 \\ c_{12}A\boldsymbol{x}_1 + c_{22}A\boldsymbol{x}_2 \end{bmatrix} \tag{20}$$

This vector $\boldsymbol{y}$ is exactly the left side of the identity (19): **vec** (ABC) with $B = [\boldsymbol{x}_1 \;\; \boldsymbol{x}_2]$ is

$$\textbf{vec}\left(\begin{bmatrix} A\boldsymbol{x}_1 & A\boldsymbol{x}_2 \end{bmatrix} \begin{bmatrix} c_{11} & c_{12} \\ c_{21} & c_{22} \end{bmatrix}\right) = \textbf{vec}\,\begin{bmatrix} c_{11}A\boldsymbol{x}_1 + c_{21}A\boldsymbol{x}_2 & c_{12}A\boldsymbol{x}_1 + c_{22}A\boldsymbol{x}_2 \end{bmatrix}$$

So if we have a linear system $K\boldsymbol{x} = \boldsymbol{b}$ with a Kronecker matrix K, (19) will reduce its size.

The pixel values in a $2D$ image (n by n) are stacked by **vec** into a column vector (length n^2 or $3n^2$ with RGB color). A video with T frames has a sequence of images at T different times. Then **vec** stacks the video into a column vector of length Tn^2.

Reference

C. Van Loan, *The ubiquitous Kronecker product*, J. Comp. Appl. Math. **123** (2000) 85–100.
Also *The Kronecker Product*, https://www.cs.cornell.edu/cv/ResearchPDF/KPhist.pdf

Problem Set IV.3

1 A matrix person might prefer to see all n eigenvectors of A in an eigenvector matrix: $AX = X\Lambda_A$. Similarly $BY = Y\Lambda_B$. Then the Kronecker product $X \otimes Y$ of size nN is the eigenvector matrix for both $A \otimes B$ and $A \oplus B$:

$$(A\otimes B)(X\otimes Y) = (X\otimes Y)(\Lambda_A\otimes\Lambda_B) \quad \text{and} \quad (A\oplus B)(X\otimes Y) = (X\otimes Y)(\Lambda_A\oplus\Lambda_B).$$

The eigenvalues of $\boldsymbol{A \otimes B}$ are the nN products $\lambda_i\mu_j$ (every λ *times* every μ). The eigenvalues of $\boldsymbol{A \oplus B}$ are the nN sums $\boldsymbol{\lambda_i + \mu_j}$ (every λ *plus* every μ).

If A and B are n by n, when is $A \oplus B$ invertible? What if (eigenvalue of A) = $-$ (eigenvalue of B)? Find 2×2 matrices so that $A \oplus B$ has rank 3 and not rank 4.

2 Prove: If A and B are symmetric positive definite, so are $A \otimes B$ and $A \oplus B$.

3 Describe a permutation P so that $P(A \otimes B) = (B \otimes A)P$. Do $A \otimes B$ and $B \otimes A$ have the same eigenvalues?

4 Suppose we want to compute $\boldsymbol{y} = (F \otimes G)\,\boldsymbol{x}$ where $\boldsymbol{x} = \text{vec}(X)$. The matrix F is m by n and G is p by q. The matrix $F \otimes G$ is mp by nq, the matrix X is q by n, the vector $\boldsymbol{x}$ is nq by 1. Show that this code finds the correct $\boldsymbol{y} = (F \otimes G)\,\boldsymbol{x}$:

$$Y = B \otimes X \otimes A^{\mathrm{T}}$$
$$\boldsymbol{y} = \text{reshape}(\boldsymbol{Y}, \boldsymbol{mp}, \boldsymbol{1})$$

5 Suppose we want to solve $(F \otimes G)\,\boldsymbol{x} = \boldsymbol{b}$ when F and G are invertible matrices (n by n). Then $\boldsymbol{b}$ and $\boldsymbol{x}$ are n^2 by 1. Show that this is equivalent to computing

$$X = G^{-1}B(F^{-1})^{\mathrm{T}} \quad \text{with} \quad \boldsymbol{x} = \textbf{vec}(X) \quad \text{and} \quad \boldsymbol{b} = \textbf{vec}(B).$$

In reality those inverse matrices are never computed. Instead we solve two systems:

Find Z from $GZ = B$

Find X from $XF^{\mathrm{T}} = Z$ or $FX^{\mathrm{T}} = Z^{\mathrm{T}}$

Show that the cost is now $O(n^3)$. The larger system $(F \otimes G)\,\boldsymbol{x} = \boldsymbol{b}$ costs $O(n^6)$.

6 What would an image look like if its pixels produced a Kronecker product $A \otimes B$?

7 How would you create a two-dimensional FFT? For an n by n image, how many operations will your 2D Fast Fourier Transform need?

IV.4 Sine and Cosine Transforms from Kronecker Sums

This section of the book presents an outstanding example of a Kronecker sum $\boldsymbol{K} = \boldsymbol{I} \otimes \boldsymbol{D} + \boldsymbol{D} \otimes \boldsymbol{I}$. The $1, -2, 1$ matrix D approximates the second derivative d^2/dx^2 in one dimension (along a line). The Kronecker sum raises it to two dimensions (in a square). This big matrix K of size N^2 approximates the Laplacian $\partial^2/\partial x^2 + \partial^2/\partial y^2$.

For Kronecker, all is normal—a very convenient way to create an important matrix K. The next steps make it practical to work with this large matrix. We are approximating Laplace's partial differential equation by N^2 difference equations:

$$\textbf{Laplace} \quad \frac{\partial^2 u}{\partial x^2} + \frac{\partial^2 u}{\partial y^2} = f(x, y) \qquad \textbf{Discrete Laplace} \quad \frac{1}{h^2} KU = F \qquad (1)$$

The N^2 components of U will be close to the true $u(x, y)$ at the N^2 points $x = ph$, $y = qh$ inside a square grid. Here $p = 1$ to N along a row of the grid and $q = 1$ to N up a column, with $h = \Delta x = \Delta y =$ meshwidth of the grid.

The difficulty is the size of the matrix K. **The solution to this difficulty is to know and use the N^2 eigenvectors of this matrix.**

It is exceptional that we know those eigenvectors. It is even more exceptional (this is virtually the only important example we have seen) that linear equations $KU = h^2 F$ are quickly solved by writing $h^2 F$ and U as combinations of the eigenvectors v_i of K:

$$\boxed{h^2 F = b_1 v_1 + b_2 v_2 + \cdots \qquad U = \frac{b_1}{\lambda_1} v_1 + \frac{b_2}{\lambda_2} v_2 + \cdots \qquad \text{Then } KU = h^2 F.} \qquad (2)$$

When K multiplies U, the λ's cancel because each $K v_i = \lambda_i v_i$. So KU matches $h^2 F$.

The eigenvectors u_j of the one-dimensional $1, -2, 1$ matrix D happen to be discrete **sine vectors**. They are sample points taken from the eigenfunctions $\sin j\pi x$ of d^2/dx^2:

$$\boxed{\begin{array}{c}\textbf{Continuous}\\ \textbf{and}\\ \textbf{Discrete}\end{array} \quad \frac{d^2}{dx^2} \sin j\pi x = -j^2\pi^2 \sin j\pi x \quad D u_j = D \begin{bmatrix} \sin j\pi h/(N+1) \\ \vdots \\ \sin j\pi N h/(N+1) \end{bmatrix} = \Lambda_j u_j} \qquad (3)$$

Eigenvectors v of the large matrix $\boldsymbol{K} = \boldsymbol{I} \otimes \boldsymbol{D} + \boldsymbol{D} \otimes \boldsymbol{I}$ were found in the Kronecker section IV.3. **Eigenvectors of K are Kronecker products $v_{jk} = u_j \otimes u_k$ of the sine eigenvectors of D.** This is true in the continuous case (for eigenfunctions). It remains true in the discrete case (for eigenvectors). The components of v_{jk} are components of u_j times components of u_k:

$$\begin{array}{c}\textbf{Continuous}\\ \textbf{eigenfunctions}\end{array} \quad \begin{array}{c} v_{jk}(x,y) = \\ (\sin j\pi x)(\sin k\pi y) \end{array} \qquad \begin{array}{c}\textbf{Discrete}\\ \textbf{eigenvectors}\end{array} \quad v_{jk}(p,q) = \left(\sin \frac{\pi j p}{N+1}\right)\left(\sin \frac{\pi k q}{N+1}\right)$$

The final step to success is the most crucial of all. **We can compute with these eigenvectors using the Fast Fourier Transform.** The Fourier matrix in Section IV.1 had complex exponentials. Its real part has cosines (which enter the Discrete Cosine Transform).

The imaginary part of F produces the Discrete Sine Transform. That DST matrix contains the eigenvectors $\boldsymbol{u}_j$ of D. The two-dimensional DST matrix contains the eigenvectors $\boldsymbol{v}_{jk}$ of the Laplace difference matrix K.

The FFT executes the computational steps of equation (2) in $O(N^2 \log_2 N)$ operations:

(1) A fast 2D transform FFT $\otimes$ FFT to find the sine coefficients b_{jk} of $h^2\boldsymbol{F}$

(2) A division of each b_{jk} by the eigenvalue $\lambda_{jk} = \lambda_j(D)\, \lambda_k(D)$

(3) A fast 2D inverse transform to find $\boldsymbol{U}$ from its sine coefficients b_{jk}/λ_{jk}.

This algorithm solves Laplace's equation in a square. Boundary values are given on the four edges. It is coded in FISHPACK. That name is a terrible pun on the translation of Poisson—who added a source term to Laplace's equation: $\partial^2 u/\partial x^2 + \partial^2 u/\partial y^2 = f(x,y)$. FISHPACK also allows the "free" or "natural" or "Neumann" boundary condition $\partial u/\partial n = 0$ with cosine eigenvectors, as well as the "fixed" or "essential" or "Dirichlet" boundary condition $u = u_0$ with sine eigenvectors.

The difference between fixed and free appears in the first and last rows of D in one dimension—and then in all the boundary rows of the Kronecker sum $K = D \oplus D$ in two dimensions:

$$D_{\text{fixed}} = \begin{bmatrix} \mathbf{-2} & 1 & & \\ 1 & -2 & 1 & \\ & \cdot & \cdot & \cdot \\ & & 1 & \mathbf{-2} \end{bmatrix} \qquad D_{\text{free}} = \begin{bmatrix} \mathbf{-1} & 1 & & \\ 1 & -2 & 1 & \\ & \cdot & \cdot & \cdot \\ & & 1 & \mathbf{-1} \end{bmatrix} \tag{4}$$

The eigenvectors of D_{fixed} led us to the Discrete Sine Transform. The eigenvectors of D_{free} would lead us to the Discrete Cosine Transform. But a different application is too interesting to miss. Instead of solving Laplace's difference equation with free boundary conditions, we will apply the two-dimensional Discrete Cosine Transform to image compression.

In that world the DCT-based algorithm is known as JPEG (jaypeg).

The Discrete Cosine Transform in JPEG

The letters JPEG stand for Joint Photographic Experts Group. This group established a family of algorithms that start with pixel values (grayscale numbers from 0 to 255, or the Red-Blue-Green intensities for each pixel). The code produces a compressed image file in the .jpg format. The algorithm can be seen in two stages:

> **Step 1** is a linear transformation of the matrix of pixel values. At the start, grayscale values are highly correlated—nearby pixels tend to have nearby values. The transform produces numbers with more independent information.

Example: Take the average and difference of two neighboring values. When the difference is small, we can transmit fewer bits and the human visual system will never know.

> **Step 2** is a nonlinear compression and quantization of the transformed signal. The compression keeps only numbers that are visually significant. The quantization converts the compressed signal into a sequence of bits—to get ready for fast transmission. Then the receiver uses those bits to reconstruct an image that is very close to the original.

Step 1 most often uses a Discrete Cosine Transform. It acts separately on 8×8 blocks of the image. (JPEG2000 offers a wavelet transform but this option has not been widely adopted.) **Each block of 64 grayscale values leads to a block of 64 cosine coefficients**—a lossless transform. The inverse transform will recover the original blocks of the image.

But before inverting we will compress and quantize those 64 numbers. This step loses information that our eyes can't see anyway.

Discrete Cosine Transforms are described in *SIAM Review* for 1999 (volume **41**, pages 135-147). They differ in their boundary conditions. The most popular choice is DCT-2 with these 8 orthogonal basis vectors in each dimension:

DCT-2 The jth component of the kth vector is $\cos\left(j + \frac{1}{2}\right) k\frac{\pi}{8} \quad j, k = 1, \ldots, 8$

Those 64 numbers go into an 8 by 8 matrix C. Then the matrix for the $2D$ cosine transform is the Kronecker product $C \otimes C$ of size 8^2. Its columns are orthogonal. It acts on each 8×8 block of the image to give an 8×8 block of Fourier cosine coefficients—which tell us the right combination of cosine basis vectors to reconstruct the original block in the image.

But we don't aim for perfect reconstruction. Step 2 discards information that we don't need. Step 1 has produced cosine coefficients c_{jk} of very different sizes—usually smaller numbers c for higher frequencies j, k. The file **https://cs.stanford.edu/people/eroberts/courses/soco/projects/data-compression/lossy/jpeg/dct.htm** shows a typical 8×8 block before and after the DCT step. The cosine coefficients are ready for Step 2, which is compression and quantization.

First, each block of 64 coefficients goes into a 64×1 vector—but not by the vec command. A zig-zag sequence is better at keeping larger coefficients first. If we end with a run of near-zeros, they can be compressed to zero (and we only transmit the length of the run: the number of zeros). Here is the start of the zig-zag order for the 64 coefficients:

$$\begin{array}{cccc} 1 & 2 & 6 & 7 \\ 3 & 5 & 8 & \\ 4 & 9 & & \\ 10 & & & \end{array}$$

Higher frequencies come later in the zig-zag order and usually with smaller coefficients. Often we can safely rescale those numbers before the rounding step gives q_{jk}:

Quantization example $\quad q_{jk} = \dfrac{c_{jk}}{j + k + 3}$ rounded to nearest integer

Now each block is represented by 64 integers q_{jk}. We transmit those numbers after encoding for efficiency. The receiver reconstructs each block of the image—approximately.

In a color image, 64 becomes 192. Every pixel combines three colors. But Red-Green-Blue are probably not the best coordinates. A better first coordinate tells us the brightness. The other two coordinates tell us "chrominance". The goal is to have three statistically independent numbers.

When the blocks are assembled, the full reconstruction often shows two unwanted artifacts. One is **blocking**: the blocks don't meet smoothly. (You may have seen this in an overcompressed printed image.) The second artifact is **ringing**: fast oscillations along edges inside the image. These oscillations are famous in the "Gibbs phenomenon" when eight cosines approximate a step function.

For a high quality picture after DCT compression, we can mostly cancel those blocking and ringing artifacts. For high definition TV, compression is essential (too many bits to keep up).

The DCT standard was set by the JPEG experts group. Then equipment was built and software was created to make image processing into an effective system.

Problem Set IV.4

1 What are the eigenvalues Λ_j of D in equation (3) ?

2 What are the eigenvalues λ_i and eigenvectors $\boldsymbol{v}_i$ of $K = I \otimes D + D \otimes I = I \oplus D$?

3 What would be the Laplace operator $K3$ on a cubic grid in 3D ?

4 What would be the N^3 by N^3 Fourier matrix $F3$ in 3D ? In 2D it was $F \otimes F$.

IV.5 Toeplitz Matrices and Shift Invariant Filters

A Toeplitz matrix has constant diagonals. The first row and column tell you the rest of the matrix, because they contain the first entry of every diagonal. **Circulant matrices** are Toeplitz matrices that satisfy the extra "wraparound" condition that makes them periodic. Effectively c_{-3} is the same as c_1 (for 4×4 circulants) :

Toeplitz matrix $A = \begin{bmatrix} a_0 & a_{-1} & a_{-2} & a_{-3} \\ a_1 & a_0 & a_{-1} & a_{-2} \\ a_2 & a_1 & a_0 & a_{-1} \\ a_3 & a_2 & a_1 & a_0 \end{bmatrix}$ **Circulant matrix** $C = \begin{bmatrix} c_0 & c_3 & c_2 & c_1 \\ c_1 & c_0 & c_3 & c_2 \\ c_2 & c_1 & c_0 & c_3 \\ c_3 & c_2 & c_1 & c_0 \end{bmatrix}$

Circulant matrices are perfect for the Discrete Fourier Transform. Always $CD = DC$. Their eigenvectors are *exactly* the columns of the Fourier matrix in Section IV.2. Their eigenvalues are exactly the values of $C(\theta) = \sum c_k e^{ik\theta}$ at the n equally spaced angles $\theta = 0, 2\pi/n, 4\pi/n, \ldots$ (where $e^{in\theta} = e^{2\pi i} = 1$).

Toeplitz matrices are nearly perfect. They are the matrices we use in signal processing and in convolutional neural nets (**CNNs**). They don't wrap around, so the analysis of A is based on the two-sided polynomial $A(\theta)$ with coefficients $a_{1-n} \ldots a_0 \ldots a_{n-1}$:

Frequency response = symbol of A	$A(\theta) = \sum a_k \, e^{ik\theta}$
$A(\theta)$ is real when A is symmetric	$a_k \, e^{ik\theta} + a_k \, e^{-ik\theta} = 2a_k \cos k\theta$
$C(\theta)$ is nonzero when C is invertible	The symbol for C^{-1} is $1/C(\theta)$

$A(\theta) \neq 0$ is not correct as a test for the invertibility of A! And A^{-1} is not Toeplitz (triangular matrices are the exception). Circulants C are cyclic convolutions. But Toeplitz matrices are noncyclic convolutions with $\boldsymbol{a} = (a_{1-n} \ldots a_{n-1})$ followed by projections:

$\boldsymbol{x}$–space $\quad Ax \quad$ = convolve $\boldsymbol{a} * \boldsymbol{x}$, then keep components 0 to $n-1$

$\boldsymbol{\theta}$–space $\quad Ax(\theta) =$ multiply $A(\theta)x(\theta)$, then project back to n coefficients

We want to use the simple polynomial $A(\theta)$ to learn about the Toeplitz matrix A.

In many problems the Toeplitz matrix is **banded**. The matrix only has w diagonals above and below the main diagonal. Only the coefficients from a_{-w} to a_w can be nonzero. Then the "bandwidth" is w, with a total of **$2w + 1$ nonzero diagonals**:

Tridiagonal Toeplitz Bandwidth $w = 1$ $\quad A = \begin{bmatrix} a_0 & a_{-1} & & \\ a_1 & a_0 & a_{-1} & \\ & a_1 & a_0 & a_{-1} \\ & & a_1 & a_0 \end{bmatrix}$

We understand tridiagonal Toeplitz matrices (and their eigenvalues) for large size n by studying the symbol $A(\theta) = a_{-1}e^{-i\theta} + a_0 + a_1 e^{i\theta}$. It is built from a_{-1}, a_0, a_1.

Toeplitz Matrices : Basic Ideas

In signal processing, a Toeplitz matrix is a **filter**. The matrix multiplication Ax in the time domain translates into ordinary multiplication $A(\theta)x(\theta)$ in the frequency domain. That is the fundamental idea *but it is not exactly true*. So Toeplitz theory is on the edge of simple Fourier analysis (one frequency at a time), but boundaries interfere.

A finite length signal $x = (x_0, \ldots, x_n)$ has boundaries at 0 and n. Those boundaries destroy a simple response to each separate frequency. We cannot just multiply by $A(\theta)$. In many applications—but not all—this inconvenient fact can be suppressed. Then a Toeplitz matrix essentially produces a convolution. If we want a **bandpass filter** that preserves frequencies $a \leq \theta \leq b$ and removes all other frequencies, then we construct $A(\theta)$ to be near 1 in that band and near zero outside the band. (Again, an ideal filter with A exactly 1 and 0 is impossible until $n = \infty$.)

Linear finite difference equations with *constant coefficients* produce Toeplitz matrices. The equations don't change as time goes forward (**LTI** = Linear Time Invariant). They don't change in space (**LSI** = Linear Shift Invariant). The familiar $-1, 2, -1$ second difference matrix is an important example:

$$\text{Tridiagonal } -1, 2, -1 \text{ matrix with symbol } \boldsymbol{A(\theta)} = -e^{-i\theta} + 2 - e^{i\theta} = \mathbf{2 - 2\cos\theta}.$$

The fact that $A(\theta) \geq 0$ tells us that A is symmetric positive semidefinite or definite. The fact that $A = 2 - 2 = 0$ when $\theta = 0$ tells us that $\lambda_{\text{min}}(A)$ will approach zero as n increases. The finite Toeplitz matrix A is barely positive definite and the infinite Toeplitz matrix is singular, all because the symbol has $A(\theta) = 0$ at $\theta = 0$.

The inverse of a Toeplitz matrix A is usually not Toeplitz. For example,

$$A^{-1} = \begin{bmatrix} 2 & -1 & 0 \\ -1 & 2 & -1 \\ 0 & -1 & 2 \end{bmatrix}^{-1} = \frac{1}{4}\begin{bmatrix} 3 & 2 & 1 \\ 2 & 4 & 2 \\ 1 & 2 & 3 \end{bmatrix} \quad \text{is not Toeplitz}$$

But Levinson found a way to use the Toeplitz pattern in a recursion—reducing the usual $O(n^3)$ solution steps for $Ax = b$ to $O(n^2)$. Superfast algorithms were proposed later, but the "Levinson-Durbin recursion" is better for moderate n. Superfast algorithms give accurate answers (not exact) for large n—one way is a circulant preconditioner.

One more general comment. It frequently happens that the first and last rows do not fit the Toeplitz pattern of "shift invariance". The entries in those boundary rows can be different from a_0 and a_1 and a_{-1}. This change can come from *boundary conditions* in a differential equation or in a filter. When A multiplies a vector $x = (x_1, \ldots, x_n)$, the Toeplitz matrix (sharp cutoff) is assuming $x_0 = 0$ and $x_{n+1} = 0$ (*zero-padding*).

Zero padding may give a poor approximation at the endpoints. Often we change the boundary rows for a better approximation. A close analysis of those changes can be difficult, because the constant-diagonal pattern in the Toeplitz matrix is perturbed.

Fast Multiplication by the FFT

For tridiagonal matrices, we don't need a special algorithm to multiply Ax. This requires only $3n$ separate multiplications. But for a full dense matrix, the Toeplitz and circulant properties allow major speedup in θ-space by using the Fast Fourier Transform.

Circulants are cyclic convolutions. In θ-space, the matrix-vector multiplication Cx becomes $(\sum c_k e^{ik\theta})(\sum x_k e^{ik\theta})$. This product will give powers of $e^{i\theta}$ that go outside the range from $k = 0$ to $k = n-1$. **Those higher frequencies and negative frequencies are "aliases" of standard frequencies**—they are equal at every $\theta = p2\pi/n$:

$$\textbf{Aliasing} \qquad e^{in\theta} = 1 \text{ and } e^{i(n+1)\theta} = e^{i\theta} \text{ and } e^{i(n+p)\theta} = e^{ip\theta} \text{ at } \theta = \frac{2\pi}{n}, \frac{4\pi}{n}, \ldots$$

Cyclic convolution $c \circledast x$ brings every term of $(\sum c_k e^{ik\theta})(\sum x_k e^{ik\theta})$ back to a term with $0 \leq k < n$. So a circulant multiplication Cx needs only $O(n \log_2 n)$ steps because of convolution by the Fast Fourier Transform.

$$\textbf{Cyclic convolution} \qquad \boldsymbol{c \circledast x \text{ and } c \circledast d \text{ give the entries of } Cx \text{ and } CD.}$$

A Toeplitz matrix multiplication Ax is not cyclic. Higher frequencies in $A(\theta)x(\theta)$ don't fold back perfectly into lower frequencies. But we can use cyclic multiplication and a circulant by a doubling trick: **embed A into a circulant matrix C.**

$$A = \begin{bmatrix} a_0 & a_{-1} & \cdot & \cdot & a_{1-n} \\ a_1 & \ddots & & & \cdot \\ \cdot & & \ddots & & \cdot \\ \cdot & & & \ddots & a_{-1} \\ a_{n-1} & \cdot & \cdot & a_1 & a_0 \end{bmatrix} \qquad C = \begin{bmatrix} A & * \\ * & * \end{bmatrix} = \begin{bmatrix} a_0 & a_{-1} & \cdot & \cdot & a_{1-n} & a_{n-1} & \cdot & \cdot & a_1 \\ a_1 & \ddots & & & \cdot & \ddots & & & \cdot \\ \cdot & & & & \cdot & & & & \cdot \\ \cdot & & & & a_{-1} & & & & a_{n-1} \\ a_{n-1} & \cdot & \cdot & a_1 & a_0 & & & & a_{1-n} \\ a_{1-n} & \ddots & & & & \ddots & & & \cdot \\ \cdot & & & & & & & & \cdot \\ \cdot & & & & & & & & a_{-1} \\ a_{-1} & \cdot & \cdot & a_{1-n} & a_{n-1} & \cdot & \cdot & a_1 & a_0 \end{bmatrix}$$

To compute Ax of size n, we can use this circulant matrix C of size $2n+1$.

1. Add $n-1$ zeros to extend x to a vector X of size $2n-1$.
2. Multiply CX using the Fast Fourier Transform (cyclic convolution).
3. Then the first n components of CX produce the desired Ax.

If size $2n$ is preferred for C, a diagonal of a_0's can go between a_{1-n} and a_{n-1}.

Toeplitz Eigenvalues and Szegö's Theorem

The exact eigenvalues of circulant matrices were found in Section IV.2. The eigenvectors are known in advance—always the same! They are the columns of the Fourier matrix F. The eigenvalues of C are the components of the vector Fc. They are the values of $C(\theta)$ at n equally spaced points. In other words, the eigenvalues are the discrete Fourier transform of column 0 of the circulant matrix C.

As always, circulant formulas are exact and Toeplitz formulas are close. You will see integrals instead of point values of $A(\theta)$. *The formulas only become exact as the size of the Toeplitz matrix becomes infinite.* So there is a limit as $n \to \infty$ included in Szegö's Theorem about the eigenvalues of A. Two special cases and then the full theorem.

Szegö As $n \to \infty$, the trace and the log determinant of a Toeplitz matrix A satisfy

$$\frac{1}{n}\,\text{trace}\,(A) = \frac{1}{2\pi}\int_0^{2\pi} A(\theta)\,d\theta = a_0 \tag{1}$$

$$\lim_{n\to\infty}\frac{1}{n}\log(\det A) = \frac{1}{2\pi}\int_0^{2\pi} \log(A(\theta))\,d\theta \tag{2}$$

The trace and determinant of any matrix are the sum and product of the eigenvalues. We are seeing their arithmetic mean in (1) and their geometric mean in (2): log determinant = sum of $\log \lambda_k$. Those two limits are the most important cases $F(\lambda)=\lambda$ and $F(\lambda)=\log\lambda$ of the full theorem, which allows any continuous function F of the eigenvalues of A:

Szegö's Theorem $$\lim_{n\to\infty}\frac{1}{n}\sum_{k=0}^{n-1} F(\lambda_k) = \frac{1}{2\pi}\int_0^{2\pi} F(A(\theta))\,d\theta. \tag{3}$$

The control on $A(\theta)$ comes from Wiener's condition that $\sum |a_k| < \infty$. Excellent notes are posted on the Stanford website **ee.stanford.edu/~gray/toeplitz.pdf**.

Those notes by Professor Gray also develop a major application to discrete time random processes. When the process is weakly stationary, the statistics stay the same at every time step. *The covariance matrix between outputs at times t and T depends only on the difference $T - t$.* **Then the covariance matrix is Toeplitz = shift invariant.**

Before applications, we mention three more key topics in Toeplitz matrix theory:

- The Gohberg-Semencul formula for A^{-1} (Gohberg was truly remarkable)
- The Wiener-Hopf factorization of infinite systems $A_\infty \boldsymbol{x} = \boldsymbol{b}$ (so was Wiener!)
- The test for invertibility of A_∞ is $A_\infty(\theta) \neq 0$ and winding number = 0

An **infinite one-step shift** models the difficulties that come with $n = \infty$. It has a diagonal of 1's above or below the main diagonal. **Its symbol is just $e^{i\theta}$ or $e^{-i\theta}$** (never zero). But $e^{i\theta}$ winds around zero as θ goes from 0 to 2π, and the infinite shift matrix $S(x_0, x_1, \ldots) = (x_1, x_2, \ldots)$ or $S\boldsymbol{x} = (0, x_0, x_1, \ldots)$ is not invertible.

Lowpass Filters in Signal Processing

A filter is a convolution: Multiply by a Toeplitz matrix. Going through a "lowpass filter", the constant vector $\boldsymbol{x} = (\ldots, 1, 1, 1, 1, \ldots)$ can come out unchanged: $A\boldsymbol{x} = \boldsymbol{x}$. But an oscillating high-frequency signal like $\boldsymbol{y} = (\ldots, -1, 1, -1, 1, \ldots)$ has $A\boldsymbol{y} \approx \mathbf{0}$. Those are the outputs if $A(0) = \Sigma a_k = 1$ and $A(\pi) \approx 0$. Here is a lowpass example.

$$\textbf{Lowpass averaging filter} \quad (A\boldsymbol{x})_n = \frac{\mathbf{1}}{\mathbf{4}} x_{n+1} + \frac{\mathbf{1}}{\mathbf{2}} x_n + \frac{\mathbf{1}}{\mathbf{4}} x_{n-1}.$$

That Toeplitz matrix A is symmetric. Its three diagonals have entries $\frac{1}{4}, \frac{2}{4}, \frac{1}{4}$. Its symbol $A(\theta)$ (frequency response) is real and $A(\theta) = 1$ at frequency $\theta = 0$: lowpass filter.

$$\textbf{Frequency response} \qquad A(\theta) = \frac{1}{4}\left(e^{-i\theta} + 2 + e^{i\theta}\right) = \frac{1}{2}(1 + \cos\theta) \geq 0.$$

The highest frequency $\theta = \pi$ produces the infinitely long plus-minus signal $\boldsymbol{y} = (\ldots, -1, 1, -1, 1, \ldots)$. That signal has $A\boldsymbol{y} = \mathbf{0}$. It is filtered out. A typical component of $A\boldsymbol{y}$ is $-\frac{1}{4} + \frac{1}{2} - \frac{1}{4} = 0$. And we see this also from the symbol: $\boldsymbol{A(\theta) = \frac{1}{2}(1 + \cos\theta)}$ **is zero at $\boldsymbol{\theta = \pi}$.**

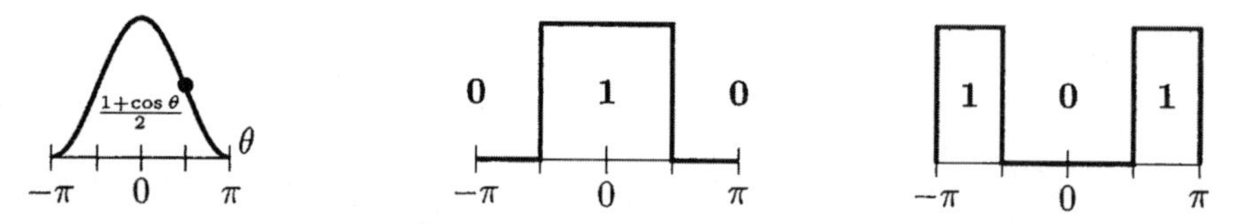

Figure IV.3: Frequency responses $A(\theta)$: Short lowpass filter, ideal lowpass, ideal highpass.

You see two *ideal filters* in Figure IV.3. They are achievable only with infinitely many nonzero diagonals in the matrix A because an ordinary polynomial can't stay constant. The numbers a_k down the diagonals of A are the Fourier coefficients of $A(\theta)$.

In practice, filters are a compromise between short and ideal. *Equiripple filters* are a natural favorite—they oscillate around 1 and around 0. The ripples (oscillations around the ideal) all have the same height. That height decreases as we use more coefficients a_k. The filter gets sharper—the drop from 1 to 0 is steeper—but computing $A\boldsymbol{x}$ takes longer.

Averages and Differences and Wavelets

Lowpass filters are *running averages*. Highpass filters are *running differences*. That word "running" means that a window moves along the vector $\boldsymbol{x}$. Suppose the window lets only 3 components of the signal $\boldsymbol{x}$ show through. The lowpass filter multiplies them by $\frac{1}{4}, \frac{1}{2}, \frac{1}{4}$ and adds, to find the averaged signal $A\boldsymbol{x}$. A highpass filter could alternate those signs to produce $-\frac{1}{4}, \frac{1}{2}, -\frac{1}{4}$. Then A is taking second differences instead of averages of averages.

The moving window creates a convolution = Toeplitz matrix A or D:

$$(Ax)_n = \frac{1}{4}x_{n-1} + \frac{1}{2}x_n + \frac{1}{4}x_{n+1} \qquad (Dx)_n = -\frac{1}{4}x_{n-1} + \frac{1}{2}x_n - \frac{1}{4}x_{n+1}$$

The idea of "wavelets" is to use both filters: averages A and differences D.
Downsample the outputs: Delete $(Ax)_n$ and $(Dx)_n$ for odd n. Keep half-signals.
Now send the half-length signal $(Ax)_{2n}$ through both filters A, D. **Downsample again.**

Wavelet transform with downsampling **A for low frequencies, D for high frequencies**	$(x)_n \longrightarrow (Ax)_{2n} \longrightarrow (AAx)_{4n}$ $\searrow (Dx)_{2n} \searrow (DAx)_{4n}$

The total signal length is unchanged by its wavelet transform $(A^2x)_{4n}, (DAx)_{4n}, (Dx)_{2n}$. But the signal is separated into frequency blocks low-low, high-low, and high. We compress those pieces separately—and the high frequencies are compressed most.

To invert the wavelet transform, go backwards in the flow chart. Upsample each piece of the transform by inserting zeros. Then reverse the arrows and assemble the original signal $(x)_n$ from the blocks AAx, DAx, Dx of its transform:

Inverse wavelet transform	$(AAx)_{4n} \longrightarrow (Ax)_{2n} \longrightarrow (x)_n$ $(DAx)_{4n} \nearrow (Dx)_{2n} \nearrow$

This process uses carefully chosen filters A and D to reduce the length of a signal without losing the information that we want to see and hear. The best wavelet transforms are adjusted so that A and D produce orthogonal matrices or symmetric matrices. The Daubechies 4-coefficient filters are a favorite, with these diagonals in A and D:

$$a_k = 1+\sqrt{3} \quad 3+\sqrt{3} \quad 3-\sqrt{3} \quad 1-\sqrt{3} \quad d_k = \sqrt{3}-1 \quad 3-\sqrt{3} \quad -3-\sqrt{3} \quad \sqrt{3}+1$$

As always, finite Toeplitz matrices with these diagonal entries have to be adjusted at the boundaries. Toeplitz has *zeros* outside the matrix, but good wavelets often use *reflections*.

Problem Set IV.5

1 Show that equations (2) and (3) are the same when $F(\lambda) = \log \lambda$.

2 Suppose $F(\lambda) = \lambda^2$. Then Szegö's Theorem (3) gives the limit of the average eigenvalue of A^2.

(a) Show by squaring A with diagonals a_{-1}, a_0, a_1 that the symbol of A^2 is $(A(\theta))^2$.

(b) Integrating that polynomial $(A(\theta))^2$ from 0 to 2π produces its constant term. What is that term for the $-1, 2, -1$ matrix ?

The $A = LU$ factorization of the $-1, 2, -1$ symmetric second difference matrix ($n = 4$) is

$$\begin{bmatrix} 2 & -1 & & \\ -1 & 2 & -1 & \\ & -1 & 2 & -1 \\ & & -1 & 2 \end{bmatrix} = \begin{bmatrix} 1 & & & \\ -1/2 & 1 & & \\ & -2/3 & 1 & \\ & & -3/4 & 1 \end{bmatrix} \begin{bmatrix} 2/1 & -1 & & \\ & 3/2 & -1 & \\ & & 4/3 & -1 \\ & & & 5/4 \end{bmatrix}$$

3 Verify that $(LU)_{44} = 2$ as required and $\det A = 5 = n + 1$.

4 These factors of the $-1, 2, -1$ Toeplitz matrix are not Toeplitz matrices. But as n increases the last row of L and the last column of U nearly end with $-1, 1$. Verify that the limit symbols $(-e^{-i\theta}+1)$ and $(-e^{i\theta}+1)$ multiply to give the correct symbol of A.

5 The symbol $S - 2e^{i\theta} - 2e^{-i\theta}$ factors into $2 - e^{i\theta}$ times $2 - e^{-i\theta}$. When the symmetric Toeplitz matrix S with diagonals $-2, 5, -2$ is factored into $S = A^{\mathrm{T}}A$ with upper triangular A, what would you expect the last column to approach as $n \to \infty$? (A will only have two nonzero diagonals.)

6 Use the Cholesky command A = chol(S) with the $-2, 5 - 2$ matrix S in Problem 5, to verify that the last row and column of A^{T} and A approach the predicted limits.

IV.6 Graphs and Laplacians and Kirchhoff's Laws

A graph consists of a set of nodes and a set of edges between those nodes. This is the most important model for discrete applied mathematics simple, useful, and general. That word *discrete* is used in contrast to *continuous*: we have vectors instead of functions, we take differences and sums instead of derivatives and integrals, we depend on linear algebra instead of calculus.

Start with the **incidence matrix** of the graph. With m edges and n nodes, the incidence matrix A is m by n. Row i of A corresponds to edge i in the graph. If that edge goes from node j to node k, then *row i of A has -1 in column j and $+1$ in column k.* So each row of A adds to zero and $(1, 1, \ldots, 1)$ is in the nullspace.

The nullspace of A contains all constant vectors $\boldsymbol{x} = (c, c, \ldots, c)$.

We assume the graph is connected—if there is no edge from node j to node k, there is at least a path of edges connecting them. Here are the dimensions of the four subspaces:

$$\textbf{dim N}(A) = 1 \quad \textbf{dim C}(A) = \textbf{dim C}(A^{\mathrm{T}}) = n - 1 \quad \textbf{dim N}(A^{\mathrm{T}}) = m - n + 1$$

The constant vector $\mathbf{1} = (1, 1, \ldots, 1)$ is the simplest choice for the nullspace basis. Then the row space contains all vectors $\boldsymbol{x}$ with $x_1 + x_2 + \cdots + x_n = 0$ (so $\boldsymbol{x}$ is orthogonal to $\mathbf{1}$). To find **bases** for all four subspaces, we can use trees and loops:

- $\mathbf{C}(A^{\mathrm{T}})$ $n - 1$ rows of A that produce a tree in the graph (a tree has no loops)
- $\mathbf{C}(A)$ the first $n - 1$ columns of A (or any $n - 1$ columns of A)
- $\mathbf{N}(A^{\mathrm{T}})$ flows around the $m - n + 1$ small loops in the graph: see equation (3)

If orthogonal bases are desired, choose the right and left singular vectors in $A = U\Sigma V^{\mathrm{T}}$.

Example from Section I.3 $m = 5$ **edges and** $n = 4$ **nodes**

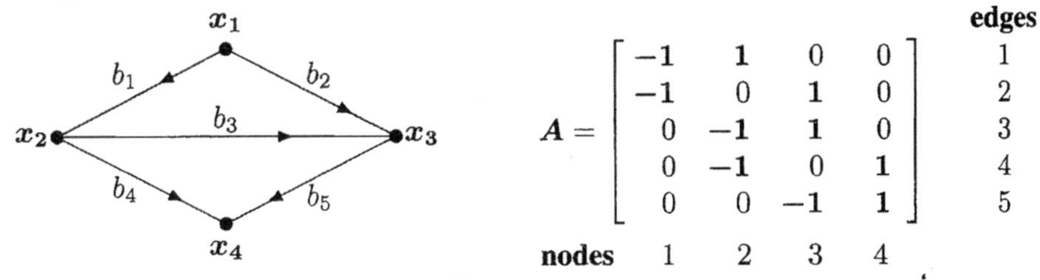

$$A = \begin{bmatrix} -1 & 1 & 0 & 0 \\ -1 & 0 & 1 & 0 \\ 0 & -1 & 1 & 0 \\ 0 & -1 & 0 & 1 \\ 0 & 0 & -1 & 1 \end{bmatrix} \begin{matrix} 1 \\ 2 \\ 3 \\ 4 \\ 5 \end{matrix} \quad \text{edges; nodes } 1\ 2\ 3\ 4$$

The **graph Laplacian matrix $L = A^{\mathrm{T}} A$** is square and symmetric and *positive semidefinite.*

$A^{\mathrm{T}} A$ has $n - 1$ positive eigenvalues $\lambda = \sigma^2$ and one zero eigenvalue (because $A\mathbf{1} = \mathbf{0}$). The special form $\boldsymbol{A^{\mathrm{T}} A = D - B}$ stands out in our example with 5 edges and 4 nodes:

Laplacian **One missing edge**	$$A^{\mathrm{T}} A = \begin{bmatrix} \mathbf{2} & -1 & -1 & \mathbf{0} \\ -1 & \mathbf{3} & -1 & -1 \\ -1 & -1 & \mathbf{3} & -1 \\ \mathbf{0} & -1 & -1 & \mathbf{2} \end{bmatrix} = D - B$$

The **degree matrix is** $D = \text{diag}(2,3,3,2)$. It counts the edges into nodes $1,2,3,4$. The **adjacency matrix** B has entries 0 and 1. An edge from j to k produces $b_{jk}=1$. A **complete graph** (all edges present) has $D = (n-1)\,I$ and $B =$ all ones minus I.

If every pair of nodes is connected by an edge, the graph is **complete**. It will have $m = (n-1) + (n-2) + \cdots + 1 = \frac{1}{2}n(n-1)$ edges. B is the all-ones matrix (minus I). All degrees are $n-1$, so $D = (n-1)I$. At the other extreme, there are only $n-1$ edges. In this case there are *no loops* in the connected graph: the graph is a *tree*. The number m of edges in any connected graph is between $m = n-1$ and $m = \frac{1}{2}n(n-1)$.

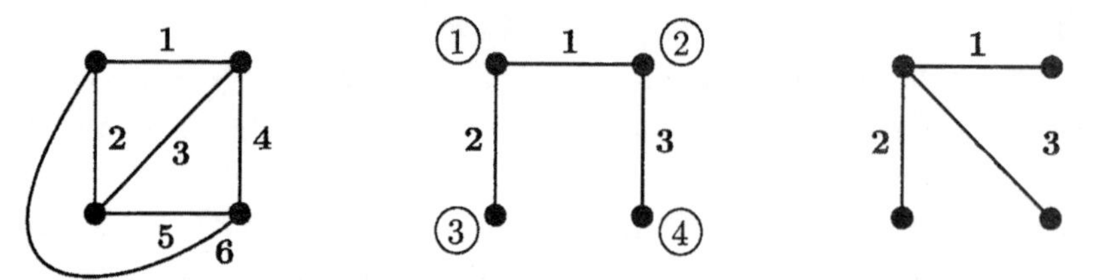

Figure IV.4: Complete graph and two trees, all with $n = 4$ nodes: $m = 6$ or 3 edges. The middle figure shows the numbering of the four nodes in all three graphs.

Those graphs are connected to linear algebra by their m by n incidence matrices A, A_2, A_3

$$\begin{bmatrix} -1 & 1 & 0 & 0 \\ -1 & 0 & 1 & 0 \\ 0 & -1 & 1 & 0 \\ 0 & -1 & 0 & 1 \\ 0 & 0 & -1 & 1 \\ -1 & 0 & 0 & 1 \end{bmatrix} \qquad \begin{bmatrix} -1 & 1 & 0 & 0 \\ -1 & 0 & 1 & 0 \\ 0 & -1 & 0 & 1 \end{bmatrix} \qquad \begin{bmatrix} -1 & 1 & 0 & 0 \\ -1 & 0 & 1 & 0 \\ -1 & 0 & 0 & 1 \end{bmatrix}$$

first tree incidence matrix A_2   **second tree incidence matrix A_3**

Our convention is that -1 comes before $+1$ in each row. But these are not directed graphs! Flows on the edges (the *currents* y_1 *to* y_m) can be positive or negative. And all information about a graph (its nodes and edges) is revealed by its incidence matrix.

The incidence matrix has n columns when the graph has n nodes. Those column vectors add up to the zero vector. Say that in a different way: The all-ones vector $\boldsymbol{x} = (1,1,1,1)$ is in the nullspace of all three incidence matrices. The nullspace of A is a single line through that all-ones vector. $A\boldsymbol{x} = \mathbf{0}$ requires $x_1 = x_2 = x_3 = x_4$ so that $\boldsymbol{x} = (c,c,c,c)$.

$$\boxed{A\boldsymbol{x} = \mathbf{0}} \qquad \begin{array}{rrrrl} -x_1 & +x_2 & & & = 0 \quad \textbf{(then } \boldsymbol{x_1 = x_2}\textbf{)} \\ -x_1 & & +x_3 & & = 0 \quad \textbf{(then } \boldsymbol{x_1 = x_3}\textbf{)} \\ & -x_2 & +x_3 & & = 0 \\ & -x_2 & & +x_4 & = 0 \quad \textbf{(then } \boldsymbol{x_2 = x_4}\textbf{)} \\ & & -x_3 & +x_4 & = 0 \\ -x_1 & & & +x_4 & = 0 \end{array} \qquad (1)$$

Kirchhoff's Current Law

The equation $Ax = \mathbf{0}$ is not very interesting. All its solutions are constant vectors. The equation $A^{\mathrm{T}}\boldsymbol{y} = \mathbf{0}$ is extremely interesting: a central equation of applied mathematics. We need to see what it means and to find a full set of $n - m + 1$ **independent solutions**.

Kirchhoff's Current Law

KCL is $A^{\mathrm{T}}\boldsymbol{y} = \mathbf{0}$

$$\begin{bmatrix} -1 & -1 & 0 & 0 & 0 & -1 \\ 1 & 0 & -1 & -1 & 0 & 0 \\ 0 & 1 & 1 & 0 & -1 & 0 \\ 0 & 0 & 0 & 1 & 1 & 1 \end{bmatrix} \begin{bmatrix} y_1 \\ y_2 \\ y_3 \\ y_4 \\ y_5 \\ y_6 \end{bmatrix} = \begin{bmatrix} 0 \\ 0 \\ 0 \\ 0 \end{bmatrix}. \quad (2)$$

First, count the solutions. With $m = 6$ unknown y's and $r = 3$ independent equations, there are $6 - 3$ **independent solutions.** The nullspace of A^{T} has dimension $m - r = \mathbf{3}$. We want to identify a basis for that subspace of $\mathbf{R}^6$.

What is the meaning of the four equations in Kirchhoff's Current Law $A^{\mathrm{T}}\boldsymbol{y} = \mathbf{0}$? Each equation is a balance law for currents going in or out of a node:

KCL = Balance of currents: Flow into each node equals flow out from that node.

At node 4, the last equation in (2) is $y_4 + y_5 + y_6 = 0$. The total net flow into node 4 is zero (or electrons would pile up). This balance of currents or forces or money occurs everywhere in engineering and science and economics. It is the balance equation of equilibrium.

The key to solving $A^{\mathrm{T}}\boldsymbol{y} = \mathbf{0}$ is to look at the small loops in the graph. A loop is a "cycle" of edges—a path that comes back to the start. The first graph in Figure IV.4 has three small loops. Going around those loops are these edges:

loop 1: Forward on edge 2, backward on edges 3 and 1
loop 2: Forward on edges 3 and 5, backward on edge 4
loop 3: Forward on edge 6, backward on edges 5 and 2

Flow around a loop automatically satisfies Kirchhoff's Current Law. At each node in the loop, the flow into the node goes out to the next node. The three loops in the graph produce three independent solutions to $A^{\mathrm{T}}\boldsymbol{y} = \mathbf{0}$. Each $\boldsymbol{y}$ gives six edge currents around a loop:

$$A^{\mathrm{T}}\boldsymbol{y} = \mathbf{0} \quad \text{for} \quad \boldsymbol{y}_1 = \begin{bmatrix} -1 \\ 1 \\ -1 \\ 0 \\ 0 \\ 0 \end{bmatrix} \quad \text{and} \quad \boldsymbol{y}_2 = \begin{bmatrix} 0 \\ 0 \\ 1 \\ -1 \\ 1 \\ 0 \end{bmatrix} \quad \text{and} \quad \boldsymbol{y}_3 = \begin{bmatrix} 0 \\ -1 \\ 0 \\ 0 \\ -1 \\ 1 \end{bmatrix}. \quad (3)$$

There are no more independent solutions even if there are more (larger) loops! The large loop around the whole graph is exactly the sum of the three small loops. So the solution $\boldsymbol{y} = (-1, 0, 0, -1, 0, 1)$ for that outer loop is exactly the sum $\boldsymbol{y}_1 + \boldsymbol{y}_2 + \boldsymbol{y}_3$.

The subspace dimensions lead to a fundamental identity in topology (discovered by Euler):

$$\boxed{\begin{array}{c}\textbf{(Number of nodes)} - \textbf{(Number of edges)} + \textbf{(Number of loops)} = \\ (n) - (m) + (m - n + 1) = 1.\end{array}} \tag{4}$$

The reader will already know that **a tree has no loops**. Our second and third graphs were trees with 4 nodes and 3 edges. Then Euler's count is $(4) - (3) + (0) = 1$. And $A^{\mathrm{T}}\boldsymbol{y} = \mathbf{0}$ has only the solution $\boldsymbol{y} = \mathbf{0}$. The rows of A are independent for every tree.

$$A_2 = \begin{bmatrix} -1 & 1 & 0 & 0 \\ -1 & 0 & 1 & 0 \\ 0 & -1 & 0 & 1 \end{bmatrix} \quad \text{and} \quad A_3 = \begin{bmatrix} -1 & 1 & 0 & 0 \\ -1 & 0 & 1 & 0 \\ -1 & 0 & 0 & 1 \end{bmatrix}.$$

The $A^{\mathrm{T}}CA$ Framework in Applied Mathematics

Graphs are perfect examples for three equations that I see everywhere in engineering and science and economics. Those equations describe a system in *steady state equilibrium.* For flows in an electrical network (currents along the six edges of our first graph) the three equations connect the *voltages* $\boldsymbol{x} = (x_1, x_2, x_3, x_4)$ *at the four nodes* and the *direct currents* $\boldsymbol{y} = (y_1, y_2, y_3, y_4, y_5, y_6)$ *along the six edges.*

Voltage differences across edges	$\boldsymbol{e} = A\boldsymbol{x}$	e_1 = voltage at end node 2 − voltage at start node 1
Ohm's Law on each edge	$\boldsymbol{y} = C\boldsymbol{e}$	current $y_1 = c_1$ times e_1 = (conductance)(voltage)
Kirchhoff's Law with current sources	$\boldsymbol{f} = A^{\mathrm{T}}\boldsymbol{y}$	current sources $\boldsymbol{f}$ into nodes balance the internal currents $\boldsymbol{y}$

Those three equations $\boldsymbol{e} = A\boldsymbol{x}$ and $\boldsymbol{y} = C\boldsymbol{e}$ and $\boldsymbol{f} = A^{\mathrm{T}}\boldsymbol{y}$ combine into one equilibrium equation $\boldsymbol{A^{\mathrm{T}}CAx = f}$. This is the form of so many fundamental laws. The beauty is in the appearance of both A and A^{T}. The result is that the governing matrix $A^{\mathrm{T}}CA$ is symmetric. **$A^{\mathrm{T}}CA$ is positive semidefinite** because $A\boldsymbol{x} = \mathbf{0}$ has the all-ones solution $\boldsymbol{x} = (1, \ldots, 1)$.

With a boundary condition such as $x_4 = 0$ (which grounds node 4 and removes the last column of A) **the reduced matrix $A^{\mathrm{T}}CA$ becomes symmetric positive definite.**

The grounded network has $n - 1 = 3$ unknown voltages ($x_4 = 0$ is known)

The reduced incidence matrix A is now 6 by 3: full rank 3.

The system matrix $A^{\mathrm{T}}CA$ is $(3 \times 6)(6 \times 6)(6 \times 3) = \mathbf{3 \times 3}$

The energy is positive: $\boldsymbol{x}^{\mathrm{T}}A^{\mathrm{T}}CA\boldsymbol{x} = (A\boldsymbol{x})^{\mathrm{T}}C(A\boldsymbol{x}) > 0$ if $\boldsymbol{x} \neq 0$

Now $A^{\mathrm{T}}CA$ is symmetric and invertible and positive definite.

This $A^{\mathrm{T}}CA$ framework is the foundation of my MIT course 18.085 on Computational Science and Engineering. It is the point where linear algebra has an important message for large-scale computations (like the finite element method). The video lectures and textbook emphasize the applications of $A^{\mathrm{T}}A$ and $A^{\mathrm{T}}CA$.

By preserving the symmetric positive definite structure of the governing equations—which are often partial differential equations—the format of $A^{\mathrm{T}}CA$ fits the laws of science. Kirchhoff's Current Law $A^{\mathrm{T}}\boldsymbol{y} = \mathbf{0}$ becomes a model for all balance laws: conservation of charge, balance of forces, zero net income in economics, conservation of mass and energy, continuity of every kind.

The same $A^{\mathrm{T}}CA$ matrix enters in linear regression (least squares applied to $A\boldsymbol{x} = \boldsymbol{b}$).

$A^{\mathrm{T}}A\widehat{\boldsymbol{x}} = A^{\mathrm{T}}\boldsymbol{b}$	Normal equation for the vector $\widehat{\boldsymbol{x}}$ that best fits the data $\boldsymbol{b}$
$A^{\mathrm{T}}CA\widehat{\boldsymbol{x}} = A^{\mathrm{T}}C\boldsymbol{b}$	Least squares weighted by the inverse covariance matrix $C = V^{-1}$
$\min \lVert\boldsymbol{b} - A\boldsymbol{x}\rVert_C^2$	Minimum squared error $(\boldsymbol{b} - A\boldsymbol{x})^{\mathrm{T}}C(\boldsymbol{b} - A\boldsymbol{x})$

Deep learning in the final chapter will again be an optimization problem. Find the weights between each layer of neurons so the learning function F correctly classifies the training data. In the 20th century, when F was linear, this was not so successful. In this century, each neuron also applies a nonlinear activation function like **ReLU**(x) (the larger of 0 and x). Deep learning has now become amazingly powerful.

The overall function F that classifies the data is *continuous and piecewise linear.* Its graph has an astonishing number of small flat pieces. Every application of **ReLU**(x) adds a fold to the graph of F. That fold crosses the other folds to divide feature space into many many pieces. **Please see Section VII.1 on deep neural nets.**

Constructing all these flat pieces provides the mathematical power for deep learning.

The Graph Laplacian Matrix

$K = A^{\mathrm{T}}CA$ is a weighted graph Laplacian—the weights are in C. The standard Laplacian matrix is $G = A^{\mathrm{T}}A$, with unit weights $(C = I)$. Both Laplacians are crucial matrices for theory and applications. The main facts about $A^{\mathrm{T}}A$ and $A^{\mathrm{T}}CA$ apply to every connected graph. K can be a stiffness matrix or a conductance matrix in engineering.

1	Every row and column of G and K adds to zero because $\boldsymbol{x} = (1, \ldots, 1)$ has $A\boldsymbol{x} = \mathbf{0}$.
2	$G = A^{\mathrm{T}}A$ is symmetric because edges go both ways (**undirected graph**).
3	The diagonal entry $(A^{\mathrm{T}}A)_{ii}$ counts the edges meeting at node i : **the degree.**
4	The off-diagonal entry is $(A^{\mathrm{T}}A)_{ij} = -1$ when an edge connects nodes i and j.
5	G and K are **positive semidefinite** but not positive definite (because $A\boldsymbol{x} = \mathbf{0}$ in **1**).

$$A^{\mathrm{T}}A = \text{diagonal} + \text{off-diagonal} = \textbf{degree matrix} - \textbf{adjacency matrix} = D - B.$$

Problem Set IV.6

1 What are the Laplacian matrices $A^{\mathrm{T}}A$ for a triangle graph and a square graph? The incidence matrix A reverses sign if all arrows are reversed—but signs in $A^{\mathrm{T}}A$ don't depend on arrows.

2 What is $A^{\mathrm{T}}A$ for a complete graph (all ten edges between $n = 5$ nodes)?

3 For a triangle graph with weights c_1, c_2, c_3 on edges $1 \to 2,\ 1 \to 3,\ 2 \to 3$, show by matrix multiplication that

$$K = A^{\mathrm{T}}CA = \begin{bmatrix} c_1 + c_2 & -c_1 & -c_2 \\ -c_1 & c_1 + c_3 & -c_3 \\ -c_2 & -c_3 & c_2 + c_3 \end{bmatrix}$$

4 That matrix $K = A^{\mathrm{T}}CA$ is the sum of $m = 3$ **"element matrices"**:

$$K = c_1 \begin{bmatrix} 1 & -1 & 0 \\ -1 & 1 & 0 \\ 0 & 0 & 0 \end{bmatrix} + c_2 \begin{bmatrix} 1 & 0 & -1 \\ 0 & 0 & 0 \\ -1 & 0 & 1 \end{bmatrix} + c_3 \begin{bmatrix} 0 & 0 & 0 \\ 0 & 1 & -1 \\ 0 & -1 & 1 \end{bmatrix}$$

Show that those rank-1 matrices come from $K = A^{\mathrm{T}}(CA) =$ **columns times rows.**

5 Draw a tree with $n = 4$ nodes and $m = 3$ edges. There should be $m - n + 1 = 0$ solutions to the current law $A^{\mathrm{T}}\boldsymbol{w} = \mathbf{0}$. Explain this conclusion: Rows of A that correspond to a tree in the graph are *independent.*

6 A complete graph with $n = 4$ nodes and $m = 6$ edges apparently can't be drawn in a plane. Can you prove (after experiment) that edges will intersect?

7 (a) For that complete graph with 4 nodes and 6 edges, find the matrix $A^{\mathrm{T}}A$.

(b) Also find $6 - 4 + 1$ solutions (from loops) to Kirchhoff's Law $A^{\mathrm{T}}\boldsymbol{w} = \mathbf{0}$.

8 Explain Euler's formula (the beginning of topology) for any graph in a plane:

$$(\text{number of nodes}) - (\text{number of edges}) + (\text{number of small loops}) = 1$$

9 For a triangle graph, find the eigenvalues and eigenvectors of $G = A^{\mathrm{T}}A$. The eigenvectors are not completely determined because G has a repeated __________. Find one choice for the SVD: $A = U\Sigma V^{\mathrm{T}}$.

IV.7 Clustering by Spectral Methods and k-means

How to understand a graph with many nodes (and not all possible edges)? An important starting point is to separate the nodes into two or more **clusters**—like groups of friends. Edges are more likely within the clusters than between different clusters. We hope the clusters have similar size so that we are not just picking off a few loners. Identifying these clusters will give a first rough understanding of our graph.

A key step in decoding genetic data is to *cluster genes* that show highly correlated (and sometimes anti-correlated!) expression levels. Clustered genes may lie in the same cellular pathway. The great achievement of the Human Genome project was to tell us the pieces in the puzzle of life: the rows of G. We now face the greater problem of fitting those pieces together to produce *function*: such as creating proteins.

An Example with Two Clusters

The figure below shows $n = 5$ nodes. Those nodes are separated into $k = 2$ clusters. The points marked by $*$ are the centroids $(2,1)$ and $(-1,2/3)$ of the two clusters. The first centroid is the average $\frac{1}{2}(1,1) + \frac{1}{2}(3,1)$ of the two points, and the second centroid is the average $\frac{1}{3}[(0,0) + (-3,0) + (0,2)]$. Those centroids c_1 and c_2 minimize the sum of squared distances $||c - a_j||^2$ to the points a_j in the clusters.

These clusters were produced by the famous k-means algorithm with $k = 2$. This is a simple way to cluster the nodes—but not the only way. (And probably not the fastest or best way for a large set of nodes.) Before using eigenvalues and cuts to produce clusters, we show how k-means is one more excellent example of a central theme in this book:

$$\textbf{Approximate an } m \times n \textbf{ matrix } A \textbf{ by } CR = (m \times k)(k \times n) \tag{1}$$

The rank of CR is low because C has only k columns and R has k rows. In the k-means approximation, *the columns of C are the centroids of the clusters*. What is R?

Each column of R has a single 1 and $k-1$ zeros. More exactly, $R_{ij} = 1$ (or 0) if centroid i is closest (or not) to the point x_j. Then the 1's in row i of R tell us the cluster of nodes around the centroid (marked by $*$) in column i of C.

For 5 nodes and 2 clusters, R has only two different columns (centroids) in $A \approx CR$.

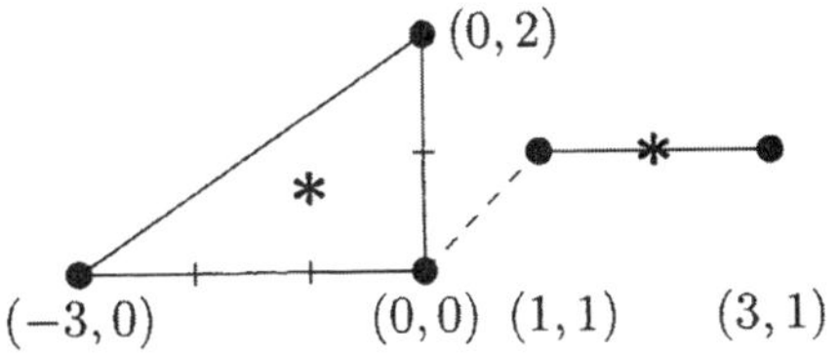

$(0,2)$ $(-3,0)$ $(0,0)$ $(1,1)$ $(3,1)$

$$\begin{bmatrix} 0 & 1 & 3 & 0 & -3 \\ 0 & 1 & 1 & 2 & 0 \end{bmatrix} \approx \begin{bmatrix} -1 & 2 & 2 & -1 & -1 \\ 2/3 & 1 & 1 & 2/3 & 2/3 \end{bmatrix}$$

$$A \approx CR = \begin{bmatrix} -1 & 2 \\ 2/3 & 1 \end{bmatrix} \begin{bmatrix} 1 & 0 & 0 & 1 & 1 \\ 0 & 1 & 1 & 0 & 0 \end{bmatrix}$$

Four Methods for Clustering

Out of many applications, we start with this one: **to break a graph in two pieces.** Those pieces are clusters of nodes. Most edges should be inside one of the clusters.

1. Each cluster should contain roughly half of the nodes.

2. The number of edges between clusters should be relatively small.

For load balancing in high performance computing, we are assigning equal work to two processors (with small communication between them). For social networks we are identifying two distinct groups. We are segmenting an image. We are reordering rows and columns of a matrix to make the off-diagonal blocks sparse.

Many algorithms have been and will be invented to partition a graph. I will focus on four successful methods that extend to more difficult problems: **Spectral clustering** (using the graph Laplacian or the modularity matrix), **minimum cut, and weighted k-means.** Here are those four methods:

I. Find the **Fiedler vector** $\boldsymbol{z}$ that solves $A^{\mathrm{T}}CA\boldsymbol{z} = \lambda D\boldsymbol{z}$. The matrix $A^{\mathrm{T}}CA$ is the graph Laplacian. Its diagonal D contains the total weights on edges into each of the nodes. *D normalizes the Laplacian.* The Fiedler vector has

The eigenvector for $\lambda_1 = 0$ is $(1, \ldots, 1)$. The Fiedler eigenvalue comes next: $\lambda = \lambda_2$.

Positive and negative components of its eigenvector indicate the two clusters of nodes.

II. Replace the graph Laplacian matrix $A^{\mathrm{T}}CA$ by the **modularity matrix** M. Choose the eigenvector that comes with the largest eigenvalue of M. Again it will be the positive and negative components that indicate the two clusters:

$$\textbf{Modularity matrix} \qquad M = (\text{adjacency matrix}) - \frac{1}{2m}\boldsymbol{d}\boldsymbol{d}^{\mathrm{T}}$$

The vector $\boldsymbol{d}$ gives the degrees of the n nodes (the number of edges adjacent to the nodes). Each row and column of $M = M^{\mathrm{T}}$ adds to zero, so one eigenvector of M is again $(1, 1, \ldots, 1)$. If its eigenvalue $\lambda = 0$ happens to be the largest—so M has no positive eigenvalues— then all nodes will and should go into one cluster.

The article by Mark Newman in PNAS **103** (2006) 8577-8582 makes a strong case for the modularity matrix in clustering the nodes.

III. Find the **minimum normalized cut** that separates the nodes in two clusters P and Q. The unnormalized measure of a cut is the **sum of edge weights w_{ij}** across that cut. Those edges connect a node in P to a node outside P:

$$\textbf{Weight across cut} \quad \textit{links}(P) = \sum w_{ij} \ \ \textbf{for } i \textbf{ in } P \textbf{ and } j \textbf{ not in } P. \qquad (2)$$

By this measure, a minimum cut could have no nodes in P. *So we normalize by the sizes of P and Q.* These are sums of weights inside clusters:

$$\textbf{Size of cluster} \qquad \textit{size}(P) = \sum w_{ij} \ \ \textbf{for } i \textbf{ in } P. \qquad (3)$$

Note that an edge inside P is counted twice, as w_{ij} and w_{ji}. The unweighted size would just count the nodes, and lead to "ratio cut." Here we divide weight across the cut by the weighted sizes of P and Q, to normalize the key quantity *Ncut*:

Normalized cut weight $$Ncut(P,Q) = \frac{links(P)}{size(P)} + \frac{links(Q)}{size(Q)}. \tag{4}$$

Shi and Malik found that minimizing *Ncut*(P, Q) gives a good partitioning of the graph. That application was to segmentation of images. They uncovered the connection to the Laplacian L.

The definition of *Ncut* extends from two clusters P and Q to k clusters $P_1, \ldots, P_k$:

Normalized K-cut $$Ncut(P_1, \ldots, P_k) = \sum_{i=1}^{K} \frac{links(P_i)}{size(P_i)} \tag{5}$$

We are coming close to **k-means clustering**. Start with $k = 2$ clusters (P and Q).

IV. k-means Represent the nodes in the graph as vectors $\boldsymbol{a}_1, \ldots, \boldsymbol{a}_n$. The clusters P and Q have centers $\boldsymbol{c}_P$ and $\boldsymbol{c}_Q$. We minimize the total squared distance from nodes to those "centroids".

2-means clustering
$\boldsymbol{c}_P, \boldsymbol{c}_Q$ = centroids **Minimize** $$E = \sum_{i \text{ in } P} ||\boldsymbol{a}_i - \boldsymbol{c}_P||^2 + \sum_{i \text{ in } Q} ||\boldsymbol{a}_i - \boldsymbol{c}_Q||^2 \tag{6}$$

The centroid is the *average* $\boldsymbol{c}_P = (\sum \boldsymbol{a}_i)/|P|$ of the vectors in cluster P.

The vector $\boldsymbol{a}_i$ may or may not represent the physical location of node i. So the clustering objective E is not restricted to Euclidean distance. The more general **kernel k-means** algorithm works entirely with a kernel matrix K that assigns inner products $K_{ij} = \boldsymbol{a}_i^{\mathrm{T}} \boldsymbol{a}_j$. Distances and means are computed from a weighted K.

The distance measure E will also be weighted, to improve the clusters P and Q.

The Normalized Laplacian Matrix

The first step to L is $A^{\mathrm{T}}A$. This A is the m by n incidence matrix of the graph. Off the diagonal, the i, j entry of $A^{\mathrm{T}}A$ is -1 if an edge connects nodes i and j. The diagonal entries make all row sums zero. Then $(A^{\mathrm{T}}A)_{ii}$ = number of edges into node i = degree of node i. With all weights equal to one, $A^{\mathrm{T}}A$ = *degree matrix* − *adjacency matrix*.

The edge weights in C can be conductances or spring constants or edge lengths. They appear on and off the diagonal of $\boldsymbol{A^{\mathrm{T}}CA = D - W}$ = **node weight matrix − edge weight matrix**. Off the diagonal, the entries of $-W$ are minus the weights w_{ij}. The diagonal entries d_i still make all row sums zero: $\boldsymbol{D} = \textbf{diag}(\textbf{sum}(\boldsymbol{W}))$.

The all-ones vector $\mathbf{1} = \mathbf{ones}(n,1)$ is in the nullspace of $A^{\mathrm{T}}CA$, because $A\mathbf{1} = \mathbf{0}$. Each row of A has 1 and -1. Equivalently, $D\mathbf{1}$ cancels $W\mathbf{1}$ (all row sums are zero). The next eigenvector is like the lowest vibration mode of a drum, with $\lambda_2 > 0$.

For the **normalized weighted Laplacian**, multiply $A^{\mathrm{T}}CA$ on the left and right by $D^{-1/2}$, preserving symmetry. Row i and column j are divided by $\sqrt{d_i}$ and $\sqrt{d_j}$, so the i,j entry of $A^{\mathrm{T}}CA$ is divided by $\sqrt{d_i d_j}$. Then L has $d_i/d_i = 1$—along its main diagonal.

Normalized Laplacian L **Normalized weights n_{ij}**	$\boldsymbol{L = D^{-1/2}\, A^{\mathrm{T}}CA\, D^{-1/2} = I - N \quad n_{ij} = \dfrac{w_{ij}}{\sqrt{d_i d_j}}}$

(7)

A triangle graph has $n = 3$ nodes and $m = 3$ edge weights $c_1, c_2, c_3 = w_{12}, w_{13}, w_{23}$:

$$\underset{\boldsymbol{A^{\mathrm{T}}CA = D - W}}{\begin{bmatrix} w_{12}+w_{13} & -w_{12} & -w_{13} \\ -w_{21} & w_{21}+w_{23} & -w_{23} \\ -w_{31} & -w_{32} & w_{31}+w_{32} \end{bmatrix}} \qquad \underset{\boldsymbol{L = D^{-1/2}A^{\mathrm{T}}CA\,D^{-1/2}}}{L = \begin{bmatrix} 1 & -n_{12} & -n_{13} \\ -n_{21} & 1 & -n_{23} \\ -n_{31} & -n_{32} & 1 \end{bmatrix}} \tag{8}$$

The normalized Laplacian $L = I - N$ is like a *correlation matrix* in statistics, with unit diagonal. Three of its properties are crucial for clustering:

1. L is symmetric positive semidefinite: orthogonal eigenvectors, eigenvalues $\lambda \geq 0$.

2. The eigenvector for $\lambda = 0$ is $\mathbf{u} = (\sqrt{d_1}, \ldots, \sqrt{d_n})$. Then $L\mathbf{u} = D^{-1/2}A^{\mathrm{T}}CA\mathbf{1} = 0$.

3. The second eigenvector $\boldsymbol{v}$ of L minimizes the **Rayleigh quotient** on a subspace:

λ_2 = smallest nonzero eigenvalue of L **Minimize subject to $x^{\mathrm{T}}u = 0$**	$\boldsymbol{\min \dfrac{x^{\mathrm{T}}Lx}{x^{\mathrm{T}}x} = \dfrac{v^{\mathrm{T}}Lv}{v^{\mathrm{T}}v} = \lambda_2 \ \text{at} \ x = v}$

(9)

The quotient $\boldsymbol{x}^{\mathrm{T}}L\boldsymbol{x}/\boldsymbol{x}^{\mathrm{T}}\boldsymbol{x}$ gives an upper bound for λ_2, for any vector $\boldsymbol{x}$ orthogonal to the first eigenvector $D^{1/2}\mathbf{1}$. A good lower bound on λ_2 is more difficult to find.

Normalized versus Unnormalized

The algorithms of clustering could use the unnormalized matrix $A^{\mathrm{T}}CA$. But L usually gives better results. The connection between them is $Lv = D^{-1/2}A^{\mathrm{T}}CAD^{-1/2}v = \lambda v$. With $z = D^{-1/2}v$ this has the simple and important form $A^{\mathrm{T}}CAz = \lambda Dz$:

Normalized Fiedler vector z	$A^{\mathrm{T}}CAz = \lambda Dz$ **with** $\mathbf{1}^{\mathrm{T}}Dz = 0.$

(10)

For this "generalized" eigenvalue problem, the eigenvector for $\lambda = 0$ is still the all-ones vector $\mathbf{1} = (1, \ldots, 1)$. The next eigenvector $\boldsymbol{z}$ is D-orthogonal to $\mathbf{1}$, which means $\mathbf{1}^{\mathrm{T}}D\,z =$

0 (Section I.10). By changing $\boldsymbol{x}$ to $D^{1/2}\boldsymbol{y}$, the Rayleigh quotient will find that second eigenvector $\boldsymbol{z}$:

$$\textbf{Same eigenvalue } \lambda_2 \quad \textbf{Fiedler } z = D^{-1/2}\,v \qquad \min_{\mathbf{1}^{\mathrm{T}} D\boldsymbol{y}=0} \frac{\boldsymbol{y}^{\mathrm{T}} A^{\mathrm{T}} C A \boldsymbol{y}}{\boldsymbol{y}^{\mathrm{T}} D \boldsymbol{y}} = \frac{\sum\sum w_{ij}(y_i - y_j)^2}{\sum d_i\, y_i^2} = \lambda_2 \text{ at } \boldsymbol{y} = \boldsymbol{z}. \tag{11}$$

In $A\boldsymbol{y}$, the incidence matrix A gives the differences $y_i - y_j$. C multiplies them by w_{ij}.

Note *For some authors, the Fiedler vector* $\boldsymbol{v}$ is an eigenvector of $A^{\mathrm{T}}CA$. We prefer $\boldsymbol{z} = D^{-1/2}\boldsymbol{v}$. Then $A^{\mathrm{T}}CA\boldsymbol{z} = \lambda_2 D\boldsymbol{z}$. Experiments seem to give similar clusters from $\boldsymbol{v}$ and $\boldsymbol{z}$. Those weighted degrees d_i (the sum of edge weights into node i) have normalized the ordinary $A^{\mathrm{T}}CA$ eigenvalue problem, to improve the clustering.

Why would we solve an eigenvalue problem $L\boldsymbol{v} = \lambda\boldsymbol{v}$ (usually expensive) as a first step in reordering a linear system $A\boldsymbol{x}=\boldsymbol{b}$? One answer is that we don't need an accurate eigenvector $\boldsymbol{v}$. A "hierarchical" *multilevel method* combines nodes to give a smaller L and a satisfactory $\boldsymbol{v}$. The fastest k-means algorithms coarsen the graph level by level, and then adjust the coarse clustering during the refinement phase.

Example 1 A 20-node graph has two built-in clusters P and Q (to find from $\boldsymbol{z}$). The MATLAB code creates edges within P and within Q, with probability 0.7. Edges between nodes in P and Q have smaller probability 0.1. All edges have weights $w_{ij} = 1$, so $C = I$. P and Q are obvious from the graph but not from its adjacency matrix W.

With $G = A^{\mathrm{T}}A$, the eigenvalue command $[V, E] = \textbf{eig}(G, D)$ solves $A^{\mathrm{T}}A\boldsymbol{x} = \lambda D\boldsymbol{x}$. Sorting the λ's leads to λ_2 and its Fiedler vector $\boldsymbol{z}$. Des Higham's third graph shows how the components of $\boldsymbol{z}$ fall into two clusters (plus and minus), to give a good reordering. He provided this MATLAB code.

```
N = 10; W = zeros(2*N, 2*N);                % Generate 2N nodes in two clusters
rand('state', 100)                          % rand repeats to give the same graph
for i = 1:2*N-1
   for j = i+1:2*N
      p = 0.7-0.6 * mod(j-i, 2);            % p = 0.1 when j - i is odd, 0.7 else
      W(i, j) = rand < p;                   % Insert edges with probability p
   end                                      % The weights are w_ij = 1 (or zero)
end                                         % So far W is strictly upper triangular
W = W + W'; D = diag(sum(W));               % Adjacency matrix W, degrees in D
G = D-W; [V, E] = eig(G, D);                % Eigenvalues of Gx = λDx in E
[a, b] = sort(diag(E)); z = V(:, b(2));     % Fiedler eigenvector z for λ2
plot(sort(z), '.-');                        % Show + - groups of Fiedler components
```

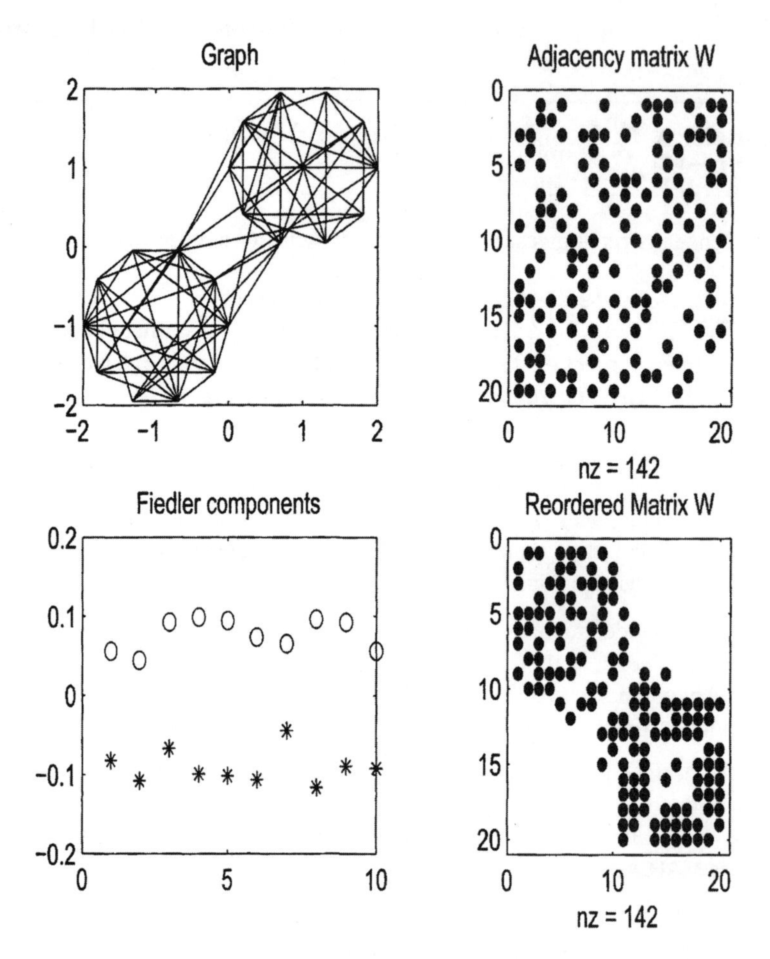

Application to Microarray Data

Microarray data comes as a matrix M from m genes and n samples. Its entries m_{ij} record the activity (expression level) of gene i in sample j. The n by n weight matrix $M^{\mathrm{T}}M$ measures the similarity between samples (the nodes in a complete graph).

The off-diagonal entries of $M^{\mathrm{T}}M$ enter W. The row sums of W go into D. Then $D - W$ is the weighted Laplacian matrix $A^{\mathrm{T}}CA$. We solve $A^{\mathrm{T}}CAz = \lambda Dz$.

Higham, Kalna, and Kibble report tests on three data sets. Those involve leukemia ($m = 5000$ genes, $n = 38$ patients), brain tumors ($m = 7129$, $n = 40$), and lymphoma. "The normalized spectral algorithm is far superior to the unnormalized version at revealing biologically relevant information."

The experiments also show how the *next eigenvector after Fiedler* helps to produce $k = 3$ clusters. The k lowest eigenvalues provide eigenvectors to identify k clusters.

Cuts Connected to Eigenvectors

How is the graph cut separating P from Q related to the Fiedler eigenvector in $A^{\mathrm{T}}CAz = \lambda Dz$? The crucial link comes from comparing *Ncut*(P,Q) in (5) with the Rayleigh quotient $y^{\mathrm{T}}A^{\mathrm{T}}CAy/y^{\mathrm{T}}Dy$ in (11). The perfect indicator of a cut would be a vector y with all components equal to p or $-q$ (two values only):

Two values Node i goes in P if $y_i = p$ Node i goes in Q if $y_i = -q$

$1^{\mathrm{T}}Dy$ will multiply one group of d_i by p and the other group by $-q$. The first d_i add to *size*(P) = sum of w_{ij} (i in P) = sum of d_i (i in P). The second group of d_i adds to *size*(Q). **The constraint $1^{\mathrm{T}}Dy = 0$ becomes $p\ size(P) = q\ size(Q)$.**

When we substitute this y into the Rayleigh quotient, we get exactly $Ncut(P,Q)$! The differences $y_i - y_j$ are zero inside P and Q. They are $p+q$ across the cut:

$$\textbf{Numerator} \quad y^{\mathrm{T}}A^{\mathrm{T}}CAy = \sum\sum w_{ij}\,(y_i - y_j)^2 = (p+q)^2\, links(P,Q) \tag{12}$$

$$\textbf{Denominator} \quad y^{\mathrm{T}}Dy = p^2\, size(P) + q^2\, size(Q) = p\,(p\, size(P)) + q\,(p\, size(P)). \tag{13}$$

That last step used $p\ size(P) = q\ size(Q)$. Cancel $p+q$ in the quotient:

$$\begin{matrix}\text{Rayleigh}\\ \text{quotient}\end{matrix} \quad \frac{(p+q)\ links(P,Q)}{p\ size(P)} = \frac{p\ links(P,Q)}{p\ size(P)} + \frac{q\ links(P,Q)}{q\ size(Q)} = Ncut(P,Q). \tag{14}$$

The *Ncut* problem is the same as the eigenvalue problem, with the extra constraint that y has only two values. (This problem is NP-hard: there are so many choices of P and Q.) The Fiedler vector z will not satisfy this two-value condition. But its components in that specially good example clearly separated into two groups. Clustering by z is a success if we can make it efficient.

Clustering by k-means

The most basic problem begins with n points $a_1, \ldots, a_n$ in d-dimensional space. The goal is to partition those points into k clusters. The clusters $P_1, \ldots P_k$ have centroids $c_1, \ldots, c_k$. *Each centroid* c *minimizes the total distance* $\sum ||c - a_j||^2$ to points a_j in its cluster. The centroid is the mean (the average) of those n_j points:

$$\textbf{Centroid of } P_j \quad c_j = \frac{\text{sum of } a\text{'s}}{\text{number of } a\text{'s}} \quad \text{minimizes } \sum ||c - a||^2 \text{ for } a\text{'s in cluster } P_j.$$

The goal is to find the partition $P_1, \ldots, P_k$ with **minimum total distance D to centroids**:

$$\boxed{\textbf{Clustering} \quad \text{Minimize } D = D_1 + \cdots + D_k = \sum ||c_j - a_i||^2 \text{ for } a_i \text{ in } P_j.} \tag{15}$$

Key idea Each clustering into $P_1, \ldots, P_k$ of the nodes produces k centroids (step 1). **Each set of centroids produces a clustering** (step 2), where $\boldsymbol{a}$ moves into P_j if $\boldsymbol{c}_j$ is the closest centroid to $\boldsymbol{a}$. (In case of equally close centroids, choose one arbitrarily.) The classical "*batch k-means algorithm*" iterates from a clustering to its centroids to a new clustering. In eqn. (1) it is a factorization $A = CR$ by alternating least squares!

***k*-means**	**1.** Find the centroids $\boldsymbol{c}_j$ of the (old) clustering $P_1, \ldots, P_k$.
	2. Find the (new) clustering that puts $\boldsymbol{a}$ in P_j if $\boldsymbol{c}_j$ is the closest centroid.

Each step reduces the total distance D. We reset the centroids $\boldsymbol{c}_j$ for each P_j, and then we improve to new P_j around those $\boldsymbol{c}_j$. Since D decreases at both steps, the k-means algorithm converges. But it might not converge to the global minimum.

It is hard to know much about the limit clusters. Non-optimal partitions can give local minima. Better partitions come from weighted distances.

Step 1 is the more expensive, to compute all the distances $||\boldsymbol{c}_j - \boldsymbol{a}_i||^2$. The complexity is normally $O(n^2)$ per iteration. When the algorithm is extended below to kernel k-means, generating a kernel matrix K from the data can cost $O(n^2 d)$.

Step 2 is the "Voronoidal idea" of finding the set closest to each centroid.

Weights and the Kernel Method

When we introduce weights in the distances, they appear in the centroids:

$$\textbf{Distances} \quad d(\boldsymbol{x}, \boldsymbol{a}_i) = w_i \, ||\boldsymbol{x} - \boldsymbol{a}_i||^2 \quad \textbf{Centroid of } P_j \quad \boldsymbol{c}_j = \frac{\sum w_i \, \boldsymbol{a}_i}{\sum w_i} \quad (\boldsymbol{a}_i \text{ in } P_j) \quad (16)$$

The weighted distance $D_j = \sum w_i \, ||\boldsymbol{x} - \boldsymbol{a}_i||^2$ is minimized by $\boldsymbol{x} = \boldsymbol{c}_j$ in step 1. To reduce the total $D = D_1 + \cdots + D_k$, step 2 resets the clusters. Each $\boldsymbol{a}_i$ goes with the closest centroid. Then iterate step 1 (new centroids) and step 2 (new clusters).

A key point is that distances to centroids only require dot products $\boldsymbol{a}_i \cdot \boldsymbol{a}_j$:

$$\textbf{Each } \boldsymbol{i} \textbf{ in } \boldsymbol{P_j} \qquad ||\boldsymbol{c}_j - \boldsymbol{a}_i||^2 = \boldsymbol{c}_j \cdot \boldsymbol{c}_j - 2\boldsymbol{c}_j \cdot \boldsymbol{a}_i + \boldsymbol{a}_i \cdot \boldsymbol{a}_i \qquad (17)$$

Kernel method The weighted kernel matrix K has entries $\boldsymbol{a}_i \cdot \boldsymbol{a}_\ell$. Those vectors $\boldsymbol{a}_i$ need not be actual positions in space. Each application can map the nodes of the graph to vectors $\boldsymbol{a}_i$ in a linear or *nonlinear* way, by its own rule. When the nodes are points $\boldsymbol{x}_i$ in input space, their representing vectors $\boldsymbol{a}_i = \phi(\boldsymbol{x}_i)$ can be points in a high-dimensional **feature space**. Three kernels are commonly used:

In vision	**Polynomial**	$K_{i\ell} = (\boldsymbol{x}_i \cdot \boldsymbol{x}_\ell + c)^d$
In statistics	**Gaussian**	$K_{i\ell} = \exp(-||\boldsymbol{x}_i - \boldsymbol{x}_\ell||^2 / 2\sigma^2)$
In neural networks	**Sigmoid**	$K_{i\ell} = \tanh(c\,\boldsymbol{x}_i \cdot \boldsymbol{x}_\ell + \theta)$

The distance in (17) needs only the kernel matrix because of the centroid formula (16).

$$\textbf{Sum over nodes in } P_j \quad \sum \|c_j - a_i\|^2 = \frac{\sum\sum w_i w_\ell K_{i\ell}}{(\sum w_i)^2} - 2\frac{\sum w_i K_{i\ell}}{\sum w_i} + \sum K_{ii}. \quad (18)$$

The **kernel batch k-means algorithm** uses the matrix K to compute this total distance.

For large data sets, k-means and **eig**$(A^{\mathrm{T}}CA, D)$ will be expensive. Here are two approaches that create a sequence of more manageable problems. **Random sampling** finds the best partition for a sample of the nodes. Use its centroids to partition all nodes, by assignment to the nearest centroid. Sampling has become a major research direction, aiming to prove that with high probability the partition is good.

Dhillon's **graclus** code uses **multilevel clustering**: graph coarsening, then clustering at the base level, and then refinement. Coarsening forms supernodes with the sum of edge weights. For the small supergraph at base level, spectral clustering or recursive 2-means will be fast. **This multilevel approach is like algebraic multigrid.**

Applications of Clustering

The reason for this section is the wide variety of applications. Here is a collection that goes far beyond clustering. This part of applied mathematics has grown very quickly.

1. Learning theory, training sets, neural networks, Hidden Markov Models
2. Classification, regression, pattern recognition, Support Vector Machines
3. Statistical learning, maximum likelihood, Bayesian statistics, spatial statistics, kriging, time series, ARMA models, stationary processes, prediction
4. Social networks, small world networks, six degrees of separation, organization theory, probability distributions with heavy tails
5. Data mining, document indexing, semantic indexing, word-document matrix, image retrieval, kernel-based learning, Nystrom method, low rank approximation
6. Bioinformatics, microarray data, systems biology, protein homology detection
7. Cheminformatics, drug design, ligand binding, pairwise similarity, decision trees
8. Information theory, vector quantization, rate distortion theory, Bregman divergences
9. Image segmentation, computer vision, texture, min cut, normalized cuts
10. Predictive control, feedback samples, robotics, adaptive control, Riccati equations.

Problem Set IV.7

1 If the graph is a line of 4 nodes, and all weights are 1 $(C = I)$, the best cut is down the middle. Find this cut from the $\pm$ components of the Fiedler vector $\boldsymbol{z}$:

$$A^{\mathrm{T}}CA\,\boldsymbol{z} = \begin{bmatrix} 1 & -1 & & \\ -1 & 2 & -1 & \\ & -1 & 2 & -1 \\ & & -1 & 1 \end{bmatrix} \begin{bmatrix} z_1 \\ z_2 \\ z_3 \\ z_4 \end{bmatrix} = \lambda_2 \begin{bmatrix} 1 & & & \\ & 2 & & \\ & & 2 & \\ & & & 1 \end{bmatrix} \begin{bmatrix} z_1 \\ z_2 \\ z_3 \\ z_4 \end{bmatrix} = \lambda_2 D\boldsymbol{z}.$$

Here $\boldsymbol{\lambda_2 = \frac{1}{2}}$. Solve for $\boldsymbol{z}$ by hand, and check $[\,1 \;\; 1 \;\; 1 \;\; 1\,]\,D\boldsymbol{z} = 0$.

2 For the same 4-node tree, compute $links(P)$ and $size(P)$ and $Ncut(P,Q)$ for the cut down the middle.

3 Starting from the same four points $1,2,3,4$ find the centroids $\boldsymbol{c_P}$ and $\boldsymbol{c_Q}$ and the total distance D for the clusters $P = \{1,2\}$ and $Q = \{3,4\}$. The k-means algorithm will not change P and Q when it assigns the four points to nearest centroids.

4 Start the k-means algorithm with $P = \{1,2,4\}$ and $Q = \{3\}$. Find the two centroids and reassign points to the nearest centroid.

5 For the clusters $P = \{1,2,3\}$ and $Q = \{4\}$, the centroids are $c_P = 2$ and $c_Q = 4$. Resolving a tie the wrong way leaves this partition with no improvement. But find its total distance D.

6 If the graph is a 2 by 4 mesh of 8 nodes, with weights $C = I$, use $\mathsf{eig}(A^{\mathrm{T}}A, D)$ to find the Fiedler vector $\boldsymbol{z}$. The incidence matrix A is 10 by 8 and $D = \mathsf{diag}(\mathsf{diag}(A^{\mathrm{T}}A))$. What clusters come from the $\pm$ components of $\boldsymbol{z}$?

7 Use the Fiedler code with probabilities narrowed from $p = 0.1$ and 0.7 to $p = 0.5$ and 0.6. Compute z and plot the graph and its partition.

Problems 8-11 are about the graph with nodes $\boldsymbol{(0,0), (1,0), (3,0), (0,4), (0,8)}$.

8 Which clusters P and Q maximize the minimum distance D^* between them?

9 Find those best clusters by the *greedy algorithm*. Start with five clusters, and combine the two closest clusters. What are the best k clusters for $k = 4,3,2$?

10 The **minimum spanning tree** is the shortest group of edges that connects all nodes. There will be $n-1$ edges and no loops, or the total length will not be minimal.

Dijkstra's algorithm Start with any node like $(0,0)$. At each step, include the shortest edge that connects a new node to the partial tree already created.

11 The minimum spanning tree can also be found by *greedy inclusion*. With the edges in increasing order of length, *keep each edge unless it completes a loop*.

IV.8 Completing Rank One Matrices

Filling up missing entries in a rank-1 matrix is directly connected to finding (or not finding) cycles in a graph. This theory of rank-1 completion developed from the following question:

We are given $m + n - 1$ nonzero entries in an m by n matrix A.

When does the requirement rank $(A) = 1$ *determine all the other entries?*

The answer depends on the *positions* of those $m+n-1$ nonzeros. Here are three examples:

$$A_1 = \begin{bmatrix} \times & \times & \times \\ \times & & \\ \times & & \end{bmatrix} \qquad A_2 = \begin{bmatrix} \times & \times & \\ \times & \times & \\ & & \times \end{bmatrix} \qquad A_3 = \begin{bmatrix} \times & \times & & \\ & \times & \times & \\ \times & & \times & \\ & & & \times \end{bmatrix}$$

success **failure** **failure**

In A_1, we are given column 1 with no zeros. Columns 2 and 3 must be multiples of column 1, if A has rank 1. Since we are given the first entries in columns 2 and 3, those columns are completely determined.

Here is another approach to A_1. In any rank 1 matrix, *every* 2 *by* 2 *determinant must be zero.* So the 2, 2 entry of A_1 is decided by $a_{22}a_{11} = a_{12}a_{21}$.

In A_2, the first four entries might not satisfy determinant $= 0$. Then we are doomed to failure. If that determinant *is* zero, we could choose any $a_{31} \neq 0$ in column 1, and complete A_2 to rank 1. This is the usual situation: no solution or infinitely many solutions.

That example shows failure whenever we know all four entries of a 2 by 2 submatrix. Does every failure occur this way? No, A_3 *has a different failure.*

In A_3, that leading 3 by 3 submatrix has too many specified entries. There are 6 instead of $3 + 3 - 1 = 5$. For most choices of those 6 entries, rank 1 is impossible for a 3 by 3:

$$\begin{bmatrix} 1 & 1 & \\ & 1 & 1 \\ 1 & & 2 \end{bmatrix} \quad \text{leads to} \quad \begin{bmatrix} 1 & 1 & \mathbf{1} \\ \mathbf{1} & 1 & 1 \\ 1 & \mathbf{2} & 2 \end{bmatrix} \quad \text{and the rank is 2.}$$

Spanning Trees in a Graph

Alex Postnikov explained the right way to look at this problem of rank-1 completion. He constructed a graph with m nodes for the m rows and n nodes for the n columns. For each prescribed entry A_{ij}, the graph has an edge connecting row node i to column node j. Then the pattern of $\times$'s in the matrices above becomes a pattern of edges in their row-column graphs (next page).

The A_1 and A_3 examples produce these two graphs:

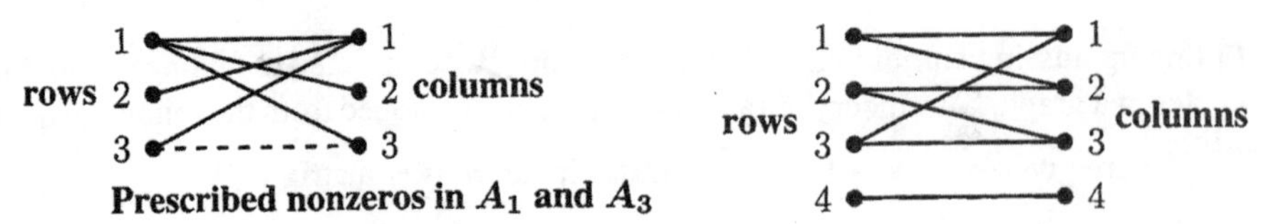

These are **bipartite graphs** because all their edges go from one part to the other part. Success for A_1 and failure for A_3 can be explained by these graphs.

Graph A_1 The 5 edges form a **spanning tree**. It is a *tree* because that graph has no closed loop (**no cycle**). The tree is *spanning* because it connects all 6 nodes. If we want to discover the $3,3$ entry of A_1 (in a rank-1 completion), we add that dotted line edge to complete a cycle. Then A_{33} is determined by A_{11}, A_{13}, and A_{31} in the cycle. Those four numbers produce a zero determinant.

Graph A_3 The 7 edges do *not* form a spanning tree. It is not a tree because there is a cycle (in the top six edges). The cycle imposes a requirement on those six entries of A_3:

If $A_3 = \boldsymbol{uv}^{\mathrm{T}}$ has rank 1, its entries must satisfy $\dfrac{A_{11}A_{22}A_{33}}{A_{12}A_{23}A_{31}} = \dfrac{(u_1v_1)(u_2v_2)(u_3v_3)}{(u_1v_2)(u_2v_3)(u_3v_1)} = 1.$

If this condition happens to hold, there are infinitely many ways to complete A_3 with rank 1.

Conclusion *The partial matrix A has a unique rank-1 completion if and only if the $m+n-1$ prescribed entries A_{ij} produce $m+n-1$ edges (row i to column j) that form a* ***spanning tree*** *in the row-column graph.* **The tree reaches all nodes with no loops.**

Open problem Which $(m+n-2)2$ entries of A can be specified, to allow a unique completion to a rank-2 matrix? Appendix C to this book confirms $(m+n-2)2$ as the correct number of independent parameters for a rank 2 matrix.

Problem Set IV.8

1 Draw the bipartite graph with 3 row and column nodes for the example matrix A_2. Do the 5 edges from A_2 make up a spanning tree?

2 Construct a 5 by 5 matrix A_4 with $5+5-1=9$ nonzeros in a cycle of length 8. What equation like $A_{11}A_{22}A_{33} = A_{12}A_{23}A_{31}$ must hold for completion to a rank-1 matrix?

3 For a connected graph with M edges and N nodes, what requirement on M and N comes from each of the words *spanning tree*?

4 How do you know that a graph with N nodes and $N-1$ edges is a spanning tree?

IV.9 The Orthogonal Procrustes Problem

Here is a neat (and useful) application of the SVD. It starts with vectors $x_1, \ldots, x_n$ and $y_1, \ldots, y_n$. **Which orthogonal matrix Q will multiply the y's to come as close as possible to the x's ?** This question turns up in a surprising number of applications.

Notice the limitation to an orthogonal matrix Q. That doesn't allow for translation and it doesn't allow for rescaling. If the mean of the y's equals the mean of the x's, then no translation is needed. Otherwise most codes will subtract those mean values to achieve new x's and y's with $\sum x_i = \sum y_i =$ **zero vector**. Then the two sets are centered with mean zero. And if we allow rescaling of the vectors to equalize their lengths in advance, that leads to the "generalized Procrustes problem".

> *I have to tell you about Procrustes and the myth.* Procrustes himself was famous for rescaling. He invited passing strangers to spend a comfortable night in his special bed. He claimed that the bed would adjust its length to match the visitor. But the reality was that Procrustes adjusted the length of the visitor to fit the bed. (Short visitors were stretched on the rack, tall visitors had their legs chopped off. I am not sure what this myth tells us about the Greeks, but it isn't good.) Theseus was up to the challenge, and he adjusted Procrustes to fit his own bed. Unfortunately fatal.

They solved the generalized problem and we solve the standard problem: orthogonal Q.

Solution	**1.** Construct matrices X and Y with columns $x_1, \ldots, x_n$ and $y_1, \ldots, y_n$
	2. Form the square matrix $Y^{\mathrm{T}}X$
	3. Find the singular value decomposition $Y^{\mathrm{T}}X = U\Sigma V^{\mathrm{T}}$
	4. The orthogonal matrix $Q = V^{\mathrm{T}}U$ minimizes $\|\|X - YQ\|\|_F^2$

Discussion

The distance between the columns x_k and $y_k Q$ is the usual Euclidean length $||x_k - y_k Q||$. For the (squared) distance from all the x's to all the yQ's, it is natural to add up the squares of those column lengths. This produces the (squared) **Frobenius norm** $||X - YQ||_F^2$. We must show that **$Q = V^{\mathrm{T}}U$ in Step 4 minimizes this norm**— it is the best possible Q.

Three small observations will be helpful in the proof:

(i) The squared Frobenius norm $||A||_F^2$ is the trace of $A^{\mathrm{T}}A$.

(ii) The trace of $A^{\mathrm{T}}B$ equals the trace of $B^{\mathrm{T}}A$ and also the trace of BA^{T}.

(iii) The squared norm $||A||_F^2$ is the same as $||AQ||_F^2$ and $||A^{\mathrm{T}}||_F^2$.

The trace of a square matrix is the sum of the entries on the main diagonal. It is also the sum of the eigenvalues. Then the three observations are easily explained.

(i) The diagonal entries of $A^{\mathrm{T}}A$ are the squared column lengths—so they add to $||A||_F^2$.

(ii) $A^{\mathrm{T}}B$ and its transpose $B^{\mathrm{T}}A$ have the same diagonal—and therefore the same trace. $A^{\mathrm{T}}B$ and BA^{T} have the same nonzero eigenvalues. Their diagonals include all $a_{ij}b_{ij}$.

(iii) AQ has the same column lengths as A, because Q is an orthogonal matrix. Both $||A||_F^2$ and $||A^{\mathrm{T}}||_F^2$ add up the squares of all the entries of A, so they are equal.

Proof that $Q = V^{\mathrm{T}}U$ is the orthogonal matrix that minimizes $||X - YQ||_F^2$.
We minimize trace $(X-YQ)^{\mathrm{T}}(X-YQ) = \text{trace}\,(X^{\mathrm{T}}X)+\text{trace}\,(Y^{\mathrm{T}}Y)-2\,\textbf{trace}\,(\boldsymbol{Q^{\mathrm{T}}Y^{\mathrm{T}}X})$.

$X^{\mathrm{T}}X$ and $Y^{\mathrm{T}}Y$ are fixed. So we *maximize* that last trace. This is the moment for the SVD: $\boldsymbol{Y^{\mathrm{T}}X = U\Sigma V^{\mathrm{T}}}$.

$$\textbf{trace}\,(\boldsymbol{Q^{\mathrm{T}}Y^{\mathrm{T}}X}) = \text{trace}\,(Q^{\mathrm{T}}U\Sigma V^{\mathrm{T}}) = \text{trace}\,(V^{\mathrm{T}}Q^{\mathrm{T}}U\Sigma) = \textbf{trace}\,(\boldsymbol{Z\Sigma}). \qquad (1)$$

The matrix $Z = V^{\mathrm{T}}Q^{\mathrm{T}}U$ is a product of orthogonal matrices and therefore orthogonal. Σ is a diagonal matrix of positive numbers $\sigma_1, \ldots, \sigma_r$. So the trace of that matrix $Z\Sigma$ is $z_{11}\sigma_1 + \cdots + z_{rr}\sigma_r$. To maximize that number, **the best choice is $Z = I$.**

$\boldsymbol{Z = V^{\mathrm{T}}Q^{\mathrm{T}}U = I}$ **means that $Q = UV^{\mathrm{T}}$ solves the Procrustes problem**

In case the original X and Y were orthogonal matrices, the perfect rotation to produce $X = YQ$ (with zero error $X - YQ$) is clearly $Q = Y^{\mathrm{T}}X$. This agrees with the answer $Q = UV^{\mathrm{T}}$. The singular values of an orthogonal matrix are all 1, so the decomposition $Y^{\mathrm{T}}X = U\Sigma V^{\mathrm{T}}$ in equation (1) is exactly UV^{T}.

Notes The Procrustes problem was originally solved in Schöneman's 1964 thesis. *Procrustes Problems* by Gower and Dijksterhuis (Oxford University Press, 2004) develops many of its applications. We have borrowed the proof given above from Golub and Van Loan (*Matrix Computations* 4th edition, page 328). That page also begins the important applications of the SVD to **angles between subspaces.**

Problem Set IV.9

1 Which orthogonal matrix Q minimizes $||X - YQ||_F^2$? Use the solution $Q = UV^{\mathrm{T}}$ above and also minimize that norm as a function of θ (set the θ-derivative to zero):

$$X = \begin{bmatrix} 1 & 2 \\ 2 & 1 \end{bmatrix} \quad Y = \begin{bmatrix} 1 & 0 \\ 0 & 1 \end{bmatrix} \quad Q = \begin{bmatrix} \cos\theta & -\sin\theta \\ \sin\theta & \cos\theta \end{bmatrix}$$

IV.10 Distance Matrices

Suppose n points are at positions x_1 to x_n in d-dimensional space. The n by n distance matrix D contains the squared distances $D_{ij} = ||x_i - x_j||^2$ between pairs of points. Thus D is symmetric. The main diagonal of D has zeros from $||x_i - x_i||^2 = 0$. Here is the key question about extracting information from D:

Can we recover the positions $x_1, \ldots, x_n$ from the Euclidean distance matrix D?

The immediate answer is *no*. Suppose we find one solution—one set of possible positions x_i. Then we could shift those positions by a constant x_0. We could multiply all x's by any orthogonal matrix. Those are **rigid motions** and they don't change the distances $||x_i - x_j||$. Recognizing that one solution will lead to this family of equivalent solutions, the key question remains (and we answer it):

Are there always positions x_1 to x_n consistent with the distance matrix D? *Yes.*

There is always a position matrix X (with columns x_1 to x_n) that produces the given distances in D. The matrix X has d rows when the points are in d-dimensional space ($d = 2$ for a map, $d = 3$ for our world, $d > 3$ is allowed and it happens). One problem is to determine the minimum dimension d.

This problem of finding X from D has a long history. At first this was a purely mathematical question. But applications soon appeared. We mention just three of them:

1. **Wireless sensor networks**: Measuring the travel times between pairs of sensors yields D. Then we solve for the sensor positions X (the network topology).

2. **Shapes of molecules**: Nuclear magnetic resonance gives distances between atoms. Then we know the matrix D. We solve for position matrices X. This example and many others will involve **noise** (errors in D) and even **missing entries**.

3. **Machine learning**: The examples in a training set are converted to *feature vectors* in high dimensions. Those vectors might lie close to a plane or a curved surface that has much lower dimension. Finding that surface (approximately) is a giant step toward understanding the data and classifying new examples. Then the *kernel trick* reduces the dimension. Again this can involve a very noisy environment.

In preparing this section of the book, we relied on a wonderful paper "*Euclidean Distance Matrices*" posted in 2015 to the arXiv: 1502.07541v2 [cs.OH]. Its authors are Ivan Dokmanic, Reza Parhizkar, Juri Ranieri, and Martin Vetterli. This paper is full of ideas, algorithms, codes, and many applications. They are clearly explained! Please find it on the Web to learn more.

Positions X from Distances D

Here is the key observation that simplifies the problem. It connects each (distance)^2 in D to four entries in the matrix $X^{\mathrm{T}}X$ (dot products of the desired vectors $\boldsymbol{x}_i$ and $\boldsymbol{x}_j$):

$$||\boldsymbol{x}_i - \boldsymbol{x}_j||^2 = (\boldsymbol{x}_i - \boldsymbol{x}_j)^{\mathrm{T}}(\boldsymbol{x}_i - \boldsymbol{x}_j) = \boldsymbol{x}_i^{\mathrm{T}}\boldsymbol{x}_i - \boldsymbol{x}_i^{\mathrm{T}}\boldsymbol{x}_j - \boldsymbol{x}_j^{\mathrm{T}}\boldsymbol{x}_i + \boldsymbol{x}_j^{\mathrm{T}}\boldsymbol{x}_j \quad (1)$$

The first term $\boldsymbol{x}_i^{\mathrm{T}}\boldsymbol{x}_i$ produces a matrix with constant rows (no dependence on j). The last term $\boldsymbol{x}_j^{\mathrm{T}}\boldsymbol{x}_j$ produces a matrix with constant columns (no dependence on i). The numbers $||\boldsymbol{x}_i||^2$ and $||\boldsymbol{x}_j||^2$ in both of those matrices are on the main diagonal of $\boldsymbol{G} = \boldsymbol{X}^{\mathrm{T}}\boldsymbol{X}$. Those are the numbers in the column vector $\mathrm{diag}(\boldsymbol{G})$.

The middle terms $-2\boldsymbol{x}_i^{\mathrm{T}}\boldsymbol{x}_j$ in (1) are exactly the numbers in $-2G = -2X^{\mathrm{T}}X$. So we can rewrite (1) as an equation for the matrix D, using the symbol $\mathbf{1}$ for the column vector of n ones. That gives constant columns and $\mathbf{1}^{\mathrm{T}}$ gives constant rows.

$$\boxed{\boldsymbol{D} = \mathbf{1}\,\mathrm{diag}(\boldsymbol{G})^{\mathrm{T}} - 2\boldsymbol{G} + \mathrm{diag}(\boldsymbol{G})\,\mathbf{1}^{\mathrm{T}}.} \quad (2)$$

Our problem is to recover G from D. Then the positions in X will come from $X^{\mathrm{T}}X = G$. You see that a solution X can be multiplied by any orthogonal matrix Q, and still $(QX)^{\mathrm{T}}(QX) = G$. Rotations are expected and allowed by Q.

Solving $X^{\mathrm{T}}X = G$ for the d by n matrix X will tell us that the points $\boldsymbol{x}_1$ to $\boldsymbol{x}_n$ can be placed in the space $\mathbf{R}^d$. The rank of G will be the *spatial dimension*—this is the smallest dimension consistent with the given distance matrix D.

Since two terms in equation (2) are rank 1 matrices, we learn that D has rank at most $d + 2$. Note that our points could all be shifted by a constant vector without changing squared distances in D. So D is an *affine* dimension (shift allowed) instead of a *subspace* dimension.

Now we solve equation (2) for $\boldsymbol{G} = \boldsymbol{X}^{\mathrm{T}}\boldsymbol{X}$. Place the first point at the origin: $\boldsymbol{x}_1 = \mathbf{0}$. Then every $||\boldsymbol{x}_i - \boldsymbol{x}_1||^2$ is just $||\boldsymbol{x}_i||^2$. The first column $\boldsymbol{d}_1$ of D (which is given) is exactly the same as $\mathrm{diag}(X^{\mathrm{T}}X) = \mathrm{diag}(G) = (||\boldsymbol{x}_1||^2, ||\boldsymbol{x}_2||^2, \ldots, ||\boldsymbol{x}_n||^2)$.

$$\mathrm{diag}(G) = \boldsymbol{d}_1 \quad \text{and} \quad \mathrm{diag}(G)\,\mathbf{1}^{\mathrm{T}} = \boldsymbol{d}_1\mathbf{1}^{\mathrm{T}}. \quad (3)$$

Now G comes from D. G will be positive semidefinite provided the distances in D obey the triangle inequality (see Menger: *Amer. J. Math.* **53**; Schoenberg: *Annals Math.* **36**):

$$\boxed{\boldsymbol{X}^{\mathrm{T}}\boldsymbol{X} = \boldsymbol{G} = -\tfrac{1}{2}(\boldsymbol{D} - \mathbf{1}\boldsymbol{d}_1^{\mathrm{T}} - \boldsymbol{d}_1\mathbf{1}^{\mathrm{T}}).} \quad (4)$$

Once we know G we find X from $X^{\mathrm{T}}X = G$. Use elimination on G or use its eigenvalues and eigenvectors. Both of those give a position matrix X with $\boldsymbol{x}_1 = \mathbf{0}$:

If $G = Q\Lambda Q^{\mathrm{T}}$ (eigenvalues and eigenvectors) then X can be $\sqrt{\Lambda}Q^{\mathrm{T}}$

If $G = U^{\mathrm{T}}U$ (elimination = Cholesky factorization) then X can be U (upper triangular)

In both cases we can keep only rank(G) rows in X. The other rows are all zero, coming from zero eigenvalues in Λ or from an early end to elimination.

We remove zero rows from X to see the dimension d of the point set of $\boldsymbol{x}$'s. If the squared distances in D include measurement noise, set small eigenvalues to zero.

This is the classical MDS algorithm: MultiDimensional Scaling with first point $\boldsymbol{x}_1 = \mathbf{0}$.

Centering X or Rotating to Match Anchor Points

Centering: Often we might prefer to place the **centroid** of the $\boldsymbol{x}$'s at the origin. The centroid of $\boldsymbol{x}_1, \ldots, \boldsymbol{x}_n$ is just the average of those vectors:

$$\textbf{Centroid} \qquad c = \frac{1}{n}(\boldsymbol{x}_1 + \cdots + \boldsymbol{x}_n) = \frac{1}{n} X \mathbf{1}. \tag{5}$$

Just multiply any position matrix X by the matrix $I - \frac{1}{n}\mathbf{1}\mathbf{1}^{\mathrm{T}}$ to put the centroid at $\mathbf{0}$.

Anchor points: Possibly a few of the positions have been selected in advance: **N anchor points $\boldsymbol{y}_i$ in a matrix Y**. Those positions may not agree with the computed $\boldsymbol{x}_i$. So we choose the corresponding N columns from the computed position matrix X, and then find the rotation Q that moves those columns closest to Y.

The best orthogonal matrix Q solves the Procrustes problem in Section III.9. It is found from the singular value decomposition $X_N Y^{\mathrm{T}} = U\Sigma V^{\mathrm{T}}$. The orthogonal matrix Q that moves the N positions in X_N closest to the anchor points in Y is $Q = VU^{\mathrm{T}}$.

J. C. Gower, Properties of Euclidean and non-Euclidean distance matrices, *Linear Algebra and Its Applications* **67** (1985) 81-97.

Problem Set IV.10

1 $||\boldsymbol{x}_1 - \boldsymbol{x}_2||^2 = 1$ and $||\boldsymbol{x}_2 - \boldsymbol{x}_3||^2 = 1$ and $||\boldsymbol{x}_1 - \boldsymbol{x}_3||^2 = 6$ will violate the triangle inequality. Construct G and confirm that it is not positive semidefinite: no solution X to $G = X^{\mathrm{T}}X$.

2 $||\boldsymbol{x}_1 - \boldsymbol{x}_2||^2 = 9$ and $||\boldsymbol{x}_2 - \boldsymbol{x}_3||^2 = 16$ and $||\boldsymbol{x}_1 - \boldsymbol{x}_3||^2 = 25$ do satisfy the triangle inequality $3 + 4 > 5$. Construct G and find points $\boldsymbol{x}_1, \boldsymbol{x}_2, \boldsymbol{x}_3$ that match these distances.

3 If all $||\boldsymbol{x}_i - \boldsymbol{x}_j||^2 = 1$ for $\boldsymbol{x}_1, \boldsymbol{x}_2, \boldsymbol{x}_3, \boldsymbol{x}_4$, find G and then X. The points lie in $\mathbf{R}^d$ for which dimension d?

Part V

Probability and Statistics

V.1 Mean, Variance, and Probability

V.2 Probability Distributions

V.3 Moments, Cumulants, and Inequalities of Statistics

V.4 Covariance Matrices and Joint Probabilities

V.5 Multivariate Gaussian and Weighted Least Squares

V.6 Markov Chains

Part V : Probability and Statistics

These subjects have jumped forward in importance. When an output is predicted, we need its probability. When that output is measured, we need its statistics. Those computations were restricted in the past to simple equations and small samples. Now we can solve differential equations for probability distributions (*master equations*). We can compute the statistics of large samples. This chapter aims at a basic understanding of these key ideas :

1 Mean m and variance σ^2 : Expected value and sample value

2 Probability distribution and cumulative distribution

3 Covariance matrix and joint probabilities

4 Normal (Gaussian) distribution : single variable and multivariable

5 Standardized random variable $(x - m)/\sigma$

6 The Central Limit Theorem

7 Binomial distribution and uniform distribution

8 Markov and Chebyshev inequalities (distance from mean)

9 Weighted least squares and Kalman filter : $\widehat{x}$ and its variance

10 Markov matrix and Markov chain

V.1 Mean, Variance, and Probability

We are starting with the three fundamental words of this chapter: *mean, variance, and probability.* Let me give a rough explanation of their meaning before I write any formulas:

The **mean** is the *average value* or expected value

The **variance** σ^2 measures the average *squared distance* from the mean m

The **probabilities** of n different outcomes are positive numbers $p_1, \ldots, p_n$ adding to 1.

Certainly the mean is easy to understand. We will start there. But right away we have two different situations that you have to keep straight. On the one hand, we may have the results (*sample values*) from a completed trial. On the other hand, we may have the expected results (*expected values*) from future trials. Let me give examples of both:

Sample values Five random freshmen have ages $\mathbf{18, 17, 18, 19, 17}$

Sample mean $\frac{1}{5}(18 + 17 + 18 + 19 + 17) = \mathbf{17.8}$

Probabilities The ages in a freshmen class are 17 (**20%**), 18 (**50%**), 19 (**30%**)

A random freshman has **expected age E** [x] $= (0.2)\,17 + (0.5)\,18 + (0.3)\,19 = \mathbf{18.1}$

Both numbers 17.8 and 18.1 are correct averages. The sample mean starts with N samples $x_1, \ldots, x_N$ from a completed trial. Their mean is the *average* of the N observed samples:

$$\textbf{Sample mean} \qquad m = \mu = \frac{1}{N}(x_1 + x_2 + \cdots + x_N) \tag{1}$$

The **expected value of** x starts with the probabilities $p_1, \ldots, p_n$ of the ages $x_1, \ldots, x_n$:

$$\textbf{Expected value} \quad m = \mathrm{E}[x] = p_1 x_1 + p_2 x_2 + \cdots + p_n x_n. \tag{2}$$

This is $\boldsymbol{p} \cdot \boldsymbol{x}$. Notice that $m = \mathrm{E}[x]$ tells us what to expect, $m = \mu$ tells us what we got.

A fair coin has probability $p_0 = \frac{1}{2}$ of tails and $p_1 = \frac{1}{2}$ of heads. Then $\mathrm{E}[x] = \left(\frac{1}{2}\right) 0 + \frac{1}{2}(1)$. The fraction of heads in N flips of the coin is the sample mean, expected to approach $\mathrm{E}[x] = \frac{1}{2}$. By taking many samples (large N), the sample results will come close to the probabilities. The **"Law of Large Numbers"** says that with probability 1, the sample mean will converge to its expected value $\mathrm{E}[x]$ as the sample size N increases.

This does *not* mean that if we have seen more tails than heads, the next sample is likely to be heads. The odds remain 50-50. The first 100 or 1000 flips do affect the sample mean. *But* 1000 *flips will not affect its limit*—because you are dividing by $N \to \infty$.

Variance (around the mean)

The **variance** σ^2 measures expected distance (squared) from the expected mean $\mathrm{E}[x]$. The **sample variance** S^2 measures actual distance (squared) from the actual sample mean. The square root is the **standard deviation** $\boldsymbol{\sigma}$ **or** $\boldsymbol{S}$. After an exam, I email μ and S to the class. I don't know the expected mean and variance because I don't know the probabilities p_1 to p_{100} for each score. (After teaching for 50 years, I still have no idea what to expect.)

The deviation is always deviation *from the mean*—sample or expected. We are looking for the size of the "spread" around the mean value $x = m$. Start with N samples.

$$\textbf{Sample variance} \qquad \boldsymbol{S^2 = \frac{1}{N-1}\left[(x_1 - m)^2 + \cdots + (x_N - m)^2\right]} \tag{3}$$

The sample ages $x = 18, 17, 18, 19, 17$ have mean $m = 17.8$. That sample has variance 0.7 :

$$S^2 = \frac{1}{4}\left[(.2)^2 + (-.8)^2 + (.2)^2 + (1.2)^2 + (-.8)^2\right] = \frac{1}{4}(2.8) = \mathbf{0.7}$$

The minus signs disappear when we compute squares. Please notice ! Statisticians divide by $N - 1 = 4$ (and not $N = 5$) so that S^2 is an unbiased estimate of σ^2. One degree of freedom is already accounted for in the sample mean.

An important identity comes from splitting each $(x - m)^2$ into $x^2 - 2mx + m^2$:

$$\begin{aligned} \text{sum of } (x_i - m)^2 &= (\text{sum of } x_i^2) - 2m(\text{sum of } x_i) + (\text{sum of } m^2) \\ &= (\text{sum of } x_i^2) - 2m(Nm) + Nm^2 \\ \textbf{sum of } \boldsymbol{(x_i - m)^2} &= \boldsymbol{(\textbf{sum of } x_i^2) - Nm^2}. \end{aligned} \tag{4}$$

This is an equivalent way to find $(x_1 - m)^2 + \cdots + (x_N - m^2)$ by adding $x_1^2 + \cdots + x_N^2$. To find the sample variance S^2, divide this by $N - 1$.

Now start with probabilities p_i (never negative !) instead of samples. We find expected values instead of sample values. The variance σ^2 is the crucial number in statistics.

$$\textbf{Variance} \qquad \boldsymbol{\sigma^2 = \mathrm{E}\left[(x - m)^2\right] = p_1(x_1 - m)^2 + \cdots + p_n(x_n - m)^2.} \tag{5}$$

We are squaring the distance from the expected value $m = \mathrm{E}[x]$. We don't have samples, only expectations. We know probabilities but we don't know experimental outcomes.

Example 1 Find the variance σ^2 of the ages of college freshmen.

Solution The probabilities of ages $x_i = 17, 18, 19$ were $p_i = 0.2$ and 0.5 and 0.3. The expected value was $\boldsymbol{m = \sum p_i x_i = 18.1}$. The variance uses those same probabilities :

$$\begin{aligned} \boldsymbol{\sigma^2} &= (0.2)(\mathbf{17 - 18.1})^2 + (0.5)(\mathbf{18 - 18.1})^2 + (0.3)(\mathbf{19 - 18.1})^2 \\ &= (0.2)(\mathbf{1.21}) + (0.5)(\mathbf{0.01}) + (0.3)(\mathbf{0.81}) = \mathbf{0.49}. \end{aligned}$$

The **standard deviation** is the square root $\boldsymbol{\sigma = 0.7}$.
This measures the spread of 17, 18, 19 around $\mathrm{E}[x]$, weighted by probabilities 0.2, 0.5, 0.3.

Equation (4) gives another way to compute the variance σ^2 :

$$\boxed{\sigma^2 = \mathrm{E}\left[x^2\right] - (\mathrm{E}\left[x\right])^2 = \boldsymbol{\sum p_i\, x_i^2 - \left(\sum p_i\, x_i\right)^2}}$$

Continuous Probability Distributions

Up to now we have allowed for n possible outcomes $x_1, \ldots, x_n$. With ages 17, 18, 19, we only had $n = 3$. If we measure age in days instead of years, there will be a thousand possible ages (too many). Better to allow *every number between* 17 *and* 20—a continuum of possible ages. Then the probabilities p_1, p_2, p_3 for ages x_1, x_2, x_3 have to move to a **probability distribution** $\boldsymbol{p(x)}$ for a whole continuous range of ages $17 \le x \le 20$.

The best way to explain probability distributions is to give you two examples. They will be the **uniform distribution** and the **normal distribution**. The first (uniform) is easy. The normal distribution is all-important.

Uniform distribution Suppose ages are uniformly distributed between 17.0 and 20.0. All ages between those numbers are "equally likely". Of course any one exact age has no chance at all. There is zero probability that you will hit the exact number $x = 17.1$ or $x = 17 + \sqrt{2}$. What you can truthfully provide (assuming our uniform distribution) is **the chance** $\boldsymbol{F(x)}$ **that a random freshman has age less than** $\boldsymbol{x}$:

The chance of age less than $x = 17$ is $F(17) = 0$	$x \le 17$ won't happen
The chance of age less than $x = 20$ is $F(20) = 1$	$x \le 20$ will happen
The chance of age less than x is $\boldsymbol{F(x) = \frac{1}{3}(x - 17)}$	$\boldsymbol{F}$ **goes from 0 to 1**

That formula $F(x) = \frac{1}{3}(x - 17)$ gives $F = 0$ at $x = 17$; then $x \le 17$ won't happen. It gives $F(x) = 1$ at $x = 20$; then $x \le 20$ is sure. Between 17 and 20, the graph of the **cumulative distribution** $\boldsymbol{F(x)}$ increases linearly for this uniform model. Let me draw the graphs of $F(x)$ and its derivative $p(x) =$ "probability density function".

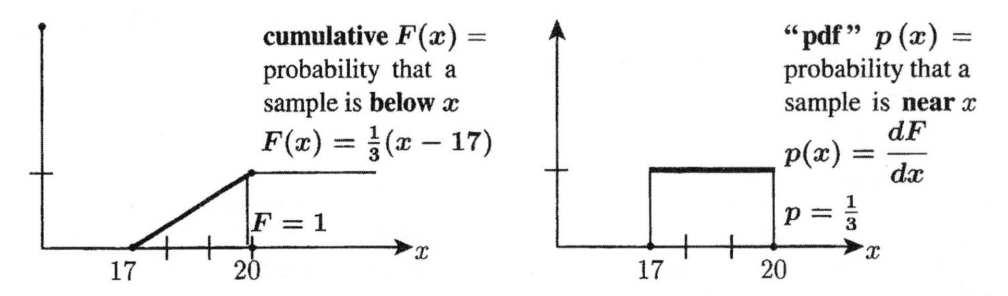

Figure V.1: $F(x)$ is the cumulative distribution and its derivative $p(x) = dF/dx$ is the **probability density function (pdf)**. For this **uniform distribution**, $p(x)$ is constant between 17 and 20. The total area under the graph of $p(x)$ is the total probability $F = 1$.

You could say that $p(x)\,dx$ is the probability of a sample falling in between x and $x + dx$. This is "infinitesimally true": $p(x)\,dx$ is $F(x + dx) - F(x)$. Here is the full connection of $F(x)$ to $p(x)$:

$$\boldsymbol{F = \text{integral of } p} \qquad \textbf{Probability of } \boldsymbol{a \le x \le b = \int_a^b p(x)\,dx = F(b) - F(a)} \qquad (6)$$

$F(b)$ is the probability of $x \le b$. I subtract $F(a)$ to keep $x \ge a$. That leaves $a \le x \le b$.

Mean and Variance of $p(x)$

What are the mean m and variance σ^2 for a probability distribution? Previously we added $p_i x_i$ to get the mean (expected value). With a continuous distribution we **integrate** $xp(x)$:

$$\textbf{Mean} \qquad m = \mathrm{E}[x] = \int x\, p(x)\, dx = \int_{x=17}^{20} (x) \left(\frac{1}{3}\right) dx = 18.5$$

For this uniform distribution, the mean m is halfway between 17 and 20. Then the probability of a random value x below this halfway point $m = 18.5$ is $F(m) = \frac{1}{2}$.

In MATLAB, $x = \text{rand}(1)$ chooses a random number uniformly between 0 and 1. Then the expected mean is $m = \frac{1}{2}$. The interval from 0 to x has probability $F(x) = x$. The interval below the mean m always has probability $F(m) = \frac{1}{2}$.

The variance is the average squared distance to the mean. With N outcomes, σ^2 is the sum of $p_i(x_i - m)^2$. For a continuous random variable x, the sum changes to an **integral**.

$$\textbf{Variance} \qquad \sigma^2 = \mathrm{E}\left[(x-m)^2\right] = \int p(x)\,(x-m)^2\, dx \tag{7}$$

When ages are uniform between $17 \le x \le 20$, the integral can shift to $0 \le x \le 3$:

$$\sigma^2 = \int_{17}^{20} \frac{1}{3}(x-18.5)^2\, dx = \int_{0}^{3} \frac{1}{3}(x-1.5)^2\, dx = \frac{1}{9}(x-1.5)^3\Bigg]_{x=0}^{x=3} = \frac{2}{9}(1.5)^3 = \frac{3}{4}.$$

That is a typical example, and here is the complete picture for a uniform $p(x)$, 0 to a.

Uniform distribution for $0 \le x \le a$

Density $p(x) = \dfrac{1}{a}$ **Cumulative** $F(x) = \dfrac{x}{a}$

$$\textbf{Mean } m = \frac{a}{2} \textbf{ halfway} \qquad \textbf{Variance } \sigma^2 = \int_0^a \frac{1}{a}\left(x - \frac{a}{2}\right)^2 dx = \frac{a^2}{12} \tag{8}$$

The mean is a multiple of a, the variance is a multiple of a^2. For $a = 3$, $\sigma^2 = \frac{9}{12} = \frac{3}{4}$. For one random number between 0 and 1 $\left(\text{mean } \frac{1}{2}\right)$ the variance is $\sigma^2 = \frac{1}{12}$.

Normal Distribution: Bell-shaped Curve

The normal distribution is also called the "Gaussian" distribution. It is the most important of all probability density functions $p(x)$. The reason for its overwhelming importance comes from repeating an experiment and averaging the outcomes. The experiments have their own distribution (like heads and tails). *The average approaches a normal distribution.*

Central Limit Theorem (informal) The average of N samples of "any" probability distribution approaches a normal distribution as $N \to \infty$ (proved in Section V.3).

Start with the "standard normal distribution". It is symmetric around $x = 0$, so its mean value is $m = 0$. It is chosen to have a standard variance $\sigma^2 = 1$. It is called $\mathbf{N}(0, 1)$.

$$\textbf{Standard normal distribution} \quad p(x) = \frac{1}{\sqrt{2\pi}} e^{-x^2/2}. \tag{9}$$

The graph of $p(x)$ is the **bell-shaped curve** in Figure V.2. The standard facts are

$$\textbf{Total probability} = \mathbf{1} \qquad \int_{-\infty}^{\infty} p(x)\, dx = \frac{1}{\sqrt{2\pi}} \int_{-\infty}^{\infty} e^{-x^2/2}\, dx = \mathbf{1}$$

$$\textbf{Mean E}\,[x] = \mathbf{0} \qquad m = \frac{1}{\sqrt{2\pi}} \int_{-\infty}^{\infty} x e^{-x^2/2}\, dx = \mathbf{0}$$

$$\textbf{Variance E}\,[x^2] = \mathbf{1} \qquad \sigma^2 = \frac{1}{\sqrt{2\pi}} \int_{-\infty}^{\infty} (x-0)^2 e^{-x^2/2}\, dx = \mathbf{1}$$

The zero mean was easy because we are integrating an odd function. Changing x to $-x$ shows that "integral $= -$ integral". So that integral must be $m = 0$.

The other two integrals apply the idea in Problem 12 to reach 1. Figure V.2 shows a graph of $p(x)$ for the normal distribution $\mathbf{N}(0, \sigma)$ and also its cumulative distribution $F(x) =$ integral of $p(x)$. From the symmetry of $p(x)$ you see *mean = zero*. From $F(x)$ you see a very important practical approximation for opinion polling:

The probability that a random sample falls between $-\sigma$ and σ is $\boldsymbol{F(\sigma) - F(-\sigma) \approx \frac{2}{3}}$.

This is because $\int_{-\sigma}^{\sigma} p(x)\, dx$ equals $\int_{-\infty}^{\sigma} p(x)\, dx - \int_{-\infty}^{-\sigma} p(x)\, dx = F(\sigma) - F(-\sigma)$.

Similarly, the probability that a random x lies between -2σ and 2σ ("*less than two standard deviations from the mean*") is $F(2\sigma) - F(-2\sigma) \approx 0.95$. If you have an experimental result further than 2σ from the mean, it is fairly sure to be not accidental: chance $= 0.05$. Drug tests may look for a tighter confirmation, like probability 0.001. Searching for the Higgs boson used a hyper-strict test of 5σ deviation from pure accident.

The normal distribution with any mean m and standard deviation σ comes by shifting and stretching the standard $\mathbf{N}(0, 1)$. **Shift x to $x - m$. Stretch $x - m$ to $(x - m)/\sigma$.**

$$\begin{array}{l}\textbf{Gaussian density } p(x) \\ \textbf{Normal distribution N}(m, \sigma)\end{array} \qquad p(x) = \frac{1}{\sigma\sqrt{2\pi}} e^{-(x-m)^2/2\sigma^2} \tag{10}$$

The integral of $p(x)$ is $F(x)$—the probability that a random sample will fall below x. The differential $p(x)\,dx = F(x+dx) - F(x)$ is the probability that a random sample will fall between x and $x + dx$. There is no simple formula to integrate $e^{-x^2/2}$, so this cumulative distribution $F(x)$ is computed and tabulated very carefully.

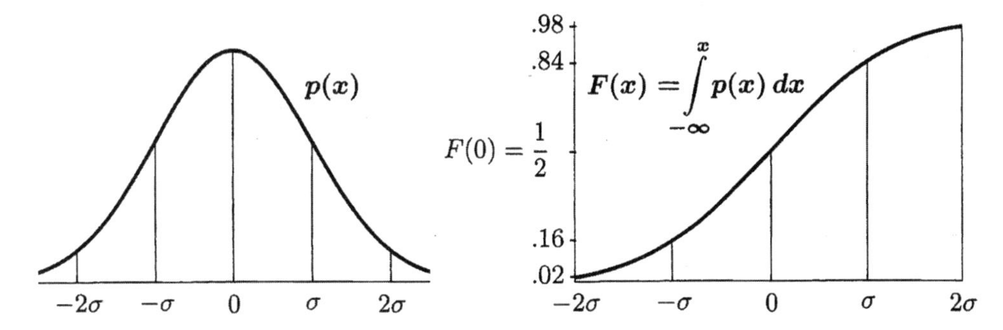

Figure V.2: The standard normal distribution $p\,(x)$ has mean $m = 0$ and $\sigma = 1$.

N Coin Flips and $N \to \infty$

Example 2 **Suppose x is 1 or -1 with equal probabilities $p_1 = p_{-1} = \frac{1}{2}$.**

The mean value is $m = \frac{1}{2}(1) + \frac{1}{2}(-1) = 0$. The variance is $\sigma^2 = \frac{1}{2}(1)^2 + \frac{1}{2}(-1)^2 = 1$.

The key question is the *average* $A_N = (x_1 + \cdots + x_N)/N$. The independent x_i are ± 1 and we are dividing their sum by N. The expected mean of A_N is still zero. The law of large numbers says that this sample average approaches zero with probability 1. How fast does A_N approach zero? **What is its variance σ_N^2 ?**

$$\text{By linearity} \quad \sigma_N^2 = \frac{\sigma^2}{N^2} + \frac{\sigma^2}{N^2} + \cdots + \frac{\sigma^2}{N^2} = N\frac{\sigma^2}{N^2} = \frac{1}{N} \quad \text{since } \sigma^2 = 1. \tag{11}$$

Example 3 **Change outputs from 1 or -1 to $x = 1$ or $x = 0$. Keep $p_1 = p_0 = \frac{1}{2}$.**

The new mean value $m = \frac{1}{2}$ falls halfway between 0 and 1. The variance moves to $\sigma^2 = \frac{1}{4}$:

$$m = \frac{1}{2}(1) + \frac{1}{2}(0) = \frac{1}{2} \quad \text{and} \quad \sigma^2 = \frac{1}{2}\left(1 - \frac{1}{2}\right)^2 + \frac{1}{2}\left(0 - \frac{1}{2}\right)^2 = \frac{1}{4}.$$

$$\text{The average } A_N \text{ now has mean } \frac{1}{2} \text{ and variance } \frac{1}{4N^2} + \cdots + \frac{1}{4N^2} = \frac{1}{4N} = \sigma_N^2. \tag{12}$$

This σ_N is half the size of σ_N in Example 2. This must be correct because the new range 0 to 1 is half as long as -1 to 1. Examples 2-3 are showing a law of linearity.

The new 0 − 1 variable x_{new} is $\frac{1}{2}\,x_{\text{old}} + \frac{1}{2}$. So the mean m is increased to $\frac{1}{2}$ and the variance is *multiplied* by $\left(\frac{1}{2}\right)^2$. A shift changes m and the rescaling changes σ^2.

Linearity $\quad x_{\text{new}} = a x_{\text{old}} + b$ has $m_{\text{new}} = a m_{\text{old}} + b$ and $\sigma^2{}_{\text{new}} = a^2 \sigma^2{}_{\text{old}}$ (13)

Here are the results from three numerical tests: random 0 or 1 averaged over N trials.
[**48** 1's from $N =$ **100**] [**5035** 1's from $N =$ **10000**] [**19967** 1's from $N =$ **40000**].
The standardized X $= (x - m)/\sigma = \left(A_N - \frac{1}{2}\right) / \, 2\sqrt{N}$ was [**−.40**] [**.70**] [**−.33**].

The Central Limit Theorem says that the average of many coin flips will approach a normal distribution. Let us begin to see how that happens: **binomial approaches normal.**

The "binomial" probabilities $p_0, \ldots, p_N$ count the number of heads in N coin flips.

For each (fair) flip, the probability of heads is $\frac{1}{2}$. For $N = 3$ flips, the probability of heads all three times is $\left(\frac{1}{2}\right)^3 = \frac{1}{8}$. The probability of heads twice and tails once is $\frac{3}{8}$, from three sequences HHT and HTH and THH. These numbers $\frac{1}{8}$ and $\frac{3}{8}$ are pieces of $\left(\frac{1}{2} + \frac{1}{2}\right)^3 = \frac{1}{8} + \frac{3}{8} + \frac{3}{8} + \frac{1}{8} = 1$. *The average number of heads in 3 flips is* 1.5.

$$\textbf{Mean}\ m = (3 \text{ heads})\frac{1}{8} + (2 \text{ heads})\frac{3}{8} + (1 \text{ head})\frac{3}{8} + 0 = \frac{3}{8} + \frac{6}{8} + \frac{3}{8} = \mathbf{1.5\ heads}$$

With N flips, Example 3 (or common sense) gives a mean of $m = \Sigma x_i p_i = \frac{1}{2}N$ heads.

The variance σ^2 is based on the *squared distance* from this mean $N/2$. With $N = 3$ the variance is $\sigma^2 = \frac{3}{4}$ (*which is* $N/4$). To find σ^2 we add $(x_i - m)^2\,p_i$ with $m = 1.5$:

$$\sigma^2 = (3-1.5)^2\,\frac{1}{8} + (2-1.5)^2\,\frac{3}{8} + (1-1.5)^2\,\frac{3}{8} + (0-1.5)^2\,\frac{1}{8} = \frac{9+3+3+9}{32} = \frac{3}{4}.$$

For any N, the variance for a binomial distribution is $\sigma_N^2 = N/4$. Then $\sigma_N = \sqrt{N}/2$.

Figure V.3 shows how the probabilities of 0, 1, 2, 3, 4 heads in $N = 4$ flips come close to a bell-shaped Gaussian. That Gaussian is centered at the mean value $m = N/2 = 2$. To reach the standard Gaussian (mean 0 and variance 1) we shift and rescale that graph. If x is the number of heads in N flips—the average of N zero-one outcomes—then x is shifted by its mean $m = N/2$ and rescaled by $\sigma = \sqrt{N}/2$ to produce the standard X :

Shifted and scaled $\quad X = \dfrac{x - m}{\sigma} = \dfrac{x - \frac{1}{2}N}{\sqrt{N}/2} \quad (N = 4 \text{ has } X = x - 2)$

Subtracting m is "centering" or "detrending". The mean of X is zero.

Dividing by σ is "normalizing" or "standardizing". The variance of X is 1.

It is fun to see the Central Limit Theorem giving the right answer at the center point $X = 0$. At that point, the factor $e^{-X^2/2}$ equals 1. We know that the variance for N coin flips is $\sigma^2 = N/4$. The center of the bell-shaped curve has height $1/\sqrt{2\pi\sigma^2} = \sqrt{\mathbf{2/N\pi}}$.

What is the height at the center of the coin-flip distribution p_0 to p_N (the binomial distribution)? For $N = 4$, the probabilities for $0, 1, 2, 3, 4$ heads come from $\left(\frac{1}{2} + \frac{1}{2}\right)^4$.

$$\textbf{Center probability } \frac{\mathbf{6}}{\mathbf{16}} \qquad \left(\frac{1}{2} + \frac{1}{2}\right)^4 = \frac{1}{16} + \frac{4}{16} + \frac{\mathbf{6}}{\mathbf{16}} + \frac{4}{16} + \frac{1}{16} = 1.$$

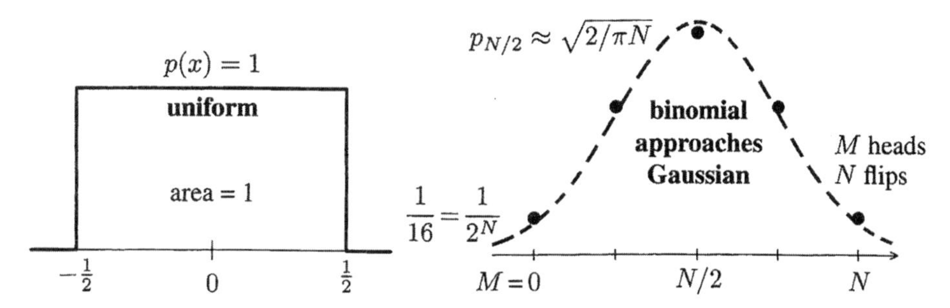

Figure V.3: The probabilities $p = (1, 4, 6, 4, 1)/16$ for the number of heads in 4 flips. These p_i approach a Gaussian distribution with variance $\sigma^2 = N/4$ centered at $m = N/2$. For X, the Central Limit Theorem gives convergence to the normal distribution $\mathbf{N}(0, 1)$.

The binomial theorem in Problem 8 tells us the center probability $p_{N/2}$ for any even N:

$$\boxed{\text{The center probability } \left(\frac{N}{2} \text{ heads, } \frac{N}{2} \text{ tails}\right) \text{ is } \quad \frac{1}{2^N} \frac{N!}{(N/2)!\,(N/2)!}}$$

For $N = 4$, those factorials produce $4!/2!\,2! = 24/4 = 6$. For large N, Stirling's formula $\sqrt{2\pi N}(N/e)^N$ is a close approximation to $N!$. Use this formula for N and twice for $N/2$:

$$\begin{array}{c}\textbf{Limit of coin-flip}\\ \textbf{Center probability}\end{array} \qquad p_{N/2} \approx \frac{1}{2^N} \frac{\sqrt{2\pi N}(N/e)^N}{\pi N (N/2e)^N} = \frac{\sqrt{2}}{\sqrt{\pi N}} = \frac{1}{\sqrt{2\pi}\sigma}. \tag{14}$$

The last step used the variance $\sigma^2 = N/4$ for coin-tossing. The result $1/\sqrt{2\pi}\sigma$ matches the center value (above) for the Gaussian. The Central Limit Theorem is true:

The centered binomial distribution approaches the normal distribution $p(x)$ as $N \to \infty$.

Monte Carlo Estimation Methods

Scientific computing has to work with errors in the data. Financial computing has to work with unsure numbers and predictions. So much of applied mathematics meets this problem: **accepting uncertainty in the inputs and estimating the variance in the outputs.**

How to estimate that variance? Often probability distributions $p(x)$ are not known. What we can do is to try different inputs $\boldsymbol{b}$ and compute the outputs x and take an average. This is the simplest form of a **Monte Carlo method** (named after the gambling palace on the Riviera, where I once saw a fight about whether the bet was placed in time). *Monte Carlo approximates an expected value* $\mathrm{E}[x]$ *by a sample average* $(x_1+\cdots+x_N)/N$.

Please understand that every x_k can be expensive to compute. We are not flipping coins. Each sample comes from a set of data b_k. *Monte Carlo randomly chooses this data* b_k*, it computes the outputs* x_k*, and then it averages those* x*'s.* Decent accuracy for $\mathrm{E}[x]$ often requires many samples b and huge computing cost. The error in approximating $\mathrm{E}[x]$ by $(x_1+\cdots+x_N)/N$ is usually of order $1/\sqrt{N}$. *Slow improvement as* N *increases.*

That $1/\sqrt{N}$ estimate came for coin flips in equation (11). Averaging N independent samples x_k of variance σ^2 reduces the variance to σ^2/N.

"Quasi-Monte Carlo" can sometimes reduce this variance to σ^2/N^2: a big difference! The inputs b_k are selected very carefully—not just randomly. This QMC approach is surveyed in the journal *Acta Numerica* 2013. The newer idea of "Multilevel Monte Carlo" is outlined by Michael Giles in *Acta Numerica* 2015. Here is how it works.

Suppose it is simpler to simulate another variable $y(b)$ which is close to $x(b)$. Then use N computations of $y(b_k)$ and only $N^* < N$ computations of $x(b_k)$ to estimate $\mathrm{E}[x]$.

2-level Monte Carlo $\quad \mathrm{E}[x] \approx \dfrac{1}{N}\displaystyle\sum_1^N y(b_k) + \dfrac{1}{N^*}\sum_1^{N^*}[x(b_k)-y(b_k)]$

The idea is that $x-y$ has a smaller variance σ^* than the original x. Therefore N^* can be smaller than N, with the same accuracy for $\mathrm{E}[x]$. We do N cheap simulations to find the y's. Those cost C each. We only do N^* expensive simulations involving x's. Those cost C^* each. The total computing cost is $NC + N^*C^*$.

Calculus minimizes the overall variance for a fixed total cost. The optimal ratio N^*/N is $\sqrt{C/C^*}\ \sigma^*/\sigma$. Three-level Monte Carlo would simulate x, y, and z:

$$\mathrm{E}[x] \approx \frac{1}{N}\sum_1^N z(b_k) + \frac{1}{N^*}\sum_1^{N^*}[y(b_k)-z(b_k)] + \frac{1}{N^{**}}\sum_1^{N^{**}}[x(b_k)-y(b_k)].$$

Giles optimizes $N, N^*, N^{**}, \ldots$ to keep $\mathrm{E}[x] \le$ fixed E_0, and provides a MATLAB code.

Review : Three Formulas for the Mean and the Variance

The formulas for m and σ^2 are the starting point for all of probability and statistics. There are three different cases to keep straight : **sample** values X_i, **expected** values (**discrete** p_i), **expected** values (**continuous** $p(x)$). Here are the mean m and the variance S^2 or σ^2 :

Samples X_1 to X_N	$m = \dfrac{X_1 + \cdots + X_N}{N}$	$S^2 = \dfrac{(X_1 - m)^2 + \cdots + (X_N - m)^2}{N-1}$
Sum of outputs x_i times probabilities p_i	$m = \sum_1^n p_i x_i$	$\sigma^2 = \sum_1^n p_i (x_i - m)^2$
Integral of outputs x with probability density	$m = \int x\, p(x)\, dx$	$\sigma^2 = \int (x-m)^2\, p(x)\, dx$

Problem Set V.1

1 The previous table has no probabilities p on line 1. How can these formulas be parallel ? Answer : *We expect a fraction p_i of the samples to be $X = x_i$.* If this is exactly true, $X = x_i$ is repeated _____ times. Then lines 1 and 2 give the same m.

When we work with samples, we just include each output X as often as it comes. We get the "empirical" mean (line 1) instead of the expected mean.

2 Add 7 to every output x. What happens to the mean and the variance ? What are the new sample mean, the new expected mean, and the new variance ?

3 We know: $\frac{1}{3}$ of all integers are divisible by 3 and $\frac{1}{7}$ of integers are divisible by 7. What fraction of integers will be divisible by 3 or 7 or both ?

4 Suppose you sample from the numbers 1 to 1000 with equal probabilities $1/1000$. What are the probabilities p_0 to p_9 that the last digit of your sample is $0, \ldots, 9$? What is the expected mean m of that last digit? What is its variance σ^2 ?

5 Sample again from 1 to 1000 but look at the last digit of the sample *squared.* That square could end with $x = 0, 1, 4, 5, 6$, or 9. What are the probabilities $p_0, p_1, p_4, p_5, p_6, p_9$? What are the (expected) mean m and variance σ^2 of that number x?

6 (a little tricky) Sample again from 1 to 1000 with equal probabilities and let x be the *first* digit ($x = 1$ if the number is 15). What are the probabilities p_1 to p_9 (adding to 1) of $x = 1, \ldots, 9$? What are the mean and variance of x?

7 Suppose you have $N = 4$ samples $157, 312, 696, 602$ in Problem 5. What are the first digits x_1 to x_4 of the squares? What is the sample mean μ ? What is the sample variance S^2 ? Remember to divide by $N - 1 = 3$ and not $N = 4$.

8 Equation (4) gave a second equivalent form for S^2 (the variance using samples):

$$S^2 = \frac{1}{N-1} \text{ sum of } (x_i - m)^2 = \frac{1}{N-1}\left[\left(\text{sum of } x_i^2\right) - Nm^2\right].$$

Verify the matching identity for the expected variance σ^2 (using $m = \Sigma\, p_i\, x_i$):

$$\boldsymbol{\sigma^2 = \textbf{sum of } p_i\,(x_i - m)^2 = \left(\textbf{sum of } p_i\, x_i^2\right) - m^2.}$$

9 If all 24 samples from a population produce the same age $x = 20$, what are the sample mean μ and the sample variance S^2? What if $x = 20$ or 21, 12 times each?

10 Computer experiment: Find the average $A_{1000000}$ of a million random 0-1 samples! What is your value of the standardized variable $X = \left(A_N - \frac{1}{2}\right)/2\sqrt{N}$?

11 The probability p_i to get i heads in N coin flips is the *binomial number* $b_i = \binom{N}{i}$ divided by 2^N. The b_i add to $(1+1)^N = 2^N$ so the probabilities p_i add to 1.

$$p_0 + \cdots + p_N = \left(\frac{1}{2} + \frac{1}{2}\right)^N = \frac{1}{2^N}(b_0 + \cdots + b_N) \text{ with } b_i = \frac{N!}{i!\,(N-i)!}$$

$$N{=}4 \text{ leads to } b_0 = \frac{24}{24},\; b_1 = \frac{24}{(1)(6)} = 4,\; b_2 = \frac{24}{(2)(2)} = 6,\; p_i = \frac{1}{16}(1, 4, 6, 4, 1).$$

Notice $b_i = b_{N-i}$. *Problem*: Confirm that the mean $m = 0p_0 + \cdots + Np_N$ equals $\frac{N}{2}$.

12 For any function $f(x)$ the expected value is $\mathrm{E}[f] = \sum p_i\, f(x_i)$ or $\int p(x)\, f(x)\, dx$ (discrete or continuous probability). The function can be x or $(x-m)^2$ or x^2. If the mean is $\mathrm{E}[x] = m$ and the variance is $\mathrm{E}[(x-m)^2] = \sigma^2$ **what is** $\mathbf{E}[\boldsymbol{x^2}]$?

13 Show that the standard normal distribution $p(x)$ has total probability $\int p(x)\, dx = 1$ as required. A famous trick multiplies $\int p(x)\, dx$ by $\int p(y)\, dy$ and computes the integral over all x and all y ($-\infty$ to ∞). The trick is to replace $dx\, dy$ in that double integral by $r\, dr\, d\theta$ (polar coordinates with $x^2 + y^2 = r^2$). Explain each step:

$$2\pi \int_{-\infty}^{\infty} p(x)\, dx \int_{-\infty}^{\infty} p(y)\, dy = \iint_{-\infty}^{\infty} e^{-(x^2+y^2)/2} dx\, dy = \int_{\theta=0}^{2\pi} \int_{r=0}^{\infty} e^{-r^2/2}\, r\, dr\, d\theta = \mathbf{2\pi}.$$

V.2 Probability Distributions

The applications of probability begin with the numbers $p_0, p_1, p_2 \ldots$ that give the probability of each possible outcome. For continuous probability we have a density function $p(x)$. The total probability is always $\sum p_i = 1$ or $\int p(x)\,dx = 1$.

There are dozens of famous and useful possibilities for p. We have chosen seven that are specially important: two with discrete probabilities $p_0, p_1, p_2 \ldots$ and five with continuous probabilities $p(x)$ or $p(x, y)$.

Binomial	Tossing a coin n times
Poisson	Rare events
Exponential	Forgetting the past
Gaussian = Normal	Averages of many tries
Log-normal	Logarithm has normal distribution
Chi-squared	Distance squared in n dimensions
Multivariable Gaussian	Probabilities for a vector (in V.5)

Each of those has a mean, and a variance around that mean. You will want to know those fundamental numbers μ and σ^2. On this topic, Wikipedia is organized in a useful way—with graphs of $p(x)$ and its integral $\Phi(x)$ (the cumulative distribution = total probability up to x). There is a systematic list of other properties of those seven special probability distributions, and others too.

1. Binomial distribution

For each trial the outcome is 1 or 0 (success or failure, heads or tails). The probability of success is $p_{1,1} = p$. The probability of failure is $p_{0,1} = 1 - p = q$. A fair coin has $p = \frac{1}{2}$.

The probabilities of $0, 1, 2$ successes in $n = 2$ trials are

$$p_{0,2} = (1-p)^2 \qquad p_{1,2} = 2p(1-p) \qquad p_{2,2} = p^2 \tag{1}$$

The probability of exactly k successes in n trials involves the binomial coefficient $\binom{n}{k}$:

$$\boldsymbol{p_{k,n} = \binom{n}{k} p^k (1-p)^{n-k}} \quad \text{with} \quad \binom{n}{k} = \frac{n\,!}{k\,!\,(n-k)\,!} \quad \text{and} \quad 0\,! = 1 \tag{2}$$

For $n = 2$ those binomial coefficients are $1, 2, 1$ as shown in equation (1):

$$\binom{2}{0} = \frac{2\,!}{0\,!\;2\,!} = 1 \qquad \binom{2}{1} = \frac{2\,!}{1\,!\;1\,!} = \mathbf{2} \qquad \binom{2}{2} = \frac{2\,!}{2\,!\;0\,!} = 1$$

For a fair coin, $k = 0, 1, 2$ successes in $n=2$ trials have $p_{0,2} = \frac{1}{4}$ and $p_{1,2} = \frac{1}{2}$ and $p_{2,2} = \frac{1}{4}$.

Mean value in one trial $\mu = (0)p_0 + (1)p_1 = (0)(1-p) + (1)p$	$\mu = \mathrm{E}[x] = p$	(3)
Binomial distribution Mean value μ_n **in** n **trials is** $n\mu$	$\mu_n = np$	(4)
Variance in one trial $\sigma^2 = \mathrm{E}[(X-\mu)^2] = (1-p)(0-\mu)^2 + p(1-\mu)^2$ $= (1-p)p^2 + p(1-p)^2 = p(1-p)(p+1-p)$	$\sigma^2 = p(1-p)$	(5)
Binomial distribution Variance σ_n^2 **in** n **trials is** $n\sigma^2$	$\sigma_n^2 = np(1-p)$	(6)

The answers for n independent trials came quickly: *Multiply by* n. The same answers come more slowly from the binomial probabilities $p_{k,n}$ in equation (2) for k successes. The sum of $kp_{k,n}$ is still $\mu_n = np$ and the sum of $(k-\mu_n)^2 p_{k,n}$ is $\sigma_n^2 = np(1-p)$.

2. Poisson Distribution

The *Poisson distribution* is fascinating, because there are different ways to approach it. One way is to connect it directly to the *binomial distribution* (probability p of success in each trial and $p_{k,n}$ for k successes in n trials). Then Poisson explains this limiting situation of **rare events but many trials with λ successes:**

$p \to 0$ The success probability p is small for each trial (going to zero)
$n \to \infty$ The number of trials n is large (going to infinity)
$np = \lambda$ The average (expected) number of successes in n trials is $\lambda = np =$ **constant**

What are the probabilities of $0, 1, 2$ successes in n trials when $p \to 0$ and $np = \lambda$?

n failures, 0 successes Probability $p_{0,n} = (1-p)^n = \left(1 - \dfrac{\lambda}{n}\right)^n \to e^{-\lambda}$

$n-1$ failures, 1 success Probability $p_{1,n} = np(1-p)^{n-1} = \dfrac{\lambda}{1-p}\left(1 - \dfrac{\lambda}{n}\right)^n \to \lambda e^{-\lambda}$

$n-2$ failures, 2 successes Probability $p_{2,n} = \dfrac{1}{2}n(n-1)\,p^2(1-p)^{n-2}$

$$= \frac{1}{2}\frac{(\lambda^2 - \lambda p)}{(1-p)^2}\left(1 - \frac{\lambda}{n}\right)^n \to \frac{1}{2}\lambda^2 e^{-\lambda}$$

At every step we applied the same key facts from a calculus course:

$$\left(1 + \frac{1}{n}\right)^n \to e \qquad \left(1 + \frac{\lambda}{n}\right)^n \to e^{\lambda} \qquad \left(1 - \frac{\lambda}{n}\right)^n \to e^{-\lambda}$$

For k successes in n trials with probability $p = \lambda/n$ each time, the binomial probability $p_{n,k}$ approaches the Poisson probability $P_k = \lambda^k e^{-\lambda}/k\,!$

$$\textbf{Poisson probability} \quad P_k = \frac{\lambda^k}{k\,!}\,e^{-\lambda} \tag{7}$$

The sum of those Poisson probabilities P_k is verified to equal 1 :

$$P_0 + P_1 + P_2 + P_3 + \cdots = e^{-\lambda}\left(1 + \lambda + \frac{\lambda^2}{2!} + \frac{\lambda^3}{3!} + \cdots\right) = e^{-\lambda}e^{\lambda} = 1.$$

Now comes the calculation of the Poisson mean (λ) and the variance (also λ).

The mean of the Poisson distribution is λ. The slow way to find it is

$$\mu_P = 0P_0 + 1P_1 + 2P_2 + \cdots = e^{-\lambda}\left(0 + \lambda + \frac{\lambda^2}{1!} + \frac{\lambda^3}{2!} + \cdots\right) = e^{-\lambda}(\lambda e^{\lambda}) = \lambda$$

Fast way: **Poisson mean** $\boldsymbol{\mu_P}$ = limit of binomial mean np. So $\boldsymbol{\mu_P = \lambda}$

Variance $\boldsymbol{\sigma_P^2}$ = limit of binomial variance $np(1-p)$. So $\boldsymbol{\sigma_P^2 = \lambda}$

Applications of Poisson

We associate Poisson with **rare events** (they have $p \to 0$). But we wait a long time, or we include a large population, to produce a decent chance that one or more events will actually occur. Poisson is not for the probability that you will be struck by lightning : say one in a million, over your lifetime. It is for the probability that someone in a city of 100, 000 will be struck. In this case $\boldsymbol{\lambda = pn} = 100,000/1,000,000 = \frac{1}{10}$. This is the expected number of lightning strikes.

The Poisson distribution is often applied to counting rare events over a long time :

The number of big meteors striking the Earth

The number of campers attacked by mountain lions

The number of failures of big banks

Poisson assumes independent events ! Those examples might not fit. One of the most difficult aspects of applied probability is to estimate the dependence of one event on other events. One bank failure may mean more bank failures.

The assumption **iid** means **i**ndependent and **i**dentically **d**istributed—not always true.

3. Exponential distribution

The exponential distribution is continuous (not discrete). It describes the **waiting time in a Poisson process**. How long until lightning strikes your city ? The key assumption is :

The waiting time is independent of the time you have already waited.

The future is independent of the past. Is this true for a television set to fail ? It is true for a computer to fail ? If failure depends on a random disaster, the waiting time has exponential distribution. The failure rate λ is constant. If failure comes from slow decay, the independence requirement will not hold.

The probability density function (pdf) for an exponential distribution is

$$\boldsymbol{p(x) = \lambda e^{-\lambda x}} \quad \text{for} \quad x \geq 0 \tag{8}$$

The cumulative distribution (probability of an event before time t) **is the integral of** $\boldsymbol{p}$:

$$\boldsymbol{F(t)} = \int_0^t \lambda e^{-\lambda x}\, dx = \Big[-e^{-\lambda x}\Big]_{x=0}^{x=-t} = \boldsymbol{1 - e^{-\lambda t}} \tag{9}$$

The **mean** of $p(x)$ is the **average waiting time** :

$$\boldsymbol{\mu} = \int_0^\infty x\, p(x)\, dx = \int_0^\infty x\lambda e^{-\lambda x}\, dx = \frac{\boldsymbol{1}}{\boldsymbol{\lambda}} \tag{10}$$

The **variance** of $p(x)$ is the expected value of $(x-\mu)^2$:

$$\boldsymbol{\sigma^2} = \int_0^\infty \left(x - \frac{1}{\lambda}\right)^2 \lambda e^{-\lambda x}\, dx = \frac{\boldsymbol{1}}{\boldsymbol{\lambda^2}} \tag{11}$$

That equation for the mean $\mu = 1/\lambda$ is not a surprise. If tables at a restaurant open up at an average rate of 3 tables per hour (night and day) then the average wait (expected waiting time) is 20 minutes.

Note that this number $\beta = \dfrac{\boldsymbol{1}}{\boldsymbol{\lambda}}$ is often used instead of λ itself.

$$\textbf{Exponential distribution} \quad p(x) = \frac{1}{\beta} e^{-\lambda/\beta} \qquad\qquad \textbf{Mean} \quad \mu = \beta. \tag{12}$$

The exponential distribution has no memory : The probability of waiting at least y hours more for a table is unfortunately not affected by already having waited x hours :

$$\textbf{No memory} \quad \textbf{Prob}\{t > x + y \textbf{ given that } t > x\} = \textbf{Prob}\{t > y\}. \tag{13}$$

This reduces to a simple statement about integrals of $p(t) = \lambda e^{-\lambda t}$:

$$\int_{x+y}^\infty \lambda e^{-\lambda t}\, dt \Big/ \int_x^\infty \lambda e^{-\lambda t}\, dt = \int_y^\infty \lambda e^{-\lambda t}\, dt \quad \text{is } e^{-\lambda(x+y)}/e^{-\lambda x} = e^{-\lambda y}.$$

Another remarkable fact. What if your phone and computer have failure rates λ_p and λ_c? Assume that their failures are independent. What is the probability distribution $p(t_{\text{min}})$ of the time t_{min} of the first failure? Answer: This distribution is **exponential with failure rate $\lambda_p + \lambda_c$**. Look at survival instead of failure in equation (9), and multiply:

$$\text{Prob}\{t_{\text{first}} > t\} = \left[\text{Prob}\{t_{\textbf{phone}} > t\}\right]\left[\text{Prob}\{t_{\textbf{computer}} > t\}\right] = e^{-\lambda_p t}\, e^{-\lambda_c t} = e^{-(\lambda_p+\lambda_c)t}.$$

If failures only occur at integer times $t = 0, 1, 2, 3, \ldots$ then the exponential distribution (which has continuous time) is replaced by the geometric distribution (discrete time).

$$\textbf{The probability of failure at time } n \textbf{ is } p_n = (1-\alpha)\alpha^n. \qquad (14)$$

The factor $1 - \alpha$ is included so that $p_0 + p_1 + \cdots = (1 - \alpha) + (\alpha - \alpha^2) + \cdots = 1$. The total probability is 1.

4. Normal Distribution (Gaussian Distribution)

The normal distribution is right at the center of probability theory. It is produced when we average the results from another distribution. At the same time it leads to new distributions that we will see next: **log-normal and multivariate normal and "chi-squared"**.

Nice formulas are often possible, even though the integral of e^{-x^2} is not an elementary function. That integral (from 0 to x) cannot be expressed in terms of exponentials and powers of x. The saving grace is that integrals from $-\infty$ to $+\infty$ do have simple forms. The integral of $e^{-x^2/2}$ is $\sqrt{2\pi}$, so we divide by that number.

Mean 0, variance 1 The standard normal distribution $\mathbf{N}(0,1)$ has $p(x) = \dfrac{1}{\sqrt{2\pi}} e^{-x^2/2}$

Mean μ, variance σ^2 The distribution $\mathbf{N}(\mu, \sigma^2)$ has $p(x) = \dfrac{1}{\sqrt{2\pi}\sigma} e^{-(x-\mu)^2/2\sigma^2}$ (15)

These distributions are symmetric around the point $x = \mu$. The function $p(x)$ increases up to that point and decreases for $x > \mu$. The second derivative is negative between $x = \mu - \sigma$ and $\mu + \sigma$. Those are points of inflection where $d^2p/dx^2 = 0$. Outside that interval the function is convex ($d^2p/dx^2 > 0$). 67% of the area under $p(x)$ is in that interval. So the probability of $|x - \mu| < \sigma$ is 0.67.

The probability of $|x - \mu| < 2\sigma$ is 95% (x is less than 2 standard deviations from its mean value μ). This is the famous bell-shaped curve (Figure V.2). It is *not* heavy-tailed.

Question If $p(x) = e^{-ax^2+bx+c}$ is a probability distribution, what are μ and σ?

Answer Complete that exponent $-ax^2 + bx + c$ to a square $-a\left(x - \frac{b}{2a}\right)^2$ plus a constant. In this form we identify the mean μ as $b/2a$. The number a outside the parentheses is $1/(2\sigma^2)$. The role of the constant c is to make $\int p(x)\,dx = 1$.

Normal distributions are symmetric around their center point $x = \mu$. So they are not suitable for variables x that are never negative. A log-normal distribution might succeed.

The cumulative distribution is the integral from $-\infty$ to x of $p(x)$. When $p(x)$ is standard normal ($\mu = 0$ and $\sigma^2 = 1$), its integral from $-\infty$ is often written $\Phi(x)$:

$$\textbf{Cumulative distribution} \qquad \Phi(x) = \frac{1}{\sqrt{2\pi}} \int_{-\infty}^{x} e^{-t^2/2}\, dt \tag{16}$$

This is very closely related to the well-tabulated "error function" $\text{erf}(x)$:

$$\textbf{Error function} \quad \textbf{erf}(x) = \frac{2}{\sqrt{\pi}} \int_{0}^{x} e^{-s^2}\, ds \tag{17}$$

We can change that variable s to $t/\sqrt{2}$, and add $\frac{1}{2}$ for the integral from $-\infty$ to 0. Then

$$\text{Integral of } p(x) = \Phi(x) = \frac{1}{2}\left[1 + \text{erf}\left(\frac{x}{\sqrt{2}}\right)\right] = \text{shifted and scaled error function}$$

For the normal distribution $\mathbf{N}(\mu, \sigma)$ just change x to $(x - \mu)/\sigma$.

The next pages will describe two more probability distributions—important in their own right—that come directly from this one-variable normal distribution $p(x)$:

1. When x has a normal distribution, the exponential e^x has a **log-normal distribution.**
2. If x has the standard $\mathbf{N}(0, 1)$ distribution, then x^2 has a **chi-squared distribution.**

If $x_1, \ldots, x_n$ are independent normals with mean zero and variance 1, then

3. Their sum $x_1 + \cdots + x_n$ has a **normal distribution** with mean zero and variance n
4. $x_1^2 + \cdots + x_n^2$ has the **chi-squared distribution** χ_n^2 with n degrees of freedom.

Ratios of sums of squares have the ***F*-distribution** (not studied here).

What we want most is to allow random variables that are not independent. In this situation we will have covariances as well as variances. The mean values are $\mu_1, \ldots, \mu_n$:

Variances	**= expected values of** $(x_i - \mu_i)^2$
Covariances	**= expected values of** $(x_i - \mu_i)(x_j - \mu_j)$

Those numbers fill the variance-covariance matrix C in Section V.5. When the variables $x_1, \ldots, x_n$ are independent, the covariances are zero and C is diagonal. When the variables are not independent, like stocks and bonds, C is symmetric and positive definite (in an extreme case, semidefinite). **Then the joint distribution of Gaussian variables $x_1, x_2, \ldots, x_n$ is "multivariate Gaussian":**

$$p(\boldsymbol{x}) = p(x_1, \ldots, x_n) = \frac{1}{(\sqrt{2\pi})^n \sqrt{\det C}}\, e^{-(\boldsymbol{x}-\boldsymbol{\mu})^{\mathrm{T}} C^{-1} (\boldsymbol{x}-\boldsymbol{\mu})} \tag{18}$$

For $n = 1$, the 1 by 1 covariance matrix is $C = [\sigma^2]$ and $p(x)$ is the usual Gaussian.

5. Log-normal Distribution

The distribution of x is *log-normal* when the distribution of $y = \log x$ is normal. This requires $x > 0$. The log-normal distribution only enters for positive random variables.

The normal distribution appears (from the central limit theorem) when you average many independent samples of a random variable y. You center and rescale by $(x - \mu)$, to approach the standard normal $\mathbf{N}(0, 1)$ with $\mu = 0$ and $\sigma^2 = 1$. Similarly the log-normal distribution appears when you take the *product* of many positive sample values. The arithmetic mean for normal compares with the geometric mean for log-normal.

To see the formula for $p(x)$, start with a normal distribution for $y = \log x$. The total probability must be 1. Change variables and remember that $dy = dx/x$. This brings a factor $1/x$ into the log-normal distribution $p(x)$:

$$1 = \int_{-\infty}^{\infty} \frac{1}{\sigma\sqrt{2\pi}} e^{-(y-\mu)^2/2\sigma^2}\, dy = \int_0^{\infty} \frac{1}{\sigma\sqrt{2\pi}\, x} e^{-(\log x-\mu)^2/2\sigma^2}\, dx = \int_0^{\infty} p(x)\, dx. \tag{19}$$

The applications of log-normal distributions always involve the requirement that $x > 0$.

6. Chi-squared Distribution

Start with this question: If x has a standard normal distribution with $\mu = 0$ and $\sigma^2 = 1$, what is the **probability distribution of $s = x^2$**? This will be the χ_1^2 distribution of s, where the Greek letter *chi* and the subscript 1 tell us that we are squaring one standard normal variable x. Certainly $s = x^2 \geq 0$.

We need the probability that s is between y and $y + dy$. This happens two ways. Either $\sqrt{s}$ is between $\sqrt{y}$ and $\sqrt{y + dy}$ or $\sqrt{s}$ is between $-\sqrt{y + dy}$ and $-\sqrt{y}$. Those are equally likely since the standard normal is symmetric across zero. And $\sqrt{y + dy} = \sqrt{y} + dy/2\sqrt{y}$ + terms in $(dy)^2$:

$$\textbf{Prob}\,\{y < s < y + dy\} = \textbf{2 Prob}\left\{\sqrt{y} < \sqrt{s} < \sqrt{y} + \frac{dy}{2\sqrt{y}}\right\} = \frac{2}{\sqrt{2\pi}} e^{-(\sqrt{y})^2/2} \frac{dy}{2\sqrt{y}}$$

This answers our first question. **The distribution of probabilities for $s = y^2$ is $p_1(s)$:**

$$\boldsymbol{\chi_1^2} \textbf{ distribution} \qquad p_1(s) = \frac{1}{\sqrt{2\pi s}} e^{-s/2}, \quad s > 0 \tag{20}$$

Note: We mention that the cumulative distribution for $s = x^2$ connects directly to the standard normal cumulative distribution $\Phi(x)$ in equation (16):

$$\text{Prob}\{s < y\} = \text{Prob}\{-\sqrt{y} < x < \sqrt{y}\} = \Phi(\sqrt{y}) - (1 - \Phi(\sqrt{y})).$$

By definition the derivative at $y = s$ is the χ_1^2 probability distribution, agreeing with (20):

$$\begin{array}{l}\text{New approach}\\ \text{Same formula}\end{array} \qquad p_1(s) = \frac{d}{dy}\left[2\Phi(\sqrt{y}) - 1\right] = \frac{1}{\sqrt{s}}\frac{1}{\sqrt{2\pi}} e^{-s/2}.$$

Move now to the **sum of two squares. This is $s_2 = x_1^2 + x_2^2$**. The standard normal variables x_1 and x_2 are independent. We have to look at all combinations $x_1^2 = s$ and $x_2^2 = s_2 - s$ that add to s_2. The probability distribution p_2 for s_2 is the integral over all those combinations:

$$p_2(s_2) = \int_0^{s_2} p_1(s)\, p_1(s_2 - s)\, ds = \text{"convolution of } p_1 \text{ with } p_1\text{".} \quad (21)$$

Then formula (20) for $p_1(s)$ leads to (22) for $p_2(s_2)$—it is an exponential distribution!

$$\textbf{Chi-squared with } n = 2 \textbf{ is exponential: } s_2 = x_1^2 + x_2^2 \quad p_2(s_2) = \frac{1}{2} e^{-s_2/2} \quad (22)$$

This convolution gives a successful approach to χ_n^2 for every $n = 3, 4, 5, \ldots$ The variable s_n is the sum of squares of n **independent standard normals**. We can think of s_n as $s_1 + s_{n-1}$ = one square plus $n - 1$ squares. Then use the same idea as in equation (21):

$$p_n(s_n) = \int_0^{s_n} p_1(s)\, p_{n-1}(s_n - s)\, ds = \text{"convolution of } p_1 \text{ with } p_{n-1}\text{"}$$

This integral p_n has the form $C s_n^{(n-2)/2}\, e^{-s_n/2}$. The number C must make the total probability $\int p_n ds_n$ equal to 1. Integration by parts will steadily reduce the exponent $(n-2)/2$ until we reach $-1/2$ or 0. Those are the two cases we have completed (for $n = 1$ and $n = 2$). So we know how to find C—and we have the χ_n^2 probability distribution for $s_n = x_1^2 + \cdots + x_n^2$:

$s_n = \chi_n^2$ = **sum of squares of n standard normals**	$p_n(s_n) = C s_n^{(n-2)/2} e^{-s_n/2}$	(23)
Integral of p_n must be 1	$C = \dfrac{1}{2^{n/2}\Gamma(n/2)} = \dfrac{1}{2^{n/2}\left(\frac{n}{2} - 1\right)!}$	

The Gamma function $\Gamma(n) = (n-1)\Gamma(n-1)$ is $(n-1)!$ and for $n = \frac{1}{2}$ this is $\Gamma\left(\frac{1}{2}\right) = \sqrt{\pi}$.

Typical Application of χ^2

We manufacture something. Then we test it. We have sample values $x_1, \ldots, x_n$ of its thickness. We compute the average thickness $\overline{x}$ and the sample variance S^2:

$$\overline{x} = \frac{1}{n} \sum_1^n x_i \quad \text{and} \quad S^2 = \frac{1}{n-1} \sum_1^n (x_i - \overline{x})^2.$$

S^2 is a sum of squares with $n - 1$ degrees of freedom. One degree of freedom was used by the mean $\overline{x}$. For $n = 1$, $\overline{x}$ is x_1 and $S = 0$. For $n = 2$, $\overline{x}$ is $\frac{1}{2}(x_1 + x_2)$ and $S^2 = \frac{1}{2}(x_1 - x_2)^2$. S^2 has the probability distribution p_{n-1} for χ_{n-1}^2 given by (23).

Problem Set V.2

1 If $p_1(x) = \dfrac{1}{\sqrt{2\pi}\,\sigma_1} e^{-x^2/2\sigma_1^2}$ and $p_2(x) = \dfrac{1}{\sqrt{2\pi}\,\sigma_2} e^{-x^2/2\sigma_2^2}$ show that $p_1 p_2$ is also a normal distribution: mean = zero and variance = $\sigma^2 = \sigma_1^2\sigma_2^2/(\sigma_1^2+\sigma_2^2)$. The product of Gaussians is Gaussian. If p_1 and p_2 have means m_1 and m_2 then $p_1 p_2$ will have mean $(m_1\sigma_2^2 + m_2\sigma_1^2)/(\sigma_1^2+\sigma_2^2)$.

2 Important: The **convolution** of those Gaussians $p_1(x)$ and $p_2(x)$ is also Gaussian:

$$(\boldsymbol{p_1} * \boldsymbol{p_2})(x) = \int_{-\infty}^{\infty} p_1(t)\, p_2(x-t)\, dt = \frac{1}{\sqrt{2\pi(\sigma_1^2+\sigma_2^2)}} e^{-x^2/2(\sigma_1^2+\sigma_2^2)}$$

A good proof takes Fourier transforms F and uses the convolution theorem:

$$F[p_1(x) * p_2(x)] = F(p_1(x)) \;\; F(p_2(x)).$$

This is a success because the transforms $F(p_i(x))$ are multiples of $\exp(-\sigma_i^2 k^2/2)$. Multiplying those transforms gives a product as in Problem 1. *That product involves* $\sigma_1^2 + \sigma_2^2$. Then the inverse Fourier transform produces $p_1 * p_2$ as another Gaussian.

Question What is the variance σ_n^2 for the convolution of n identical Gaussians $\mathrm{N}(0, \sigma^2)$?

3 Verify that the convolution $P(x) = \displaystyle\int p(t)\, p(x-t)\, dt$ has $\displaystyle\int P(x)\, dx = 1$:

$$\int_{x=-\infty}^{\infty} P(x)\, dx = \int_x \int_t p(t)\, p(x-t)\, dt = \int_t \int_x p(t)\, p(x-t)\, dt = \underline{\qquad}.$$

4 Explain why the probability distribution $P(x)$ for the sum of two random variables is the convolution $P = p_1 * p_2$ of their separate probability distributions. An example comes from rolling two dice and adding the results:

$$\text{Probabilities for sum} \qquad \left(\frac{1}{6},\frac{1}{6},\frac{1}{6},\frac{1}{6},\frac{1}{6},\frac{1}{6}\right) * \left(\frac{1}{6},\frac{1}{6},\frac{1}{6},\frac{1}{6},\frac{1}{6},\frac{1}{6}\right) =$$

$$\text{Probabilities of 2 to 12} \qquad \left(\frac{1}{36},\frac{2}{36},\frac{3}{36},\frac{4}{36},\frac{5}{36},\frac{6}{36},\frac{5}{36},\frac{4}{36},\frac{3}{36},\frac{2}{36},\frac{1}{36}\right)$$

V.3 Moments, Cumulants, and Inequalities of Statistics

Suppose we know the average value (the mean $\overline{X} = \mathrm{E}[X]$) of a random variable X. What we want to know is something about its probabilities: the probability that X is greater than or equal to a. A larger cutoff a will make that probability smaller.

Markov found a simple bound $\overline{X}/a$ on the probability that $X \geq a$, for any *nonnegative* random variable X. As an example we could choose $a = 2\overline{X}$. Then Markov says: *A random sample will be larger than or equal to $2\overline{X}$ with probability not greater than $\frac{1}{2}$.*

Markov's inequality assumes that $X \geq 0$: no samples are negative
Then the probability of $X(s) \geq a$ is at most $\dfrac{\mathrm{E}[X]}{a} = \dfrac{\text{mean of } X}{a} = \dfrac{\overline{X}}{a}$

An example will show why Markov's inequality is true (and what it means). Suppose the numbers $0, 1, 2, \ldots$ (*all nonnegative*!) appear with probabilities $p_0, p_1, p_2, \ldots$ And suppose that this distribution has mean $\mathrm{E}[X] = \overline{X} = 1$:

$$\textbf{Mean value} \qquad \overline{X} = 0p_0 + 1p_1 + 2p_2 + 3p_3 + 4p_4 + 5p_5 + \cdots = 1. \tag{1}$$

Then Markov's inequality with $a = 3$ says that the probability of $X \geq 3$ is at most $\frac{1}{3}$:

$$\textbf{Markov} \qquad \boldsymbol{p_3 + p_4 + p_5 + \cdots \leq \frac{1}{3}}. \tag{2}$$

Proof of Markov: Write equation (1) in a more revealing way:

$$0p_0 + 1p_1 + 2p_2 + \boldsymbol{3(p_3 + p_4 + p_5 + \cdots)} + p_4 + 2p_5 + \cdots = 1. \tag{3}$$

Every term in equation (3) is greater than or equal to zero. Therefore the bold term could not be larger than 1:

$$\boldsymbol{3(p_3 + p_4 + p_5 + \cdots) \leq 1} \textbf{ which is Markov's inequality (2).} \tag{4}$$

Equation (3) tells us even more. **When could $\boldsymbol{p_3 + p_4 + p_5 + \cdots}$ be equal to $\frac{1}{3}$?** Now the bold term in equation (3) is equal to 1, so every other term must be zero:

$$p_1 = 0 \qquad 2p_2 = 0 \qquad p_4 = 0 \qquad 2p_5 = 0 \ldots$$

This leaves only p_0 and p_3. So it forces $p_3 = \frac{1}{3}$ and $p_0 = \frac{2}{3}$ because the p's must add to 1.

Conclusions First, Markov is correct. Second, if there is equality and $\mathrm{Prob}\{x \geq a\}$ is equal to $\mathbf{E}[x]/a$, then all probabilities are actually zero except for these two :

$$\mathrm{Prob}\{x = a\} = \frac{\mathbf{E}[x]}{a} \quad \text{and} \quad \mathrm{Prob}\{x = 0\} = 1 - \frac{\mathbf{E}[x]}{a}.$$

Now we give a formal proof, allowing continuous as well as discrete probabilities.

The proof of Markov's inequality takes four quick steps.

$$\begin{aligned}
\overline{X} = \mathbf{E}[X] &= \sum_{\text{all } s} X(s) \text{ times (Probability of } X(s)) \\
&\geq \sum_{X(s) \geq a} X(s) \text{ times (Probability of } X(s)) \\
&\geq \sum_{X(s) \geq a} a \text{ times (Probability of } X(s)) \\
&= \boldsymbol{a} \textbf{ times (Probability that } \boldsymbol{X(s) \geq a})
\end{aligned}$$

Dividing by a produces Markov's inequality: **(Probability that $X(s) \geq a$) $\leq \overline{X}/a$.**

The second useful inequality is Chebyshev's. This applies to all random variables $X(s)$, not just to nonnegative functions. It provides an estimate for the probability of events that are *far from the mean* $\overline{X}$—so we are looking at the "tail" of the probability distribution. For "heavy-tailed" distributions a large deviation $|X(s) - \overline{X}|$ has higher than usual probability. The probability of $|X - \overline{X}| \geq a$ will decrease as the number a increases.

Chebyshev's inequality for any probability distribution $X(s)$

The probability of $|X(s) - \overline{X}| \geq a$ is at most $\dfrac{\sigma^2}{a^2}$

The proof is to apply Markov's inequality to the nonnegative function $Y(s) = (X(s) - \overline{X})^2$. By the definition of variance, **the mean of that function Y is σ^2** ! We are interested in the events with $|X(s) - \overline{X}| \geq a$. Square both sides to get $Y(s) \geq a^2$. Then use Markov:

$$\textbf{Chebyshev from Markov} \qquad \text{Prob}\left(Y(s) \geq a^2\right) \leq \frac{\text{mean of } Y}{a^2} = \frac{\sigma^2}{a^2}. \tag{5}$$

Those are easy inequalities. They could be improved. Sometimes they are enough to establish that a particular randomized algorithm will succeed. When they are not enough, you generally have to go beyond the mean and variance to higher "moments" and "cumulants". Often the key idea (as in Chernoff's inequality) is to use a **generating function**, which connects all those moments.

Connecting a list of numbers to a function is a powerful idea in mathematics. One real function $f(x) = \Sigma\, a_n x^n$ or one complex function $F(x) = \Sigma\, a_n e^{inx}$ contains (in a different form) the information in all the numbers a_n. Those numbers could be probabilities or "moments" or "cumulants". Then f will be their *generating function.*

Moments and Central Moments

The mean and variance are fundamental numbers, for a discrete set of probabilities p_i and for probability density functions $p(x)$. But the basic theory of statistics goes further. For every n, **the moments are** $m_n = \mathrm{E}[x^n]$. So far we know m_0, m_1, m_2:

Zeroth moment = 1 $\quad \Sigma\, p_i = 1$ or $\int p(x)\,dx = 1$

First moment = mean = $\mathrm{E}[x]$ $\quad \Sigma\, ip_i = m$ or $\int xp(x)\,dx = m$

Second moment (around 0) $\quad \Sigma\, i^2 p_i$ or $\int x^2 p(x)\,dx = \sigma^2 + m^2 = \mathrm{E}[x^2]$

Second central moment (around m) $\quad \Sigma\,(i-m)^2\,p_i = \sigma^2$ or $\int (x-m)^2\,p(x)\,dx = \sigma^2$

The nth moment m_n is $\Sigma\; i^n p_i$ or $\int x^n\,p(x)\,dx$. But the **central moments** (around m) are more useful. They are μ_n = expected value of $(x-m)^n$.

$$\textbf{nth central moment} \quad \mu_n = \Sigma\,(i-m)^n\,p_i \text{ or } \int (x-m)^n\,p(x)\,dx \tag{6}$$

$$\textbf{nth normalized central moment} \quad = \mu_n/\sigma^n \tag{7}$$

Every symmetric distribution with $p_i = p_{-i}$ or $p(x) = p(-x)$ will have mean zero. *All odd moments will be zero*, because the terms left of zero and right of zero will cancel.

When $p(x)$ is *symmertic around its mean value*, the odd *central* moments $\mu_1, \mu_5, \ldots$ will all be zero. [Best example = normal distribution.] The normalized third central moment $\gamma = \mu_3/\sigma^3$ is called the **skewness** of the distribution.

Question: What is the skewness of Bernoulli's coin flip probabilities $p_0 = 1 - p$ and $p_1 = p$?

Answer: First, the mean is p. The variance is $\mu_2 = \sigma^2 = (1-p)p^2 + p(1-p)^2 = p(1-p)$. The third central moment is μ_3, and it depends on distances from the mean p:

$$\mu_3 = (1-p)(0-p)^3 + p(1-p)^3 = p(1-p)(1-2p) \qquad \textbf{skewness} \;\; \gamma = \frac{\mu_3}{\sigma^3} = \frac{1-2p}{\sqrt{p(1-p)}}.$$

Moments are larger when you are far from the "center". For a seesaw, that is the center of mass. For probability, the center is the mean m. A heavy-tailed distribution will have a large value of μ_4. We generally use the fourth moment of a normal distribution (it is $3\sigma^4$) as a basis for comparison. Then the **kurtosis** of any distribution is called kappa:

$$\textbf{Kurtosis} \quad \kappa = \frac{\mu_4 - 3\sigma^4}{\sigma^4} = \frac{\mu_4}{\sigma^4} - 3. \tag{8}$$

Generating Functions and Cumulants

Four key functions are created from the probabilities and the moments and the cumulants. We start with a discrete random variable X. The event $X = n$ has probability p_n. Those numbers p_n lead to the first three functions. Then $\boldsymbol{K(t) = \log M(t)}$.

Probability generating function	$G(z) = \sum_0^\infty p_n z^n$	(9)
Characteristic function	$\phi(t) = \sum_0^\infty p_n e^{itn}$	(10)
Moment generating function	$M(t) = \sum_0^\infty m_n \frac{t^n}{n!}$	(11)
Cumulant generating function	$K(t) = \sum_0^\infty \kappa_n \frac{t^n}{n!}$	(12)

All probabilities p_n and moments m_n and cumulants κ_n can be recovered from the nth derivatives of the functions G, M, K at $z = 0$ and $t = 0$:

$$p_n = \frac{1}{n!}\frac{d^nG}{dz^n}(0) \qquad m_n = \frac{d^nM}{dt^n}(0) \qquad \kappa_n = \frac{d^nK}{dt^n}(0) \tag{13}$$

And there is a very revealing way to connect the four generating functions to expectations:

$$\boxed{G(z) = \mathrm{E}\left[z^X\right] \quad \phi(t) = \mathrm{E}\left[e^{itX}\right] \quad M(t) = \mathrm{E}\left[e^{tX}\right] \quad K(t) = \log \mathrm{E}\left[e^{tX}\right]} \tag{14}$$

Of course there must be a purpose behind these four functions. We are capturing an infinite set of coefficients in each function. The key point of cumulants is that they lead to this property of $K(t) = \log M(t)$:

$$K_{X+Y}(t) = K_X(t) + K_Y(t) \quad \textbf{for independent random variables } X \textbf{ and } Y.$$

Examples One coin flip gives the **Bernoulli distribution** with probabilities $p_0 = 1 - p$ and $p_1 = p$. Then $p = \mathrm{E}[x] = \mathrm{E}[x^2] = \mathrm{E}[x^3] = \cdots$ and $\boldsymbol{M(t) = 1 - p + pe^t}$.

Cumulative generating function $\boldsymbol{K(t) = \log(1 - p + pe^t)}$.

The first cumulant is $\kappa_1 = \dfrac{dK}{dt} = \left[\dfrac{pe^t}{1 - p + pe^t}\right]_{t=0} = p.$

For the **binomial distribution** (N independent coin flips) multiply every cumulant by N.

For the **Poisson distribution** $p_n = e^{-\lambda}\lambda^n/n!$ the function $K(t)$ is $\lambda(e^t - 1)$. All $\kappa_n = \lambda$.

Generating Functions for Continuous Distributions

We cannot leave out the normal distribution! For a continuous distribution, all the generating functions have integrals involving $p(x)$ instead of sums involving p_n:

$$\phi(t) = \int_{-\infty}^{\infty} p(x)\, e^{itx}\, dx \qquad M(t) = \int_{-\infty}^{\infty} p(x)\, e^{tx}\, dx \qquad K(t) = \log M(t). \quad (15)$$

The moment generating function for a normal distribution (mean μ, variance σ^2) is $M(t) = e^{\mu t} e^{\sigma^2 t^2/2}$. The cumulant generating function is its logarithm $\boldsymbol{K(t) = \mu t + \sigma^2 t^2/2}$. So the normal distribution is highly exceptional with only those two nonzero cumulants:

Normal distribution	$\kappa_1 = \mu$	$\kappa_2 = \sigma^2$	$\kappa_3 = \kappa_4 = \ldots = 0$

Key facts about 3 cumulants $\kappa_1 =$ **mean** $\kappa_2 =$ **variance** $\kappa_3 =$ **third central moment**

Because the cumulants of independent processes can be added, they appear throughout combinatorics and statistics and physics. Higher cumulants are more complicated than κ_3.

The Central Limit Theorem

In a few lines, we can justify the great limit theorem of probability. It concerns the standardized averages $Z_n = \sum (X_k - m)/\sigma\sqrt{N}$ of N independent samples $X_1, \ldots, X_N$ with mean m and variance σ^2. The central limit theorem says: **The distribution of Z_n approaches the standard normal distribution (mean zero, variance 1) as $N \to \infty$.**

The proof uses the characteristic function of the standardized variable $Y = (X - m)/\sigma$:

$$\mathrm{E}\left[e^{itY}\right] = \mathrm{E}\left[1 + itY - \frac{1}{2} t^2 Y^2 + O(t^3)\right] = 1 + 0 - \frac{1}{2} t^2 + O(t^3) \qquad (16)$$

Certainly $Z_N = (Y_1 + Y_2 + \cdots + Y_N)/\sqrt{N}$. So its characteristic function is a product of N identical characteristic functions of $Y/\sqrt{N}$: the number t is fixed.

$$\left[\mathrm{E}\left(e^{itY/\sqrt{N}}\right)\right]^N = \left[1 - \frac{1}{2}\left(\frac{t}{\sqrt{N}}\right)^2 + O\left(\frac{t}{\sqrt{N}}\right)^3\right]^N \to e^{-t^2/2} \quad \text{as} \quad N \to \infty. \quad (17)$$

That limit $e^{-t^2/2}$ is the characteristic function of a standard normal distribution $\mathrm{N}(0, 1)$.

Chernoff's Inequality for Sums

The mean value of a sum $X = X_1 + \cdots + X_n$ is always $\overline{X} = \overline{X}_1 + \cdots + \overline{X}_n$. If those variables are *independent*, then also the variance σ^2 of X is $\sigma^2 = \sigma_1^2 + \cdots + \sigma_n^2$.

At the start of this section, Chebyshev's inequality gave a bound on the probability of samples X_i that are far from their means $\overline{X}_i$. It applies also to their sums X and $\overline{X}$.

$$\textbf{Chebyshev for a sum} \quad \textbf{Prob}\,(|X - \overline{X}| \geq a) \leq \frac{\sigma_1^2 + \cdots + \sigma_n^2}{a^2} \tag{18}$$

The inequality is tight for $n = 1$. Can the inequality be improved for a sum ? I would have guessed *no*. Chernoff says *yes*. There is a subtle point here. Chernoff assumed not just independence of each pair X_i and X_j, but *joint independence* of all the X_i together:

$$\text{Multiply probabilities} \qquad p(x_1, \ldots, x_n) = p_1(x_1) \cdots p_n(x_n). \tag{19}$$

A diagonal covariance matrix only needs pairwise independence. Equation (19) says more. And it leads to much stronger bounds (20) for a sum than Chebyshev's inequality (18).

The key point about Chernoff's inequality is its *exponential bound*. Sums X that are far from their mean $\overline{X}$ are exponentially unlikely. That is because an exceptional sum $X = X_1 + \cdots + X_n$ usually needs *several* X_i to be far from their means $\overline{X}_i$.

An example is the number of heads in tossing n coins. Then $\overline{X}_i = p_i$ for each coin. The total number X of heads will have mean value $\overline{X} = p_1 + \cdots + p_n$.

$$\begin{array}{ll} \textbf{Upper Chernoff} & \textbf{Prob}\,(X \geq (1+\delta)\overline{X}) \leq e^{-\overline{X}\delta^2/(2+\delta)} \\ \textbf{Lower Chernoff} & \textbf{Prob}\,(X \leq (1-\delta)\overline{X}) \leq e^{-\overline{X}\delta^2/2} \end{array} \tag{20}$$

In his online notes for the MIT class 18.310, Michel Goemans points out how strong this is. Suppose we have n flips of a fair coin. Then $\overline{X} = n/2$ (half heads and half tails). Now take a small $\delta^2 = (4 \log n)/n$, and compare the Chernoff bounds (20) for δ and 2δ.

The probability that $X \leq (1 - 2\delta)\overline{X}$ is very much smaller than the probability of $X \leq (1-\delta)\overline{X}$. By doubling δ, the δ^2 in Chernoff's exponent changes to $4\delta^2$. The probability drops from $1/n$ to $1/n^4$:

$$\begin{aligned} \overline{X}\delta^2/2 &= \log n \ \text{ gives a bound } e^{-\log n} = 1/n \\ \overline{X}(2\delta)^2/2 &= 4\log n \ \text{ gives a bound } e^{-4\log n} = 1/n^4 \end{aligned}$$

Chebyshev would have had $\mathbf{1/4n}$ where Chernoff has $\mathbf{1/n^4}$: an exponential difference ! We can point to the key step in proving Chernoff bounds. First center X so that $\overline{X} = 0$:

$$\begin{array}{llll} \textbf{Usual Chebyshev} & \textbf{Prob}\,(|X| \geq a) & = \text{Prob}\,(X^2 \geq a^2) & \leq \mathbf{E}\,[X^2]/a^2 \\ \textbf{Upper Chernoff} & \textbf{Prob}\,(X \geq a) & = \text{Prob}\,(e^{sX} \geq e^{sa}) & \leq \mathbf{E}\left[e^{sX}\right]/e^{sa} \\ \textbf{Lower Chernoff} & \textbf{Prob}\,(X \leq a) & = \text{Prob}\,(e^{-sX} \geq e^{-sa}) & \leq \mathbf{E}\left[e^{-sX}\right]/e^{-sa} \end{array}$$

We need both exponentials (plus and minus). Especially we need the moment generating function. This tells us $M(s) = \mathrm{E}\,[e^{sX}]$ and $M(-s) = \mathrm{E}\,[e^{-sX}]$. Then Chernoff will follow from a careful choice of s. It has many applications in randomized linear algebra.

To work with non-independent samples we need **covariances** as well as variances. And we also need Markov-Chebyshev-Chernoff inequalities for matrices. Those come now.

Markov and Chebyshev Inequalities for Matrices

Stochastic gradient descent (Section VI.5) is the established algorithm for optimizing the weights in a neural net. Stochastic means *random*—and those weights go into matrices. So an essential step in the statistics of deep learning is to develop inequalities for the *eigenvalues and the trace of random matrices.*

We will not aim for an exhaustive presentation: just the basic facts. We will write $\boldsymbol{X \leq A}$ when $A - X$ is positive semidefinite: energy ≥ 0 and all eigenvalues ≥ 0. Otherwise $\boldsymbol{X \not\leq A}$. In this case $A - X$ has a negative eigenvalue.

Markov's inequality Suppose $X \geq 0$ is a semidefinite or definite random matrix with mean $\mathrm{E}[X] = \overline{X}$. If A is any positive definite matrix, then

$$\textbf{Prob}\,\{\boldsymbol{X \not\leq A}\} = \textbf{Prob}\,\{A - X \text{ is not positive semidefinite }\} \leq \textbf{Trace of } \boldsymbol{\overline{X} A^{-1}}. \quad (21)$$

If X and A are scalars x and a, compare with Markov's inequality $\text{Prob}\,\{x \geq a\} \leq \overline{x}/a$.

Proof If $A^{1/2}$ is the positive definite square root of A, here is the key step:

$$(\text{Trace of } A^{-1/2} X A^{-1/2}) > 1 \text{ if } X \not\leq A. \quad (22)$$

When $A - X$ is not positive semidefinite, there must be a vector $\boldsymbol{v}$ with negative energy $\boldsymbol{v}^{\mathrm{T}}(A - X)\boldsymbol{v} < 0$. Set $\boldsymbol{w} = A^{1/2}\boldsymbol{v}$ so that $\boldsymbol{w}^{\mathrm{T}}\boldsymbol{w} < \boldsymbol{w}^{\mathrm{T}} A^{-1/2} X A^{-1/2} \boldsymbol{w}$. Then the largest eigenvalue of $A^{-1/2} X A^{-1/2}$ is $\lambda_{\max} > 1$:

$$\textbf{Rayleigh quotient} \qquad \lambda_{\max} = \max \frac{\boldsymbol{y}^{\mathrm{T}} A^{-1/2} X A^{-1/2} \boldsymbol{y}}{\boldsymbol{y}^{\mathrm{T}} \boldsymbol{y}} > 1$$

No eigenvalues of $A^{-1/2} X A^{-1/2}$ are negative, so its trace is larger than 1. This is (22). Then taking expectations of both sides of (22) will produce Markov's inequality (21):

$$\text{Prob}\,\{X \not\leq A\} \leq \mathrm{E}[\text{trace}\,(A^{-1/2} X A^{-1/2})] = \text{trace}\,(A^{-1/2} \overline{X} A^{-1/2}) = \text{trace}\,(\overline{X} A^{-1}).$$

Now we turn to Chebyshev's inequality $\text{Prob}\,\{|\,X - \overline{X}\,| \geq a\} \leq \sigma^2/a^2$. For a symmetric matrix $A = Q\Lambda Q^{\mathrm{T}}$, we will use a fact about its absolute value $|A| = Q|\Lambda|Q^{\mathrm{T}}$:

If $\boldsymbol{A^2 - B^2}$ *is positive semidefinite then* $|\boldsymbol{A}| - |\boldsymbol{B}|$ *is also positive semidefinite.*

The proof is not completely simple. The opposite statement is not true. We save examples for the Problem Set and use this fact now:

> **Chebyshev's inequality for a random matrix $\boldsymbol{X}$ with mean 0**
>
> **If $\boldsymbol{A}$ is positive definite then Prob $\{|\boldsymbol{X}| \not\leq \boldsymbol{A}\} <$ trace $(\mathbf{E}\,[\boldsymbol{X^2}]\boldsymbol{A^{-2}})$.** (23)

If $A - |X|$ is not positive semidefinite then $A^2 - X^2$ is not positive semidefinite (as above):

$$\text{Prob}\,\{|X| \not\leq A\} \leq \text{Prob}\,\{X^2 \not\leq A^2\} < \text{trace}\,(\mathrm{E}\,[X^2] A^{-2}).$$

That last step was Markov's inequality for the positive semidefinite matrix X^2.

Wikipedia offers a multidimensional Chebyshev inequality for a random $\boldsymbol{x}$ in $\mathbf{R}^N$. Suppose its mean is $\mathrm{E}[\boldsymbol{x}] = \boldsymbol{m}$ and its covariance matrix is $V = \mathrm{E}[(\boldsymbol{x}-\boldsymbol{m})(\boldsymbol{x}-\boldsymbol{m})^{\mathrm{T}}]$:

If V is positive definite and $t > 0$ then **Prob** $\{(\boldsymbol{x}-\boldsymbol{m})^{\mathrm{T}} V^{-1}(\boldsymbol{x}-\boldsymbol{m}) > t^2\} \leq N/t^2$.

Matrix Chernoff Inequalities

Chernoff inequalities deal with a sum Y of random variables X_k—previously scalars and now n by n positive semidefinite (or positive definite) matrices. We look at the smallest and largest eigenvalues of that sum. The key point of Chernoff is that for a sum to be far from its mean value, it normally needs *several terms* in the sum to be fairly far off—and that combination of unusual events is exponentially unlikely.

So we get exponentially small bounds for tail probabilities (distances from the mean).

Matrix Chernoff Suppose each matrix X_k in $Y = \Sigma X_k$ has eigenvalues $0 \leq \lambda \leq C$. Let $\mu_{\min}$ and $\mu_{\max}$ be the extreme eigenvalues of the average sum $\overline{Y} = \Sigma \overline{X}_k$. Then

$$\mathbf{E}[\lambda_{\min}(Y)] \geq \left(1 - \frac{1}{e}\right)\mu_{\min} - C\log n \tag{24}$$

$$\mathbf{E}[\lambda_{\max}(Y)] \leq (e-1)\ \mu_{\max} + C\log n \tag{25}$$

Eigenvalues of the sum Y far from their mean are exponentially unlikely:

$$\textbf{Prob}\{\lambda_{\min}(Y) \leq t\,\mu_{\min}\} \leq n\,e^{-(1-t)^2\mu_{\min}/2C} \tag{26}$$

$$\textbf{Prob}\{\lambda_{\max}(Y) \geq t\,\mu_{\max}\} \leq n\left(\frac{e}{t}\right)^{t\mu_{\max}/C} \quad \text{for } t \geq e \tag{27}$$

Joel Tropp's online textbook is an excellent presentation of matrix inequalities. His book proves sharper estimates for the expectations and exponential bounds in (24-27).

Joel Tropp, *An Introduction to Matrix Concentration Inequalities*, arXiv:1501.01591.

There are many more inequalities after Chernoff! We stop with an application of the inequality (26) to the probability that a graph with edges at random is connected.

Erdős-Renyi Random Graphs

This is a chance to define random graphs. Start with n nodes. **Each edge is present with probability p.** Then the question is: **For which p is the graph likely to be connected?**

The n by n adjacency matrix has $M_{jk} = M_{kj} = 1$ when nodes j and k are connected. The random variable x_{jk} for that two-way edge is 0 or 1, with probability $1-p$ and p:

Adjacency matrix $\quad M = \text{sum of random matrices} = \sum_{j<k} x_{jk}(E_{jk} + E_{kj}). \quad (28)$

E_{jk} is the matrix with a single 1 in position (j,k). The Laplacian matrix is $\boldsymbol{L} = \boldsymbol{D} - \boldsymbol{M}$.

Row sums minus M $\quad \boldsymbol{L} = \sum_{j<k} x_{jk}(E_{jj} + E_{kk} - E_{jk} - E_{kj}) \quad (29)$

For every edge (meaning $x_{jk} = 1$), the degree matrix D has two 1's from $E_{jj} + E_{kk}$. The adjacency matrix M has two 1's from $E_{jk} + E_{kj}$. Together those four entries have $\lambda = 2$ and 0. So L is positive semidefinite and $\lambda = 0$ has the eigenvector $(1, \ldots, 1)$.

The second smallest eigenvalue of L will be positive when the graph is connected. With two pieces, we can number nodes in one part before nodes in the other part. Then L has separate blocks from the two parts. Each block has an all-ones eigenvector, and L has $\lambda_1 = \lambda_2 = 0$ (two zero eigenvalues with separate eigenvectors of 1's).

We need a random matrix Y of size $n-1$ whose smallest eigenvalue is $\boldsymbol{\lambda_1(Y) = \lambda_2(L)}$. Then a connected graph will have $\lambda_1(Y) > 0$. A suitable Y is ULU^{T}, where the $n-1$ rows of U are orthogonal unit vectors that are perpendicular to the all-ones vector $\mathbf{1}$. Y is like L, but with that all-ones eigenvector and its zero eigenvalue removed.

We now apply the matrix Chernoff theorem to discover when $\lambda_{\min}(Y) > 0$ and the graph is connected. Y is the sum of random matrices X_{jk} (for all $j < k$):

$$\boldsymbol{Y} = ULU^{\mathrm{T}} = \sum x_{jk}\, U\,(E_{jj} + E_{kk} - E_{jk} - E_{kj})\, U^{\mathrm{T}} = \sum X_{jk} \tag{30}$$

Each X_{jk} is semidefinite with eigenvalues ≤ 2 as above, since $||U|| = ||U^{\mathrm{T}}|| = 1$.

Then $C = 2$ in the Chernoff theorem. We also need the smallest eigenvalue $\mu_{\min}$ of the average matrix $\overline{Y}$. The expected value of each random number x_{jk} is p—the probability of including that edge in the graph. The expected value of $Y = \sum X_{jk}$ is $p\,n\,I_{n-1}$:

$$\overline{\boldsymbol{Y}} = pU\left[\sum_{j<k}(E_{jj} + E_{kk} - E_{jk} - E_{kj})\right]U^{\mathrm{T}}$$

$$= pU[(n-1)I_n - (\mathbf{1}\mathbf{1}^{\mathrm{T}} - I_n)]U^{\mathrm{T}} = \boldsymbol{p\,n\,I_{n-1}} \tag{31}$$

That term $(n-1)I_n$ came from adding all the diagonal matrices E_{jj} and E_{kk}. The off-diagonal matrices E_{jk} and E_{kj} add to $\mathbf{1}\mathbf{1}^{\mathrm{T}} - I_n$ = all-ones matrix with zero diagonal. $UU^{\mathrm{T}} = I_{n-1}$ produced $p\,n\,I_{n-1}$, and we have found the smallest eigenvalue $\boldsymbol{\mu_{\min} = p\,n}$.

Now apply the inequality (26) in Chernoff's theorem with $C = 2$:

$$\textbf{Prob}\,\{\lambda_2(L) \leq t\,p\,n\} = \textbf{Prob}\,\{\lambda_1(Y) \leq t\,p\,n\} \leq (n-1)\,e^{-(1-t)^2 pn/4}. \tag{32}$$

As $t \to 0$, the crucial quantity is $(n-1)e^{-pn/4}$. This is below 1 if its logarithm is below zero.

$$\log(n-1) - \frac{1}{4}pn < 0 \quad \text{or} \quad p > \frac{4\log(n-1)}{n}. \tag{33}$$

Edges in the random graph are included with probability p. *If p is large enough to satisfy* (33), *the graph is probably connected.* A tighter argument would remove the factor 4. That produces the optimal cutoff value of p for connected random graphs.

Problem Set V.3

1 Find the probability generating function $G(z)$ for Poisson's $p_n = e^{-\lambda}\lambda^n/n\,!$

2 Independent random variables x and y have $p(x,y) = p(x)\,p(y)$. Derive the special property $K_{X+Y}(t) = K_X(t) + K_Y(t)$ of their cumulant generating functions.

3 A fair coin flip has outcomes $X = 0$ and $X = 1$ with probabilities $\frac{1}{2}$ and $\frac{1}{2}$. What is the probability that $X \geq 2\overline{X}$? Show that Markov's inequality gives the exact probability $\overline{X}/2$ in this case.

4 Throwing two ordinary dice has an outcome between $X = 2$ and $X = 12$ (two 1's or two 6's). The mean value is 7. What is the actual probability p that $X \geq 12$? Show that Markov's $\overline{X}/12$ overestimates that probability p in this case where $a = 12$.

5 Here is another proof of Markov's basic inequality for a nonnegative variable X.

For $a > 0$, the random variable $Y = \begin{cases} 0 & \text{if } X \leq a \\ a & \text{if } X > a \end{cases}$ has $Y \leq X$. Why?

Explain the final step a **Prob** $[\boldsymbol{X} \geq \boldsymbol{t}] = \mathbf{E}\,[\boldsymbol{Y}] \leq \mathbf{E}\,[\boldsymbol{X}]$ to Markov's bound $\mathrm{E}\,[X]/a$.

6 Show that the largest eigenvalue of a random $Y = Y^{\mathrm{T}}$ is a convex function of Y:

$$\boldsymbol{\lambda}_{\mathbf{max}}(\overline{\boldsymbol{Y}}) = \lambda_{\max}(p_1Y_1 + \cdots + p_nY_n) \leq p_1\lambda_{\max}(Y_1) + \cdots + p_n\lambda_{\max}(Y_n) = \mathrm{E}[\boldsymbol{\lambda}_{\mathbf{max}}(\boldsymbol{Y})].$$

7 Show that $A - B$ is positive semidefinite but $A^2 - B^2$ is not:

$$A = \begin{bmatrix} 2 & 1 \\ 1 & 1 \end{bmatrix} \qquad B = \begin{bmatrix} 1 & 1 \\ 1 & 1 \end{bmatrix}$$

8 Prove this amazing identity when random samples $0 < x_1 < x_2 < \cdots < x_n$ have probabilities p_1 to p_n:

$$\textbf{mean} = \mathbf{E}[x] = \sum_1^n p_i x_i = \int_{t=0}^{\infty} (\textbf{Probability that } \boldsymbol{x} > \boldsymbol{t})\, dt.$$

Hint: That probability is $\sum p_i = 1$ up to $t = x_1$ and then it is $1 - p_1$ as far as $t = x_2$.

V.4 Covariance Matrices and Joint Probabilities

Linear algebra enters when we run M different experiments at once. We might measure age and height and weight ($M = 3$ measurements of N people). Each experiment has its own mean value. So we have a vector $\boldsymbol{m} = (m_1, m_2, m_3)$ containing the M mean values. Those could be *sample means* of age and height and weight. Or m_1, m_2, m_3 could be *expected values* of age, height, weight based on known probabilities.

A matrix becomes involved when we look at variances. Each experiment will have a sample variance S_i^2 or an expected $\sigma_i^2 = \mathrm{E}\left[(x_i - m_i)^2\right]$ based on the squared distance from its mean. Those M numbers $\sigma_1^2, \ldots, \sigma_M^2$ will go on the main diagonal of the "variance-covariance matrix". So far we have made no connection between the M parallel experiments. They measure different random variables, but the experiments are not necessarily independent!

If we measure age and height and weight $(\boldsymbol{a}, \boldsymbol{h}, \boldsymbol{w})$ for children, the results will be strongly correlated. Older children are generally taller and heavier. Suppose the means m_a, m_h, m_w are known. Then $\sigma_a^2, \sigma_h^2, \sigma_w^2$ are the separate variances in age, height, weight. **The new numbers are the covariances like σ_{ah}, which measures the connection of age to height.**

$$\textbf{Covariance} \quad \sigma_{ah} = \mathrm{E}\left[(\textbf{age} - \textbf{mean age})\,(\textbf{height} - \textbf{mean height})\right]. \tag{1}$$

This definition needs a close look. To compute σ_{ah}, it is not enough to know the probability of each age and the probability of each height. We have to know the **joint probability of each pair (age and height)**. This is because age is connected to height.

p_{ah} = probability that a random child has age $= \boldsymbol{a}$ **and** height $= \boldsymbol{h}$: both at once

p_{ij} = **probability that experiment 1 produces x_i and experiment 2 produces y_j**

Suppose experiment 1 (age) has mean m_1. Experiment 2 (height) has mean m_2. The covariance in equation (1) between experiments 1 and 2 looks at **all pairs** of ages x_i and heights y_j. We multiply by the joint probability p_{ij} of that pair.

Expected value of
$(x - m_1)(y - m_2)$

$$\textbf{Covariance} \quad \sigma_{12} = \sum_{\textbf{all } i,j}\sum p_{ij}(x_i - m_1)(y_j - m_2) \tag{2}$$

To capture this idea of "joint probability p_{ij}" we begin with two small examples.

Example 1 **Flip two coins separately**. With 1 for heads and 0 for tails, the results can be $(1,1)$ or $(1,0)$ or $(0,1)$ or $(0,0)$. Those four outcomes all have probability $\left(\frac{1}{2}\right)^2 = \frac{1}{4}$. For independent experiments we multiply probabilities:

$$p_{ij} = \textbf{Probability of } (i,j) = (\textbf{Probability of } i) \textbf{ times } (\textbf{Probability of } j).$$

Example 2 *Glue the coins together*, facing the same way. The only possibilities are $(1,1)$ and $(0,0)$. Those have probabilities $\frac{1}{2}$ and $\frac{1}{2}$. The probabilities p_{10} and p_{01} are zero. $(1,0)$ and $(0,1)$ won't happen because the coins stick together: both heads or both tails.

Joint probability matrices for Examples 1 and 2 $\quad P = \begin{bmatrix} \frac{1}{4} & \frac{1}{4} \\ \frac{1}{4} & \frac{1}{4} \end{bmatrix}$ and $P = \begin{bmatrix} \frac{1}{2} & 0 \\ 0 & \frac{1}{2} \end{bmatrix}$.

Let me stay longer with P, to show it in good matrix notation. The matrix shows the probability p_{ij} of each pair (x_i, y_j)—starting with $(x_1, y_1) =$ (heads, heads) and $(x_1, y_2) =$ (heads, tails). Notice the row sums p_1, p_2 and column sums P_1, P_2 and the total sum $= 1$.

$$\textbf{Probability matrix} \quad P = \begin{bmatrix} p_{11} & p_{12} \\ p_{21} & p_{22} \end{bmatrix} \quad \begin{matrix} p_{11} + p_{12} = \boldsymbol{p_1} \\ p_{21} + p_{22} = \boldsymbol{p_2} \end{matrix} \quad \begin{pmatrix} \text{first} \\ \text{coin} \end{pmatrix}$$

(second coin) column sums $\quad \boldsymbol{P_1} \quad \boldsymbol{P_2} \qquad$ 4 entries add to 1

Those sums p_1, p_2 and P_1, P_2 are the **marginals** of the joint probability matrix P:

$p_1 = p_{11} + p_{12} =$ chance of heads from **coin 1** (coin 2 can be heads or tails)
$P_1 = p_{11} + p_{21} =$ chance of heads from **coin 2** (coin 1 can be heads or tails)

Example 1 showed *independent* random variables. Every probability p_{ij} equals p_i times p_j ($\frac{1}{2}$ times $\frac{1}{2}$ gave $p_{ij} = \frac{1}{4}$ in that example). In this case **the covariance $\boldsymbol{\sigma_{12}}$ will be zero.** Heads or tails from the first coin gave no information about the second coin.

Zero covariance σ_{12} for independent trials	$V = \begin{bmatrix} \sigma_1^2 & 0 \\ 0 & \sigma_2^2 \end{bmatrix}$	$=$ **diagonal covariance matrix V.**

Independent experiments have $\sigma_{12} = 0$ because every p_{ij} equals $(p_i)(p_j)$ in equation (2):

$$\sigma_{12} = \sum_i \sum_j (p_i)(p_j)(x_i - m_1)(y_j - m_2) = \left[\sum_i (p_i)(x_i - m_1)\right]\left[\sum_j (p_j)(y_j - m_2)\right] = [0][0].$$

Example 3 The glued coins show perfect correlation. Heads on one means heads on the other. The covariance σ_{12} moves from 0 to $\boldsymbol{\sigma_1}$ **times** $\boldsymbol{\sigma_2}$. This is the largest possible value of σ_{12}. Here it is $\left(\frac{1}{2}\right)\left(\frac{1}{2}\right) = \sigma_{12} = \left(\frac{1}{4}\right)$, as a separate computation confirms:

$$\textbf{Means} = \frac{1}{2} \qquad \sigma_{12} = \frac{1}{2}\left(1 - \frac{1}{2}\right)\left(1 - \frac{1}{2}\right) + 0 + 0 + \frac{1}{2}\left(0 - \frac{1}{2}\right)\left(0 - \frac{1}{2}\right) = \frac{1}{4}$$

Heads or tails from coin 1 gives complete information about heads or tails from the glued coin 2:

Glued coins give largest possible covariances
Singular covariance matrix: determinant = 0 $\quad V_{\text{glue}} = \begin{bmatrix} \sigma_1^2 & \sigma_1\sigma_2 \\ \sigma_1\sigma_2 & \sigma_2^2 \end{bmatrix}$

Always $\sigma_1^2\sigma_2^2 \geq (\sigma_{12})^2$. Thus σ_{12} is *between* $-\sigma_1\sigma_2$ *and* $\sigma_1\sigma_2$. The matrix V is **positive definite** (or in this singular case of glued coins, V is **positive semidefinite**). Those are important facts about all M by M covariance matrices V for M experiments.

Note that the **sample covariance matrix** S from N trials is certainly semidefinite. Every new sample $X =$ (age, height, weight) contributes to the **sample mean** $\overline{\boldsymbol{X}}$ (a vector). Each rank-one term $(X_i - \overline{X})(X_i - \overline{X})^{\mathrm{T}}$ is positive semidefinite and we just add to reach the matrix S. No probabilities in S, just actual outcomes:

$$\overline{\boldsymbol{X}} = \frac{X_1 + \cdots + X_N}{N} \qquad S = \frac{(X_1 - \overline{X})(X_1 - \overline{X})^{\mathrm{T}} + \cdots + (X_N - \overline{X})(X_N - \overline{X})^{\mathrm{T}}}{N-1} \tag{3}$$

The Covariance Matrix V is Positive Semidefinite

Come back to the *expected* covariance σ_{12} between two experiments 1 and 2 (two coins):

$$\begin{aligned} \sigma_{12} &= \text{expected value of } [(\textit{output}\,1 - \textit{mean}\,1) \text{ times } (\textit{output}\,2 - \textit{mean}\,2)] \\ \boldsymbol{\sigma_{12}} &= \boldsymbol{\sum\sum p_{ij}\,(x_i - m_1)\,(y_j - m_2)}. \text{ The sum includes all } i, j. \end{aligned} \tag{4}$$

$p_{ij} \geq 0$ is the probability of seeing outputs x_i in experiment 1 **and** y_j in experiment 2. Some pair of outputs must appear. Therefore the N^2 joint probabilities p_{ij} add to 1.

$$\textbf{Total probability (all pairs) is 1} \qquad \sum_{\text{all } i,j}\sum p_{ij} = 1. \tag{5}$$

Here is another fact we need. *Fix on one particular output* x_i in experiment 1. Allow *all outputs* y_j in experiment 2. Add the probabilities of $(x_i, y_1), (x_i, y_2), \ldots, (x_i, y_n)$:

$$\textbf{Row sum } \boldsymbol{p_i} \textbf{ of } \boldsymbol{P} \qquad \sum_{j=1}^{n} \boldsymbol{p_{ij}} = \textbf{probability } \boldsymbol{p_i} \textbf{ of } \boldsymbol{x_i} \textbf{ in experiment 1.} \tag{6}$$

Some y_j must happen in experiment 2 ! Whether the two coins are completely separate or glued, we get the same answer $\frac{1}{2}$ for the probability $p_H = p_{HH} + p_{HT}$ that coin 1 is heads:

$$\text{(separate) } P_{HH} + P_{HT} = \frac{1}{4} + \frac{1}{4} = \boldsymbol{\frac{1}{2}} \qquad \text{(glued) } P_{HH} + P_{HT} = \frac{1}{2} + 0 = \boldsymbol{\frac{1}{2}}.$$

That basic reasoning allows us to write one matrix formula that includes the covariance σ_{12} along with the separate variances σ_1^2 and σ_2^2 for experiment 1 and experiment 2. We get the whole covariance matrix V by adding the matrices V_{ij} for each pair (i, j):

$$\begin{array}{l}\textbf{Covariance matrix}\\ \boldsymbol{V} \textbf{ = sum of all } \boldsymbol{V_{ij}}\end{array} \qquad V = \sum_{\text{all } i,j}\sum p_{ij} \begin{bmatrix} (x_i - m_1)^2 & (x_i - m_1)(y_j - m_2) \\ (x_i - m_1)(y_j - m_2) & (y_j - m_2)^2 \end{bmatrix} \tag{7}$$

Off the diagonal, this is equation (2) for the covariance σ_{12}. On the diagonal, we are getting the ordinary variances σ_1^2 and σ_2^2. I will show in detail how we get $V_{11} = \sigma_1^2$ by using equation (6). Allowing all j just leaves the probability p_i of x_i in experiment 1:

$$V_{11} = \sum_{\text{all } i,j}\sum p_{ij}(x_i - m_1)^2 = \sum_{\text{all } i} (\text{probability of } x_i)\,(x_i - m_1)^2 = \boldsymbol{\sigma_1^2}. \quad (8)$$

Please look at that twice. It is the key to producing the whole covariance matrix by one formula (7). The beauty of that formula is that it combines 2 by 2 matrices V_{ij}. And the matrix V_{ij} in (7) for each pair of outcomes i, j is **positive semidefinite**:

V_{ij} has diagonal entries $p_{ij}(x_i - m_1)^2 \geq 0$ and $p_{ij}(y_j - m_2)^2 \geq 0$ and $\det(V_{ij}) = 0$.

That matrix V_{ij} has rank 1. Equation (7) multiplies p_{ij} *times column* $\boldsymbol{U}$ *times row* $\boldsymbol{U}^{\mathrm{T}}$:

$$\begin{bmatrix} (x_i - m_1)^2 & (x_i - m_1)(y_j - m_2) \\ (x_i - m_1)(y_j - m_2) & (y_j - m_2)^2 \end{bmatrix} = \begin{bmatrix} x_i - m_1 \\ y_j - m_2 \end{bmatrix} \begin{bmatrix} x_i - m_1 & y_j - m_2 \end{bmatrix} \quad (9)$$

Every matrix $\boldsymbol{p_{ij}UU^{\mathrm{T}}}$ *is positive semidefinite.* So the whole matrix V (the sum of those rank 1 matrices) is **at least semidefinite**—and probably V is definite.

The covariance matrix V is positive definite unless the experiments are dependent.

Now we move from two variables x and y to M variables like age-height-weight. The output from each trial is a vector X with M components. (Each child has an age-height-weight vector $\boldsymbol{X}$ with 3 components.) The covariance matrix V is now M by M. The matrix $\boldsymbol{V}$ is created from the output vectors $\boldsymbol{X}$ and their average $\overline{\boldsymbol{X}} = \mathbf{E}[\boldsymbol{X}]$:

Covariance matrix $$\boldsymbol{V} = \mathbf{E}\left[(\boldsymbol{X} - \overline{\boldsymbol{X}})\,(\boldsymbol{X} - \overline{\boldsymbol{X}})^{\mathrm{T}}\right] = \sum \boldsymbol{p_{ij}}\,(\boldsymbol{X} - \overline{\boldsymbol{X}})\,(\boldsymbol{X} - \overline{\boldsymbol{X}})^{\mathrm{T}} \quad (10)$$

Remember that $\boldsymbol{X}\boldsymbol{X}^{\mathrm{T}}$ and $\overline{\boldsymbol{X}}\,\overline{\boldsymbol{X}}^{\mathrm{T}}$ = (column) (row) are M by M matrices.

For $M = 1$ (one variable) you see that $\overline{X}$ is the mean m and V is the variance σ^2. For $M = 2$ (two coins) you see that $\overline{X}$ is (m_1, m_2) and V matches equation (7). The expectation always adds up outputs times their probabilities. For age-height-weight the output could be X = (5 years, 31 inches, 48 pounds) and its probability is $\boldsymbol{p_{5,31,48}}$.

Now comes a new idea. *Take any linear combination* $\mathbf{c}^{\mathrm{T}}\boldsymbol{X} = c_1X_1 + \cdots + c_MX_M$. With $\mathbf{c} = (6, 2, 5)$ this would be $\mathbf{c}^{\mathrm{T}}\boldsymbol{X} = 6 \times \text{age} + 2 \times \text{height} + 5 \times \text{weight}$. By linearity we know that its expected value $\mathrm{E}[\mathbf{c}^{\mathrm{T}}\boldsymbol{X}]$ is $\mathbf{c}^{\mathrm{T}}\mathrm{E}[\boldsymbol{X}] = \mathbf{c}^{\mathrm{T}}\overline{\boldsymbol{X}}$:

$\mathbf{E}[\mathbf{c}^{\mathrm{T}}\boldsymbol{X}] = \mathbf{c}^{\mathrm{T}}\mathbf{E}[\boldsymbol{X}]$ = 6 (expected age) + 2 (expected height) + 5 (expected weight).

More than the mean of $c^{\mathrm{T}}X$, we also know its *variance* $\sigma^2 = c^{\mathrm{T}}Vc$:

$$\begin{aligned}\text{Variance of } c^{\mathrm{T}}X &= \mathbf{E}\left[\left(c^{\mathrm{T}}X - c^{\mathrm{T}}\overline{X}\right)\left(c^{\mathrm{T}}X - c^{\mathrm{T}}\overline{X}\right)^{\mathrm{T}}\right] \\ &= c^{\mathrm{T}}\mathbf{E}\left[\left(X - \overline{X}\right)\left(X - \overline{X}\right)^{\mathrm{T}}\right]c = c^{\mathrm{T}}Vc\end{aligned} \tag{11}$$

Now the key point: *The variance of* $c^{\mathrm{T}}X$ *can never be negative.* So $c^{\mathrm{T}}Vc \geq 0$. New proof: *The covariance matrix V is positive semidefinite by the energy test* $c^{\mathrm{T}}Vc \geq 0$.

Covariance matrices V open up the link between probability and linear algebra: **V equals $Q\Lambda Q^{\mathrm{T}}$** with eigenvalues $\lambda_i \geq 0$ and orthonormal eigenvectors q_1 to q_M.

Diagonalizing the covariance matrix V means finding M *independent* experiments as combinations of the original M experiments.

Confession I am not entirely happy with that proof based on $c^{\mathrm{T}}Vc \geq 0$. The expectation symbol $\mathbf{E}$ is hiding the key idea of joint probability. Allow me to show directly that the covariance matrix V is positive semidefinite (at least for the age-height-weight example). The proof is simply that **V is the sum of the joint probability p_{ahw} of each combination (age, height, weight) times the positive semidefinite matrix UU^{T}**. Here U is $X - \overline{X}$:

$$V = \sum_{\text{all } a,h,w} p_{ahw}\, U U^{\mathrm{T}} \quad \text{with} \quad U = \begin{bmatrix}\text{age}\\ \text{height}\\ \text{weight}\end{bmatrix} - \begin{bmatrix}\text{mean age}\\ \text{mean height}\\ \text{mean weight}\end{bmatrix} \tag{12}$$

This is exactly like the 2 by 2 coin flip matrix V in equation (7). Now $M = 3$.

The value of the expectation symbol E is that it also allows *pdf*'s: probability density functions like $p(x, y, z)$ for continuous random variables x and y and z. If we allow all numbers as ages and heights and weights, instead of age $i = 0, 1, 2, 3 \ldots$, then we need $p(x, y, z)$ instead of p_{ijk}. The sums in this section of the book would all change to integrals. But we still have $V = \mathrm{E}\left[UU^{\mathrm{T}}\right]$:

$$\textbf{Covariance matrix } V = \iiint p(x,y,z)\, UU^{\mathrm{T}}\, dx\, dy\, dz \quad \text{with} \quad U = \begin{bmatrix} x - \overline{x}\\ y - \overline{y}\\ z - \overline{z}\end{bmatrix}. \tag{13}$$

Always $\iiint p = 1$. Examples 1–2 emphasized how p can give diagonal V or singular V:

Independent variables x, y, z $\quad p(x, y, z) = p_1(x)\, p_2(y)\, p_3(z)$.

Dependent variables x, y, z $\quad p(x, y, z) = 0$ except when $cx + dy + ez = 0$.

The Mean and Variance of $z = x + y$

Start with the sample mean. We have N samples of x. Their mean (= average) is the number m_x. We also have N samples of y and their mean is m_y. **The sample mean of $z = x + y$ is clearly $m_z = m_x + m_y$:**

$$\textbf{Mean of sum} = \textbf{Sum of means} \qquad \frac{1}{N}\sum_{1}^{N}(x_i + y_i) = \frac{1}{N}\sum_{1}^{N} x_i + \frac{1}{N}\sum_{1}^{N} y_i. \tag{14}$$

Nice to see something that simple. The *expected* mean of $z = x+y$ doesn't look so simple, but it must come out as $\mathbf{E}[z] = \mathbf{E}[x] + \mathbf{E}[y]$. Here is one way to see this.

The joint probability of the pair (x_i, y_j) is p_{ij}. Its value depends on whether the experiments are independent, which we don't know. But for the expectation of the sum $z = x+y$, dependence or independence of x and y doesn't matter. *Expected values still add*:

$$\mathbf{E}[x + y] = \sum_i \sum_j p_{ij}(x_i + y_j) = \sum_i \sum_j p_{ij} x_i + \sum_i \sum_j p_{ij} y_j. \tag{15}$$

All the sums go from 1 to N. We can add in any order. For the first term on the right side, add the p_{ij} along row i of the probability matrix P to get p_i. That double sum gives $\mathrm{E}[x]$:

$$\sum_i \sum_j p_{ij} x_i = \sum_i (p_{i1} + \cdots + p_{iN}) x_i = \sum_i p_i x_i = \mathbf{E}[x].$$

For the last term, add p_{ij} down column j of the matrix to get the probability P_j of y_j. Those pairs (x_1, y_j) and (x_2, y_j) and ... and (x_N, y_j) are all the ways to produce y_j:

$$\sum_i \sum_j p_{ij} y_j = \sum_j (p_{1j} + \cdots + p_{Nj}) y_j = \sum_j P_j y_j = \mathbf{E}[y].$$

Now equation (15) says that $\mathrm{E}[x + y] = \mathrm{E}[x] + \mathrm{E}[y]$. Mean of sum = Sum of means.

What about the variance of $z = x + y$? The joint probabilities p_{ij} and the covariance σ_{xy} will be involved. Let me separate the variance of $x + y$ into three simple pieces:

$$\begin{aligned}\sigma_z^2 &= \sum\sum p_{ij}(x_i + y_j - m_x - m_y)^2 \\ &= \sum\sum p_{ij}(x_i - m_x)^2 + \sum\sum p_{ij}(y_j - m_y)^2 + 2\sum\sum p_{ij}(x_i - m_x)(y_j - m_y)\end{aligned}$$

The first piece is $\boldsymbol{\sigma_x^2}$. The second piece is $\boldsymbol{\sigma_y^2}$. The last piece is $\boldsymbol{2\sigma_{xy}}$.

$$\textbf{The variance of } z = x + y \quad \textbf{is} \quad \boldsymbol{\sigma_z^2 = \sigma_x^2 + \sigma_y^2 + 2\sigma_{xy}}. \tag{16}$$

The Covariance Matrix for $Z = AX$

Here is a good way to see σ_z^2 when $z = x + y$. Think of (x, y) as a column vector $\boldsymbol{X}$. Think of the 1 by 2 matrix $\boldsymbol{A} = \begin{bmatrix} 1 & 1 \end{bmatrix}$ multiplying that vector $\boldsymbol{X} = (x, y)$. Then $\boldsymbol{AX}$ is the sum $z = x + y$. The variance $\boldsymbol{\sigma_z^2}$ in equation (16) goes into matrix notation as

$$\boldsymbol{\sigma_z^2} = \begin{bmatrix} 1 & 1 \end{bmatrix} \begin{bmatrix} \sigma_x^2 & \sigma_{xy} \\ \sigma_{xy} & \sigma_y^2 \end{bmatrix} \begin{bmatrix} 1 \\ 1 \end{bmatrix} \quad \text{which is} \quad \boldsymbol{\sigma_z^2 = AVA^{\mathrm{T}}}. \qquad (17)$$

You can see that $\sigma_z^2 = AVA^{\mathrm{T}}$ in (17) agrees with $\sigma_x^2 + \sigma_y^2 + 2\sigma_{xy}$ in (16).

Now for the main point. The vector $\boldsymbol{X}$ could have M components coming from M experiments (instead of only 2). Those experiments will have an M by M covariance matrix $\boldsymbol{V_X}$. The matrix A could be K by M. Then AX is a vector with K combinations of the M outputs (instead of one combination $x + y$ of 2 two outputs).

That vector $\boldsymbol{Z} = \boldsymbol{AX}$ of length K has a K by K covariance matrix $V_{\boldsymbol{Z}}$. Then the great rule for covariance matrices—of which equation (17) was only a 1 by 2 example—is this beautiful formula: The covariance matrix of $\boldsymbol{AX}$ is **$\boldsymbol{A}$ (covariance matrix of $\boldsymbol{X}$) $\boldsymbol{A}^{\mathrm{T}}$**:

$$\boxed{\textbf{The covariance matrix of } \boldsymbol{Z = AX} \textbf{ is } \quad \boldsymbol{V_Z = AV_X A^{\mathrm{T}}}} \qquad (18)$$

To me, this neat formula shows the beauty of matrix multiplication. I won't prove this formula, just admire it. It is constantly used in applications.

The Correlation ρ

Correlation ρ_{xy} is closely related to covariance σ_{xy}. They both measure dependence or independence. Start by rescaling or "standardizing" the random variables x and y **The new $\boldsymbol{X = x/\sigma_x}$ and $\boldsymbol{Y = y/\sigma_y}$ have variance $\boldsymbol{\sigma_X^2 = \sigma_Y^2 = 1}$.** This is just like dividing a vector $\boldsymbol{v}$ by its length to produce a unit vector $\boldsymbol{v}/||\boldsymbol{v}||$ of length 1.

The correlation of $\boldsymbol{x}$ and $\boldsymbol{y}$ is the covariance of $\boldsymbol{X}$ and $\boldsymbol{Y}$. If the original covariance of x and y was σ_{xy}, then rescaling to X and Y will divide by σ_x and σ_y:

$$\boxed{\textbf{Correlation } \boldsymbol{\rho_{xy}} = \frac{\boldsymbol{\sigma_{xy}}}{\boldsymbol{\sigma_x \sigma_y}} = \textbf{covariance of } \frac{\boldsymbol{x}}{\boldsymbol{\sigma_x}} \textbf{ and } \frac{\boldsymbol{y}}{\boldsymbol{\sigma_y}}} \quad \textbf{Always } \boldsymbol{-1 \le \rho_{xy} \le 1}$$

Zero covariance gives zero correlation. *Independent random variables* produce $\rho_{xy} = 0$.

We know that always $(\rho_{xy})^2 \le \sigma_x^2 \sigma_y^2$ (the covariance matrix V is at least positive semidefinite). Then $\boldsymbol{(\sigma_{xy})^2 \le 1}$. Correlation near $\rho = +1$ means strong dependence in the same direction: often voting the same. Negative correlation means that y tends to be below its mean when x is above its mean: Voting in opposite directions when ρ is near -1.

Example 4 *Suppose that y is just $-x$.* A coin flip has outputs $x = 0$ or 1. The same flip has outputs $y = 0$ or -1. The mean m_x is $\frac{1}{2}$ for a fair coin, and m_y is $-\frac{1}{2}$. The covariance of x and y is $\sigma_{xy} = -\sigma_x\sigma_y$. The correlation divides by $\sigma_x\sigma_y$ to get $\boldsymbol{\rho_{xy} = -1}$. In this case the correlation matrix R has determinant zero (singular and only semidefinite):

$$\textbf{Correlation matrix} \quad \boldsymbol{R} = \begin{bmatrix} \mathbf{1} & \boldsymbol{\rho_{xy}} \\ \boldsymbol{\rho_{xy}} & \mathbf{1} \end{bmatrix} \qquad \boldsymbol{R} = \begin{bmatrix} \mathbf{1} & \mathbf{-1} \\ \mathbf{-1} & \mathbf{1} \end{bmatrix} \text{ when } \boldsymbol{y = -x}$$

R always has 1's on the diagonal because we normalized to $\sigma_X = \sigma_Y = 1$. R is the correlation matrix for x and y, and it is also the covariance matrix for $X = x/\sigma_x$ and $Y = y/\sigma_y$.

That number ρ_{xy} is also called the Pearson coefficient.

Example 5 Suppose the random variables x, y, z are *independent. What matrix is R?*

Answer *R is the identity matrix.* All three correlations $\rho_{xx}, \rho_{yy}, \rho_{zz}$ are 1 by definition. All three cross-correlations $\rho_{xy}, \rho_{xz}, \rho_{yz}$ are zero by independence.

The correlation matrix R comes from the covariance matrix V, when we rescale every row and every column. Divide each row i and column i by the ith standard deviation σ_i.

(a) $\boldsymbol{R = DVD}$ for the diagonal matrix $D = \text{diag}\,[1/\sigma_1, \ldots, 1/\sigma_M]$. Then every $R_{ii} = 1$.

(b) If covariance V is positive definite, correlation $R = DVD$ is also positive definite.

■ WORKED EXAMPLE ■

Suppose x and y are independent random variables with mean 0 and variance 1. Then the covariance matrix $\boldsymbol{V_X}$ for $\boldsymbol{X} = (x, y)$ is the 2 by 2 identity matrix. What are the mean $\boldsymbol{m_Z}$ and the covariance matrix $\boldsymbol{V_Z}$ for the 3-component vector $\boldsymbol{Z} = (x, y, ax + by)$?

Solution

$$\textbf{Z is connected to X by A} \qquad Z = \begin{bmatrix} x \\ y \\ ax + by \end{bmatrix} = \begin{bmatrix} 1 & 0 \\ 0 & 1 \\ a & b \end{bmatrix} \begin{bmatrix} x \\ y \end{bmatrix} = A\boldsymbol{X}.$$

The vector $\boldsymbol{m_X}$ contains the means of the M components of $\boldsymbol{X}$. The vector $\boldsymbol{m_Z}$ contains the means of the K components of $\boldsymbol{Z} = A\boldsymbol{X}$. The matrix connection between the means of $\boldsymbol{X}$ and $\boldsymbol{Z}$ has to be linear: $\boldsymbol{m_Z} = A\boldsymbol{m_X}$. The mean of $ax + by$ is $am_x + bm_y$.

The covariance matrix for $\boldsymbol{Z}$ is $\boldsymbol{V_Z} = AA^{\mathrm{T}}$, when $\boldsymbol{V_X}$ is the 2 by 2 identity matrix:

$$\boldsymbol{V_Z} = \begin{array}{c}\textbf{covariance matrix for}\\ \boldsymbol{Z = (x, y, ax+by)}\end{array} = \begin{bmatrix} 1 & 0 \\ 0 & 1 \\ a & b \end{bmatrix} \begin{bmatrix} 1 & 0 & a \\ 0 & 1 & b \end{bmatrix} = \begin{bmatrix} 1 & 0 & a \\ 0 & 1 & b \\ a & b & a^2+b^2 \end{bmatrix}.$$

Interpretation: x and y are independent with covariance $\sigma_{xy} = 0$. Then the covariance of x with $ax + by$ is a and the covariance of y with $ax + by$ is b. Those just come from the two independent parts of $ax + by$. Finally, equation (18) gives the variance of $ax + by$:

$$\textbf{Use } \boldsymbol{V_Z = AV_XA^{\mathrm{T}}} \qquad \sigma^2_{ax+by} = \sigma^2_{ax} + \sigma^2_{by} + 2\sigma_{ax,by} = a^2 + b^2 + 0.$$

The 3 by 3 matrix $\boldsymbol{V_Z}$ is *singular*. Its determinant is $a^2 + b^2 - a^2 - b^2 = 0$. The third component $z = ax + by$ is completely dependent on x and y. The rank of $\boldsymbol{V_Z}$ is only 2.

GPS Example The signal from a GPS satellite includes its departure time. The receiver clock gives the arrival time. The receiver multiplies the travel time by the speed of light. Then it knows the distance from that satellite. Distances from four or more satellites will pinpoint the receiver position (using least squares !).

One problem: The speed of light changes in the ionosphere. But the correction will be almost the same for all nearby receivers. If one receiver stays in a known position, we can take differences from that one. **Differential GPS** reduces the error variance by fixing one receiver:

$$\begin{array}{ll}\textbf{Difference matrix} & \textbf{Covariance matrix}\\ \boldsymbol{A = [\,1 \quad -1\,]} & \boldsymbol{V_Z = AV_XA^{\mathrm{T}}}\end{array} \qquad \begin{aligned} \boldsymbol{V_Z} &= \begin{bmatrix} 1 & -1 \end{bmatrix} \begin{bmatrix} \sigma_1^2 & \sigma_{12} \\ \sigma_{12} & \sigma_2^2 \end{bmatrix} \begin{bmatrix} 1 \\ -1 \end{bmatrix} \\ &= \boldsymbol{\sigma_1^2 - 2\sigma_{12} + \sigma_2^2} \end{aligned}$$

Errors in the speed of light are gone. Then centimeter positioning accuracy is achievable. (The key ideas are on page 320 of *Algorithms for Global Positioning* by Borre and Strang.) The GPS world is all about time and space and amazing accuracy.

Problem Set V.4

1 (a) Compute the variance σ^2 when the coin flip probabilities are p and $1 - p$ (tails $= 0$, heads $= 1$).

(b) The sum of N independent flips (0 or 1) is the count of heads after N tries. The rule (16-17-18) for the variance of a sum gives $\sigma^2 =$ ______.

2 What is the covariance σ_{kl} between the results $x_1, \ldots, x_n$ of Experiment 3 and the results $y_1, \ldots, y_n$ of Experiment 5 ? Your formula will look like σ_{12} in equation (2). Then the $(3, 5)$ and $(5, 3)$ entries of the covariance matrix V are $\sigma_{35} = \sigma_{53}$.

3 For $M = 3$ experiments, the variance-covariance matrix V will be 3 by 3. There will be a probability p_{ijk} that the three outputs are x_i and y_j and z_k. Write down a formula like equation (7) for the matrix V.

4 What is the covariance matrix V for $M = 3$ independent experiments with means m_1, m_2, m_3 and variances $\sigma_1^2, \sigma_2^2, \sigma_3^2$?

5 When the covariance matrix for outputs $\boldsymbol{X}$ is $\boldsymbol{V}$, the covariance matrix for outputs $\boldsymbol{Z} = \boldsymbol{AX}$ is $\boldsymbol{AVA}^{\mathrm{T}}$. Explain this neat formula by linearity:

$$Z = \mathrm{E}\left[(AX - \overline{AX})(AX - \overline{AX})^{\mathrm{T}}\right] = A\mathrm{E}\left[(X - \overline{X})(X - \overline{X})^{\mathrm{T}}\right] A^{\mathrm{T}}.$$

Problems 6–10 are about the conditional probability that $\boldsymbol{Y = y_j}$ when we know $\boldsymbol{X = x_i}$. Notation: **Prob ($\boldsymbol{Y = y_j | X = x_i}$)** = probability of the outcome y_j given that $X = x_i$.

Example 1 *Coin* 1 *is glued to coin* 2. Then Prob (Y = heads when X = heads) is **1**.
Example 2 *Independent coin flips*: X gives no information about Y. Useless to know X.

Then Prob (Y = heads $|X$ = heads) is the same as Prob (Y = heads).

6 Explain the **sum rule** of conditional probability:

$$\text{Prob}\,(Y = y_j) = \text{ sum over all outputs } x_i \text{ of Prob}\,(Y = y_j | X = x_i).$$

7 The n by n matrix P contains **joint probabilities** p_{ij} = Prob ($X = x_i$ **and** $Y = y_j$).

Explain why the conditional Prob ($Y = y_j | X = x_i$) equals $\dfrac{p_{ij}}{p_{i1} + \cdots + p_{in}} = \dfrac{p_{ij}}{p_i}$.

8 For this joint probability matrix with Prob (x_1, y_2) = 0.3, find Prob ($y_2 | x_1$) and Prob (x_1).

$$P = \begin{bmatrix} p_{11} & p_{12} \\ p_{21} & p_{22} \end{bmatrix} = \begin{bmatrix} 0.1 & 0.3 \\ 0.2 & 0.4 \end{bmatrix}$$

The entries p_{ij} add to 1.
Some i, j must happen.

9 Explain the **product rule** of conditional probability:

p_{ij} = Prob ($X = x_i$ **and** $Y = y_j$) equals Prob ($Y = y_j | X = x_i$) times Prob ($X = x_i$).

10 Derive this **Bayes Theorem** for p_{ij} from the product rule in Problem 8:

$$\text{Prob}\,(Y = y_j \textbf{ and } X = x_i) = \frac{\text{Prob}\,(X = x_i | Y = y_j)\ \text{Prob}\,(Y = y_j)}{\text{Prob}\,(X = x_i)}$$

"Bayesians" use prior information. "Frequentists" only use sampling information.

V.5 Multivariate Gaussian and Weighted Least Squares

The normal probability density $p(x)$ (the Gaussian) depends on only two numbers:

$$\textbf{Mean } m \textbf{ and variance } \sigma^2 \qquad p(x) = \frac{1}{\sqrt{2\pi}\,\sigma}\, e^{-(x-m)^2/2\sigma^2}. \tag{1}$$

The graph of $p(x)$ is a bell-shaped curve centered at $x = m$. The continuous variable x can be anywhere between $-\infty$ and ∞. With probability close to $\frac{2}{3}$, that random x will lie between $m - \sigma$ and $m + \sigma$ (less than one standard deviation σ from its mean value m).

$$\int_{-\infty}^{\infty} p(x)\,dx = 1 \quad \text{and} \quad \int_{m-\sigma}^{m+\sigma} p(x)\,dx = \frac{1}{\sqrt{2\pi}} \int_{-1}^{1} e^{-X^2/2}\,dX \approx \frac{2}{3}. \tag{2}$$

That integral has a change of variables from x to $X = (x-m)/\sigma$. This simplifies the exponent to $-X^2/2$ and it simplifies the limits of integration to -1 and 1. Even the $1/\sigma$ from $p(x)$ disappears because dX equals dx/σ. Every Gaussian becomes a **standard Gaussian** with mean $m = 0$ and variance $\sigma^2 = 1$:

$$\textbf{The standard normal distribution } N(0,1) \quad \textbf{has} \quad p(x) = \frac{1}{\sqrt{2\pi}}\, e^{-x^2/2}. \tag{3}$$

Integrating $p(x)$ from $-\infty$ to x gives the cumulative distribution $F(x)$: the probability that a random sample is below x. That probability will be $F = \frac{1}{2}$ at $x = 0$ (the mean).

Two-dimensional Gaussians

Now we have $M = 2$ Gaussian random variables x and y. They have means m_1 and m_2. They have variances σ_1^2 and σ_2^2. If they are *independent*, then their probability density $p(x,y)$ is just $\mathbf{p_1(x)}$ **times** $\mathbf{p_2(y)}$. Multiply probabilities when variables are independent:

$$\textbf{Independent } x \textbf{ and } y \quad p(x,y) = \frac{1}{2\pi\sigma_1\sigma_2}\, e^{-(x-m_1)^2/2\sigma_1^2}\, e^{-(y-m_2)^2/2\sigma_2^2} \tag{4}$$

The covariance of x and y will be $\boldsymbol{\sigma_{12} = 0}$. The covariance matrix V will be *diagonal*. The variances σ_1^2 and σ_2^2 are always on the main diagonal of V. The exponent in $p(x,y)$ is just the sum of the x-exponent and the y-exponent. Good to notice that the two exponents can be combined into $-\frac{1}{2}\,(\boldsymbol{x}-\boldsymbol{m})^{\mathrm{T}}\, \boldsymbol{V}^{-1}\,(\boldsymbol{x}-\boldsymbol{m})$ with the inverse covariance matrix V^{-1} in the middle. This exponent is $-[\boldsymbol{x}-\boldsymbol{m}]^{\mathrm{T}} V^{-1} [\boldsymbol{x}-\boldsymbol{m}]/2$:

$$-\frac{(x-m_1)^2}{2\sigma_1^2} - \frac{(y-m_2)^2}{2\sigma_2^2} = -\frac{1}{2}\begin{bmatrix} x-m_1 & y-m_2 \end{bmatrix} \begin{bmatrix} \sigma_1^2 & 0 \\ 0 & \sigma_2^2 \end{bmatrix}^{-1} \begin{bmatrix} x-m_1 \\ y-m_2 \end{bmatrix} \tag{5}$$

Non-independent x and y

We are ready to give up independence. The exponent (5) with V^{-1} is still correct when V is no longer a diagonal matrix. **Now the Gaussian depends on a vector $\boldsymbol{m}$ and a matrix V.**

When $M = 2$, the first variable x may give partial information about the second variable y (and vice versa). Maybe part of y is decided by x and part is truly independent. It is the M by M covariance matrix V that accounts for dependencies between the M variables $\boldsymbol{x} = x_1, \ldots, x_M$. **The inverse covariance matrix V^{-1} goes into $p(\boldsymbol{x})$:**

Multivariate Gaussian probability distribution

$$p(\boldsymbol{x}) = \frac{1}{(\sqrt{2\pi})^M \sqrt{\det V}}\, e^{-(\boldsymbol{x}-\boldsymbol{m})^{\mathrm{T}} V^{-1} (\boldsymbol{x}-\boldsymbol{m})/2} \tag{6}$$

The vectors $\boldsymbol{x} = (x_1, \ldots, x_M)$ and $\boldsymbol{m} = (m_1, \ldots, m_M)$ contain the random variables and their means. The M square roots of 2π and the determinant of V are included to make the total probability equal to 1. Let me check by linear algebra. I use the eigenvalues λ and the orthonormal eigenvectors $\boldsymbol{q}$ of the symmetric matrix $V = Q\Lambda Q^{\mathrm{T}}$. So $\boldsymbol{V^{-1} = Q\Lambda^{-1}Q^{\mathrm{T}}}$:

$$\boldsymbol{X} = \boldsymbol{x} - \boldsymbol{m} \qquad (\boldsymbol{x} - \boldsymbol{m})^{\mathrm{T}} \boldsymbol{V}^{-1} (\boldsymbol{x} - \boldsymbol{m}) = \boldsymbol{X}^{\mathrm{T}} \boldsymbol{Q} \boldsymbol{\Lambda}^{-1} \boldsymbol{Q}^{\mathrm{T}} \boldsymbol{X} = \boldsymbol{Y}^{\mathrm{T}} \boldsymbol{\Lambda}^{-1} \boldsymbol{Y}$$

Notice! The combinations $\boldsymbol{Y} = \boldsymbol{Q}^{\mathrm{T}} \boldsymbol{X} = \boldsymbol{Q}^{\mathrm{T}} (\boldsymbol{x} - \boldsymbol{m})$ are statistically independent. *Their covariance matrix Λ is diagonal.*

> This step of diagonalizing V by its eigenvector matrix Q is the same as "uncorrelating" the random variables. Covariances are zero for the new variables $Y_1, \ldots Y_M$. This is the point where linear algebra helps calculus to compute multidimensional integrals.

The integral of $p(\boldsymbol{x})$ is not changed when we center the variable $\boldsymbol{x}$ by subtracting $\boldsymbol{m}$ to reach $\boldsymbol{X}$, and rotate that variable to reach $\boldsymbol{Y} = \boldsymbol{Q}^{\mathrm{T}} \boldsymbol{X}$. The matrix $\boldsymbol{\Lambda}$ is diagonal! So the integral we want splits into M separate one-dimensional integrals that we know:

$$\int \cdots \int e^{-\boldsymbol{Y}^{\mathrm{T}} \boldsymbol{\Lambda}^{-1} \boldsymbol{Y}/2}\, d\boldsymbol{Y} = \left(\int_{-\infty}^{\infty} e^{-y_1^2/2\lambda_1}\, dy_1 \right) \cdots \left(\int_{-\infty}^{\infty} e^{-y_M^2/2\lambda_M}\, dy_M \right)$$

$$= \left(\sqrt{2\pi\lambda_1}\right) \cdots \left(\sqrt{2\pi\lambda_M}\right) = \left(\sqrt{2\pi}\right)^M \sqrt{\det V}. \tag{7}$$

The determinant of V (also the determinant of Λ) is the product $(\lambda_1) \ldots (\lambda_M)$ of the eigenvalues. Then (7) gives the correct number to divide by so that $p(x_1, \ldots, x_M)$ in equation (6) has integral $= 1$ as desired.

The mean and variance of $p(\boldsymbol{x})$ are also M-dimensional integrals. The same idea of diagonalizing V by its eigenvectors $\boldsymbol{q}_1$ to $\boldsymbol{q}_M$ and introducing $\boldsymbol{Y} = \boldsymbol{Q}^{\mathrm{T}} \boldsymbol{X}$ will find those integrals:

Vector $\boldsymbol{m}$ of means
$$\int \cdots \int \boldsymbol{x}\, p(\boldsymbol{x})\, d\boldsymbol{x} = (m_1, m_2, \ldots) = \boldsymbol{m} \tag{8}$$

Covariance matrix V
$$\int \cdots \int (\boldsymbol{x} - \boldsymbol{m})\, p(\boldsymbol{x}) (\boldsymbol{x} - \boldsymbol{m})^{\mathrm{T}}\, d\boldsymbol{x} = \boldsymbol{V} \tag{9}$$

Conclusion: Formula (6) for the probability density $p(\boldsymbol{x})$ has all the properties we want.

Weighted Least Squares

In Chapter 4, least squares started from an unsolvable system $A\boldsymbol{x} = \boldsymbol{b}$. We chose $\widehat{\boldsymbol{x}}$ to minimize the error $||\boldsymbol{b} - A\boldsymbol{x}||^2$. That led us to the least squares equation $A^{\mathrm{T}}A\widehat{\boldsymbol{x}} = A^{\mathrm{T}}\boldsymbol{b}$. The best $A\widehat{\boldsymbol{x}}$ is the projection of $\boldsymbol{b}$ onto the column space of A. But is this squared distance $E = ||\boldsymbol{b} - A\boldsymbol{x}||^2$ the right error measure to minimize?

If the measurement errors in $\boldsymbol{b}$ are independent random variables, with mean $m = 0$ and variance $\sigma^2 = 1$ and a normal distribution, Gauss would say **yes**: *Use least squares.* If the errors are not independent or their variances are not equal. Gauss would say **no**: *Use **weighted** least squares,*

This section will show that the good measure of error is $\boldsymbol{E = (b - Ax)^{\mathrm{T}} V^{-1}(b - Ax)}$. The equation for the best $\widehat{\boldsymbol{x}}$ uses the covariance matrix V:

$$\textbf{Weighted least squares} \qquad \boldsymbol{A^{\mathrm{T}}V^{-1}A\widehat{x} = A^{\mathrm{T}}V^{-1}b}. \tag{10}$$

The most important examples have m *independent* errors in $\boldsymbol{b}$. Those errors have variances $\sigma_1^2, \ldots, \sigma_m^2$. By independence, V is a diagonal matrix. The good weights $1/\sigma_1^2, \ldots, 1/\sigma_m^2$ come from V^{-1}. *We are weighting the errors in* $\boldsymbol{b}$ *to have* **variance** $= \mathbf{1}$ (and **covariance** $= \mathbf{0}$).

$$\textbf{Weighted least squares} \quad \textbf{Independent errors in } \boldsymbol{b} \qquad \text{Minimize} \quad E = \sum_{i=1}^{m} \frac{(\boldsymbol{b} - A\boldsymbol{x})_i^2}{\sigma_i^2} \tag{11}$$

By weighting the errors, we are "whitening" the noise. **White noise** is a quick description of independent errors based on the standard Gaussian $\mathbf{N}(0, 1)$ with mean zero and $\sigma^2 = 1$.

Let me write down the steps to equations (10) and (11) for the best $\widehat{\boldsymbol{x}}$:

Start with $A\boldsymbol{x} = \boldsymbol{b}$ (m equations, n unknowns, $m > n$, no solution)

Each right side b_i has mean zero and variance σ_i^2. The b_i are independent

Divide the ith equation by σ_i to have variance $= 1$ for every b_i/σ_i

That division turns $A\boldsymbol{x} = \boldsymbol{b}$ into $V^{-1/2}A\boldsymbol{x} = V^{-1/2}\boldsymbol{b}$ with $V^{-1/2} = \mathrm{diag}\,(1/\sigma_1, \ldots, 1/\sigma_m)$

Ordinary least squares on those weighted equations has $A \to V^{-1/2}A$ and $\boldsymbol{b} \to V^{-1/2}\boldsymbol{b}$

$$(V^{-1/2}A)^{\mathrm{T}}(V^{-1/2}A)\widehat{\boldsymbol{x}} = (V^{-1/2}A)^{\mathrm{T}}V^{-1/2}\boldsymbol{b} \;\text{ is }\; \boldsymbol{A^{\mathrm{T}}V^{-1}A\widehat{x} = A^{\mathrm{T}}V^{-1}b}. \tag{12}$$

Because of $1/\sigma^2$ in V^{-1}, more reliable equations (*smaller* σ) get heavier weights. This is the main point of weighted least squares.

Those diagonal weightings (uncoupled equations) are the most frequent and the simplest. They apply to *independent errors in the* b_i. When these measurement errors are not independent, V is no longer diagonal—but (12) is still the correct weighted equation.

In practice, finding all the covariances can be serious work. Diagonal V is simpler.

The Variance in the Estimated $\widehat{x}$

One more point: Often the important question is not the best $\widehat{x}$ for a particular b. This is only one sample! The real goal is to know the **reliability of the whole experiment.** This is measured (as reliability always is) by the **variance in the estimate** $\widehat{x}$. First, zero mean in b gives zero mean in $\widehat{x}$. Then the formula connecting variance V in the inputs b to variance W in the outputs $\widehat{x}$ turns out to be beautiful:

$$\textbf{Variance-covariance matrix for } \widehat{x} \quad W = \mathrm{E}[(\widehat{x}-x)(\widehat{x}-x)^{\mathrm{T}}] = (A^{\mathrm{T}}V^{-1}A)^{-1}. \tag{13}$$

That smallest possible variance comes from the best possible weighting, which is V^{-1}.

This key formula is a perfect application of Section V.4. **If b has covariance matrix V, then $\widehat{x} = Lb$ has covariance matrix LVL^{T}.** The matrix L is $(A^{\mathrm{T}}V^{-1}A)^{-1}A^{\mathrm{T}}V^{-1}$, because $\widehat{x} = Lb$ solves the weighted equation $(A^{\mathrm{T}}V^{-1}A)\,\widehat{x} = A^{\mathrm{T}}V^{-1}b$. Substitute this into LVL^{T} and watch equation (13) appear:

$$LVL^{\mathrm{T}} = (A^{\mathrm{T}}V^{-1}A)^{-1}A^{\mathrm{T}}V^{-1} \quad V \quad V^{-1}A\,(A^{\mathrm{T}}V^{-1}A)^{-1} = (A^{\mathrm{T}}V^{-1}A)^{-1}.$$

This is the covariance W of the output $\widehat{x}$. It is time for an example.

Example 1 Suppose a doctor measures your heart rate x three times $(m=3, n=1)$:

$$\begin{matrix} x = b_1 \\ x = b_2 \\ x = b_3 \end{matrix} \quad \text{is} \quad Ax = b \quad \text{with} \quad A = \begin{bmatrix} 1 \\ 1 \\ 1 \end{bmatrix} \quad \text{and} \quad V = \begin{bmatrix} \sigma_1^2 & 0 & 0 \\ 0 & \sigma_2^2 & 0 \\ 0 & 0 & \sigma_3^2 \end{bmatrix}$$

The variances could be $\sigma_1^2 = 1/9$ and $\sigma_2^2 = 1/4$ and $\sigma_3^2 = 1$. The weights are 3 then 2 then 1. You are getting more nervous as measurements are taken: b_3 is less reliable than b_2 and b_1. All three measurements contain some information, so they all go into the best (weighted) estimate $\widehat{x}$:

$$V^{-1/2}A\widehat{x} = V^{-1/2}b \quad \text{is} \quad \begin{matrix} 3x = 3b_1 \\ 2x = 2b_2 \\ 1x = 1b_3 \end{matrix} \quad \text{leading to} \quad A^{\mathrm{T}}V^{-1}A\widehat{x} = A^{\mathrm{T}}V^{-1}b$$

$$\begin{bmatrix} 1 & 1 & 1 \end{bmatrix} \begin{bmatrix} 9 & & \\ & 4 & \\ & & 1 \end{bmatrix} \begin{bmatrix} 1 \\ 1 \\ 1 \end{bmatrix} \widehat{x} = \begin{bmatrix} 1 & 1 & 1 \end{bmatrix} \begin{bmatrix} 9 & & \\ & 4 & \\ & & 1 \end{bmatrix} \begin{bmatrix} b_1 \\ b_2 \\ b_3 \end{bmatrix}$$

$$\widehat{x} = \frac{9b_1 + 4b_2 + b_3}{14} \quad \textbf{is the best weighted average of } b_1, b_2, b_3$$

Most weight is on b_1 since its variance σ_1 is smallest. The variance of $\widehat{x}$ has the beautiful formula $\boldsymbol{W} = (\boldsymbol{A}^{\mathrm{T}}\boldsymbol{V}^{-1}\boldsymbol{A})^{-1}$. That variance $W = \frac{1}{14}$ went down from $\frac{1}{9}$ by including b_2 and b_3:

$$\textbf{Variance of } \widehat{x} \qquad \left(\begin{bmatrix} 1 & 1 & 1 \end{bmatrix} \begin{bmatrix} 9 & & \\ & 4 & \\ & & 1 \end{bmatrix} \begin{bmatrix} 1 \\ 1 \\ 1 \end{bmatrix} \right)^{-1} = \frac{1}{14} \quad \text{is smaller than} \quad \frac{1}{9}$$

The BLUE theorem of Gauss (proved on the website) says that our $\widehat{x} = Lb$ is the Best Linear Unbiased Estimate of the solution to $Ax = b$. For any other unbiased choice $x^* = L^*b$, the variance $W^* = L^*VL^{*\mathrm{T}}$ will be greater than our $W = LVL^{\mathrm{T}}$.
Note: Greater means that $W^* - W$ will be positive semidefinite. Unbiased means $L^*A = I$. So an exact $Ax = b$ will produce the right answer $x = L^*b = L^*Ax$.

I must add that there are reasons not to minimize squared errors in the first place. One reason: This $\widehat{x}$ often has many small components. The squares of small numbers are very small, and they appear when we minimize. It is easier to make sense of *sparse* vectors—only a few nonzeros. Statisticians often prefer to minimize **unsquared errors**: **the sum of** $|(b - Ax)_i|$. *This error measure is* ℓ^1 *instead of* ℓ^2. Because of the absolute values, the equation for $\widehat{x}$ using the ℓ^1 norm becomes nonlinear.

Fast new algorithms are computing a sparse $\widehat{x}$ quickly and **the future belongs to** $\boldsymbol{\ell^1}$. Section IV.4 on compressive sensing was an impressive application of regression in ℓ^1.

The Kalman Filter

The "Kalman filter" is the great algorithm in dynamic least squares. That word *dynamic* means that new measurements b_k keep coming. So the best estimate $\widehat{x}_k$ keeps changing (based on all of $b_0, \ldots, b_k$). More than that, the matrix A is changing. So $\widehat{x}_2$ will be our best least squares estimate of the latest solution to the **whole history of observation equations and update equations (state equations).** Up to time 2, there are 3 observations and 2 state equations:

$$A_0x_0 = b_0 \qquad x_1 = F_0x_0 \qquad A_1x_1 = b_1 \qquad x_2 = F_1x_1 \qquad A_2x_2 = b_2 \tag{14}$$

The Kalman idea is to introduce one equation at a time. There will be errors in each equation. With every new equation, we update the best estimate $\widehat{x}_k$ for the current x_k. But history is not forgotten! This new estimate $\widehat{x}_k$ uses all the past observations b_0 to b_{k-1} and all the state equations $x_{\text{new}} = F_{\text{old}}\, x_{\text{old}}$. A large and growing least squares problem.

One more important point about (14). Each least squares equation is **weighted** using the covariance matrix V_k for the error in b_k. There is even a covariance matrix C_k for errors in the update equations $x_{k+1} = F_kx_k$. The best $\widehat{x}_1$ then depends on b_0, b_1 and F_1 and V_0, V_1 and C_1. The good way to write $\widehat{x}_1$ is as an update to the previous $\widehat{x}_1$.

Let me concentrate on a simplified problem, without the matrices F_k and the covariances C_k. We are estimating the same true $\boldsymbol{x}$ at every step. How do we get $\widehat{\boldsymbol{x}}_1$ from $\widehat{\boldsymbol{x}}_0$?

OLD $\quad A_0\,\boldsymbol{x}_0 = \boldsymbol{b}_0$ leads to the weighted equation $A_0^{\mathrm{T}}\,V_0^{-1}\,A_0\,\widehat{\boldsymbol{x}}_0 = A_0^{\mathrm{T}}\,V_0^{-1}\,\boldsymbol{b}_0. \qquad (15)$

NEW $\quad \begin{bmatrix} A_0 \\ A_1 \end{bmatrix} \widehat{\boldsymbol{x}}_1 = \begin{bmatrix} \boldsymbol{b}_0 \\ \boldsymbol{b}_1 \end{bmatrix}$ leads to the following weighted equation for $\widehat{\boldsymbol{x}}_1$:

$$\begin{bmatrix} A_0^{\mathrm{T}} & A_1^{\mathrm{T}} \end{bmatrix} \begin{bmatrix} V_0^{-1} & \\ & V_1^{-1} \end{bmatrix} \begin{bmatrix} A_0 \\ A_1 \end{bmatrix} \widehat{\boldsymbol{x}}_1 = \begin{bmatrix} A_0^{\mathrm{T}} & A_1^{\mathrm{T}} \end{bmatrix} \begin{bmatrix} V_0^{-1} & \\ & V_1^{-1} \end{bmatrix} \begin{bmatrix} \boldsymbol{b}_0 \\ \boldsymbol{b}_1 \end{bmatrix}. \qquad (16)$$

Yes, we could just solve that new problem and forget the old one. But the old solution $\widehat{\boldsymbol{x}}_0$ needed work that we hope to reuse in $\widehat{\boldsymbol{x}}_1$. What we look for is **an update to $\widehat{\boldsymbol{x}}_0$** :

Kalman update gives $\widehat{\boldsymbol{x}}_1$ from $\widehat{\boldsymbol{x}}_0$	$\widehat{\boldsymbol{x}}_1 = \widehat{\boldsymbol{x}}_0 + K_1(\boldsymbol{b}_1 - A_1\,\widehat{\boldsymbol{x}}_0).$	(17)

The update correction is the mismatch $\boldsymbol{b}_1 - A_1\widehat{\boldsymbol{x}}_0$ between the old state $\widehat{\boldsymbol{x}}_0$ and the new measurements $\boldsymbol{b}_1$—multiplied by the *Kalman gain matrix* K_1. The formula for K_1 comes from comparing the solutions $\widehat{\boldsymbol{x}}_1$ and $\widehat{\boldsymbol{x}}_0$ to (15) and (16). And when we update $\widehat{\boldsymbol{x}}_0$ to $\widehat{\boldsymbol{x}}_1$ based on new data $\boldsymbol{b}_1$, **we also update the covariance matrix $\boldsymbol{W}_0$ to $\boldsymbol{W}_1$.**

Remember $W_0 = (A_0^{\mathrm{T}}\,V_0^{-1}\,A_0)^{-1}$ from equation (13). Update its inverse from W_0^{-1} to W_1^{-1}:

Covariance $\boldsymbol{W}_1$ of errors in $\widehat{\boldsymbol{x}}_1$	$\boldsymbol{W}_1^{-1} = \boldsymbol{W}_0^{-1} + \boldsymbol{A}_1^{\mathrm{T}}\,\boldsymbol{V}_1^{-1}\,\boldsymbol{A}_1$	(18)
Kalman gain matrix $\boldsymbol{K}_1$	$\boldsymbol{K}_1 = \boldsymbol{W}_1\,\boldsymbol{A}_1^{\mathrm{T}}\,\boldsymbol{V}_1^{-1}$	(19)

This is the heart of the Kalman filter. Notice the importance of the covariance matrices W_k. Those matrices measure the reliability of the whole process, where the vector $\widehat{\boldsymbol{x}}_k$ estimates the current state based on the particular measurements $\boldsymbol{b}_0$ to $\boldsymbol{b}_k$.

Whole chapters and whole books are written to explain the dynamic Kalman filter, when the states $\boldsymbol{x}_k$ are also changing (based on the matrices F_k). There is a *prediction* of $\boldsymbol{x}_k$ using F, followed by a *correction* using the new data $\boldsymbol{b}$. Perhaps best to stop here!

This page was about **recursive least squares**: adding new data $\boldsymbol{b}_k$ and updating the best current estimate $\widehat{\boldsymbol{x}}_k$ based on all the data—and updating its covariance matrix W_k. The updating idea began with the Sherman-Morrison-Woodbury formula for $(A-UV^{\mathrm{T}})^{-1}$ in Section III.1. Numerically that is the key to Kalman's success—exchanging inverse matrices of size n for inverse matrices of size k.

Problem Set V.5

1 Two measurements of the same variable x give two equations $x = b_1$ and $x = b_2$. Suppose the means are zero and the variances are σ_1^2 and σ_2^2, with independent errors: V is diagonal with entries σ_1^2 and σ_2^2. Write the two equations as $A\boldsymbol{x} = \boldsymbol{b}$ (A is 2 by 1). As in the text Example 1, find this best estimate $\widehat{\boldsymbol{x}}$ based on b_1 and b_2:

$$\widehat{\boldsymbol{x}} = \frac{b_1/\sigma_1^2 + b_2/\sigma_2^2}{1/\sigma_1^2 + 1/\sigma_2^2} \qquad \mathrm{E}\left[\widehat{\boldsymbol{x}}\,\widehat{\boldsymbol{x}}^{\mathrm{T}}\right] = \left(\frac{1}{\sigma_1^2} + \frac{1}{\sigma_2^2}\right)^{-1}.$$

2 (a) In Problem 1, suppose the second measurement b_2 becomes super-exact and its variance $\sigma_2 \to 0$. What is the best estimate $\widehat{x}$ when σ_2 reaches zero?

(b) The opposite case has $\sigma_2 \to \infty$ and no information in b_2. What is now the best estimate $\widehat{x}$ based on b_1 and b_2 ?

3 If x and y are independent with probabilities $p_1(x)$ and $p_2(y)$, then $p(x,y) = p_1(x)\,p_2(y)$. By separating double integrals into products of single integrals ($-\infty$ to ∞) show that $\iint p(x,y)\,dx\,dy = \mathbf{1}$ and $\iint (x+y)\,p(x,y)\,dx\,dy = \boldsymbol{m_1 + m_2}$.

4 Continue Problem 3 for independent x, y to show that $p(x,y) = p_1(x)\,p_2(y)$ has

$$\iint (x - m_1)^2\,p(x,y)\,dx\,dy = \boldsymbol{\sigma_1^2} \qquad \iint (x - m_1)(y - m_2)\,p(x,y)\,dx\,dy = \mathbf{0}.$$

So the 2 by 2 covariance matrix V is diagonal and its entries are ______.

5 Suppose $\widehat{x}_k$ is the average of $b_1, \ldots, b_k$. A new measurement b_{k+1} arrives. The Kalman update equation (17) gives the new average $\widehat{x}_{k+1}$:

Verify that $\widehat{x}_{k+1} = \widehat{x}_k + \dfrac{1}{k+1}(b_{k+1} - \widehat{x}_k)$ *is the correct average of* $b_1 \ldots, b_{k+1}$.

Also check the update equation (18) for the variance $W_{k+1} = \sigma^2/(k+1)$ of this average $\widehat{x}$ assuming that $W_k = \sigma^2/k$ and b_{k+1} has variance $V = \sigma^2$.

6 **(Steady model)** Problem 5 was *static* least squares. All the sample averages $\widehat{x}_k$ were estimates of the same x. To make the Kalman filter *dynamic*, include also a *state equation* $x_{k+1} = Fx_k$ *with its own error variance* s^2. The dynamic least squares problem allows x to "drift" as k increases:

$$\begin{bmatrix} 1 & \\ -F & 1 \\ & 1 \end{bmatrix} \begin{bmatrix} x_0 \\ x_1 \end{bmatrix} = \begin{bmatrix} b_0 \\ 0 \\ b_1 \end{bmatrix} \quad \text{with variances} \quad \begin{bmatrix} \sigma^2 \\ s^2 \\ \sigma^2 \end{bmatrix}.$$

With $F = 1$, divide both sides of those three equations by σ, s, and σ. Find $\widehat{x_0}$ and $\widehat{x_1}$ by least squares, which gives more weight to the recent b_1. The Kalman filter is developed in *Algorithms for Global Positioning* (Borre and Strang).

V.6 Markov Chains

The key facts about Markov chains are illustrated by rental cars! Start with 100 cars in Chicago. Every month, cars move between Chicago and Denver.

80% of the Chicago cars stay in Chicago	30% of the Denver cars move to Chicago
20% of the Chicago cars move to Denver	70% of the Denver cars stay in Denver

In matrix language, the movement of cars from month n to $n+1$ is given by $\boldsymbol{y}_{n+1} = \boldsymbol{P}\boldsymbol{y}_n$:

$$\boxed{y_{n+1} = \begin{bmatrix} \textbf{Chicago cars} \\ \textbf{Denver cars} \end{bmatrix}_{n+1} = \begin{bmatrix} 0.8 & 0.3 \\ 0.2 & 0.7 \end{bmatrix} \begin{bmatrix} \textbf{Chicago cars} \\ \textbf{Denver cars} \end{bmatrix}_{n} = \boldsymbol{P}\boldsymbol{y}_n} \quad (1)$$

Every month we multiply by that **"Markov matrix"** $\boldsymbol{P}$. Both columns add to 1. After n months the distribution of cars is $\boldsymbol{y}_n = \boldsymbol{P}^n\boldsymbol{y}_0$. Our example has $\boldsymbol{y}_0 = (100, 0)$ since all cars start in Chicago:

$$y_0 = \begin{bmatrix} 100 \\ 0 \end{bmatrix} \quad y_1 = \begin{bmatrix} 80 \\ 20 \end{bmatrix} \quad y_2 = \begin{bmatrix} 70 \\ 30 \end{bmatrix} \quad y_3 = \begin{bmatrix} 65 \\ 35 \end{bmatrix} \cdots \quad y_\infty = \begin{bmatrix} \mathbf{60} \\ \mathbf{40} \end{bmatrix}.$$

Suppose that all 100 cars start in Denver instead of Chicago:

$$y_0 = \begin{bmatrix} 0 \\ 100 \end{bmatrix} \quad y_1 = \begin{bmatrix} 30 \\ 70 \end{bmatrix} \quad y_2 = \begin{bmatrix} 45 \\ 55 \end{bmatrix} \quad y_3 = \begin{bmatrix} 52.5 \\ 47.5 \end{bmatrix} \cdots \quad y_\infty = \begin{bmatrix} \mathbf{60} \\ \mathbf{40} \end{bmatrix}.$$

Both ways lead to the same 60-40 limiting distribution. It doesn't matter where the cars start. Since we are looking at powers P^n of the matrix P, this is a problem for the eigenvalues and eigenvectors of P:

$$\textbf{Eigenvalues} \quad \det \begin{bmatrix} .8-\lambda & .3 \\ .2 & .7-\lambda \end{bmatrix} = \lambda^2 - 1.5\lambda + 0.5 = (\lambda - \mathbf{1})(\lambda - \mathbf{0.5}) \quad \begin{matrix} \lambda = \mathbf{1} \\ \lambda = \mathbf{0.5} \end{matrix}$$

$$\textbf{Eigenvectors} \quad \begin{bmatrix} 0.8 & 0.3 \\ 0.2 & 0.7 \end{bmatrix} \begin{bmatrix} .60 \\ .40 \end{bmatrix} = \begin{bmatrix} \mathbf{.60} \\ \mathbf{.40} \end{bmatrix} \qquad \begin{bmatrix} 0.8 & 0.3 \\ 0.2 & 0.7 \end{bmatrix} \begin{bmatrix} 1 \\ -1 \end{bmatrix} = \frac{1}{2} \begin{bmatrix} \mathbf{1} \\ \mathbf{-1} \end{bmatrix}$$

Those explain everything. The limiting 60-40 distribution of cars is the **steady state**: **eigenvalue** $\boldsymbol{\lambda_1 = 1}$. So $\boldsymbol{y}_n = (60, 40)$ gives $\boldsymbol{y}_{n+1} = (60, 40)$. Then $\lambda_2 = \frac{1}{2}$ means:

Every month the distance to steady state is multiplied by $\frac{1}{2}$.

You see that in the numbers above: $100, 80, 70, 65$ in Chicago has multiplied the distance to 60 (steady state) by $\frac{1}{2}$ every month. Similarly $0, 20, 30, 35$ in Denver is halving the distance to 40 (steady state). In matrix notation, P is diagonalized as $X\Lambda X^{-1}$ by using its eigenvectors and eigenvalues. Then $\boldsymbol{P}^n = (X\Lambda X^{-1}) \dots (X\Lambda X^{-1}) = \boldsymbol{X\Lambda^n X^{-1}}$:

$$P^n = X\Lambda^n X^{-1} = \begin{bmatrix} .6 & 1 \\ .4 & -1 \end{bmatrix} \begin{bmatrix} 1^n & \\ & (\frac{1}{2})^n \end{bmatrix} \begin{bmatrix} 1 & 1 \\ .4 & -.6 \end{bmatrix} = \begin{bmatrix} \mathbf{.6} & \mathbf{.6} \\ \mathbf{.4} & \mathbf{.4} \end{bmatrix} + \left(\frac{\mathbf{1}}{\mathbf{2}}\right)^n \begin{bmatrix} \mathbf{.4} & \mathbf{-.6} \\ \mathbf{-.4} & \mathbf{.6} \end{bmatrix}$$

For $n = 1$ we have P. For $n = \infty$ the limiting matrix P^∞ has .6, .4 in both columns. The 100 cars could start in the Chicago column or the Denver column—always a 60-40 split as $n \to \infty$. This is the outstanding feature of positive Markov matrices like P.

The requirements for a **positive Markov matrix** are:

> **All $p_{ij} > 0$ and each column of P adds to 1 (so no car is lost). Then $\mathbf{1}^{\mathrm{T}}P = \mathbf{1}^{\mathrm{T}}$**
>
> The matrix P has $\lambda_1 = 1$ (largest eigenvalue) and $\boldsymbol{x}_1 > 0$ (positive eigenvector).

It is the Perron-Frobenius Theorem for positive matrices that guarantees $\lambda_1 > 0$ and $\boldsymbol{x}_1 > 0$. Then the fact that columns add to 1 tells us that $P^{\mathrm{T}}\mathbf{1} = \mathbf{1}$ and $\lambda_1 = 1$. And this produces the steady state $\boldsymbol{y}_\infty$ as a multiple of $\boldsymbol{x}_1$. Remember that row 1 of X^{-1} is the left eigenvector $\begin{bmatrix} \mathbf{1} & \mathbf{1} & \ldots & \mathbf{1} \end{bmatrix}$:

Convergence

$$P^n = X\Lambda^n X^{-1} = \begin{bmatrix} \boldsymbol{x}_1 & \boldsymbol{x}_2 \cdots \boldsymbol{x}_n \end{bmatrix} \begin{bmatrix} 1 & & \\ & \lambda_2^n & \\ & & \ddots \end{bmatrix} \begin{bmatrix} & X^{-1} & \end{bmatrix} \to \begin{bmatrix} \boldsymbol{x}_1 & \boldsymbol{x}_1 \cdots \boldsymbol{x}_1 \end{bmatrix}$$

As $n \to \infty$, only the 1 in that diagonal matrix Λ^n will survive. Columns times rows become **column $\boldsymbol{x}_1$ of X times** $\begin{bmatrix} 1 & 1 & \ldots & 1 \end{bmatrix}$. The limiting matrix P^∞ has $\boldsymbol{x}_1$ in every column! The steady state $\boldsymbol{y}_\infty = P^\infty \boldsymbol{y}_0$ has to be a multiple of that column vector $\boldsymbol{x}_1$.

The multiple was $(60, 40)$ in our example because we started with 100 cars. A Markov chain doesn't destroy old cars or add new cars—it eventually distributes them according to the leading eigenvector $\boldsymbol{x}_1$ of P.

Now we look at P as a **matrix of probabilities**. Then comes Perron-Frobenius.

Transition Probabilities

Markov chains are perfect examples of linear algebra within probability theory. The fundamental numbers are the probabilities p_{ij} of moving from state j at time n to state i at time $n + 1$:

Transition probabilities $\;p_{ij} =$ **Probability that** $x(n+1) = i$ **if** $x(n) = j$. (2)

There are two key points hidden in that simple statement. First, the probability p_{ij} does not depend on n. The rules stay the same at all times. Second, *the probabilities $\boldsymbol{y}_{n+1}$ for the new state $x(n+1)$ depend only on the current state $x(n)$*—not on any earlier history.

One question is still to answer: What are the possible "states" of the Markov chain? In the example the states are Chicago and Denver (*sorry, cities*). Here are three options:

Finite Markov Chain	Each state $x(n)$ is one of the numbers $1, 2, \ldots, N$
Infinite State Markov Chain	Each state $x(n)$ is an integer $n = 0, 1, 2, \ldots$
Continuous Markov Chain	Each state $x(n)$ is a real number.

We mostly choose finite chains with N possible states. The initial state $x(0)$ could be given. Or we may only know the **vector y_0 of probabilities** for that initial state. Unlike differential equations, $x(0)$ *does not determine* $x(1)$. If $x(0) = j$, that only determines the N probabilities in y_1 for the new state $x(1)$. That new state is a number from 1 to N.

The probabilities for those new states are the numbers $p_{1j}, p_{2j}, \ldots, p_{Nj}$. **Those probabilities must add to 1:**

$$\textbf{Column } j \textbf{ of } P \qquad p_{1j} + p_{2j} + \cdots + p_{Nj} = 1. \tag{3}$$

Those numbers go naturally into a matrix P = N by N matrix of probabilities p_{ij}. Those are called transition probabilities and **P is the transition matrix.** It tells us everything we can know (only probabilities, not facts!) about the transition from state $x(0) = j$ to state $x(1) = i$. And this same matrix P applies to the transition from $x(n)$ to $x(n+1)$ at every future time.

$$\begin{array}{c}\textbf{Transition}\\ \textbf{matrix } P\\ y_{n+1} = Py_n\end{array} \qquad y_{n+1} = \begin{bmatrix} \text{Prob}\{x(n+1)=1\} \\ \vdots \\ \text{Prob}\{x(n+1)=N\} \end{bmatrix} = \begin{bmatrix} p_{11} & \cdots & p_{1N} \\ \vdots & & \vdots \\ p_{N1} & \cdots & p_{NN} \end{bmatrix} \begin{bmatrix} \text{Prob}\{x(n)=1\} \\ \vdots \\ \text{Prob}\{x(n)=N\} \end{bmatrix} \tag{4}$$

All entries of the matrix P have $0 \leq p_{ij} \leq 1$. By equation (3), each column adds to 1:

$$\mathbf{1}^{\mathrm{T}} = \textbf{row vector of } N \textbf{ ones} \qquad \mathbf{1}^{\mathrm{T}}P = \mathbf{1}^{\mathrm{T}} \text{ and } P^{\mathrm{T}}\mathbf{1} = \mathbf{1} \qquad \mathbf{1} = \textbf{column vector of } N \textbf{ ones}$$

So P^{T} is a nonnegative matrix with eigenvalue $\lambda = 1$ and eigenvector $\mathbf{1} = (1, 1, \ldots, 1)$. And P is also a nonnegative matrix with eigenvalue $\lambda = 1$. But we must find the eigenvector with $Pv = v$:

Example 1 $\quad P = \begin{bmatrix} \mathbf{0.8} & \mathbf{0.3} \\ \mathbf{0.2} & \mathbf{0.7} \end{bmatrix} \quad \begin{bmatrix} 1 & 1 \end{bmatrix} P = \begin{bmatrix} 1 & 1 \end{bmatrix} \quad Pv = \begin{bmatrix} 0.8 & 0.3 \\ 0.2 & 0.7 \end{bmatrix} \begin{bmatrix} \mathbf{0.6} \\ \mathbf{0.4} \end{bmatrix} = v$

Thus $v = (0.6, 0.4)$. The trace is $0.8 + 0.7 = 1.5$. So the second eigenvalue is $\lambda_2 = \mathbf{0.5}$. The second eigenvector v_2 of P is always orthogonal to the first eigenvector $(1, 1)$ of P^{T}.

We return to the transition equation $y_{n+1} = Py_n$ for the probabilities. Those vectors y_n and y_{n+1} contain the probabilities for the N different states, at time n and at time $n+1$. At all times, **the columns of P^n add to 1** (and so do the probabilities in y_n):

$$\mathbf{1}^{\mathrm{T}} y_{n+1} = \mathbf{1}^{\mathrm{T}}(Py_n) = (\mathbf{1}^{\mathrm{T}}P)y_n = \mathbf{1}^{\mathrm{T}}y_n = 1. \tag{5}$$

Here is the fundamental question for a Markov chain. The transition matrix P is fixed and known. The starting state $x(0)$ or the vector y_0 of probabilities for that state may be known or unknown. Key question: **Do the probability vectors $y_n = P^n y_0$ have a limit y_∞ as $n \to \infty$?** We expect y_∞ to be independent of the initial probabilities in y_0. We will see that y_∞ often exists, but not for every P. When it exists, **y_∞ tells us how often we expect to be in each state.**

Example 2 The transition matrix can be $P = \begin{bmatrix} 0 & 1 \\ 1 & 0 \end{bmatrix}$ = **switching matrix.**

This means: The system changes its state at every time step. State 1 at time n leads to State 2 at time $n+1$. If the initial probabilities were in $\boldsymbol{y}_0 = \left(\frac{1}{3}, \frac{2}{3}\right)$ then $\boldsymbol{y}_1 = P\boldsymbol{y}_0 = \left(\frac{2}{3}, \frac{1}{3}\right)$. The probabilities go back and forth between those two vectors. *No steady state.*

Our matrices P always have the eigenvalue $\lambda = 1$. Its eigenvector would be a steady state (nothing changes when we multiply by P). But this particular P also has the eigenvalue $\lambda = -1$. Its effect will not die out as $n \to \infty$. The only steadiness will be seen in $\boldsymbol{y}_0 = \boldsymbol{y}_2 = \boldsymbol{y}_4 \ldots$ and separately in $\boldsymbol{y}_1 = \boldsymbol{y}_3 = \boldsymbol{y}_5 \ldots$ The powers of this matrix P oscillate between P and I.

Other matrices P do have a steady state: $\boldsymbol{P^n} \to \boldsymbol{P^\infty}$ and $\boldsymbol{y}_n \to \boldsymbol{y}_\infty$. Notice that the actual states x are still changing—based on the probabilities p_{ij} in P. **The vector $\boldsymbol{y}_\infty$ tells us the fraction $(y_{1\infty}, \ldots, y_{N\infty})$ of time that the system is eventually in each state.**

Positive P or Nonnegative P

There is a clear difference between our two examples: $P_1 > 0$ and $P_2 \geq 0$.

$P_1 = \begin{bmatrix} 0.8 & 0.3 \\ 0.2 & 0.7 \end{bmatrix}$ has eigenvalues **1** and $\frac{\mathbf{1}}{\mathbf{2}}$. The powers $\left(\dfrac{1}{2}\right)^n$ approach zero.

$P_2 = \begin{bmatrix} 0 & 1 \\ 1 & 0 \end{bmatrix}$ has eigenvalues **1** and $\mathbf{-1}$. The powers $(-1)^n$ don't approach zero.

Every column of P adds to 1. Then $\lambda = 1$ is an eigenvalue, because the rows of $P - I$ add to the zero row. *And no entry of P is negative*. The two properties together ensure that no eigenvalue can have $|\lambda| > 1$. But there is an important difference between P_1 and P_2.

P_2 has zero entries. This opens the possibility that the magnitude of λ_2 could be 1.

P_1 has strictly positive entries. This guarantees that the magnitude of λ_2 has $|\lambda_2| < 1$.

The Perron-Frobenius Theorem for $\boldsymbol{P_1} > \mathbf{0}$ (strictly positive entries) guarantees success:

1. **The largest eigenvalue λ_1 of P and its eigenvector v_1 are strictly positive.**
2. **All other eigenvalues $\lambda_2, \ldots, \lambda_N$ have $|\lambda| < \lambda_1$. Markov matrices have $\lambda_1 = 1$.**

A third example P_3 shows that zeros in P don't always ruin the approach to steady state:

Example 3 $\boldsymbol{P_3} = \begin{bmatrix} 1 & \frac{1}{2} \\ 0 & \frac{1}{2} \end{bmatrix}$ has $\lambda = 1, \frac{1}{2}$ with $\boldsymbol{v}_1 = \begin{bmatrix} 1 \\ 0 \end{bmatrix}$ and $\boldsymbol{v}_2 = \begin{bmatrix} 1 \\ -1 \end{bmatrix}$.

Even with that zero in P_3, all columns of $(P_3)^n$ approach $\boldsymbol{v}_1$ = first eigenvector:

$$(\boldsymbol{P_3})^n = \begin{bmatrix} 1 & \frac{1}{2} \\ 0 & \frac{1}{2} \end{bmatrix}^n = \begin{bmatrix} 1 & 1-\left(\frac{1}{2}\right)^n \\ 0 & \left(\frac{1}{2}\right)^n \end{bmatrix} \to \begin{bmatrix} \mathbf{1} & \mathbf{1} \\ \mathbf{0} & \mathbf{0} \end{bmatrix}. \tag{6}$$

This is the steady state that we want and expect. Then $\boldsymbol{y}_n = (P_3)^n \boldsymbol{y}_0$ approaches that same eigenvector $\boldsymbol{v}_1 = \begin{bmatrix} 1 & 0 \end{bmatrix}$. This Markov chain moves everybody to state 1 as $n \to \infty$.

Convergence to Steady State as $n \to \infty$

For strictly positive Markov matrices, the best way to see the convergence $P^n \to P^\infty$ is by diagonalizing P. *Assume for now* that p has n independent eigenvectors. The eigenvalue matrix Λ starts with $\lambda_1 = 1$. Its eigenvector matrix X has the leading eigenvector $\boldsymbol{v}_1$ in its first column. As n increases, that eigenvector $\boldsymbol{v}_1$ will appear in every column of P^n.

$$P = X\Lambda X^{-1} \text{ means that } P^n = (X\Lambda X^{-1})\ldots(X\Lambda X^{-1}) = \boldsymbol{X\Lambda^n X^{-1}}$$

The columns of X are the eigenvectors $\boldsymbol{v}_1, \ldots, \boldsymbol{v}_n$ of P. The rows of X^{-1} are the eigenvectors of P^{T} (starting with the all-ones vector $\mathbf{1}^{\mathrm{T}}$). Because $\lambda_1 = 1$ and all other eigenvalues have $|\lambda| < 1$, the diagonal matrix Λ^n will approach Λ^∞ with just a single "1" in the top corner:

$$\boldsymbol{P^n} = X\Lambda^n X^{-1} \text{approaches } \boldsymbol{P^\infty} = \begin{bmatrix} \boldsymbol{v}_1 & \boldsymbol{v}_2 & \cdots \end{bmatrix} \begin{bmatrix} \mathbf{1} & & \\ & \mathbf{0} & \\ & & \mathbf{0} \\ & & & \cdot \end{bmatrix} \begin{bmatrix} \mathbf{1}^{\mathrm{T}} \\ \cdot \\ \cdot \\ \cdot \end{bmatrix} = \begin{bmatrix} \boldsymbol{v}_1 & \boldsymbol{v}_1 & \cdots \end{bmatrix}$$

Here are $P, P^2, P^3, \ldots$ converging to the rank one matrix $P^\infty = \boldsymbol{v}_1\mathbf{1}^{\mathrm{T}}$ with $\boldsymbol{v}_1$ in all columns:

$$\begin{bmatrix} .80 & .30 \\ .20 & .70 \end{bmatrix} \quad \begin{bmatrix} .70 & .45 \\ .30 & .55 \end{bmatrix} \quad \begin{bmatrix} .65 & .525 \\ .35 & .475 \end{bmatrix} \text{ approach } \begin{bmatrix} \mathbf{.60} & \mathbf{.60} \\ \mathbf{.40} & \mathbf{.40} \end{bmatrix} = \begin{bmatrix} \mathbf{.6} \\ \mathbf{.4} \end{bmatrix} \begin{bmatrix} 1 & 1 \end{bmatrix}$$

At this point we state and prove the Perron-Frobenius Theorem. Actually we prove Perron's part (strictly positive matrices). Then Frobenius allows zeros in P. This brings the *possibility* that $|\lambda_2|$ equals λ_1. In that case P^n will not converge (unless $P = I$) to the usual $P^\infty = \boldsymbol{v}_1\mathbf{1}^{\mathrm{T}}$.

Perron-Frobenius Theorem

One matrix theorem dominates this subject. The Perron-Frobenius Theorem applies when all $a_{ij} \geq 0$. There is no requirement that all columns add to 1. We prove the neatest form, when all $a_{ij} > 0$. Then the largest eigenvalue $\lambda_{\max}$ and also its eigenvector $\boldsymbol{x}$ are positive.

Perron-Frobenius for $A > 0$ ***All numbers in $A\boldsymbol{x} = \lambda_{\max}\boldsymbol{x}$ are strictly positive.***

Proof Start with $A > 0$. The key idea is to look at all numbers t such that $A\boldsymbol{x} \geq t\boldsymbol{x}$ for some nonnegative vector $\boldsymbol{x}$ (other than $\boldsymbol{x} = \mathbf{0}$). We are allowing inequality in $A\boldsymbol{x} \geq t\boldsymbol{x}$ in order to have many small positive candidates t. For the **largest value $t_{\max}$** (which is attained), we will show that ***equality holds***: $A\boldsymbol{x} = t_{\max}\boldsymbol{x}$. Then $t_{\max}$ is our eigenvalue $\lambda_{\max}$ and $\boldsymbol{x}$ is the eigenvector—which we now prove.

If $Ax \geq t_{\max} x$ is not an equality, multiply both sides by A. Because $A > 0$, that produces a strict inequality $A^2 x > t_{\max} Ax$. Therefore the positive vector $y = Ax$ satisfies $Ay > t_{\max}\, y$. This means that $t_{\max}$ could be increased. This contradiction forces the equality $Ax = t_{\max} x$, and *we have an eigenvalue*. Its eigenvector x is positive because on the left side of that equality, Ax is sure to be positive.

To see that no eigenvalue can be larger than $t_{\max}$, suppose $Az = \lambda z$. Since λ and z may involve negative or complex numbers, we take absolute values: $|\lambda||z| = |Az| \leq A|z|$ by the "triangle inequality." This $|z|$ is a nonnegative vector, so this $|\lambda|$ is one of the possible candidates t. Therefore $|\lambda|$ cannot exceed $t_{\max}$—which must be $\lambda_{\max}$.

Finer Points of Markov Theory

Returning to Markov, we left two cases unresolved in proving $P^n \to P^\infty = [\boldsymbol{v}_1\ \boldsymbol{v}_1 \ldots \boldsymbol{v}_1]$.

1. $P > 0$ could be strictly positive, but it might not have n independent eigenvectors.
2. $P \geq 0$ might have zero entries. Then $|\lambda_2| = 1$ becomes a possibility—not a certainty.

Case 1 is a technical problem. The important fact $P^n \to P^\infty$ is still true even if we don't have an invertible eigenvector matrix X. We do have separation between $\lambda_1 = 1$ and all other eigenvalues. As long as the eigenvector with $P\boldsymbol{v}_1 = \boldsymbol{v}_1$ goes in the first column of X, the first column of $X^{-1}PX$ will still be $(1, 0, \ldots, 0)$. The submatrix A in the last rows and columns of $X^{-1}PX$ has the other eigenvalues of P. They all have $|\lambda| < 1$ by Perron-Frobenius. **We will prove that $\boldsymbol{A^n \to 0}$.**

Main point from algebra: $X^{-1}PX$ can always be made triangular (usually diagonal).

If $|\lambda_2| < 1$ then $P^n \to P^\infty$

We want to prove that $\boldsymbol{P^n \to P^\infty = [\ v_1\ \ v_1\ \ldots\ v_1\ ]}$ **whenever** $\boldsymbol{|\lambda_2| < 1}$. The matrix $P \geq 0$ could contain zeros, as in Example 3. The matrix P might not have n independent eigenvectors, so we can't diagonalize P. But we can separate $\lambda_1 = 1$ from the rest of the matrix with $|\lambda| < 1$. This will be enough to prove that P^n approaches P^∞:

$$X^{-1}PX = \begin{bmatrix} 1 & 0 \\ 0 & A \end{bmatrix} \quad \text{and the eigenvalues of } A \text{ have } |\lambda_2| < 1, \ldots, |\lambda_n| < 1.$$

By isolating that matrix A, we get a clean result with many applications.

$$\boldsymbol{P^n} = X(X^{-1}PX)^n X^{-1} = \boldsymbol{X} \begin{bmatrix} \mathbf{1} & \mathbf{0} \\ \mathbf{0} & \boldsymbol{A^n} \end{bmatrix} \boldsymbol{X^{-1}} \text{ converges to } \boldsymbol{v}_1 \mathbf{1}^{\mathrm{T}} \text{ and } \boldsymbol{A^n \to 0}.$$

If all eigenvalues of A have $|\lambda| < 1$, then $A^n \to 0$ as $n \to \infty$

Step 1 Find an upper triangular matrix $S = M^{-1}AM$ that has small norm $||S|| < 1$.

Step 2 Then $A^n = (MSM^{-1})^n = MS^nM^{-1}$ has $||A^n|| \leq ||M||\,||S||^n\,||M^{-1}|| \to \mathbf{0}$.

We need to find that triangular matrix S—then the proof is complete. Since $S = M^{-1}AM$ is similar to A, S will have the same eigenvalues as A. But the eigenvalues of a triangular matrix are seen on its main diagonal:

$$S = \begin{bmatrix} \lambda_2 & a & b \\ 0 & \lambda_3 & c \\ 0 & 0 & \lambda_4 \end{bmatrix} \quad \text{has } ||S|| < 1 \text{ if } a, b, c \text{ are very small}$$

Key point: The largest eigenvalue is not a norm! If a, b, c are large then the powers S^2, S^3, S^4 will start to grow. Eventually $\lambda_2^n, \lambda_3^n, \lambda_4^n$ do their part and S^n falls back toward the zero matrix. If we want to guarantee no growth at the beginning then we want the *norm* of S to be $||S|| < 1$. Then $||S^n||$ will stay below $||S||^n$ and go directly toward zero.

We know that $|\lambda_2| < 1, |\lambda_3| < 1, |\lambda_4| < 1$. If a, b, c are small the norm is below 1:

$$||S|| \leq ||\text{diagonal part}|| + ||\text{off-diagonal part}|| < 1.$$

We reach this triangular $S = M^{-1}AM$ in two steps. First, every square matrix A is similar to some upper triangular matrix $T = Q^{-1}AQ$. This is Schur's Theorem with an orthogonal matrix Q. Its proof is straightforward, on page 343 of *Introduction to Linear Algebra*. Then we reduce A, B, C in T by a diagonal matrix D to reach small a, b, c in S:

$$D^{-1}TD = \begin{bmatrix} 1 & & \\ & 1/d & \\ & & 1/d^2 \end{bmatrix} \begin{bmatrix} \lambda_2 & A & B \\ 0 & \lambda_3 & C \\ 0 & 0 & \lambda_4 \end{bmatrix} \begin{bmatrix} 1 & & \\ & d & \\ & & d^2 \end{bmatrix} = \begin{bmatrix} \lambda_2 & dA & d^2B \\ 0 & \lambda_3 & dC \\ 0 & 0 & \lambda_4 \end{bmatrix} = S.$$

For small d, the off-diagonal numbers dA and d^2B and dC become as small as we want. Then S is $D^{-1}(Q^{-1}AQ)D = M^{-1}AM$, as required for Step 1 and Step 2.

If $P \geq 0$ is *not strictly positive*, everything depends on the eigenvalues of P. We face the possibility that $|\lambda_2| = 1$ and the powers P^n do not converge. Here are examples:

$$P = \begin{bmatrix} 1 & 0 & 0 \\ 0 & 0.5 & 0.5 \\ 0 & 0.5 & 0.5 \end{bmatrix} \quad (\lambda = 1, 1, 0) \qquad P = \begin{bmatrix} 0 & 0 & 1 \\ 1 & 0 & 0 \\ 0 & 1 & 0 \end{bmatrix} \quad (\lambda^3 = 1)$$

"Gambler's Ruin"

This matrix P is a classic Markov example. Two gamblers have \$3 between them. The system has four states $(3, 0), (2, 1), (1, 2)$, and $(0, 3)$. The *absorbing* states are $(3, 0)$ and $(0, 3)$, when a player has won all the money—the game is over and there is no way to leave either of those states. With those two steady states we must expect $\lambda = 1$ twice.

The transient states are $(2, 1)$ and $(1, 2)$, when a \$1 game is played—with probability p that Player 1 will win and probability $q = 1 - p$ that Player 2 will win. The 4 by 4 transition matrix P has those numbers $\boldsymbol{p}$ and $\boldsymbol{q}$ in its middle columns, where Player 1 has \$2 or \$1.

$$P = \begin{bmatrix} 1 & p & 0 & 0 \\ 0 & 0 & p & 0 \\ 0 & q & 0 & 0 \\ 0 & 0 & q & 1 \end{bmatrix}$$

Question: What are the four eigenvalues of P?
Answer: $\lambda = 1, 1, \sqrt{pq}$, and $-\sqrt{pq}$. So $|\lambda_2| = 1$ and there is no unique steady state.

Question: What is the probability that the game will continue forever with no winner?
Answer: **Zero.** With probability 1 this game will end.

Question: **If the game starts at $(2, 1)$, what is the probability p^* that Player 1 wins?**
Answer: Good question! Player 1 will win immediately in round 1 with probability p. The probability is $q = 1 - p$ that Player 2 will win round 1 and change the status to $(\$1, \$2)$. Then the probability is p that Player 1 will win round 2 and return the status back to $(\$2, \$1)$. From there Player 1 eventually wins with probability p^*. From this we can find p^*:

$$p^* = p + qpp^* \quad \text{and} \quad p^* = \frac{p}{1 - qp}$$

Master Equations: Continuous Markov Processes

Master equations are blessed with an impressive name. They are linear differential equations $d\boldsymbol{p}/dt = A\boldsymbol{p}$ for a probability vector $\boldsymbol{p}(t)$ (nonnegative components that sum to 1). The matrix A is special: **negative or zero** on the diagonal, **positive or zero** off the diagonal, **columns add to zero**. This continuous Markov process has probabilities $e^{At}\boldsymbol{p}(0)$.

The probability of being in state j at time t is $p_j(t)$. The probability for the state to change from j to i in a small time interval dt is $a_{ij}\,dt$. Given $\boldsymbol{p}(0)$, the solution comes from a matrix exponential $\boldsymbol{p}(t) = e^{tA}\boldsymbol{p}(0)$. That matrix e^{At} will be a Markov matrix.

Proof. If n is large, $I + (tA/n)$ is an ordinary Markov matrix. Its columns add to 1. Then $\left(I + \frac{tA}{n}\right)^n$ converges to $P = e^{tA}$ which is also Markov. And as $t \to \infty$, e^{tA} converges in the usual way to a limit P^∞.

An example is the matrix A with diagonals $1, -2, 1$, except that $A_{11} = A_{NN} = -1$. This is minus the *graph Laplacian* on a line of nodes. Finite difference approximations to the heat equation with Neumann boundary conditions use that matrix.

This A appears in the master equation for the *bimolecular reaction* $\boldsymbol{A + B \to C}$. A molecule of A chemically combines with a molecule of B to form a molecule of C.

$$A = \begin{array}{c} \\ ⓪ \\ ① \\ ② \\ ③ \\ ④ \end{array} \begin{array}{c} \begin{matrix} ⓪ & ① & ② & ③ & ④ \end{matrix} \\ \begin{bmatrix} -16 & 1 & 0 & 0 & 0 \\ 16 & -10 & 2 & 0 & 0 \\ 0 & 9 & -6 & 3 & 0 \\ 0 & 0 & 4 & -4 & 4 \\ 0 & 0 & 0 & 1 & -4 \end{bmatrix} \end{array}$$

Columns of A add to zero
Columns of $I + \frac{tA}{n}$ add to one
Then $P = e^{At}$ is Markov

Problem Set V.6

1 Find a Markov matrix $P \geq 0$ that has a zero entry but P^2 is strictly positive. The columns of P add to 1 so $\lambda_{\text{max}} = 1$. How do you know that the other eigenvalues of P have $|\lambda| < 1$? Then P^n approaches P^∞ with $\boldsymbol{v}_1$ in every column.

2 If A has all positive entries then $A^{\mathrm{T}}A$ and AA^{T} have all positive entries. Use Perron's theorem to show: The rank 1 matrix $\sigma_1 \boldsymbol{u}_1 \boldsymbol{v}_1^{\mathrm{T}}$ closest to A is also positive.

3 These matrices have $||A|| > 1$ and $||M|| > 1$. Find matrices B and C so that $||BAB^{-1}|| < 1$ and $||CMC^{-1}|| < 1$. This is surely possible because the _____ of A and M are below 1 in absolute value. Why is it impossible if M is Markov?

$$A = \begin{bmatrix} \frac{1}{2} & 1 \\ 0 & \frac{1}{2} \end{bmatrix} \qquad M = \begin{bmatrix} .8 & .1 \\ .8 & .1 \end{bmatrix}$$

4 Why is $||BZB^{-1}|| \leq 1$ impossible for any B but $||CYC^{-1}|| \leq 1$ is possible?

$$Z = \begin{bmatrix} 1 & 1 \\ 0 & 1 \end{bmatrix} \qquad Y = \begin{bmatrix} 1 & 1 \\ 0 & -1 \end{bmatrix}$$

5 If you take powers of A, what is the limit of A^n as $n \to \infty$?

$$A = \begin{bmatrix} 2/3 & 1/3 \\ 1/3 & 2/3 \end{bmatrix} \quad \text{and also} \quad A = \frac{1}{4}\begin{bmatrix} 2 & 1 & 1 \\ 1 & 2 & 1 \\ 1 & 1 & 2 \end{bmatrix}$$

6 Suppose that every year 99% of the people in New York and Florida go to Florida and New York—but 1% die off (I am sorry about this question). Can you create a 3 by 3 Markov matrix P corresponding to the three states New York-Florida-dead. What is the limit of the matrices P^n as $n \to \infty$?

Part VI

Optimization

VI.1 Minimum Problems : Convexity and Newton's Method

VI.2 Lagrange Multipliers = Derivatives of the Cost

VI.3 Linear Programming, Game Theory, and Duality

VI.4 Gradient Descent Toward the Minimum

VI.5 Stochastic Gradient Descent and ADAM

Part VI : Optimization

The goal of optimization is to minimize a function $F(x_1, \ldots, x_n)$—often with many variables. This subject must begin with the most important equation of calculus: *Derivative = Zero at the minimum point* $\boldsymbol{x}^*$. With n variables, F has n partial derivatives $\partial F/\partial x_i$. If there are no "constraints" that $\boldsymbol{x}$ must satisfy, we have n equations $\partial F/\partial x_i = 0$ for n unknowns $x_1^*, \ldots, x_n^*$.

At the same time, there are often conditions that the vector $\boldsymbol{x}$ must satisfy. These **constraints on** $\boldsymbol{x}$ could be equations $A\boldsymbol{x} = \boldsymbol{b}$ or inequalities $\boldsymbol{x} \geq \boldsymbol{0}$. The constraints will enter the equations through Lagrange multipliers $\lambda_1, \ldots, \lambda_m$. Now we have $m + n$ unknowns (x's and λ's) and $m + n$ equations (derivatives $= 0$). So this subject combines linear algebra and multivariable calculus. We are often in high dimensions.

This introduction ends with key facts of calculus: the approximation of $\boldsymbol{F}(\boldsymbol{x} + \boldsymbol{\Delta x})$ by $\boldsymbol{F}(\boldsymbol{x}) + \boldsymbol{\Delta x}^{\mathrm{T}} \boldsymbol{\nabla F} + \frac{1}{2}\boldsymbol{x}^{\mathrm{T}} \boldsymbol{H x}$. Please see that important page.

Evidently this is part of mathematics. Yet optimization has its own ideas and certainly its own algorithms. It is not always presented as an essential course in the mathematics department. But departments in the social sciences and the physical sciences—economics, finance, psychology, sociology and every field of engineering—use and teach this subject because they need it.

We have organized this chapter to emphasize the key ideas of optimization:

VI.1 The central importance of **convexity**, which replaces linearity. Convexity involves second derivatives—the graph of $F(\boldsymbol{x})$ will bend upwards. A big question computationally is whether we can find and use all those second partial derivatives $\partial^2 F/\partial x_i \partial x_j$. The choice is between **"Newton methods"** that use them and **"gradient methods"** that don't: second-order methods or first-order methods.

Generally neural networks for deep learning involve very many unknowns. Then gradient methods (first order) are chosen. *Often F is not convex!* The last sections of this chapter describe those important algorithms—they move along the gradient (the vector of first derivatives) toward the minimum of $F(\boldsymbol{x})$.

VI.2 The meaning of **Lagrange multipliers**, which build the constraints into the equation *derivative = zero*. Most importantly, those multipliers give the **derivatives of the cost with respect to the constraints**. They are the Greeks of mathematical finance.

VI.3 The classical problems **LP, QP, SDP** of "mathematical programming". The unknowns are vectors or matrices. The inequalities ask for nonnegative vectors $\boldsymbol{x} \geq \mathbf{0}$ or positive semidefinite matrices $\boldsymbol{X} \geq \mathbf{0}$. Each minimization problem has a **dual problem**—a *maximization*. The multipliers in one problem become the unknowns in the dual problem. They both appear in a **2-person game**.

VI.4 First-order algorithms begin with **gradient descent**. The derivative of the cost in the search direction is negative. The choice of direction to move and how far to move— this is the art of computational optimization. You will see crucial decisions to be made, like adding "momentum" to descend more quickly.

Levenberg-Marquardt combines gradient descent with Newton's method. The idea is to get near $\boldsymbol{x}^*$ with a first order method and then converge quickly with (almost) second order. This is a favorite for nonlinear least squares.

VI.5 **Stochastic gradient descent.** In neural networks, the function to minimize is a sum of many terms—the losses in all samples in the training data. The learning function F depends on the "weights". Computing its gradient is expensive. So each step learns only a **minibatch** of B training samples—chosen randomly or "stochastically". One step accounts for a part of the data but not all. *We hope and expect that the part is reasonably typical of the whole.*

Stochastic gradient descent—often with speedup terms from "ADAM" to account for earlier step directions—has become the workhorse of deep learning. The partial derivatives we need from F are computed by **backpropagation**. This key idea will be explained separately in the final chapter of the book.

Like so much of applied mathematics, optimization has a discrete form and a continuous form. Our unknowns are vectors and our constraints involve matrices. For the *calculus of variations* the unknowns are functions and the constraints involve integrals. The vector equation "first derivative = zero" becomes the Euler-Lagrange differential equation "first variation = zero".

That is a parallel (and continuous) world of optimization but we won't go there.

The Expression "argmin"

The minimum of the function $F(x) = (x-1)^2$ is zero: *min* $F(x) = 0$. That tells us how low the graph of F goes. But it does not tell us which number $\boldsymbol{x}^*$ gives the minimum. In optimization, that "argument" $\boldsymbol{x}^*$ is the number we usually solve for. The minimizing x for $F = (x-1)^2$ is $\boldsymbol{x}^* = \textbf{argmin}\, \boldsymbol{F}(\boldsymbol{x}) = \mathbf{1}$.

$$\textbf{argmin}\, \boldsymbol{F}(\boldsymbol{x}) = \textbf{value(s) of } x \textbf{ where } F \textbf{ reaches its minimum.}$$

For strictly convex functions, $\operatorname{argmin} F(x)$ is **one point** x^*: an isolated minimum.

Multivariable Calculus

Machine learning involves functions $F(x_1, \ldots, x_n)$ of many variables. We need basic facts about the first and second derivatives of F. These are "partial derivatives" when $n > 1$.

The important facts are in equations (1)-(2)-(3). I don't believe you need a whole course (too much about integrals) to use these facts in optimizing a deep learning function $F(x)$.

One function F
One variable x

$$\boldsymbol{F(x+\Delta x) \approx F(x) + \Delta x \frac{dF}{dx}(x) + \frac{1}{2}(\Delta x)^2 \frac{d^2F}{dx^2}(x)} \quad (1)$$

This is the beginning of a Taylor series—and we don't often go beyond that second-order term. The first terms $F(x)+(\Delta x)(dF/dx)$ give a *first order* approximation to $F(x+\Delta x)$, using information at x. Then the $(\Delta x)^2$ term makes it a *second order* approximation.

The Δx term gives a point on the tangent line—tangent to the graph of $F(x)$. The $(\Delta x)^2$ term moves from the tangent line to the "tangent parabola". The function F will be **convex**—its slope increases and its graph bends upward, as in $y = x^2$—when the second derivative of $F(x)$ is positive: $\boldsymbol{d^2F/dx^2 > 0}$. **Equation (2) lifts (1) into $\boldsymbol{n}$ dimensions.**

One function F
Variables x_1 to x_n

$$\boldsymbol{F(x+\Delta x) \approx F(x) + (\Delta x)^{\mathrm{T}} \nabla F + \frac{1}{2}(\Delta x)^{\mathrm{T}} H (\Delta x)} \quad (2)$$

This is the important formula! **The vector $\boldsymbol{\nabla F}$ is the gradient of $\boldsymbol{F}$**—the column vector of n partial derivatives $\partial F/\partial x_1$ to $\partial F/\partial x_n$. **The matrix $\boldsymbol{H}$ is the Hessian matrix.** H is the symmetric matrix of second derivatives $H_{ij} = \partial^2 F/\partial x_i\, \partial x_j = \partial^2 F/\partial x_j\, \partial x_i$.

The graph of $y = F(x_1, \ldots, x_n)$ is now a surface in $(n+1)$-dimensional space. The tangent line becomes a tangent plane at x. When the second derivative matrix H is positive definite, F is a *strictly convex function*: *it stays above its tangents.* A convex function F has a **minimum** at $\boldsymbol{x^*}$ if $\boldsymbol{f = \nabla F(x^*) = 0}$: n equations for $\boldsymbol{x^*}$.

Sometimes we meet m different functions $f_1(\boldsymbol{x})$ to $f_m(\boldsymbol{x})$: a *vector function* $\boldsymbol{f}$:

m functions $\boldsymbol{f} = (f_1, \ldots, f_m)$
n variables $\boldsymbol{x} = (x_1, \ldots, x_n)$

$$\boldsymbol{f(x+\Delta x) \approx f(x) + J(x)\,\Delta x} \quad (3)$$

The symbol $\boldsymbol{J(x)}$ represents the m by n **Jacobian matrix of $\boldsymbol{f(x)}$** at the point $\boldsymbol{x}$. The m rows of J contain the gradient vectors of the m functions $f_1(\boldsymbol{x})$ to $f_m(\boldsymbol{x})$.

$$\textbf{Jacobian matrix } \boldsymbol{J} = \begin{bmatrix} (\nabla f_1)^{\mathrm{T}} \\ \vdots \\ (\nabla f_m)^{\mathrm{T}} \end{bmatrix} = \begin{bmatrix} \dfrac{\partial f_1}{\partial x_1} & \cdots & \dfrac{\partial f_1}{\partial x_n} \\ \vdots & & \vdots \\ \dfrac{\partial f_m}{\partial x_1} & \cdots & \dfrac{\partial f_m}{\partial x_n} \end{bmatrix} \quad (4)$$

The **Hessian matrix $\boldsymbol{H}$ is the Jacobian $\boldsymbol{J}$ of the gradient $\boldsymbol{f = \nabla F}$**! The determinant of J (when $m = n$) appears in n-dimensional integrals. It is the r in the area formula $\iint r\, dr\, d\theta$.

VI.1 Minimum Problems : Convexity and Newton's Method

This part of the book will focus on problems of minimization, for functions $F(\boldsymbol{x})$ with many variables: $F(\boldsymbol{x}) = F(x_1, \ldots, x_n)$. There will often be constraints on the vectors $\boldsymbol{x}$:

Linear constraints	$A\boldsymbol{x} = \boldsymbol{b}$	(the set of these $\boldsymbol{x}$ is convex)
Inequality constraints	$\boldsymbol{x} \geq \boldsymbol{0}$	(the set of these $\boldsymbol{x}$ is convex)
Integer constraints	**Each x_i is 0 or 1**	(the set of these $\boldsymbol{x}$ is not convex)

The problem statement could be unstructured or highly structured. Then the algorithms to find the minimizing $\boldsymbol{x}$ range from very general to very specific. Here are examples:

Unstructured Minimize $F(\boldsymbol{x})$ for vectors $\boldsymbol{x}$ in a subset K of $\mathbf{R}^n$

Structured Minimize a quadratic cost $F(\boldsymbol{x}) = \frac{1}{2}\boldsymbol{x}^{\mathrm{T}}S\boldsymbol{x}$ constrained by $A\boldsymbol{x} = \boldsymbol{b}$

Minimize a linear cost $F(\boldsymbol{x}) = \boldsymbol{c}^{\mathrm{T}}\boldsymbol{x}$ constrained by $A\boldsymbol{x} = \boldsymbol{b}$ and $\boldsymbol{x} \geq \boldsymbol{0}$

Minimize with a binary constraint: each x_i is 0 or 1

We cannot go far with those problems until we recognize the crucial role of **convexity**. We hope the function $F(\boldsymbol{x})$ is convex. We hope the constraint set K is convex. In the theory of optimization, we have to live without linearity. **Convexity** will take control when linearity is lost. Here is convexity for a function $F(\boldsymbol{x})$ and a constraint set K:

K is a convex set	If $\boldsymbol{x}$ and $\boldsymbol{y}$ are in K, so is the line from $\boldsymbol{x}$ to $\boldsymbol{y}$
F is a convex function	The set of points on and above the graph of F is convex
F is smooth and convex	$F(\boldsymbol{x}) \geq F(\boldsymbol{y}) + (\boldsymbol{\nabla} \boldsymbol{F}(\boldsymbol{y}), \boldsymbol{x} - \boldsymbol{y})$

That last inequality says that the graph of a convex F *stays above its tangent lines*.

A triangle in the plane is certainly a convex set in $\mathbf{R}^2$. What about the union of two triangles ? Right now I can see only two ways for the union to be convex:

1) One triangle contains the other triangle. The union is the bigger triangle.

2) They share a complete side and their union has no inward-pointing angles.

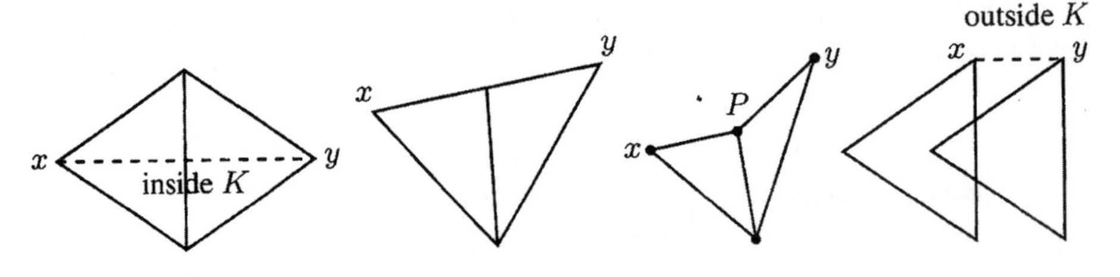

Figure VI.1: Two convex sets and two non-convex sets in $\mathbf{R}^2$: Inward-pointing at P.

For functions F there is a direct way to define convexity. Look at all points $px+(1-p)\,y$ between x and y. The graph of F stays on or goes below a straight line graph.

$$\boxed{\textbf{F is convex} \quad F(px+(1-p)y) \le pF(x)+(1-p)F(y) \text{ for } 0<p<1.} \tag{1}$$

For a **strictly convex** function, this holds with strict inequality (replace $\le$ by $<$). Then the graph of F goes strictly below the chord that connects the point $x, F(x)$ to $y, F(y)$.
Interesting that the graph stays above its tangent lines and below its chords.

Here are three examples of convex functions. Only F_2 is strictly convex:

$$F_1 = ax+b \qquad F_2 = x^2 \text{ (but not } -x^2) \qquad F_3 = \max(F_1, F_2)$$

The convexity of that function F_3 is an important fact. This is where linearity fails but convexity succeeds! The maximum of two or more linear functions is rarely linear. But **the maximum $F(x)$ of two or more convex functions $F_i(x)$ is always convex.** For any $z = px+(1-p)y$ between x and y, each function F_i has

$$\boxed{F_i(z) \le pF_i(x)+(1-p)F_i(y) \le p\,F(x)+(1-p)\,F(y).} \tag{2}$$

This is true for each i. Then $F(z) = \max F_i(z) \le pF(x)+(1-p)F(y)$, as required.

The maximum of any family of convex functions (in particular any family of linear functions) is convex. Suppose we have *all the linear functions that stay below a convex function* F. Then the maximum of those linear functions below F is exactly equal to F.

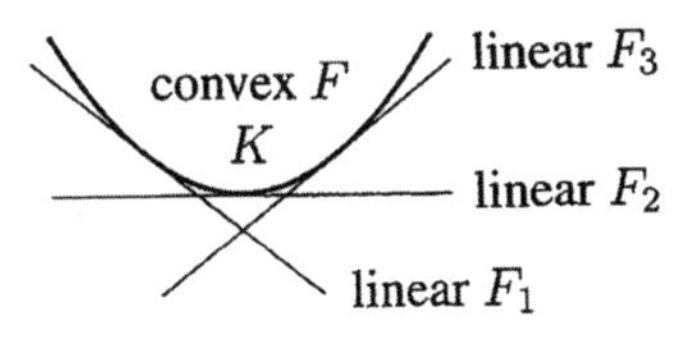

Figure VI.2: **A convex function F is the maximum of all its tangent functions.**

A convex set K is the intersection of all half-spaces that contain it. And the *intersection* of any family of convex sets is convex. But the *union* of convex sets is not always convex.

Similarly the *maximum* of any family of convex functions will be convex. But the *minimum* of two convex functions is generally not convex—it can have a "double well".

Here are two useful facts about matrices, based on pos def + pos def = pos def:

The set of *positive definite n by n matrices* is convex.

The set of *positive semidefinite n by n matrices* is convex.

The first set is "open". The second set is "closed". It contains its semidefinite limit points.

The Second Derivative Matrix

An ordinary function $f(x)$ is convex if $d^2f/dx^2 \geq 0$. Reason: The slope df/dx is increasing. *The curve bends upward* (like the parabola $f = x^2$ with second derivative $= 2$). The extension to n variables involves the **n by n matrix $H(x)$** of second derivatives. If $F(x)$ is a smooth function then there is an almost perfect test for convexity:

> $F(x_1, \ldots, x_n)$ is convex if and only if its second derivative matrix $H(x)$ is **positive semidefinite at all x**. That **Hessian matrix** is symmetric because $\partial^2 F/\partial x_i \partial x_j = \partial^2 F/\partial x_j \partial x_i$. The function F is **strictly** convex if $H(x)$ is **positive definite at all x**.
>
> $$H(x) = \begin{bmatrix} \partial^2 F/\partial x_1^2 & \partial^2 F/\partial x_1 \partial x_2 & \bullet \\ \partial^2 F/\partial x_2 \partial x_1 & \partial^2 F/\partial x_2^2 & \bullet \\ \bullet & \bullet & \bullet \end{bmatrix} \qquad H_{ij} = \frac{\partial^2 F}{\partial x_i \partial x_j} = H_{ji}$$

A linear function $F = c^{\mathrm{T}}x$ is convex (but not strictly convex). Above its graph is a half-space: flat boundary. Its second derivative matrix is $H = 0$ (very semidefinite).

A quadratic $F = \frac{1}{2}x^{\mathrm{T}}Sx$ has gradient Sx. Its symmetric second derivative matrix is S. Above its graph is a bowl, when S is positive definite. This function F is strictly convex.

Convexity Prevents Two Local Minima

We minimize a convex function $F(x)$ for x in a convex set K. That double convexity has a favorable effect: *There will not be two isolated solutions.* If x and y are in K and they give the same minimum, then all points z on the line between them are also in K and give that minimum. Convexity avoids the truly dangerous situation when F has its minimum value at an unknown number of separate points in K.

This contribution of convexity is already clear for ordinary functions $F(x)$ with one variable x. Here is the graph of a non-convex function with minima at x and y and z.

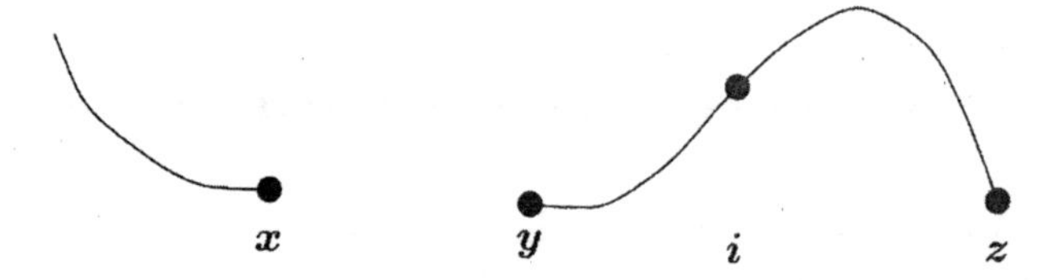

F is not convex. It is concave after the inflection point i, where $\partial^2 F/\partial x^2$ goes negative. And F is not defined on a convex set K (because of the gap between x and y). To fix both problems, we could connect x to y by a straight line, and end the graph at i.

For a convex problem to have multiple solutions x and y, the interval between them must be filled with solutions. Never two isolated minima, usually just a single point. **The set of minimizing points x in a convex problem is convex.**

The **CVX** system provides MATLAB software for "disciplined convex programming". The user chooses least squares, linear and quadratic programming,... See **cvxr.com/cvx**.

The ℓ^1 and ℓ^2 and ℓ^∞ Norms of x

Norms $F(x) = ||x||$ are convex functions of x. The unit ball where $||x|| \leq 1$ is a convex set K of vectors x. That first sentence is exactly the triangle inequality:

$$\textbf{Convexity of } ||x|| \quad ||p\,x + (1-p)\,y|| \leq p||x|| + (1-p)||y||$$

There are three favorite vector norms $\ell^1, \ell^2, \ell^\infty$. We draw the unit balls $||x|| \leq 1$ in $\mathbf{R}^2$:

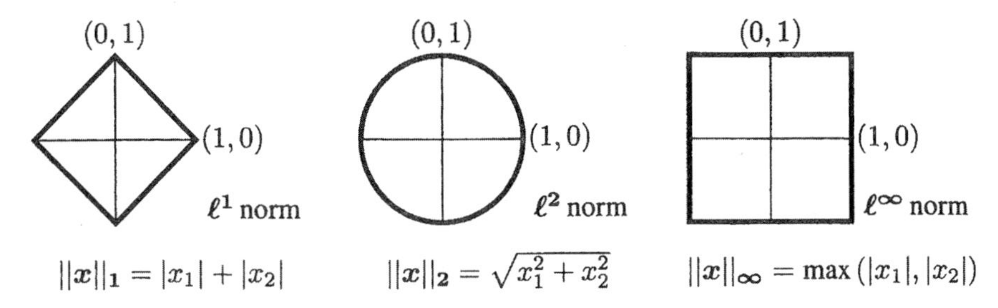

$||x||_1 = |x_1| + |x_2|$ $\qquad$ $||x||_2 = \sqrt{x_1^2 + x_2^2}$ $\qquad$ $||x||_\infty = \max(|x_1|, |x_2|)$

Figure VI.3: For all norms, the convex "unit ball" where $||x|| \leq 1$ is centered at $x = 0$.

Newton's Method

We are looking for the point x^* where $F(x)$ has a minimum and its gradient $\nabla F(x^*)$ is the zero vector. We have reached a nearby point x_k. We aim to move to a new point x_{k+1} that is closer than x_k to $x^* = \text{argmin}\, F(x)$. What is a suitable step $x_{k+1} - x_k$ to reach that new point x_{k+1} ?

Calculus provides an answer. Near our current point x_k, the gradient ∇F is often well estimated by using its first derivatives—which are the second derivatives of $F(x)$. Those second derivatives $\partial^2 F/\partial x_i \partial x_j$ are in the Hessian matrix H:

$$\nabla F(x_{k+1}) \approx \nabla F(x_k) + H(x_k)(x_{k+1} - x_k). \tag{3}$$

We want that left hand side to be zero. So the natural choice for x_{k+1} comes when the *right side is zero*: we have n linear equations for the step $\Delta x_k = x_{k+1} - x_k$:

$$\boxed{\textbf{Newton's Method} \qquad H(x_k)(\Delta x_k) = -\nabla F(x_k) \quad \text{and} \quad x_{k+1} = x_k + \Delta x_k} \tag{4}$$

Newton's method is also producing the minimizer of a quadratic function built from F and its derivatives ∇F and its second derivatives H at the point x_k:

$$x_{k+1} \text{ minimizes } F(x_k) + \nabla F(x_k)^{\mathrm{T}}(x - x_k) + \frac{1}{2}(x - x_k)^{\mathrm{T}} H(x_k)(x - x_k). \tag{5}$$

Newton's method is *second order*. It uses second derivatives (in H). There will still be an error in the new $\boldsymbol{x}_{k+1}$. But that error is proportional to the **square** of the error in $\boldsymbol{x}_k$:

$$\textbf{Quadratic convergence} \qquad ||\boldsymbol{x}_{k+1} - \boldsymbol{x}^*|| \leq C\,||\boldsymbol{x}_k - \boldsymbol{x}^*||^2. \tag{6}$$

If $\boldsymbol{x}_k$ is close to $\boldsymbol{x}^*$, then $\boldsymbol{x}_{k+1}$ will be much closer. An example is the computation of $\boldsymbol{x}^* = \sqrt{4} = 2$ in one dimension. Newton is solving $\boldsymbol{x}^2 - 4 = 0$:

$$\textbf{Minimize} \quad \boldsymbol{F(x) = \tfrac{1}{3}x^3 - 4x} \quad \textbf{with} \quad \boldsymbol{\nabla F(x) = x^2 - 4} \quad \textbf{and} \quad \boldsymbol{H(x) = 2x}$$

One step of Newton's method: $H(x_k)(\Delta x_k) = 2x_k(x_{k+1} - x_k) = -x_k^2 + 4.$

Then $2x_k x_{k+1} = x_k^2 + 4.$ So Newton chooses $\boldsymbol{x_{k+1} = \dfrac{1}{2}\left(x_k + \dfrac{4}{x_k}\right)}$.

Guess the square root, divide into 4, and average the two numbers. We can start from 2.5:

$$x_0 = \mathbf{2.5} \qquad x_1 = \mathbf{2.05} \qquad x_2 = \mathbf{2.0006} \qquad x_3 = \mathbf{2.000000009}$$

The wrong decimal is twice as far out at each step. **The error $\boldsymbol{x_k - 2}$ is squared**:

$$x_{k+1} - 2 = \frac{1}{2}\left(x_k + \frac{4}{x_k}\right) - 2 = \frac{1}{2x_k}(x_k - 2)^2 \qquad\qquad ||\boldsymbol{x}_{k+1} - \boldsymbol{x}^*|| \approx \frac{1}{4}||\boldsymbol{x}_k - \boldsymbol{x}^*||^2$$

Squaring the error explains the speed of Newton's method—*provided $\boldsymbol{x}_k$ is close.*

How well does Newton's method work in practice? At the start, $\boldsymbol{x}_0$ may not be close to $\boldsymbol{x}^*$. We cannot trust the second derivative at $\boldsymbol{x}_0$ to be useful. So we compute the Newton step $\Delta\boldsymbol{x}_0 = \boldsymbol{x}_1 - \boldsymbol{x}_0$, but then we allow **backtracking**:

> Choose $\alpha < \frac{1}{2}$ and $\beta < 1$ and reduce the step $\Delta\boldsymbol{x}$ by the factor β until we know that the new $\boldsymbol{x}_{k+1} = \boldsymbol{x}_k + t\Delta\boldsymbol{x}$ is producing a sufficient drop in $F(\boldsymbol{x})$:

$$\textbf{Reduce } \boldsymbol{t} \textbf{ until the drop in } \boldsymbol{F} \textbf{ satisfies} \quad F(\boldsymbol{x}_k + t\boldsymbol{\Delta x}) \leq F(\boldsymbol{x}_k) + \alpha\, t\, \boldsymbol{\nabla F}^{\mathrm{T}} \boldsymbol{\Delta x}. \tag{7}$$

We return to backtracking in Section VI.4. It is a safety step in any search direction, to know a safe choice of $\boldsymbol{x}_{k+1}$ after the direction from $\boldsymbol{x}_k$ has been set. Or we can fix a small stepsize—this is the hyperparameter—and skip the search in training a large neural network.

Summary Newton's method is eventually fast, because it uses the second derivatives of $F(\boldsymbol{x})$. But those can be too expensive to compute—especially in high dimensions. *Quasi-Newton methods* in Section III.1 allow the Hessian H to be built gradually from information gained as the algorithm continues. Often neural networks are simply too large to use H. *Gradient descent* is the algorithm of choice, to develop in VI.4-5.

The next two pages describe a compromise that gets better near $\boldsymbol{x}^*$.

Levenberg-Marquardt for Nonlinear Least Squares

Least squares begins with a set of m data points (t_i, y_i). It aims to fit those m points as well as possible by choosing the parameters $\boldsymbol{p} = (p_1, \ldots p_n)$ in a fitting function $\widehat{Y}(t, \boldsymbol{p})$. Suppose the parameters are the usual $\boldsymbol{p} = (C, D)$ for a straight line fit by $\widehat{y} = C + Dt$. Then the sum of squared errors depends on C and D:

$$E(C, D) = (y_1 - C - Dt_1)^2 + \cdots + (y_m - C - Dt_m)^2. \tag{8}$$

The minimum error E is at the values $\widehat{C}$ and $\widehat{D}$ where $\partial E/\partial C = 0$ and $\partial E/\partial D = 0$. Those equations are linear since C and D appear linearly in the fitting function $\widehat{y} = C + Dt$. We are finding the best least squares solution to m linear equations $\boldsymbol{J p} = \boldsymbol{y}$:

m equations
2 unknowns
no solution

$$J\begin{bmatrix} C \\ D \end{bmatrix} = \begin{bmatrix} 1 & t_1 \\ \cdot & \cdot \\ \cdot & \cdot \\ 1 & t_m \end{bmatrix}\begin{bmatrix} C \\ D \end{bmatrix} = \begin{bmatrix} y_1 \\ \cdot \\ \cdot \\ y_m \end{bmatrix} = \boldsymbol{y}$$

2 equations
2 unknowns

$$J^{\mathrm{T}} J\begin{bmatrix} \widehat{C} \\ \widehat{D} \end{bmatrix} = J^{\mathrm{T}}\boldsymbol{y} \text{ for the best parameters } \widehat{\boldsymbol{p}} = \begin{bmatrix} \widehat{C} \\ \widehat{D} \end{bmatrix}.$$

This is linear least squares. The fitting function $\widehat{\boldsymbol{y}}$ is linear in C and D. J would normally be called A. But for **nonlinear** least squares the fitting function $\widehat{\boldsymbol{y}}(\boldsymbol{p})$ depends in a nonlinear way on the n parameters $\boldsymbol{p} = (p_1, \ldots, p_n)$. When we minimize the total error E = sum of squares, we expect $\boldsymbol{n}$ **nonlinear equations** to determine the best parameters:

$$\begin{aligned} \boldsymbol{E}(\boldsymbol{p}) &= \sum_{i=1}^{m}(y_i - \widehat{y}_i)^2 = (\boldsymbol{y} - \widehat{\boldsymbol{y}}(\boldsymbol{p}))^{\mathrm{T}}(\boldsymbol{y} - \widehat{\boldsymbol{y}}(\boldsymbol{p})) \\ &= \boldsymbol{y}^{\mathrm{T}}\boldsymbol{y} - 2\boldsymbol{y}^{\mathrm{T}}\widehat{\boldsymbol{y}}(\boldsymbol{p}) + \widehat{\boldsymbol{y}}(\boldsymbol{p})^{\mathrm{T}}\widehat{\boldsymbol{y}}(\boldsymbol{p}). \end{aligned} \tag{9}$$

This is the "square loss" error function to minimize by choosing the best parameters $\widehat{\boldsymbol{p}}$.

Applications can include a weighting matrix or whitening matrix W. Often W is a diagonal matrix of inverse variances $1/\sigma_1^2, \ldots, 1/\sigma_m^2$. Those enter the total error $\boldsymbol{E} = (\boldsymbol{y} - \widehat{\boldsymbol{y}})^{\mathrm{T}} W (\boldsymbol{y} - \widehat{\boldsymbol{y}})$. Therefore they enter the normal equations $J^{\mathrm{T}} W J \widehat{\boldsymbol{p}} = J^{\mathrm{T}} W \boldsymbol{y}$. This gives more weight to data with smaller variance σ_i: the data that is more reliable. For simplicity we go forward with $W = I$: unit weights.

Our problem is to minimize $\boldsymbol{E}(\boldsymbol{p})$ in equation (9). So we compute its gradient vector $\partial E/\partial \boldsymbol{p} = \boldsymbol{\nabla E}$. This gradient is constant for linear least squares—but $\boldsymbol{\nabla E}$ depends on $\boldsymbol{p}$ for our nonlinear problem. The next page describes an algorithm to minimize E—approximating Newton but avoiding second derivatives of E.

$$\boxed{\nabla E = 2J^{\mathrm{T}}(y - \widehat{y}\,(p_n)) = \mathbf{0} \text{ with } m \text{ by } n \text{ Jacobian matrix } J = \frac{\partial y}{\partial p} \text{ at } \widehat{p}.} \tag{10}$$

J was a constant m by 2 matrix when the fitting function $\widehat{y} = C + Dt$ was linear in the parameters $p = (C, D)$. *The least squares equation for minimum error is* $\nabla E = \mathbf{0}$. In the linear case this was $J^{\mathrm{T}}J\widehat{p} = J^{\mathrm{T}}\widehat{y}$. In the nonlinear case we have a first order method (**gradient descent**) and an **approximate Newton's method** to solve (10):

$$\textbf{Gradient descent} \qquad p_{n+1} - p_n = -sJ^{\mathrm{T}}\,(y - \widehat{y}\,(p_n)) \tag{11}$$

$$\textbf{Newton (approximate)} \qquad J^{\mathrm{T}}J\,(p_{n+1} - p_n) = J^{\mathrm{T}}\,(y - \widehat{y}\,(p_n)) \tag{12}$$

That symmetric matrix $J^{\mathrm{T}}J$ is an approximation to the second derivative matrix $\frac{1}{2}H$ (the Hessian of the function E). To see this, substitute the first order approximation $\widehat{y}\,(p + \Delta p) \approx \widehat{y}\,(p) + J\Delta p$ into the loss function E in (9):

$$E(p + \Delta p) \approx (y - \widehat{y}\,(p) - J\Delta p)^{\mathrm{T}}(y - \widehat{y}\,(p) - J\Delta p). \tag{13}$$

The second order term is $\Delta p^{\mathrm{T}}J^{\mathrm{T}}J\Delta p$. So $2J^{\mathrm{T}}J$ is acting like a Hessian matrix.

The key idea of Levenberg and Marquardt was to combine the gradient descent and Newton update rules (11)-(12) into one rule. It has a parameter λ. Small values of λ will lean toward Newton, large values of λ will lean more toward gradient descent. Here is a favorite (and flexible) method for nonlinear least squares problems:

$$\boxed{\textbf{Levenberg-Marquardt} \qquad (J^{\mathrm{T}}J + \lambda I)\,(p_{n+1} - p_n) = J^{\mathrm{T}}(y - \widehat{y}\,(p_n)).} \tag{14}$$

You start with a fairly large λ. The starting p_0 is probably not close to the best choice p^*. At this distance from the minimizing point p^*, you cannot really trust the accuracy of $J^{\mathrm{T}}J$—when the problem is nonlinear and the Hessian of E depends on p.

As the approximations $p_1, p_2, \ldots$ get closer to the correct value p^*, the matrix $J^{\mathrm{T}}J$ becomes trustworthy. Then we bring λ toward zero. The goal is to enjoy fast convergence (nearly Newton) to the solution of $\nabla E\,(p^*) = \mathbf{0}$ and the minimum of $E(p)$.

A useful variant is to multiply λI in (14) by the diagonal matrix $\operatorname{diag}(J^{\mathrm{T}}J)$. That makes λ dimensionless. As with all gradient descents, the code must check that the error E decreases at each step—and adjust the stepsize as needed. A good decrease signals that λ can be reduced and the next iteration moves closer to Newton.

Is it exactly Newton when $\lambda = 0$ *? I am sorry but I don't think it is. Look back at* (13).

The quadratic term suggests that the second derivative matrix (Hessian) is $2J^{\mathrm{T}}J$. But (13) is only a first order approximation. For linear least squares, first order was exact. In a nonlinear problem that cannot be true. The official name here is *Gauss-Newton*.

To say this differently, we cannot compute a second derivative by squaring a first derivative.

Nevertheless Levenberg-Marquardt is an enhanced first order method, extremely useful for nonlinear least squares. It is one way to train neural networks of moderate size.

Problem Set VI.1

1 When is the union of two circular discs a convex set? Or two squares?

2 The text proposes only two ways for the union of two triangles in $\mathbf{R}^2$ to be convex. Is this test correct? What if the triangles are in $\mathbf{R}^3$?

3 The **"convex hull"** of any set S in $\mathbf{R}^n$ is the smallest convex set K that contains S. From the set S, how could you construct its convex hull K?

4 (a) Explain why the intersection $K_1 \cap K_2$ of two convex sets is a convex set.

(b) How does this prove that the maximum F_3 of two convex functions F_1 and F_2 is a convex function? Use the text definition: F is convex when the set of points on and above its graph is convex. What set is above the graph of F_3?

5 Suppose K is convex and $F(x) = 1$ for x in K and $F(x) = 0$ for x not in K. Is F a convex function? What if the 0 and 1 are reversed?

6 From their second derivatives, show that these functions are convex:

(a) Entropy $x \log x$

(b) $\log(e^x + e^y)$

(c) ℓ^p norm $||\boldsymbol{x}||_p = (|x_1|^p + |x_2|^p)^{1/p}, \; p \geq 1$

(d) $\lambda_{\max}(S)$ as a function of the symmetric S

7 $R(\boldsymbol{x}) = \dfrac{\boldsymbol{x}^{\mathrm{T}} S \boldsymbol{x}}{\boldsymbol{x}^{\mathrm{T}} \boldsymbol{x}} = \dfrac{x^2 + 2y^2}{x^2 + y^2}$ is *not* convex. It has a maximum as well as a minimum. At a maximum point, the second derivative matrix H is ______.

8 This chapter includes statements of the form $\underset{x}{\min}\ \underset{y}{\max}\ K(x, y) = \underset{y}{\max}\ \underset{x}{\min}\ K(x, y)$.

But minimax = maximin is not always true! Explain this example:

$$\underset{x}{\min}\ \underset{y}{\max}\ (x + y) \text{ and } \underset{y}{\max}\ \underset{x}{\min}\ (x + y) \text{ are } +\infty \text{ and } -\infty.$$

9 Suppose $f(x, y)$ is a smooth convex function and $f(0,0) = f(1,0) = f(0,1) = 0$.

(a) What do you know about $f\left(\frac{1}{2}, \frac{1}{2}\right)$?

(b) What do you know about the derivatives $a = \partial^2 f / \partial x^2, \quad b = \partial^2 f / \partial x \partial y, \; c = \partial^2 f / \partial y^2$?

10 Could any smooth function $f(x, y)$ in the circle $x^2 + y^2 \leq 1$ be written as the difference $g(x, y) - h(x, y)$ of two convex functions g and h? Probably yes.

The next four problems are about Newton's method.

11 Show that equation (5) is correct: **Newton's Δx minimizes that quadratic.**

12 What is Newton's method to solve $x^2 + 1 = 0$? Since there is no (real) solution, the method can't converge. (The iterations give a neat example of "chaos".)

13 What is Newton's method to solve $\sin x = 0$? Since this has many solutions, it may be hard to predict the limit x^* of the Newton iterations.

14 From $\boldsymbol{x}_0 = (u_0, v_0)$ find $\boldsymbol{x}_1 = (u_1, v_1)$ in Newton's method for the equations $u^3 - v = 0$ and $v^3 - u = 0$. Newton converges to the solution $(0,0)$ or $(1,1)$ or $(-1,-1)$ or it goes off to infinity. If you use four colors for the starting points (u_0, v_0) that lead to those four limits, the printed picture is beautiful.

15 If f is a convex function, we know that $f(x/2 + y/2) \leq \frac{1}{2}f(x) + \frac{1}{2}f(y)$. If this "halfway test" holds for every x and y, show that the "quarterway test" $f(3x/4 + y/4) \leq \frac{3}{4}f(x) + \frac{1}{4}f(y)$ is also passed. This is a test halfway between x and $x/2 + y/2$. So two halfway tests give the quarterway test.

The same reasoning will eventually give $f(px + (1-p)y) \leq p\,f(x) + (1-p)\,f(y)$ for any fraction $p = m/2^n \leq 1$. These fractions are dense in the whole interval $0 \leq p \leq 1$. If f is a continuous function, then the halfway test for all x, y leads to the $px + (1-p)y$ test for all $0 \leq p \leq 1$. *So the halfway test proves f is convex.*

16 Draw the graph of any strictly convex function $f(x)$.

Draw the chord between any two points on the graph.

Draw the tangent lines at those same two points.

Between x and y, verify that *tangent lines* $< f(x) <$ *chord*.

VI.2 Lagrange Multipliers = Derivatives of the Cost

Unstructured problems deal with convex functions $F(x)$ on convex sets K. This section starts with highly structured problems, to see how Lagrange multipliers deal with constraints. We want to bring out the meaning of the multipliers $\lambda_1, \ldots, \lambda_m$. After introducing them and using them, it is a big mistake to discard them.

Our first example is in two dimensions. The function F is quadratic. The set K is linear.

$$\textbf{Minimize } F(x) = x_1^2 + x_2^2 \textbf{ on the line } K: a_1 x_1 + a_2 x_2 = b$$

On the line K, we are looking for the point that is nearest to $(0, 0)$. The cost $F(x)$ is distance squared. In Figure VI.4, the constraint line is **tangent to the circle** at the winning point $x^* = (x_1^*, x_2^*)$. We discover this from simple calculus, *after we bring the constraint equation* $a_1x_1 + a_2x_2 = b$ *into the function* $F = x_1^2 + x_2^2$.

This was Lagrange's beautiful idea.

Multiply $a_1x_1 + a_2x_2 - b$ by an unknown multiplier λ and add it to $F(x)$

$$\textbf{Lagrangian } L(x, \lambda) = F(x) + \lambda(a_1x_1 + a_2x_2 - b) = x_1^2 + x_2^2 + \lambda(a_1x_1 + a_2x_2 - b) \quad (1)$$

Set the derivatives $\partial L/\partial x_1$ and $\partial L/\partial x_2$ and $\partial L/\partial \lambda$ to zero.

Solve those three equations for x_1, x_2, λ.

$$\partial L/\partial x_1 = 2x_1 + \lambda a_1 = 0 \quad (2a)$$

$$\partial L/\partial x_2 = 2x_2 + \lambda a_2 = 0 \quad (2b)$$

$$\partial L/\partial \lambda = a_1x_1 + a_2x_2 - b = 0 \quad \text{(the constraint !)} \quad (2c)$$

The first equations give $x_1 = -\frac{1}{2}\lambda a_1$ and $x_2 = -\frac{1}{2}\lambda a_2$. Substitute into $a_1x_1 + a_2x_2 = b$:

$$-\frac{1}{2}\lambda a_1^2 - \frac{1}{2}\lambda a_2^2 = b \quad \text{and} \quad \lambda = \frac{-2b}{a_1^2 + a_2^2}. \quad (3)$$

Substituting λ into (2a) and (2b) reveals the closest point (x_1^*, x_2^*) and the minimum cost $(x_1^*)^2 + (x_2^*)^2$:

$$x_1^* = -\frac{1}{2}\lambda a_1 = \frac{a_1 b}{a_1^2 + a_2^2} \qquad x_2^* = -\frac{1}{2}\lambda a_2 = \frac{a_2 b}{a_1^2 + a_2^2} \qquad (x_1^*)^2 + (x_2^*)^2 = \frac{b^2}{a_1^2 + a_2^2}$$

The derivative of the minimum cost with respect to the constraint level b is minus the Lagrange multiplier :

$$\frac{d}{db}\left(\frac{b^2}{a_1^2 + a_2^2}\right) = \frac{2b}{a_1^2 + a_2^2} = -\lambda. \quad (4)$$

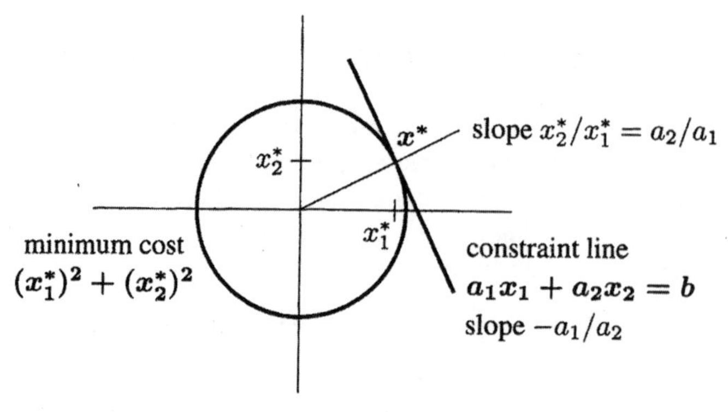

Figure VI.4: The constraint line is tangent to the minimum cost circle at the solution $\boldsymbol{x}^*$.

Minimizing a Quadratic with Linear Constraints

We will move that example from the plane $\mathbf{R}^2$ to the space $\mathbf{R}^n$. Instead of one constraint on $\boldsymbol{x}$ we have m **constraints $A^{\mathrm{T}}\boldsymbol{x} = \boldsymbol{b}$. The matrix A^{T} will be m by n.** There will be m Lagrange multipliers $\lambda_1, \ldots, \lambda_m$: one for each constraint. The cost function $F(\boldsymbol{x}) = \frac{1}{2}\boldsymbol{x}^{\mathrm{T}}S\boldsymbol{x}$ allows any symmetric positive definite matrix S.

$$\textbf{Problem: Minimize } F = \tfrac{1}{2}\boldsymbol{x}^{\mathrm{T}}S\boldsymbol{x} \textbf{ subject to } A^{\mathrm{T}}\boldsymbol{x} = \boldsymbol{b}. \qquad (5)$$

With m constraints there will be m Lagrange multipliers $\boldsymbol{\lambda} = (\lambda_1, \ldots, \lambda_m)$. They build the constraints $A^{\mathrm{T}}\boldsymbol{x} = \boldsymbol{b}$ into the Lagrangian $L(\boldsymbol{x}, \boldsymbol{\lambda}) = \frac{1}{2}\boldsymbol{x}^{\mathrm{T}}S\boldsymbol{x} + \boldsymbol{\lambda}^{\mathrm{T}}(A^{\mathrm{T}}\boldsymbol{x} - \boldsymbol{b})$. The $n + m$ derivatives of L give $n + m$ equations for a vector $\boldsymbol{x}$ in $\mathbf{R}^n$ and $\boldsymbol{\lambda}$ in $\mathbf{R}^m$:

$$\begin{array}{ll} \boldsymbol{x}\text{-derivatives of } L: & S\boldsymbol{x} + A\boldsymbol{\lambda} = \mathbf{0} \\ \boldsymbol{\lambda}\text{-derivatives of } L: & A^{\mathrm{T}}\boldsymbol{x} \qquad = \boldsymbol{b} \end{array} \qquad (6)$$

The first equations give $\boldsymbol{x} = -S^{-1}A\boldsymbol{\lambda}$. Then the second equations give $-A^{\mathrm{T}}S^{-1}A\boldsymbol{\lambda} = \boldsymbol{b}$. This determines the optimal $\boldsymbol{\lambda}^*$ and therefore the optimal $\boldsymbol{x}^*$:

$$\textbf{Solution } \boldsymbol{\lambda}^*, \boldsymbol{x}^* \qquad \boldsymbol{\lambda}^* = -(A^{\mathrm{T}}S^{-1}A)^{-1}\boldsymbol{b} \qquad \boldsymbol{x}^* = S^{-1}A(A^{\mathrm{T}}S^{-1}A)^{-1}\boldsymbol{b}. \quad (7)$$

$$\text{Minimum cost } F^* = \frac{1}{2}(\boldsymbol{x}^*)^{\mathrm{T}}S\boldsymbol{x}^* = \frac{1}{2}\boldsymbol{b}^{\mathrm{T}}(A^{\mathrm{T}}S^{-1}A)^{-1}A^{\mathrm{T}}S^{-1}SS^{-1}A(A^{\mathrm{T}}S^{-1}A)^{-1}\boldsymbol{b}.$$

This simplifies a lot !

$$\begin{array}{l} \textbf{Minimum cost } F^* = \dfrac{1}{2}\boldsymbol{b}^{\mathrm{T}}(A^{\mathrm{T}}S^{-1}A)^{-1}\boldsymbol{b} \\ \textbf{Gradient of cost } \dfrac{\partial F^*}{\partial \boldsymbol{b}} = (A^{\mathrm{T}}S^{-1}A)^{-1}\boldsymbol{b} = -\boldsymbol{\lambda}^* \end{array} \qquad (8)$$

This is truly a model problem. When the constraint changes to an inequality $A^{\mathrm{T}}\boldsymbol{x} \leq \boldsymbol{b}$, the multipliers become $\lambda_i \geq 0$ and the problem becomes harder.

May I return to the "saddle point matrix" or "KKT matrix" in equation (6):

$$M\begin{bmatrix} \boldsymbol{x} \\ \boldsymbol{\lambda} \end{bmatrix} = \begin{bmatrix} S & A \\ A^{\mathrm{T}} & 0 \end{bmatrix}\begin{bmatrix} \boldsymbol{x} \\ \boldsymbol{\lambda} \end{bmatrix} = \begin{bmatrix} 0 \\ \boldsymbol{b} \end{bmatrix} \tag{9}$$

That matrix M is not positive definite or negative definite. Suppose you multiply the first block row $[S \quad A]$ by $A^{\mathrm{T}}S^{-1}$ to get $[A^{\mathrm{T}} \quad A^{\mathrm{T}}S^{-1}A]$. Subtract from the second block row to see a zero block:

$$\begin{bmatrix} S & A \\ 0 & -A^{\mathrm{T}}S^{-1}A \end{bmatrix}\begin{bmatrix} \boldsymbol{x} \\ \boldsymbol{\lambda} \end{bmatrix} = \begin{bmatrix} 0 \\ \boldsymbol{b} \end{bmatrix}. \tag{10}$$

This is just elimination on the 2 by 2 block matrix M. That new block in the 2, 2 position is called the **Schur complement** (named after the greatest linear algebraist of all time).

We reached the same equation $-A^{\mathrm{T}}S^{-1}A\boldsymbol{\lambda} = \boldsymbol{b}$ as before. Elimination is just an organized way to solve linear equations. The first n pivots were positive because S is **positive** definite. Now there will be m negative pivots because $-A^{\mathrm{T}}S^{-1}A$ is **negative** definite. This is the unmistakable sign of a **saddle point in the Lagrangian** $L(\boldsymbol{x}, \boldsymbol{\lambda})$.

That function $L = \frac{1}{2}\boldsymbol{x}^{\mathrm{T}}S\boldsymbol{x} + \boldsymbol{\lambda}^{\mathrm{T}}(A^{\mathrm{T}}\boldsymbol{x} - \boldsymbol{b})$ is **convex in $\boldsymbol{x}$ and concave in $\boldsymbol{\lambda}$**!

Minimax = Maximin

There is more to learn from this problem. The $\boldsymbol{x}$-derivative of L and the $\boldsymbol{\lambda}$-derivative of L were set to zero in equation (6). We solved those two equations for $\boldsymbol{x}^*$ and $\boldsymbol{\lambda}^*$. That pair $(\boldsymbol{x}^*, \boldsymbol{\lambda}^*)$ is a saddle point of L in equation (7). By solving (7) we found the minimum cost and its derivative in (8).

Suppose we separate this into two problems: a minimum and a maximum problem. First minimize $L(\boldsymbol{x}, \boldsymbol{\lambda})$ for each fixed λ. The minimizing $\boldsymbol{x}^*$ depends on $\boldsymbol{\lambda}$. Then find the $\boldsymbol{\lambda}^*$ that maximizes $L(\boldsymbol{x}^*(\boldsymbol{\lambda}), \boldsymbol{\lambda})$.

Minimize L at $\boldsymbol{x}^* = -S^{-1}A\boldsymbol{\lambda}$ At that point $\boldsymbol{x}^*$, $\min L = -\dfrac{1}{2}\boldsymbol{\lambda}^{\mathrm{T}}A^{\mathrm{T}}S^{-1}A\boldsymbol{\lambda} - \boldsymbol{\lambda}^{\mathrm{T}}\boldsymbol{b}$

Maximize that minimum $\boldsymbol{\lambda}^* = -(A^{\mathrm{T}}S^{-1}A)^{-1}\boldsymbol{b}$ gives $L = \dfrac{1}{2}\boldsymbol{b}^{\mathrm{T}}(A^{\mathrm{T}}S^{-1}A)^{-1}\boldsymbol{b}$

$$\boxed{\max_{\boldsymbol{\lambda}} \; \min_{\boldsymbol{x}} \; L = \frac{1}{2}\boldsymbol{b}^{\mathrm{T}}(A^{\mathrm{T}}S^{-1}A)^{-1}\boldsymbol{b}}$$

This *maximin* was $\boldsymbol{x}$ first and $\boldsymbol{\lambda}$ second. The reverse order is *minimax*: $\boldsymbol{\lambda}$ first, $\boldsymbol{x}$ second.

The maximum over $\boldsymbol{\lambda}$ of $L(\boldsymbol{x}, \boldsymbol{\lambda}) = \dfrac{1}{2}\boldsymbol{x}^{\mathrm{T}}S\boldsymbol{x} + \boldsymbol{\lambda}^{\mathrm{T}}(A^{\mathrm{T}}\boldsymbol{x} - \boldsymbol{b})$ is $\begin{cases} +\infty \text{ if } A^{\mathrm{T}}\boldsymbol{x} \neq \boldsymbol{b} \\ \frac{1}{2}\boldsymbol{x}^{\mathrm{T}}S\boldsymbol{x} \text{ if } A^{\mathrm{T}}\boldsymbol{x} = \boldsymbol{b} \end{cases}$

The minimum over $\boldsymbol{x}$ of that maximum over $\boldsymbol{\lambda}$ is our answer $\frac{1}{2}\boldsymbol{b}^{\mathrm{T}}(A^{\mathrm{T}}S^{-1}A)^{-1}\boldsymbol{b}$.

$$\boxed{\min_{\boldsymbol{x}} \; \max_{\boldsymbol{\lambda}} \; L = \frac{1}{2}\boldsymbol{b}^{\mathrm{T}}(A^{\mathrm{T}}S^{-1}A)^{-1}\boldsymbol{b}}$$

At the saddle point $(\boldsymbol{x}^*, \boldsymbol{\lambda}^*)$ we have $\dfrac{\partial L}{\partial \boldsymbol{x}} = \dfrac{\partial L}{\partial \boldsymbol{\lambda}} = 0$ **and** $\displaystyle\max_{\boldsymbol{\lambda}} \min_{\boldsymbol{x}} L = \min_{\boldsymbol{x}} \max_{\boldsymbol{\lambda}} L.$

Dual Problems in Science and Engineering

Minimizing a quadratic $\frac{1}{2}\boldsymbol{x}^{\mathrm{T}}S\boldsymbol{x}$ with a linear constraint $A^{\mathrm{T}}\boldsymbol{x} = \boldsymbol{b}$ is not just an abstract exercise. It is the central problem in physical applied mathematics–when a linear differential equation is made discrete. Here are two major examples.

1 Network equations for electrical circuits

Unknowns: Voltages at nodes, currents along edges

Equations: Kirchhoff's Laws and Ohm's Law

Matrices: $A^{\mathrm{T}}S^{-1}A$ is the **conductance matrix.**

2 Finite element method for structures

Unknowns: Displacements at nodes, stresses in the structure

Equations: Balance of forces and stress-strain relations

Matrices: $A^{\mathrm{T}}S^{-1}A$ is the **stiffness matrix.**

The full list would extend to every field of engineering. The stiffness matrix and the conductance matrix are symmetric positive definite. Normally the constraints are equations and not inequalities. Then mathematics offers three approaches to the modeling of the physical problem:

(i) Linear equations with the stiffness matrix or conductance matrix or system matrix

(ii) Minimization with currents or stresses as the unknowns $\boldsymbol{x}$

(iii) Maximization with voltages or displacements as the unknowns $\boldsymbol{\lambda}$

In the end, the linear equations (i) are the popular choice. We reduce equation (9) to equation (10). Those network equations account for Kirchhoff and Ohm together. The structure equations account for force balance and properties of the material. All electrical and mechanical laws are built into the final system.

For problems of fluid flow, that system of equations is often in its saddle point form. The unknowns $\boldsymbol{x}$ and $\boldsymbol{\lambda}$ are velocities and pressures. The numerical analysis is well described in *Finite Elements and Fast Iterative Solvers* by Elman, Silvester, and Wathen.

For network equations and finite element equations leading to conductance matrices and stiffness matrices $A^{\mathrm{T}}CA$, one reference is my textbook on *Computational Science and Engineering*. The video lectures for 18.085 are on **ocw.mit.edu.**

In statistics and least squares (linear regression), the matrix $A^{\mathrm{T}}\Sigma^{-1}A$ includes $\Sigma =$ covariance matrix. We divide by variances σ^2 to whiten the noise.

For nonlinear problems, the energy is no longer a quadratic $\frac{1}{2}\boldsymbol{x}^{\mathrm{T}}S\boldsymbol{x}$. *Geometric nonlinearities* appear in the matrix A. *Material nonlinearities* (usually simpler) appear in the matrix C. Large displacements and large stresses are a typical source of nonlinearity.

Prob 1 Set VI.2

1 Mi. imize $F(\boldsymbol{x}) = \frac{1}{2}\boldsymbol{x}^{\mathrm{T}}S\boldsymbol{x} = \frac{1}{2}x_1^2 + 2x_2^2$ subject to $A^{\mathrm{T}}\boldsymbol{x} = x_1 + 3x_2 = b$.

(a) What is the Lagrangian $L(\boldsymbol{x}, \lambda)$ for this problem?

(b) What are the three equations "derivative of L = zero"?

(c) Solve those equations to find $\boldsymbol{x}^* = (x_1^*, x_2^*)$ and the multiplier λ^*.

(d) Draw Figure VI.4 for this problem with constraint line tangent to cost circle.

(e) Verify that the derivative of the minimum cost is $\partial F^*/\partial b = -\lambda^*$.

2 Minimize $F(\boldsymbol{x}) = \frac{1}{2}\left(x_1^2 + 4x_2^2\right)$ subject to $2x_1 + x_2 = 5$. Find and solve the three equations $\partial L/\partial x_1 = 0$ and $\partial L/\partial x_2 = 0$ and $\partial L/\partial \lambda = 0$. Draw the constraint line $2x_1 + x_2 = 5$ tangent to the ellipse $\frac{1}{2}\left(x_1^2 + 4x_2^2\right) = F_{\text{min}}$ at the minimum point (x_1^*, x_2^*).

3 The saddle point matrix in Problem 1 is

$$M = \begin{bmatrix} S & A \\ A^{\mathrm{T}} & 0 \end{bmatrix} = \begin{bmatrix} 1 & 0 & 1 \\ 0 & 4 & 3 \\ 1 & 3 & 0 \end{bmatrix}.$$

Reduce M to a triangular matrix U by elimination, and verify that

$$U = \begin{bmatrix} S & A \\ 0 & -A^{\mathrm{T}}S^{-1}A \end{bmatrix}.$$

How many positive pivots for M? How many positive eigenvalues for M?

4 For any invertible symmetric matrix S, *the number of positive pivots equals the number of positive eigenvalues.* The pivots appear in $S = LDL^{\mathrm{T}}$ (triangular L) The eigenvalues appear in $S = Q\Lambda Q^{\mathrm{T}}$ (orthogonal Q). A nice proof sends L and Q to either I or $-I$ *without becoming singular part way* (see Problem Set III.2). The eigenvalues stay real and don't cross zero. So their signs are the same in D and Λ.

Prove this "Law of Inertia" for any 2 by 2 invertible symmetric matrix S: **S has 0 or 1 or 2 positive eigenvalues when it has 0 or 1 or 2 positive pivots.**

1. Take the determinant of $LDL^{\mathrm{T}} = Q\Lambda Q^{\mathrm{T}}$ to show that $\det D$ and $\det \Lambda$ have the same sign. If the determinant is negative then S has __ positive eigenvalue in Λ and __ positive pivot in D.
2. If the determinant is positive then S could be positive definite or negative definite. Show that both pivots are positive when both eigenvalues are positive.

5 Find the minimum value of $F(\boldsymbol{x}) = \frac{1}{2}\left(x_1^2 + x_2^2 + x_3^2\right)$ with one constraint $x_1 + x_2 + x_3 = 3$ and then with an additional constraint $x_1 + 2x_2 + 3x_3 = 12$. The second minimum value should be less than the first minimum value: Why? The first problem has a __ tangent to a sphere in $\mathbf{R}^3$. The second problem has a __ tangent to a sphere in $\mathbf{R}^3$.

VI.3 Linear Programming, Game Theory, and Duality

This section is about highly structured optimization problems. Linear programming comes first—linear cost and linear constraints (including inequalities). It was also historically first, when Dantzig invented the simplex algorithm to find the optimal solution. Our approach here will be to see the "duality" between a minimum problem and a maximum—*two linear programs that are solved at the same time.*

An inequality constraint $\boldsymbol{x_k} \geq \mathbf{0}$ has two states—active and inactive. If the minimizing solution ends up with $x_k^* > 0$, then that requirement was inactive—it didn't change anything. Its Lagrange multiplier will have $\lambda_k^* = 0$. The minimum cost is not affected by that constraint on x_k. But if the constraint $x_k \geq 0$ actively forces the best $\boldsymbol{x}^*$ to have $x_k^* = 0$, then the multiplier will have $\lambda_k^* > 0$. So the optimality condition is $\boldsymbol{x_k^* \lambda_k^* = 0}$ for each k.

One more point about linear programming. It solves all **2-person zero sum games.** Profit to one player is loss to the other player. The optimal strategies produce a saddle point.

Inequality constraints are still present in quadratic programming (**QP**) and semidefinite programming (**SDP**). The constraints in SDP involve symmetric matrices. The inequality $S \geq 0$ means that the matrix is positive semidefinite (or definite). If the best S is actually positive definite, then the constraint $S \geq 0$ was not active and the Lagrange multiplier (now also a matrix) will be zero.

Linear Programming

Linear programming starts with a cost vector $\boldsymbol{c} = (c_1, \ldots, c_n)$. The problem is to minimize the cost $F(\boldsymbol{x}) = c_1x_1 + \cdots + c_nx_n = \boldsymbol{c}^{\mathrm{T}}\boldsymbol{x}$. The constraints are m linear equations $A\boldsymbol{x} = \boldsymbol{b}$ and n inequalities $x_1 \geq 0, \ldots, x_n \geq 0$. We just write $\boldsymbol{x} \geq 0$ to include all n components:

$$\textbf{Linear Program} \qquad \boxed{\textbf{Minimize } \boldsymbol{c}^{\mathrm{T}}\boldsymbol{x} \textbf{ subject to } A\boldsymbol{x} = \boldsymbol{b} \textbf{ and } \boldsymbol{x} \geq 0} \qquad (1)$$

If A is 1 by 3, $A\boldsymbol{x} = \boldsymbol{b}$ gives a plane like $x_1 + x_2 + 2x_3 = 4$ in 3-dimensional space. That plane will be chopped off by the constraints $x_1 \geq 0, x_2 \geq 0, x_3 \geq 0$. This leaves a triangle on a plane, with corners at $(x_1, x_2, x_3) = (4, 0, 0)$ and $(0, 4, 0)$ and $(0, 0, 2)$. Our problem is to find the point $\boldsymbol{x}^*$ in this triangle that minimizes the cost $\boldsymbol{c}^{\mathrm{T}}\boldsymbol{x}$.

Because the cost is linear, **its minimum will be reached at one of those corners.** Linear programming has to find that minimum cost corner. Computing all corners is exponentially impractical when m and n are large. So the *simplex method* finds one starting corner that satisfies $A\boldsymbol{x} = \boldsymbol{b}$ and $\boldsymbol{x} \geq \mathbf{0}$. Then it moves along an edge of the constraint set K to another (lower cost) corner. The cost $\boldsymbol{c}^{\mathrm{T}}\boldsymbol{x}$ drops at every step.

It is a linear algebra problem to find the steepest edge and the next corner (where that edge ends). The simplex method repeats this step many times, from corner to corner.

New starting corner, new steepest edge, new lower cost corner in the simplex method. In practice the number of steps is polynomial (but in theory it could be exponential).

Our interest here is to identify the **dual problem**—a maximum problem for $\boldsymbol{y}$ in $\mathbf{R}^m$. It is standard to use $\boldsymbol{y}$ instead of λ for the dual unknowns—the Lagrange multipliers.

Dual Problem | **Maximize** $\boldsymbol{y}^{\mathrm{T}}\boldsymbol{b}$ **subject to** $A^{\mathrm{T}}\boldsymbol{y} \le \boldsymbol{c}$. (2)

This is another linear program for the simplex method to solve. It has the same inputs $A, \boldsymbol{b}, \boldsymbol{c}$ as before. When the matrix A is m by n, the matrix A^{T} is n by m. So $A^{\mathrm{T}}\boldsymbol{y} \le \boldsymbol{c}$ has n constraints. A beautiful fact: $\boldsymbol{y}^{\mathrm{T}}\boldsymbol{b}$ in the maximum problem is never larger than $\boldsymbol{c}^{\mathrm{T}}\boldsymbol{x}$ in the minimum problem.

Weak duality $\quad \boldsymbol{y}^{\mathrm{T}}\boldsymbol{b} = \boldsymbol{y}^{\mathrm{T}}(A\boldsymbol{x}) = (A^{\mathrm{T}}\boldsymbol{y})^{\mathrm{T}}\boldsymbol{x} \le \boldsymbol{c}^{\mathrm{T}}\boldsymbol{x}$ $\quad$ **Maximum (2) $\le$ minimum (1)**

Maximizing pushes $\boldsymbol{y}^{\mathrm{T}}\boldsymbol{b}$ upward. Minimizing pushes $\boldsymbol{c}^{\mathrm{T}}\boldsymbol{x}$ downward. The great duality theorem (**the minimax theorem**) says that they meet at the best $\boldsymbol{x}^*$ and the best $\boldsymbol{y}^*$.

Duality $\quad$ **The maximum of $\boldsymbol{y}^{\mathrm{T}}\boldsymbol{b}$ equals the minimum of $\boldsymbol{c}^{\mathrm{T}}\boldsymbol{x}$.**

The simplex method will solve both problems at once. For many years that method had no competition. Now this has changed. Newer algorithms go directly *through* the set of allowed $\boldsymbol{x}$'s instead of traveling around its edges (from corner to corner). *Interior point methods* are competitive because they can use calculus to achieve steepest descent.

The situation right now is that either method could win—along the edges or inside.

Max Flow-Min Cut

Here is a special linear program. The matrix A will be the incidence matrix of a graph. That means that flow in equals flow out at every node. Each edge of the graph has a *capacity* M_j—which the flow y_j along that edge cannot exceed.

The maximum problem is to send the greatest possible flow from the source node s to the sink node t. This flow is returned from t to s on a special edge with unlimited capacity—drawn on the next page. The constraints on $\boldsymbol{y}$ are Kirchhoff's Current Law $A^{\mathrm{T}}\boldsymbol{y} = 0$ and the capacity bounds $|y_j| \le M_j$ on each edge of the graph. The beauty of this example is that you can solve it by common sense (for a small graph). In the process, you discover and solve the dual minimum problem, which is **min cut**.

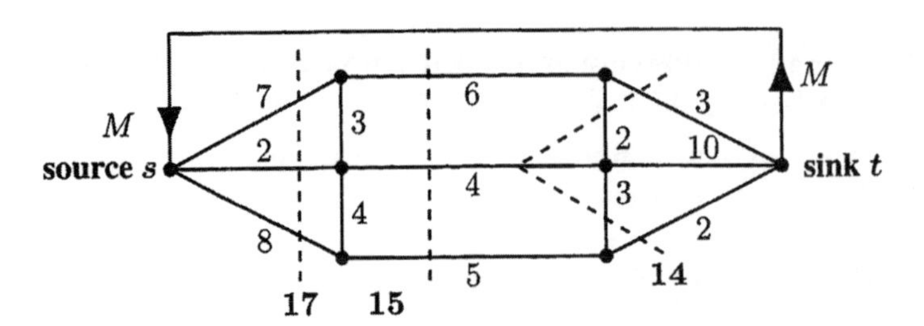

Max flow problem
Maximize M
with flow $y_j \leq M_j$
The capacity M_j is shown on every edge

Figure VI.5: The max flow M is bounded by the capacity of any cut (dotted line). *By duality, the capacity of the* ***minimum cut*** *equals the* ***maximum flow***: $M = \mathbf{14}$.

Begin by sending flow out of the source. The three edges from s have capacity $7 + 2 + 8 = \mathbf{17}$. Is there a tighter bound than $M \leq 17$?

Yes, a cut through the three middle edges only has capacity $6 + 4 + 5 = \mathbf{15}$. Therefore 17 cannot reach the sink. Is there a tighter bound than $M \leq 15$?

Yes, a cut through five later edges only has capacity $3 + 2 + 4 + 3 + 2 = \mathbf{14}$. The total flow M cannot exceed 14. Is that flow of 14 achievable and is this the tightest cut?

Yes, this is the min cut (*it is an* ℓ^1 *problem*!) and by duality 14 is the max flow.

Wikipedia shows a list of faster and faster algorithms to solve this important problem. It has many applications. If the capacities M_j are integers, the optimal flows y_j are integers. Normally integer programs are extra difficult, but not here.

A special max flow problem has all capacities $M_j = 1$ or 0. The graph is **bipartite** (all edges go from a node in part 1 to a node in part 2). We are matching people in part 1 to jobs in part 2 (at most one person per job and one job per person). Then the maximum matching is $M =$ max flow in the graph $=$ max number of assignments.

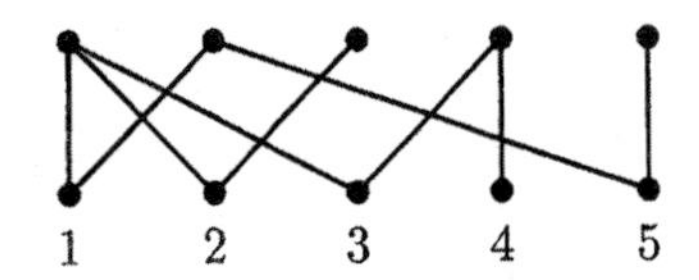

This bipartite graph allows a perfect matching: $M = 5$. Remove the edge from 2 down to 1. Now only $M = 4$ assignments are possible, because 2 and 5 will only be qualified for one job (5).

For bipartite graphs, max flow $=$ min cut is König's theorem and Hall's marriage theorem.

Two Person Games

Games with three or more players are very difficult to solve. Groups of players can combine against the others, and those alliances are unstable. New teams will often form. It took John Nash to make good progress, leading to his Nobel Prize (in economics!). But two-person zero-sum games were completely solved by von Neumann. We will see their close connection to linear programming and duality.

The players are X and Y. There is a payoff matrix A. At every turn, player X chooses a row of A and player Y chooses a column. The number in that row and column of A is the payoff, and then the players take another turn.

To match with linear programming, I will make the payment go from player X to Y. Then X wants to minimize and Y wants to maximize. Here is a very small payoff matrix. It has two rows for X to choose and three columns for Y.

Payoff matrix

	y_1	y_2	y_3
x_1	**1**	**0**	**2**
x_2	**3**	**−1**	**4**

Y likes those large numbers in column 3. X sees that the smallest number in that column is 2 (in row 1). Both players have no reason to move from this simple strategy of column 3 for Y and row 1 for X. The payoff of 2 is from X to Y:

2 is smallest in its column and largest in its row

This is a **saddle point**. Y cannot expect to win more than 2. X cannot expect to lose less than 2. Every play of the game will be the same because no player has an incentive to change. The optimal strategies $\boldsymbol{x}^*$ and $\boldsymbol{y}^*$ are clear: row 1 for X and column 3 for Y.

But a change in column 3 will require new thinking by both players.

New payoff matrix

	y_1	y_2	y_3
x_1	**1**	**0**	**4**
x_2	**3**	**−1**	**2**

X likes those small and favorable numbers in column 2. But Y will never choose that column. Column 3 looks best (biggest) for Y, and X should counter by choosing row 2 (to avoid paying 4). But then column 1 becomes better than column 3 for Y, because winning 3 in column 1 is better than winning 2.

You are seeing that Y still wants column 3 but must go sometimes to column 1. Similarly X must have a *mixed strategy*: choose rows 1 and 2 with probabilities x_1 and x_2. The choice at each turn must be unpredictable, or the other player will take advantage. So the decision for X is two probabilities $x_1 \geq 0$ and $x_2 \geq 0$ that add to $x_1 + x_2 = 1$. The payoff matrix has a new row from this mixed strategy:

row 1	1	0	4
row 2	3	−1	2
$\boldsymbol{x_1(\textbf{row 1}) + x_2(\textbf{row 2})}$	$\boldsymbol{x_1 + 3x_2}$	$\boldsymbol{-x_2}$	$\boldsymbol{4x_1 + 2x_2}$

X will choose fractions x_1 and x_2 to make the worst (*largest*) payoff as small as possible. Remembering $x_2 = 1 - x_1$, this will happen when the two largest payoffs are equal:

$$x_1 + 3x_2 = 4x_1 + 2x_2 \quad \textbf{means} \quad \boldsymbol{x_1 + 3(1 - x_1) = 4x_1 + 2(1 - x_1)}.$$

That equation gives $\boldsymbol{x_1^* = \frac{1}{4}}$ **and** $\boldsymbol{x_2^* = \frac{3}{4}}$. **The new mixed row is** $\boldsymbol{2.5, -.75, 2.5}$.

Similarly Y will choose columns $1, 2, 3$ with probabilities y_1, y_2, y_3. Again they add to 1. That mixed strategy combines the three columns of A into a new column for Y.

column 1	column 2	column 3	**mix 1, 2, 3**
1	0	4	$\boldsymbol{y_1 + 4y_3}$
3	−1	2	$\boldsymbol{3y_1 - y_2 + 2y_3}$

Y will choose the fractions $y_1 + y_2 + y_3 = 1$ to make the worst (*smallest*) payoff as large as possible. That happens when $y_2 = 0$ and $y_3 = 1 - y_1$. The two mixed payoffs are equal:

$$y_1 + 4(1 - y_1) = 3y_1 + 2(1 - y_1) \quad \text{gives} \quad -3y_1 + 4 = y_1 + 2 \quad \text{and} \quad y_1^* = y_3^* = \tfrac{1}{2}.$$

The new mixed column has 2.5 in both components. These optimal strategies identify 2.5 as the value of the game. With the mixed strategy $x_1^* = \frac{1}{4}$ and $x_2^* = \frac{3}{4}$, Player X can guarantee to pay no more than 2.5. Player Y can guarantee to receive no less than 2.5. We have found the saddle point (best mixed strategies, with minimax payoff from X = maximin payoff to $Y = \mathbf{2.5}$) of this two-person game.

	y_1	y_2	y_3	$\frac{1}{2}$**col 1** + $\frac{1}{2}$**col 2**
row 1	1	0	4	**2.5**
row 2	3	−1	2	**2.5**
$\frac{1}{4}$ **row 1** + $\frac{3}{4}$**row 2**	**2.5**	−.75	**2.5**	

Von Neumann's **minimax theorem** for games gives a solution for every payoff matrix. It is equivalent to the duality theorem *min* $\boldsymbol{c}^{\mathrm{T}}\boldsymbol{x} =$ *max* $\boldsymbol{y}^{\mathrm{T}}\boldsymbol{b}$ for linear programming.

Semidefinite Programming (SDP)

The cost to minimize is still $\boldsymbol{c}^{\mathrm{T}}\boldsymbol{x}$: linear cost. But now the constraints on $\boldsymbol{x}$ involve symmetric matrices S. We are given S_0 to S_n and $S(\boldsymbol{x}) = S_0 + x_1S_1 + \cdots + x_nS_n$ is required to be *positive semidefinite (or definite)*. Fortunately this is a convex set of $\boldsymbol{x}$'s—the average of two semidefinite matrices is semidefinite. (Just average the two energies $\boldsymbol{v}^{\mathrm{T}}S\boldsymbol{v} \geq 0$.)

Now the set of allowed $\boldsymbol{x}$'s could have curved sides instead of flat sides:

$$S_0 + x_1S_1 + x_2S_2 = \begin{bmatrix} x_1 & 1 \\ 1 & x_2 \end{bmatrix} \text{ is positive semidefinite when } \boldsymbol{x_1 \geq 0} \text{ and } \boldsymbol{x_1x_2 \geq 1}.$$

Minimizing the maximum eigenvalue of $S(\boldsymbol{x})$ is also included with an extra variable t:

$$\text{Minimize } t \text{ so that } tI - S(\boldsymbol{x}) \text{ is positive semidefinite.}$$

And SDP could also minimize the largest singular value—the L^2 norm of $S(\boldsymbol{x})$:

$$\text{Minimize } t \text{ so that } \begin{bmatrix} tI & S(\boldsymbol{x}) \\ S(\boldsymbol{x})^{\mathrm{T}} & tI \end{bmatrix} \text{ is positive semidefinite.}$$

For those and most semidefinite problems, interior-point methods are the best. We don't travel around the boundary of the constraint set (from corner to corner, as the simplex method does for linear programs). Instead we travel through the interior of the set. Essentially we are solving a least squares problem at each iteration—usually 5 to 50 iterations.

As in linear programming, there is a **dual problem** (a maximization). The value of this dual is always *below* the value $\boldsymbol{c}^{\mathrm{T}}\boldsymbol{x}$ of the original. When we maximize in the dual and minimize $\boldsymbol{c}^{\mathrm{T}}\boldsymbol{x}$ in the primal, we hope to make those answers equal. But this might not happen for semidefinite programs with matrix inequalities.

SDP gives a solution method for matrix problems that previously looked too difficult.

Problem Set VI.3

1 Is the constraint $\boldsymbol{x} \geq \mathbf{0}$ needed in equation (3) for weak duality? Is the inequality $A^{\mathrm{T}}\boldsymbol{y} \leq \boldsymbol{c}$ already enough to prove that $(A^{\mathrm{T}}\boldsymbol{y})^{\mathrm{T}}\boldsymbol{x} \leq \boldsymbol{c}^{\mathrm{T}}\boldsymbol{x}$? I don't think so.

2 Suppose the constraints are $x_1 + x_2 + 2x_3 = 4$ and $x_1 \geq 0, x_2 \geq 0, x_3 \geq 0$. Find the three corners of this triangle in $\mathbf{R}^3$. Which corner minimizes the cost $\boldsymbol{c}^{\mathrm{T}}\boldsymbol{x} = 5x_1 + 3x_2 + 8x_3$?

3 What maximum problem for y is the dual to Problem 2? One constraint in the primal problem means one unknown y in the dual problem. Solve this dual problem.

4 Suppose the constraints are $\boldsymbol{x} \geq \mathbf{0}$ and $x_1 + 2x_3 + x_4 = 4$ and $x_2 + x_3 - x_4 = 2$. Two equality constraints on four unknowns, so a corner like $\boldsymbol{x} = (0, 6, 0, 4)$ has $4 - 2 = 2$ zeros. Find another corner with $\boldsymbol{x} = (x_1, x_2, 0, 0)$ and show that it costs more than the first corner.

5 Find the optimal (minimizing) strategy for X to choose rows. Find the optimal (maximizing) strategy for Y to choose columns. What is the payoff from X to Y at this optimal minimax point $\boldsymbol{x}^*, \boldsymbol{y}^*$?

$$\textbf{Payoff matrices} \quad \begin{bmatrix} 1 & 2 \\ 4 & 8 \end{bmatrix} \qquad \begin{bmatrix} 1 & 4 \\ 8 & 2 \end{bmatrix}$$

6 If $A^{\mathrm{T}} = -A$ (antisymmetric payoff matrix), why is this a fair game for X and Y with minimax payoff equal to zero?

7 Suppose the payoff matrix is a diagonal matrix Σ with entries $\sigma_1 > \sigma_2 > \ldots > \sigma_n$. What strategies are optimal for X and Y?

8 Convert $||(x_1, x_2, x_3)||_1 \leq 2$ in the $\boldsymbol{\ell}^1$ **norm** to eight linear inequalities $Ax \leq \boldsymbol{b}$. The constraint $||x|| \leq 2$ in the $\boldsymbol{\ell}^\infty$ **norm** also produces eight linear inequalities.

9 In the $\boldsymbol{\ell}^2$ **norm,** $||x|| \leq 2$ is a quadratic inequality $x_1^2 + x_2^2 + x_3^2 \leq 4$. But in semidefinite programming (SDP) this becomes one matrix inequality $\boldsymbol{X}\boldsymbol{X}^{\mathrm{T}} \leq 4I$.

Why is this constraint $\boldsymbol{X}\boldsymbol{X}^{\mathrm{T}} \leq 4I$ equivalent to $\boldsymbol{x}^{\mathrm{T}}\boldsymbol{x} \leq 4$?

Note Duality offers an important option: Solve the primal *or the dual*. That applies to optimization in machine learning, as this paper shows:

F. Bach, *Duality between subgradient and conditional gradient methods*, SIAM Journal of Optimization **25** (2015) 115-129; arXiv: 1211.6302.

VI.4 Gradient Descent Toward the Minimum

This section of the book is about a fundamental problem: **Minimize a function** $\boldsymbol{f(x_1,\ldots,x_n)}$. Calculus teaches us that all the first derivatives $\partial f/\partial x_i$ are zero at the minimum (when f is smooth). If we have $n = 20$ unknowns (a small number in deep learning) then minimizing one function f produces 20 equations $\partial f/\partial x_i = 0$. *"Gradient descent" uses the derivatives* $\partial f/\partial x_i$ *to find a direction that reduces* $f(\boldsymbol{x})$. The steepest direction, in which $f(\boldsymbol{x})$ decreases fastest, is given by the gradient $-\boldsymbol{\nabla f}$:

$$\textbf{Gradient descent} \qquad \boldsymbol{x}_{k+1} = \boldsymbol{x}_k - s_k \boldsymbol{\nabla f}(\boldsymbol{x}_k) \tag{1}$$

The symbol $\boldsymbol{\nabla f}$ represents the vector of n partial derivatives of f: its **gradient**. So (1) is a vector equation for each step $k = 1, 2, 3, \ldots$ and s_k is the *stepsize* or the *learning rate*. We hope to move toward the point $\boldsymbol{x}^*$ where the graph of $f(\boldsymbol{x})$ hits bottom.

We are willing to assume for now that 20 first derivatives exist and can be computed. We are not willing to assume that those 20 functions also have 20 convenient derivatives $\partial/\partial x_j(\partial f/\partial x_i)$. Those are the 210 **second derivatives** of f—which go into a 20 by 20 symmetric matrix H. (Symmetry reduces $n^2 = 400$ to $\frac{1}{2}n^2 + \frac{1}{2}n = 210$ computations.) The second derivatives would be very useful extra information, but in many problems we have to go without.

You should know that 20 first derivatives and 210 second derivatives don't multiply the computing cost by 20 and 210. The neat idea of **automatic differentiation**—rediscovered and extended as **backpropagation** in machine learning—makes those cost factors much smaller in practice. This idea is described in Section VII.2.

Return for a moment to equation (1). The step $-s_k\boldsymbol{\nabla f}(\boldsymbol{x}_k)$ includes a minus sign (to descend) and a factor s_k (to control the the stepsize) and the gradient vector $\boldsymbol{\nabla f}$ (containing the first derivatives of f computed at the current point $\boldsymbol{x}_k$). A lot of thought and computational experience has gone into the choice of stepsize and search direction.

We start with the main facts from calculus about derivatives and gradient vectors $\boldsymbol{\nabla f}$.

The Derivative of $f(x)$: $n = 1$

The derivative of $f(x)$ involves a *limit*—this is the key difference between calculus and algebra. We are comparing the values of f at two nearby points x and $x + \Delta x$, as Δx approaches zero. More accurately, we are watching the slope $\Delta f/\Delta x$ between two points on the graph of $f(x)$:

$$\textbf{Derivative of } \boldsymbol{f} \textbf{ at } x \qquad \frac{df}{dx} = \textbf{limit of } \frac{\Delta f}{\Delta x} = \textbf{limit of } \left[\frac{f(x+\Delta x) - f(x)}{\Delta x}\right]. \tag{2}$$

This is a forward difference when Δx is positive and a backward difference when $\Delta x < 0$. When we have the same limit from both sides, that number is the slope of the graph at x.

The ramp function $\text{ReLU}(x) = f(x) = \max(0, x)$ is heavily involved in deep learning (see VII.1). It has unequal slopes **1** to the right and **0** to the left of $x = 0$. So the derivative df/dx does not exist at that corner point in the graph. For $n = 1$, df/dx is the gradient $\boldsymbol{\nabla f}$.

$$\text{ReLU} = \begin{matrix} x \text{ for } x \geq 0 \\ 0 \text{ for } x \leq 0 \end{matrix} \qquad \textbf{slope } \frac{\Delta f}{\Delta x} = \frac{f(0+\Delta x) - f(0)}{\Delta x} = \begin{matrix} \boldsymbol{\Delta x / \Delta x = 1} \text{ if } \Delta x > 0 \\ \boldsymbol{0 / \Delta x = 0} \text{ if } \Delta x < 0 \end{matrix}$$

For the smooth function $\boldsymbol{f(x) = x^2}$, the ratio $\Delta f/\Delta x$ will safely approach the derivative df/dx from both sides. But the approach could be slow (just first order). Look again at the point $x = 0$, where the true derivative $df/dx = 2x$ is now zero:

$$\text{The ratio } \frac{\Delta f}{\Delta x} \text{ at } x = 0 \text{ is } \frac{f(\Delta x) - f(0)}{\Delta x} = \frac{(\Delta x)^2 - 0}{\Delta x} = \boldsymbol{\Delta x} \text{ Then limit} = \text{slope} = 0.$$

In this case and in almost all cases, we get a better ratio (closer to the limiting slope df/dx) by averaging the **forward difference** (where $\Delta x > 0$) with the **backward difference** (where $\Delta x < 0$). The average is the more accurate **centered difference**.

$$\begin{matrix}\textbf{Centered} \\ \textbf{at } x\end{matrix} \quad \frac{1}{2}\left[\frac{f(x+\Delta x) - f(x)}{\Delta x} + \frac{f(x-\Delta x) - f(x)}{-\Delta x}\right] = \frac{\boldsymbol{f(x+\Delta x) - f(x-\Delta x)}}{\boldsymbol{2\,\Delta x}}$$

For the example $f(x) = x^2$ this centering will produce the exact derivative $df/dx = 2x$. In the picture we are averaging plus and minus slopes to get the correct slope 0 at $x = 0$. For all smooth functions, the centered differences reduce the error to size $(\Delta x)^2$. This is a big improvement over the error of size Δx for uncentered differences $f(x+\Delta x) - f(x)$.

$f(x) = 0$ $f(x) = x$
slope 0 **slope 1**

centered $f(x) = x^2$
backward **forward**

Figure VI.6: ReLU function = ramp from deep learning. Centered slope of $f = x^2$ is exact.

Most finite difference approximations are centered for extra accuracy. But we are still dividing by a small number $2\,\Delta x$. And for a multivariable function $F(x_1, x_2, \ldots, x_n)$ we will need ratios $\Delta F/\Delta x_i$ in n different directions—possibly for large n. Those ratios approximate the n partial derivatives that go into the **gradient vector grad** $\boldsymbol{F} = \boldsymbol{\nabla F}$.

The gradient of $\boldsymbol{F(x_1, \ldots, x_n)}$ is the column vector $\boldsymbol{\nabla F = \left(\frac{\partial F}{\partial x_1}, \ldots, \frac{\partial F}{\partial x_n}\right)}$.

Its components are the n partial derivatives of F. $\boldsymbol{\nabla F}$ points in the steepest direction.

Examples 1-3 will show the value of vector notation ($\boldsymbol{\nabla F}$ is always a column vector).

Example 1 For a constant vector $\boldsymbol{a} = (a_1, \ldots, a_n)$, $F(\boldsymbol{x}) = \boldsymbol{a}^{\mathrm{T}}\boldsymbol{x}$ has gradient $\boldsymbol{\nabla F} = \boldsymbol{a}$. The partial derivatives of $F = a_1x_1 + \cdots + a_nx_n$ are the numbers $\partial F/\partial x_k = a_k$.

Example 2 For a symmetric matrix S, the gradient of $F(\boldsymbol{x}) = \boldsymbol{x}^{\mathrm{T}} S \boldsymbol{x}$ is $\boldsymbol{\nabla F} = \boldsymbol{2\, S x}$. To see this, write out the function $F(x_1, x_2)$ when $n = 2$. The matrix S is 2 by 2:

$$F = \begin{bmatrix} x_1 & x_2 \end{bmatrix} \begin{bmatrix} a & b \\ b & c \end{bmatrix} \begin{bmatrix} x_1 \\ x_2 \end{bmatrix} = \begin{matrix} ax_1^2 + cx_2^2 \\ +2bx_1x_2 \end{matrix} \quad \begin{bmatrix} \partial f/\partial x_1 \\ \partial f/\partial x_2 \end{bmatrix} = 2 \begin{bmatrix} ax_1 + bx_2 \\ bx_1 + cx_2 \end{bmatrix} = \boldsymbol{2S} \begin{bmatrix} \boldsymbol{x_1} \\ \boldsymbol{x_2} \end{bmatrix}.$$

Example 3 For a positive definite symmetric S, the minimum of a quadratic $F(\boldsymbol{x}) = \frac{1}{2}\boldsymbol{x}^{\mathrm{T}} \boldsymbol{S x} - \boldsymbol{a}^{\mathrm{T}} \boldsymbol{x}$ is the negative number $\boldsymbol{F_{\min}} = -\frac{1}{2}\,\boldsymbol{a}^{\mathrm{T}} \boldsymbol{S a}$ at $\boldsymbol{x}^* = \boldsymbol{S^{-1} a}$. This is an important example! The minimum occurs where first derivatives of F are zero:

$$\boldsymbol{\nabla F} = \begin{bmatrix} \partial F/\partial x_1 \\ : \\ \partial F/\partial x_n \end{bmatrix} = \boldsymbol{Sx - a = 0 \ \text{at} \ x^* = S^{-1}a = \arg\min F.} \tag{3}$$

As always, **that notation arg min F stands for the point x^* where the minimum of $F(x) = \frac{1}{2}x^{\mathrm{T}} Sx - a^{\mathrm{T}}x$ is reached.** Often we are more interested in this minimizing $\boldsymbol{x}^*$ than in the actual minimum value $F_{\min} = F(\boldsymbol{x}^*)$ at that point:

$$\boldsymbol{F_{\min}} \text{ is } \frac{1}{2}(S^{-1}\boldsymbol{a})^{\mathrm{T}} S (S^{-1}\boldsymbol{a}) - \boldsymbol{a}^{\mathrm{T}}(S^{-1}\boldsymbol{a}) = \frac{1}{2}\boldsymbol{a}^{\mathrm{T}} S^{-1}\boldsymbol{a} - \boldsymbol{a}^{\mathrm{T}} S^{-1}\boldsymbol{a} = -\frac{\boldsymbol{1}}{\boldsymbol{2}}\boldsymbol{a}^{\mathrm{T}}\boldsymbol{S}^{-1}\boldsymbol{a}.$$

The graph of F is a bowl passing through zero at $\boldsymbol{x} = \boldsymbol{0}$ and dipping to its minimum at $\boldsymbol{x}^*$.

Example 4 The **determinant** $F(\boldsymbol{x}) = \det \boldsymbol{X}$ is a function of all n^2 variables x_{ij}. In the formula for $\det \boldsymbol{X}$, each x_{ij} along a row is multiplied by its "cofactor" C_{ij}. This cofactor is a determinant of size $n-1$, using all rows of $\boldsymbol{X}$ except row i and all columns except column j—and multiplied by $(-1)^{i+j}$:

$$\text{The partial derivatives } \frac{\partial(\det X)}{\partial x_{ij}} = C_{ij} \text{ in the matrix of cofactors of } X \text{ give } \boldsymbol{\nabla F}.$$

Example 5 The **logarithm of the determinant** is a most remarkable function:

$$\boldsymbol{L(X)} = \log(\det \boldsymbol{X}) \text{ has partial derivatives } \frac{\partial L}{\partial x_{ij}} = \frac{C_{ij}}{\det X} = \boldsymbol{j, i} \textbf{ entry of } \boldsymbol{X^{-1}}.$$

The chain rule for $L = \log F$ is $(\partial L/\partial F)(\partial F/\partial x_{ij}) = (1/F)(\partial F/\partial x_{ij}) = (1/\det X)\, C_{ij}$. Then this ratio of cofactors to determinant gives the j, i entries of the inverse matrix X^{-1}.

It is neat that X^{-1} contains the n^2 first derivatives of $L = \log\det X$. The second derivatives of L are remarkable too. We have n^2 variables x_{ij} and n^2 first derivatives in $\boldsymbol{\nabla L} = (X^{-1})^{\mathrm{T}}$. This means n^4 second derivatives! What is amazing is that the matrix of second derivatives is **negative definite** when $X = S$ is symmetric positive definite. So we reverse the sign of L: **positive definite second derivatives $\Rightarrow$ convex function.**

– log (det S) is a convex function of the entries of the positive definite matrix S.

The Geometry of the Gradient Vector ∇f

Start with a function $f(x, y)$. It has $n=2$ variables. Its gradient is $\nabla f = (\partial f/\partial x, \partial f/\partial y)$. This vector changes length as we move the point x, y where the derivatives are computed:

$$\nabla f = \left(\frac{\partial f}{\partial x}, \frac{\partial f}{\partial y}\right) \quad \textbf{Length} = ||\nabla f|| = \sqrt{\left(\frac{\partial f}{\partial x}\right)^2 + \left(\frac{\partial f}{\partial y}\right)^2} = \text{steepest slope of } f$$

That length $||\nabla f||$ tells us the steepness of the graph of $z = f(x, y)$. The graph is normally a curved surface—like a mountain or a valley in xyz space. At each point there is a slope $\partial f/\partial x$ in the x-direction and a slope $\partial f/\partial y$ in the y-direction. **The steepest slope is in the direction of ∇f = grad f. The magnitude of that steepest slope is $||\nabla f||$.**

Example 6 The graph of a linear function $f(x, y) = ax + by$ is the plane $z = ax + by$. The gradient is the vector $\nabla f = \begin{bmatrix} a \\ b \end{bmatrix}$ of partial derivatives. The length of that vector is $||\nabla f|| = \sqrt{a^2 + b^2}$ = slope of the roof. The slope is steepest in the direction of ∇f.

That steepest direction is perpendicular to the level direction. The level direction z = constant has $ax + by$ = constant. It is the safe direction to walk, perpendicular to ∇f. The component of ∇f in that flat direction is zero. Figure VI.7 shows the two perpendicular directions (level and steepest) on the plane $z = x + 2y = f(x, y)$.

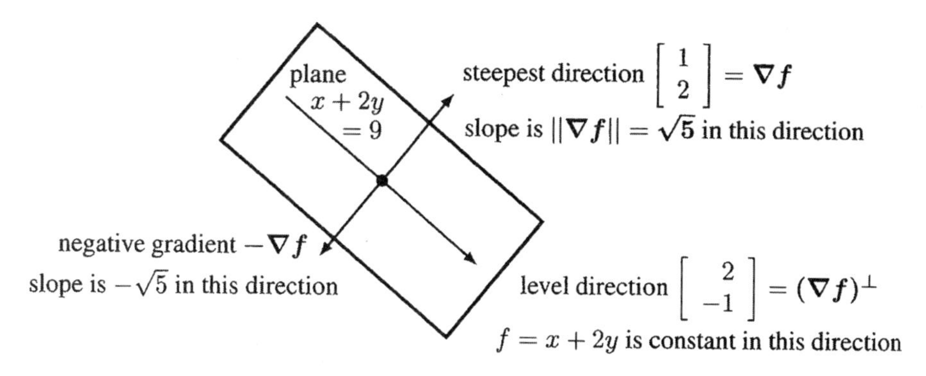

Figure VI.7: The negative gradient $-\nabla f$ gives the direction of steepest descent.

For the nonlinear function $f(x, y) = ax^2 + by^2$, the gradient is $\nabla f = \begin{bmatrix} 2ax \\ 2by \end{bmatrix}$. That tells us the steepest direction, changing from point to point. We are on a curved surface (a bowl opening upward). The bottom of the bowl is at $x = y = 0$ where the gradient vector is zero. The slope in the steepest direction is $||\nabla f||$. At the minimum, $\nabla f = (2ax, 2by) = (0, 0)$ and *slope = zero*.

The level direction has $z = ax^2 + by^2 =$ constant height. That plane $z =$ constant cuts through the bowl in a level curve. In this example the level curve $ax^2 + by^2 = c$ is an ellipse. The direction of the ellipse (level direction) is perpendicular to the gradient vector (steepest direction). But there is a serious difficulty for steepest descent:

The steepest direction changes as you go down! The gradient doesn't point to the bottom!

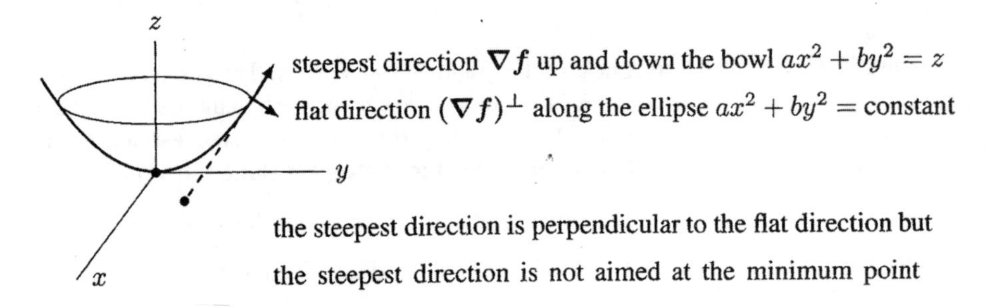

Figure VI.8: **Steepest descent moves down the bowl in the gradient direction** $\begin{bmatrix} -2ax \\ -2by \end{bmatrix}$.

Let me repeat. At the point x_0, y_0 the gradient direction for $f = ax^2 + by^2$ is along $\nabla f = (2ax_0, 2by_0)$. The steepest line through x_0, y_0 is $2ax_0(y - y_0) = 2by_0(x - x_0)$. But then the lowest point $(x, y) = (0, 0)$ does not lie on the line! **We will not find that minimum point in one step of "gradient descent". The steepest direction does not lead to the bottom of the bowl**—except when $b = a$ and the bowl is circular.

Water changes direction as it goes down a mountain. Sooner or later, we must change direction too. In practice we keep going in the gradient direction and stop when our cost function f is not decreasing quickly. At that point Step 1 ends and we recompute the gradient ∇f. This gives a new descent direction for Step 2.

An Important Example with Zig-Zag

The example $f(x, y) = \frac{1}{2}(x^2 + by^2)$ is extremely useful for $0 < b \le 1$. Its gradient ∇f has two components $\partial f/\partial x = x$ and $\partial f/\partial y = by$. The minimum value of f is zero. That minimum is reached at the point $(x^*, y^*) = (0, 0)$. Best of all, steepest descent with exact line search produces a simple formula for each point (x_k, y_k) in the slow progress down the bowl toward $(0, 0)$. Starting from $(x_0, y_0) = (b, 1)$ we find these points:

$$\boxed{x_k = b\left(\frac{b-1}{b+1}\right)^k \quad y_k = \left(\frac{1-b}{1+b}\right)^k \quad f(x_k, y_k) = \left(\frac{1-b}{1+b}\right)^{2k} f(x_0, y_0)} \quad (4)$$

If $b = 1$, you see immediate success in one step. The point (x_1, y_1) is $(0,0)$. The bowl is perfectly circular with $f = \frac{1}{2}(x^2 + y^2)$. The negative gradient direction goes exactly through $(0,0)$. Then the first step of gradient descent finds that correct minimizing point where $f = 0$.

The real purpose of this example is seen when b is small. The crucial ratio in equation (4) is $r = (b-1)/(b+1)$. For $b = \frac{1}{10}$ this ratio is $r = -9/11$. For $b = \frac{1}{100}$ the ratio is $-99/101$. The ratio is approaching -1 and the progress toward $(0,0)$ has virtually stopped when b is very small.

Figure VI.9 shows the frustrating zig-zag pattern of the steps toward $(0,0)$. Every step is short and progress is very slow. This is a case where the stepsize s_k in $\boldsymbol{x}_{k+1} = \boldsymbol{x}_k - s_k \boldsymbol{\nabla} \boldsymbol{f}(\boldsymbol{x}_k)$ was exactly chosen to minimize f (an exact line search). But the direction of $-\boldsymbol{\nabla} \boldsymbol{f}$, even if steepest, is pointing far from the final answer $(x^*, y^*) = (0,0)$.

The bowl has become a narrow valley when b is small, and we are uselessly crossing the valley instead of moving down the valley to the bottom.

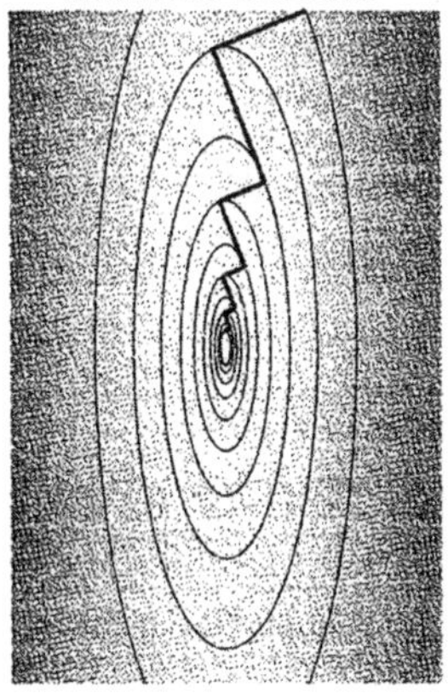

The first descent step starts out perpendicular to the level set. As it crosses through lower level sets, the function $f(x,y)$ is decreasing. **Eventually its path is tangent to a level set L.** Descent has stopped. Going further will increase f. The first step ends. The next step is perpendicular to L. So the zig-zag path took a 90 ° turn.

Figure VI.9: Slow convergence on a zig-zag path to the minimum of $f = x^2 + by^2$.

For b close to 1, this gradient descent is faster. First-order convergence means that the distance to $(x^*, y^*) = (0,0)$ is reduced by the constant factor $(1-b)/(1+b)$ at every step. The following analysis will show that linear convergence extends to all strongly convex functions f—first when each line search is exact, and then (more realistically) when the search at each step is close to exact.

Machine learning often uses **stochastic gradient descent**. The next section will describe this variation (especially useful when n is large and $\boldsymbol{x}$ has many components). And we recall that **Newton's method** uses second derivatives to produce quadratic convergence—the reduction factor $(1-b)/(1+b)$ drops to zero and the error is squared at every step. (Our model problem of minimizing a quadratic $\frac{1}{2}\boldsymbol{x}^{\mathrm{T}} S \boldsymbol{x}$ is solved in one step.) This is a gold standard that approximation algorithms don't often reach in practice.

Convergence Analysis for Steepest Descent

On this page we are following the presentation by Boyd and Vandenberghe in *Convex Optimization* (published by Cambridge University Press). From the specific choice of $f(x,y) = \frac{1}{2}(x^2 + by^2)$, we move to any strongly convex $f(\boldsymbol{x})$ in n dimensions. Convexity is tested by the positive definiteness of the symmetric matrix $H = \nabla^2 \boldsymbol{f}$ of second derivatives (the Hessian matrix). In one dimension this is the number d^2f/dx^2:

Strongly convex $\boldsymbol{m > 0}$ $\qquad \boldsymbol{H_{ij}} = \dfrac{\boldsymbol{\partial^2 f}}{\boldsymbol{\partial x_i \partial x_j}}$ has eigenvalues between $\boldsymbol{m} \leq \boldsymbol{\lambda} \leq \boldsymbol{M}$ at all $\boldsymbol{x}$

The quadratic $f = \frac{1}{2}\left(x^2 + by^2\right)$ has second derivatives 1 and b. The mixed derivative $\partial^2 f/\partial x \partial y$ is zero. So the matrix is diagonal and its two eigenvalues are $m = b$ and $M = 1$. We will now see that **the ratio $\boldsymbol{m/M}$ controls the speed of steepest descent.**

The gradient descent step is $\boldsymbol{x}_{k+1} = \boldsymbol{x}_k - s\nabla \boldsymbol{f}_k$. We estimate f by its Taylor series:

$$f(\boldsymbol{x}_{k+1}) \leq f(\boldsymbol{x}_k) + \nabla \boldsymbol{f}^{\mathrm{T}}(\boldsymbol{x}_{k+1} - \boldsymbol{x}_k) + \frac{M}{2}||\boldsymbol{x}_{k+1} - \boldsymbol{x}_k||^2 \tag{5}$$

$$= f(\boldsymbol{x}_k) - s||\nabla \boldsymbol{f}||^2 + \frac{Ms^2}{2}||\nabla \boldsymbol{f}||^2 \tag{6}$$

The best s minimizes the left side (exact line search). The minimum of the right side is at $s = 1/M$. Substituting that number for s, the next point $\boldsymbol{x}_{k+1}$ has

$$f(\boldsymbol{x}_{k+1}) \leq f(\boldsymbol{x}_k) - \frac{1}{2M}||\nabla \boldsymbol{f}(\boldsymbol{x}_k)||^2. \tag{7}$$

A parallel argument uses m instead of M to reverse the inequality sign in (5).

$$f(\boldsymbol{x}^*) \geq f(\boldsymbol{x}_k) - \frac{1}{2m}||\nabla \boldsymbol{f}(\boldsymbol{x}_k)||^2. \tag{8}$$

Multiply (7) by M and (8) by m and subtract to remove $||\nabla \boldsymbol{f}(\boldsymbol{x}_k)||^2$. Rewrite the result as

Steady drop in f

$$\boxed{f(\boldsymbol{x}_{k+1}) - f(\boldsymbol{x}^*) \leq \left(1 - \frac{m}{M}\right)\left(f(\boldsymbol{x}_k) - f(\boldsymbol{x}^*)\right)} \tag{9}$$

This says that every step reduces the height above the bottom of the valley by at least $c = 1 - \frac{m}{M}$. That is **linear convergence: very slow when $\boldsymbol{b = m/M}$ is small.**

Our zig-zag example had $m = b$ and $M = 1$. The estimate (9) guarantees that the height $f(\boldsymbol{x}_k)$ above $f(\boldsymbol{x}^*) = 0$ is reduced by at least $1 - b$. The exact formula in that totally computable problem produced the reduction factor $(1-b)^2/(1+b)^2$. When b is small this is about $1 - 4b$. So the actual improvement was only 4 times better than the rough estimate $1 - b$ in (9). **This gives us considerable faith that (9) is realistic.**

Inexact Line Search and Backtracking

That ratio m/M appears throughout approximation theory and numerical linear algebra. This is the point of mathematical analysis—to find numbers like m/M that control the rate of descent to the minimum value $f(\boldsymbol{x}^*)$.

Up to now all line searches were exact: $\boldsymbol{x}_{k+1}$ exactly minimized $f(\boldsymbol{x})$ along the line $\boldsymbol{x} = \boldsymbol{x}_k - s\boldsymbol{\nabla f}_k$. Choosing s is a one-variable minimization. The line moves from $\boldsymbol{x}_k$ in the direction of steepest descent. But we can't expect an exact formula for minimizing a general function $f(\boldsymbol{x})$, even just along a line. So we need a fast sensible way to find an approximate minimum (and the analysis needs a bound on this additional error).

One sensible way is **backtracking**. Start with the full step $s = 1$ to $\boldsymbol{X} = \boldsymbol{x}_k - \boldsymbol{\nabla f}_k$.

Test If $f(\boldsymbol{X}) \le f(\boldsymbol{x}_k) - \dfrac{s}{3}||\boldsymbol{\nabla f}_k||^2$, with $s = 1$, stop and accept $\boldsymbol{X}$ as $\boldsymbol{x}_{k+1}$.

Otherwise backtrack: Reduce s to $\frac{1}{2}$ and try the test on $\boldsymbol{X} = \boldsymbol{x}_k - \frac{1}{2}\boldsymbol{\nabla f}_k$.

If the test fails again, try the stepsize $s = \frac{1}{4}$. Since $\boldsymbol{\nabla f}$ is a descent direction, the test is eventually passed. The factors $\frac{1}{3}$ and $\frac{1}{2}$ could be any numbers $\alpha < \frac{1}{2}$ and $\beta < 1$.

Boyd and Vandenberghe show that the convergence analysis for exact line search extends also to this backtracking search. Of course the guaranteed reduction factor $1 - (m/M)$ for each step toward the minimum is now not so large. But the new factor $1 - \min(2m\alpha, 2m\alpha\beta/M)$ is still below 1. Steepest descent with backtracking search still has linear convergence—a constant factor (or better) at every step.

Momentum and the Path of a Heavy Ball

The slow zig-zag path of steepest descent is a real problem. We have to improve it. Our model example $f = \frac{1}{2}(x^2 + by^2)$ has only two variables x, y and its second derivative matrix H is diagonal—constant entries $f_{xx} = 1$ and $f_{yy} = b$. But it shows the zig-zag problem very clearly when $\boldsymbol{b = \lambda_{\min}/\lambda_{\max} = m/M}$ **is small.**

Key idea: Zig-zag would not happen for a heavy ball rolling downhill. Its momentum carries it through the narrow valley—bumping the sides but moving mostly forward. So we **add momentum with coefficient β to the gradient** (Polyak's important idea). This gives one of the most convenient and powerful ideas in deep learning.

The direction $\boldsymbol{z}_k$ of the new step remembers the previous direction $\boldsymbol{z}_{k-1}$.

Descent with momentum
$$\boxed{\boldsymbol{x}_{k+1} = \boldsymbol{x}_k - s\boldsymbol{z}_k \text{ with } \boldsymbol{z}_k = \boldsymbol{\nabla f}(\boldsymbol{x}_k) + \beta\boldsymbol{z}_{k-1}} \quad (10)$$

Now we have two coefficients to choose—the stepsize s and also β. Most important, **the step to $\boldsymbol{x}_{k+1}$ in equation (10) involves $\boldsymbol{z}_{k-1}$.** Momentum has turned a one-step method (gradient descent) into a two-step method. To get back to one step, we have to rewrite equation (10) as **two coupled equations** (one vector equation) for the state at time $k+1$:

Descent with momentum
$$\boxed{\begin{aligned} \boldsymbol{x}_{k+1} \qquad\qquad &= \boldsymbol{x}_k - s\boldsymbol{z}_k \\ \boldsymbol{z}_{k+1} - \boldsymbol{\nabla f}(\boldsymbol{x}_{k+1}) &= \qquad \beta\boldsymbol{z}_k \end{aligned}} \quad (11)$$

With those two equations, we have recovered a one-step method. This is exactly like reducing a single second order differential equation to a system of two first order equations. Second order reduces to first order when dy/dt becomes a second unknown along with y.

$$\textbf{2nd order equation} \quad \textbf{1st order system} \qquad \frac{d^2y}{dt^2} + b\frac{dy}{dt} + ky = 0 \text{ becomes } \frac{d}{dt}\begin{bmatrix} y \\ dy/dt \end{bmatrix} = \begin{bmatrix} 0 & 1 \\ -k & -b \end{bmatrix}\begin{bmatrix} y \\ dy/dt \end{bmatrix}.$$

Interesting that this b is damping the motion while β adds momentum to encourage it.

The Quadratic Model

When $f(\boldsymbol{x}) = \frac{1}{2}\boldsymbol{x}^{\mathrm{T}}S\boldsymbol{x}$ is quadratic, its gradient $\boldsymbol{\nabla f} = S\boldsymbol{x}$ is linear. This is the model problem to understand: S is symmetric positive definite and $\boldsymbol{\nabla f}(\boldsymbol{x}_{k+1})$ becomes $S\boldsymbol{x}_{k+1}$ in equation (11). Our 2 by 2 supermodel is included, when the matrix S is diagonal with entries 1 and b. For a bigger matrix S, you will see that its largest and smallest eigenvalues determine the best choices for β and the stepsize s—so the 2 by 2 case actually contains the essence of the whole problem.

To follow the steps of accelerated descent, we track each eigenvector of S. Suppose $S\boldsymbol{q} = \lambda\boldsymbol{q}$ and $\boldsymbol{x}_k = c_k\,\boldsymbol{q}$ and $\boldsymbol{z}_k = d_k\,\boldsymbol{q}$ and $\boldsymbol{\nabla f}_k = S\boldsymbol{x}_k = \lambda\, c_k\,\boldsymbol{q}$. Then our equation (11) connects the numbers c_k and d_k at step k to c_{k+1} and d_{k+1} at step $k+1$.

$$\begin{array}{l}\textbf{Following the} \\ \textbf{eigenvector } \boldsymbol{q}\end{array} \quad \begin{array}{rl} c_{k+1} & = c_k - s\,d_k \\ -\lambda\,c_{k+1} + d_{k+1} = & \beta\,d_k \end{array} \quad \begin{bmatrix} 1 & 0 \\ -\lambda & 1 \end{bmatrix}\begin{bmatrix} c_{k+1} \\ d_{k+1} \end{bmatrix} = \begin{bmatrix} 1 & -s \\ 0 & \beta \end{bmatrix}\begin{bmatrix} c_k \\ d_k \end{bmatrix} \tag{12}$$

Finally we invert the first matrix ($-\lambda$ becomes $+\lambda$) to see each descent step clearly:

$$\begin{array}{c}\textbf{Descent step} \\ \textbf{multiplies by } \boldsymbol{R}\end{array} \quad \boxed{\begin{bmatrix} c_{k+1} \\ d_{k+1} \end{bmatrix} = \begin{bmatrix} 1 & 0 \\ \lambda & 1 \end{bmatrix}\begin{bmatrix} 1 & -s \\ 0 & \beta \end{bmatrix}\begin{bmatrix} c_k \\ d_k \end{bmatrix} = \begin{bmatrix} 1 & -s \\ \lambda & \beta - \lambda s \end{bmatrix}\begin{bmatrix} c_k \\ d_k \end{bmatrix} = \boldsymbol{R}\begin{bmatrix} c_k \\ d_k \end{bmatrix}} \tag{13}$$

After k steps the starting vector is multiplied by R^k. For fast convergence to zero (which is the minimum of $f = \frac{1}{2}\boldsymbol{x}^{\mathrm{T}}S\boldsymbol{x}$) we want both eigenvalues e_1 and e_2 of R to be as small as possible. Clearly those eigenvalues of R depend on the eigenvalue λ of S. That eigenvalue λ could be anywhere between $\lambda_{\min}(S)$ and $\lambda_{\max}(S)$. Our problem is:

$$\textbf{Choose } s \textbf{ and } \beta \textbf{ to minimize } \max\Big[|e_1(\lambda)|,\ |e_2(\lambda)|\Big] \text{ for } \lambda_{\min}(S) \le \lambda \le \lambda_{\max}(S). \tag{14}$$

It seems a miracle that this problem has a beautiful solution. The optimal s and β are

$$\boxed{s = \left(\frac{2}{\sqrt{\lambda_{\max}} + \sqrt{\lambda_{\min}}}\right)^2 \quad \text{and} \quad \beta = \left(\frac{\sqrt{\lambda_{\max}} - \sqrt{\lambda_{\min}}}{\sqrt{\lambda_{\max}} + \sqrt{\lambda_{\min}}}\right)^2.} \tag{15}$$

Think of the 2 by 2 supermodel, when S has eigenvalues $\lambda_{\max} = 1$ and $\lambda_{\min} = b$:

$$s = \left(\frac{2}{1+\sqrt{b}}\right)^2 \quad \text{and} \quad \beta = \left(\frac{1-\sqrt{b}}{1+\sqrt{b}}\right)^2 \tag{16}$$

These choices of stepsize and momentum give a convergence rate that looks like the rate in equation (4) for ordinary steepest descent (no momentum). But there is a crucial difference: **b is replaced by $\sqrt{b}$.**

$$\boxed{\textbf{Ordinary descent factor} \quad \left(\frac{1-b}{1+b}\right)^2 \qquad \textbf{Accelerated descent factor} \quad \left(\frac{1-\sqrt{b}}{1+\sqrt{b}}\right)^2} \tag{17}$$

So similar but so different. The real test comes when b is very small. Then the ordinary descent factor is essentially $1 - 4b$, very close to 1. The accelerated descent factor is essentially $1 - 4\sqrt{b}$, *much further from* 1.

To emphasize the improvement that momentum brings, suppose $b = 1/100$. Then $\sqrt{b} = 1/10$ (ten times larger than b). The convergence factors in equation (17) are

$$\textbf{Steepest descent} \quad \left(\frac{.99}{1.01}\right)^2 = .96 \qquad \textbf{Accelerated descent} \quad \left(\frac{.9}{1.1}\right)^2 = .67$$

Ten steps of ordinary descent multiply the starting error by 0.67. This is matched by a single momentum step. Ten steps with the momentum term multiply the error by 0.018.

Notice that $\lambda_{\max}/\lambda_{\min} = 1/b = \kappa$ is the **condition number of S**. This controls everything. For the non-quadratic problems studied next, the condition number is still the key. That number κ becomes L/μ as you will see.

A short editorial *This is not part of the expository textbook.* It concerns the rate of convergence for gradient descent. We know that one step methods (computing $\boldsymbol{x}_{k+1}$ from $\boldsymbol{x}_k$) can multiply the error by $1 - O(1/\kappa)$. Two step methods that use $\boldsymbol{x}_{k-1}$ in the momentum term can achieve $1 - O(\sqrt{1/\kappa})$. The condition number is $\kappa = \lambda_{\max}/\lambda_{\min}$ for the convex quadratic model $f = \frac{1}{2}\boldsymbol{x}^{\mathrm{T}} S\boldsymbol{x}$. Our example had $\mathbf{1}/\boldsymbol{\kappa} = \boldsymbol{b}$.

It is natural to hope that $\mathbf{1} - c\kappa^{-1/n}$ can be achieved by using n known values $\boldsymbol{x}_k, \boldsymbol{x}_{k-1}, \ldots, \boldsymbol{x}_{k-n+1}$. *This might be impossible*—even if it is exactly parallel to finite difference formulas for $d\boldsymbol{x}/dt = \boldsymbol{f}(\boldsymbol{x})$. Stability there requires a stepsize bound on Δt. Stability in descent requires a bound on the stepsize s. MATLAB's low order code ODE15S is often chosen for stiff equations and ODE45 is the workhorse for smoother solutions. Those are predictor-corrector combinations, not prominent so far in optimization.

The speedup from momentum is like "**overrelaxation**" in the 1950's. David Young's thesis brought the same improvement for iterative methods applied to a class of linear equations $A\boldsymbol{x} = \boldsymbol{b}$. In those early days of numerical analysis, A was separated into $S - T$ and the iterations were $S\boldsymbol{x}_{k+1} = T\boldsymbol{x}_k + \boldsymbol{b}$. They took forever. Now overrelaxation is virtually forgotten, replaced by faster methods (multigrid). Will accelerated steepest descent give way to completely different ideas for minimizing $f(\boldsymbol{x})$?

This is possible but hard to imagine.

Nesterov Acceleration

Another way to bring $\boldsymbol{x}_{k-1}$ into the formula for $\boldsymbol{x}_{k+1}$ is due to Yuri Nesterov. Instead of evaluating the gradient $\nabla \boldsymbol{f}$ at $\boldsymbol{x}_k$, he shifted that evaluation point to $\boldsymbol{x}_k + \gamma_k(\boldsymbol{x}_k - \boldsymbol{x}_{k-1})$. And choosing $\gamma = \beta$ (the momentum coefficient) combines both ideas.

Gradient Descent	Stepsize s	$\beta = 0$	$\gamma = 0$
Heavy Ball	Stepsize s	Momentum β	$\gamma = 0$
Nesterov Acceleration	Stepsize s	Momentum β	shift $\nabla \boldsymbol{f}$ by $\gamma \Delta \boldsymbol{x}$

Accelerated descent involves all three parameters $\boldsymbol{s}, \boldsymbol{\beta}, \boldsymbol{\gamma}$:

$$\boldsymbol{x}_{k+1} = \boldsymbol{x}_k + \boldsymbol{\beta}\,(\boldsymbol{x}_k - \boldsymbol{x}_{k-1}) - \boldsymbol{s}\,\nabla \boldsymbol{f}\,(\boldsymbol{x}_k + \boldsymbol{\gamma}\,(\boldsymbol{x}_k - \boldsymbol{x}_{k-1})) \tag{18}$$

To analyze the convergence rate for Nesterov with $\gamma = \beta$, reduce (18) to first order:

$$\textbf{Nesterov} \qquad \boldsymbol{x}_{k+1} = \boldsymbol{y}_k - s\nabla \boldsymbol{f}(\boldsymbol{y}_k) \text{ and } \boldsymbol{y}_{k+1} = \boldsymbol{x}_{k+1} + \beta(\boldsymbol{x}_{k+1} - \boldsymbol{x}_k). \tag{19}$$

Suppose $f(\boldsymbol{x}) = \frac{1}{2}\boldsymbol{x}^{\mathrm{T}} S\boldsymbol{x}$ and $\nabla \boldsymbol{f} = S\boldsymbol{x}$ and $S\boldsymbol{q} = \lambda \boldsymbol{q}$ as before. To track this eigenvector set $\boldsymbol{x}_k = c_k\boldsymbol{q}$ and $\boldsymbol{y}_k = d_k\boldsymbol{q}$ and $\nabla \boldsymbol{f}(\boldsymbol{y}_k) = \lambda d_k \boldsymbol{q}$ in (19):

$c_{k+1} = (1 - s\lambda)d_k$ and $d_{k+1} = (1+\beta)c_{k+1} - \beta c_k = (1+\beta)\,(1-s\lambda)d_k - \beta c_k$ becomes

$$\begin{bmatrix} c_{k+1} \\ d_{k+1} \end{bmatrix} = \begin{bmatrix} 0 & 1 - s\lambda \\ -\beta & (1+\beta)(1-s\lambda) \end{bmatrix} \begin{bmatrix} c_k \\ d_k \end{bmatrix} = R \begin{bmatrix} c_k \\ d_k \end{bmatrix} \tag{20}$$

Every Nesterov step is a multiplication by $\boldsymbol{R}$. Suppose R has eigenvalues e_1 and e_2, depending on s and β and λ. We want the larger of $|e_1|$ and $|e_2|$ to be as small as possible for all λ between $\lambda_{\text{min}}(S)$ and $\lambda_{\text{max}}(S)$. These choices for s and β give small e's:

$$s = \frac{1}{\lambda_{\text{max}}} \text{ and } \beta = \frac{\sqrt{\lambda_{\text{max}}} - \sqrt{\lambda_{\text{min}}}}{\sqrt{\lambda_{\text{max}}} + \sqrt{\lambda_{\text{min}}}} \text{ give } \max(|e_1|, |e_2|) = \frac{\sqrt{\lambda_{\text{max}}} - \sqrt{\lambda_{\text{min}}}}{\sqrt{\lambda_{\text{max}}}} \tag{21}$$

When S is the 2 by 2 matrix with eigenvalues $\lambda_{\text{max}} = 1$ and $\lambda_{\text{min}} = b$, that convergence factor (the largest eigenvalue of R) is $\mathbf{1} - \sqrt{\boldsymbol{b}}$.

This shows the same highly important improvement (*from b to $\sqrt{b}$*) as the momentum (heavy ball) formula. The complete analysis by Lessard, Recht, and Packard discovered that Nesterov's choices for s and β can be slightly improved. Su, Boyd, and Candès developed a deeper link between a particular Nesterov optimization and this equation:

$$\textbf{Model for descent} \qquad \frac{d^2y}{dt^2} + \frac{3}{t}\frac{dy}{dt} + \nabla f(t) = 0.$$

Functions to Minimize : The Big Picture

The function $f(x)$ can be strictly convex or barely convex or non-convex. Its gradient can be linear or nonlinear. Here is a list of function types in increasing order of difficulty.

1. $\boldsymbol{f(x,y)} = \frac{1}{2}(\boldsymbol{x}^2 + \boldsymbol{by}^2)$. This has only 2 variables. Its gradient $\nabla \boldsymbol{f} = (x, by)$ is linear. Its Hessian $\boldsymbol{H}$ is a diagonal 2 by 2 matrix with entries 1 and b. Those are the eigenvalues of H. The condition number is $\boldsymbol{\kappa = 1/b}$ when $0<b<1$. *Strictly convex.*

2. $\boldsymbol{f(x_1, \ldots, x_n)} = \frac{1}{2}\boldsymbol{x}^{\mathrm{T}}S\boldsymbol{x} - \boldsymbol{c}^{\mathrm{T}}\boldsymbol{x}$. Here S is a symmetric positive definite matrix. The gradient $\nabla \boldsymbol{f} = \boldsymbol{Sx} - \boldsymbol{c}$ is linear. The Hessian is $\boldsymbol{H} = \boldsymbol{S}$. Its eigenvalues are λ_1 to λ_n. Its condition number is $\boldsymbol{\kappa} = \boldsymbol{\lambda_{\max}/\lambda_{\min}}$. *Strictly convex.*

3. $\boldsymbol{f(x_1, \ldots, x_n)}$ = **smooth strictly convex function**. Its Hessian $\boldsymbol{H(x)}$ is positive definite at all $\boldsymbol{x}$ (H varies with $\boldsymbol{x}$). The eigenvalues of H are $\lambda_1(\boldsymbol{x})$ to $\lambda_n(\boldsymbol{x})$, always positive. The condition number is the maximum over all $\boldsymbol{x}$ of $\lambda_{\max}/\lambda_{\min}$.

   An essentially equivalent condition number is the ratio $L/\lambda_{\min}(\boldsymbol{x})$:

   $$\boldsymbol{L} = \textbf{"Lipschitz constant" in } ||\nabla \boldsymbol{f(x)} - \nabla \boldsymbol{f(y)}|| \leq \boldsymbol{L}\, ||\boldsymbol{x} - \boldsymbol{y}||. \tag{22}$$

   This allows corners in the gradient $\nabla \boldsymbol{f}$ and jumps in the second derivative matrix H.

4. $\boldsymbol{f(x_1, \ldots, x_n)}$ = **convex but not strictly convex**. The Hessian can be only semidefinite, with $\boldsymbol{\lambda_{\min}} = \mathbf{0}$. A small example is the ramp function f = ReLU(x) = max $(0, x)$. The gradient $\nabla \boldsymbol{f}$ becomes a "**subgradient**" that has multiple values at a corner point. The subgradient of ReLU at $x = 0$ has all values from 0 to 1. *The lines through* $(0,0)$ *with those slopes stay below the ramp function ReLU* $(\boldsymbol{x})$.

Positive definite H is allowed but so is $\lambda_{\min} = 0$. The condition number can be infinite.

The simplest example with $\lambda_{\min} = 0$ has its minimum along the whole line $x + y = 0$:

$$f(x,y) = (\boldsymbol{x}+\boldsymbol{y})^2 = \begin{bmatrix} x & y \end{bmatrix} \begin{bmatrix} \mathbf{1} & \mathbf{1} \\ \mathbf{1} & \mathbf{1} \end{bmatrix} \begin{bmatrix} x \\ y \end{bmatrix} \quad \text{with a } \textbf{semidefinite matrix } \boldsymbol{S}$$

This degeneracy is very typical of deep learning. The number of weights used by the network often far exceeds the number of training samples that determine those weights. (The "MNIST" data set can have $60,000$ training samples and $300,000$ weights.) Those weights are underdetermined but still gradient descent succeeds. *Why do its weights generalize well—to give good answers for unseen test data?*

When strict convexity is lost (Case 4), a convergence proof is still possible. But the condition number is infinite. And the rate of convergence can become sublinear.

Note. We speak about linear convergence when the error $\boldsymbol{x}_k - \boldsymbol{x}^*$ (the distance to the minimizing point) is reduced by an approximately constant factor $C < 1$ at each step:

$$\textbf{Linear convergence} \qquad ||\boldsymbol{x}_{k+1} - \boldsymbol{x}^*|| \approx C\,||\boldsymbol{x}_k - \boldsymbol{x}^*||. \tag{23}$$

This means that the error decreases **exponentially** (like C^k or $e^{k \log C}$ with $\log C < 0$). Exponential sounds fast, but it can be very slow when C is near 1.

In minimizing quadratics, **non-adaptive methods generally converge to minimum norm solutions**. Those solutions (like $\boldsymbol{x}^+ = \boldsymbol{A}^+\boldsymbol{b}$ from the pseudoinverse A^+) have zero component in the row space of A. They have the *largest margin*.

Here are good textbook references for gradient descent, including the stochastic version that is coming in VI.5:

1. D. Bertsekas, *Convex Analysis and Optimization*, Athena Scientific (2003).
2. S. Boyd and L. Vandenberghe, *Convex Optimization*, Cambridge Univ. Press (2004).
3. Yu. Nesterov, *Introductory Lectures on Convex Optimization*, Springer (2004).
4. J. Nocedal and S. Wright , *Numerical Optimization*, Springer (1999).

Four articles coauthored by Ben Recht have brought essential new ideas to the analysis of gradient methods. Papers 1 and 2 study accelerated descent (this section). Papers 3 and 4 study stochastic descent and adaptive descent. Video 5 is excellent.

1. L. Lessard, B. Recht, and A. Packard. *Analysis and design of optimization algorithms via integral quadratic constraints*, arXiv: 1408.3595v7, 28 Oct 2015.
2. A. C. Wilson, B. Recht, and M. Jordan, *A Lyapunov analysis of momentum methods in optimization*, arXiv: 1611.02635v3, 31 Dec 2016.
3. A. C. Wilson, R. Roelofs, M. Stern , N. Srebro, and B. Recht, *The marginal value of adaptive gradient methods in machine learning*, arXiv :1705.08292v1, 23 May 2017.
4. C. Zhang, S. Bengio, M. Hardt, B. Recht, and O. Vinyals, *Understanding deep learning requires rethinking generalization*, arXiv :1611.03530v2, 26 Feb 2017, International Conference on Learning Representations (2017).
5. **https://simons.berkeley.edu/talks/ben-recht-2013-09-04**

Section VI.5 on stochastic gradient descent returns to these papers for this message: Adaptive methods can possibly converge to undesirable weights in deep learning. Gradient descent from $\boldsymbol{x}_0 = \boldsymbol{0}$ (and SGD) finds the minimum norm solution to least squares.

Those adaptive methods (variants of *Adam*) are popular. The formula for $\boldsymbol{x}_{k+1}$ that stopped at $\boldsymbol{x}_{k-1}$ will go much further back—to include *all earlier points* starting at $\boldsymbol{x}_0$. In many problems that leads to faster training. As with momentum, *memory can help*.

When there are more weights to determine than samples to use (underdetermined problems), we can have multiple minimum points for $f(\boldsymbol{x})$ and multiple solutions to $\boldsymbol{\nabla} f = \boldsymbol{0}$. A crucial question in VI.5 is whether improved adaptive methods find good solutions.

Constraints and Proximal Points

How does steepest descent deal with a constraint restricting $\boldsymbol{x}$ to a convex set K? The **projected gradient** and **proximal gradient** methods use four fundamental ideas.

1 Projection onto K *The projection* $\boldsymbol{\Pi x}$ *of* $\boldsymbol{x}$ *onto* $\boldsymbol{K}$ *is the point in* $\boldsymbol{K}$ *nearest to* $\boldsymbol{x}$.

If K is curved then Π is not linear. Think of projection onto the unit ball $||\boldsymbol{x}|| \leq 1$. In this case $\boldsymbol{\Pi x} = \text{proj}_K(\boldsymbol{x}) = \boldsymbol{x}/||\boldsymbol{x}||$ for points outside the ball. A key property is that **$\boldsymbol{\Pi}$ is a contraction**. Projecting two points onto K reduces the distance between them:

$$\textbf{Projection } \boldsymbol{\Pi} = \textbf{proj}_K \qquad ||\boldsymbol{\Pi}\, \boldsymbol{x} - \boldsymbol{\Pi}\, \boldsymbol{z}|| \leq ||\boldsymbol{x} - \boldsymbol{z}|| \quad \text{for all } \boldsymbol{x} \text{ and } \boldsymbol{z} \text{ in } \mathbf{R}^n. \tag{24}$$

2 Proximal mapping $\textbf{Prox}_f(\boldsymbol{x})$ *is the vector* $\boldsymbol{z}$ *that minimizes* $\frac{1}{2}||\boldsymbol{x} - \boldsymbol{z}||^2 + f(\boldsymbol{z})$. In case $f = 0$ inside K and $f = \infty$ outside K, $\text{Prox}_f(\boldsymbol{x})$ is exactly the projection $\boldsymbol{\Pi x}$.

Important example If $f(\boldsymbol{x}) = c||\boldsymbol{x}||_1$ then Prox_f is the **shrinkage function** in statistics.

The ith component of $\boldsymbol{x}$ leads to
the ith component of $\text{Prox}_f(\boldsymbol{x})$
This is **soft thresholding** $\boldsymbol{S(x)}$
$S(x_i) = \text{sign}\,(x_i) \cdot \max\,(x_i - c,\, 0)$

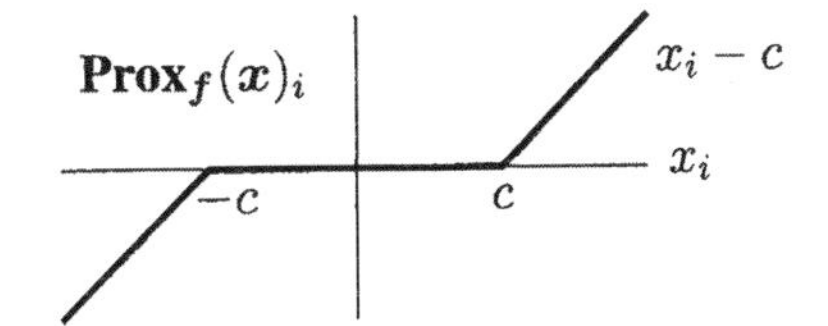

We are denoising and regularizing, as in the ℓ^1 LASSO construction of Section III.4. The graph shows how small components are set to zero—producing zeros is the effect that the $\boldsymbol{\ell^1}$ norm already achieves compared to $\boldsymbol{\ell^2}$.

3 Projected gradient descent takes a normal descent step (which may go outside the constraint set K). Then it projects the result back onto K : a basic idea.

$$\boxed{\textbf{Projected descent} \qquad \boldsymbol{x}_{k+1} = \text{proj}_K(\boldsymbol{x}_k - s_k \boldsymbol{\nabla f}(\boldsymbol{x}_k))} \tag{25}$$

4 Proximal gradient descent also starts with a normal step. Now the projection onto K is replaced by the **proximal map** to determine $s = s_k$: a subtle idea.

$$\boxed{\textbf{Proximal descent} \qquad \boldsymbol{x}_{k+1} = \text{prox}_s(\boldsymbol{x}_k - s\boldsymbol{\nabla f}(\boldsymbol{x}_k))} \tag{26}$$

The LASSO function to minimize is $f(\boldsymbol{x}) = \frac{1}{2}||\boldsymbol{b} - A\boldsymbol{x}||_2^2 + \lambda||\boldsymbol{x}||_1$. The proximal mapping decides the soft thresholding and the stepsize s at $\boldsymbol{x}$ by

$$\begin{aligned} \text{prox}_s(\boldsymbol{x}) &= \underset{z}{\text{argmin}}\ \frac{1}{2s}||\boldsymbol{x} - \boldsymbol{z}||^2 + \lambda||\boldsymbol{x}||_1 \\ &= \underset{z}{\text{argmin}}\ \frac{1}{2}||\boldsymbol{x} - \boldsymbol{z}||^2 + \lambda s||\boldsymbol{x}||_1 = S_{s\lambda}(\boldsymbol{x}) \end{aligned} \tag{27}$$

That produces the soft-thresholding function S in the graph above as the update $\boldsymbol{x}_{k+1}$.

$$\boxed{\textbf{Proximal gradients for LASSO (fast descent)} \quad \boldsymbol{x}_{k+1} = S_{\lambda s}(\boldsymbol{x}_k + sA^{\mathrm{T}}(\boldsymbol{b} - A\boldsymbol{x}_k))}$$

Problem Set VI.4

1 For a 1 by 1 matrix in Example 3, the determinant is just $\det X = x_{11}$. Find the first and second derivatives of $F(X) = -\log(\det X) = -\log x_{11}$ for $x_{11} > 0$. Sketch the graph of $F = -\log x$ to see that this function F is convex.

2 The determinant of a 2 by 2 matrix is $\det(X) = ad - bc$. Its first derivatives are $d, -c, -b, a$ in $\boldsymbol{\nabla F}$. After dividing by $\det X$, those fill the inverse matrix X^{-1}. That division by $\det X$ makes them the four derivatives of $\log(\det X)$:

$$\textbf{Derivatives of } \boldsymbol{F = \det X} \quad \boldsymbol{\nabla F} = \begin{bmatrix} d & -c \\ -b & a \end{bmatrix} \qquad \textbf{Inverse of } \boldsymbol{X} = \frac{1}{\det \boldsymbol{X}} \begin{bmatrix} d & -b \\ -c & a \end{bmatrix} = \frac{(\boldsymbol{\nabla F})^{\mathrm{T}}}{\det \boldsymbol{X}}$$

Symmetry gives $b = c$. Then $F = -\log(ad - b^2)$ is a convex function of a, b, d. *Show that the* 3 *by* 3 *second derivative matrix of this function* F *is positive definite.*

3 Show how equations (7) and (8) lead to the basic estimate (9) for linear convergence of steepest descent. (This extends to backtracking for another choice of c.)

4 A non-quadratic example with its minimum at $x = 0$ and $y = +\infty$ is

$$f(x, y) = \frac{1}{2}x^2 + e^{-y} \qquad \nabla f = \begin{bmatrix} x \\ -e^{-y} \end{bmatrix} \quad H = \begin{bmatrix} 1 & 0 \\ 0 & e^{-y} \end{bmatrix} \quad \kappa = \frac{1}{e^{-y}}$$

5 Explain why projection onto a convex set K is a *contraction* in equation (24). Why is the distance $||\boldsymbol{x} - \boldsymbol{y}||$ never increased when $\boldsymbol{x}$ and $\boldsymbol{y}$ are projected onto K?

6 What is the gradient descent equation $\boldsymbol{x}_{k+1} = \boldsymbol{x}_k - s_k \nabla f(\boldsymbol{x}_k)$ for the least squares problem of minimizing $f(\boldsymbol{x}) = \frac{1}{2}||A\boldsymbol{x} - \boldsymbol{b}||^2$?

VI.5 Stochastic Gradient Descent and ADAM

Gradient descent is fundamental in training a deep neural network. It is based on a step of the form $\boldsymbol{x}_{k+1} = \boldsymbol{x}_k - s_k \nabla L(\boldsymbol{x}_k)$. That step should lead us downhill toward the point $\boldsymbol{x}^*$ where the loss function $L(\boldsymbol{x})$ is minimized for the test data $\boldsymbol{v}$. But for large networks with many samples in the training set, this algorithm (as it stands) is not successful!

It is important to recognize two different problems with classical steepest descent:

1. Computing ∇L at every descent step—the derivatives of the total loss L with respect to all the weights $\boldsymbol{x}$ in the network—is too expensive. That total loss adds the individual losses $\ell(\boldsymbol{x}, \boldsymbol{v}_i)$ *for every sample* $\boldsymbol{v}_i$ *in the training set*—potentially millions of separate losses are computed and added in every computation of $\boldsymbol{L}$.

2. The number of weights is even larger. So $\nabla_{\boldsymbol{x}} \boldsymbol{L} = \mathbf{0}$ for many different choices $\boldsymbol{x}^*$ of the weights. **Some of those choices can give poor results on unseen test data.** The learning function $\boldsymbol{F}$ can fail to "generalize". But **stochastic gradient descent (SGD)** does find weights $\boldsymbol{x}^*$ that generalize—weights that will succeed on unseen vectors $\boldsymbol{v}$ from a similar population.

Stochastic gradient descent uses only a "minibatch" of the training data at each step. B samples will be chosen randomly. Replacing the full batch of all the training data by a minibatch changes $L(\boldsymbol{x}) = \frac{1}{n} \sum \ell_i(\boldsymbol{x})$ to a sum of only B losses. This resolves both difficulties at once. The success of deep learning rests on these two facts:

1. Computing $\nabla \ell_i$ by backpropagation on B samples is much faster. Often $B = 1$.

2. The stochastic algorithm produces weights $\boldsymbol{x}^*$ that also succeed on unseen data.

The first point is clear. The calculation per step is greatly reduced. The second point is a miracle. Generalizing well to new data is a gift that researchers work hard to explain.

We can describe the big picture of weight optimization in a large neural network. The weights are determined by the training data. That data often consists of thousands of samples. We know the "*features*" of each sample—maybe its height and weight, or its shape and color, or the number of nouns and verbs and commas (for a sample of text). Those features go into a vector $\boldsymbol{v}$ for each sample. We use a minibatch of samples.

And for each sample in the training set, we know if it is "a cat or a dog"—or if the text is "poetry or prose". We look for a **learning function** $\boldsymbol{F}$ that assigns good weights. Then for $\boldsymbol{v}$ in a similar population, F outputs the correct classification "cat" or "poetry".

We use this function F for unidentified **test data**. The features of the test data are the new inputs $\boldsymbol{v}$. The output from F will be the correct (?) classification—provided the function has learned the training data in a way that generalizes.

Here is a remarkable observation from experience. *We don't want to fit the training data too perfectly.* That would often be **overfitting**. The function F becomes oversensitive. It memorized everything but it hasn't learned anything. **Generalization by SGD** is the ability to give the correct classification for unseen test data $\boldsymbol{v}$, based on the weights $\boldsymbol{x}$ that were learned from the training data.

I compare overfitting with choosing a polynomial of degree 60 *that fits exactly to* 61 *data points.* Its 61 coefficients a_0 to a_{60} will perfectly learn the data. But that high degree polynomial will oscillate wildly between the data points. For test data at a nearby point, the perfect-fit polynomial gives a *completely wrong answer.*

So a fundamental strategy in training a neural network (which means finding a function that learns from the training data and generalizes well to test data) is **early stopping**. Machine learning needs to know when to quit! Possibly this is true of human learning too.

The Loss Function and the Learning Function

Now we establish the optimization problem that the network will solve. We need to define the "loss" $\boldsymbol{L(x)}$ that our function will (approximately) minimize. This is the sum of the errors in classifying each of the training data vectors $\boldsymbol{v}$. And we need to describe the form of the learning function $\boldsymbol{F}$ that classifies each data vector $\boldsymbol{v}$.

At the beginning of machine learning the function F was *linear*—a severe limitation. Now F is certainly nonlinear. Just the inclusion of one particular nonlinear function at each neuron in each layer has made a dramatic difference. It has turned out that with thousands of samples, the function F can be correctly trained.

It is the processing power of the computer that makes for fast operations on the data. In particular, we depend on the speed of GPU's (the Graphical Processing Units that were originally developed for computer games). They make deep learning possible.

We first choose a loss function to minimize. Then we describe stochastic gradient descent. The gradient is determined by the network architecture—the "feedforward" steps whose weights we will optimize. Our goal in this section is to find *optimization algorithms that apply to very large problems.* Then Chapter VII will describe how the architecture and the algorithms have made the learning functions successful.

Here are three loss functions—cross-entropy is a favorite for neural nets. Section VII.4 will describe the advantages of cross-entropy loss over square loss (as in least squares).

1 **Square loss** $L(\boldsymbol{x}) = \dfrac{1}{N}\sum_{1}^{N} ||F(\boldsymbol{x}, \boldsymbol{v}_i) - \text{true}||^2$: sum over the training samples $\boldsymbol{v}_i$

2 **Hinge loss** $L(\boldsymbol{x}) = \dfrac{1}{N}\sum_{1}^{N} \textbf{max}\,(0, 1 - t\,F(\boldsymbol{x}))$ for **classification** $t = 1$ or -1

3 **Cross-entropy loss** $L(\boldsymbol{x}) = -\dfrac{1}{N}\sum_{1}^{N} [y_i \log \widehat{y}_i + (1-y_i)\log(1-\widehat{y}_i)]$ for $y_i = 0$ or 1

Cross-entropy loss or "logistic loss" is preferred for *logistic regression* (with two choices only). The true label $y_i = 0$ or 1 could be -1 or 1 ($\widehat{y}_i$ is a computed label).

For a minibatch of size B, replace N by B. And choose the B samples randomly.

This section was enormously improved by Suvrit Sra's lecture in 18.065 on 20 April 2018.

Stochastic Descent Using One Sample Per Step

To simplify, suppose each minibatch contains only one sample $\boldsymbol{v}_k$ (so $B = 1$). That sample is chosen randomly. The theory of stochastic descent usually assumes that the sample is replaced after use—in principle the sample could be chosen again at step $k+1$. But replacement is expensive compared to starting with a random ordering of the samples. In practice, we often omit replacement and work through samples in a random order.

Each pass through the training data is **one epoch** of the descent algorithm. Ordinary gradient descent computes one epoch per step (batch mode). Stochastic gradient descent needs many steps (for minibatches). The online advice is to choose $B \leq 32$.

Stochastic descent began with a seminal paper of Herbert Robbins and Sutton Monro in the *Annals of Mathematical Statistics* **22** (1951) 400-407: A Stochastic Approximation Method. Their goal was a fast method that converges to $\boldsymbol{x}^*$ in probability:

$$\text{To prove} \qquad \text{Prob}\,(||\boldsymbol{x}_k - \boldsymbol{x}^*|| > \epsilon) \text{ approaches zero as } k \to \infty.$$

Stochastic descent is more sensitive to the stepsizes s_k than full gradient descent. If we randomly choose sample $\boldsymbol{v}_i$ at step k, then the kth descent step is familiar:

$$\boxed{\boldsymbol{x}_{k+1} = \boldsymbol{x}_k - s_k \nabla_{\boldsymbol{x}}\, \ell(\boldsymbol{x}_k, \boldsymbol{v}_i)} \qquad \boxed{\nabla_{\boldsymbol{x}}\, \ell = \text{derivative of the loss term from sample } \boldsymbol{v}_i}$$

We are doing much less work per step (B inputs instead of all inputs from the training set). But we do not necessarily converge more slowly. A typical feature of stochastic gradient descent is **"semi-convergence"**: fast convergence at the start.

Early steps of SGD often converge more quickly than GD toward the solution $\boldsymbol{x}^*$.

This is highly desirable for deep learning. Section VI.4 showed zig-zag for full batch mode. This was improved by adding momentum from the previous step (which we may also do for SGD). Another improvement frequently comes by using **adaptive methods** like some variation of ADAM. Adaptive methods look further back than momentum—now all previous descent directions are remembered and used. Those come later in this section.

Here we pause to look at semi-convergence: Fast start by stochastic gradient descent. We admit immediately that later iterations of SGD are frequently erratic. **Convergence at the start changes to large oscillations near the solution**. Figure VI.10 will show this. One response is to stop early. And thereby we avoid overfitting the data.

In the following example, the solution $\boldsymbol{x}^*$ is in a specific interval I. If the current approximation $\boldsymbol{x}_k$ is outside I, the next approximation $\boldsymbol{x}_{k+1}$ is closer to I (or inside I). That gives semiconvergence—a good start. *But eventually the $\boldsymbol{x}_k$ bounce around inside I.*

Fast Convergence at the Start : Least Squares with $n = 1$

We learned from Suvrit Sra that the simplest example is the best. The vector $\boldsymbol{x}$ has only one component x. The ith loss is $\ell_i = \frac{1}{2}(a_i x - b_i)^2$ with $a_i > 0$. The gradient of ℓ_i is its derivative $a_i(a_i x - b_i)$. It is zero and ℓ_i is minimized at $\boldsymbol{x} = \boldsymbol{b_i / a_i}$. The total loss over all N samples is $L(x) = \frac{1}{2N}\sum (a_i x - b_i)^2$: Least squares with N equations, 1 unknown.

$$\text{The equation to solve is } \nabla \boldsymbol{L} = \frac{1}{N}\sum_{1}^{N} a_i(a_i x - b_i) = 0. \text{ The solution is } \boldsymbol{x}^* = \frac{\sum \boldsymbol{a_i b_i}}{\sum \boldsymbol{a_i^2}} \quad (1)$$

Important If B/A is the largest ratio b_i/a_i, then the true solution $\boldsymbol{x}^*$ **is below $\boldsymbol{B/A}$**. This follows from a row of four inequalities :

$$\text{all } \frac{\boldsymbol{b_i}}{\boldsymbol{a_i}} \le \frac{\boldsymbol{B}}{\boldsymbol{A}} \qquad A\, a_i\, b_i \le B\, a_i^2 \qquad A\left(\sum a_i b_i\right) \le B\left(\sum a_i^2\right) \qquad \boldsymbol{x}^* = \frac{\sum a_i b_i}{\sum a_i^2} \le \frac{\boldsymbol{B}}{\boldsymbol{A}}. \quad (2)$$

Similarly $\boldsymbol{x}^*$ is above the smallest ratio $\boldsymbol{\beta/\alpha}$. Conclusion: If $\boldsymbol{x}_k$ is outside the interval $\boldsymbol{I}$ from β/α to B/A, then the kth gradient descent step will move *toward that interval $\boldsymbol{I}$* containing $\boldsymbol{x}^*$. Here is what we can expect from stochastic gradient descent:

If $\boldsymbol{x_k}$ is outside $\boldsymbol{I}$, then $\boldsymbol{x_{k+1}}$ moves toward the interval $\boldsymbol{\beta/\alpha \le x \le B/A}$.
If $\boldsymbol{x_k}$ is inside $\boldsymbol{I}$, then so is $\boldsymbol{x_{k+1}}$. The iterations can bounce around inside $\boldsymbol{I}$.

A typical sequence $\boldsymbol{x}_0, \boldsymbol{x}_1, \boldsymbol{x}_2, \ldots$ from minimizing $||A\boldsymbol{x} - \boldsymbol{b}||^2$ by stochastic gradient descent is graphed in Figure VI.10. *You see the fast start and the oscillating finish.* This behavior is a perfect signal to think about early stopping or averaging (page 365) when the oscillations start.

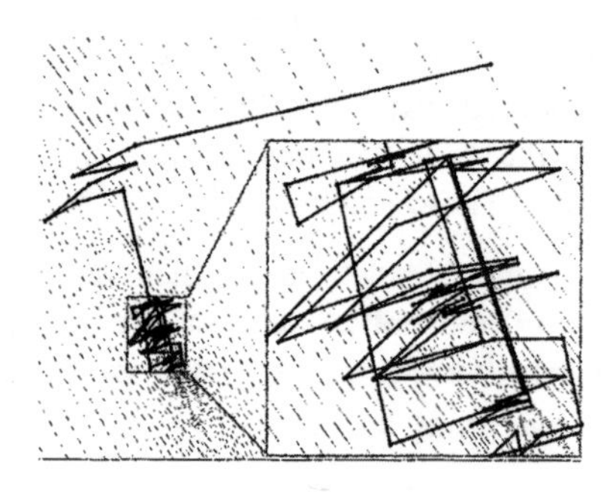

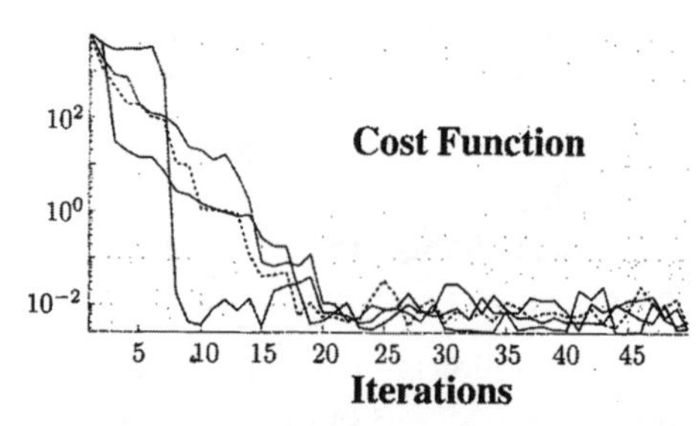

Figure VI.10: The left figure shows a trajectory of stochastic gradient descent with two unknowns. The early iterations succeed but later iterations oscillate (as shown in the inset). On the right, the quadratic cost function decreases quickly at first and then fluctuates instead of converging. The four paths start from the same x_0 with random choices of i in equation (3). The condition number of the 40 by 2 matrix A is only 8.6.

Randomized Kaczmarz is Stochastic Gradient Descent for $Ax = b$

Kaczmarz for $Ax = b$ with random $i(k)$
$$\boldsymbol{x}_{k+1} = \boldsymbol{x}_k + \frac{b_i - \boldsymbol{a}_i^{\mathrm{T}} \boldsymbol{x}_k}{||\boldsymbol{a}_i||^2} \boldsymbol{a}_i \qquad (3)$$

We are randomly selecting row i of A at step k. We are adjusting $\boldsymbol{x}_{k+1}$ to solve equation i in $A\boldsymbol{x} = \boldsymbol{b}$. (Multiply equation (3) by $\boldsymbol{a}_i^{\mathrm{T}}$ to verify that $\boldsymbol{a}_i^{\mathrm{T}} \boldsymbol{x}_{k+1} = b_i$. This is equation i in $A\boldsymbol{x} = \boldsymbol{b}$.) Geometrically, $\boldsymbol{x}_{k+1}$ is the projection of $\boldsymbol{x}_k$ onto one of the hyperplanes $\boldsymbol{a}_i^{\mathrm{T}} \boldsymbol{x} = b_i$ that meet at $\boldsymbol{x}^* = A^{-1}\boldsymbol{b}$.

This algorithm resisted a close analysis for many years. The equations $\boldsymbol{a}_1^{\mathrm{T}} \boldsymbol{x} = b_1$, $\boldsymbol{a}_2^{\mathrm{T}} \boldsymbol{x} = b_2 \ldots$ were taken in cyclic order with step $s = 1$. Then Strohmer and Vershynin proved fast convergence for random Kaczmarz. They used SGD with *norm-squared sampling (importance sampling)* as in Section II.4: Choose row i of A with probability p_i proportional to $||\boldsymbol{a}_i^{\mathrm{T}}||^2$.

The previous page described the Kaczmarz iterations for $A\boldsymbol{x} = \boldsymbol{b}$ when A was N by 1. The sequence $x_0, x_1, x_2, \ldots$ moved toward the interval I. The least squares solution x^* was in that interval. For an N by K matrix A, we expect the K by 1 vectors $\boldsymbol{x}_i$ to move into a K-dimensional box around $\boldsymbol{x}^*$. Figure VI.10 showed this for $K = 2$.

The next page will present numerical experiments for stochastic gradient descent:

A variant of random Kaczmarz was developed by Gower and Richtarik, with no less than six equivalent randomized interpretations. Here are references that connect the many variants from the original by Kaczmarz in the 1937 Bulletin de l'Académie Polonaise.

1 T. Strohmer and R. Vershynin , A randomized Kaczmarz algorithm with exponential convergence, *Journal of Fourier Analysis and Applications* **15** (2009) 262-278.

2 A. Ma, D. Needell, and A. Ramdas, Convergence properties of the randomized extended Gauss-Seidel and Kaczmarz methods, arXiv : 1503.08235v3 1 Feb 2018.

3 D. Needell, N. Srebro, and R. Ward, Stochastic gradient descent, weighted sampling, and the randomized Kaczmarz algorithm, *Math. Progr.* **155** (2015) 549-573.

4 R. M. Gower and P. Richtarik, Randomized iterative methods for linear systems, *SIAM J. Matrix Analysis.* **36** (2015) 1660-1690; arXiv : 1506.03296v5 6 Jan 2016.

5 L. Bottou et al, in *Advances in Neural Information Processing Systems*, NIPS **16** (2004) and NIPS **20** (2008), MIT Press.

6 S. Ma, R. Bassily, and M. Belkin, *The power of interpolation: Understanding the effectiveness of SGD in modern over-parametrized learning*, arXiv : 1712.06559.

7 S. Reddi, S. Sra, B. Poczos, and A. Smola, *Fast stochastic methods for nonsmooth nonconvex optimization*, arXiv: 1605.06900, 23 May 2016.

Random Kaczmarz and Iterated Projections

Suppose $A\boldsymbol{x}^* = \boldsymbol{b}$. A typical step of random Kaczmarz projects the current error $\boldsymbol{x}_k - \boldsymbol{x}^*$ onto the hyperplane $\boldsymbol{a}_i^{\mathrm{T}}\boldsymbol{x} = \boldsymbol{b}_i$. Here i is chosen randomly at step k (often with importance sampling using probabilities proportional to $||a_i||^2$). To see that projection matrix $\boldsymbol{a}_i\boldsymbol{a}_i^{\mathrm{T}}/\boldsymbol{a}_i^{\mathrm{T}}\boldsymbol{a}_i$, substitute $b_i = \boldsymbol{a}_i^{\mathrm{T}}\boldsymbol{x}^*$ into the update step (3):

$$\boldsymbol{x}_{k+1} - \boldsymbol{x}^* = \boldsymbol{x}_k - \boldsymbol{x}^* + \frac{b_i - \boldsymbol{a}_i^{\mathrm{T}}\boldsymbol{x}_k}{||\boldsymbol{a}_i||^2}\,\boldsymbol{a}_i = (\boldsymbol{x}_k - \boldsymbol{x}^*) - \frac{\boldsymbol{a}_i\boldsymbol{a}_i^{\mathrm{T}}}{\boldsymbol{a}_i^{\mathrm{T}}\boldsymbol{a}_i}(\boldsymbol{x}_k - \boldsymbol{x}^*) \qquad (4)$$

Orthogonal projection never increases length. The error can only decrease. The error norm $||\boldsymbol{x}_k - \boldsymbol{x}^*||$ decreases steadily, even if the cost function $||A\boldsymbol{x}_k - \boldsymbol{b}||$ does not. *But convergence is usually slow*! Strohmer-Vershynin estimate the expected error:

$$\mathrm{E}\left[||\boldsymbol{x}_k - \boldsymbol{x}^*||^2\right] \le \left(1 - \frac{1}{c^2}\right)^k ||\boldsymbol{x}_0 - \boldsymbol{x}^*||^2, \; c = \text{ condition number of } A. \qquad (5)$$

This is slow compared to gradient descent (there c^2 is replaced by c, and then $\sqrt{c}$ with momentum in VI.4). But (5) is independent of the size of A: attractive for large problems.

The theory of alternating projections was initiated by von Neumann (in Hilbert space). See books and papers by Bauschke-Borwein, Escalante-Raydan, Diaconis, Xu,...

Our experiments converge slowly! The 100 by 10 matrix A is random with $c \approx 400$. The figures show random Kaczmarz for $600,000$ steps. We measure convergence by the angle θ_k between $\boldsymbol{x}_k - \boldsymbol{x}^*$ and the row $\boldsymbol{a}_i$ chosen at step k. The error equation (4) is

$$||\boldsymbol{x}_{k+1} - \boldsymbol{x}^*||^2 = (1 - \cos^2\theta_k)\,||\boldsymbol{x}_k - \boldsymbol{x}^*||^2 \qquad (6)$$

The graph shows that those numbers $1 - \cos^2\theta_k$ *are very close to* 1: **slow convergence.** But the second graph confirms that convergence does occur. The Strohmer-Vershynin bound (5) becomes $\mathrm{E}\left[\cos^2\theta_k\right] \ge 1/c^2$. Our example matrix has $1/c^2 \approx 10^{-5}$ and often $\cos^2\theta_k \approx 2 \cdot 10^{-5}$, confirming that bound.

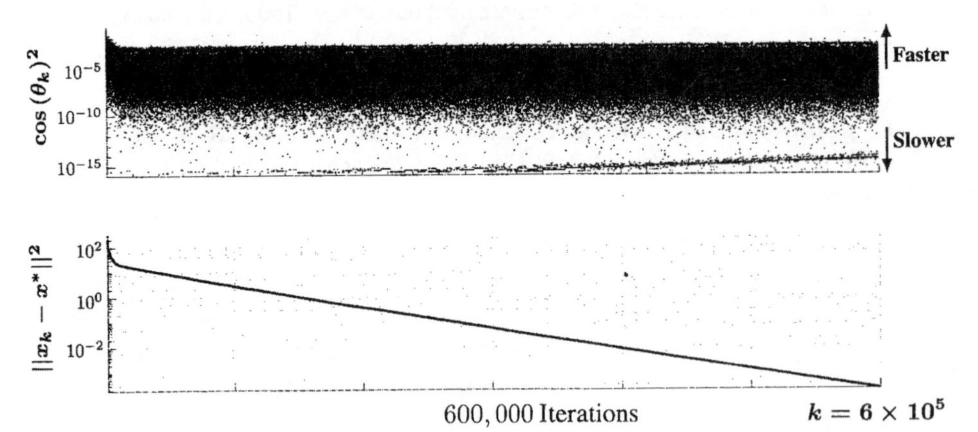

Figure VI.11: Convergence of the squared error for random Kaczmarz. Equation (6) with $1 - \cos^2\theta_k$ close to $1 - 10^{-5}$ produces the slow convergence in the lower graph.

Convergence in Expectation

For a stochastic algorithm like SGD, we need a convergence proof that accounts for randomness—in the assumptions and also in the conclusions. Suvrit Sra provided us with such a proof, and we reproduce it here. The function $f(\boldsymbol{x})$ is a sum $\frac{1}{n}\Sigma f_i(\boldsymbol{x})$ of n terms. The sampling chooses $i(k)$ at step k uniformly from the numbers 1 to n (with replacement!) and the stepsize is s = constant$/\sqrt{T}$. First come assumptions on $f(\boldsymbol{x})$ and $\boldsymbol{\nabla f}(\boldsymbol{x})$, followed by a standard requirement (no bias) for the random sampling.

1	**Lipschitz smoothness of $\boldsymbol{\nabla f}(\boldsymbol{x})$**	$\|\|\boldsymbol{\nabla f}(\boldsymbol{x}) - \boldsymbol{\nabla f}(\boldsymbol{y})\|\| \leq L\,\|\|\boldsymbol{x}-\boldsymbol{y}\|\|$
2	**Bounded gradients**	$\|\|\boldsymbol{\nabla f}_{i(k)}(\boldsymbol{x})\|\| \leq G$
3	**Unbiased stochastic gradients**	$\mathbf{E}\left[\boldsymbol{\nabla f}_{i(k)}(\boldsymbol{x}) - \boldsymbol{\nabla f}(\boldsymbol{x})\right] = 0$

From Assumption 1 it follows that

$$f(\boldsymbol{x}_{k+1}) \leq f(\boldsymbol{x}_k) + (\boldsymbol{\nabla f}(\boldsymbol{x}_k), \boldsymbol{x}_{k+1} - \boldsymbol{x}_k) + \frac{1}{2}Ls^2||\boldsymbol{\nabla f}_{i(k)}(\boldsymbol{x}_k)||^2$$

$$f(\boldsymbol{x}_{k+1}) \leq f(\boldsymbol{x}_k) + (\boldsymbol{\nabla f}(\boldsymbol{x}_k), -s\,\boldsymbol{\nabla f}_{i(k)}(\boldsymbol{x}_k)) + \frac{1}{2}Ls^2||\boldsymbol{\nabla f}_{i(k)}(\boldsymbol{x}_k)||^2$$

Now take expectations of both sides and use Assumptions 2-5:

$$\mathbf{E}\left[f(\boldsymbol{x}_{k+1})\right] \leq \mathbf{E}\left[f(\boldsymbol{x}_k)\right] - s\,\mathbf{E}\left[\,||\boldsymbol{\nabla f}(\boldsymbol{x}_k)||^2\,\right] + \frac{1}{2}Ls^2G^2$$

$$\Rightarrow \mathbf{E}\left[\,||\boldsymbol{\nabla f}(\boldsymbol{x}_k)||^2\,\right] \leq \frac{1}{s}\mathbf{E}\left[f(\boldsymbol{x}_k) - f(\boldsymbol{x}_{k+1})\right] + \frac{1}{2}Ls^2G^2. \tag{7}$$

Choose the stepsize $s = c/\sqrt{T}$, and add up (7) from $k = 1$ to T. The sum telescopes:

$$\frac{1}{T}\sum_{k=1}^{T}\mathbf{E}\left[\,||\boldsymbol{\nabla f}(\boldsymbol{x}_k)||^2\,\right] \leq \frac{1}{\sqrt{T}}\left(\frac{f(\boldsymbol{x}_1) - f(\boldsymbol{x}^*)}{c} + \frac{Lc}{2}G^2\right) = \frac{C}{\sqrt{T}}. \tag{8}$$

Here $f(\boldsymbol{x}^*)$ is the global minimum. The smallest term in (8) is below the average:

$$\min_{1\leq k\leq T}\ \mathbf{E}\left[\,||\boldsymbol{\nabla f}(\boldsymbol{x}_k)||^2\,\right] \leq C/\sqrt{T}. \tag{9}$$

The conclusion of Sra's theorem is convergence in expectation at a sublinear rate.

Weight Averaging Inside SGD

The idea of **averaging the outputs** from several steps of stochastic gradient descent looks promising. The learning rate (stepsize) can be constant or cyclical over each group of outputs. Gordon Wilson et al have named this method *Stochastic Weight Averaging* (SWA). They emphasize that this gives promising results for training deep networks, with better generalization and almost no overhead. It seems natural and effective.

P. Izmailov, D. Podoprikhin, T. Garipov, D. Vetrov, A. Gordon Wilson, *Averaging weights leads to wider optima and better generalization*, arXiv : 1803.05407.

Adaptive Methods Using Earlier Gradients

For faster convergence of gradient descent and stochastic gradient descent, adaptive methods have been a major direction. The idea is *to use gradients from earlier steps.* That "memory" guides the choice of search direction $\boldsymbol{D}$ and the all-important stepsize s. We are searching for the vector $\boldsymbol{x}^*$ that minimizes a specified loss function $L(\boldsymbol{x})$. In the step from $\boldsymbol{x}_k$ to $\boldsymbol{x}_{k+1}$, we are free to choose $\boldsymbol{D}_k$ and s_k:

$$\boldsymbol{D}_k = \boldsymbol{D}(\boldsymbol{\nabla L}_k, \boldsymbol{\nabla L}_{k-1}, \ldots, \boldsymbol{\nabla L}_0) \quad \text{and} \quad s_k = s(\boldsymbol{\nabla L}_k, \boldsymbol{\nabla L}_{k-1}, \ldots, \boldsymbol{\nabla L}_0). \qquad (10)$$

For a standard SGD iteration, $\boldsymbol{D}_k$ depends only on the current gradient $\boldsymbol{\nabla L_k}$ (and s_k might be $s/\sqrt{k}$). That gradient $\boldsymbol{\nabla L_k}(\boldsymbol{x}_k, B)$ is evaluated only on a random minibatch B of the test data. Now, deep networks often have the option of using some or all of the earlier gradients (computed on earlier random minibatches):

$$\textbf{Adaptive Stochastic Gradient Descent} \qquad \boldsymbol{x}_{k+1} = \boldsymbol{x}_k - s_k \boldsymbol{D}_k \qquad (11)$$

Success or failure will depend on $\boldsymbol{D}_k$ and s_k. The first adaptive method (called ADAGRAD) chose the usual search direction $\boldsymbol{D}_k = \boldsymbol{\nabla L}(\boldsymbol{x}_k)$ but computed the stepsize from all previous gradients [Duchi-Hazan-Singer]:

$$\textbf{ADAGRAD stepsize} \qquad s_k = \left(\frac{\alpha}{\sqrt{k}}\right)\left[\frac{1}{k}\,\text{diag}\left(\sum_{1}^{k} ||\boldsymbol{\nabla L}_i||^2\right)\right]^{1/2} \qquad (12)$$

$\alpha/\sqrt{k}$ is a typical decreasing stepsize in proving convergence of stochastic descent. It is often omitted when it slows down convergence in practice. The "memory factor" in (12) led to real gains in convergence speed. Those gains made adaptive methods a focus for more development.

Exponential moving averages in ADAM [Kingma-Ba] have become the favorites. Unlike (12), recent gradients $\boldsymbol{\nabla L}$ have greater weight than earlier gradients in both s_k and the step direction $\boldsymbol{D}_k$. The exponential weights in $\boldsymbol{D}$ and s come from $\delta < 1$ and $\beta < 1$:

$$\boldsymbol{D}_k = (1-\delta)\sum_{i=1}^{k} \delta^{k-i}\, \boldsymbol{\nabla L}(\boldsymbol{x}_i) \qquad s_k = \left(\frac{\alpha}{\sqrt{k}}\right)\left[(1-\beta)\,\text{diag}\sum_{i=1}^{k} \beta^{k-i} ||\boldsymbol{\nabla L}(\boldsymbol{x}_i)||^2\right]^{1/2} \qquad (13)$$

Typical values are $\delta = 0.9$ and $\beta = 0.999$. Small values of δ and β will effectively kill off the moving memory and lose the advantages of adaptive methods for convergence speed. That speed is important in the total cost of gradient descent! The look-back formula for the direction $\boldsymbol{D}_k$ is like including momentum in the heavy ball method of Section VI.4.

The actual computation of $\boldsymbol{D}_k$ and s_k will be a recursive combination of old and new:

$$\boldsymbol{D}_k = \delta \boldsymbol{D}_{k-1} + (1-\delta)\boldsymbol{\nabla L}(\boldsymbol{x}_k) \qquad s_k^2 = \beta s_{k-1}^2 + (1-\beta)\,||\boldsymbol{\nabla L}(\boldsymbol{x}_k)||^2 \qquad (14)$$

For several class projects, this adaptive method clearly produced faster convergence.

ADAM is highly popular in practice (this is written in 2018). But several authors have pointed out its defects. Hardt, Recht, and Singer constructed examples to show that its limit x_∞ for the weights in deep learning could be problematic: Convergence may fail or (worse) the limiting weights may *generalize poorly* in applications to unseen test data.

Equations (13)–(14) follow the recent conference paper of Reddi, Kale, and Kumar. Those authors prove non-convergence of ADAM, with simple examples in which *the stepsize s_k increases in time*—an undesired outcome. In their example, ADAM takes the wrong direction twice and the right direction once in every three steps. The exponential decay scales down that good step and overall the stepsizes s_k do not decrease. A large β (near 1) is needed and used, but there are always convex optimization problems on which ADAM will fail. *The idea is still good.*

One approach is to use an increasing minibatch size B. The NIPS 2018 paper proposes a new adaptive algorithm YOGI, which better controls the learning rate (the stepsize). Compared with ADAM, a key change is to an additive update; other steps are unchanged. At this moment, experiments are showing improved results with YOGI.

And after fast convergence to weights that nearly solve $\nabla L(x) = \mathbf{0}$ there is still the crucial issue: **Why do those weights generalize well to unseen test data?**

References

1. S. Ruder, *An overview of gradient descent optimization algorithms*, arXiv:1609.04747.
2. J. Duchi, E. Hazan, and Y. Singer. Adaptive subgradient methods for online learning and stochastic optimization. *J. of Machine Learning Research* **12** (2011) 2121–2159.
3. P. Kingma and J. Ba, ADAM: A method for stochastic optimization, ICLR, **2015**.
4. M. Hardt, B. Recht, and Y. Singer, *Train faster, generalize better: Stability of stochastic gradient descent*, arXiv:1509.01240v2, 7 Feb 2017, Proc. ICML (2016).
5. A. Wilson, R. Roelofs, M. Stern, N. Srebro, and B. Recht, *The marginal value of adaptive gradient methods in machine learning*, arXiv: 1705.08292, 23 May 2017.
6. S. Reddi, S. Kale, and S. Kumar, On the convergence of ADAM and beyond, ICLR **2018**: *Proc. Intl. Conference on Learning Representations.*
7. S. Reddi, M. Zaheer, D. Sachan, S. Kale, and S. Kumar, *Adaptive methods for nonconvex optimization*, NIPS (2018).

We end this chapter by emphasizing: Stochastic gradient descent is now the leading method to find weights x that minimize the loss $L(x)$ and solve $\nabla L(x^*) = \mathbf{0}$. Those weights from SGD normally succeed on unseen test data.

Generalization : Why is Deep Learning So Effective ?

We end Chapter VI—and connect to Chapter VII—with a short discussion of a central question for deep learning. The issue here is **generalization**. This refers to the behavior of a neural network on test data that it has not seen. If we construct a function $F(\boldsymbol{x}, \boldsymbol{v})$ that successfully classifies the known training data $\boldsymbol{v}$, will F continue to give correct results when $\boldsymbol{v}$ is outside the training set ?

The answer must lie in the stochastic gradient descent algorithm that chooses weights. Those weights $\boldsymbol{x}$ minimize a loss function $L(\boldsymbol{x}, \boldsymbol{v})$ over the training data. The question is : **Why do the computed weights do so well on the test data ?**

Often we have more free parameters in $\boldsymbol{x}$ than data in $\boldsymbol{v}$. In that case we can expect many sets of weights (many vectors $\boldsymbol{x}$) to be equally accurate on the training set. Those weights could be good or bad. They could generalize well or poorly. Our algorithm chooses a particular $\boldsymbol{x}$ and applies those weights to new data $\boldsymbol{v}_{\text{test}}$.

An unusual experiment produced unexpectedly positive results. The components of each input vector $\boldsymbol{v}$ were randomly shuffled. So the individual features represented by $\boldsymbol{v}$ suddenly had no meaning. Nevertheless the deep neural net learned those randomized samples. The learning function $\boldsymbol{F}(\boldsymbol{x}, \boldsymbol{v})$ still classified the test data correctly. Of course $\boldsymbol{F}$ could not succeed with unseen data, when the components of $\boldsymbol{v}$ are reordered.

It is a common feature of optimization that smooth functions are easier to approximate than irregular functions. But here, for completely randomized input vectors, stochastic gradient descent needed only three times as many epochs (triple the number of iterations) to learn the training data. This random labeling of the training samples (the experiment has become famous) is described in arXiv : 1611.03530.

The Kaczmarz Method in Tomographic Imaging (CT)

A key property of Kaczmarz is its quick success in early iterations. This is called *semi-convergence* in tomography (where solving $A\boldsymbol{x} = \boldsymbol{b}$ constructs a CT image, and the method produces a regularized solution when the data is noisy). Quick semi-convergence for noisy data is an excellent property for such a simple method. The first steps all approach the correct interval from α/β to A/B (for one scalar unknown). But inside that interval, Kaczmarz jumps around unimpressively.

We are entering here the enormous topic of ill-conditioned inverse problems (see the books of P. C. Hansen). In this book we can do no more than open the door.

Problem Set VI.5

1 The rank-one matrix $P = \boldsymbol{a}\boldsymbol{a}^{\mathrm{T}}/\boldsymbol{a}^{\mathrm{T}}\boldsymbol{a}$ is an orthogonal projection onto the line through $\boldsymbol{a}$. Verify that $P^2 = P$ (projection) and that $P\boldsymbol{x}$ is on that line and that $\boldsymbol{x} - P\boldsymbol{x}$ is always perpendicular to $\boldsymbol{a}$ (*why is* $\boldsymbol{a}^{\mathrm{T}}\boldsymbol{x} = \boldsymbol{a}^{\mathrm{T}}P\boldsymbol{x}$?)

2 Verify that equation (4) which shows that $\boldsymbol{x}_{k+1} - \boldsymbol{x}^*$ is exactly $P(\boldsymbol{x}_k - \boldsymbol{x}^*)$.

3 If A has only two rows $\boldsymbol{a}_1$ and $\boldsymbol{a}_2$, then Kaczmarz will produce the *alternating projections* in this figure. Starting from any error vector $e_0 = \boldsymbol{x}_0 - \boldsymbol{x}^*$, why does $\boldsymbol{e}_k$ approach zero ? How fast ?

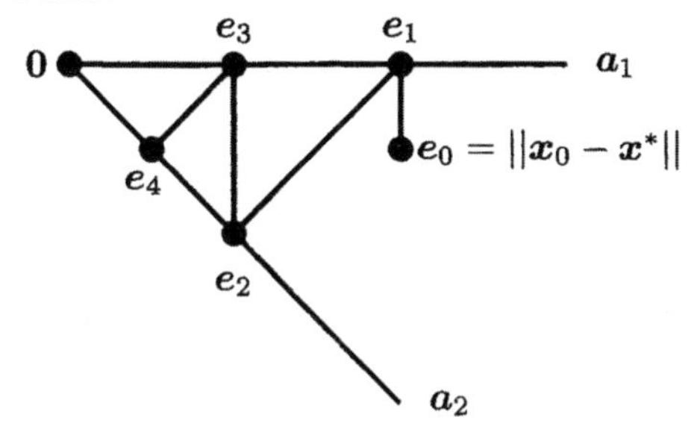

4 Suppose we want to minimize $F(x, y) = y^2 + (y - x)^2$. The actual minimum is $F = 0$ at $(x^*, y^*) = (0, 0)$. Find the gradient vector $\boldsymbol{\nabla F}$ at the starting point $(x_0, y_0) = (1, 1)$. For full gradient descent (*not stochastic*) with step $s = \frac{1}{2}$, where is (x_1, y_1) ?

5 In minimizing $F(\boldsymbol{x}) = ||A\boldsymbol{x} - \boldsymbol{b}||^2$, stochastic gradient descent with minibatch size $B = 1$ will solve one equation $\boldsymbol{a}_i^{\mathrm{T}}\boldsymbol{x} = b_i$ at each step. Explain the typical step for minibatch size $B = 2$.

6 (Experiment) For a random A and $\boldsymbol{b}$ (20 by 4 and 20 by 1), try stochastic gradient descent with minibatch sizes $B = 1$ and $B = 2$. Compare the convergence rates— the ratios $r_k = ||\boldsymbol{x}_{k+1} - \boldsymbol{x}^*|| \,/\, ||\boldsymbol{x}_k - \boldsymbol{x}^*||$.

7 (Experiment) Try the weight averaging on page 365 proposed in arXiv : 1803.05407. Apply it to the minimization of $||A\boldsymbol{x} - \boldsymbol{b}||^2$ with randomly chosen A (20 by 10) and $\boldsymbol{b}$ (20 by 1), and minibatch $B = 1$.

Do averages in stochastic descent converge faster than the usual iterates $\boldsymbol{x}_k$?

Part VII

Learning from Data

VII.1 The Construction of Deep Neural Networks

VII.2 Convolutional Neural Nets

VII.3 Backpropagation and the Chain Rule

VII.4 Hyperparameters : The Fateful Decisions

VII.5 The World of Machine Learning

Part VII : Learning from Data

This part of the book is a great adventure—hopefully for the reader, certainly for the author, and it involves the whole science of thought and intelligence. You could call it Machine Learning (ML) or Artificial Intelligence (AI). Human intelligence created it (but we don't fully understand what we have done). Out of some combination of ideas and failures, attempting at first to imitate the neurons in the brain, a successful approach has emerged to finding **patterns in data**.

What is important to understand about deep learning is that those data-fitting computations, of almost unprecedented size, are often heavily underdetermined. There are a great many points in the training data, but there are far more weights to be computed in a deep network. The art of deep learning is to find, among many possible solutions, one that will **generalize to new data.**

It is a remarkable observation that learning on deep neural nets with many weights leads to a successful tradeoff: F is accurate on the training set *and* the unseen test set. This is the good outcome from minibatch gradient descent with momentum and the hyperparameters from Section VII.4 (including stepsize selection and early stopping).

This chapter is organized in an irregular order. **Deep learning comes first**. Earlier models like Support Vector Machines and Kernel Methods are briefly described in VII.5. The order is anhistorical, and the reader will know why. Neural nets have become the primary architecture for the most interesting (and the most difficult) problems of machine learning. That multi-layer architecture often succeeds, but by no means always! This book has been preparing for deep learning and we simply give it first place.

Sections VII.1-2 describe the learning function $\boldsymbol{F}(\boldsymbol{x}, \boldsymbol{v})$ for *fully connected nets and convolutional nets*. The training data is given by a set of feature vectors $\boldsymbol{v}$. The weights that allow $\boldsymbol{F}$ to classify that data are in the vector $\boldsymbol{x}$. To optimize $\boldsymbol{F}$, gradient descent needs its derivatives $\boldsymbol{\partial F / \partial x}$. The weights $\boldsymbol{x}$ are the matrices $A_1, \ldots, A_L$ and bias vectors $\boldsymbol{b}_1, \ldots, \boldsymbol{b}_L$ that take the sample data $\boldsymbol{v} = \boldsymbol{v}_0$ to the output $\boldsymbol{w} = \boldsymbol{v}_L$.

Formulas for $\partial F / \partial A$ and $\partial F / \partial b$ are not difficult. Those formulas are useful to see. But real codes use *automatic differentiation (AD) for backpropagation* (Section VII.3). Each hidden layer with its optimized weights learns more about the data and the population from which it comes—in order to classify new and unseen data from the same population.

The Functions of Deep Learning

Suppose one of the digits $0, 1, \ldots, 9$ is drawn in a square. How does a person recognize which digit it is? That neuroscience question is not answered here. How can a computer recognize which digit it is? This is a machine learning question. Probably both answers begin with the same idea: *Learn from examples.*

So we start with M different images (the training set). An image will be a set of p small pixels—or a vector $\boldsymbol{v} = (v_1, \ldots, v_p)$. The component v_i tells us the "grayscale" of the ith pixel in the image: how dark or light it is. So we have M images each with p features: M vectors $\boldsymbol{v}$ in p-dimensional space. For every $\boldsymbol{v}$ in that training set we know the digit it represents.

In a way, we know a function. We have M inputs in $\mathbf{R}^p$ each with an output from 0 to 9. But we don't have a "rule". We are helpless with a new input. Machine learning proposes to create a rule that succeeds on (most of) the training images. But "succeed" means much more than that: The rule should give the correct digit for a much wider set of test images, taken from the same population. This essential requirement is called *generalization.*

What form shall the rule take? Here we meet the fundamental question. Our first answer might be: $F(\boldsymbol{v})$ could be a linear function from $\mathbf{R}^p$ to $\mathbf{R}^{10}$ (a 10 by p matrix). The 10 outputs would be probabilities of the numbers 0 to 9. We would have $10p$ entries and M training samples to get mostly right.

The difficulty is: Linearity is far too limited. Artistically, two zeros could make an 8. 1 and 0 could combine into a handwritten 9 or possibly 6. Images don't add. In recognizing faces instead of numbers, we will need a lot of pixels—and the input-output rule is nowhere near linear.

Artificial intelligence languished for a generation, waiting for new ideas. There is no claim that the absolutely best class of functions has now been found. That class needs to allow a great many parameters (called weights). And it must remain feasible to compute all those weights (in a reasonable time) from knowledge of the training set.

The choice that has succeeded beyond expectation—and has turned shallow learning into deep learning—is *Continuous Piecewise Linear (CPL) functions.* **Linear** for simplicity, **continuous** to model an unknown but reasonable rule, and **piecewise** to achieve the nonlinearity that is an absolute requirement for real images and data.

This leaves the crucial question of computability. What parameters will quickly describe a large family of CPL functions? Linear finite elements start with a triangular mesh. But specifying many individual nodes in $\mathbf{R}^p$ is expensive. Much better if those nodes are the *intersections* of a smaller number of lines (or hyperplanes). Please know that a regular grid is too simple.

Here is a first construction of a piecewise linear function of the data vector $\boldsymbol{v}$. Choose a matrix A_1 and vector $\boldsymbol{b}_1$. Then set to zero (this is the nonlinear step) all negative components of $A_1\boldsymbol{v} + \boldsymbol{b}_1$. Then multiply by a matrix A_2 to produce 10 outputs in $\boldsymbol{w} = F(\boldsymbol{v}) = A_2(A_1\boldsymbol{v}+\boldsymbol{b}_1)_+$. That vector $(A_1\boldsymbol{v}+\boldsymbol{b}_1)_+$ forms a "hidden layer" between the input $\boldsymbol{v}$ and the output $\boldsymbol{w}$.

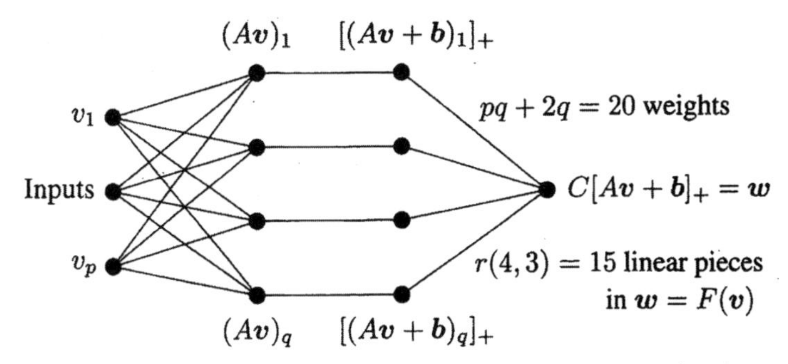

Actually the nonlinear function called ReLU $(x) = x_+ = \max(x,0)$ was originally smoothed into a logistic curve like $1/(1+e^{-x})$. It was reasonable to think that continuous derivatives would help in optimizing the weights A_1, b_1, A_2. That proved to be wrong.

The graph of each component of $(A_1v + b_1)_+$ has two halfplanes (one is flat, from the zeros where $A_1v + b_1$ is negative). If A_1 is q by p, the input space $\mathbf{R}^p$ is sliced by q hyperplanes into r pieces. We can count those pieces ! This measures the "expressivity" of the overall function $F(v)$. The formula from combinatorics uses the binomial coefficients (see Section VII.1):

$$r(q,p) = \binom{q}{0} + \binom{q}{1} + \cdots + \binom{q}{p}$$

This number gives an impression of the graph of F. But our function is not yet sufficiently expressive, and one more idea is needed.

Here is the indispensable ingredient in the learning function F. The best way to create complex functions from simple functions is by **composition**. Each F_i is linear (or affine) followed by the nonlinear ReLU : $F_i(v) = (A_iv + b_i)_+$. Their composition is $F(v) = F_L(F_{L-1}(\ldots F_2(F_1(v))))$. We now have $L-1$ hidden layers before the final output layer. The network becomes deeper as L increases. That depth can grow quickly for convolutional nets (with banded Toeplitz matrices A).

The great optimization problem of deep learning is to compute weights A_i and b_i that will make the outputs $F(v)$ nearly correct—close to the digit $w(v)$ that the image v represents. This problem of minimizing some measure of $F(v) - w(v)$ is solved by following a gradient downhill. The gradient of this complicated function is computed by *backpropagation*—the workhorse of deep learning that executes the chain rule.

A historic competition in 2012 was to identify the 1.2 million images collected in ImageNet. The breakthrough neural network in AlexNet had 60 million weights. Its accuracy (after 5 days of stochastic gradient descent) cut in half the next best error rate. Deep learning had arrived.

Our goal here was to identify continuous piecewise linear functions as powerful approximators. That family is also convenient—closed under addition and maximization and composition. The magic is that the learning function $F(A_i, b_i, v)$ gives accurate results on images v that F has never seen.

This two-page essay was written for SIAM News (December 2018).

Bias vs. Variance : Underfit vs. Overfit

A **training set** contains N vectors $\boldsymbol{v}_1, \ldots, \boldsymbol{v}_N$ with m components each (the m features of each sample). For each of those N points in $\mathbf{R}^m$, we are given a value y_i. We assume there is an unknown function $f(\boldsymbol{x})$ so that $y_i = f(\boldsymbol{x}_i) + \epsilon_i$, where the noise ϵ has zero mean and variance σ^2. That is the function $f(\boldsymbol{x})$ that our algorithms try to learn.

Our learning algorithm actually finds a function $F(\boldsymbol{x})$ close to $f(\boldsymbol{x})$. For example, F from our learning algorithm could be linear (not that great) or piecewise linear (much better) – this depends on the algorithm we use. We fervently hope that $F(\boldsymbol{x})$ will be close to the correct $f(\boldsymbol{x})$ **not only on the training samples but also for later test samples**.

The warning is often repeated, and always the same: **Don't overfit the data**. The option is there, to reproduce all known observations. It is more important to prevent large swings in the learning function (which is built from the weights). This function is going to be applied to new data. Implicitly or explicitly, we need to **regularize** this function F.

Ordinarily, we regularize by adding a penalty term like $\lambda||\boldsymbol{x}||$ to the function that we are minimizing. This gives a smoother and more stable solution as the minimum point. For deep learning problems this isn't always necessary! We don't fully understand why steepest descent or stochastic steepest descent will find a near minimum that generalizes well to unseen test data—with no penalty term. Perhaps the success comes from following this rule: *Stop the minimization early before you overfit.*

If F does poorly on the training samples with large error (**bias**), that is *underfitting*

If F does well on the training samples but not well on test samples, that is *overfitting*.

This is the **bias-variance tradeoff.** High bias from underfitting, high variance from overfitting. Suppose we scale f and F so that $\mathrm{E}\,[F(\boldsymbol{x})] = 1$.

$$\textbf{Bias} = \mathrm{E}\,[f(\boldsymbol{x}) - F(\boldsymbol{x})] \qquad\qquad \textbf{Variance} = \mathrm{E}\,[(F(\boldsymbol{x}))^2] - (\mathrm{E}\,[F(\boldsymbol{x})])^2$$

We are forced into this tradeoff by the following identity for (Bias)^2 + (Variance) + (Noise)^2 :

$$\mathrm{E}\,[(y - F(\boldsymbol{x}))^2] = (\mathrm{E}\,[f(\boldsymbol{x}) - F(\boldsymbol{x})])^2 + \mathrm{E}\,[(F(\boldsymbol{x}))^2] - (\mathrm{E}\,[F(\boldsymbol{x})])^2 + \mathrm{E}\,[(y - f(\boldsymbol{x}))^2]$$

Again, *bias* comes from allowing less freedom and using fewer parameters (weights). *Variance* is large when we provide too much freedom and too many parameters for F. Then the learned function F can be super-accurate on the training set but out of control on an unseen test set. **Overfitting produces an F that does not generalize.**

Here are links to six sites that support codes for machine learning:

Caffe : arXiv:1408.5093	Keras : http://keras.io/
MatConvNet : www.vlfeat.org/matconvnet	Theano : arXiv : 1605.02688
Torch : torch.ch	TensorFlow : www.tensorflow.org

VII.1 The Construction of Deep Neural Networks

Deep neural networks have evolved into a major force in machine learning. Step by step, the structure of the network has become more resilient and powerful—and more easily adapted to new applications. One way to begin is to describe essential pieces in the structure. Those pieces come together into a **learning function $F(x, v)$** *with weights x that capture information from the training data v*—to prepare for use with new test data.

Here are important steps in creating that function F:

1	Key operation	**Composition $F = F_3(F_2(F_1(x, v)))$**
2	Key rule	**Chain rule for x-derivatives of F**
3	Key algorithm	**Stochastic gradient descent to find the best weights x**
4	Key subroutine	**Backpropagation to execute the chain rule**
5	Key nonlinearity	**$\text{ReLU}(y) = \max(y, 0) =$ ramp function**

Our first step is to describe the pieces $F_1, F_2, F_3, \ldots$ for one layer of neurons at a time. The weights x that connect the layers v are optimized in creating F. The vector $v = v_0$ comes from the training set, and the function F_k produces the vector v_k at layer k. The whole success is to build the power of F from those pieces F_k in equation (1).

F_k is a Piecewise Linear Function of v_{k-1}

The input to F_k is a vector v_{k-1} of length N_{k-1}. The output is a vector v_k of length N_k, ready for input to F_{k+1}. This function F_k has two parts, first linear and then nonlinear:

1. The linear part of F_k yields $A_k v_{k-1} + b_k$ (that bias vector b_k makes this "affine")

2. A fixed nonlinear function like ReLU is applied to *each component* of $A_k v_{k-1} + b_k$

$$\boxed{v_k = F_k(v_{k-1}) = \text{ReLU}\,(A_k v_{k-1} + b_k)} \qquad (1)$$

The training data for each sample is in a feature vector v_0. The matrix A_k has shape N_k by N_{k-1}. The column vector b_k has N_k components. **These A_k and b_k are weights** constructed by the optimization algorithm. Frequently *stochastic gradient descent* computes optimal weights $x = (A_1, b_1, \ldots, A_L, b_L)$ in the central computation of deep learning. It relies on backpropagation to find the x-derivatives of F, to solve $\nabla F = 0$.

The activation function $\text{ReLU}(y) = \max(y, 0)$ gives flexibility and adaptability. Linear steps alone were of limited power and ultimately they were unsuccessful.

ReLU is applied to every "neuron" in every internal layer. There are N_k neurons in layer k, containing the N_k outputs from $A_k v_{k-1} + b_k$. Notice that ReLU itself is continuous and piecewise linear, as its graph shows. (The graph is just a ramp with slopes 0 and 1. Its derivative is the usual step function.) When we choose ReLU, the composite function $F = F_L(F_2(F_1(x, v)))$ has an important and attractive property:

The learning function F is continuous and piecewise linear in v.

One Internal Layer ($L = 2$)

Suppose we have measured $m = 3$ features of one sample point in the training set. Those features are the 3 components of the input vector $\boldsymbol{v} = \boldsymbol{v}_0$. Then the first function F_1 in the chain multiplies $\boldsymbol{v}_0$ by a matrix A_1 and adds an offset vector $\boldsymbol{b}_1$ (bias vector). If A_1 is 4 by 3 and the vector $\boldsymbol{b}_1$ is 4 by 1, we have 4 components of $A_0\boldsymbol{v}_0 + \boldsymbol{b}_0$.

That step found 4 combinations of the 3 original features in $\boldsymbol{v} = \boldsymbol{v}_0$. The 12 weights in the matrix A_1 were optimized over many feature vectors $\boldsymbol{v}_0$ in the training set, to choose a 4 by 3 matrix (and a 4 by 1 bias vector) that would find 4 insightful combinations.

The final step to reach $\boldsymbol{v}_1$ is to apply the nonlinear "activation function" to each of the 4 components of $A_1\boldsymbol{v}_0 + \boldsymbol{b}_1$. Historically, the graph of that nonlinear function was often given by a smooth "*S-curve*". Particular choices then and now are in Figure VII.1.

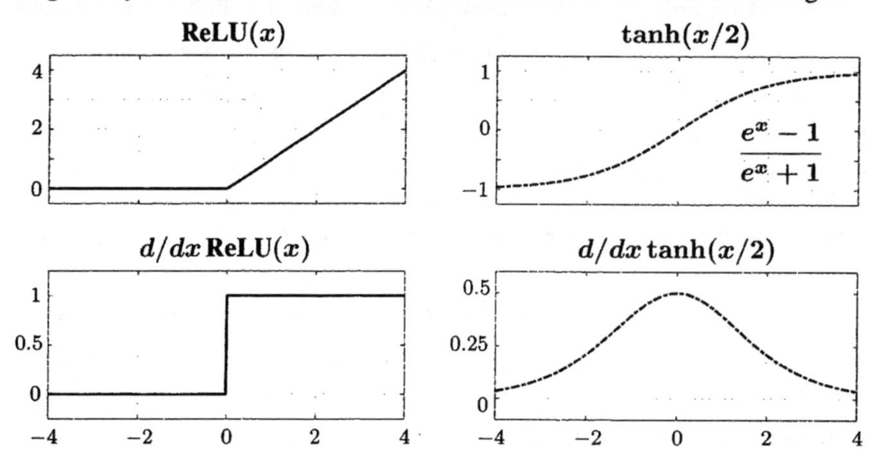

Figure VII.1: The **R**ectified **L**inear **U**nit and a sigmoid option for nonlinearity.

Previously it was thought that a sudden change of slope would be dangerous and possibly unstable. But large scale numerical experiments indicated otherwise! A better result was achieved by the **ramp function** $\text{ReLU}(y) = \max(y, 0)$. We will work with ReLU:

$$\textbf{Substitute } A_1\boldsymbol{v}_0 + \boldsymbol{b}_1 \textbf{ into ReLU to find } \boldsymbol{v}_1 \qquad (\boldsymbol{v}_1)_k = \max((A_1\boldsymbol{v}_0 + \boldsymbol{b}_1)_k, 0). \tag{2}$$

Now we have the components of $\boldsymbol{v}_1$ at the four "neurons" in layer 1. The input layer held the three components of this particular sample of training data. We may have thousands or millions of samples. The optimization algorithm found A_1 and $\boldsymbol{b}_1$, possibly by stochastic gradient descent using backpropagation to compute gradients of the overall loss.

Suppose our neural net is shallow instead of deep. It only has this first layer of 4 neurons. Then the final step will multiply the 4-component vector $\boldsymbol{v}_1$ by a 1 by 4 matrix A_2 (a row vector). It can add a single number $\boldsymbol{b}_2$ to reach the value $\boldsymbol{v}_2 = A_2\boldsymbol{v}_1 + \boldsymbol{b}_2$. The nonlinear function ReLU is not applied to the output.

$$\begin{array}{l}\text{Overall we compute } \boldsymbol{v}_2 = F(\boldsymbol{x}, \boldsymbol{v}_0) \text{ for each feature vector } \boldsymbol{v}_0 \text{ in the training set.} \\ \text{The steps are } \boldsymbol{v}_2 = A_2\boldsymbol{v}_1 + \boldsymbol{b}_2 = A_2\,(\text{ReLU}\,(A_1\boldsymbol{v}_0 + \boldsymbol{b}_1)) + \boldsymbol{b}_2 = F(\boldsymbol{x}, \boldsymbol{v}_0).\end{array} \tag{3}$$

The goal in optimizing $x = A_1, b_1, A_2, b_2$ is that the output values $v_\ell = v_2$ at the last layer $\ell = 2$ should correctly capture the important features of the training data v_0.

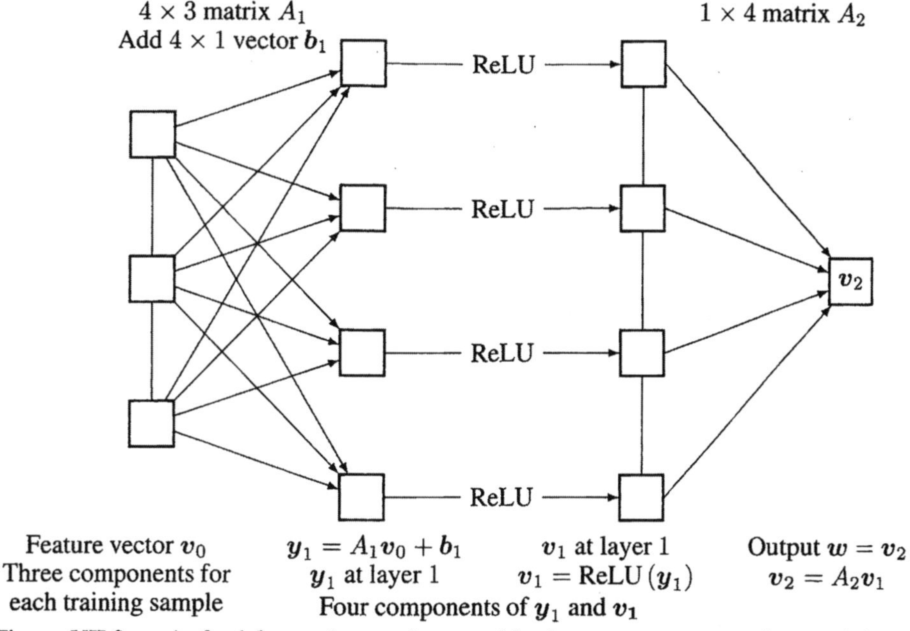

Figure VII.2: A feed-forward neural net with 4 neurons on **one internal layer**. The output v_2 (plus or minus) classifies the input v_0 (dog or cat). Then v_2 is a composite measure of the 3-component feature vector v_0. This net has 20 weights in A_k and b_k.

> For a **classification problem** each sample v_0 of the training data is assigned **1 or −1**. We want the output v_2 to have that correct sign (most of the time). For a **regression problem** we use the numerical value (**not just the sign**) of v_2. We do not choose enough weights A_k and b_k to get every sample correct. And we do not necessarily want to ! That would probably be **overfitting the training data**. It could give erratic results when F is applied to new and unknown test data.

> Depending on our choice of loss function $L(x, v_2)$ to minimize, this problem can be like least squares or entropy minimization. We are choosing x = weight matrices A_k and bias vectors b_k to minimize L. Those two loss functions—square loss and cross-entropy loss—are compared in Section VII.4.

Our hope is that **the function F has "learned" the data**. This is machine learning. We don't want to choose so many weights in x that every input sample is sure to be correctly classified. *That is not learning*. That is simply fitting (overfitting) the data.

We want a balance where the function F has learned what is important in recognizing *dog versus cat—or identifying an oncoming car versus a turning car.*

Machine learning doesn't aim to capture every detail of the numbers $0, 1, 2 \ldots, 9$. It just aims to capture enough information to decide correctly *which number it is.*

The Initial Weights x_0 in Gradient Descent

The architecture in a neural net decides the form of the learning function $F(\boldsymbol{x}, \boldsymbol{v})$. The training data goes into $\boldsymbol{v}$. Then we *initialize* the weights $\boldsymbol{x}$ in the matrices A and vectors $\boldsymbol{b}$. From those initial weights $\boldsymbol{x}_0$, the optimization algorithm (normally a form of gradient descent) computes weights $\boldsymbol{x}_1$ and $\boldsymbol{x}_2$ and onward, aiming to minimize the total loss.

The question is: *What weights $\boldsymbol{x}_0$ to start with?* Choosing $\boldsymbol{x}_0 = 0$ would be a disaster. Poor initialization is an important cause of failure in deep learning. A proper choice of the net and the initial $\boldsymbol{x}_0$ has random (and independent) weights that meet two requirements:

1. $\boldsymbol{x}_0$ has a carefully chosen variance σ^2.

2. The hidden layers in the neural net have enough neurons (not too narrow).

Hanin and Rolnick show that the initial variance σ^2 controls the mean of the computed weights. The layer widths control the variance of the weights. The key point is this: *Many-layered depth can reduce the loss on the training set. But if σ^2 is wrong or width is sacrificed, then gradient descent can lose control of the weights. They can explode to infinity or implode to zero.*

The danger controlled by the variance σ^2 of $\boldsymbol{x}_0$ is exponentially large or exponentially small weights. The good choice is $\sigma^2 = 2/\text{fan-in}$. The fan-in is the maximum number of inputs to neurons (Figure VII.2 has fan-in $= 4$ at the output). The initialization "He uniform" in Keras makes this choice of σ^2.

The danger from narrow hidden layers is exponentially large variance of $\boldsymbol{x}$ for deep nets. If layer j has n_j neurons, the quantity to control is the sum of 1/(layer widths n_j).

Looking ahead, convolutional nets (ConvNets) and residual networks (ResNets) can be very deep. Exploding or vanishing weights is a constant danger. Ideas from physics (*mean field theory*) have become powerful tools to explain and also avoid these dangers. Pennington and coauthors proposed a way to stay on the edge between fast growth and decay, even for $10,000$ layers. A key is to use orthogonal transformations: Exactly as in matrix multiplication $Q_1Q_2Q_3$, orthogonality leaves the size unchanged.

For ConvNets, fan-in becomes the number of features times the kernel size (and not the full size of A). For ResNets, a correct σ^2 normally removes both dangers. Very deep networks can produce very impressive learning.

The key point: Deep learning can go wrong if it doesn't start right.

K. He, X.Zhang, S. Ren, and J. Sun, *Delving deep into rectifiers*, arXiv: 1502.01852.

B. Hanin and D. Rolnick, *How to start training: The effect of initialization and architecture*, arXiv: 1803.01719, 19 Jun 2018.

L. Xiao, Y. Bahri, J. Sohl-Dickstein, S. Schoenholz, and J. Pennington, *Dynamical isometry and a mean field theory of CNNs: How to train* $10,000$ *layers*, arXiv: 1806.05393, 2018.

Stride and Subsampling

Those words represent two ways to achieve the same goal: *Reduce the dimension.* Suppose we start with a 1D signal of length 128. We want to filter that signal—multiply that vector by a weight matrix A. We also want to reduce the length to 64. Here are two ways to reach that goal.

In two steps Multiply the 128-component vector $\boldsymbol{v}$ by A, and then discard the odd-numbered components of the output. This is filtering followed by subsampling. The output is $(\downarrow 2)\, A\boldsymbol{v}$.

In one step Discard the odd-numbered rows of the matrix A. The new matrix A_2 becomes short and wide: 64 rows and 128 columns. The "**stride**" of the filter is now 2. Now multiply the 128-component vector $\boldsymbol{v}$ by A_2. Then $A_2\boldsymbol{v}$ is the same as $(\downarrow 2)\, A\boldsymbol{v}$. A stride of 3 would keep every third component.

Certainly the one-step striding method is more efficient. If the stride is 4, the dimension is divided by 4. In two dimensions (for images) it is reduced by 16.

The two-step method makes clear that half or three-fourths of the information is lost. Here is a way to reduce the dimension from 128 to 64 as before, but to run less risk of destroying important information: *Max-pooling.*

Max-pooling

Multiply the 128-component vector $\boldsymbol{v}$ by A, as before. Then from each even-odd pair of outputs like $(A\boldsymbol{v})_2$ and $(A\boldsymbol{v})_3$, *keep the maximum.* Please notice right away: Max-pooling is simple and fast, **but taking the maximum is not a linear operation.** It is a sensible route to dimension reduction, pure and simple.

For an image (a 2-dimensional signal) we might use max-pooling over every 2 by 2 square of pixels. Each dimension is reduced by 2, The image dimension is reduced by 4. This speeds up the training, when the number of neurons on a hidden layer is divided by 4.

Normally a max-pooling step is given its own separate place in the overall architecture of the neural net. Thus a part of that architecture might look like this:

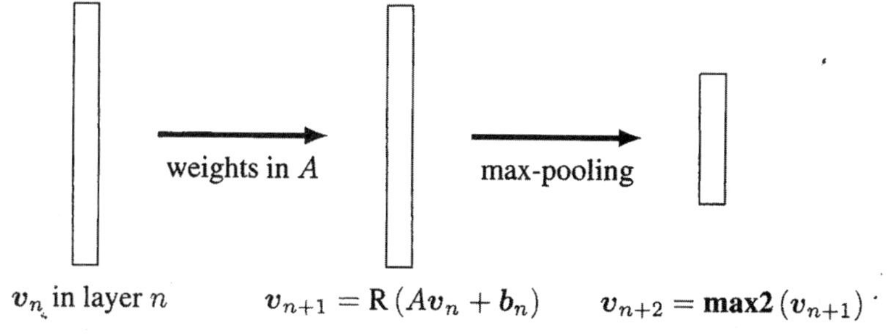

$\boldsymbol{v}_n$ in layer n $\qquad \boldsymbol{v}_{n+1} = \mathrm{R}\,(A\boldsymbol{v}_n + \boldsymbol{b}_n) \qquad \boldsymbol{v}_{n+2} = \mathbf{max2}\,(\boldsymbol{v}_{n+1})$

Dimension reduction has another important advantage, in addition to reducing the computation. *Pooling also reduces the possibility of overfitting.* Average pooling would keep the *average* of the numbers in each pool: now the pooling layer is linear.

The Graph of the Learning Function $F(v)$

The graph of $F(v)$ is a surface made up of many, many flat pieces—they are planes or hyperplanes that fit together along all the folds where ReLU produced a change of slope. This is like origami except that this graph has flat pieces going to infinity. And the graph might not be in $\mathbf{R}^3$—the feature vector $v = v_0$ has $N_0 = m$ components.

Part of the *mathematics of deep learning* is to estimate the number of flat pieces and to visualize how they fit into one piecewise linear surface. That estimate comes after an example of a neural net with one internal layer. Each feature vector v_0 contains m measurements like height, weight, age of a sample in the training set.

In the example, F had three inputs in v_0 and one output v_2. Its graph will be a piecewise flat surface in 4-dimensional space. The height of the graph is $v_2 = F(v_0)$, over the point v_0 in 3-dimensional space. Limitations of space in the book (and severe limitations of imagination in the author) prevent us from drawing that graph in $\mathbf{R}^4$. Nevertheless we can try to count the flat pieces, based on 3 inputs and 4 neurons and 1 output.

Note 1 With only $m = 2$ inputs (2 features for each training sample) the graph of F is a surface in 3D. We can and will make an attempt to describe it.

Note 2 You actually see points on the graph of F when you run examples on **playground.tensorflow.org**. This is a very instructive website.

That website offers four options for the training set of points v_0. You choose the number of layers and neurons. Please choose the ReLU activation function! Then the program counts epochs as gradient descent optimizes the weights. (An *epoch* sees all samples on average once.) If you have allowed enough layers and neurons to correctly classify the blue and orange training samples, you will see a polygon separating them. **That polygon shows where $F = 0$.** It is the cross-section of the graph of $z = F(v)$ at height $z = 0$.

That polygon separating blue from orange (or *plus* from *minus*: this is classification) is the analog of a separating hyperplane in a Support Vector Machine. If we were limited to linear functions and a straight line between a blue ball and an orange ring around it, separation would be impossible. But for the deep learning function F this is not difficult...

We will discuss experiments on this **playground.tensorflow** site in the Problem Set.

Important Note: Fully Connected versus Convolutional

We don't want to mislead the reader. Those "fully connected" nets are often not the most effective. If the weights around one pixel in an image can be repeated around all pixels (why not?), then one row of A is all we need. The row can assign zero weights to faraway pixels. Local **convolutional neural nets** (**CNN's**) are the subject of Section VII.2.

You will see that the count grows exponentially with the number of neurons and layers. That is a useful insight into the power of deep learning. We badly need insight because the size and depth of the neural network make it difficult to visualize in full detail.

Counting Flat Pieces in the Graph : One Internal Layer

It is easy to count entries in the weight matrices A_k and the bias vectors $\boldsymbol{b}_k$. Those numbers determine the function F. But it is far more interesting to count the number of flat pieces in the graph of F. This number measures the **expressivity** of the neural network. $F(\boldsymbol{x}, \boldsymbol{v})$ is a more complicated function than we fully understand (at least so far). The system is deciding and acting on its own, without explicit approval of its "thinking". For driverless cars we will see the consequences fairly soon.

Suppose $\boldsymbol{v}_0$ has m components and $A_1\boldsymbol{v}_0+\boldsymbol{b}_1$ has N components. We have N functions of $\boldsymbol{v}_0$. Each of those linear functions is zero along a hyperplane (dimension $m-1$) in $\mathbf{R}^m$. When we apply ReLU to that linear function it becomes piecewise linear, with a fold along that hyperplane. On one side of the fold its graph is sloping, on the other side the function changes from negative to zero.

Then the next matrix A_2 combines those N piecewise linear functions of v_0, so we now have folds along *N different hyperplanes in $\mathbf{R}^m$*. This describes each piecewise linear component of the next layer $A_2(\text{ReLU}(A_1\boldsymbol{v}_0+\boldsymbol{b}_1))$ in the typical case.

You could think of N straight folds in the plane (the folds are actually along N hyperplanes in m-dimensional space). The first fold separates the plane in two pieces. The next fold from ReLU will leave us with four pieces. The third fold is more difficult to visualize, but the following figure shows that there are seven (*not eight*) pieces.

In combinatorial theory, we have a **hyperplane arrangement**—and a theorem of Tom Zaslavsky counts the pieces. The proof is presented in Richard Stanley's great textbook on *Enumerative Combinatorics* (2001). But that theorem is more complicated than we need, because it allows the fold lines to meet in all possible ways. Our task is simpler because we assume that the fold lines are in "general position"—$m+1$ folds don't meet. For this case we now apply the neat counting argument given by Raghu, Poole, Kleinberg, Gangul, and Dickstein : *On the Expressive Power of Deep Neural Networks*, arXiv : 1606.05336v6 : See also *The Number of Response Regions* by Pascanu, Montufar, and Bengio on arXiv 1312.6098.

Theorem For $\boldsymbol{v}$ in $\mathbf{R}^m$, suppose the graph of $F(\boldsymbol{v})$ has folds along N hyperplanes $H_1,\ldots,H_N$. Those come from N linear equations $\boldsymbol{a}_i^{\mathrm{T}}\boldsymbol{v}+b_i=0$, in other words from ReLU at N neurons. Then the number of linear pieces of F and regions bounded by the N hyperplanes is $r(N,m)$:

$$\boldsymbol{r}(N,\boldsymbol{m})=\sum_{i=0}^{m}\binom{N}{i}=\binom{N}{0}+\binom{N}{1}+\cdots+\binom{N}{m}. \tag{4}$$

These binomial coefficients are

$$\binom{N}{i}=\frac{N!}{i!(N-i)!}\quad\text{with } 0!=1 \text{ and } \binom{N}{0}=1 \text{ and } \binom{N}{i}=0 \text{ for } i>N.$$

Example The function $F(x,y,z)=\text{ReLU}(x)+\text{ReLU}(y)+\text{ReLU}(z)$ has 3 folds along the 3 planes $x=0, y=0, z=0$. Those planes divide $\mathbf{R}^3$ into $r(3,3)=8$ pieces where $F=x+y+z$ and $x+z$ and x and 0 (and 4 more). Adding $\text{ReLU}(x+y+z-1)$ gives a fourth fold and $r(4,3)=15$ pieces of $\mathbf{R}^3$. Not 16 because the new fold plane $x+y+z=1$ does not meet the 8th original piece where $x<0, y<0, z<0$.

George Polya's famous YouTube video *Let Us Teach Guessing* cut a cake by 5 planes. He helps the class to find $r(5,3) = 26$ pieces. Formula (4) allows m-dimensional cakes.

One hyperplane in $\mathbf{R}^m$ produces $\binom{1}{0} + \binom{1}{1} = 2$ regions. And $N = 2$ hyperplanes will produce $r(2,m) = 1 + 2 + 1 = 4$ regions provided $m > 1$. When $m = 1$ we have two folds in a line, which only separates the line into $r(2,1) = 3$ pieces.

The count r of linear pieces will follow from the recursive formula

$$\boxed{r(N,m) = r(N-1,m) + r(N-1,m-1).} \tag{5}$$

To understand that recursion, start with $N-1$ hyperplanes in $\mathbf{R}^m$ and $r(N-1,m)$ regions. *Add one more hyperplane* H (dimension $m-1$). The established $N-1$ hyperplanes cut H into $r(N-1,m-1)$ regions. Each of those pieces of H divides one existing region into two, adding $r(N-1,m-1)$ regions to the original $r(N-1,m)$; see Figure VII.3. So the recursion is correct, and we now apply equation (5) to compute $r(N,m)$.

The count starts at $r(1,0) = r(0,1) = \mathbf{1}$. Then (4) is proved by induction on $N+m$:

$$\begin{aligned} r(N-1,m) + r(N-1,m-1) &= \sum_0^m \binom{N-1}{i} + \sum_0^{m-1} \binom{N-1}{i} \\ &= \binom{N-1}{0} + \sum_0^{m-1} \left[\binom{N-1}{i} + \binom{N-1}{i+1}\right] \\ &= \binom{N}{0} + \sum_0^{m-1} \binom{N}{i+1} = \sum_0^m \binom{N}{i}. \end{aligned} \tag{6}$$

The two terms in brackets (second line) became one term because of a useful identity:

$$\binom{N-1}{i} + \binom{N-1}{i+1} = \binom{N}{i+1} \quad \text{and the induction is complete.}$$

Mike Giles made that presentation clearer, and he suggested Figure VII.3 to show the effect of the last hyperplane H. There are $r = 2^N$ linear pieces of $F(\boldsymbol{v})$ for $N \leq m$ and $r \approx N^m/m!$ pieces for $N >> m$, when the hidden layer has many neurons.

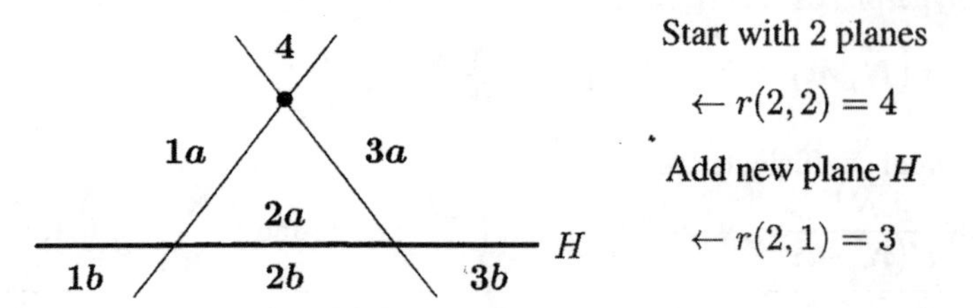

Figure VII.3: The $r(2,1) = 3$ pieces of H create 3 new regions. Then the count becomes $r(3,2) = 4 + 3 = 7$ flat regions in the continuous piecewise linear surface $\boldsymbol{v}_2 = F(\boldsymbol{v}_0)$. A fourth fold will cross all 3 existing folds and create 4 new regions, so $r\,(4,2) = 11$.

Flat Pieces of $F(v)$ with More Hidden Layers

Counting the linear pieces of $F(\boldsymbol{v})$ is much harder with 2 internal layers in the network. Again $\boldsymbol{v}_0$ and $\boldsymbol{v}_1$ have m and N_1 components. Now $A_1\boldsymbol{v}_1 + \boldsymbol{b}_1$ will have N_2 components before ReLU. Each one is like the function F for one layer, described above. Then application of ReLU will create new folds in its graph. Those folds are along the lines where a component of $A_1\boldsymbol{v}_1 + \boldsymbol{b}_1$ is zero.

Remember that each component of $A_1\boldsymbol{v}_1 + \boldsymbol{b}_1$ is piecewise linear, not linear. So it crosses zero (if it does) along a piecewise linear surface, not a hyperplane. The straight lines in Figure VII.3 for the folds in $\boldsymbol{v}_1$ will change to *piecewise straight lines* for the folds in $\boldsymbol{v}_2$. In m dimensions they are connected pieces of hyperplanes. So the count becomes variable, depending on the details of $\boldsymbol{v}_0, A_1, \boldsymbol{b}_1, A_2$, and $\boldsymbol{b}_2$.

Still we can estimate the number of linear pieces. We have N_2 piecewise straight lines (or piecewise hyperplanes in $\mathbf{R}^m$) from N_2 ReLU's at the second hidden layer. If those lines were actually straight, we would have a total of $N_1 + N_2$ folds in each component of $\boldsymbol{v}_3 = F(\boldsymbol{v}_0)$. *Then the formula* (4) *to count the pieces would have* $N_1 + N_2$ *in place of* N. This is our estimate (open for improvement) with two layers between $\boldsymbol{v}_0$ and $\boldsymbol{v}_3$.

Composition $F_3(F_2(F_1(v)))$

The word "composition" would simply represent "matrix multiplication" if all our functions were linear: $F_k(\boldsymbol{v}) = A_k\boldsymbol{v}$. Then $F(\boldsymbol{v}_0) = A_3A_2A_1\boldsymbol{v}_0$: just one matrix. For nonlinear F_k the meaning is the same: Compute $\boldsymbol{v}_1 = F_1(\boldsymbol{v}_0)$, then $\boldsymbol{v}_2 = F_2(\boldsymbol{v}_1)$, and finally $\boldsymbol{v}_3 = F_3(\boldsymbol{v}_2)$. This operation of **composition** $\boldsymbol{F_3(F_2(F_1(v_0)))}$ is far more powerful in creating functions than addition!

For a neural network, composition produces continuous piecewise linear functions $F(\boldsymbol{v}_0)$. The 13th problem on Hilbert's list of 23 unsolved problems in 1900 asked a question about *all* continuous functions. A famous generalization of his question was this:

> Is every continuous function $F(x, y, z)$ of three variables the composition of continuous functions $G_1, \ldots, G_N$ of two variables? The answer is *yes*.

Hilbert seems to have expected the answer *no*. But a positive answer was given in 1957 by Vladimir Arnold (age 19). His teacher Andrey Kolmogorov had previously created multivariable functions out of 3-variable functions.

Related questions have negative answers. If $F(x, y, z)$ has continuous derivatives, it may be impossible for all the 2-variable functions to have continuous derivatives (Vitushkin). And to construct 2-variable continuous functions $F(x, y)$ as compositions of 1-variable continuous functions (the ultimate 13th problem) you must allow *addition*. The 2-variable functions xy and x^y use 1-variable functions exp, log, and log log:

$$\boldsymbol{xy = \exp(\log x + \log y)} \quad \text{and} \quad \boldsymbol{x^y = \exp(\exp(\log y + \log\log x))}. \qquad (7)$$

So much to learn from the Web. A chapter of *Kolmogorov's Heritage in Mathematics* (Springer, 2007) connects these questions explicitly to neural networks.

Is the answer to Hilbert still yes for continuous piecewise linear functions on $\mathbf{R}^m$?

Neural Nets Give Universal Approximation

The previous paragraphs wandered into the analysis of functions $f(\boldsymbol{v})$ of several variables. For deep learning a key question is the approximation of f by a neural net—when the weights $\boldsymbol{x}$ are chosen to bring $F(\boldsymbol{x},\boldsymbol{v})$ close to $f(\boldsymbol{v})$.

There is a qualitative question and also a quantitative question:

1 For any continuous function $f(\boldsymbol{v})$ with $\boldsymbol{v}$ in a cube in $\mathbf{R}^d$, can a net with enough layers and neurons and weights $\boldsymbol{x}$ give uniform approximation to f within any desired accuracy $\epsilon > 0$? This property is called **universality**.

$$\textbf{If } f(\boldsymbol{v}) \textbf{ is continuous there exists } \boldsymbol{x} \textbf{ so that } |F(\boldsymbol{x},\boldsymbol{v}) - f(\boldsymbol{v})| < \epsilon \textbf{ for all } \boldsymbol{v}. \tag{8}$$

2 If $f(\boldsymbol{v})$ belongs to a normed space S of smooth functions, how quickly does the approximation error improve as the net has more weights?

$$\textbf{Accuracy of approximation to } f \qquad \min_{\boldsymbol{x}} ||F(\boldsymbol{x},\boldsymbol{v}) - f(\boldsymbol{v})|| \leq C||f||_S \tag{9}$$

Function spaces S often use the $\boldsymbol{L^2}$ or $\boldsymbol{L^1}$ or $\boldsymbol{L^\infty}$ norm of the function f and its partial derivatives up to order r. Functional analysis gives those spaces a meaning even for non-integer r. C usually decreases as the smoothness parameter r is increased. For continuous piecewise linear approximation over a uniform grid with meshwidth h we often find $C = O(h^2)$.

The response to Question 1 is *yes*. Wikipedia notes that one hidden layer (with enough neurons!) is sufficient for approximation within ϵ. The 1989 proof by George Cybenko used a sigmoid function rather than ReLU, and the theorem is continually being extended. Ding-Xuan Zhou proved that we can require the A_k to be convolution matrices (the structure becomes a CNN). Convolutions have many fewer weights than arbitrary matrices—and universality allows many convolutions.

The response to Question **2** by Mhaskar, Liao, and Poggio begins with the degree of approximation to functions $f(v_1,\ldots,v_d)$ with continuous derivatives of order r. *For n weights the usual error bound is $Cn^{-r/d}$.* The novelty is their introduction of **composite functions** built from 2-variable functions, as in $f(v_1,v_2,v_3,v_4) = f_3(f_1(v_1,v_2),f_2(v_3,v_4))$. For a composite function, the approximation by a hierarchical net is much more accurate. *The error bound becomes* $\boldsymbol{Cn^{-r/2}}$.

The proof applies the standard result for $d = 2$ variables to each function f_1, f_2, f_3. A difference of composite functions is a composite of 2-variable differences.

1 G. Cybenko, Approximation by superpositions of a sigmoidal function, *Mathematics of Control, Signals, and Systems* **7** (1989) 303-314.

2 K. Hornik, Approximation capabilities of multilayer feedforward networks, *Neural Networks* **4** (1991) 251-257.

3 H. Mhaskar, Q. Liao, and T. Poggio, *Learning functions: When is deep better than shallow*, arXiv: 01603.00988v4; 29 May 2016.

4 D.-X. Zhou, *Universality of deep convolutional neural networks*, arXiv : 1805.10769, 20 Jul 2018.

5 D. Rolnick and M. Tegmark, *The power of deeper networks for expressing natural functions*, arXiv : 1705.05502, 27 Apr 2018.

Problem Set VII.1

1 In the example $F = \text{ReLU}(x) + \text{ReLU}(y) + \text{ReLU}(z)$ that follows formula (4) for $r(N, m)$, suppose the 4th fold comes from ReLU $(x + y + z)$. Its fold plane $x + y + z = 0$ now meets the 3 original fold planes $x = 0, y = 0, z = 0$ at a single point $(0, 0, 0)$—*an exceptional case*. Describe the 16 (not 15) linear pieces of F = sum of these four ReLU's.

2 Suppose we have $m = 2$ inputs and N neurons on a hidden layer, so $F(x, y)$ is a linear combination of N ReLU's. Write out the formula for $r(N, 2)$ to show that the count of linear pieces of F has leading term $\frac{1}{2}N^2$.

3 Suppose we have $N = 18$ lines in a plane. If 9 are vertical and 9 are horizontal, how many pieces of the plane ? Compare with $r(18, 2)$ when the lines are in general position and no three lines meet.

4 What weight matrix A_1 and bias vector $\boldsymbol{b}_1$ will produce ReLU $(x + 2y - 4)$ and ReLU $(3x - y + 1)$ and ReLU $(2x + 5y - 6)$ as the $N = 3$ components of the first hidden layer ? (The input layer has 2 components x and y.) If the output w is the sum of those three ReLU's, how many pieces in $w(x, y)$?

5 Folding a line four times gives $r(4, 1) = 5$ pieces. Folding a plane four times gives $r(4, 2) = 11$ pieces. According to formula (4), how many flat subsets come from folding $\mathbf{R}^3$ four times ? The flat subsets of $\mathbf{R}^3$ meet at 2D planes (like a door frame).

6 The binomial theorem finds the coefficients $\binom{N}{k}$ in $(a + b)^N = \sum_{0}^{N} \binom{N}{k} a^k b^{N-k}$.

For $a = b = 1$ what does this reveal about those coefficients and $r(N, m)$ for $m \geq N$?

7 In Figure VII.3, one more fold will produce 11 flat pieces in the graph of $z = F(x, y)$. Check that formula (4) gives $r(4, 2) = 11$. How many pieces after five folds ?

8 Explain with words or show with graphs why each of these statements about Continuous Piecewise Linear functions (CPL functions) is true :

M The maximum $M(x, y)$ of two CPL functions $F_1(x, y)$ and $F_2(x, y)$ is CPL.

S The sum $S(x, y)$ of two CPL functions $F_1(x, y)$ and $F_2(x, y)$ is CPL.

C If the one-variable functions $y = F_1(x)$ and $z = F_2(y)$ are CPL, so is the composition $C(x) = z = (F_2(F_1(x))$.

9 How many weights and biases are in a network with $m = N_0 = 4$ inputs in each feature vector $\boldsymbol{v}_0$ and $N = 6$ neurons on each of the 3 hidden layers? How many activation functions (ReLU) are in this network, before the final output?

10 (Experimental) In a neural network with two internal layers and a total of 10 neurons, should you put more of those neurons in layer **1** or layer **2**?

Problems 11–13 use the blue ball, orange ring example on playground.tensorflow.org with one hidden layer and activation by ReLU (not Tanh). When learning succeeds, a white polygon separates blue from orange in the figure that follows.

11 Does learning succeed for $N = 4$? What is the count $r(N, 2)$ of flat pieces in $F(\boldsymbol{x})$? The white polygon shows where flat pieces in the graph of $F(\boldsymbol{x})$ change sign as they go through the base plane $z = 0$. How many sides in the polygon?

12 Reduce to $N = 3$ neurons in one layer. Does F still classify blue and orange correctly? How many flat pieces $r(3, 2)$ in the graph of $F(\boldsymbol{v})$ and how many sides in the separating polygon?

13 Reduce further to $N = 2$ neurons in one layer. Does learning still succeed? What is the count $r(2, 2)$ of flat pieces? How many folds in the graph of $F(\boldsymbol{v})$? How many sides in the white separator?

14 Example 2 has blue and orange in two quadrants each. With one layer, do $N = 3$ neurons and even $N = 2$ neurons classify that training data correctly? How many flat pieces are needed for success? Describe the unusual graph of $F(\boldsymbol{v})$ when $N = 2$.

15 Example 4 with blue and orange spirals is much more difficult! With one hidden layer, can the network learn this training data? Describe the results as N increases.

16 Try that difficult example with two hidden layers. Start with $4 + 4$ and $6 + 2$ and $2 + 6$ neurons. Is $2 + 6$ better or worse or more unusual than $6 + 2$?

17 How many neurons bring complete separation of the spirals with two hidden layers? Can three layers succeed with fewer neurons than two layers?

I found that $4 + 4 + 2$ and $4 + 4 + 4$ neurons give very unstable iterations for that spiral graph. There were spikes in the training loss until the algorithm stopped trying. playground.tensorflow.org (on our back cover!) was a gift from Daniel Smilkov.

18 What is the smallest number of pieces that 20 fold lines can produce in a plane?

19 How many pieces are produced from 10 vertical and 10 horizontal folds?

20 What is the maximum number of pieces from 20 fold lines in a plane?

VII.2 Convolutional Neural Nets

This section is about networks with a different architecture. Up to now, each layer was fully connected to the next layer. If one layer had n neurons and the next layer had m neurons, then the matrix A connecting those layers is m by n. *There were mn independent weights in A.* The weights from all layers were chosen to give a final output that matched the training data. The derivatives needed in that optimization were computed by backpropagation. Now we might have only 3 or 9 independent weights per layer.

That fully connected net will be extremely inefficient for image recognition. First, the weight matrices A will be huge. If one image has 200 by 300 pixels, then its input layer has $60,000$ components. The weight matrix A_1 for the first hidden layer has $60,000$ columns. The problem is: We are looking for connections between faraway pixels. Almost always, the important connections in an image are **local**.

Text and music have a 1D local structure: a time series

Images have a 2D local structure: 3 copies for red-green-blue

Video has a 3D local structure: Images in a time series

More than this, the search for structure is essentially the same everywhere in the image. There is normally no reason to process one part of a text or image or video differently from other parts. We can use the same weights in all parts: *Share the weights.* The neural net of local connections between pixels is **shift-invariant**: the same everywhere.

The result is a big reduction in the number of independent weights. Suppose each neuron is connected to only E neurons on the next layer, and those connections are the same for all neurons. Then the matrix A between those layers has only E independent weights x. The optimization of those weights becomes enormously faster. In reality we have time to create several different channels with their own E or E^2 weights. They can look for edges in different directions (horizontal, vertical, and diagonal).

In one dimension, a banded shift-invariant matrix is a **Toeplitz matrix** or a **filter**. Multiplication by that matrix A is a **convolution** $x * v$. The network of connections between all layers is a **Convolutional Neural Net** (**CNN** or **ConvNet**). Here $E = 3$.

$$A = \begin{bmatrix} x_1 & x_0 & x_{-1} & 0 & 0 & 0 \\ 0 & x_1 & x_0 & x_{-1} & 0 & 0 \\ 0 & 0 & x_1 & x_0 & x_{-1} & 0 \\ 0 & 0 & 0 & x_1 & x_0 & x_{-1} \end{bmatrix} \qquad \begin{array}{l} \boldsymbol{v} = (v_0, v_1, v_2, v_3, v_4, v_5) \\ \boldsymbol{y} = A\boldsymbol{v} = \quad (y_1, y_2, y_3, y_4) \\ N+2 \text{ inputs and } N \text{ outputs} \end{array}$$

It is valuable to see A as *a combination of shift matrices* $\boldsymbol{L}, \boldsymbol{C}, \boldsymbol{R}$: *Left, Center, Right.*

$$\textbf{Each shift has a diagonal of 1's} \qquad \boldsymbol{A = x_1 L + x_0 C + x_{-1} R}$$

Then the derivatives of $\boldsymbol{y} = A\boldsymbol{v} = \boldsymbol{x_1 L v + x_0 C v + x_{-1} R v}$ are exceptionally simple:

$$\boxed{\frac{\partial \boldsymbol{y}}{\partial x_1} = L\boldsymbol{v} \qquad \frac{\partial \boldsymbol{y}}{\partial x_0} = C\boldsymbol{v} \qquad \frac{\partial \boldsymbol{y}}{\partial x_{-1}} = R\boldsymbol{v}} \tag{1}$$

Convolutions in Two Dimensions

When the input $\boldsymbol{v}$ is an image, the convolution with $\boldsymbol{x}$ becomes two-dimensional. The numbers x_{-1}, x_0, x_1 change to $E^2 = 3^2$ independent weights. The inputs v_{ij} have two indices and $\boldsymbol{v}$ represents $(N+2)^2$ pixels. The outputs have only N^2 pixels unless we pad with zeros at the boundary. The 2D convolution $\boldsymbol{x} * \boldsymbol{v}$ is a *linear combination of* 9 *shifts.*

$$\textbf{Weights} \begin{bmatrix} x_{11} & x_{01} & x_{-11} \\ x_{10} & x_{00} & x_{-10} \\ x_{1-1} & x_{0-1} & x_{-1-1} \end{bmatrix}$$

Input image v_{ij} i,j from$(0,0)$ to$(N+1,N+1)$
Output image y_{ij} i,j from $(1,1)$ to (N,N)
Shifts L, C, R, U, D = Left, Center, **R**ight, **U**p, **D**own

$$A = x_{11}LU + x_{01}CU + x_{-11}RU + x_{10}L + x_{00}C + x_{-10}R + x_{1-1}LD + x_{0-1}CD + x_{-1-1}RD$$

This expresses the convolution matrix A as a combination of 9 shifts. The derivatives of the output $\boldsymbol{y} = A\boldsymbol{v}$ are again exceptionally simple. We use these nine derivatives to create the gradients $\boldsymbol{\nabla F}$ and $\boldsymbol{\nabla L}$ that are needed in stochastic gradient descent to improve the weights $\boldsymbol{x}_k$. The next iteration $\boldsymbol{x}_{k+1} = \boldsymbol{x}_k - s\boldsymbol{\nabla L}_k$ has weights that better match the correct outputs from the training data.

These nine derivatives of $\boldsymbol{y} = A\boldsymbol{v}$ are computed inside backpropagation:

$$\frac{\partial \boldsymbol{y}}{\partial x_{11}} = LU\boldsymbol{v} \quad \frac{\partial \boldsymbol{y}}{\partial x_{01}} = CU\boldsymbol{v} \quad \frac{\partial \boldsymbol{y}}{\partial x_{-11}} = RU\boldsymbol{v} \quad \cdots \quad \frac{\partial \boldsymbol{y}}{\partial x_{-1-1}} = RD\boldsymbol{v} \qquad (2)$$

CNN's can readily afford to have $\boldsymbol{B}$ **parallel channels** (and that number B can vary as we go deeper into the net). The count of weights in $\boldsymbol{x}$ is so much reduced by weight sharing and weight locality, that we don't need and we can't expect one set of $E^2 = 9$ weights to do all the work of a convolutional net.

Let me highlight the operational meaning of convolution. In 1 dimension, the formal algebraic definition $y_j = \sum x_i v_{j-i} = \sum x_{j-k} v_k$ involves a "flip" of the v's or the x's. This is a source of confusion that we do not need. We look instead at left-right shifts L and R of the whole signal (in 1D) and also up-down shifts U and D in two dimensions. Each shift is a matrix with a diagonal full of 1's. That saves us from the complication of remembering flipped subscripts.

A convolution is a combination of shift matrices (producing a filter or Toeplitz matrix)
A cyclic convolution is a combination of cyclic shifts (producing a circulant matrix)
A continuous convolution is a continuous combination (an integral) of shifts
In deep learning, the coefficients in the combination will be the "weights" to be learned.

Two-dimensional Convolutional Nets

Now we come to the real success of CNN's: **Image recognition**. ConvNets and deep learning have produced a small revolution in computer vision. The applications are to self-driving cars, drones, medical imaging, security, robotics—there is nowhere to stop. Our interest is in the algebra and geometry and intuition that makes all this possible.

In two dimensions (for images) the matrix A is **block Toeplitz**. Each small block is E by E. This is a familiar structure in computational engineering. The count E^2 of independent weights to be optimized is far smaller than for a fully connected network.

The same weights are used around all pixels (*shift-invariance*). The matrix produces a 2D convolution $\boldsymbol{x} * \boldsymbol{v}$. Frequently A is called a **filter**.

To understand an image, look to see where it changes. *Find the edges.* Our eyes look for sharp cutoffs and steep gradients. Our computer can do the same by creating a filter. The dot products between a smooth function and a moving filter window will be smooth. But when an edge in the image lines up with a diagonal wall, *we see a spike.* Those dot products (fixed image) · (moving image) are exactly the **"convolution"** of the two images.

The difficulty with two or more dimensions is that edges can have many directions. We will need horizontal and vertical and diagonal filters for the test images. And filters have many purposes, including smoothing, gradient detection, and edge detection.

1 Smoothing For a 2D function f, the natural smoother is *convolution with a Gaussian*:

$$Gf(x,y) = \frac{1}{2\pi\sigma^2}\, e^{-(x^2+y^2)/2\sigma^2} * f = \frac{1}{\sqrt{2\pi}\,\sigma}\, e^{-x^2/2\sigma^2} * \frac{1}{\sqrt{2\pi}\,\sigma}\, e^{-y^2/2\sigma^2} * f(x,y)$$

This shows G as a product of 1D smoothers. The Gaussian is everywhere positive, so it is *averaging*: Gf cannot have a larger maximum than f. The filter removes noise (at a price in sharp edges). For small variance σ^2, details become clearer.

For a 2D vector (a matrix f_{ij} instead of a function $f(x,y)$) the Gaussian must become discrete. The perfection of radial symmetry will be lost because the matrix G is square. Here is a 5 by 5 discrete Gaussian G $(E=5)$:

$$G = \frac{1}{273}\begin{bmatrix} 1 & 4 & 7 & 4 & 1 \\ 4 & 16 & 26 & 16 & 4 \\ 7 & 26 & 41 & 26 & 7 \\ 4 & 16 & 26 & 16 & 4 \\ 1 & 4 & 7 & 4 & 1 \end{bmatrix} \approx \frac{1}{289}\begin{bmatrix} 1 \\ 4 \\ 7 \\ 4 \\ 1 \end{bmatrix}\begin{bmatrix} 1 & 4 & 7 & 4 & 1 \end{bmatrix} \qquad (3)$$

We also lost our exact product of 1D filters. To come closer, use a larger matrix $G = \boldsymbol{x}\boldsymbol{x}^{\mathrm{T}}$ with $\boldsymbol{x} = (.006, .061, .242, .383, .242, .061, .006)$ and discard the small outside pixels.

2 Gradient detection Image processing (as distinct from learning by a CNN) needs filters that detect the gradient. They contain specially chosen weights. We mention some simple filters just to indicate how they can find gradients—the first derivatives of f.

One dimension $(x_1, x_0, x_{-1}) = \left(-\frac{1}{2}, 0, \frac{1}{2}\right)$ $\left[\left(\frac{1}{2}, 0, -\frac{1}{2}\right) \text{ in convolution form}\right]$
$E = 3$

In this case the components of Av are centered differences: $(Av)_i = \frac{1}{2}v_{i+1} - \frac{1}{2}v_{i-1}$. When the components of v are increasing linearly from left to right, as in $v_i = 3i$, the output from the filter is $\frac{1}{2}3(i+1) - \frac{1}{2}3(i-1) = 3 =$ *correct gradient.*

The flip to $\left(\frac{1}{2}, 0, -\frac{1}{2}\right)$ comes from the definition of convolution as $\sum x_{i-k}v_k$.

Two dimensions These 3×3 *Sobel operators* approximate $\partial/\partial x$ and $\partial/\partial y$:

$$E = 3 \qquad \frac{\partial}{\partial x} \approx \frac{1}{2}\begin{bmatrix} -1 & 0 & 1 \\ -2 & 0 & 2 \\ -1 & 0 & 1 \end{bmatrix} \qquad \frac{\partial}{\partial y} \approx \frac{1}{2}\begin{bmatrix} -1 & -2 & -1 \\ 0 & 0 & 0 \\ 1 & 2 & 1 \end{bmatrix} \tag{4}$$

For functions, the gradient vector $g = \text{grad}\, f$ has $||g||^2 = |\partial f/\partial x|^2 + |\partial f/\partial y|^2$.

Those weights were created for image processing, to locate the most important features of a typical image: *its edges.* These would be candidates for E by E filters inside a 2D convolutional matrix A. But remember that in deep learning, weights like $\frac{1}{2}$ and $-\frac{1}{2}$ are not chosen by the user. They are created from the training data.

Sections IV.2 and IV.5 of this book studied cyclic convolutions and Toeplitz matrices. Shift-invariance led to the application of discrete Fourier transforms. But in a CNN, ReLU is likely to act on each neuron. The network may include zero-padding—as well as max-pooling layers. So we cannot expect to apply the full power of Fourier analysis.

3 Edge detection After the gradient direction is estimated, we look for edges—the most valuable features to know. "*Canny Edge Detection*" is a highly developed process. Now we don't want smoothing, which would blur the edge. The good filters become *Laplacians of Gaussians*:

$$E\, f(x,y) = \nabla^2[g(x,y) * f(x,y)] = [\nabla^2 g(x,y)] * f(x,y). \tag{5}$$

The Laplacian $\nabla^2 G$ of a Gaussian is $(x^2 + y^2 - 2\sigma^2)\, e^{-(x^2+y^2)/2\sigma^2}/\pi\sigma^4$.

The Stride of a Convolutional Filter

Important The filters described so far all have a *stride* $S = 1$. For a larger stride, the *moving window takes longer steps* as it moves across the image. Here is the matrix A for a 1-dimensional 3-weight filter with a stride of 2. Notice especially that the length of the output $y = Av$ is reduced by that factor of 2 (previously four outputs and now two):

$$\textbf{Stride } S = 2 \qquad A = \begin{bmatrix} x_1 & x_0 & x_{-1} & 0 & 0 \\ 0 & 0 & x_1 & x_0 & x_{-1} \end{bmatrix} \tag{6}$$

Now the nonzero weights like x_1 in L are two columns apart (S columns apart for stride S). In 2D, a stride $S = 2$ reduces each direction by 2 and the whole output by 4.

Extending the Signal

Instead of losing neurons at the edges of the image when A is not square, we can *extend the input layer.* We are "inventing" components beyond the image boundary. Then the output $y = Av$ fits the image block: equal dimensions for input and output.

The simplest and most popular approach is **zero-padding**: Choose all additional components to be *zeros.* The extra columns on the left and right of A multiply those zeros. In between, we have a square Toeplitz matrix as in Section IV.5. It is still determined by a much smaller set of weights than the number of entries in A.

For periodic signals, zero-padding is replaced by *wraparound.* The Toeplitz matrix becomes a circulant (Section IV.2). The Discrete Fourier Transform tells its eigenvalues. The eigenvectors are always the columns of the Fourier matrix. The multiplication Av is a *cyclic convolution* and the Convolution Rule applies.

A more accurate choice is to go beyond the boundary by *reflection.* If the last component of the signal is v_N, and the matrix is asking for v_{N+1} and v_{N+2}, we can reuse v_N and v_{N-1} (or else v_{N-1} and v_{N-2}). Whatever the length of v and the size of A, all the matrix entries in A come from the same E weights x_{-1} to x_1 or x_{-2} to x_2 (and E^2 weights in 2D).

Note Another idea. We might accept the original dimension (128 in our example) and use the reduction to 64 as a way to apply **two filters C_1 and C_2**. Each filter output is downsampled from 128 to 64. The total sample count remains 128. If the filters are suitably independent, no information is lost and the original 128 values can be recovered.

This process is linear. Two 64 by 128 matrices are combined into 128 by 128: square. If that matrix is invertible, as we intend, the filter bank is *lossless.*

This is what CNN's usually do: Add more channels of weight matrices A in order to capture more features of the training sample. The neural net has a bank of B filters.

Filter Banks and Wavelets

The idea in those last paragraphs produces a **filter bank**. This is just a set of B different filters (convolutions). In signal processing, an important case combines a lowpass filter C_1 with a highpass filter C_2. The output of $C_1 v$ is a smoothed signal (dominated by low frequencies). The output $C_2 v$ is dominated by high frequencies. A perfect cutoff by ideal filters cannot be achieved by finite matrices C_1 and C_2.

From two filters we have a total of 256 output components. Then both outputs are subsampled. The result is 128 components, separated approximately into *averages and differences*—low frequencies and high frequencies. The matrix is 128 by 128.

Wavelets The wavelet idea is to repeat the same steps on the 64 components of the lowpass output $(\downarrow 2)\, C_1 x$. *Then* $(\downarrow 2)\, C_1\, (\downarrow 2)\, C_1 x$ *is an average of averages.* Its frequencies are concentrated in the lowest quarter ($|\omega| \leq \pi/4$) of all frequencies. The mid-frequency output $(\downarrow 2)\, C_2\, (\downarrow 2)\, C_1 x$ with 32 components will not be subdivided. Then $128 = 64 + 32 + 16 + 16$.

In the limit of infinite subdivision, *wavelets* enter. This low-high frequency separation is an important theme in signal processing. It has not been so important for deep learning. But with multiple channels in a CNN, frequency separation could be effective.

Counting the Number of Inputs and Outputs

In a one-dimensional problem, suppose a layer has N neurons. We apply a convolutional matrix with E nonzero weights. The stride is S, and we pad the input signal by P zeros at each end. How many outputs (M numbers) does this filter produce ?

$$\boxed{\textbf{Karpathy's formula} \qquad M = \frac{N-E+2P}{S}+1} \tag{7}$$

In a 2D or 3D problem, this 1D formula applies in each direction.

Suppose $E = 3$ and the stride is $S = 1$. If we add one zero ($P = 1$) at each end, then

$$M = N - 3 + 2 + 1 = N \qquad \text{(input length = output length)}$$

This case $2P = E - 1$ with stride $S = 1$ is the most common architecture for CNN's.

If we don't pad the input with zeros, then $P = 0$ and $M = N - 2$ (as in the 4 by 6 matrix A at the start of this section). In 2 dimensions this becomes $M^2 = (N-2)^2$. We lose neurons this way, but we avoid zero-padding.

Now suppose the stride is $S = 2$. Then $N - E$ must be an even number. Otherwise the formula (4) produces a fraction. Here are two examples of success for stride $S = 2$, with $N - E = 5 - 3$ and padding $P = 0$ or $P = 1$ at both ends of the five inputs :

$$\textbf{Stride } \textbf{2} \qquad \begin{bmatrix} x_{-1} & x_0 & x_1 & 0 & 0 \\ 0 & 0 & x_{-1} & x_0 & x_1 \end{bmatrix} \begin{bmatrix} x_{-1} & x_0 & x_1 & 0 & 0 & 0 & 0 \\ 0 & 0 & x_{-1} & x_0 & x_1 & 0 & 0 \\ 0 & 0 & 0 & 0 & x_{-1} & x_0 & x_1 \end{bmatrix}$$

Again, our counts apply in each direction to an image in 2D or a tensor.

A Deep Convolutional Network

Recognizing images is a major application of deep learning (and a major success). The success came with the creation of AlexNet and the development of convolutional nets. This page will describe a deep network of local convolutional matrices for image recognition. We follow the prize-winning paper of Simonyan and Zisserman from ICLR 2015. That paper recommends a deep architecture of $L = 16$–19 layers with small (3×3) filters. The network has a breadth of B parallel channels (B images on each layer).

If the breadth B were to stay the same at all layers, and all filters had E by E local weights, a straightforward formula would estimate the number W of weights in the net:

$$\boxed{W \approx LBE^2 \qquad \textbf{\textit{L} layers, \textit{B} channels, \textit{E} by \textit{E} local convolutions}} \tag{8}$$

Notice that W does not depend on the count of neurons on each layer. This is because A has E^2 weights, whatever its size. Pooling will change that size without changing E^2.

But the count of B channels can change—*and it is very common to end a CNN with fully-connected layers*. This will radically change the weight count W !

It is valuable to discuss the decisions taken by Simonyan and Zisserman, together with other options. Their choices led to $W \approx 135,000,000$ weights. The computations were on four NVIDIA GPU's, and training one net took 2-3 weeks. The reader may have less computing power (and smaller problems). So the network hyperparameters L and B will be reduced. We believe that the important principles remain the same.

A key point here is the recommendation to reduce the size E of the local convolutions. 5 by 5 and 7 by 7 filters were rejected. In fact a 1 by 1 convolutional layer can be a way to introduce an extra bank of ReLU's—as in the ResNets coming next.

The authors compare three convolution layers, each with 3 by 3 filters, to a single layer of less local 7 by 7 convolutions. They are comparing 27 weights with 49 weights, and three nonlinear layers with one. In both cases the influence of a single data point spreads to three neighbors vertically and horizontally in the image or the RGB images ($B = 3$). Preference goes to the 3 by 3 filters with extra nonlinearities from more neurons per layer.

Softmax Outputs for Multiclass Networks

In recognizing digits, we have 10 possible outputs. For letters and other symbols, 26 or more. With multiple output classes, we need an appropriate way to decide the very last layer (the output layer $\boldsymbol{w}$ in the neural net that started with $\boldsymbol{v}$). "Softmax" replaces the two-output case of logistic regression. **We are turning n numbers into probabilities.**

The outputs $w_1, \ldots, w_n$ are converted to probabilities $p_1, \ldots, p_n$ that add to 1 :

$$\textbf{Softmax} \qquad p_j = \frac{1}{S}\, e^{w_j} \quad \text{where} \quad S = \sum_{k=1}^{n} e^{w_k} \tag{9}$$

Certainly softmax assigns the largest probability p_j to the largest output w_j. But e^w is a nonlinear function of w. So the softmax assignment is not invariant to scale: If we double all the outputs w_j, softmax will produce different probabilities p_j. For small w's softmax actually deemphasizes the largest number $w_{\max}$.

In the CNN example of **teachyourmachine.com** to recognize digits, you will see how softmax produces the probabilities displayed in a pie chart—an excellent visual aid.

CNN We need a lot of weights to fit the data, and we are proud that we can compute them (with the help of gradient descent). But there is no justification for the number of weights to be uselessly large—*if weights can be reused.* For long signals in 1D and especially images in 2D, we may have no reason to change the weights from pixel to pixel.

1. cs231n.github.io/convolutional-networks/ (karpathy@cs.stanford.edu)

2. K. Simonyan and A. Zisserman, *Very deep convolutional networks for large-scale image recognition*, ICLR (2015), arXiv: 1409.1556v6, 10 Apr 2015.

3. A. Krizhevsky, I. Sutskever, and G. Hinton, *ImageNet classification with deep convolutional neural networks*, NIPS (2012) 1106–1114.

4. Y. LeCun and Y. Bengio, *Convolutional networks for images, speech, and time-series, Handbook of Brain Theory and Neural Networks*, MIT Press (1998).

Support Vector Machine in the Last Layer

For CNN's in computer vision, the final layer often has a special form. If the previous layers used ReLU and max-pooling (both piecewise linear), the last step can become a difference-of-convex program, and eventually a multiclass Support Vector Machine (SVM). Then optimization of the weights in a piecewise linear CNN can be one layer at a time.

L. Berrada, A. Zisserman, and P. Kumar, *Trusting SVM for piecewise linear CNNs*, arXiv : 1611.02185, 6 Mar 2017.

The World Championship at the Game of Go

A dramatic achievement by a deep convolutional network was to defeat the (human) world champion at Go. This is a difficult game played on a 19 by 19 board. In turn, two players put down "stones" in attempting to surround those of the opponent. When a group of one color has no open space beside it (left, right, up, or down), those stones are removed from the board. Wikipedia has an animated game.

AlphaGo defeated the leading player Lee Sedol by 4 games to 1 in 2016. It had trained on thousands of human games. This was a convincing victory, but not overwhelming. Then the neural network was deepened and improved. Google's new version AlphaGo Zero learned to play without any human intervention—simply by playing against itself. Now it defeated its former self AlphaGo by 100 to 0.

The key point about the new and better version is that **the machine learned by itself**. It was told the rules and nothing more. The first version had been fed earlier games, aiming to discover why winners had won and losers had lost. The outcome from the new approach was parallel to the machine translation of languages. To master a language, special cases from grammar seemed essential. How else to learn all those exceptions ? The translation team at Google was telling the system what it needed to know.

Meanwhile another small team was taking a different approach : Let the machine figure it out. In both cases, playing Go and translating languages, success came with a deeper neural net and more games and no coaching.

It is the depth and the architecture of AlphaGo Zero that interest us here. The hyperparameters will come in Section VII.4 : the fateful decisions. The parallel history of Google Translate must wait until VII.5 because Recurrent Neural Networks (**RNN's**) are needed—to capture the sequential structure of text.

It is interesting that the machine often makes opening moves that have seldom or never been chosen by humans. The input to the network is a board position and its history. The output vector gives the probability of selecting each move—and also a scalar that estimates the probability of winning from that position. Every step communicates with a Monte Carlo tree search, to produce *reinforcement learning*.

Residual Networks (ResNets)

Networks are becoming seriously deeper with more and more hidden layers. Mostly these are convolutional layers with a moderate number of independent weights. But depth brings dangers. Information can jam up and never reach the output. The problem of "vanishing gradients" can be serious: so many multiplications in propagating so far, with the result that computed gradients are exponentially small. When it is well designed, depth is a good thing—**but you must create paths for learning to move forward.**

The remarkable thing is that those fast paths can be very simple: "*skip connections*" that go directly to the next layer—bypassing the usual step $\boldsymbol{v}_n = (A_n\boldsymbol{v}_{n-1} + \boldsymbol{b}_n)_+$. An efficient proposal of Veit, Wilber, and Belongie is to allow either a *skip* or a normal convolution, with a ReLU step every time. If the net has L layers, there will be 2^L possible routes— fast or normal from each layer to the next.

One result is that entire layers can be removed without significant impact. The nth layer is reached by 2^{n-1} possible paths. Many paths have length well below n, not counting the skips.

It is hard to predict whether deep ConvNets will be replaced by ResNets.

K. He, X. Zhang, S. Ren, and J. Sun, *Deep residual learning for image recognition*, arXiv: 1512.03385, 10 Dec 2015. This paper works with extremely deep neural nets by adding shortcuts that skip layers, with weights $A = I$. Otherwise depth can degrade performance.

K. He, X. Zhang, S. Ren, and J. Sun, *Identity mappings in deep residual networks*, arXiv: 1603.05027, 25 Jul 2016.

A. Veit, M. Wilber, and S. Belongie, *Residual networks behave like ensembles of relatively shallow networks*, arXiv: 1605.06431, 27 Oct 2016.

A Simple CNN: Learning to Read Letters

One of the class projects at MIT was a convolutional net. The user begins by drawing multiple copies (not many) of A and B. On this training set, the correct classification is part of the input from the user. Then comes the mysterious step of *learning this data*—creating a continuous piecewise linear function $F(\boldsymbol{v})$ that gives high probability to the correct answer (the letter that was intended).

For learning to read digits, 10 probabilities appear in a pie chart. You quickly discover that too small a training set leads to frequent errors. If the examples had centered numbers or letters, and the test images are not centered, the user understands why those errors appear.

One purpose of **teachyourmachine.com** is education in machine learning at all levels (schools included). It is accessible to every reader.

These final references apply highly original ideas from signal processing to CNN's:

R. Balestriero and R. Baraniuk, *Mad Max: Affine spline insights into deep learning*, arXiv: 1805.06576.

S. Mallat, *Understanding deep convolutional networks*, Phil. Trans. Roy. Soc. **374** (2016); arXiv: 1601.04920.

C.-C. J. Kuo, *The CNN as a guided multilayer RECOS transform*, IEEE Signal Proc. Mag. **34** (2017) 81-89; arXiv: 1701.08481.

Problem Set VII.2

1 Wikipedia proposes a 5×5 matrix (different from equation (3)) to approximate a Gaussian. Compare the two filters acting on a horizontal edge (all 1's above all 0's) and a diagonal edge (lower triangle of 1's, upper triangle of 0's).

2 What matrix—corresponding to the Sobel matrices in equation (4)—would you use to find gradients in the 45° diagonal direction?

3 (Recommended) For image recognition, remember that the input sample $\boldsymbol{v}$ is a matrix (say 3 by 3). Pad it with zeros on all sides to be 5 by 5. Now apply a convolution as in the text (before equation (2)) to produce a 3 by 3 output $A\boldsymbol{v}$. What are the $1,1$ and $2,2$ entries of $A\boldsymbol{v}$?

4 Here are two matrix approximations L to the Laplacian $\partial^2 u/\partial x^2 + \partial^2 u/\partial y^2 = \nabla^2 u$:

$$\begin{bmatrix} 0 & 1 & 0 \\ 1 & -4 & 1 \\ 0 & 1 & 0 \end{bmatrix} \quad \text{and} \quad \begin{bmatrix} 1 & 4 & 1 \\ 4 & -20 & 4 \\ 1 & 4 & 1 \end{bmatrix}.$$

What are the responses $\boldsymbol{LV}$ and $\boldsymbol{LD}$ to a vertical or diagonal step edge?

$$V = \begin{bmatrix} 2 & 2 & 2 & 6 & 6 & 6 \\ 2 & 2 & 2 & 6 & 6 & 6 \\ 2 & 2 & 2 & 6 & 6 & 6 \\ 2 & 2 & 2 & 6 & 6 & 6 \end{bmatrix} \qquad D = \begin{bmatrix} 0 & 0 & 0 & 0 & 1 & 1 \\ 0 & 0 & 0 & 1 & 1 & 1 \\ 0 & 0 & 1 & 1 & 1 & 1 \\ 0 & 1 & 1 & 1 & 1 & 1 \end{bmatrix}$$

5 *Could a convolutional net learn calculus*? Start with the derivatives of fourth degree polynomials $p(x)$. The inputs could be graphs of $p = a_0 + a_1 x + \cdots + a_4 x^4$ for $0 \le x \le 1$ and a training set of a's. The correct outputs would be the coefficients $0, a_1, 2a_2, 3a_3, 4a_4$ from dp/dx. Using softmax with 5 classes, could you design and create a CNN to learn differential calculus?

6 Would it be easier or harder to learn integral calculus? With the same inputs, the six outputs would be $0, a_0, \frac{1}{2}a_1, \frac{1}{3}a_2, \frac{1}{4}a_3, \frac{1}{5}a_4$.

7 How difficult is addition of polynomials, with two graphs as inputs? The training set outputs would be the correct sums $a_0 + b_0, \ldots, a_4 + b_4$ of the coefficients. Is multiplication of polynomials difficult with 9 outputs $a_0 b_0, a_0 b_1 + a_1 b_0, \ldots, a_4 b_4$?

The inputs in 5-7 are pictures of the graphs. Cleve Moler reported on experiments:

https://blogs.mathworks.com/cleve/2018/08/06/teaching-calculus-to-a-deep-learner

Also .. **2018/10/22/teaching-a-newcomer-about-teaching-calculus-to-a-deep-learner**

A theory of deep learning for hidden physics is emerging: for example see arXiv: 1808.04327.

VII.3 Backpropagation and the Chain Rule

Deep learning is fundamentally a giant problem in optimization. We are choosing numerical "weights" to minimize a loss function L (which depends on those weights). $L(\boldsymbol{x})$ adds up all the losses $\ell(\boldsymbol{w} - \textbf{true}) = \ell(F(\boldsymbol{x}, \boldsymbol{v}) - \text{true})$ between the computed outputs $\boldsymbol{w} = F(\boldsymbol{x}, \boldsymbol{v})$ and the true classifications of the inputs $\boldsymbol{v}$. Calculus tells us the system of equations to solve for the weights that minimize L:

The partial derivatives of L with respect to the weights x should be zero.

Gradient descent in all its variations needs to *compute derivatives* (components of the gradient of F) at the current values of the weights. The derivatives $\partial F/\partial \boldsymbol{x}$ lead to $\partial L/\partial \boldsymbol{x}$. From that information we move to new weights that give a smaller loss. Then we recompute derivatives of F and L at the new values of the weights, and repeat.

Backpropagation is a method to compute derivatives quickly, using the chain rule:

Chain rule
$$\frac{dF}{dx} = \frac{d}{dx}(F_3(F_2(F_1(x)))) = \left(\frac{dF_3}{dF_2}(F_2(F_1(x)))\right)\left(\frac{dF_2}{dF_1}(F_1(x))\right)\left(\frac{dF_1}{dx}(x)\right)$$

One goal is a way to visualize how the function F is computed from the weights $\boldsymbol{x}_1, \boldsymbol{x}_2, \ldots, \boldsymbol{x}_N$. A neat way to do this is a **computational graph**. It separates the big computation into small steps, and we can find the derivative of each step (each computation) on the graph. Then the chain rule from calculus gives the derivatives of the final output $\boldsymbol{w} = F(\boldsymbol{x}, \boldsymbol{v})$ with respect to all the weights $\boldsymbol{x}$. For a standard net, the steps in the chain rule can correspond to layers in the neural net.

This is an incredibly efficient improvement on the separate computation of each derivative $\partial F/\partial x_i$. At first it seems unbelievable, that reorganizing the computations can make such an enormous difference. In the end (the doubter might say) you have to compute derivatives for each step and multiply by the chain rule. But the method does work—and N derivatives are computed in far less than N times the cost of one derivative $\partial F/\partial x_1$.

Backpropagation has been discovered many times. Another name is **automatic differentiation (AD)**. You will see that the steps can be arranged in two basic ways: **forward-mode** and **backward-mode**. The right choice of mode can make a large difference in the cost (a factor of thousands). That choice depends on whether you have many functions F depending on a few inputs, or few functions F depending on many inputs.

Deep learning has basically one loss function depending on many weights. The right choice is "backward-mode AD". This is what we call **backpropagation**. It is the computational heart of deep learning. We will illustrate computational graphs and backpropagation by a small example.

The computational graphs were inspired by the brilliant exposition of Christopher Olah, posted on his blog (colah.github.io). Since 2017 he has published on (https://distill.pub). And the new paper by Catherine and Desmond Higham (arXiv: 1801.05894, to appear in *SIAM Review*) gives special attention to backpropagation, with very useful codes.

Derivatives $\partial F/\partial \boldsymbol{x}$ of the Learning Function $F(\boldsymbol{x}, \boldsymbol{v})$

The weights $\boldsymbol{x}$ consist of all the matrices $A_1, \ldots, A_L$ and the bias vectors $\boldsymbol{b}_1, \ldots, \boldsymbol{b}_L$. The inputs $\boldsymbol{v} = \boldsymbol{v}_0$ are the training data. The outputs $\boldsymbol{w} = F(\boldsymbol{x}, \boldsymbol{v}_0)$ appear in layer L. Thus $\boldsymbol{w} = \boldsymbol{v}_L$ is the last step in the neural net, after $\boldsymbol{v}_1, \ldots, \boldsymbol{v}_{L-1}$ in the hidden layers.

Each new layer $\boldsymbol{v}_n$ comes from the previous layer by $\mathrm{R}\,(\boldsymbol{b}_n + A_n \boldsymbol{v}_{n-1})$. Here R is the nonlinear activation function (usually ReLU) applied one component at a time.

Thus deep learning carries us from $\boldsymbol{v} = \boldsymbol{v}_0$ to $\boldsymbol{w} = \boldsymbol{v}_L$. Then we substitute $\boldsymbol{w}$ into the loss function to measure the error for that sample $\boldsymbol{v}$. It may be a classification error: 0 instead of 1, or 1 instead of 0. It may be a least squares regression error $||\boldsymbol{g} - \boldsymbol{w}||^2$, with $\boldsymbol{w}$ instead of a desired output $\boldsymbol{g}$. Often it is a "cross-entropy". The total loss $L(\boldsymbol{x})$ is the sum of the losses on all input vectors $\boldsymbol{v}$.

The giant optimization of deep learning aims to find the weights $\boldsymbol{x}$ that minimize L. For full gradient descent the loss is $L(\boldsymbol{x})$. For stochastic gradient descent the loss at each iteration is $\ell(\boldsymbol{x})$—from a single input or a minibatch of inputs. **In all cases we need the derivatives $\partial \boldsymbol{w}/\partial \boldsymbol{x}$ of the outputs $\boldsymbol{w}$** (the components of the last layer) with respect to the weights $\boldsymbol{x}$ (the A's and $\boldsymbol{b}$'s that carry us from layer to layer).

This is one reason that deep learning is so expensive and takes so long—even on GPU's. For convolutional nets the derivatives were found quickly and easily in Section VII.2.

Computation of $\partial F/\partial \boldsymbol{x}$: Explicit Formulas

We plan to compute the derivatives $\partial \boldsymbol{F}/\partial \boldsymbol{x}$ in two ways. The first way is to present the explicit formulas: the derivative with respect to each and every weight. The second way is to describe the **backpropagation algorithm** that is constantly used in practice.

Start with the last bias vector $\boldsymbol{b}_L$ and weight matrix A_L that produce the final output $\boldsymbol{v}_L = \boldsymbol{w}$. There is no nonlinearity at this layer, and we drop the layer index L:

$$\boxed{\boldsymbol{v}_L = \boldsymbol{b}_L + A_L \boldsymbol{v}_{L-1} \quad \text{or simply} \quad \boldsymbol{w} = \boldsymbol{b} + A\boldsymbol{v}.} \tag{1}$$

Our goal is to find the derivatives $\partial w_i/\partial b_j$ and $\partial w_i/\partial A_{jk}$ for all components of $\boldsymbol{b} + A\boldsymbol{v}$. When j is different from i, *the ith output w_i is not affected by b_j or A_{jk}*. Multiplying A times $\boldsymbol{v}$, row j of A produces w_j and not w_i. We introduce the symbol $\boldsymbol{\delta}$ which is 1 or 0:

$$\boldsymbol{\delta_{ij} = 1} \text{ if } i = j \qquad \boldsymbol{\delta_{ij} = 0} \text{ if } i \neq j \qquad \text{The identity matrix } I \text{ has entries } \boldsymbol{\delta_{ij}}.$$

Columns of I are **1-hot vectors!** The derivatives are 1 or 0 or $\boldsymbol{v}_k$ (Section I.12):

Fully connected layer
Independent weights A_{jk}

$$\boxed{\frac{\partial w_i}{\partial b_j} = \delta_{ij} \quad \text{and} \quad \frac{\partial w_i}{\partial A_{jk}} = \delta_{ij} v_k} \tag{2}$$

Example There are six b's and a's in $\begin{bmatrix} w_1 \\ w_2 \end{bmatrix} = \begin{bmatrix} b_1 \\ b_2 \end{bmatrix} + \begin{bmatrix} a_{11}v_1 + a_{12}v_2 \\ a_{21}v_1 + a_{22}v_2 \end{bmatrix}$.

Derivatives of w_1 $\quad \dfrac{\partial w_1}{\partial b_1} = 1, \dfrac{\partial w_1}{\partial b_2} = 0, \dfrac{\partial w_1}{\partial a_{11}} = v_1, \dfrac{\partial w_1}{\partial a_{12}} = v_2, \dfrac{\partial w_1}{\partial a_{21}} = \dfrac{\partial w_1}{\partial a_{22}} = 0.$

Combining Weights b and A into M

It is often convenient to combine the bias vector b and the matrix A into one matrix M:

$$\textbf{Matrix of weights} \qquad M = \begin{bmatrix} 1 & 0^{\mathrm{T}} \\ b & A \end{bmatrix} \quad \text{has} \quad M\begin{bmatrix} 1 \\ v \end{bmatrix} = \begin{bmatrix} 1 \\ b + Av \end{bmatrix}. \qquad (3)$$

For each layer of the neural net, the top entry (*the zeroth entry*) is now **fixed at 1**. After multiplying that layer by M, *the zeroth component on the next layer is still* 1. Then $\mathrm{ReLU}(1) = 1$ preserves that entry in every hidden layer.

At the beginning of this book (page **v**), the big image of a neural net had squares for zeroth entries and circles for all other entries. *Every square now contains a* 1.

This block matrix M produces a compact derivative formula for the last layer $w = Mv$.

$$M = \begin{bmatrix} 1 & 0^{\mathrm{T}} \\ b & A \end{bmatrix} \quad \text{and} \quad \frac{\partial w_i}{\partial M_{jk}} = \delta_{ij} v_k \quad \text{for} \quad i > 0. \qquad (4)$$

Both v and w begin with 1's. Then $k = 0$ correctly gives $\partial w_0 / \partial M_{j0} = 0$ for $j > 0$.

Derivatives for Hidden Layers

Now suppose there is one hidden layer, so $L = 2$. The output is $w = v_L = v_2$, the hidden layer contains v_1, and the input is $v_0 = v$. The nonlinear R is probably ReLU.

$$v_1 = \mathrm{R}\,(b_1 + A_1 v_0) \quad \text{and} \quad w = b_2 + A_2 v_1 = b_2 + A_2 \mathrm{R}\,(b_1 + A_1 v_0).$$

Equation (2) still gives the derivatives of w with respect to the last weights b_2 and A_2. The function R is absent at the output and v is v_1. But the derivatives of w with respect to b_1 and A_1 do involve the nonlinear function R acting on $b_1 + A_1 v_0$.

So the derivatives in $\partial w / \partial A_1$ need the chain rule $\partial f / \partial x = (\partial f / \partial g)(\partial g / \partial x)$:

$$\textbf{Chain rule} \quad \frac{\partial w}{\partial A_1} = \frac{\partial [A_2 \mathrm{R}\,(b_1 + A_1 v_0)]}{\partial A_1} = A_2\,\mathrm{R}'(b_1 + A_1 v_0)\,\frac{\partial (b_1 + A_1 v_0)}{\partial A_1}. \qquad (5)$$

That chain rule has three factors. Starting from v_0 at layer $L - 2 = 0$, the weights b_1 and A_1 bring us toward the layer $L - 1 = 1$. The derivatives of that step are exactly like equation (2). But the output of that partial step is not v_{L-1}. To find that hidden layer we first have to apply R. So the chain rule includes its derivative R'. Then the final step (to w) multiplies by the last weight matrix A_2.

The Problem Set extends these formulas to L layers. They could be useful. But with pooling and batch normalization, automatic differentiation seems to defeat hard coding.

Very important Notice how formulas like (2) and (5) go **backwards from w to v**. Automatic backpropagation will do this too. "**Reverse mode**" starts with the output.

Details of the Derivatives $\partial w/\partial A_1$

We feel some responsibility to look more closely at equation (5). Its nonlinear part R' comes from the derivative of the nonlinear activation function. The usual choice is the ramp function ReLU $(x) = (x)_+$, and we see ReLU as the limiting case of an S-shaped sigmoid function. Here is ReLU together with its first two derivatives:

$$\text{ReLU}(x) = \max(0, x) = (x)_+ \qquad \textbf{Ramp function } R(x)$$

$$dR/dx = \begin{cases} 0 & x < 0 \\ 1 & x > 0 \end{cases} \qquad \begin{array}{l}\textbf{Step function} \\ H(x) = dR/dx\end{array}$$

$$d^2R/dx^2 = \begin{cases} 0 & x \neq 0 \\ 1 & \text{integral over all } x \end{cases} \qquad \begin{array}{l}\textbf{Delta function} \\ \delta(x) = d^2R/dx^2\end{array}$$

The delta function represents an *impulse*. It models a finite change in an infinitesimal time. It is physically impossible but mathematically convenient. It is defined, not at every point x, but by its effect on integrals from $-\infty$ to ∞ of a continuous function $g(x)$:

$$\int \delta(x)\, dx = 1 \qquad \int \delta(x)\, g(x)\, dx = g(0) \qquad \int \delta(x-a)\, g(x)\, dx = g(a)$$

With ReLU, a neuron could stay at zero through all the steps of deep learning. This "dying ReLU" can be avoided in several ways—it is generally not a major problem. One way that firmly avoids it is to change to a Leaky ReLU with a nonzero gradient:

$$\textbf{Leaky ReLU}\,(x) = \begin{cases} \boldsymbol{x} & x \geq 0 \\ \boldsymbol{.01x} & x \leq 0 \end{cases} \quad \text{Always ReLU}\,(ax) = a\text{ReLU}\,(x) \tag{6}$$

Geoffrey Hinton pointed out that if all the bias vectors $\boldsymbol{b}_1, \ldots, \boldsymbol{b}_L$ are set to zero at every layer of the net, the scale of the input $\boldsymbol{v}$ passes straight through to the output $\boldsymbol{w} = F(\boldsymbol{v})$. Thus $F(A\boldsymbol{v}) = aF(\boldsymbol{v})$. (A final softmax would lose this scale invariance.)

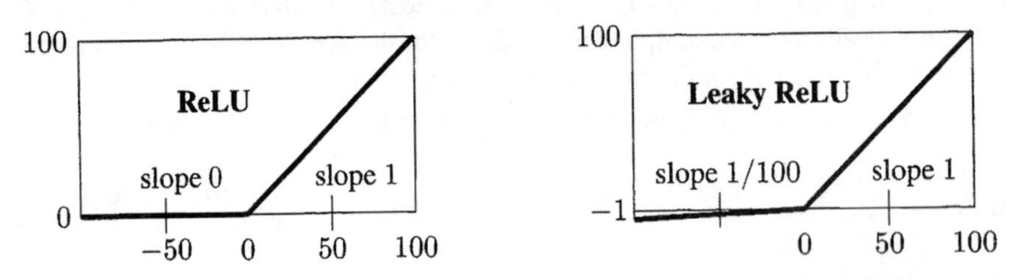

Figure VII.4: The graphs of ReLU and Leaky ReLU (two options for nonlinear activation).

Returning to formula (5), write A and $\boldsymbol{b}$ for the matrix A_{L-1} and the vector $\boldsymbol{b}_{L-1}$ that produce the last hidden layer. Then ReLU and A_L and $\boldsymbol{b}_L$ produce the final output $\boldsymbol{w} = \boldsymbol{v}_L$. Our interest is in $\partial w/\partial A$, the dependence of $\boldsymbol{w}$ on the next to last matrix of weights.

$$\boxed{\boldsymbol{w} = A_L(R(A\boldsymbol{v} + \boldsymbol{b})) + \boldsymbol{b}_L \quad \text{and} \quad \frac{\partial w}{\partial A} = A_L R'(A\boldsymbol{v} + \boldsymbol{b}) \frac{\partial(A\boldsymbol{v} + \boldsymbol{b})}{\partial A}} \tag{7}$$

We think of R as a diagonal matrix of ReLU functions acting component by component on $Av + b$. Then $J = R'(Av + b)$ is a diagonal matrix with 1's for positive components and 0's for negative components. (Don't ask about zeros.) Formula (7) has become (8):

$$w = A_L\, R\,(Av + b) \quad \text{and} \quad \frac{\partial w}{\partial A} = A_L\, J\, \frac{\partial (Av + b)}{\partial A} \tag{8}$$

We know every component (v_k or zero) of the third factor from the derivatives in (2).

When the sigmoid function R_a replaces the ReLU function, the diagonal matrix $J = R_a'(Av + b)$ no longer contains 1's and 0's. Now we evaluate the derivative dR_a/dx at each component of $Av + b$.

In practice, backpropagation finds the derivatives with respect to all A's and b's. It creates those derivatives automatically (and effectively).

Computational Graphs

Suppose $F(x, y)$ is a function of two variables x and y. Those inputs are the first two nodes in the computational graph. A typical step in the computation—**an edge in the graph**—is one of the operations of arithmetic (addition, subtraction, multiplication,...). The final output is the function $F(x, y)$. Our example will be $\boldsymbol{F = x^2(x + y)}$.

Here is the graph that computes F with intermediate nodes $c = x^2$ and $s = x + y$:

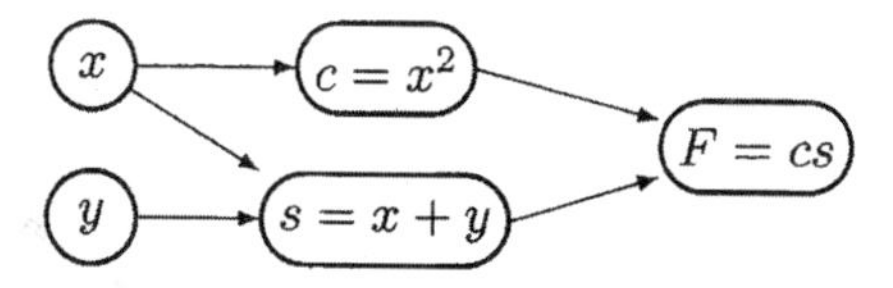

When we have inputs x and y, for example $\boldsymbol{x = 2}$ and $\boldsymbol{y = 3}$, the edges lead to $\boldsymbol{c = 4}$ and $\boldsymbol{s = 5}$ and $\boldsymbol{F = 20}$. This agrees with the algebra that we normally crowd into one line: $F = x^2(x + y) = 2^2(2 + 3) = 4(5) = 20$.

$x = 2$ → $c = x^2 = 4$ → $F = cs = 20$
$y = 3$ → $s = x + y = 5$ → $F = cs = 20$

Now we compute the derivative of each step—each edge in the graph. Begin with the x-derivative. At first we choose *forward-mode*, starting with the input x and moving toward the output function $x^2(x + y)$. So the first steps use the power rule for $c = x^2$ and the sum rule for $s = x + y$. The last step applies the product rule to $F = c$ times s.

$$\frac{\partial c}{\partial x} = 2x \qquad \frac{\partial s}{\partial x} = 1 \qquad \frac{\partial F}{\partial c} = s \qquad \frac{\partial F}{\partial s} = c$$

Moving through the graph produces the chain rule!

$$\frac{\partial \boldsymbol{F}}{\partial \boldsymbol{x}} = \frac{\partial F}{\partial c}\frac{\partial c}{\partial x} + \frac{\partial F}{\partial s}\frac{\partial s}{\partial x}$$
$$= (s)(2x) + (c)(1) = (5)(4) + (4)(1) = \mathbf{24}$$

The result is to compute the derivative of the output F with respect to *one input* x. You can see those x-derivatives on the computational graph.

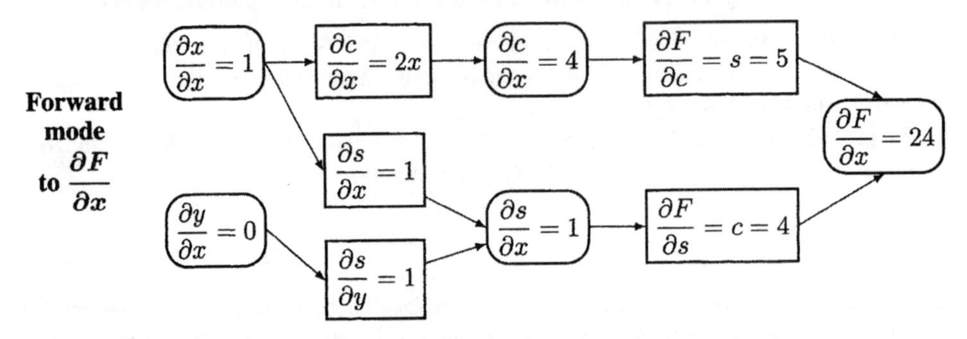

There will be a similar graph for y-derivatives—the forward mode leading to $\partial F/\partial y$. Here is the chain rule and the numbers that would appear in that graph for $x = 2$ and $y = 3$ and $c = x^2 = 2^2$ and $s = x + y = 2 + 3$ and $F = cs$:

$$\frac{\partial \boldsymbol{F}}{\partial \boldsymbol{y}} = \frac{\partial F}{\partial c}\frac{\partial c}{\partial y} + \frac{\partial F}{\partial s}\frac{\partial s}{\partial y}$$
$$= (s)(0) + (c)(1) = (5)(0) + (4)(1) = \mathbf{4}$$

The computational graph for $\partial F/\partial y$ is not drawn but the point is important: *Forward mode requires a new graph for each input x_i, to compute the partial derivative $\partial F/\partial x_i$.*

Reverse Mode Graph for One Output

The reverse mode starts with the output F. **It computes the derivatives with respect to both inputs**. The computations go **backward** through the graph.

That means it does not follow the empty line that started with $\partial y/\partial x = 0$ in the forward graph for x-derivatives. And it would not follow the empty line $\partial x/\partial y = 0$ in the forward graph (not drawn) for y-derivatives. A larger and more realistic problem with N inputs will have N forward graphs, each with $N - 1$ empty lines (because the N inputs are independent). The derivative of x_i with respect to every other input x_j is $\partial x_i/\partial x_j = 0$.

Instead of N forward graphs from N inputs, we will have **one backward graph from one output**. Here is that reverse-mode computational graph. It finds the derivative of F with respect to every node. It starts with $\partial F/\partial F = 1$ and goes in reverse.

A computational graph executes the chain rule to find derivatives. The reverse mode finds all derivatives $\partial F/\partial x_i$ by following all chains **backward from output to input.** Those chains all appear as paths **on one graph**—not as separate chain rules for exponentially many possible paths. This is the success of reverse mode.

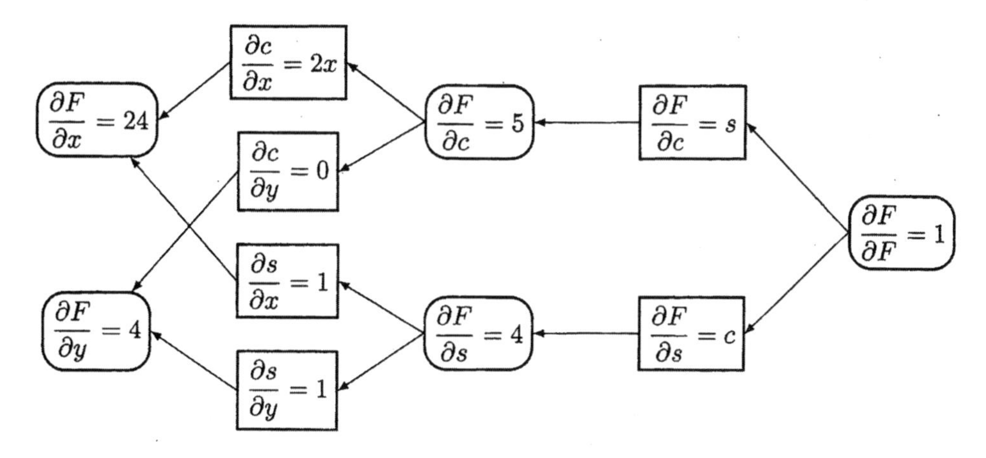

Figure VII.5: Reverse-mode computation of the gradient $\left(\frac{\partial F}{\partial x}, \frac{\partial F}{\partial y}\right)$ at $x = 2, y = 3$.

Product of Matrices ABC: Which Order?

The decision between forward and reverse order also appears in matrix multiplication! If we are asked to multiply A times B times C, the associative law offers two choices for the multiplication order:

AB first or BC first? **Compute $(AB)\,C$ or $A(BC)$?**

The result is the same but the number of individual multiplications can be very different. Suppose the matrix A is m by n, and B is n by p, and C is p by q.

First way

$$AB = (m \times n)\,(n \times p) \text{ has } \boldsymbol{mnp} \text{ multiplications}$$
$$(AB)C = (m \times p)\,(p \times q) \text{ has } \boldsymbol{mpq} \text{ multiplications}$$

Second way

$$BC = (n \times p)\,(p \times q) \text{ has } \boldsymbol{npq} \text{ multiplications}$$
$$A(BC) = (m \times n)\,(n \times q) \text{ has } \boldsymbol{mnq} \text{ multiplications}$$

So the comparison is between $mp(n+q)$ and $nq(m+p)$. Divide both numbers by $mnpq$:

> The first way is faster when $\dfrac{1}{q} + \dfrac{1}{n}$ is smaller than $\dfrac{1}{m} + \dfrac{1}{p}$.

Here is an extreme case (extremely important). *Suppose C is a column vector*: p by 1. Thus $q = 1$. Should you multiply BC to get another column vector (n by 1) and then $A(BC)$ to find the output (m by 1)? Or should you multiply AB first?

The question almost answers itself. The correct $A(BC)$ produces a vector at each step. The matrix-vector multiplication BC has np steps. The next matrix-vector multiplication $A(BC)$ has mn steps. Compare those $np + mn$ steps to the cost of starting with the matrix-matrix option AB (mnp steps!). Nobody in their right mind would do that.

But if A is a row vector, $(AB)C$ is better. Row times matrix each time.

This will match the central computation of deep learning: **Training the network = optimizing the weights.** The output $F(\boldsymbol{v})$ from a deep network is a chain starting with $\boldsymbol{v}$:

$$F(\boldsymbol{v}) = A_L\boldsymbol{v}_{L-1} = A_L(\mathrm{R}\,A_{L-1}(\ldots(\mathrm{R}\,A_2(\mathbf{R}\,A_1\boldsymbol{v})))) \quad \text{is forward through the net.}$$

The derivatives of F with respect to the matrices A (and the bias vectors $\boldsymbol{b}$) are easiest for the *last matrix* A_L in $A_L\boldsymbol{v}_{L-1}$. The derivative of $A\boldsymbol{v}$ with respect to A contains v's:

$$\frac{\partial F_i}{\partial A_{jk}} = \delta_{ij}\, v_k. \quad \text{Next is the derivative of } A_L \operatorname{ReLU}(A_{L-1}v_{L-1}) \text{ with respect to } A_{L-1}.$$

We can explain how that same reverse mode also appears in the comparison of direct methods *versus* adjoint methods for optimization (choosing good weights).

Adjoint Methods

The same question of the best order for matrix multiplication ABC comes up in a big class of optimization problems. We are solving a square system of N linear equations $E\boldsymbol{v} = \boldsymbol{b}$. The vector $\boldsymbol{b}$ depends on **design variables $\boldsymbol{p} = (p_1, \ldots, p_M)$**. Therefore the solution vector $\boldsymbol{v} = E^{-1}\boldsymbol{b}$ depends on $\mathbf{p}$. **The matrix $\partial \boldsymbol{v}/\partial \boldsymbol{p}$ containing the derivatives $\partial v_i/\partial p_j$ will be N by M.**

To repeat: We are minimizing $F(\boldsymbol{v})$. The vector $\boldsymbol{v}$ depends on the design variables $\boldsymbol{p}$. So we need a chain rule that multiplies derivatives $\partial F/\partial v_i$ times derivatives $\partial v_i/\partial p_j$. Let me show how this becomes a product of **three matrices**—and the multiplication order is decisive. **Three sets of derivatives** control how F depends on the input variables p_i:

$A = \boldsymbol{\partial F/\partial v_i}$	The derivatives of F with respect to $v_1, \ldots, v_N$
$B = \boldsymbol{\partial v_i/\partial b_k}$	The derivative of each v_i with respect to each b_k
$C = \boldsymbol{\partial b_k/\partial p_j}$	The derivative of each b_k with respect to each p_j

To see $\partial v_i/\partial p_j$ we take derivatives of the equation $E\boldsymbol{v} = \boldsymbol{b}$ with respect to the p_j:

$$E\,\frac{\partial \boldsymbol{v}}{\partial p_j} = \frac{\partial \boldsymbol{b}}{\partial p_j}, \quad j = 1, \ldots, M \quad \text{so} \quad \frac{\partial \boldsymbol{x}}{\partial \boldsymbol{p}} = E^{-1}\frac{\partial \boldsymbol{b}}{\partial \boldsymbol{p}}. \tag{9}$$

It seems that we have M linear systems of size N. Those will be expensive to solve over and over, as we search for the choice of $\boldsymbol{p}$ that minimizes $F(\boldsymbol{v})$. The matrix $\partial \boldsymbol{v}/\partial \boldsymbol{p}$ contains the derivatives of $v_1, \ldots, v_N$ with respect to the design variables $p_1, \ldots, p_M$.

Suppose for now that the cost function $F(\boldsymbol{v}) = \boldsymbol{c}^{\mathrm{T}}\boldsymbol{v}$ is linear (so $\partial F/\partial \boldsymbol{v} = \mathbf{c}^{\mathrm{T}}$). Then what optimization actually needs is the gradient of $F(\boldsymbol{v})$ with respect to the p's. The first set of derivatives $\partial F/\partial \boldsymbol{v}$ is just the vector $\boldsymbol{c}^{\mathrm{T}}$:

$$\boxed{\frac{\partial F}{\partial \boldsymbol{p}} = \frac{\partial F}{\partial \boldsymbol{v}}\frac{\partial \boldsymbol{v}}{\partial \boldsymbol{p}} = \mathbf{c}^{\mathrm{T}}\, E^{-1}\,\frac{\partial \boldsymbol{b}}{\partial \boldsymbol{p}} \quad \text{has three factors to be multiplied.}} \tag{10}$$

This is the key equation. It ends with a product of a **row vector c^{T}** times an **N by N matrix** E^{-1} times an **N by M matrix** $\partial b/\partial p$. How should we compute that product?

Again the question almost answers itself. We do *not* want to multiply two matrices. So we are *not* computing $\partial v/\partial p$ after all. Instead the good first step is to find $c^{\mathrm{T}}E^{-1}$. This produces a row vector λ^{T}. In other words we solve the adjoint equation $E^{\mathrm{T}}\lambda = c$:

$$\textbf{Adjoint equation} \qquad E^{\mathrm{T}}\lambda = c \quad \text{gives} \quad \lambda^{\mathrm{T}}E = c^{\mathrm{T}} \quad \text{and} \quad \lambda^{\mathrm{T}} = c^{\mathrm{T}}E^{-1}. \tag{11}$$

Substituting λ^{T} for $c^{\mathrm{T}}E^{-1}$ in equation (10), the final step multiplies that row vector times the derivatives of the vector b (its gradient):

$$\textbf{Gradient of the cost } F \qquad \frac{\partial F}{\partial p} = \lambda^{\mathrm{T}}\frac{\partial b}{\partial p} \qquad (1 \text{ by } N \text{ times } N \text{ by } M). \tag{12}$$

The optimal order is $(AB)C$ because the first factor A is actually the row vector λ^{T}.

This example of an adjoint method started with $Ex = b$. The right hand side b depended on design parameters p. So the solution $x = E^{-1}b$ depended on p. Then the cost function $F(x) = c^{\mathrm{T}}x$ depended on p.

The adjoint equation $A^{\mathrm{T}}\lambda = c$ found the vector λ that efficiently combined the last two steps. "Adjoint" has a parallel meaning to "transpose" and we can apply it also to differential equations. The design variables $p_1, \ldots, p_M$ might appear in the matrix E, or in an eigenvalue problem or a differential equation.

Our point here is to emphasize and reinforce the key idea of backpropagation:

The reverse mode can order the derivative computations in a faster way.

Adjoints and Sensitivity for Deep Layers

Coming closer to the problem of deep learning, what are the derivatives $\partial w/\partial x_j$ of the outputs $w = (w_1, \ldots w_M)$ at layer L with respect to the parameters $x = (x_1, \ldots, x_N)$? That output $w = v_L$ is seen after L steps from the input v_0. We write step n as

$$v_n = F_n(v_{n-1}, x_n) \quad \text{where } F_n \text{ depends on the weights (parameters) } x_n. \tag{13}$$

(13) is a **recurrence relation**. And the same P parameters x could be used at every step. Deep learning has new parameters for each new layer—which gives it "learning power" that an ordinary recurrence relation cannot hope for. In fact a typical recurrence (13) is just a finite difference analog of a differential equation $dv/dt = f(v, x, t)$.

The analogy is not bad. In this case too we may be aiming for a desired output $v(T)$, and we are choosing parameters x to bring us close. The problem is to find the matrix of derivatives $J = \partial v_N/\partial x_M$. We have to apply the chain rule to equation (13), all the way back from N to 0. Here is a step of the chain:

$$v_N = F_N(v_{N-1}, x_N) = F_N(F_{N-1}(v_{N-2}, x_{N-1}), x_N). \tag{14}$$

Take its derivatives with respect to x_{N-1}, to see the rule over the last two layers:

$$\frac{\partial v_N}{\partial x_{N-1}} = \frac{\partial F_N}{\partial v_{N-1}} \frac{\partial v_{N-1}}{\partial x_{N-1}} \qquad \frac{\partial v_N}{\partial x_{N-2}} = \frac{\partial F_N}{\partial v_{N-1}} \frac{\partial v_{N-1}}{\partial x_{N-2}} = \frac{\partial F_N}{\partial v_{N-1}} \frac{\partial v_{N-1}}{\partial v_{N-2}} \frac{\partial v_{N-2}}{\partial x_{N-2}}$$

That last expression is a triple product ABC. The calculation requires a decision: *Start with AB or start with BC?* Both the adjoint method for optimization and the reverse mode of backpropagation would counsel: *Begin with AB.*

The last two pages developed from class notes by Steven Johnson: *Adjoint methods and sensitivity analysis for recurrence relations*, **http://math.mit.edu/~stevenj/18.336/recurrence2.pdf**. Also online: *Notes on adjoint methods for* 18.335.

For deep learning, the recurrence relation is between layers of the net.

Problem Set VII.3

1 If $\boldsymbol{x}$ and $\boldsymbol{y}$ are column vectors in $\mathbf{R}^n$, is it faster to multiply $\boldsymbol{x}(\boldsymbol{y}^{\mathrm{T}}\boldsymbol{x})$ or $(\boldsymbol{x}\boldsymbol{y}^{\mathrm{T}})\boldsymbol{x}$?

2 If A is an m by n matrix with $m > n$, is it faster to multiply $A(A^{\mathrm{T}}A)$ or $(AA^{\mathrm{T}})A$?

3 (a) If $A\boldsymbol{x} = \boldsymbol{b}$, what are the derivatives $\partial x_i/\partial b_j$ with A fixed?

(b) What are the derivatives of $\partial x_i/\partial A_{jk}$ with $\boldsymbol{b}$ fixed?

4 For $\boldsymbol{x}$ and $\boldsymbol{y}$ in $\mathbf{R}^n$, what are $\partial(\boldsymbol{x}^{\mathrm{T}}\boldsymbol{y})/\partial x_i$ and $\partial(\boldsymbol{x}\boldsymbol{y}^{\mathrm{T}})/\partial x_i$?

5 Draw a computational graph to compute the function $f(x,y) = x^3(x-y)$. Use the graph to compute $f(2,3)$.

6 Draw a reverse mode graph to compute the derivatives $\partial f/\partial x$ and $\partial f/\partial y$ for $f = x^3(x-y)$. Use the graph to find those derivatives at $x = 2$ and $x = 3$.

7 Suppose A is a Toeplitz matrix in a convolutional neural net (**CNN**). The number a_k is on diagonal $k = 1-n, \ldots, n-1$. If $\boldsymbol{w} = A\boldsymbol{v}$, what is the derivative $\partial w_i/\partial a_k$?

8 In a max-pooling layer, suppose $w_i = \max(v_{2i-1}, v_{2i})$. Find all $\partial w_i/\partial v_j$.

9 To understand the chain rule, start from this identity and let $\Delta x \to 0$:

$$\frac{f(g(x+\Delta x)) - f(g(x))}{\Delta x} = \frac{f(g(x+\Delta x)) - f(g(x))}{g(x+\Delta x) - g(x)} \; \frac{g(x+\Delta x) - g(x)}{\Delta x}$$

Then the derivative at x of $f(g(x))$ equals df/dg at $g(x)$ times dg/dx at x.

Question: Find the derivative at $x = 0$ of $\sin(\cos(\sin x))$.

Backpropagation is essentially equivalent to **AD** (automatic differentiation) in reverse mode:

A. Griewank and A. Walther, *Evaluating Derivatives*, SIAM (2008).

VII.4 Hyperparameters : The Fateful Decisions

After the loss function is chosen and the network architecture is decided, there are still critical decisions to be made. We must choose the *hyperparameters*. They govern the algorithm itself—the computation of the weights. Those weights represent what the computer has learned from the training set : how to predict the output from the features in the input. In machine learning, the decisions include those hyperparameters and the loss function and dropout and regularization.

The goal is to find patterns that distinguish 5 from 7 and 2—by looking at pixels. The hyperparameters decide how quickly and accurately those patterns are discovered. The **stepsize** s_k in gradient descent is first and foremost. That number appears in the iteration $\boldsymbol{x}_{k+1} = \boldsymbol{x}_k - s_k \, \boldsymbol{\nabla} \boldsymbol{L}\,(\boldsymbol{x}_k)$ or one of its variants : *accelerated* (by momentum) or *adaptive* (ADAM) or *stochastic* with a random minibatch of training data at each step k.

The words **learning rate** are often used in place of stepsize. Depending on the author, the two might be identical or differ by a normalizing factor. Also : η_k *often replaces* s_k. First we ask for the optimal stepsize when there is only one unknown. Then we point to a general approach. Eventually we want a faster decision.

1. **Choose** $s_k = 1/L''(x_k)$. Newton uses the second derivative of L. That choice accounts for the quadratic term in the Taylor series for $L(x)$ around the point x_k. As a result, Newton's method is *second order*: The error in x_{k+1} is proportional to the *square* of the error in x_k. Near the minimizing x^*, convergence is fast.

   In more dimensions, the second derivative becomes the Hessian matrix $H(\boldsymbol{x}) = \boldsymbol{\nabla}^2 \boldsymbol{L}(\boldsymbol{x}_k)$. Its size is the number of weights (components of $\boldsymbol{x}$). To find $\boldsymbol{x}_{k+1}$, Newton solves a large system of equations $H(\boldsymbol{x}_k)\,(\boldsymbol{x}_{k+1} - \boldsymbol{x}_k) = -\boldsymbol{\nabla} \boldsymbol{L}(\boldsymbol{x}_k)$. Gradient descent replaces H by a single number $1/s_k$.

2. **Decide** s_k **from a line search**. The gradient $\boldsymbol{\nabla} \boldsymbol{L}(\boldsymbol{x}_k)$ sets the direction of the line. The current point $\boldsymbol{x}_k$ is the start of the line. By evaluating $L(\boldsymbol{x})$ at points on the line, we find a nearly minimizing point—which becomes $\boldsymbol{x}_{k+1}$.

   Line search is a practical idea. One algorithm is *backtracking*, as described in Section VI.4. This reduces the stepsize s by a constant factor until the decrease in L is consistent with the steepness of the gradient (again within a chosen factor). Optimizing a line search is a carefully studied 1-dimensional problem.

   But no method is perfect. We look next at the effect of a poor stepsize s.

Too Small or Too Large

We need to identify the difficulties with a poor choice of learning rate :

s_k **is too small**	Then gradient descent takes too long to minimize $L(\boldsymbol{x})$ Many steps $\boldsymbol{x}_{k+1} - \boldsymbol{x}_k = -s_k \, \boldsymbol{\nabla} \boldsymbol{L}\,(\boldsymbol{x}_k)$ with small improvement
s_k **is too large**	We are overshooting the best choice $\boldsymbol{x}_{k+1}$ in the descent direction Gradient descent will jump around the minimizing $\boldsymbol{x}^*$.

Suppose the first steps s_0 and s_1 are found by line searches, and work well. We may want to stay with that learning rate for the early iterations. **Normally we reduce** s as the minimization of $L(\boldsymbol{x})$ continues.

Larger steps at the start	Get somewhere close to the optimal weights $\boldsymbol{x}^*$
Smaller steps at the end	Aim for convergence without overshoot

A learning rate schedule $s_k = s_0/\sqrt{k}$ or $s_k = s_0/k$ systematically reduces the steps.

After reaching weights $\boldsymbol{x}$ that are close to minimizing the loss function $L(\boldsymbol{x}, \boldsymbol{v})$ we may want to bring new $\boldsymbol{v}$'s from a **validation set**. This is not yet production mode. The purpose of **cross-validation** is to confirm that the computed weights $\boldsymbol{x}$ are capable of producing accurate outputs from new data.

Cross-validation

Cross-validation aims to estimate the validity of our model and the strength of our learning function. Is the model too weak or too simple to give accurate predictions and classifications? Are we overfitting the training data and therefore at risk with new test data? You could say that cross-validation works more carefully with a relatively small data set, so that testing and production can go forward quickly on a larger data set.

Note Another statistical method—for another purpose—also reuses the data. This is the *bootstrap* introduced by Brad Efron. It is used (and needed) when the sample size is small or its distribution is not known. We aim for maximum understanding by returning to the (small) sample and *reusing that data* to extract new information. Normally small data sets are not the context for applications to deep learning.

A first step in cross-validation is to divide the available data into K subsets. If $K = 2$, these would essentially be the training set and test set—but we are usually aiming for more information from smaller sets before working with a big test set. *K-fold cross-validation* uses each of K subsets separately as a test set. In every trial, the other $K - 1$ subsets form the training set. We are reworking the same data (moderate size) to learn more than one optimization can teach us.

Cross-validation can make a learning rate adaptive: changing as descent proceeds.

There are many variants, like "double cross-validation". In a standardized m by n least squares problem $A\boldsymbol{x} = \boldsymbol{b}$, Wikipedia gives the expected value $(m - n - 1)/(m + n - 1)$ for the mean square error. Higher errors normally indicate overfitting. The corresponding test in deep learning warns us to consider earlier stopping.

This section on hyperparameters was influenced and improved by Bengio's long chapter in a remarkable book. The book title is *Neural Networks : Tricks of the Trade* (2nd edition), edited by G. Montavon, G. Orr, and K.-R. Müller. It is published by Springer (2012) with substantial contributions from leaders in the field.

Batch Normalization of Each Layer

As training goes forward, the mean and variance of the original population can change at every layer of the network. This change in the distribution of inputs is "covariate shift". We often have to adjust the stepsize and other hyperparameters, due to this shift in the statistics of layers. A good plan is to *normalize the input to each layer.*

Normalization makes the training safer and faster. The need for dropout often disappears. Fewer iterations can now give more accurate weights. And the cost can be very moderate. *Often we just train two additional parameters on each layer.*

The problem is greatest when the nonlinear function is a sigmoid rather than ReLU. The sigmoid "saturates" by approaching a limit like 1 (while ReLU increases forever as $x \to \infty$). The nonlinear sigmoid becomes virtually linear and even constant when x becomes large. Training slows down because the nonlinearity is barely used.

It remains to decide the point at which inputs will be normalized. Ioffe and Szegedy avoid computing covariance matrices (far too expensive). Their normalizing transform acts on each input $\boldsymbol{v}_1, \ldots, \boldsymbol{v}_B$ in a minibatch of size B :

mean	$\boldsymbol{\mu} = (\boldsymbol{v}_1 + \cdots + \boldsymbol{v}_B) / B$
variance	$\sigma^2 = \left(\|\boldsymbol{v}_1 - \boldsymbol{\mu}\|^2 + \cdots + \|\boldsymbol{v}_B - \boldsymbol{\mu}\|^2\right) / B$
normalize	$\boldsymbol{V}_i = (\boldsymbol{v}_i - \boldsymbol{\mu}) / \sqrt{\sigma^2 + \epsilon}$ for small $\epsilon > 0$
scale/shift	$\boldsymbol{y}_i = \boldsymbol{\gamma} \boldsymbol{V}_i + \boldsymbol{\beta}$ ($\boldsymbol{\gamma}$ and $\boldsymbol{\beta}$ are trainable parameters)

The key point is to **normalize the inputs $\boldsymbol{y}_i$ to each new layer**. What was good for the original batch of vectors (at layer zero) is also good for the inputs to each hidden layer.

S. Ioffe and C. Szegedy, *Batch normalization*, arXiv : 1502.03167v3, 2 Mar 2015.

Dropout

Dropout is the removal of randomly selected neurons in the network. Those are components of the input layer $\boldsymbol{v}_0$ or of hidden layers $\boldsymbol{v}_n$ before the output layer $\boldsymbol{v}_L$. All weights in the A's and $\boldsymbol{b}$'s connected to those dropped neurons disappear from the net (Figure VII.6). Typically hidden layer neurons might be given probability $p = 0.5$ of surviving, and input components might have $p = 0.8$ or higher. *The main objective of random dropout is to avoid overfitting.* It is a relatively inexpensive averaging method compared to combining predictions from many networks.

Dropout was proposed by five leaders in the development of deep learning algorithms: N. Srivastava, G. Hinton, A. Krizhevsky, I. Sutskever, and R. Salakhutdinov. Their paper "*Dropout*" appears in: *Journal of Machine Learning Research* **15** (2014) 1929-1958. For recent connections of dropout to physics and uncertainty see arXiv : 1506.02142 and 1809.08327.

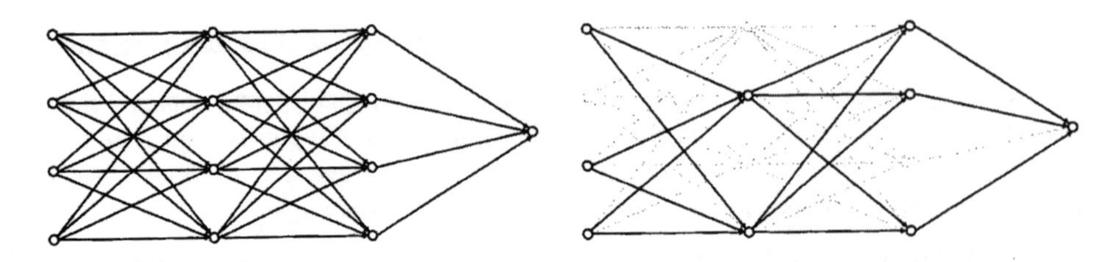

Figure VII.6: Crossed neurons have dropped out in the thinned network.

Dropout offers a way to compute with many different neural architectures at once. In training, each new $\boldsymbol{v}_0$ (the feature vector of an input sample) leads to a new thinned network. Starting with N neurons there are 2^N possible thinned networks.

At test time, we use the full network (no dropout) with weights rescaled from the training weights. The outgoing weights from an undropped neuron are multiplied by p in the rescaling. This approximate averaging at test time led the five authors to reduced generalization errors—more simply than from other regularization methods.

One inspiration for dropout was genetic reproduction—where half of each parent's genes are dropped and there is a small random mutation. That dropout for a child seems more unforgiving and permanent than dropout for deep learning—which averages over many thinned networks. (True, we see some averaging over siblings. But the authors conjecture that over time, our genes are forced to be robust in order to survive.)

The dropout model uses a zero-one random variable r (a Bernoulli variable). Then $r = 1$ with probability p and $r = 0$ with probability $1 - p$. The usual feed-forward step to layer n is $\boldsymbol{y}_n = A_n \boldsymbol{v}_{n-1} + \boldsymbol{b}_n$, followed by the nonlinear $\boldsymbol{v}_n = \mathrm{R}\boldsymbol{y}_n$. *Now a random r multiplies each component of $\boldsymbol{v}_{n-1}$* to drop that neuron when $r = 0$. Component by component, $\boldsymbol{v}_{n-1}$ is multiplied by 0 or 1 to give $\boldsymbol{v}^*_{n-1}$. Then $\boldsymbol{y}_n = A_n \boldsymbol{v}^*_{n-1} + \boldsymbol{b}_n$.

To compute gradients, use backpropagation for each training example in the minibatch. Then average those gradients. Stochastic gradient descent can still include acceleration (momentum added) and adaptive descent and weight decay. The authors highly recommend regularizing the weights, for example by a maximum norm requirement $||\boldsymbol{a}|| \leq c$ on the columns of all weight matrices A.

Exploring Hyperparameter Space

Often we optimize hyperparameters using experiments or experience. To decide the learning rate, we may try three possibilities and measure the drop in the loss function. A geometric sequence like $.1, .01, .001$ would make more sense than an arithmetic sequence $.05, .03, .01$. And if the smallest or largest choice gives the best results, then continue the experiment to the next number in the series. In this stepsize example, you would be considering computational cost as well as validation error.

LeCun emphasizes that for a multiparameter search, *random sampling* is the way to cover many possibilities quickly. Grid search is too slow in multiple dimensions.

Loss Functions

The loss function measures the difference between the correct and the computed output for each sample. The correct output—often a classification $y = 0, 1$ or $y = 1, 2, \ldots, n$—is part of the training data. The computed output at the final layer is $\boldsymbol{w} = F(\boldsymbol{x}, \boldsymbol{v})$ from the learning function with weights $\boldsymbol{x}$ and input $\boldsymbol{v}$.

Section VI.5 defined three familiar loss functions. Then this chapter turned to the structure of the neural net and the function F. Here we come back to compare square loss with cross-entropy loss.

1. Quadratic cost (square loss): $\boldsymbol{\ell(y, w) = \frac{1}{2}||y - w||^2}$.

This is the loss function for least squares—always a possible choice. But it is not a favorite choice for deep learning. One reason is the parabolic shape for the graph of $\ell(y, w)$, as we approach zero loss at $w = y$. The derivative also approaches zero.

A zero derivative at the minimum is normal for a smooth loss function, but it frequently leads to an unwanted result: *The weights A and $\boldsymbol{b}$ change very slowly near the optimum.* Learning slows down and many iterations are needed.

$$\textbf{2. Cross-entropy loss} : \boldsymbol{\ell(y, w)} = -\frac{1}{n}\sum_{1}^{n}\left[y_i \log z_i + (1 - y_i)\log(1 - z_i)\right] \qquad (1)$$

Here we allow and expect that the N outputs w_i from training the neural net have been normalized to $z(w)$, with $0 < z_i < 1$. Often those z_i are probabilities. Then $1 - z_i$ is also between 0 and 1. So both logarithms in (1) are negative, and the minus sign assures that the overall loss is positive : $\ell > 0$.

More than that, the logarithms give a different and desirable approach to $z = 0$ or 1. For this calculation we refer to Nielsen's online book *Neural Networks and Deep Learning*, which focuses on sigmoid activation functions instead of ReLU. The price of those smooth functions is that they *saturate* (lose their nonlinearity) near their endpoints.

Cross-entropy has good properties, but where do the logarithms come from? The first point is Shannon's formula for entropy (a measure of information). If message i has probability p_i, you should allow $-\log p_i$ bits for that message. Then the expected (average) number of bits per message is best possible :

$$\textbf{Entropy} = -\sum_{1}^{m} p_i \log p_i. \ \text{ For } m = 2 \text{ this is } -p\log p - (1-p)\log(1-p). \qquad (2)$$

Cross-entropy comes in when we don't know the p_i and we use $\widehat{p}_i$ instead :

$$\textbf{Cross-entropy} = -\sum_{1}^{m} p_i \log \widehat{p}_i. \ \text{ For } m = 2 \text{ this is } -p\log \widehat{p} - (1-p)\log(1-\widehat{p}). \qquad (3)$$

(3) is always larger than (2). The true p_i are not known and the $\widehat{p}_i$ cost more. The difference is a very useful but not symmetric function called **Kullback-Leibler (KL) divergence**.

Regularization : ℓ^2 or ℓ^1 (or none)

Regularization is a voluntary but well-advised decision. It adds a *penalty term* to the loss function $L(\boldsymbol{x})$ that we minimize : an $\boldsymbol{\ell}^2$ penalty in ridge regression and $\boldsymbol{\ell}^1$ in LASSO.

RR Minimize $||\boldsymbol{b}-A\boldsymbol{x}||_2^2+\lambda_2\,||\boldsymbol{x}||_2^2$ **LASSO** Minimize $||\boldsymbol{b}-A\boldsymbol{x}||_2^2+\lambda_1\sum|x_i|$

The penalty controls the size of $\boldsymbol{x}$. *Regularization is also called weight decay.*

The coefficient λ_2 or λ_1 is a hyperparameter. Its value can be based on cross-validation. The purpose of the penalty terms is to avoid overfitting (sometimes expressed as *fitting the noise*). Cross-validation for a given λ finds the minimizing $\boldsymbol{x}$ on a test set. Then it checks by using those weights on a training set. If it sees errors from overfitting, λ is increased.

A small value of λ tends to increase the variance of the error : overfitting. Large λ will increase the bias : underfitting because the fitting term $||\boldsymbol{b} - A\boldsymbol{x}||^2$ is less important.

A different viewpoint ! Recent experiments on MNIST make it unclear if explicit regularization is always necessary. The best test performance is often seen with $\lambda = 0$ (then $\boldsymbol{x}^*$ is the minimum norm solution $A^+\boldsymbol{b}$). The analysis by Liang and Rakhlin identifies matrices for which this good result can be expected—provided the data leads to fast decay of the spectrum of the sample covariance matrix and the kernel matrix.

In many cases these are the matrices of Section III.3 : *Effectively low rank.* Similar ideas are increasingly heard, that deep learning with many extra weights and good hyperparameters will find solutions that generalize, without penalty.

T. Liang and A. Rakhlin, *Just interpolate : Kernel "ridgeless" regression can generalize,* arXiv : 1808.00387, 1 Aug 2018.

The Structure of AlphaGo Zero

It is interesting to see the sequence of operations in AlphaGo Zero, learning to play Go :

1. A convolution of 256 filters of kernel size 3×3 with stride 1 : $E = 3, S = 1$
2. Batch normalization
3. ReLU
4. A convolution of 256 filters of kernel size 3×3 with stride 1
5. Batch normalization
6. A skip connection as in ResNets that adds the input to the block
7. ReLU
8. A fully connected linear layer to a hidden layer of size 256
9. ReLU

Training was by stochastic gradient descent on a fixed data set that contained the final 2 million games of self-played data from a previous run of AlphaGo Zero.

The CNN includes a fully connected layer that outputs a vector of size $19^2 + 1$. This accounts for all positions on the 19×19 board, plus a pass move allowed in Go.

VII.5 The World of Machine Learning

Fully connected nets and convolutional nets are parts of a larger world. From training data they lead to a learning function $\boldsymbol{F}(\boldsymbol{x}, \boldsymbol{v})$. That function produces a close approximation to the correct output $\boldsymbol{w}$ for each input $\boldsymbol{v}$ ($\boldsymbol{v}$ is the vector of features of that sample). But machine learning has developed a multitude of other approaches—some long established—to the problem of learning from data.

This book cannot do justice to all those ideas. It does seem useful to describe Recurrent Neural Nets (and Support Vector Machines). We also include key words to indicate the scope of machine learning. (A *glossary* is badly needed! That would be a tremendous contribution to this field.) At the end is a list of books on topics in machine learning.

Recurrent Neural Networks (RNNs)

These networks are appropriate for data that comes in a definite order. This includes time series and natural language: *speech or text or handwriting*. In the network of connections from inputs $\boldsymbol{v}$ to outputs $\boldsymbol{w}$, the new feature is the *input from the previous time* $t-1$. This recurring input is determined by the function $h(t-1)$.

Figure VII.7 shows an outline of that new step in the architecture of the network.

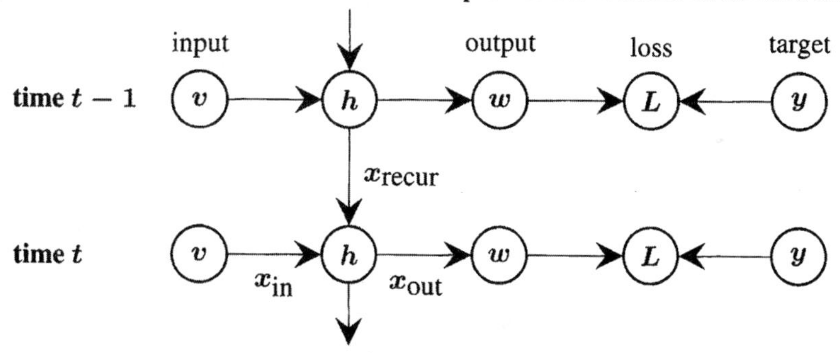

Figure VII.7: The computational graph for a recurrent network finds loss-minimizing outputs $\boldsymbol{w}$ at each time t. The inputs to $\boldsymbol{h}(t)$ are the new data $\boldsymbol{v}(t)$ and the recurrent data $\boldsymbol{h}(t-1)$ from the previous time. The weights multiplying the data are $\boldsymbol{x}_{\text{in}}$ and $\boldsymbol{x}_{\text{recur}}$ and $\boldsymbol{x}_{\text{out}}$, chosen to minimize the loss $L(\boldsymbol{y}-\boldsymbol{w})$. This network architecture is *universal*: It will compute any formula that is computable by a Turing machine.

Key Words and Ideas

1 Kernel learning (next page)

2 Support Vector Machines (next page)

3 Generative Adversarial Networks

4 Independent Component Analysis

5 Graphical models

6 Bayesian statistics

7 Random forests

8 Reinforcement learning

Support Vector Machines

Start with n points $\boldsymbol{v}_1, \ldots, \boldsymbol{v}_n$ in m-dimensional space. Each $\boldsymbol{v}_i$ comes with a *classification* $y_i = 1$ *or* $y_i = -1$. The goal proposed by Vapnik is to find a plane $\boldsymbol{w}^{\mathrm{T}}\boldsymbol{v} = b$ in m dimensions that *separates the plus points from the minus points*—if this is possible. That vector $\boldsymbol{w}$ will be perpendicular to the plane. The number b tells us the distance $|b|/||\boldsymbol{w}||$ from the line or plane or hyperplane in $\mathbf{R}^m$ to the point $(0, \ldots, 0)$.

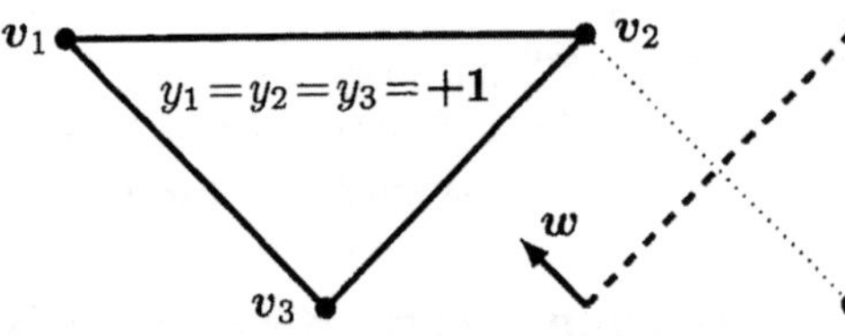

This separating line $\boldsymbol{w}^{\mathrm{T}}\boldsymbol{v} = b$ maximizes the equal distance (*the margin*) to $+$ and $-$ points. If $\boldsymbol{v}_4$ is inside the triangle, separation will be impossible

$\boldsymbol{v}_4$ has $y_4 = -1$

Problem Find $\boldsymbol{w}$ and b so that $\boldsymbol{w}^{\mathrm{T}}\boldsymbol{v}_i - b$ has the correct sign y_i for all points $i = 1, \ldots, n$.

If $\boldsymbol{v}_1, \boldsymbol{v}_2, \boldsymbol{v}_3$ are plus points $(y = +1)$ in a plane, then $\boldsymbol{v}_4$ *must be outside the triangle of* $\boldsymbol{v}_1, \boldsymbol{v}_2, \boldsymbol{v}_3$. The picture shows the line of maximum separation (*maximum margin*).

Maximum margin Minimize $\|\|\boldsymbol{w}\|\|$ under the conditions $y_i(\boldsymbol{w}^{\mathrm{T}}\boldsymbol{v}_i - b) \geq 1$.

This is a "hard margin". That inequality requires $\boldsymbol{v}_i$ to be on its correct side of the separator. If the points can't be separated, then no $\boldsymbol{w}$ and b will succeed. For a "soft margin" we go ahead to choose the best available $\boldsymbol{w}$ and b, based on hinge loss + penalty:

$$\textbf{Soft margin} \quad \text{Minimize} \left[\frac{1}{n} \sum_{1}^{n} \max\left(0, 1 - y_i(\boldsymbol{w}^{\mathrm{T}}\boldsymbol{v}_i - b)\right) \right] + \lambda\, ||\boldsymbol{w}||^2. \qquad (1)$$

That hinge loss (the maximum term) is zero when $\boldsymbol{v}_i$ is on the correct side of the separator. If separation is impossible, the penalty $\lambda||\boldsymbol{w}||^2$ balances hinge losses with margin sizes.

If we introduce a variable h_i for that hinge loss we are minimizing a quadratic function of $\boldsymbol{w}$ with linear inequalities connecting $\boldsymbol{w}, b, y_i$ and h_i. This is quadratic programming in high dimensions—well understood in theory but challenging in practice.

The Kernel Trick

SVM is linear separation. A plane separates $+$ points from $-$ points. The kernel trick allows a nonlinear separator, when feature vectors $\boldsymbol{v}$ are transformed to $N(\boldsymbol{v})$. Then the dot product of transformed vectors gives us the **kernel function** $K(\boldsymbol{v}_i, \boldsymbol{v}_j) = N(\boldsymbol{v}_i)^{\mathrm{T}} N(\boldsymbol{v}_j)$.

The key is to work entirely with K and not at all with the function N. *In fact we never see or need* N. In the linear case, this corresponds to choosing a positive definite K and not seeing the matrix A in $K = A^{\mathrm{T}}A$. The RBF kernel $\exp(-||\boldsymbol{v}_i - \boldsymbol{v}_j||^2/2\sigma^2)$ is in III.3.

M. Belkin, S. Ma, and S. Mandal, *To understand deep learning we need to understand kernel learning*, arXiv:1802.01396. "Non-smooth Laplacian kernels defeat smooth Gaussians"

T. Hofmann, B. Schölkopf, and A. J. Smola, *Kernel methods in machine learning*, Annals of Statistics **36** (2008) 1171-1220 (with extensive references).

Google Translate

An exceptional article about deep learning and the development of Google Translate appeared in the New York Times Magazine on Sunday, 14 December 2016. It tells how Google suddenly jumped from a conventional translation to a recurrent neural network. The author Gideon Lewis-Kraus describes that event as three stories in one: the work of the development team, and the group inside Google that saw what was possible, and the worldwide community of scientists who gradually shifted our understanding of how to learn: https://www.nytimes.com/2016/12/14/magazine/the-great-AI-awakening.html

The development took less than a year. Google Brain and its competitors conceived the idea in five years. The worldwide story of machine learning is an order of magnitude longer in time and space. The key point about the recent history is the earthquake it produced in the approach to learning a language:

> Instead of programming every word and grammatical rule and exception in both languages, *let the computer find the rules*. Just give it enough correct translations.
>
> If we were recognizing images, the inputs would be many examples with correct labels (the training set). The machine creates the function $F(\boldsymbol{x}, \boldsymbol{v})$.

This is closer to how children learn. And it is closer to how we learn. If you want to teach checkers or chess, the best way is to get a board and make the moves. Play the game.

The steps from this vision to neural nets and deep learning did not come easily. Marvin Minsky was certainly one of the leaders. But his book with Seymour Papert was partly about what "Perceptrons" could not do. With only one layer, the XOR function (A or B but not both) was unavailable. Depth was missing and it was needed.

The lifework of Geoffrey Hinton has made an enormous difference to this subject. For machine translation, he happened to be at Google at the right time. For image recognition, he and his students won the visual recognition challenge in 2012 (with AlexNet). Its depth changed the design of neural nets. Equally impressive is a 1986 article in *Nature*, in which Rumelhart, Hinton, and Williams foresaw that backpropagation would become crucial in optimizing the weights: *Learning representations by back-propagating errors*.

These ideas led to great work worldwide. The "cat paper" in 2011-2012 described training a face detector without labeled images. The leading author was Quoc Le: *Building high-level features using large scale unsupervised learning*: **arxiv.org/abs/1112.6209**. A large data set of 200 by 200 images was sampled from YouTube. The size was managed by localizing the receptive fields. The network had one billion weights to be trained—this is still a million times smaller than the number of neurons in our visual cortex. Reading this paper, you will see the arrival of deep learning.

A small team was quietly overtaking the big team that used rules. Eventually the paper with 31 authors arrived on **arxiv.org/abs/1609.08144**. And Google had to switch to the deep network that didn't start with rules.

Books on Machine Learning

1 Y. S. Abu-Mostafa *et al*, *Learning from Data*, AMLBook (2012).

2 C. C. Aggarwal, *Neural Networks and Deep Learning: A Textbook*, Springer (2018).

3 E. Alpaydim, *Introduction to Machine Learning*, MIT Press (2016).

4 E. Alpaydim, *Machine Learning : The New AI*, MIT Press (2016).

5 C. M. Bishop, *Pattern Recognition and Machine Learning*, Springer (2006).

6 F. Chollet, *Deep Learning with Python* and *Deep Learning with R*, Manning (2017).

7 B. Efron and T. Hastie, *Computer Age Statistical Inference*, Cambridge (2016). https://web.stanford.edu/~hastie/CASI_files/PDF/casi.pdf

8 A. Géron, *Hands-On Machine Learning with Scikit-Learn and TensorFlow*, O'Reilly (2017).

9 I. Goodfellow, Y. Bengio , and A. Courville, *Deep Learning*, MIT Press (2016).

10 T. Hastie, R. Tibshirani , and J. Friedman, *The Elements of Statistical Learning : Data Mining, Inference, and Prediction*, Springer (2011).

11 M. Mahoney, J. Duchi, and A. Gilbert, editors, *The Mathematics of Data*, American Mathematical Society (2018).

12 M. Minsky and S. Papert, *Perceptrons*, MIT Press (1969).

13 A. Moitra, *Algorithmic Aspects of Machine Learning*, Cambridge (2014).

14 G. Montavon, G. Orr, and K. R. Müller, eds, *Neural Networks : Tricks of the Trade*, 2nd edition, Springer (2012).

15 M. Nielsen, *Neural Networks and Deep Learning*, (title).com (2017).

16 B. Recht and S. Wright, *Optimization for Machine Learning*, to appear.

17 A. Rosebrock, *Deep Learning for Computer Vision with Python*, pyimagsearch (2018).

18 S. Shalev-Schwartz and S. Ben-David, *Understanding Machine Learning*: *From Theory to Algorithms*, Cambridge (2014).

19 S. Sra, S. Nowozin, and S. Wright, eds. *Optimization for Machine Learning*, MIT Press (2012).

20 G. Strang, *Linear Algebra and Learning from Data*, Wellesley-Cambridge Press (2019).

21 V. N. Vapnik, *Statistical Learning Theory*, Wiley (1998).

Image Compression by the SVD

Uncompressed image. Slider at 300.

IMAGE SIZE 600×600
#PIXELS $= 360000$

UNCOMPRESSED SIZE
proportional to number of pixels

COMPRESSED SIZE
approximately proportional to
$600 \times 300 + 300 + 300 \times 600$
$= 360300$

COMPRESSION RATIO
$360000/360300 = 1.00$

Show singular values

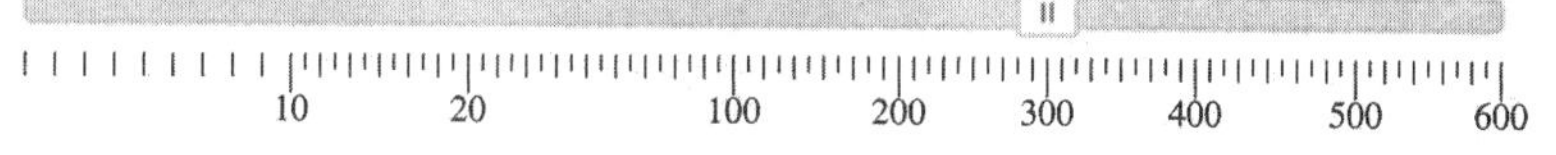

Compressed image. Slider at 20.

IMAGE SIZE 600×600
#PIXELS $= 360000$

UNCOMPRESSED SIZE
proportional to number of pixels

COMPRESSED SIZE
approximately proportional to
$600 \times 20 + 20 + 20 \times 600$
$= 24020$

COMPRESSION RATIO
$360000/24020 = 14.99$

Show singular values

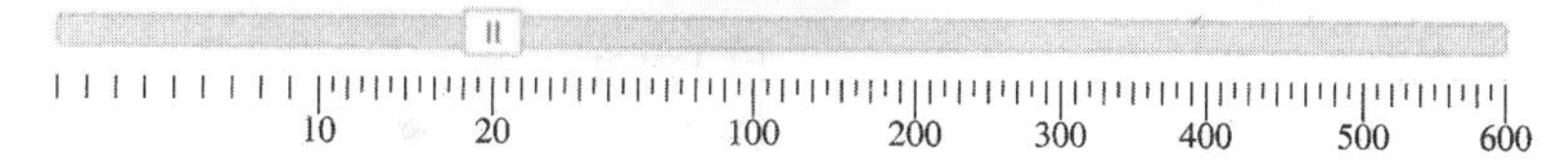

Change the number of singular values using the slider. Click on one of these images to compress it:

You can compress your own images by using the file picker or by dropping them on this page.

Codes and Algorithms for Numerical Linear Algebra

LAPACK	is the first choice for dense linear algebra codes.
ScaLAPACK	achieves high performance for very large problems.
COIN/OR	provides high quality codes for the optimization problems of operations research.

Here are sources for specific algorithms.

Direct solution of linear systems	
Basic matrix-vector operations	BLAS
Elimination with row exchanges	LAPACK
Sparse direct solvers (UMFPACK)	SuiteSparse, SuperLU
QR by Gram-Schmidt and Householder	LAPACK
Eigenvalues and singular values	
Shifted QR method for eigenvalues	LAPACK
Golub-Kahan method for the SVD	LAPACK
Iterative solutions	
Preconditioned conjugate gradients for $S\boldsymbol{x} = \boldsymbol{b}$	Trilinos
Preconditioned GMRES for $A\boldsymbol{x} = \boldsymbol{b}$	Trilinos
Krylov-Arnoldi for $A\boldsymbol{x} = \lambda\boldsymbol{x}$	ARPACK, Trilinos, SLEPc
Extreme eigenvalues of S	see also BLOPEX
Optimization	
Linear programming	CLP in COIN/OR
Semidefinite programming	CSDP in COIN/OR
Interior point methods	IPOPT in COIN/OR
Convex Optimization	CVX, CVXR
Randomized linear algebra	
Randomized factorizations via pivoted QR	users.ices.utexas.edu/
$A = CMR$ columns/mixing/rows	~pgm/main_codes.html
Interpolative decomposition (ID)	
Fast Fourier Transform	FFTW.org
Repositories of high quality codes	GAMS and Netlib.org
ACM Transactions on Mathematical Software	TOMS
Deep learning software (see also page 374)	
Deep learning in Julia	Fluxml.ai/Flux.jl/stable
Deep learning in MATLAB	Mathworks.com/learn/tutorials/deep–learning–onramp.html
Deep learning in Python and JavaScript	Tensorflow.org, Tensorflow.js
Deep learning in R	Keras, KerasR

Counting Parameters in the Basic Factorizations

$A = LU \quad A = QR \quad S = Q\Lambda Q^{\mathrm{T}} \quad A = X\Lambda X^{-1} \quad A = QS \quad A = U\Sigma V^{\mathrm{T}}$

This is a review of key ideas in linear algebra. The ideas are expressed by those factorizations and our plan is simple: *Count the parameters in each matrix.* We hope to see that in each equation like $A = LU$, the two sides have the same number of parameters.

For $A = LU$, both sides have n^2 parameters.

L: **Triangular** $n \times n$ matrix with 1's on the diagonal	$\frac{1}{2}n(n-1)$
U: **Triangular** $n \times n$ matrix with free diagonal	$\frac{1}{2}n(n+1)$
Q: **Orthogonal** $n \times n$ matrix	$\frac{1}{2}n(n-1)$
S: **Symmetric** $n \times n$ matrix	$\frac{1}{2}n(n+1)$
Λ: **Diagonal** $n \times n$ matrix	n
X: $n \times n$ matrix of independent **eigenvectors**	$n^2 - n$

Comments are needed for Q. Its first column $\boldsymbol{q}_1$ is a point on the unit sphere in $\mathbf{R}^n$. That sphere is an $\boldsymbol{n-1}$-dimensional surface, just as the unit circle $x^2 + y^2 = 1$ in $\mathbf{R}^2$ has only one parameter (the angle θ). The requirement $||\boldsymbol{q}_1|| = 1$ has used up one of the n parameters in $\boldsymbol{q}_1$. Then $\boldsymbol{q}_2$ has $n-2$ parameters—it is a unit vector and it is orthogonal to $\boldsymbol{q}_1$. The sum $(n-1) + (n-2) + \cdots + 1$ equals $\frac{1}{2}\boldsymbol{n(n-1)}$ free parameters in Q.

The eigenvector matrix X has only $n^2 - n$ parameters, not n^2. If $\boldsymbol{x}$ is an eigenvector then so is $c\boldsymbol{x}$ for any $c \neq 0$. We could require the largest component of every $\boldsymbol{x}$ to be 1. This leaves $n-1$ parameters for each eigenvector (and no free parameters for X^{-1}).

The count for the two sides now agrees in all of the first five factorizations.

For the SVD, use the reduced form $\boldsymbol{A_{m\times n} = U_{m\times r}\Sigma_{r\times r}V^{\mathrm{T}}_{r\times n}}$ (known zeros are not free parameters!) Suppose that $m \leq n$ and A is a full rank matrix with $\boldsymbol{r = m}$. The parameter count for A is $\boldsymbol{mn}$. *So is the total count for U, Σ, and V.* The reasoning for orthonormal columns in U and V is the same as for orthonormal columns in Q.

U has $\dfrac{1}{2}\boldsymbol{m(m-1)}$ $\quad \Sigma$ has $\boldsymbol{m}$ $\quad V$ has $(n-1) + \cdots + (n-m) = \boldsymbol{mn} - \dfrac{1}{2}\boldsymbol{m(m+1)}$

Finally, suppose that A is an *m by n matrix of rank r*. **How many free parameters in a rank r matrix?** We can count again for $U_{m\times r}\Sigma_{r\times r}V^{\mathrm{T}}_{r\times n}$:

U has $(m-1) + \cdots + (m-r) = \boldsymbol{mr} - \dfrac{1}{2}\boldsymbol{r(r+1)}$ $\quad V$ has $\boldsymbol{nr} - \dfrac{1}{2}\boldsymbol{r(r+1)}$ $\quad \Sigma$ has $\boldsymbol{r}$

The total parameter count for rank r is $\boldsymbol{(m+n-r)\,r}$.

We reach the same total for $A = CR$ in Section I.1. The r columns of C were taken directly from A. The row matrix R includes an r by r identity matrix (not free!). Then the count for CR agrees with the previous count for $U\Sigma V^{\mathrm{T}}$, when the rank is r:

C has $\boldsymbol{mr}$ parameters $\qquad R$ has $\boldsymbol{nr - r^2}$ parameters $\qquad$ Total $\boldsymbol{(m+n-r)\,r}$.

Index of Authors

(For most living authors, the page includes a journal or book or arXiv reference)

Abu-Mostafa, 416
Aggarwal, 416
Alpaydim, 416
Andersson, 108
Arnold, 383
Arnoldi, 115–117, 123

Ba, 366, 367
Bach, 343
Bader, 105, 108
Bahri, 378
Balestriero, 395
Banach, 91
Baraniuk, 196, 395
Bassily, 363
Bau, 115, 117
Bayes, 253, 303
Beckermann, 180, 182, 183
Belkin, 170, 363, 414
Belongie, 395
Ben-David, 416
Bengio, 356, 381, 393, 408, 416
Benner, 182
Bernoulli, 286, 287, 410
Berrada, x, 394
Bertsekas, 188, 356
Bishop, 416
Borre, 165, 302
Bottou, 363
Boyd, 185, 188, 191, 350, 354, 356
Bregman, 185, 192
Bro, 108
Buhmann, 181

Candès, 195, 196, 198, 354
Canny, 390
Carroll, 104
Cauchy, 90, 178, 200
Chang, 104
Chebyshev, 179, 263, 285, 289, 290
Chernoff, 285, 289, 291
Cholesky, 48, 54, 115, 238
Chollet, 416
Chu, 185
Combettes, 191
Cooley, 209
Courant, 174, 176
Courville, 416
Cybenko, 384

Dantzig, 338
Darbon, 193
Daubechies, 237
Davis, 152, 170
De Lathauwer, 107
de Moor, 108
Dhillon, 253
Dickstein, 381
Dijksterhuis, 258
Dokmanic, 259
Donoho, 195–198
Douglas, 191
Drineas, 108, 143
Duchi, 366, 367, 416
Durbin, 233

Eckart-Young, 58, 71, 72, 74, 179
Eckstein, 185
Efron, 408, 416
Einstein, 91
Eldridge, 170
Elman, 336
Embree, 117, 142
Erdős, 291
Euclid, 259
Euler, 242, 244, 322

Fatemi, 193
Fiedler, 246, 248, 254
Fischer, 174, 176
Fisher, 86, 87
Flyer, 181
Fornberg, 181
Fortin, 188
Fortunato, 182
Fourier, 178, 179, 204, 207, 216, 222, 391
Friedman, 416
Frobenius, 71, 73, 93, 257, 312, 315

Géron, 416
Gangul, 381
Garipov, 365
Gauss, 122, 191, 209, 268, 304, 308
Gibbs, 231
Gilbert, 416
Giles, 272, 382
Gillis, 98
Givens, 119
Glowinski, 188
Goemans, 289
Gohberg, 235
Goldfarb, 193
Goldstein, 193
Golub, 115, 120, 258
Goodfellow, 416
Gordon Wilson, 365

Gower, 258, 261, 363
Gram-Schmidt, 30, 116, 128–130
Gray, 235
Griewank, 406
Grinfeld, 108

Haar, 35
Hadamard, 30, 107, 218
Hager, 167
Hajinezhad, 188
Halko, 143, 151
Hall, 340
Hanin, 378
Hankel, 178, 183
Hansen, 368
Hardt, 356, 367
Harmon, viii
Harshman, 104, 108
Hastie, 99, 100, 198, 199, 416
Hazan, 366, 367
He, 378, 395
Hermite, 206
Hessenberg, 115, 117
Higgs, 268
Higham, 249, 250, 397
Hilbert, 78, 91, 96, 159, 178, 183, 383
Hillar, 104
Hinton, 203, 393, 409, 415
Hitchcock, 104
Hofmann, 414
Hölder, 96
Hopf, 235
Hornik, 384

Householder, 34, 131, 135

Ioffe, 409
Izmailov, 365

Jacobi, 122, 123, 191, 323
Johnson, 154
Johnson, 406
Jordan, 356

Kaczmarz, 122, 185, 193, 363, 368
Kahan, 120, 152, 170
Kale, 367
Kalman, 164, 167, 263, 308, 309
Kalna, 250
Kannan, 151, 155
Karpathy, 392
Khatri-Rao, 105, 106
Kibble, 250
Kingma, 366, 367
Kirchhoff, 18, 241, 243, 336
Kleinberg, 381
Kolda, 105, 108
König, 340
Kolmogorov, 383
Krizhevsky, 203, 393, 409
Kronecker, 105, 221, 223, 224, 226
Kruskal, 104
Krylov, 115–117, 121, 178, 181, 183
Kullback, 411
Kumar, 367, 394
Kuo, 395

Lagrange, 69, 173, 185, 322, 333
Lanczos, 115, 118
Laplace, 223, 225, 228, 239, 248
Lathauwer, 108
Le, 415
LeCun, 393, 410
Lee, 97, 108, 198
Lei, 193
Leibler, 411
Lessard, 354, 356
Levenberg, 329
Levinson, 233
Lewis-Kraus, 415
Li, 182
Liang, 412
Liao, 384
Liberty, 151
Lim, 104
Lindenstrauss, 154
Lipschitz, 355
Logan, 196
Lorentz, 91
Lustig, 196
Lyapunov, 180

Ma, 363, 414
Maggioni, 108
Mahoney, 108, 143, 146, 151, 416
Malik, 247
Mallat, 395
Mandal, 414
Markov, 253, 263, 284, 290, 293, 311, 318
Marquardt, 329
Martinsson, 130, 139, 143, 151, 155
Mazumder, 198

Menger, 260
Mhaskar, 384
Minsky, 415, 416
Mirsky, 71, 72
Moitra, 416
Moler, 396
Monro, 361
Montavon, 408, 416
Montufar, 381
Moore-Penrose, 133
Morrison, 160, 162, 309
Müller, 408, 416

Nakatsukasa, 152, 200
Nash, 340
Needell, 363
Nesterov, 354, 356
Neumann, 318
Newman, 182, 246
Newton, 165, 327, 330, 332
Nielsen, 411, 416
Nocedal, 356
Nowozin, 416
Nyquist, 195
Nystrom, 253

Ohm, 18, 242, 336
Olah, 397
Orr, 408, 416
Oseledets, 108
Osher, 193

Paatero, 108
Packard, 354, 356
Papert, 415, 416
Parhizkar, 259
Parikh, 185, 191
Pascanu, 381
Peaceman, 191

Pearson, 301
Peleato, 185
Pennington, 378
Pentland, 98
Perron, 312, 315, 319
Pesquet, 191
Pick, 183
Poczos, 363
Podoprikhin, 365
Poggio, 384
Poisson, 223, 229, 276, 287
Polya, 382
Polyak, 351
Poole, 381
Postnikov, 255
Procrustes, 67, 257
Pythagoras, 30

Rachford, 191
Raghu, 381
Ragnarsson, 108
Rakhlin, 412
Ramdas, 363
Ranieri, 259
Rao Nadakuditi, 170, 171
Rayleigh, 68, 81, 87, 173, 249, 290
Recht, 198, 354, 356, 367, 416
Reddi, 363, 367
Ren, 378, 395
Renyi, 291
Riccati, 253
Richtarik, 363
Robbins, 361
Roelofs, 356, 367
Roentgen, 196
Rokhlin, 151
Rolnick, 378, 385
Romberg, 198
Rosebrock, 416
Ruder, 367
Rudin, 193
Ruiz-Antolin, 178, 179, 182
Rumelhart, 415

Sachan, 367
Salakhutdinov, 409
Schmidt, 71
Schoenberg, 260
Schoenholz, 378
Schölkopf, 414
Schöneman, 258
Schur, 177, 335
Schwarz, 61, 90, 96, 200
Seidel, 122, 191
Semencul, 235
Seung, 97, 108
Shalev-Schwartz, 416
Shannon, 195, 411
Shepp, 196
Sherman, 160, 162, 309
Shi, 188, 247
Silvester, 336
Simonyan, 392, 393
Singer, 366, 367
Smilkov, 386
Smola, 363, 414
Sohl-Dickstein, 378
Song, 108
Sorensen, 142
Sra, viii, 360, 363, 365, 416
Srebro, 75, 198, 356, 363, 367
Srivastava, 409
Stanley, 381
Stern, 356, 367
Stewart, 74, 143
Strohmer, 122, 363
Su, 354
Sun, 378, 395
Sutskever, 203, 393, 409
Sylvester, 180–183
Szegö, 235
Szegedy, 409

Tao, 175, 195, 198
Tapper, 108
Taylor, 179, 323, 407
Tegmark, 385
Tibshirani, 99, 100, 184, 416
Toeplitz, 183, 232, 235, 387, 389
Townsend, 78, 178–183
Trefethen, 115, 117
Tropp, 143, 151, 291
Truhar, 182
Tucker, 107
Tukey, 209
Turing, 413
Turk, 98
Tygert, 151
Tyrtyshnikov, 108

Udell, 181, 182

Van Loan, 108, 115, 120, 226, 258
Vandenberghe, 350, 356
Vandermonde, 178, 180, 181
Vandewalle, 108
Vapnik, 414, 416
Veit, 395
Vempala, 151, 155
Vershynin, 122, 363
Vetrov, 365
Vetterli, 259
Vinyals, 356
Vitushkin, 383
von Neumann, 340

Wakin, 196
Walther, 406
Wang, 170
Ward, 363
Wathen, 336
Weyl, 172, 175
Wiener, 235
Wilber, 182, 395
Wilkinson, 119
Williams, 415
Wilson, 356, 367
Woodbury, 160, 162, 309
Woodruff, 108, 151
Woolfe, 151
Wright, 356, 416

Xiao, 378
Xu, 98

Yin, 193
Young, 353
Yu, 98

Zadeh, 198
Zaheer, 367
Zaslavsky, 381
Zhang, 98, 356, 378, 395
Zhong, 108
Zhou, 193, 384, 385
Zisserman, 392–394
Zolotarev, 182
Zou, 99, 100

Index

Accelerated descent, 352, 353
Accuracy, 384
Activation, iv, 375, 376
AD, 397, 406
ADAGRAD, 366
ADAM, 322, 356, 366
Adaptive, 407
Adaptive descent, 356, 361
ADI method, 182
Adjacency matrix, 203, 240, 291
Adjoint equation, 405
Adjoint methods, 404
ADMM, 99, 185, 187, 188
Affine, iii
AlexNet, ix, 373, 415
Aliasing, 234
AlphaGo Zero, 394, 412
Alternating direction, 185, 191
Alternating minimization, 97, 106, 199, 252
Antisymmetric, 52
Approximate SVD, 144, 155
Approximation, 384
Architecture, 413
Argmin, 186, 322
Arnoldi, 116, 117
Artificial intelligence, 371
Associative Law, 13, 163
Asymptotic rank, 79
Augmented Lagrangian, 185, 187
Autocorrelation, 220
Automatic differentiation, 371, 397, 406
Average pooling, 379
Averages, 236, 365

Back substitution, 25
Backpropagation, 102, 344, 371, 397
Backslash, 113, 184
Backtracking, 328, 351
Backward difference, 123
Backward-mode, 397
Banach space, 91
Banded, 203, 232
Bandpass filter, 233
Basis, 4, 5, 15, 204, 239
Basis pursuit, 184, 195
Batch mode, 361
Batch normalization, x, 409, 412
Bayes Theorem, 303
Bell-shaped curve, 279
Bernoulli, 287
BFGS (quasi-Newton), 165
Bias, iii, 375
Bias-variance, 374, 412
Bidiagonal matrix, 120
Big picture, 14, 18, 31
Binomial, 270, 271, 275, 287
Binomial theorem, 385
Bipartite graph, 256, 340
Block Toeplitz, 389
BLUE theorem, 308
Bootstrap, 408
Boundary condition, 229
Bounded variation, 193, 194
Bowl, 49
Bregman distance, 192

Caffe, viii, 374
Cake numbers, 382
Calculus, 396
Calculus of variations, 322
Canny Edge Detection, 390
Cauchy-Schwarz, 90, 96, 200
Centered difference, 345
Centering (mean 0), 75, 270
Central Limit Theorem, 267, 271, 288
Central moment, 286
Centroid, 247, 261
Chain rule, 375, 397, 406
Channels, 388
Chebyshev series, 179
Chebyshev's inequality, 285, 290
Chernoff's inequality, 285
Chi-squared, 275, 280–282
Chord, 332
Circulant, x, 213, 220, 234
Circulants $CD = DC$, 220
Classification, 377
Closest line, 136
Clustering, 245, 246
CNN, v, 203, 232, 380
Coarea formula, 194
Codes, 374
Coin flips, 269
Column pivoting, 129, 143
Column space, 1–5, 13, 14
Combinatorics, 373, 381
Companion matrix, 42
Complete graph, 240, 244
Complete spaces, 91
Complex conjugate, 205, 215
Complex matrix, 45
Composite function, 384
Composition, iv, 373, 375, 383
Compressed sensing, 146, 159, 196
Compression, 230
Computational graph, 397, 401
Computing the SVD, 120, 155
Condition number, 145, 353
Conductance matrix, 124, 336
Congruent, 53, 85, 87, 177
Conjugate gradients, 121
Connected graph, 292
Constant diagonal, 213
Continuous Piecewise Linear, 372, 375
Contraction, 357, 358
Convergence in expectation, 365
Convex, 293, 321, 324, 325
Convex hull, 331
ConvNets, 378
Convolution, 203, 214, 220, 283, 387
Convolution in 2D, 388
Convolution of functions, 219
Convolution rule, 218, 220
Convolutional net, 380, 387
Corner, 338, 343
Correlation, 300, 301
Cosine series, 212
Counting Law, 16
Courant-Fischer, 174
Covariance, 76, 289, 294
Covariance matrix, 81, 134, 295–297
CP decomposition, 97, 104
CPL, 372, 385
Cramer's Rule, 141
Cross-correlation, 219
Cross-entropy, 360
Cross-entropy loss, 411
Cross-validation, 408, 412
Cubic convergence, 119
Cumulant, 287, 288
Cumulative distribution, 266, 269
Current Law, 18, 241
CURT, 108
CVX, 326
Cycle, 256
Cyclic convolution, 214, 218, 234
Cyclic permutation, 213

Data science, vi, 11, 71
DCT, 66, 230, 231
Deep Learning, iii, vi, 371
Degree matrix, 203, 240
DEIM method, 142
Delta function, 193, 219

Derivative, 101, 344, 398, 399
Derivative $d\lambda/dt$, 169
Derivative $d\sigma/dt$, 170
Derivative of A^2, 167
Derivative of A^{-1}, 163
Descent factor, 353
Determinant, 36, 42, 47, 48, 346
DFT matrix, 205, 207
Diagonalization, 11, 43, 52, 298
Diamond, 88, 89, 184
Difference equation, 223
Difference matrix, 16, 39, 238
Digits, iii
Dimension, 4, 6
Discrete Fourier Transform, 203–207
Discrete Gaussian, 389
Discrete sines and cosines, 66
Discriminant, 86, 87
Displacement rank, 182
Distance from mean, 284
Distance matrix, 259
Document, 98
Driverless cars, 381
Dropout, 409, 410
DST matrix, 66, 229
Dual problem, 186, 190, 322, 339
Duality, ix, 96, 339, 340, 342, 343
Dying ReLU, 400

Early stopping, 360
Eckart-Young, 58, 71, 72, 74, 75
Eigenfaces, 98
Eigenfunction, 228
Eigenvalue, 1, 12, 36, 39
Eigenvalue of $A \oplus B$, 224
Eigenvalue of $A \otimes B$, 224
Eigenvalues of A^{T} and A^k, 42, 70
Eigenvalues of AB and BA, 59, 64
Eigenvector, 12, 36, 39, 216
Eigenvectors, viii, 11, 217
Element matrix, 244
Elimination, 11, 21, 23
Ellipse, 50, 62
Energy, 46, 49
Entropy, 411
Epoch, 361
Equilibrium, 242
Equiripple filter, 236
Erdős-Renyi, 291
Error equation, 116, 364
Error function, 280
Euclidean, 88, 259
Even function, 212
Expected value, 149, 264
Exploding weights, 378
Exponential distribution, 278
Expressivity, 373, 381

Factorization, 5, 11
Fan-in, 378
Fast Fourier Transform, 178
Fast multiplication, 234
Feature space, 86, 252, 375
FFT, ix, 204, 209, 211, 229, 234
Fiedler vector, 246, 248, 249, 254
Filter, 203, 233, 236, 387
Filter bank, 391
Finance, 321
Finite element, 336
Five tests, 49
Fold plane, 381
Forward mode, 401, 402
Four subspaces, 18
Fourier integral, 204, 205
Fourier matrix, ix, 35, 180, 204, 205, 216
Fourier series, 179, 204
Free parameters, 419
Frequency response, 232, 236
Frequency space, 219
Frobenius norm, 71, 257
Fully connected net, v, x, 371, 380
Function space, 91, 92
Fundamental subspaces, 14

Gain matrix, 164
Gambler's ruin, 317
Game theory, 340
GAN, 413

Gaussian, 271, 275, 389
Generalize, 359, 367, 368, 372
Generalized eigenvalue, 81
Generalized SVD, 85
Generating function, 285, 287
Geometry of the SVD, 62
Gibbs phenomenon, 231
Givens rotation, 119
GMRES, 117
Go, ix, 394, 412
Golub-Kahan, 120
Google, 394
Google Translate, 415
GPS, 165, 302
GPU, 393
Gradient, 323, 344, 345, 347
Gradient descent, 322, 344, 349
Gradient detection, 389
Gradient of cost, 334
Gram matrix, 124
Gram-Schmidt, 114, 128
Grammar, 415
Graph, 16, 203, 239
Graph Laplacian, 124, 224, 239, 243, 246
Grayscale, 229
Greedy algorithm, 254

Hölder's inequality, 96
Haar wavelet, 35
Hadamard matrix, 30
Hadamard product, 107, 218
Half-size transforms, 209
Halfway convexity test, 332
Hankel matrix, 183
Hard margin, 414
He uniform, 378
Heavy ball, 351, 366
Heavy-tailed, 285
Hermitian matrix, 206
Hessenberg matrix, 117
Hessian, 55, 323, 326
Hidden layer, 371, 372, 387, 399
Hilbert 13th problem, 383
Hilbert matrix, 78, 183
Hilbert space, 66, 91
Hinge loss, 360
Householder, 131, 135
Hyperparameters, 407–412
Hyperplane, 381, 382

iid, 277
ICA, 413
Identity for σ^2, 265, 274
Ill-conditioned, 113
Image recognition, 387, 389
ImageNet, ix, 373
Importance sampling, 363
Incidence matrix, 16, 203, 239, 240
Incoherence, 195
Incomplete LU, 122
Incomplete matrices, 159
Indefinite matrix, 50, 172
Independent, 3–5, 289
Indicator function, 189
Infinite dimensions, 91
Informative component, 171
Initialization, 378
Inner product, 9, 10, 91
Interior-point methods, 342
Interlacing, 53, 168, 170, 171, 175
Internal layer, 377
Interpolation, 180, 363
Interpolative decomposition, 139, 155
Inverse Fourier transform, 217
Inverse of $A - UV^{\mathrm{T}}$, 162
Inverse of $A \otimes B$, 221
Inverse problems, 114
Inverse transform, 204
Isolated minimum, 322

Jacobian matrix, 323
Johnson-Lindenstrauss, 154
Joint independence, 289
Joint probability, 102, 294
JPEG, 66, 229

Kaczmarz, 122, 193, 363–364
Kalman filter, 164, 167, 308–310

Karhunen-Loève, 61
Keras, viii, 374, 418
Kernel function, 414
Kernel matrix, 247
Kernel method, 181, 252
Kernel trick, 414
Khatri-Rao product, 105, 106
Kirchhoff, 241, 242, 336
KKT matrix, 173, 177, 335
Kriging, 253
Kronecker product, 105, 221, 227
Kronecker sum, 223–225, 228
Krylov, 116, 183
Kullback-Leibler, 411
Kurtosis, 286

Lagrange multiplier, 150, 321, 333
Lagrangian, 173, 333
Lanczos, 118
Laplace's equation, 229
Laplacian matrix, 203
Laplacian of Gaussian, 390
Large deviation, 285
Largest determinant, 141
Largest variance, 71
LASSO, 100, 184, 190, 357
Latent variable, 182
Law of Inertia, 53, 177, 337
Law of large numbers, 264
Leaky ReLU, 400
Learning function, iii, vi, 373, 375
Learning rate, vii, 344, 407
Least squares, iv, 109, 124, 126
Left eigenvectors, 43
Left nullspace, 14, 17
Left singular vectors, 60
Length squared, 47, 149
Level set, 194
Levenberg-Marquardt, 329, 330
Line of nodes, 223
Line search, 351
Linear convergence, 350, 356
Linear pieces, 373, 381
Linear programming, 338
Linear Time Invariance, 67, 233
Lipschitz constant, 194, 355, 365
Local structure, 387
Log-normal, 275, 280, 281
Log-rank, 181
Logan-Shepp test, 196
Logistic curve, 373
Logistic regression, 393
Loop, 17, 241
Loss function, 360, 377, 411
Low effective rank, 159, 180
Low rank approximation, 144, 155
Lowpass filter, 236, 391
LTI, 233

Machine learning, 371, 413
Machine learning codes, 374, 418
Machine translation, 394
Margin, 414
Marginals, 295
Markov chains, 311
Markov matrix, 39, 311–313, 318
Markov's inequality, 284, 290, 293
Mass matrix, 81
Master equation, 263, 318
Matricized tensor, 105
Matrix calculus, 163
Matrix Chernoff, 291
Matrix completion, viii, 159, 197, 198, 255
Matrix identities, 163
Matrix inversion lemma, 160
Matrix multiplication, 7, 10, 13
Matrix norm, 92, 94
Max-min, 174
Max-pooling, 379, 406
Maximum flow, 339, 340
Maximum of $R(x)$, 81, 172
Maximum problem, 62, 63, 68
MDS algorithm, 261
Mean, 75, 147, 264, 267
Mean field theory, 378
Mean of sum, 299
Medical norm, 95, 96

Method of multipliers, 185
Microarray, 250
Minibatch, ix, 322, 359, 367
Minimax, 174, 335, 342
Minimum, 49, 55, 338
Minimum cut, 246, 340
Minimum norm, 126, 356
Minimum variance, 150
Missing data, 197
Mixed strategy, 341
Mixing matrix, 8, 142, 143, 152, 155
MNIST, iii, 355
Modularity matrix, 246
Modulation, 208
Moments, 286
Momentum, 351, 366
Monte Carlo, 272, 394
Morrison-Woodbury, 160, 162
Moving window, 237, 390
MRI, 196
Multigrid, 122, 353
Multilevel method, 249
Multiplication, 2, 10, 214
Multiplicity, 40
Multivalued, 192
Multivariable, 275, 280, 304, 305

Netflix competition, 199
Neural net, iii, v, 377
Neuron, 375
Newton's method, 165, 327, 332
NMF, 97, 98, 190
Node, 16
Noise, 184
Nondiagonalizable, 40
Nonlinear least squares, 329
Nonnegative, 8, 97
Nonuniform DFT, 178, 182
Norm of tensor, 103
Norm of vector, 88
Norm-squared sampling, 122, 146, 149, 156, 363
Normal distribution, 268, 279, 288
Normal equation, 113, 127
Normal matrix, 180, 183
Normalize, 128, 270, 409
Normalized Laplacian, 248
NP-hard, 99
Nuclear norm, 71, 95, 100, 159, 197, 200
Nullspace, 6, 14
Nyquist-Shannon, 195

Ohm, 336
One-pixel, 196
One-sided inverse, 8
One-Zero matrices, 78
OpenCourseWare, x
Optimal strategy, 341, 343
Optimization, 321
Orthogonal, 11, 29, 52, 128
Orthogonal eigenvectors, 44
Orthogonal functions, 205
Orthogonal matrix, 29, 33, 35, 36, 257
Orthogonal subspaces, 29
Orthonormal basis, 34, 130
Outer product, 9, 10, 103
Overfitting, iii, vi, ix, 359, 360, 409
Overrelaxation, 353

Parameters, 70, 419
Payoff matrix, 341, 343
PCA, 1, 71, 75–77
Penalty, 100, 114
Perceptrons, 415
Periodic functions, 204
Permutation, 26, 28, 35
Perron-Frobenius, 314, 315
Pieces of the SVD, 57
Piecewise linear, x, 375, 381, 385
Pivot, 23, 25, 47, 48
Playground, 386
Poisson, 182, 275, 276, 287
Polar decomposition, 67
Polar form, 215
Pooling, x, 379
Positions from distances, 260
Positive definite, viii, 45–49
Positive matrix, 315

Positive semidefinite, 46, 290
Power method, 95
Preconditioner, 122, 147
Primal problem, 185
Principal axis, 51
Principal components, 71
Probability density, 266, 273, 278
Probability matrix, 295
Probability of failure, 279
Procrustes, 257, 258, 261
Product rule, 303
Projection, 32, 113, 127, 136, 153, 357
Projection matrix, 127, 153
Projects, vi, viii, 155, 366, 395
Proof of the SVD, 59, 69
Proximal, 189, 191
Proximal descent, 357
Pseudoinverse, 113, 124, 125, 132, 184
Pseudospectra, 117

Quadratic $\frac{1}{2}\boldsymbol{x}^{\mathrm{T}}S\boldsymbol{x}$, 326
Quadratic convergence, 328
Quadratic cost, 411
Quadratic formula, 38
Quadratic model, 352
Quantization, 230
Quarter circle, 79, 80
Quasi-Monte Carlo, 272
Quasi-Newton, 165, 328

Radial basis function, 181
Ramp function, 376, 400
Random forest, 413
Random graph, 291, 292
Random process, 235
Random projection, 153
Random sampling, 114, 120, 148, 253, 410
Randomization, ix, 108, 146, 155, 368
Randomized Kaczmarz, 363
Rank, 4, 5, 10, 20
Rank r, 419
Rank of $A^{\mathrm{T}}A$ and AB, 19
Rank of tensor, 103, 104
Rank one, 61, 110, 160, 255, 417
Rank revealing, 138, 143
Rank two matrix, 176
Rare events, 277
Rational approximation, 182
Rayleigh quotient, 63, 68, 81, 87, 173, 251
RBF kernel, 181, 414
Real eigenvalues, 44
Rectified Linear Unit, 376
Recurrence relation, 405
Recurrent network, 394, 413
Recursion, 382
Recursive least squares, 164, 309
Reduced form of SVD, 57
Reflection, 33, 34, 131, 237, 391
Regression, 77, 377
Regularization, 132, 410, 412
Reinforcement learning, 394, 413
Relax, 184
ReLU, iv, x, 375, 376, 400
Repeated eigenvalue, 12, 69
Rescaling, 300
Reshape, 226, 227
Residual net, 395
ResNets, 378
Restricted isometry, 196
Reverse mode, 399, 402, 403, 405
Ridge regression, 132, 184
Right singular vectors, 60
Rigid motion, 259
RNN, 394, 413
Rotation, 33, 37, 41, 62, 67
Roundoff error, 145
Row exchange, 26
Row picture, 21
Row space, 5, 14

Saddle point, ix, 50, 81, 168, 172, 174, 186, 335, 341
Sample covariance, viii, 76, 296
Sample mean, 264, 296
Sample value, 264
Sample variance, 265
Saturate, 409, 411

Scale invariance, 400
Schur complement, 177, 335
Schur's Theorem, 317
Scree plot, 78
Second derivatives, 49, 50, 326
Second difference, 123
Secular equation, 171
Semi-convergence, 361
Semidefinite, 47
Semidefinite program, 198, 342
Sensitivity, 406
Separable, 187
Separating hyperplane, 380
Separation of Variables, 225
SGD, 359, 361, 367
Share weights, 387
Sharp point, 89, 184
Sherman-Morrison-Woodbury, 162
Shift, 213, 235
Shift invariance, 203
Shift matrix, 387
Shift rule, 208
Shift-invariant, 387
Shrinkage, 189, 191, 357
SIAM News, 373
Sigmoid, iv, 252
Signal processing, 191, 211, 218
Similar matrices, 38, 43, 85, 119
Simplex method, 338
Sine Transform, 229
Singular gap, 79
Singular Value Decomposition, *see* SVD
Singular values, 56
Singular vectors, ix, 56, 59
Sketch, 151
Skewness, 286
Skip connection, 412
Skip connections, 395
Slice of tensor, 101
Smoothing, 389
Smoothness, 92
Sobel, 390, 396
Soft thresholding, 189, 192, 357
Softmax, 393
Solutions to $z^N = 1$, 206, 215
Spanning tree, 256
Sparse, 8, 89, 184, 195
Sparse PCA, 98–100
Spectral norm, 71
Spectral radius, 95
Spectral Theorem, 12, 44
Speech, 413
Spirals, 386
Spline, 395
Split algorithm, 185, 191
Split Bregman, 192, 193
Square loss, 360, 411
Square root of matrix, 67
Standard deviation, 265
Standardized, 263, 270, 288
State equation, 167
Steady state, 311
Steepest descent, 186, 347, 348, 350
Step function, 193
Stepsize, vii, 186, 344, 407
Stiffness matrix, 124, 336
Stochastic descent, viii, 359, 361, 398
Straight line fit, 136
Stretching, 62, 67
Strictly convex, 49, 323, 325, 355
Stride, 379, 390
Structured matrix, 180
Subgradient, 188, 191, 192, 355
Submatrix, 65
Subsmpling, 379
Sum of squares, 51
Support Vector Machine, 394
SVD, vi, ix, 1, 5, 11, 31, 56, 57, 60, 144
SVD for derivatives, 65
SVM, 181, 394, 413, 414
SWA, 365
Sylvester test, 180, 181, 183
Symbol, 232, 238
Symmetric matrix, 11, 36
Szegö, 235

Tangent line, 324, 325, 332
Taylor series, 323

Tensor, x, 101, 110
Tensor train, 108
Tensor unfolding, 105
Tensorflow, viii, 374, 418
Test, 47, 412
Test data, iii, ix, 359
Text mining, 98
Three bases, 138
Toeplitz matrix, 183, 232, 233, 373, 387, 406
Total probability, 268, 274
Total variance, 77
Total variation, 190, 193
Trace, 36, 77
Training, 412
Training data, iii, 359
Transition matrix, 313, 314
Tree, 17, 240, 244
Triangle inequality, 88, 260
Tridiagonal, 28, 118, 232
Tucker form, 107
Turing machine, 413
Two person game, 340

Unbiased, 308
Underfitting, 374
Unfolding, 108
Uniform distribution, 266, 267
Unit ball, 89, 96, 200
Unitarily invariant, 72
Unitary matrix, 206
Universality, 384
Unsupervised, 71
Updating, 164–166
Upper Chernoff, 289
Upper triangular, 23, 129

Vandermonde, 178, 180
Vanishing weights, 378
Variance, ix, 76, 134, 147, 150, 264, 265, 267
Variance of $\widehat{x}$, 307
Variance of sum, 299
Vector norm, 327
Vectorize (vec), 225
Video, 226
Voltage Law, 18

Wavelet, 391
Wavelet transform, 237
Weak duality, 339, 343
Weakly stationary, 235
Weight averaging, 365
Weight decay, 412
Weight sharing, 388
Weighted, 134, 243
Weighted average, 306, 307
Weights, 375
Weyl inequalities, 172, 175, 176
White noise, 134, 306
Wiener-Hopf, 235
Wikipedia, 30, 275, 394, 408
Wraparound, 391

YOGI, 367

Zero padding, 233, 391
Zig-zag, 348, 349
Zolotarev, 182

Index of Symbols

$(AB)C$ or $A(BC)$, 403
$-1, 2, -1$ matrix, 238
18.06-18.065, vi, viii, x, 155
$A = CMR$, 8, 142, 151, 156
$A = CR$, 5, 7, 245
$A = LU$, 11, 24, 27
$A = QR$, 11, 129, 143, 156
$A = QS$, 67
$A = UV$, 97
$A = U\Sigma V^{\mathrm{T}}$, 11, 57, 64, 69, 120
$A = X\Lambda X^{-1}$, 11, 39
AB, 9, 10, 64
$AV = U\Sigma$, 56
$AV_r = U_r\Sigma_r$, 57
$A \oplus B$, 223
$A \otimes B$, 221
$A^{\mathrm{T}}CA$, 242, 243
$A^+ = A^\dagger = V\Sigma^+U^{\mathrm{T}}$, 125, 132
$A^k = X\Lambda^k X^{-1}$, 39
$H_k = Q_k^{\mathrm{T}}AQ_k$, 117
M-orthogonal, 83
QQ^{T}, 32
QR algorithm, 119, 123
$Q^{\mathrm{T}}Q$, 32
$Q^{\mathrm{T}} = Q^{-1}$, 29
S-curve, 376
S-norm, 90
$S = A^{\mathrm{T}}A$, 47, 48
$S = A^{\mathrm{T}}CA$, 54
$S = Q\Lambda Q^{\mathrm{T}}$, 11, 12, 44, 51
$\boldsymbol{\nabla F}$, 323, 347
$\boldsymbol{a} * *\boldsymbol{a}$, 220
$\boldsymbol{c} * \boldsymbol{d}$, 214, 220
$\boldsymbol{c} \circledast \boldsymbol{d}$, 214, 218
$\boldsymbol{x}^{\mathrm{T}}S\boldsymbol{x}$, 46, 55
ℓ^0 norm, 89, 159
ℓ^1 vs. ℓ^2, 308
$\ell^1, \ell^2, \ell^\infty$ norms, 88, 94, 159, 327
$\boldsymbol{u}\boldsymbol{v}^{\mathrm{T}}$, 9
$\mathbf{C}(\boldsymbol{A})$, 3
$\mathrm{N}(0,1)$, 288, 304
$\mathrm{N}(m,\sigma)$, 268
$\mathbf{S} + \mathbf{T}, \mathbf{S} \cap \mathbf{T}, \mathbf{S}^\perp$, 20
Kron(A, B), 221
vec, 225–227
$\mathbf{C}(A^{\mathrm{T}})$, 9
$\mathbf{N}(A)$, 14
$\mathbf{N}(A^{\mathrm{T}}A) = \mathbf{N}(A)$, 20, 135
$\log(\det X)$, 346, 358
$\odot$, 105
$|\lambda_1| \le \sigma_1$, 61
$||A\boldsymbol{x}||/||\boldsymbol{x}||$, 62
$||\boldsymbol{f}||_C$, 92
k-means, 97, 245, 247, 251, 252, 254
MATLAB, 45, 82, 108, 221, 249, 418
Julia, 45, 82, 418
.* or $\circ$, 107, 218